Birkhäuser

Advances in Mathematical Fluid Mechanics

The *Advances in Mathematical Fluid Mechanics* series is a forum for the publication of high-quality, peer-reviewed research monographs and edited collections on the mathematical theory of fluid mechanics, with special regards to the Navier-Stokes equations and other significant viscous and inviscid fluid models. Titles in this series consider theoretical, numerical, and computational methods, as well as applications to science and engineering. Works in related areas of mathematics that have a direct bearing on fluid mechanics are also welcome. All manuscripts are peer-reviewed to meet the highest standards of scientific literature.

Grigory Panasenko • Konstantin Pileckas

Multiscale Analysis of Viscous Flows in Thin Tube Structures

Birkhäuser

Grigory Panasenko
Faculty of Mathematics and Informatics
Vilnius University
Vilnius, Lithuania

Konstantin Pileckas
Faculty of Mathematics and Informatics
Vilnius University
Vilnius, Lithuania

ISSN 2297-0320 ISSN 2297-0339 (electronic)
Advances in Mathematical Fluid Mechanics
ISBN 978-3-031-54629-7 ISBN 978-3-031-54630-3 (eBook)
https://doi.org/10.1007/978-3-031-54630-3

This book is published under the imprint Birkhäuser, www.birkhauser-science.com by the registered company Springer Nature Switzerland AG
The registered company address is: Gewerbestrasse 11, 6330 Cham, Switzerland

Preface

Mathematical models of viscous flows in thin domains have multiple applications. Such domains have one or several dimensions which are much smaller than other ones. In particular, tube structures are some unions of thin cylinders or rectangles (pipes or channels). This geometry simulates a network of blood vessels in biological applications or pipelines and cooling systems in technical applications.

Full dimension numerical computations of flows in thin domains require huge computer resources, for example, for a network of blood vessels. To reduce these resources and accelerate computations, we use asymptotic analysis where the small parameter is the ratio of thickness of pipes or channels to their length. This analysis leads to the construction of asymptotic expansions justified by error estimates. It is also implemented in some special numerical methods combining the description with reduced dimension and full dimension zooms for small zones of singular behavior of the solution.

Basically, we consider the Newtonian rheology for the fluid motion corresponding to the stationary and nonstationary Navier–Stokes or Stokes equations. These equations introduced two centuries ago are still in the spotlight of contemporary mathematics. In particular, one of the main challenges ("millennium problem") concerns the question of the global existence and uniqueness of a solution of the nonstationary Navier–Stokes equations.

The nonstationary Navier–Stokes equations have the form:

$$\begin{cases} \dfrac{\partial \mathbf{v}}{\partial t} - \nu\Delta\mathbf{v} + (\mathbf{v}\cdot\nabla)\mathbf{v} + \nabla p = 0, \\ \qquad\qquad\qquad\qquad \operatorname{div} \mathbf{v} = 0\,, \end{cases} \tag{1}$$

where the unknown functions are $\mathbf{v}$ (the n-dimensional velocity vector field) and p, the pressure (more exactly, this unknown scalar function stands for the pressure divided by the density of the fluid); these unknown functions depend on the space variable $x \in \mathbb{R}^n$ and the time $t \in [0, +\infty)$. The positive coefficient ν is the kinematic viscosity. It is related to the dynamic viscosity μ as follows: $\nu = \mu/\rho$, where ρ is the density of the fluid. This system of partial differential equations requires one initial condition for $\mathbf{v}(x,0)$. If these equations are stated in a bounded domain, then they should be completed by a boundary condition. Classical boundary condition is the given velocity $\mathbf{v}$ on the boundary (the Dirichlet type condition); however, recently other boundary conditions involving pressure were studied. We will consider also the stationary version of the Navier–Stokes equations

$$\begin{cases} -\nu\Delta\mathbf{v} + (\mathbf{v}\cdot\nabla)\mathbf{v} + \nabla p = 0, \\ \operatorname{div}\mathbf{v} = 0, \end{cases} \tag{2}$$

as well as the linearized versions: the nonstationary Stokes equations

$$\begin{cases} \dfrac{\partial\mathbf{v}}{\partial t} - \nu\Delta\mathbf{v} + \nabla p = 0, \\ \operatorname{div}\mathbf{v} = 0, \end{cases} \tag{3}$$

and the stationary Stokes equations

$$\begin{cases} -\nu\Delta\mathbf{v} + \nabla p = 0, \\ \operatorname{div}\mathbf{v} = 0. \end{cases} \tag{4}$$

The Navier–Stokes equations are often used for the description of the blood flow, although sometimes for the hemodynamical applications non-Newtonian rheologies are used. The stationary and nonstationary Navier–Stokes equations are considered in thin tube structures. The main results of the book are formulated in the form of theorems. The complete asymptotic expansions of the solutions are constructed. The estimates for the difference of the exact solution and its J-th asymptotic approximation is proved. The method of asymptotic partial decomposition of the domain (MAPDD) is formulated and justified for the stationary and nonstationary cases. It gives the asymptotically exact interface conditions of coupling of the 1D and 3D models of the flow. Namely, the geometry of the blood circulation system is presented as a big union B_ε of thin cylindrical "vessels," and the small parameter ε is the ratio of the radius to the height of the cylinders. The MAPDD simplifies the solution of the Stokes and Navier–Stokes equations in such structures by combining the description of different dimensions in one model of hybrid dimension. It provides the one-dimensional reduction in the main part of the thin structure, at some distance from the junction area, and it keeps the $3D$ (or $2D$) description within the junction area. The asymptotic analysis justifies the appropriate interface conditions between these $3D$ and $1D$ descriptions. In particular, for the steady Stokes or Navier–Stokes equations, the above $1D$ models are reduced to some Poiseuille type flows. For the nonstationary Navier–Stokes equations, the $1D$ models are reduced to some nonsteady Poiseuille type flows. This approach effectively reduces the computational burden on solvers dealing with flows within intricately shaped domains, thereby significantly expediting computations.

The structure of the monograph is as follows. It starts by the introductory Chap. 1 presenting the methods used in the book applied to the simplified settings. This chapter helps to catch the main ideas of the book with the examples of "toy problems," i.e., drastically simplified realistic models. In Chap. 1, we define the main geometrical object of the book: thin tube structure. The further chapters systematically introduce all necessary notions and recall fundamental theorems used in the book. Thus, Chap. 2 (Preliminaries) recalls the well-known inequalities in normed spaces, as well as some facts from functional analysis on linear operators in Banach and Hilbert spaces, Sobolev and Hölder spaces. A spotlight of Chap. 2 is the study of functional spaces for unbounded domains with cylindrical outlets at infinity (such domains are used for the construction of the boundary layer correctors), the analysis of the divergence equation, and the spaces of solenoidal vector-valued functions. This chapter also contains the formulations and proofs of the estimates of solutions of the divergence equation for tube structures taking into account the dependence of constants on the small parameter. Chapter 3 studies the so-called Poiseuille flows which are stationary or non-stationary flows in an infinite cylindrical tube with the given flux in the case when the pressure is a linear function of the longitudinal variable x_n, independent of the transversal variables $(x_1, \dots, x_{n-1})$; the tangential velocity of the flow vanishes while the normal velocity is independent of x_n. We prove the existence and uniqueness of such solution and the a priori estimates, i.e., the estimates of the norm of the solution via the norms of the data. We also introduce the linear operators relating the pressure slope of the Poiseuille flow to the flux. In Chap. 4, we consider the general theory of the Stokes equations in bounded and unbounded domains. For the tube structures, the existence and uniqueness theorems are proved, a priori estimates of the solution are obtained. For the problems in unbounded domains with cylindrical outlets to infinity, we prove the existence and uniqueness theorems as well as the theorems on the stabilization (exponential decay) of the solution in the outlets and for time tending to infinity. These theorems are used further for the construction of boundary layers. Chapters 5 and 6 are the spotlights of the book. In Chap. 5, we consider the stationary Stokes and Navier–Stokes equations in thin tube structures B_ε. We start with the Dirichlet conditions on the boundary of the tube structure; we construct the complete asymptotic expansion of the solution, prove the error estimates for the asymptotic approximations, and formulate and justify the method of asymptotic partial decomposition of the domain. In the second part of the chapter, we consider the Stokes equations with the given pressure at the inlets and outlets and no-slip boundary condition on the lateral surface of the tubes. We prove the existence and uniqueness theorem for this boundary value problem and construct and justify the asymptotic expansion and the MAPDD approximations. Finally, we consider the Navier–Stokes equation with the given Bernoulli pressure on the inlets and outlets and construct and justify the asymptotic expansion and the MAPDD. Chapter 6 introduces the Dirichlet boundary value problem for the nonstationary Navier–Stokes equations stated in B_ε with the initial condition $\mathbf{v}(x, 0) = 0$. The existence and uniqueness of the solution for $\varepsilon << 1$ is proved. The asymptotic expansion is constructed. It contains the boundary layers "in space" and "in time." The error estimates are proved for the asymptotic approximations and for the time-dependent MAPDD approximations. Chapter 7 is devoted to the time-periodic setting of the Navier–Stokes equations. As in the previous chapters, we prove the existence and uniqueness theorems and construct and justify the asymptotic and the MAPDD approximations.

The theoretical analysis developed in Chaps. 5 and 6 is confirmed by several numerical experiments showing that in reality the limitations of the applicability of the proposed methods are more flexible than the theoretically predicted conditions. Several numerical experiments for the flows in thin tube structures are presented. Applying the finite element method (FEM)-based codes, we provide the direct numerical computations for the full dimension and reduced dimension (MAPDD) models and compare the solutions. We evaluate the difference between these solutions and point out the limitations of the asymptotic theory and estimate the size of the boundary layer zones where the boundary layer correctors' contribution is essential. The last chapter is a brief bibliographical review on the adjacent topics.

The book is accessible for a wide range of readers: specialists in engineering, applied mathematicians working in fluid mechanics as well as in applications to biophysics and medicine, and master and PhD students in mathematics and mechanics.

The "users" of methods who are not interested in reading the proofs may pass directly to the description of algorithms.

In the book we use generally triplet numbering of formulas and propositions (number of chapter, number of section, number of formula or proposition within the section).

Vilnius, Lithuania
September, 2023

Grigory Panasenko
Konstantin Pileckas

Contents

Chapter 1
Introduction

1.1 Asymptotic Methods for Elliptic Equations in Thin Domains with Dirichlet Boundary Conditions

This section presents an overview of the methodology used to construct asymptotic expansions for solutions of elliptic equations and stationary Stokes and Navier–Stokes equations within thin tube structures, which serve as models for blood vessel networks. These structures are composed of finite unions of thin cylinders. We begin by introducing the concept of thin tube structures and formulating a model problem. It is important to note that the construction of asymptotic expansions for solutions of Stokes or Navier–Stokes equations, subject to no-slip boundary conditions, is more technically intricate compared to the case of the Laplace equation. Nevertheless, the fundamental ideas underlying the process remain similar. As a result, we focus on addressing Dirichlet's problem for the Laplace equation in the context of thin tube structures.

For our simplified model problems, we develop the asymptotic expansion of the solution and validate it by computing the residual. Furthermore, we establish the existence, uniqueness, and a priori estimation of the solution. Additionally, we evaluate the discrepancy between the exact solution and its asymptotic approximation. Throughout this work, we adopt conventional notations and draw upon well-established theorems of functional analysis. While we provide a brief introduction to these notations and theorems in this introductory section, a more comprehensive review will be presented in Chap. 2. In that chapter, we will also introduce functional spaces and formulate the primary theorems of functional analysis that will be utilized in subsequent chapters.

G. Panasenko, K. Pileckas, *Multiscale Analysis of Viscous Flows in Thin Tube Structures*, Advances in Mathematical Fluid Mechanics,
https://doi.org/10.1007/978-3-031-54630-3_1

1.1.1 Dimensional Reduction for the Poisson Equation in a Thin Rectangle

We begin by considering the simplest formulation of the boundary value problem for partial differential equations within a thin domain. Specifically, we examine the Dirichlet problem for the Laplacian in a thin rectangle denoted as $G_\varepsilon = (0, 1) \times (0, \varepsilon)$:

$$\begin{cases} -\Delta u_\varepsilon = f(x_1), \; x \in G_\varepsilon \,, \\ \quad u_\varepsilon = 0, \qquad x \in \partial G_\varepsilon, \end{cases} \tag{1.1.1}$$

where $f \in C^\infty([0, 1])$.

1.1.1.1 Construction of High-Order Asymptotic Approximations for the Case $f \in C_0^\infty([0, 1])$

We aim to construct an approximate solution for small values of ε. The process of constructing asymptotic approximations does not require any specialized mathematical prerequisites, so we present it directly. However, to establish the existence of a unique solution for the boundary value problem (1.1.1) and to evaluate the error of the constructed approximation, we will need some results from functional analysis. These necessary facts are briefly summarized below, with more extensive information available in Chap. 2.

Let us proceed with constructing the asymptotic expansion for the solution of (1.1.1) as the small parameter ε tends to zero. We assume that $f \in C_0^\infty([0, 1])$. Here, $C_0^\infty([0, 1])$ represents the space of infinitely differentiable functions defined on the interval [0, 1], which vanish in some neighborhood of the interval's ends.

We seek an asymptotic solution in the form:

$$u_\varepsilon^{(J)} = \sum_{l=0}^{J} \varepsilon^{l+2} u_l \left(x_1, \frac{x_2}{\varepsilon}\right), \tag{1.1.2}$$

where $u_l \in C^2([0, 1] \times [0, 1])$ and J is an even integer with $J \geq 0$. By substituting $u_\varepsilon^{(J)}$ into (1.1.1), we obtain the following expression:

$$
\begin{aligned}
-\Delta u_\varepsilon^{(J)} &= -\frac{\partial^2}{\partial x_1^2}u_\varepsilon^{(J)} - \frac{\partial^2}{\partial x_2^2}u_\varepsilon^{(J)} \\
&= \left\{ -\sum_{l=0}^{J} \varepsilon^{l+2}\frac{\partial^2 u_l}{\partial x_1^2}(x_1,\xi_2) - \sum_{l=0}^{J} \varepsilon^{l+2}\varepsilon^{-2}\frac{\partial^2 u_l}{\partial \xi_2^2}(x_1,\xi_2) \right\}\Bigg|_{\xi_2=\frac{x_2}{\varepsilon}} \\
&= \left\{ -\sum_{l=0}^{J} \varepsilon^{l}\frac{\partial^2 u_l}{\partial \xi_2^2}(x_1,\xi_2) - \sum_{l'=2}^{J+2} \varepsilon^{l'}\frac{\partial^2 u_{l'-2}}{\partial x_1^2}(x_1,\xi_2) \right\}\Bigg|_{\xi_2=\frac{x_2}{\varepsilon},l'=l+2} \\
&= -\sum_{l=0}^{J}\varepsilon^l\left(\frac{\partial^2 u_l}{\partial \xi_2^2} + \frac{\partial^2 u_{l-2}}{\partial x_1^2}\right) - \varepsilon^{J+1}\frac{\partial^2 u_{J-1}}{\partial x_1^2} - \varepsilon^{J+2}\frac{\partial^2 u_J}{\partial x_1^2},
\end{aligned}
$$

where $u_l = 0$ if $l < 0$. In this derivation, we utilized the evident formula for changing the subscript in the sum:

$$
\sum_{l=0}^{N} a_l = \sum_{l'=M}^{N+M} a_{l'-M} = \sum_{l=M}^{N+M} a_{l-M},
$$

where $l' = l + M$ and $l = l' - M$.

Equating this expansion to the right-hand side $f(x_1)$ leads to a series of problems for u_l:

$$
\begin{cases}
-\dfrac{\partial^2}{\partial \xi_2^2}u_l(x_1,\xi_2) = \dfrac{\partial^2}{\partial x_1^2}u_{l-2}(x_1,\xi_2) + f(x_1)\delta_{l0}, & \xi_2 \in (0,1), \\
\quad u(x_1,\xi_2) = 0, & \xi_2 = 0, \quad \xi_2 = 1,
\end{cases}
\tag{1.1.3}
$$

where δ_{ij} is the Kronecker delta. For $l = 0$, the solution is given by:

$$
u_0(x_1, x_2) = \frac{1}{2}x_2(x_2 - 1)(-f(x_1)).
$$

One can directly verify by induction that $u_l = 0$ for odd values of l. By calculating the residual on the right-hand side, we obtain the following expression:

$$
\begin{cases}
-\Delta u_\varepsilon^{(J)} = f(x_1) + r_\varepsilon(x), & x \in G_\varepsilon, \\
\quad u_\varepsilon^{(J)} = 0, & x \in \partial G_\varepsilon,
\end{cases}
\tag{1.1.4}
$$

where

$$
r_\varepsilon = \varepsilon^{J+2}\frac{\partial^2 u_J}{\partial x_1^2} = O(\varepsilon^{J+2}). \tag{1.1.5}
$$

The vanishing of $u_\varepsilon^{(J)}$ on ∂G_ε follows from the condition $f \in C_0^\infty([0, 1])$. This can be proven by induction using (1.1.3).

Thus, we have constructed the J-th order approximation of the solution. In order to prove its proximity to the exact solution, we require some prerequisites from functional analysis.

1.1.2 Justification of the Asymptotic Expansion

To establish the validity of the error estimate and prove the existence of the solution, we need to revisit certain definitions. In our analysis, we utilize the spaces $L^2(G)$ and $W^{1,2}(G)$, where G represents a domain in $\mathbb{R}^n$. The space $L^2(G)$ is a Hilbert space of real-valued functions u defined on G, which possess a finite Lebesgue integral $\int_G u^2(x)dx$. It is equipped with an inner product $(u, w)_{L^2(G)} = \int_G u(x)w(x)dx$ and a norm $\|u\|_{L^2(G)} = \sqrt{(u, u)_{L^2(G)}}$. We write $u \in C_0^\infty(G)$ if u is an infinitely differentiable function that vanishes outside some subdomain $G' \subset\subset G$ and $u \in C^\infty(G)$ if u is infinitely differentiable function in G extebdable to $\mathbb{R}^n$. Moreover, if u belongs to $L^2(G)$ and there exists a function $u_i \in L^2(G)$ satisfying the relation

$$\int\limits_G u(x)\frac{\partial v}{\partial x_i}dx = -\int\limits_G u_i(x)v(x)dx$$

for all functions $v \in C_0^\infty(G)$, then u_i is referred to as a weak partial derivative and denoted by $\dfrac{\partial u}{\partial x_i}$.

The Sobolev space $W^{1,2}(G)$ is the space of functions in $L^2(G)$ having all weak partial derivatives $\dfrac{\partial u}{\partial x_i} \in L^2(G)$, where $i = 1, \dots, n$. It is endowed with an inner product $(u, w)_{W^{1,2}(G)} = \int_G \big(u(x)w(x) + \nabla u \cdot \nabla w\big)dx$ and a norm $\|u\|_{W^{1,2}(G)} = \sqrt{(u, u)_{W^{1,2}(G)}}$. The spaces $L^2(G)$ and $W^{1,2}(G)$ can also be defined as the closures of the space $C^\infty(G)$ with respect to the norms $\|u\|_{L^2(G)}$ and $\|u\|_{W^{1,2}(G)}$, respectively. Additionally, we introduce the space $\mathring{W}^{1,2}(G)$, which is the closure of $C_0^\infty(G)$ with respect to the norm $\|\cdot\|_{W^{1,2}(G)}$.

We define a weak solution of problem (1.1.1) as a function $u_\varepsilon \in \mathring{W}^{1,2}(G_\varepsilon)$ that satisfies the following identity for any test function $v \in \mathring{W}^{1,2}(G_\varepsilon)$:

$$\int\limits_{G_\varepsilon} \nabla u_\varepsilon \cdot \nabla v dx = \int\limits_{G_\varepsilon} f(x_1)v(x)dx. \tag{1.1.6}$$

To prove the existence of the solution, we employ Poincaré inequality, which holds for any function $w \in \mathring{W}^{1,2}(G_\varepsilon)$ and is given by:

$$\|w\|_{L^2(G_\varepsilon)} \le C\varepsilon\|\nabla w\|_{L^2(G_\varepsilon)}$$

where the constant C is independent of ε.

Let H be a Hilbert space of functions in $\mathring{W}^{1,2}(G_\varepsilon)$, endowed with a new inner product

$$(w, v)_H = \int\limits_{G_\varepsilon} \nabla w(x) \cdot \nabla v(x) dx.$$

Then, (1.1.6) can be written as

$$(u_\varepsilon, v)_H = (f, v)_{L^2(G_\varepsilon)}. \tag{1.1.7}$$

Considering the linear functional $\Phi : H \to \mathbb{R}^1$ defined by $\Phi(v) = (f, v)_{L^2(G_\varepsilon)}$, we note that due to Poincaré inequality, it is a continuous functional on H. Therefore, we can apply the Riesz representation theorem, which states that there exists a unique element $u_\varepsilon \in H$ such that for all $v \in H$, $\Phi(v) = (u_\varepsilon, v)_H$. This element $u_\varepsilon \in H$ is the unique weak solution of problem (1.1.1). By taking $v = u_\varepsilon$ in (1.1.7) and applying the Cauchy-Schwarz-Bunyakovsky inequality $|(u, v)_{L^2(G_\varepsilon)}| \le \|u\|_{L^2(G_\varepsilon)}\|v\|_{L^2(G_\varepsilon)}$, followed by the Poincaré inequality, we obtain the a priori estimate

$$\|u_\varepsilon\|_{W^{1,2}(G_\varepsilon)} \le C\varepsilon\|f\|_{L^2(G_\varepsilon)}. \tag{1.1.8}$$

It is important to note that the a priori estimate (1.1.8) is valid not only for functions f dependent on one variable x_1 but also for all $f \in L^2(G_\varepsilon)$.

This estimate serves as the basis for justifying the asymptotic expansion (1.1.2). By subtracting problem (1.1.1) from (1.1.4), we find that the difference $w = u_\varepsilon - u_\varepsilon^{(J)}$ satisfies the following problem:

$$\begin{cases} -\Delta w = r_\varepsilon, \ x \in G_\varepsilon, \\ \quad w = 0, \ \ x \in \partial G_\varepsilon. \end{cases} \tag{1.1.9}$$

Applying estimate (1.1.8) and taking into account (1.1.5) and the definition of the $L^2(G_\varepsilon)$-norm, we derive the estimate:

$$\|u_\varepsilon - u_\varepsilon^{(J)}\|_{W^{1,2}(G_\varepsilon)} = O\left(\varepsilon^{J+2}\sqrt{\text{mes}G_\varepsilon}\right). \tag{1.1.10}$$

Furthermore, this estimate can be improved. For even J, replacing $u_\varepsilon^{(J)}$ by $u_\varepsilon^{(J+2)}$, we have:

$$\|u_\varepsilon - u_\varepsilon^{(J+2)}\|_{W^{1,2}(G_\varepsilon)} = O\left(\varepsilon^{J+4}\sqrt{\text{mes}G_\varepsilon}\right). \tag{1.1.11}$$

On the other hand, we find:

$$\|u_\varepsilon^{(J)} - u_\varepsilon^{(J+2)}\|_{W^{1,2}(G_\varepsilon)} = O\left(\varepsilon^{J+3}\sqrt{\text{mes}G_\varepsilon}\right). \tag{1.1.12}$$

Combining the estimates (1.1.11) and (1.1.12) and using the triangle inequality, we obtain:

$$\|u_\varepsilon - u_\varepsilon^{(J)}\|_{W^{1,2}(G_\varepsilon)} = O\left(\varepsilon^{J+3}\sqrt{\text{mes}G_\varepsilon}\right) . \tag{1.1.13}$$

This concludes the justification of the asymptotic expansion (1.1.2) and the error estimate for the problem (1.1.1).

1.1.3 Construction of High-Order Asymptotic Approximations for $f \in C^\infty([0, 1])$

In this section, we consider the scenario where the function $f(x_1)$ is infinitely differentiable on the interval [0, 1]. Additionally, let us assume that $f(x_1) = 0$ in some neighborhood of $x_1 = 1$, but $f(0) \neq 0$. In such cases, the boundary condition at $\gamma_0 = \partial G_\varepsilon \cap \{x_1 = 0\}$ is not satisfied. To rectify this, we introduce the concept of a boundary layer corrector denoted as $u_\varepsilon^{[BL,J]}$ given by the asymptotic expansion:

$$u_\varepsilon^{[BL,J]} = \sum_{l=0}^{J} \varepsilon^{l+2} u_l^{[BL]}\left(\frac{x_1}{\varepsilon}, \frac{x_2}{\varepsilon}\right) , \tag{1.1.14}$$

where $u_l^{[BL]}|_{x_1=0}$ compensates the trace $u_l|_{x_1=0}$:

$$u_l^{[BL]}|_{x_1=0} = -u_l|_{x_1=0} .$$

Thus, we encounter a chain of problems for $u_l^{[BL]}$ in the dilated variables ξ belonging to the half-strip $\Pi = \mathbb{R}_+ \times (0, 1)$ (see Fig. 1.1), expressed as:

$$\begin{cases} -\Delta_\xi u_l^{[BL]} = 0, & \xi \in \Pi, \\ u_l^{[BL]} = 0, & \xi_2 = 0, \quad \xi_2 = 1, \\ u_l^{[BL]}(0, \xi_2) = -u_l(0, \xi_2). \end{cases} \tag{1.1.15}$$

To demonstrate that the solutions of (1.1.15) decay exponentially as $\xi_1 \to +\infty$, we employ the Fourier series technique. Specifically, we assume:

Fig. 1.1 Domain of definition of boundary layer problem: half-strip

$$-u_l(0,\xi_2)=\sum_{n=1}^{\infty} b_n \sin(\xi_2\pi n),$$

and then seek $u_l^{[BL]}(\xi_1,\xi_2)$ in the form of:

$$u_l^{[BL]}(\xi_1,\xi_2)=\sum_{n=1}^{\infty} B_n(\xi_1)\sin(\xi_2\pi n)\ ,$$

which leads to the following ordinary differential equations for the coefficients B_n:

$$\begin{cases} -B_n''+(\pi n)^2 B_n &= 0, \quad \xi_1\in\mathbb{R}_+,\\ B_n(0) &= b_n,\\ B_n(\xi_1)\to 0 & \text{as } \xi_1\to+\infty. \end{cases}$$

The solution to these equations is given by $B_n=b_n e^{-\pi n\xi_1}$, resulting in the final expression for $u_l^{[BL]}$ as follows:

$$u_l^{[BL]}(\xi_1,\xi_2)=\sum_{n=1}^{\infty} b_n e^{-\pi n\xi_1}\sin(\xi_2\pi n)\ . \tag{1.1.16}$$

Hence, we have established the theorem of exponential stabilization of the solution at infinity, also known as the theorem of Phrägmen–Lindelöf type:

$$\exists C_1>0:\ \left|u_l^{[BL]}(\xi)\right|,\left|\nabla u_l^{[BL]}(\xi)\right|\le C_1 e^{-\pi\xi_1}\ .$$

To obtain an error estimate, we introduce the "corrected" asymptotic solution as:

$$u_\varepsilon^{(a,J)}=u_\varepsilon^{(J)}\left(x_1,\frac{x_2}{\varepsilon}\right)+u_\varepsilon^{[BL,J]}\left(\frac{x}{\varepsilon}\right)\eta(3x_1),$$

where $\eta \in C^2(\mathbb{R})$ is a cut-off function satisfying:

$$\eta(t) = \begin{cases} 1 \text{ if } |t| \le 1, \\ 0 \text{ if } |t| \ge 2. \end{cases}$$

Calculating the new residual

$$R_\varepsilon = r_\varepsilon - \left(\Delta u_\varepsilon^{[BL,J]} \eta(3x_1) + 2\frac{\partial u_\varepsilon^{[BL,J]}}{\partial x_1} \eta'(3x_1) + u_\varepsilon^{[BL,J]} \eta''(3x_1) \right),$$

and taking into account that the support of the terms with η belongs to the zone where the boundary layers are exponentially small, we obtain the following estimate in the $L^2(G_\varepsilon)$-norm:

$$||R_\varepsilon||_{L^2(G_\varepsilon)} = O\left(\varepsilon^{J+2}\sqrt{\text{mes } G_\varepsilon}\right).$$

As in the previous case, we obtain an error estimate:

$$\left\| u_\varepsilon - \left(u_\varepsilon^{(J)} + u_\varepsilon^{[BL,J]}\eta\right) \right\|_{W^{1,2}(G_\varepsilon)} = O\left(\varepsilon^{J+2}\sqrt{\text{mes} G_\varepsilon}\right),$$

which can be improved for even J as:

$$\left\| u_\varepsilon - \left(u_\varepsilon^{(J)} + u_\varepsilon^{[BL,J]}\eta\right) \right\|_{W^{1,2}(G_\varepsilon)} = O\left(\varepsilon^{J+3}\sqrt{\text{mes} G_\varepsilon}\right).$$

Additionally, we find a positive number δ such that for all x satisfying $\text{dist}(x, \gamma_0) \le \delta$, the boundary layer corrector $u_\varepsilon^{[BL,J]}(\frac{x}{\varepsilon})$ has a "tail" smaller than ε^J (see Fig. 1.2). We use the estimate

$$\left|u_\varepsilon^{[BL,J]}(x)\right| \le C_1 e^{-\pi x_1/\varepsilon},$$

to conclude that it is smaller than ε^J for $x_1 \ge \delta$, leading to the following formula for δ:

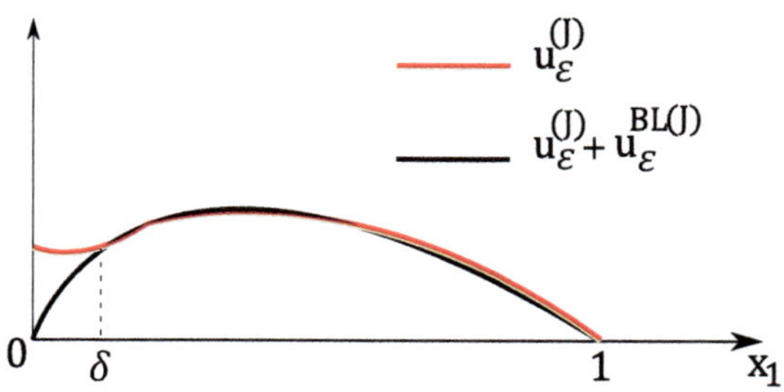

Fig. 1.2 Comparison of asymptotics with and without boundary layer

$$\delta = \text{const}\, J\varepsilon |\ln \varepsilon| \ .$$

Finally, it is important to note that the same approach can be applied when f does not vanish neither in 0 nor in 1.

1.2 Dirichlet's Problem for the Laplacian in a Thin Tube Structure

In this section, we investigate a "toy" problem related to the Laplacian in a thin tube structure. The domain consists of two regions: $B_\varepsilon = B_\varepsilon^1 \cup B_\varepsilon^2$, where $B_\varepsilon^1 = (-1, 1) \times \left(-\frac{\varepsilon}{2}, \frac{\varepsilon}{2}\right)$, and $B_\varepsilon^2 = \left(-\frac{\varepsilon}{2}, \frac{\varepsilon}{2}\right) \times (0, 1)$.

The problem under consideration is given by:

$$\begin{cases} -\Delta u_\varepsilon = 1, \ x \in B_\varepsilon, \\ u_\varepsilon = 0, \ x \in \partial B_\varepsilon. \end{cases} \tag{1.2.1}$$

This setting imitates a viscous flow in a T-shaped structure (see Fig. 1.3). While the real-life setting corresponds to the Navier–Stokes or Stokes equations, the main ideas can be effectively explained in this simplified setup.

The asymptotic expansion construction proceeds as follows:

Step 1. Define the "Poiseuille"-type solutions:

$$\begin{aligned} U_\varepsilon^1(x) &= -\frac{1}{2}\left(x_2^2 - \frac{\varepsilon^2}{4}\right) \quad \text{in } B_\varepsilon^1, \\ U_\varepsilon^2(x) &= -\frac{1}{2}\left(x_1^2 - \frac{\varepsilon^2}{4}\right) \quad \text{in } B_\varepsilon^2. \end{aligned} \tag{1.2.2}$$

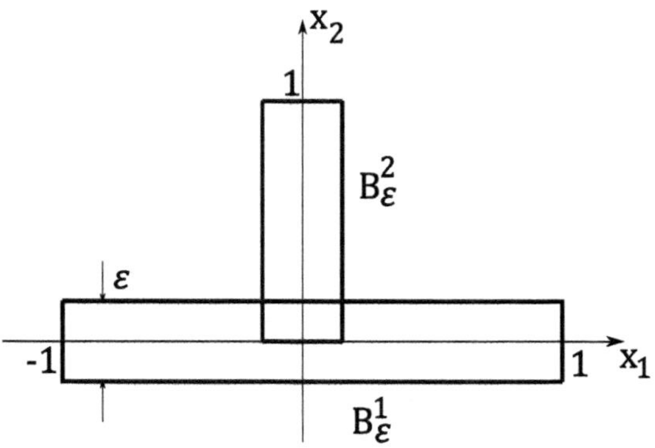

Fig. 1.3 T-shaped structure

Step 2. Multiply U_ε^1 and U_ε^2 by the cut-off functions:

$$u_\varepsilon^{12}(x) = U_\varepsilon^1(x)\left(1-\eta\left(\frac{x_1}{3\varepsilon}\right)\right) + U_\varepsilon^2(x)\left(1-\eta\left(\frac{x_2}{3\varepsilon}\right)\right) . \tag{1.2.3}$$

Substituting this expression into the equation, we obtain:

$$-\Delta u_\varepsilon^{12} = 1 + F\left(\frac{x}{\varepsilon}\right) ,$$

where the function $F(\xi)$ is defined as follows:

$$F(\xi) = \begin{cases} 1-\eta\left(\frac{\xi_1}{3}\right) - \frac{1}{18}\left(\xi_2^2 - \frac{1}{4}\right)\eta''\left(\frac{\xi_1}{3}\right) + 1- \\ -\eta\left(\frac{\xi_2}{3}\right) - \frac{1}{18}\left(\xi_1^2 - \frac{1}{4}\right)\eta''\left(\frac{\xi_2}{3}\right) - 1, & \xi \in (-6,6)^2 \cap \Pi, \\ 0, & \xi \in \Pi \setminus (-6,6)^2. \end{cases} \tag{1.2.4}$$

Here, $\Pi = \mathbb{R} \times \left(-\frac{1}{2}, \frac{1}{2}\right) \bigcup \left(\frac{1}{2}, \frac{1}{2}\right) \times \mathbb{R}_+$.

Step 3. To compensate the residual F, we introduce $u^{BL}(\xi)$, the solution to the boundary layer problem:

$$\begin{cases} -\Delta_\xi U^{BL}(\xi) = -F(\xi),\ \xi \in \Pi, \\ \quad U^{BL}(\xi) = 0, \qquad \xi \in \partial\Pi. \end{cases}$$

It is evident that there exists a unique solution in $\mathring{W}^{1,2}(\Pi)$ and $|u^{BL}|, |\nabla u^{BL}| \leq C_1 e^{-\pi|\xi|}$, where $C_1 > 0$.

Step 4. Construct boundary layer correctors as before to compensate the traces of u_ε^{12} on:

$$(a)\quad \gamma^{(-1,0)} = \left\{x_1 = -1, x_2 \in \left(-\frac{\varepsilon}{2}, \frac{\varepsilon}{2}\right)\right\} :$$

find $U^{[BL(-1,0)]}(\xi)$ defined on the semi-strip $\Pi^{(-1,0)} = \mathbb{R}_+ \times \left(-\frac{1}{2}, \frac{1}{2}\right)$ and satisfying

$$\begin{cases} -\Delta_\xi U^{[BL(-1,0)]} = 0, & \xi \in \Pi^{(-1,0)}, \\ U^{[BL(-1,0)]}(\xi) = 0, & \xi \in \partial\Pi^{(1,0)}\setminus\{\xi_1 = 0\}, \\ U^{[BL(-1,0)]}(\xi) = \frac{1}{2}(\xi_2^2 - \frac{1}{4}),\ \xi_1 = 0; \end{cases}$$

on

$$(b)\quad \gamma^{(1,0)} = \left\{x_1 = 1, x_2 \in \left(-\frac{\varepsilon}{2}, \frac{\varepsilon}{2}\right)\right\}:$$

find $U^{[BL(1,0)]}(\xi)$ defined on a semi-strip $\Pi^{(1,0)} = \mathbb{R}_- \times \left(-\frac{1}{2}, \frac{1}{2}\right)$ satisfying

$$\begin{cases} -\Delta_\xi U^{[BL(1,0)]} = 0, & \xi \in \Pi^{(-1,0)}, \\ U^{[BL(1,0)]}(\xi) = 0, & \xi \in \partial\Pi^{(1,0)} \setminus \{\xi_1 = 0\}, \\ U^{[BL(1,0)]}(\xi) = \frac{1}{2}(\xi_2^2 - \frac{1}{4}),\ \xi_1 = 0; \end{cases}$$

on

$$(c)\quad \gamma^{(0,1)} = \left\{x_1 \in \left(-\frac{\varepsilon}{2}, \frac{\varepsilon}{2}\right), x_2 = 1\right\},$$

find $U^{[BL(0,1)]}(\xi)$ defined on a semi-strip $\Pi^{(0,1)} = \left(-\frac{1}{2}, \frac{1}{2}\right) \times \mathbb{R}_-$ satisfying

$$\begin{cases} -\Delta_\xi U^{[BL(0,1)]} &= 0, & \xi \in \Pi^{(0,1)}, \\ U^{[BL(0,1)]}(\xi) &= 0, & \xi \in \partial\Pi^{(0,1)} \setminus \{\xi_2 = 0\}, \\ U^{[BL(0,1)]}(\xi) &= \frac{1}{2}(\xi_1^2 - \frac{1}{4}),\ \xi_2 = 0. \end{cases}$$

The final asymptotic approximation is given by:

$$\begin{aligned} u_\varepsilon^a(x) = {} & U_\varepsilon^1(x)\left(1 - \eta\left(\frac{x_1}{3\varepsilon}\right)\right) + U_\varepsilon^2(x)\left(1 - \eta\left(\frac{x_2}{3\varepsilon}\right)\right) \\ & + U^{BL}\left(\frac{x}{\varepsilon}\right)\eta(3x_1)\eta(3x_2) \\ & + U^{BL(-1,0)}\left(\frac{x_1+1}{\varepsilon}, \frac{x_2}{\varepsilon}\right)\eta(3(x_1+1)) \\ & + U^{BL(1,0)}\left(\frac{x_1-1}{\varepsilon}, \frac{x_2}{\varepsilon}\right)\eta(3(x_1-1)) \\ & + U^{BL(0,1)}\left(\frac{x_1}{\varepsilon}, \frac{x_2-1}{\varepsilon}\right)\eta(3(x_2-1)). \end{aligned}$$

All boundary layers decay exponentially, as proven by the Phrägmen–Lindelöf theorems, which are established by Fourier analysis for all semi-strips in the boundary layer problems.

By calculating the residual, we observe that it is of order $O(\varepsilon^{J+2}\sqrt{\text{mes } B_\varepsilon})$ in L^2-norm, and applying the a priori estimate, we obtain:

$$\|u_\varepsilon^{(a,J)} - u_\varepsilon\|_{W^{1,2}(B_\varepsilon)} = O(\varepsilon^{J+2}\sqrt{\text{mes } B_\varepsilon}).$$

1.3 Method of Asymptotic Partial Decomposition of Domain for a T-Shaped Domain

In this section, we present the method of asymptotic partial decomposition of domain, specifically designed for analyzing a T-shaped domain denoted as B_ε. The main objective is to decompose B_ε into several subdomains to efficiently solve the governing equation with appropriate boundary conditions. The decomposition is expressed in the form:

$$B_\varepsilon = B^1_{\varepsilon\delta} \cup B^2_{\varepsilon\delta} \cup B^{(0,0)}_{\varepsilon\delta} \cup B^{(-1,0)}_{\varepsilon\delta} \cup B^{(1,0)}_{\varepsilon\delta} \cup B^{(0,1)}_{\varepsilon\delta},$$

where the subdomains are defined as follows:

$$\begin{aligned}
B^1_{\varepsilon\delta} &= (-1+\delta, -\delta) \times \left(-\tfrac{\varepsilon}{2}, \tfrac{\varepsilon}{2}\right) \cup (\delta, 1-\delta) \times \left(-\tfrac{\varepsilon}{2}, \tfrac{\varepsilon}{2}\right), \\
B^2_{\varepsilon\delta} &= \left(-\tfrac{\varepsilon}{2}, \tfrac{\varepsilon}{2}\right) \times (\delta, 1-\delta), \\
B^{(0,0)}_{\varepsilon\delta} &= (-\delta, \delta) \times \left(-\tfrac{\varepsilon}{2}, \tfrac{\varepsilon}{2}\right) \cup \left(-\tfrac{\varepsilon}{2}, \tfrac{\varepsilon}{2}\right) \times (0, \delta), \\
B^{(-1,0)}_{\varepsilon\delta} &= (-1, -1+\delta) \times \left(-\tfrac{\varepsilon}{2}, \tfrac{\varepsilon}{2}\right), \\
B^{(1,0)}_{\varepsilon\delta} &= (1-\delta, 1) \times \left(-\tfrac{\varepsilon}{2}, \tfrac{\varepsilon}{2}\right), \\
B^{(0,1)}_{\varepsilon\delta} &= \left(-\tfrac{\varepsilon}{2}, \tfrac{\varepsilon}{2}\right) \times (1-\delta, 1).
\end{aligned}$$

See Fig. 1.4 for an illustration of the asymptotic domain decomposition of the T-shaped structure.

Consider the approximation u_d, which is a solution to the equation

$$-\Delta u_d = 1$$

in the various subdomains with specific boundary conditions:

(a) In $B^{(0,0)}_{\varepsilon\delta}$, the boundary conditions are:

$$\begin{aligned}
u_d &= 0, \quad x \in \partial B_\varepsilon \cap \partial B^{(0,0)}_{\varepsilon\delta}, \\
u_d &= U^1_\varepsilon, \quad x \in \partial B^{(0,0)}_{\varepsilon\delta} \cap \{x_1 = \pm\delta\}, \\
u_d &= U^2_\varepsilon, \quad x \in \partial B^{(0,0)}_{\varepsilon\delta} \cap \{x_2 = \delta\}.
\end{aligned}$$

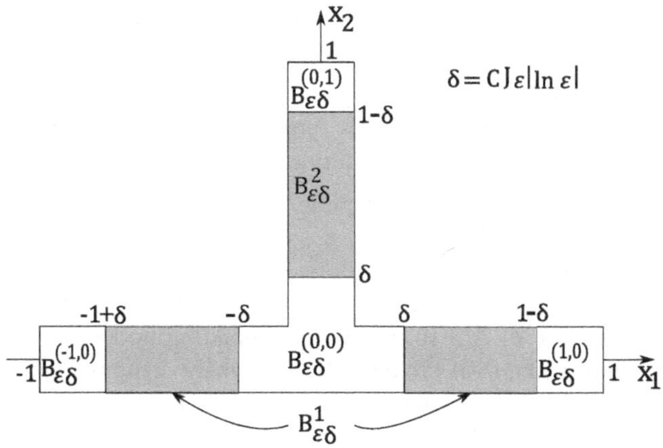

Fig. 1.4 Asymptotic domain decomposition of T-shaped structure

(b) In $B_{\varepsilon\delta}^{(\pm 1,0)}$, the boundary conditions are:

$$u_d = 0, \quad x \in \partial B_\varepsilon \cap \partial B_{\varepsilon\delta}^{(+1,0)},$$

$$u_d = U_\varepsilon^1, \quad x \in \partial B_{\varepsilon\delta}^{(\pm 1,0)} \cap \{x_1 = \pm(1-\delta)\}.$$

(c) In $B_{\varepsilon\delta}^{(0,1)}$, the boundary condition is:

$$u_d = 0, \quad x \in \partial B_\varepsilon \cap \partial B_{\varepsilon\delta}^{(0,1)},$$

$$u_d = U_\varepsilon^2, \quad x \in \{x_2 = 1-\delta\}.$$

Furthermore, we define

$$u_d = U_\varepsilon^1 \text{ in } B_{\varepsilon\delta}^1, \quad u_d = U_\varepsilon^2 \text{ in } B_{\varepsilon\delta}^2.$$

The main result of this method is an error estimate given by the following theorem:

Theorem 1.3.1 *Let $\delta = CJ\varepsilon|\ln\varepsilon|$, where C is a constant independent of ε. Then the error between the exact solution u_ε and the approximation u_d in the $W^{1,2}(B_\varepsilon)$ norm is bounded as:*

$$\|u_\varepsilon - u_d\|_{W^{1,2}(B_\varepsilon)} = O\left(\varepsilon^{J+2}\sqrt{\text{mes } B_\varepsilon}\right), \tag{1.3.1}$$

where J is a positive integer and mes B_ε *denotes the measure of the domain B_ε.*

This method effectively reduces the computational resources needed. The problem is broken down into independent subproblems in $B_{\varepsilon\delta}^{(0,0)}$, $B_{\varepsilon\delta}^{(\pm 1,0)}$, and $B_{\varepsilon\delta}^{(0,1)}$, which can be solved in parallel.

The proof of inequality (1.3.1) is predicated on the estimation of the difference $u_\varepsilon^a - u_d$, achieved by selecting an appropriate value for δ. This choice ensures that the boundary layer functions $U^{[BL]}\left(\frac{x}{\varepsilon}\right)$, $U^{[BL(-1,0)]}\left(\frac{x_1+1}{\varepsilon}, \frac{x_2}{\varepsilon}\right)$, $U^{[BL(1,0)]}\left(\frac{x_1-1}{\varepsilon}, \frac{x_2}{\varepsilon}\right)$, and $U^{BL(0,1)}\left(\frac{x_1}{\varepsilon}, \frac{x_2-1}{\varepsilon}\right)$, along with their first- and second-order derivatives, have magnitudes smaller than ε^{J+4} when evaluated at points x situated at a distance of $\delta/3$ from the corresponding nodes or vertices ($(0,0)$, $(-1,0)$, $(1,0)$, and $(0,1)$, respectively). As previously demonstrated, this value of δ is determined as $CJ\varepsilon|\ln\varepsilon|$, where C is a constant independent of ε. With this selection of δ, let us proceed to modify the factors η in the definition formula of $u_\varepsilon^{(a,J)}$ as follows:

$$
\begin{aligned}
\tilde{u}_\varepsilon^{(a,J)}(x) &= U_\varepsilon^1(x)\left(1-\eta\left(\tfrac{x_1}{3\varepsilon}\right)\right) + U_\varepsilon^2(x)\left(1-\eta\left(\tfrac{x_2}{3\varepsilon}\right)\right) \\
&+ U^{BL}\left(\tfrac{x}{\varepsilon}\right)\eta\left(\tfrac{3x_1}{\delta}\right)\eta\left(\tfrac{3x_2}{\delta}\right) \\
&+ U^{BL(-1,0)}\left(\tfrac{x_1+1}{\varepsilon}, \tfrac{x_2}{\varepsilon}\right)\eta\left(\tfrac{3(x_1+1)}{\delta}\right) \\
&+ U^{BL(1,0)}\left(\tfrac{x_1-1}{\varepsilon}, \tfrac{x_2}{\varepsilon}\right)\eta\left(\tfrac{3(x_1-1)}{\delta}\right) \\
&+ U^{BL(0,1)}\left(\tfrac{x_1}{\varepsilon}, \tfrac{x_2-1}{\varepsilon}\right)\eta\left(\tfrac{3(x_2-1)}{\delta}\right).
\end{aligned}
$$

Utilizing the aforementioned choice of δ, we can now compute the residual r_ε in the right-hand side of the Laplace equation:

$$
-\Delta\tilde{u}_\varepsilon^{(a,J)} = 1 + r_\varepsilon \tag{1.3.2}
$$

We will then prove that this residual term satisfies the following estimate:

$$
\|r_\varepsilon\|_{L^2(B_\varepsilon)} = O\left(\varepsilon^{J+2}\sqrt{\text{mes } B_\varepsilon}\right).
$$

It is important to note that $\tilde{u}_\varepsilon^{(a,J)}$ precisely satisfies Dirichlet's boundary condition on ∂B_ε.

Next, we subtract Equation (1.3.2) from $-\Delta u_d = 1$, leading to the following equation for the difference $w = \tilde{u}_\varepsilon^{(a,J)} - u_d$:

$$
-\Delta w = r_\varepsilon.
$$

Additionally, the condition $w = 0$ holds on the boundaries of small subdomains $B_{\varepsilon\delta}^{(0,0)}$, $B_{\varepsilon\delta}^{(-1,0)}$, $B_{\varepsilon\delta}^{(1,0)}$, and $B_{\varepsilon\delta}^{(0,1)}$. By applying the a priori estimate for the Laplace equation (similar to (1.1.8)), we are able to demonstrate that:

$$\|\tilde{u}_\varepsilon^{(a,J)} - u_d\|_{W^{1,2}(B_\varepsilon)} = O\left(\varepsilon^{J+2}\sqrt{\text{mes } B_\varepsilon}\right).$$

In light of the chosen value of δ, we can now evaluate the difference between the original asymptotic expansion $u_\varepsilon^{(a,J)}$ and the modified asymptotic expansion $\tilde{u}_\varepsilon^{(a,J)}$. This yields the estimate:

$$\|\tilde{u}_\varepsilon^{(a,J)} - u_\varepsilon^{(a,J)}\|_{W^{1,2}(B_\varepsilon)} = O\left(\varepsilon^{J+2}\sqrt{\text{mes } B_\varepsilon}\right).$$

Finally, taking into account the obtained estimates of order $O\left(\varepsilon^{J+2}\sqrt{\text{mes } B_\varepsilon}\right)$ for the differences $\tilde{u}_\varepsilon^{(a,J)} - u_d$, $\tilde{u}_\varepsilon^{(a,J)} - u_\varepsilon^{(a,J)}$, and $u_\varepsilon^{(a,J)} - u_\varepsilon$ in the norm $W^{1,2}(B_\varepsilon)$, we finish the proof of the theorem.

This approach can be extended to encompass the case of the Stokes and Navier–Stokes equations in thin tube structures, facilitating the modeling of blood flow in a network of vessels.

1.4 Method of Asymptotic Partial Decomposition of Domain for Flows in a Tube Structure (Applications in Hemodynamics)

Motivated by the modeling of blood flow in a network of blood vessels, we investigate the Stokes equation in a tube structure. These domains consist of connected finite unions of thin finite cylinders (in the 2D case, thin rectangles). Each tube structure can be represented schematically by its graph, where the thickness of tubes approaches zero, reducing the tubes to segments.

1.4.1 Tube Structure and Graphs

Definition 1.4.1 Consider N different points $O_1, O_2, \ldots, O_N$ in $\mathbb{R}^n$, where $n = 2, 3$, and M closed segments $e_1, e_2, \ldots, e_M$ connecting pairs of these points (i.e., $e_j = \overline{O_{i_j} O_{k_j}}$ where $i_j, k_j \in \{1, \ldots, N\}, i_j \neq k_j$). All points O_i are supposed to be the ends of some segments e_j. The segments e_j are called edges of the graph. A point O_i is a node if it is the common end of at least two edges, and it is a vertex if it is the end of only one edge. Two edges e_j and e_i can only intersect at their common nodes. The set of vertices is assumed to be non-empty.

A graph $\mathcal{B} = \bigcup\limits_{j=1}^{M} e_j$ is defined as the union of edges, and it is assumed to be a connected set (see Fig. 1.5).

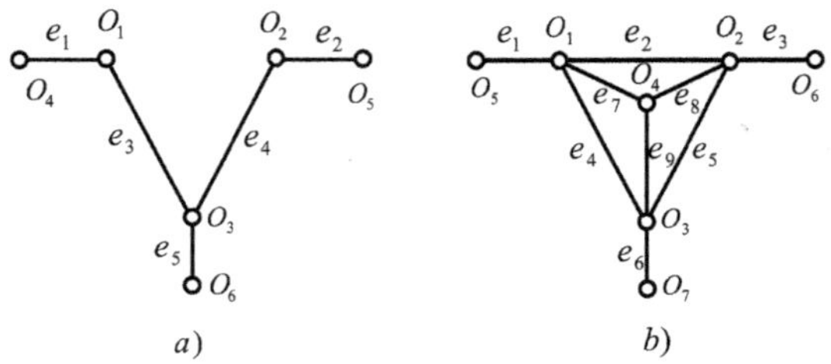

Fig. 1.5 Graphs of tube structures

Let e be an edge, denoted as $e = \overline{O_i O_j}$. Consider two Cartesian coordinate systems in $\mathbb{R}^n$. The first coordinate system has its origin at O_i, and the axis $O_i x_n^{(e)}$ aligns with the direction of the ray $[O_i O_j)$. On the other hand, the second coordinate system has its origin at O_j, with the axis $O_j \tilde{x}_n^{(e)}$ directed opposite to the ray $[O_j O_i)$.

We can choose either coordinate system, and for both cases, the local variable is denoted as x^e. It will be pointed out which end serves as the origin of the coordinate system.

Let us associate with each edge e_j a bounded domain $\sigma_j \subset \mathbb{R}^{d-1}$ with a C^2-smooth boundary $\partial\sigma_j$, where $j = 1, \dots, M$. For each edge $e_j = e$ and its corresponding domain $\sigma_j = \sigma^{(e)}$, we introduce the cylinder $\Pi_\varepsilon^{(e)}$ as follows:

$$\Pi_\varepsilon^{(e)} = \left\{ x^{(e)} \in \mathbb{R}^n : x_n^{(e)} \in (0, |e|), \frac{x^{(e)'}}{\varepsilon} \in \sigma^{(e)} \right\},$$

where $x^{(e)'} = (x_1^{(e)}, \dots, x_{n-1}^{(e)})$, $|e|$ represents the length of the edge e, and $\varepsilon > 0$ is a small parameter. Notably, the edges e_j, Cartesian coordinates of nodes and vertices O_j, as well as the domains σ_j, remain independent of ε.

Let $O_1, \dots, O_{N_1}$ be nodes and $O_{N_1+1}, \dots, O_N$ be vertices. Additionally, let $\omega^1, \dots, \omega^N$ be bounded domains in $\mathbb{R}^n$, independent of ε, and possessing Lipschitz boundaries $\partial\omega^j$. We define the nodal domains $\omega_\varepsilon^j = \{x \in \mathbb{R}^n : \dfrac{x - O_j}{\varepsilon} \in \omega^j\}$.

Definition 1.4.2 A *tube structure* refers to the following domain:

$$B_\varepsilon = \Big(\bigcup_{j=1}^{M} \Pi_\varepsilon^{(e_j)}\Big) \bigcup \Big(\bigcup_{j=1}^{N} \omega_\varepsilon^j\Big). \tag{1.4.1}$$

We assume that this domain is connected and its boundary ∂B_ε is C^2-regular.

Define the surfaces $\gamma_\varepsilon^{N_1+1}, \dots, \gamma_\varepsilon^N$ as the intersections of the entire boundary ∂B_ε with the boundaries of $\omega_\varepsilon^{N_1+1}, \dots, \omega_\varepsilon^N$, respectively. These surfaces are referred to as the inflow/outflow parts of the boundary, as in fluid flow models, they represent the fluid inlets and outlets within the domain B_ε.

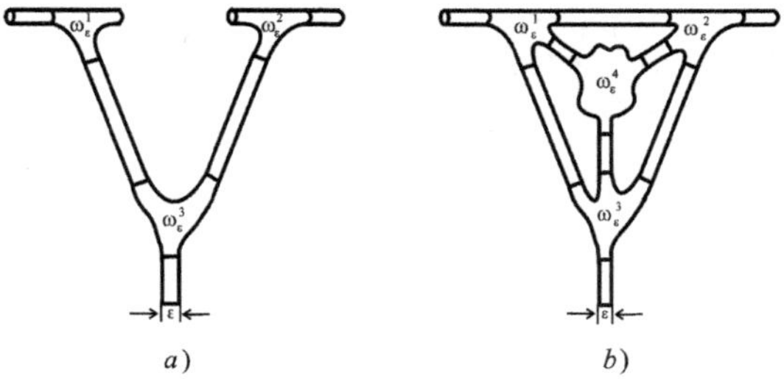

Fig. 1.6 Tube structures

Remark 1.4.1 Furthermore, it should be noted that in certain sections, instead of considering tube structures with a "completely smooth" boundary, tube structures with truncated inlets and outlets are examined. In this scenario, the smoothing domains ω_ε^j, $j = N_1 + 1, \ldots, N$ are not incorporated, and thus, B_ε is defined as:

$$B_\varepsilon = \Big(\bigcup_{j=1}^{M} \Pi_\varepsilon^{(e_j)}\Big) \bigcup \Big(\bigcup_{j=1}^{N_1} \omega_\varepsilon^j\Big) . \tag{1.4.2}$$

The surfaces $\gamma_\varepsilon^{N_1+1}, \ldots, \gamma_\varepsilon^N$ are the bases of the cylinders $\Pi_\varepsilon^{(e_{N_1}+1)}, \ldots, \Pi_\varepsilon^{(e_N)}$ associated with (i.e., containing) the vertices $O_{N_1+1}, \ldots, O_N$. As a result, the boundary ∂B_ε is generally C^2-regular, except for the corners at the junction of the lateral surfaces of the cylinders $\Pi_\varepsilon^{(e_{N_1}+1)}, \ldots, \Pi_\varepsilon^{(e_N)}$ and their bases $\gamma_\varepsilon^{N_1+1}, \ldots, \gamma_\varepsilon^N$ (see Fig. 1.6).

Remark 1.4.2 In addition, some further notations will be utilized in subsequent chapters. A semi-infinite dilated cylinder $\Pi_\infty^{(e)}$ is defined as $\Pi_\infty^{(e)} = \{x^{(e)} \in \mathbb{R}^n : x_n^{(e)} \in [0, \infty), x^{(e)\prime} \in \sigma^{(e)}\}$.

The notation for an edge e_k containing a vertex O_j as an end point is given as e_{O_j}. The domain σ^k associated with this edge is re-denoted as σ^{O_j}. It should be noted that the subscript k may differ from j. Moreover, r_1 represents the maximum diameter of domains ω^i, $i = 1, \ldots, N$, and let $r = r_1 + 1$.

Consider a node or vertex O_l and all edges e_j that have O_l as one of their end points. The union of these edges is referred to as a bundle of edges denoted as $\mathcal{B}_l$, that is, $\mathcal{B}_l = \bigcup_{j:O_l \in e_j} e_j$. A bundle of cylinders B_{O_l} refers to the union $\omega_\varepsilon^l \cup \Big(\bigcup_{j:O_l\in e_j} \Pi_\varepsilon^{(e_j)}\Big)$. Additionally, a half-bundle HB_{O_l} is defined as $HB_{O_l} = \omega_\varepsilon^l \cup \Big(\bigcup_{j:O_l\in e_j} \{x \in \Pi_\varepsilon^{(e_j)}, x_n^{(e_j)} \in [0, |e_j|/2]\}\Big)$. Furthermore, $\Omega_l = \omega^l \cup \Big(\bigcup_{j:O_l\in e_j} \Pi_\infty^{(e_j)}\Big)$ represents a bundle of dilated cylinders, i.e., a domain with cylindrical outlets to infinity.

The figure below represents the bundle of dilated cylinders (see Fig. 1.7) corresponding to the node O_4 of the structure presented in Fig. 1.6b.

Denote also $\gamma^l = \partial\omega^l \cap \partial\Omega_l$ and $\Omega_l^\varepsilon = \{x \in \mathbb{R}^n : x/\varepsilon \in \Omega_l\}$.

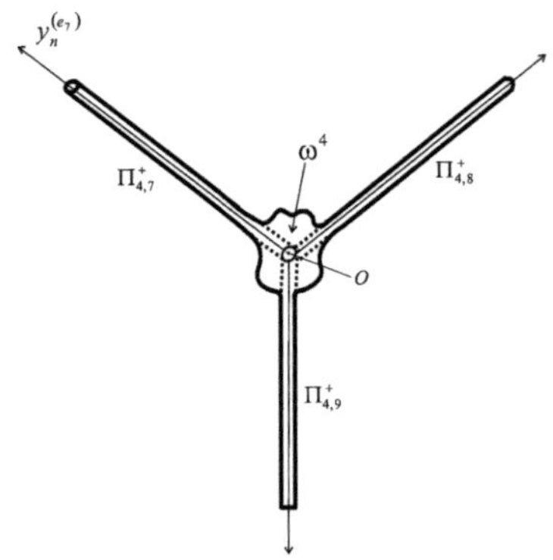

Fig. 1.7 Domain Ω_4 with three outlets to infinity corresponding to the node O_4

Moving on to the formulation of the problem, we consider the Stokes equations within the tube structure B_ε:

$$\begin{cases} -\nu\Delta\mathbf{v} + \nabla p = 0, \ x \in B_\varepsilon, \\ \operatorname{div} \mathbf{v} = 0, \ x \in B_\varepsilon, \\ \mathbf{v} = \mathbf{g}, \ x \in \partial B_\varepsilon. \end{cases} \tag{1.4.3}$$

Assume that the fluid velocity $\mathbf{g}$ at the boundary ∂B_ε has a specific structure: $\mathbf{g} = 0$ everywhere on ∂B_ε except for the sets $\gamma_\varepsilon^{N_1+1}, \ldots, \gamma_\varepsilon^N$:

$$\begin{aligned} &\mathbf{g}(x)\big|_{\sigma_\varepsilon^j} = \mathbf{g}^j\left(\tfrac{x-O_j}{\varepsilon}\right)\big|_{\gamma_\varepsilon^j}, \ j = N_1+1, \ldots, N, \\ &\mathbf{g}(x)\big|_{\partial B_\varepsilon \setminus \left(\bigcup\limits_{j=N_1+1}^{N} \gamma_\varepsilon^j\right)} = 0\,, \end{aligned} \tag{1.4.4}$$

Here, $\mathbf{g}^j$ represents a C^2-regular function with compact support in γ_ε^j. Furthermore, it is ensured that the integral $\int_{\partial B_\varepsilon} \mathbf{g} \cdot \mathbf{n} dS = 0$ and $\nu > 0$.

Now, we define a weak solution for problem (1.4.3) as a vector-valued function $\mathbf{v} \in W^{1,2}(B_\varepsilon)$ satisfying the conditions div $\mathbf{v} = 0$, $\mathbf{v}|_{\partial B_\varepsilon} = \mathbf{g}$, and such that for every $\mathbf{w} \in \mathring{W}^{1,2}(B_\varepsilon)$ with div$\mathbf{w} = 0$, the following integral identity holds:

$$\nu \int\limits_{B_\varepsilon} \nabla\mathbf{v} \cdot \nabla\mathbf{w} dx = 0.$$

Later, we present a chapter concerning the Stokes problem in various settings (see Chap. 4). In this section, we only describe the approach of the Method of Asymptotic Partial Domain Decomposition (MAPDD) for solving this problem. Its justification is similar to that of the Dirichlet boundary value problem for the Poisson equation but is more technical.

1.4.2 Partial Asymptotic Decomposition of the Domain for the Stokes Equation

Let us begin by introducing the Poiseuille flow in an infinite tube $\Pi_\varepsilon = \mathbb{R} \times \sigma_\varepsilon^j$. This flow represents an exact solution of the Stokes equations, given by the system of partial differential equations:

$$\begin{cases} -\nu\Delta \mathbf{v} + \nabla p = 0, \ x \in \Pi_\varepsilon, \\ \operatorname{div} \mathbf{v} = 0, \ x \in \Pi_\varepsilon, \\ \mathbf{v} = 0, \ x \in \partial\Pi_\varepsilon. \end{cases}$$

For the three-dimensional case ($n = 3$), the Poiseuille flow velocity $\mathbf{u}_{P_\alpha}(x)$ and the corresponding pressure $p_{P_\alpha}(x)$ are given by:

$$\mathbf{u}_{P_\alpha}(x) = \begin{pmatrix} 0 \\ 0 \\ \alpha u_0(x_1, x_2) \end{pmatrix}, \quad p_{P_\alpha}(x) = -\alpha x_3 + \beta, \tag{1.4.5}$$

where α and β are real constants and $u_0(x_1, x_2)$ satisfies:

$$\begin{cases} -\nu\Delta u_0 = 1, \ (x_1, x_2) \in \sigma_\varepsilon^j, \\ u_0 = 0, \ (x_1, x_2) \in \partial\sigma_\varepsilon^j. \end{cases}$$

For the two-dimensional case ($n = 2$), where $\sigma_\varepsilon^j = \left(-\frac{\varepsilon}{2}, \frac{\varepsilon}{2}\right)$, the expressions for $\mathbf{u}_{P_\alpha}(x)$ and $p_{P_\alpha}(x)$ are:

$$\mathbf{u}_{P_\alpha}(x) = \begin{pmatrix} 0 \\ \alpha u_0(x_1) \end{pmatrix}, \quad p_{P_\alpha}(x) = -\alpha x_2 + \beta, \quad u_0(x_1) = -\frac{1}{2\nu}\left(x_1^2 - \frac{\varepsilon^2}{4}\right). \tag{1.4.6}$$

It is important to note that in the case when σ_ε^j is a two-dimensional disk of radius $\varepsilon/2$, the expression for $u_0(x)$ becomes:

$$u_0(x) = -\frac{1}{4\nu}\left(r^2 - \frac{\varepsilon^2}{4}\right), \tag{1.4.7}$$

where r is the radial coordinate.

Let $\left(\tilde{\mathbf{u}}_P(x), \tilde{p}_P\right)$ represent the Poiseuille flow in another coordinate system, obtained by rotations, i.e., in the local coordinate system associated with e_j. It is worth mentioning that the Stokes and Navier–Stokes systems are invariant with respect to translations and rotations of the Cartesian coordinate system. Hence,

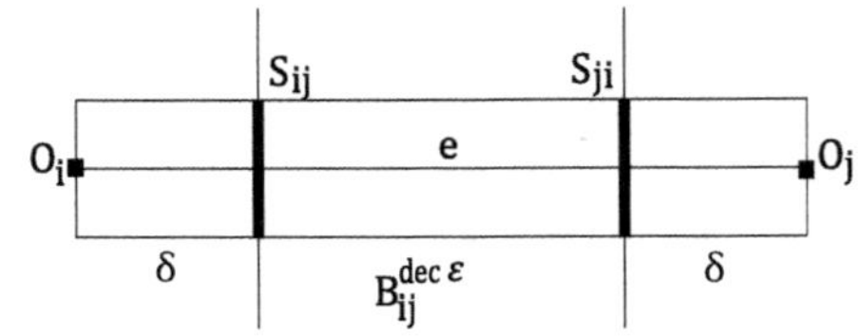

Fig. 1.8 Truncation of the cylinder $\Pi_\varepsilon^{(e)}$

when we transform to local coordinates $x^{(e)}$ with the origin O_j, we perform the change of variables $x^{(e)} = \mathcal{P}^{(e)}(x - O_j)$, where $\mathcal{P}^{(e)}$ is the orthogonal matrix relating the global coordinates x to the local ones $x^{(e)}$. Consequently, the expressions for $\tilde{\mathbf{v}}_P(x)$ and $\tilde{p}_P(x)$ are given by:

$$\tilde{\mathbf{v}}_P(x) = (\mathcal{P}^{(e)})^* \mathbf{v}_P(x^{(e)\prime}), \quad \tilde{p}_P(x) = -\alpha x_n^{(e)} + a, \tag{1.4.8}$$

where $x^{(e)} = \mathcal{P}^{(e)}(x - O_j)$.

Now, let us proceed to describe the algorithm of the method of asymptotic partial domain decomposition (MAPDD) for the Stokes problem posed in a tube structure B_ε. We introduce a small positive number δ, much greater than ε (δ to be chosen of the order $O(\varepsilon|\ln\varepsilon|)$). For each edge $e = \overline{O_i O_j}$ of the graph, we introduce two hyperplanes orthogonal to this edge, intersecting it at a distance δ from its ends.

The cross-sections of the cylinder $\Pi_\varepsilon^{(e)}$ formed by these two hyperplanes are denoted, respectively, by $S_{i,j}$ (the cross-section at a distance δ from O_i) and $S_{j,i}$ (the cross-section at a distance δ from O_j). Furthermore, we denote the part of the cylinder $\Pi_\varepsilon^{(e)}$ between these two cross-sections as $B_{ij}^{\mathrm{dec},\varepsilon}$ (see Fig. 1.8). Additionally, we define $B_i^{\varepsilon,\delta}$ as the connected, truncated part of B_ε, bounded by the cross-sections $S_{i,j}$, and containing the vertex or node O_i.

We introduce the function spaces $W_\Gamma^{1,2}(B_\varepsilon)$, consisting of all vector-valued functions from the space $W^{1,2}(B_\varepsilon)$ that vanish for $x \in \Gamma = \partial B_\varepsilon \backslash (\bigcup_{j=N_1+1}^{N} \gamma_\varepsilon^j)$ and $H_\Gamma^{1,2}(B_\varepsilon)$ which is the space of all divergence-free vector functions from $W_\Gamma^{1,2}(B_\varepsilon)$.

Now, we define the subspaces $W_{\mathrm{dec}}(B_\varepsilon, \delta)$, $W_{0,\mathrm{dec}}(B_\varepsilon, \delta)$, and $L^2_{\mathrm{dec}}(B_\varepsilon, \delta)$ as follows:

- $W_{\mathrm{dec}}(B_\varepsilon, \delta)$ consists of elements (vector-valued functions) that coincide with the Poiseuille flows $\tilde{\mathbf{v}}_P$ on each truncated cylinder $B_{ij}^{\mathrm{dec},\varepsilon}$.
- $W_{0,\mathrm{dec}}(B_\varepsilon, \delta)$ is a subspace of $W_{\mathrm{dec}}(B_\varepsilon, \delta)$ with elements that vanish on the entire boundary ∂B_ε.
- $L^2_{\mathrm{dec}}(B_\varepsilon, \delta)$ is a subspace of the space $L^2(B_\varepsilon)$ comprising affine functions of $x_1^{(e)}$ on each truncated cylinder $B_{ij}^{\mathrm{dec},\varepsilon}$.

Furthermore, we define $H_{\mathrm{dec}}(B_\varepsilon, \delta)$ as a subspace of $W_{\mathrm{dec}}(B_\varepsilon, \delta)$ consisting of divergence-free vector functions.

The method of asymptotic partial domain decomposition (MAPDD) replaces the original Stokes problem (1.4.3) by its projection on $H_{\text{dec}}(B_\varepsilon, \delta)$. Specifically, we seek $\mathbf{v}_d \in H_{\text{dec}}(B_\varepsilon, \delta)$ with the boundary condition $\mathbf{v}_d|_{\partial B_\varepsilon} = \mathbf{g}$, such that for every $\mathbf{w} \in H_{0,\text{dec}}(B_\varepsilon, \delta)$, the following integral identity holds:

$$\nu \int_{B_\varepsilon} \nabla \mathbf{v}_d \cdot \nabla \mathbf{w}\, dx = 0.$$

For numerical implementation, the weak formulation of MAPDD including the pressure is more convenient. We seek $\mathbf{v}_d \in W_{\text{dec}}(B_\varepsilon, \delta)$ with the boundary condition $\mathbf{v}_d|_{\partial B_\varepsilon} = \mathbf{g}$ and $p_d \in L^2_{\text{dec}}(B_\varepsilon, \delta)$, such that for every $\mathbf{w} \in W_{0,\text{dec}}(B_\varepsilon, \delta)$ and $q \in L^2_{\text{dec}}(B_\varepsilon, \delta)$, the following integral identity holds:

$$\nu \int_{B_\varepsilon} \nabla \mathbf{v}_d \cdot \nabla \mathbf{w}\, dx - \int_{B_\varepsilon} p_d \operatorname{div} \mathbf{w}\, dx + \int_{B_\varepsilon} q \operatorname{div} \mathbf{v}_d\, dx = 0. \tag{1.4.9}$$

It can be observed that the divergence of $\mathbf{v}_d$ is equal to zero ($\operatorname{div} \mathbf{v}_d = 0$). This is deduced from the integral identity obtained when setting $\mathbf{w} = 0$, leading to:

$$\int_{B_\varepsilon} q \operatorname{div} \mathbf{v}_d\, dx = 0 \quad \forall q \in L^2(B_\varepsilon, \delta).$$

Hence, we conclude that $\operatorname{div} \mathbf{v}_d = 0$.

It has been proven that there exists a constant $C > 0$ such that, for any given non-negative integer J, if $\delta = CJ\varepsilon|\ln \varepsilon|$, then the exact solution $(\mathbf{v}, p)$ and the MAPDD solution $(\mathbf{v}_d, p_d)$ satisfy the estimate:

$$\|\mathbf{v}_d - \mathbf{v}\|_{W^{1,2}(B_\varepsilon)} + \|p_d - p\|_{L^2(B_\varepsilon)} = O(\varepsilon^J),$$

where p_d and p are normalized so that:

$$\int_{B_\varepsilon} p_d\, dx = 0, \quad \int_{B_\varepsilon} p\, dx = 0.$$

This concludes the description of the method of asymptotic partial domain decomposition (MAPDD) for the Stokes equation in a tube structure.

Chapter 2
Preliminaries

This chapter presents the essential notations and concepts from real and functional analysis. It introduces the spaces of smooth functions, Sobolev spaces, and spaces of divergence-free vector functions, which play a fundamental role in the study of Navier–Stokes equations. The aim is not to provide a comprehensive treatment of auxiliary results (many are stated without proof) but to recall necessary prerequisites. References to specialized literature are provided for a more detailed study. However, we do include proofs for certain assertions that are deemed relatively straightforward and not commonly found in the theory of differential equations. This particularly applies to results in noncompact domains with cylindrical outlets to infinity and in the tube structure B_ε. The selection of statements to be proved is subject to our preferences.

2.1 Basic Notation, Elementary Inequalities, Auxiliary Results

2.1.1 Basic Notation

In the following, we employ standard notation commonly used in the context of partial differential equations (PDEs).

We use the symbol c to represent any constant. The precise value of c may vary from line to line in a given calculation. Sometimes, subscripts are attached to c (e.g., c_k) to denote specific constants. When we want to emphasize that the constant c depends on a fixed quantity A, we write $c(A)$.

The symbol $\mathbb{R}^n$ denotes a real n-dimensional Euclidean space, with $\mathbb{R}^1 = \mathbb{R}$. An open connected set in $\mathbb{R}^n$, usually referred to as a domain, is denoted by Ω or ω, while $\overline{\Omega}$ represents its closure. The boundary of Ω is denoted by $\partial\Omega = \overline{\Omega} \setminus \Omega$. A typical point in Ω is $x = (x_1, \dots, x_n)$. If Ω is a ball with radius r centered at some

G. Panasenko, K. Pileckas, *Multiscale Analysis of Viscous Flows in Thin Tube Structures*, Advances in Mathematical Fluid Mechanics,
https://doi.org/10.1007/978-3-031-54630-3_2

point $x_0 \in \mathbb{R}^n$, we denote it by $B_r(x_0) = \{x \in \mathbb{R}^n : |x - x_0| < r\}$. If x_0 coincides with the origin, we write $B_r(0) = B_r$.

Given two vectors $a = (a_1, \dots, a_n)$ and $b = (b_1, \dots, b_n)$ in $\mathbb{R}^n$, we define the dot product and the Euclidean norm as follows:

$$a \cdot b = \sum_{i=1}^{n} a_i b_i, \quad |a| = \left(\sum_{i=1}^{n} a_i^2 \right)^{1/2}.$$

If $f : \Omega \subset \mathbb{R}^n \to \mathbb{R}$ is a function defined in Ω, we write $f(x) = f(x_1, \cdots, x_n)$. When dealing with a vector-valued function $\mathbf{f} : \Omega \subset \mathbb{R}^n \to \mathbb{R}^m$, we represent it as $\mathbf{f}(x) = (f_1(x), \dots, f_m(x))$, where f_k denotes the kth component of $\mathbf{f}$ for $k = 1, \dots, m$. Throughout this work, we use boldface to denote vector-valued quantities, except when referring to points x in $\mathbb{R}^n$.

A function $f : \Omega \subset \mathbb{R}^n \to \mathbb{R}$ is said to be *Lipschitz continuous* in Ω if there exists a constant L such that

$$|f(x) - f(y)| \leq L|x - y| \text{ for all } x, y \in \Omega.$$

Let $\alpha = (\alpha_1, \dots, \alpha_n)$ be a multi-index with $\alpha_j \geq 0$, where $j = 1, \dots, n$. The notation $|\alpha| = \sum_{j=1}^{n} \alpha_j$ is used to denote the sum of the components of α. The expression $D^\alpha f$ represents the partial derivative

$$\frac{\partial^{|\alpha|} f}{\partial x_1^{\alpha_1} \dots \partial x_n^{\alpha_n}}.$$

For functions $f = f(x, t)$ that depend on both $x \in \mathbb{R}^n$ and $t \in \mathbb{R}^1_+$, we also use the notations:

$$\partial_t^r f = \frac{\partial^r f}{\partial t^r}, \quad \partial_x^{|\alpha|} f = \frac{\partial^{|\alpha|} f}{\partial x_1^{\alpha_1} \dots \partial x_n^{\alpha_n}}.$$

The gradient of a function f is represented by $\nabla f = \left(\frac{\partial f}{\partial x_1}, \dots, \frac{\partial f}{\partial x_n} \right)$, while

$$\nabla f \cdot \nabla g = \sum_{i=1}^{n} \frac{\partial f}{\partial x_i} \frac{\partial g}{\partial x_i}, \quad |\nabla f|^2 = \sum_{i=1}^{n} \left| \frac{\partial f}{\partial x_i} \right|^2.$$

The symbol ∇f is interpreted as either a column or row vector, depending on the context.

The Laplacian of a function f is denoted by $\Delta f = \sum_{i=1}^{n} \frac{\partial^2 f}{\partial x_i^2}$.

For vector-valued functions $\mathbf{f}(x) = (f_1(x), \dots, f_n(x))$, the gradient matrix (Jacobian matrix) is given by

$$\nabla \mathbf{f} = \begin{pmatrix} \frac{\partial f_1}{\partial x_1} & \cdots & \frac{\partial f_1}{\partial x_n} \\ \vdots & \ddots & \vdots \\ \frac{\partial f_n}{\partial x_1} & \cdots & \frac{\partial f_n}{\partial x_n} \end{pmatrix},$$

and the divergence of **f** is denoted by

$$\operatorname{div} \mathbf{f} = \nabla \cdot \mathbf{f} = \sum_{i=1}^{n} \frac{\partial f_i}{\partial x_i}.$$

Let $\mathbf{f}(x) = (f_1(x), \ldots, f_n(x))$ and $\mathbf{g}(x) = (g_1(x), \ldots, g_n(x))$ be two vector-valued functions. Then the following expressions hold:

$$\nabla \mathbf{f} \cdot \nabla \mathbf{g} = \sum_{j=1}^{n} \sum_{i=1}^{n} \frac{\partial f_j}{\partial x_i} \frac{\partial g_j}{\partial x_i}, \quad |\nabla \mathbf{f}|^2 = \sum_{j=1}^{n} \sum_{i=1}^{n} \left| \frac{\partial f_j}{\partial x_i} \right|^2,$$

$$\mathbf{g} \cdot \nabla \mathbf{f} = \begin{pmatrix} \sum_{i=1}^{n} g_i \frac{\partial f_1}{\partial x_i} \\ \vdots \\ \sum_{i=1}^{n} g_i \frac{\partial f_n}{\partial x_i} \end{pmatrix}.$$

For any matrix A, we denote the transposed matrix by A^t.

2.1.2 Elementary Inequalities

Below we list a set of elementary inequalities that are frequently utilized throughout this work and should be memorized.

1. **Young's inequality:** For $1 < q < \infty$ with $\frac{1}{q} + \frac{1}{q'} = 1$, it holds that

$$ab \leq \frac{a^q}{q} + \frac{b^{q'}}{q'}, \tag{2.1.1}$$

   for all $a, b > 0$.
2. **Young's inequality with ε:**

$$ab \leq \varepsilon a^q + \frac{(\varepsilon q)^{-q'/q}}{q'} b^{q'}, \tag{2.1.2}$$

   for all $\varepsilon > 0$ and $a, b > 0$.

3. **Cauchy's inequality with** ε: In the case where $q = q' = 2$, we have

$$ab \leq \varepsilon a^2 + \frac{1}{4\varepsilon} b^2, \tag{2.1.3}$$

for all $\varepsilon > 0$ and $a, b > 0$.

4. **Hölder's inequality:** For $1 < q < \infty$ with $\frac{1}{q} + \frac{1}{q'} = 1$, if $f \in L^q(\Omega)$ and $g \in L^{q'}(\Omega)$, then

$$\int\limits_{\Omega} |f(x)g(x)|\, dx \leq \left(\int\limits_{\Omega} |f(x)|^q\, dx \right)^{1/q} \left(\int\limits_{\Omega} |g(x)|^{q'}\, dx \right)^{1/q'}$$
$$= \|f\|_{L^q(\Omega)} \|g\|_{L^{q'}(\Omega)}. \tag{2.1.4}$$

For $1 < p_1, p_2, p_3 < \infty$ with $\frac{1}{p_1} + \frac{1}{p_2} + \frac{1}{p_3} = 1$, and if $f \in L^{p_1}(\Omega)$, $g \in L^{p_2}(\Omega)$, $h \in L^{p_3}(\Omega)$, we have

$$\int\limits_{\Omega} |f(x)g(x)h(x)|\, dx$$
$$\leq \left(\int\limits_{\Omega} |f(x)|^{p_1}\, dx \right)^{1/p_1} \left(\int\limits_{\Omega} |g(x)|^{p_2}\, dx \right)^{1/p_2} \left(\int\limits_{\Omega} |h(x)|^{p_3}\, dx \right)^{1/p_3}$$
$$= \|f\|_{L^{p_1}(\Omega)} \|g\|_{L^{p_2}(\Omega)} \|h\|_{L^{p_3}(\Omega)}. \tag{2.1.5}$$

5. **Cauchy-Schwarz-Bunyakovsky inequality:** In the case where $q = q' = 2$, the Hölder inequality becomes

$$\int\limits_{\Omega} |f(x)g(x)|\, dx \leq \left(\int\limits_{\Omega} |f(x)|^2\, dx \right)^{1/2} \left(\int\limits_{\Omega} |g(x)|^2\, dx \right)^{1/2} \tag{2.1.6}$$
$$= \|f\|_{L^2(\Omega)} \|g\|_{L^2(\Omega)}.$$

6. **Hardy's inequality:** For a measurable function f with $f(0) = 0$ and $\int_0^\infty |f'(x)|^2\, dx < \infty$, Hardy's inequality is given by

$$\int\limits_0^\infty \frac{|f(x)|^2}{x^2}\, dx \leq 4 \int\limits_0^\infty |f'(x)|^2\, dx. \tag{2.1.7}$$

The equality holds if and only if $f(x) = 0$ almost everywhere.

2.2 Some Concepts from Functional Analysis

2.2.1 Linear Operators in Banach and Hilbert Spaces

Let E be a Banach space, which is a complete normed linear space, equipped with the norm $\| \cdot \|_E$. A vector $\big(u_1, \ldots, u_n\big)$ is considered to belong to the vector space E if all its components $u_1, \ldots, u_n$ belong to E. The norm of the vector $\mathbf{u} = (u_1, \ldots, u_n) \in E$ is defined as follows:

$$\|\mathbf{u}\|_E = \sqrt{\sum_{i=1}^{n} \|u_i\|_E^2}.$$

Note that vector and scalar Banach spaces are not distinguished in the notation.

The dual space E^* of E is defined as the Banach space of all linear continuous functionals:

$$f : v \mapsto \langle f, v \rangle,$$

where $\langle f, v \rangle$ denotes the value of the functional f at v. Sometimes, we will simply write $f(v)$ without using the duality pairing notation.

A sequence $\{v_j\} \subset E$ is said to *converge strongly* to $v \in E$, denoted as $v_j \to v$, if the following condition is satisfied:

$$\lim_{j \to \infty} \|v_j - v\|_E = 0.$$

Similarly, $\{v_j\} \subset E$ *converges weakly* to $v \in E$, denoted as $v_j \rightharpoonup v$, if the following condition holds:

$$\lim_{j \to \infty} \langle f, v_j \rangle = \langle f, v \rangle \quad \forall f \in E^*.$$

For a linear subset $\mathcal{D} \subset E$, the closure (completion) of $\mathcal{D}$ in the norm $\| \cdot \|_E$ is denoted by $\overline{\mathcal{D}}^{\|\cdot\|_E}$. $\mathcal{D}$ is called *dense* in E if $\overline{\mathcal{D}}^{\|\cdot\|_E} = E$.

We refer to a Banach space E as *separable* if it contains a countable dense subset. In what follows, we consider only separable Banach and Hilbert spaces.

Theorem 2.2.1 (Hahn–Banach Theorem) *Let E be a Banach space, $E_1 \subset E$ be its linear subspace, and f_1 be a continuous linear functional defined on E_1. There exists a functional f defined on the whole of E such that*

$$\langle f, v \rangle = \langle f_1, v \rangle \quad \forall v \in E_1, \quad \|f\|_{E^*} = \|f_1\|_{E_1^*}.$$

A Banach space E is called *reflexive* if each linear continuous functional on E^* can be represented in the form:

$$f \mapsto \langle f, v\rangle \quad \forall f \in E^*,$$

for some fixed $v \in E$. Then each $v \in E$ can be identified with the functional $\langle \cdot, v\rangle$, and E can be isometrically identified with $(E^*)^*$. In this case, we write $E^{**} = E$.

Theorem 2.2.2 *Let E be reflexive. Then any weakly convergent sequence is bounded. Additionally, if $u_k \rightharpoonup u$, then*

$$\|u\|_E \leq \lim_{j\to\infty} \inf \|u_j\|_E.$$

Each bounded sequence $\{u_j\} \subset E$ contains a subsequence $\{u_{j_l}\}$ that converges weakly to some $u \in E$.

A Banach space E is called *uniformly convex* if for every $\varepsilon > 0$, there exists $\delta(\varepsilon)$ such that the following conditions

$$\|v\|_E \leq 1, \quad \|u\|_E \leq 1, \quad \|v - u\|_E \geq \varepsilon,$$

imply that

$$\|v + u\|_E < 2(1 - \delta(\varepsilon)).$$

Theorem 2.2.3 *Every uniformly convex Banach space is reflexive.*

It can be observed that if $u_k \to u$ in E, then $u_k \rightharpoonup u$. The next theorem establishes the reverse statement.

Theorem 2.2.4 *Let E be a uniformly convex Banach space. If the sequence $\{u_j\}$ converges weakly to $u \in E$ and $\|u_j\|_E \to \|u\|_E$, then $\{u_j\}$ converges strongly to u.*

Consider two Banach spaces E_1 and E_2. Let $A : E_1 \longmapsto E_2$ be a linear operator. The *range* of A is defined as $\mathcal{R}(A) := \{v \in E_2 : \; v = Au \;\text{ for some }\; u \in E_1\}$, and the *kernel* (or *null space*) of A is denoted by $\ker A := \{u \in E_1 : \; Au = 0\}$.

The linear operator $A : E_1 \longmapsto E_2$ is called *bounded* if

$$\|A\| := \sup_{\|u\|_{E_1} \leq 1} \|Au\|_{E_2} < \infty.$$

$\|A\|$ is referred to as the *norm* of A.

It can be shown that a bounded linear operator $A : E_1 \longmapsto E_2$ is *continuous*, meaning that whenever $u_j \to u$ in E_1, then $Au_j \to Au$ in E_2, and vice versa: any continuous linear operator is bounded.

An operator $A : E_1 \longmapsto E_2$ is called *completely continuous* (or *compact*) if A transforms every weakly convergent sequence $\{u_j\}$ in E_1 into a strongly convergent sequence $\{Au_j\}$ in E_2 (note that A is not necessarily a linear operator).

Let $A : E_1 \longmapsto \mathcal{R}(A) \subseteq E_2$ be a linear continuous operator. The *adjoint* operator $A^* : g \in E_2^* \longmapsto A^*g \in \mathcal{R}(A^*) \subseteq E_1^*$ is defined by the following formula:

$$\langle A^*g, v\rangle = \langle g, Av\rangle \quad \forall g \in E_2^*, \quad \forall v \in E_1.$$

The null space $\ker A^*$ of the adjoint operator A^* is called the *cokernel* of A and is denoted as $\operatorname{coker} A$.

Theorem 2.2.5 *If one of the spaces $\mathcal{R}(A)$, $\mathcal{R}(A^*)$ is closed, then both are closed, and it holds that*

$$\mathcal{R}(A) = \{w \in E_2 : \ \langle g, w\rangle = 0 \quad \forall g \in \ker A^*\},$$

$$\mathcal{R}(A^*) = \{f \in E_1^* : \ \langle f, v\rangle = 0 \quad \forall v \in \ker A\}.$$

Theorem 2.2.6 (Riesz Representation Theorem) *For each functional $f \in H^*$, there exists a unique element $u_f \in H$ such that f can be uniquely represented in the form*

$$\langle f, v\rangle = \left(u_f, v\right)_H \quad \forall v \in H.$$

Moreover, $\|f\|_{H^*} = \|u_f\|_H$. The mapping $f \mapsto u_f$ is a linear isomorphism of H^* onto H.

By virtue of the Riesz representation theorem, H^* is canonically identified with H, and we can write $H^* = H$.

Lemma 2.2.1 (Lax–Milgram Lemma) *Let H be a separable Hilbert space. Assume that $a : H \times H \to \mathbb{R}$ is a bilinear mapping, such that:*

(i) It is continuous, i.e., there exists a positive constant κ_1 such that, for any pair $(u, v) \in H \times H$

$$|a(u, v)| \le \kappa_1 \|u\| \, \|v\|,$$

and

(ii) It is coercive, i.e., there exists a positive constant κ_2 such that, for any $v \in H$,

$$|a(v, v)| \ge \kappa_2 \|v\|^2.$$

Let b be a bounded linear functional $H \to \mathbb{R}$. Then the equation

$$a(u, v) = b(v) \qquad \forall v \in H,$$

admits a unique solution $u \in H$.

Lemma is proved, for example, in [36].

Remark 2.2.1 For a symmetric bilinear form a, this theorem follows immediately from the Riesz representation theorem, considering $a(u, v)$ as a new inner product in H and applying the Riesz representation theorem to the linear bounded functional b.

Let $A : H \longmapsto H$ be a bounded, linear operator. Its adjoint $A^* : H \longmapsto H$ satisfies

$$\left(A^* u, v\right)_H = \left(u, Av\right)_H \quad \forall u, v \in H.$$

If $A = A^*$, then the operator A is called *self-adjoint*.

The number $\lambda \in \mathbb{C}$ is called an *eigenvalue* of A, and $v \in H$ is called an *eigenfunction* of A corresponding to the eigenvalue λ if v is a nontrivial solution of the equation $Av = \lambda v$. Let $\{\lambda_j\}$ be a system of eigenvalues of A and $N_j = \{v \in H : Av = \lambda_j v\}$. If $\lambda_0 = 0$ is an eigenvalue, then $N_0 = \ker A$.

Theorem 2.2.7 *Let* $A : H \longmapsto H$ *be a linear completely continuous operator.*

(*i*) *The eigenvalues of* A *form either a finite or countable set in* $\mathbb{C}$. *If it is countable, then the sequence* $\{\lambda_j\}$ *tends to zero. Moreover, if* $\lambda \neq 0$ *is an eigenvalue of* A, *then the set of all corresponding linearly independent eigenfunctions is finite-dimensional.*

(*ii*) *Let, in addition,* A *be self-adjoint. Then* A *has at least one eigenvalue. Moreover, all eigenvalues of* A *are real, and*

$$H = N_0 \oplus \sum_j N_j,$$

i.e., there is an orthonormal basis in H *consisting of eigenfunctions.*

Theorem 2.2.8 (Fredholm Alternative) *Let* $A : H \longmapsto H$ *be a linear completely continuous operator. Then either the equation*

$$u - Au = f \tag{2.2.1}$$

has a unique solution for all $f \in H$ *or the homogeneous equation*

$$u - Au = 0$$

has a nonzero solution. In the latter case, Eq. (2.2.1) *has a solution if and only if* $(f, v) = 0$ *for all* v *such that* $v - A^* v = 0$.

2.2.2 Some Results on Nonlinear Operator Equations

In this section, we present the well-known Brouwer fixed point theorem along with the more general Leray–Schauder fixed point theorem, which will serve as crucial tools for establishing results concerning nonlinear operator equations. We provide the formulations of these theorems below for future reference:

Theorem 2.2.9 (Brouwer Fixed Point Theorem) *Let K be a nonempty, convex, compact subset of an n-dimensional Euclidean space $\mathbb{R}^n$. For any continuous mapping $A : K \to K$, there exists a fixed point.*

Theorem 2.2.10 (Leray–Schauder Fixed Point Theorem) *Let H be a Hilbert space, and let $A : H \to H$ be a completely continuous nonlinear operator. If all possible solutions of the equations*

$$v^{(\lambda)} = \lambda A v^{(\lambda)}, \quad \lambda \in [0, 1], \tag{2.2.2}$$

satisfy the estimate

$$\|v^{(\lambda)}\|_H \leq \rho, \tag{2.2.3}$$

for some $\rho > 0$ independent of λ, then the equation

$$v = Av \tag{2.2.4}$$

has at least one solution v satisfying the estimate $\|v\|_H \leq \rho$.

It is worth noting that the results presented in this section are well-known and can be found in numerous classical books on functional analysis and its applications to partial differential equations (PDEs), such as those authored by W. Rudin [189], L.V. Kantorovich, G.P. Akilov [84], A.N. Kolmogorov, S.V. Fomin [94], V.A. Trenogin [205], and N.S. Papageorgiou and P. Winker [169], among others.

2.3 Hölder and Sobolev Function Spaces

2.3.1 Hölder Spaces

Let $C^k(\mathbb{R}^n)$ denote the space of k-times continuously differentiable functions, where $k \geq 0$ is an integer. Consider a bounded domain $\Omega \subset \mathbb{R}^n$. We define $C^k(\bar{\Omega})$ as the space of restrictions of functions from $C^k(\mathbb{R}^n)$ to $\bar{\Omega}$. The norm in $C^k(\bar{\Omega})$ is given by

$$\|u\|_{C^k(\bar{\Omega})} = \sum_{|\alpha| \leq k} \sup_{x \in \bar{\Omega}} |D^\alpha u(x)|,$$

where α denotes a multi-index and $D^\alpha u$ represents the α-th order partial derivative of u. For $k = 0$, we write $C^0(\bar{\Omega}) = C(\bar{\Omega})$. Additionally, we define $C_0^k(\Omega)$, for $k \geq 0$, as the subspace of functions from $C^k(\bar{\Omega})$ having compact support in Ω.

Given $\delta \in (0, 1)$, a function u satisfying the estimate

$$|u(x) - u(y)| \leq C|x - y|^\delta \quad \text{for all } x, y \in \bar{\Omega},$$

for some constant C, is said to be *Hölder continuous in $\bar{\Omega}$ with the exponent δ*. The δ-Hölder seminorm of u is defined as

$$[u]_{C^{0,\delta}(\bar{\Omega})} = \sup_{x,y\in\bar{\Omega},x\neq y} \left\{ \frac{|u(x) - u(y)|}{|x - y|^\delta} \right\}.$$

The *Hölder space* $C^{k,\delta}(\bar{\Omega})$ (where $k \geq 0$ is an integer) consists of all functions $u \in C^k(\bar{\Omega})$ for which the norm

$$\|u\|_{C^{k,\delta}(\bar{\Omega})} = \sum_{|\alpha|\leq k} \sup_{x\in\bar{\Omega}} |D^\alpha u(x)| + \sum_{|\alpha|=k} [D^\alpha u]_{C^{0,\delta}(\bar{\Omega})}$$

is finite. Thus, the space $C^{k,\delta}(\bar{\Omega})$ comprises functions that are k-times continuously differentiable in $\bar{\Omega}$ and whose k^{th}-order partial derivatives are Hölder continuous with exponent δ. It is well-known that $C^{k,\delta}(\bar{\Omega})$ is a Banach space.

2.3.2 Boundaries

Let Ω be a bounded domain in $\mathbb{R}^n$. We say that $\partial\Omega$ is C^k-regular (where k is an integer) if there exists a positive number r_0 such that, for each point $x_0 \in \partial\Omega$, the portion of $\partial\Omega$ within the ball $B_{r_0}(x_0)$ can be represented, in a local system of coordinates, as the graph of a C^k-function $\varphi^{(x_0)}$ such that $\varphi^{(x_0)}(x_0) = 0$. Specifically, there exist a coordinate system $\{y\}$ and a C^k-function $\varphi^{(x_0)} : \mathbb{R}^{n-1} \to \mathbb{R}$ such that

$$\Omega \cap B_r(x_0) = \{y \in B_r(0) : y_n > \varphi^{(x_0)}(y')\},$$

where $y' = (y_1, y_2, \ldots, y_{n-1})$, and the coordinates $\{y\}$ are related to the old coordinates $\{x\}$ by the formulas

$$y_i = \sum_{l=1}^n a_{il}(x_l - x_{0l}), \quad i = 1, 2, \ldots, n,$$

where $A = (a_{jl})_{j,l=1,2,\ldots,n}$ is a constant orthogonal $n \times n$ matrix and the axis y_n is directed along the outward normal vector to $\partial\Omega$.

If all functions $\varphi^{(x_0)}$ belong to the Hölder space $C^{k,\delta}$, we say that the boundary $\partial\Omega$ is $C^{k,\delta}$-regular ($\partial\Omega \in C^{k,\delta}$). Moreover, $\partial\Omega$ is C^∞ ($\partial\Omega \in C^\infty$) if it is C^k-regular for any $k = 1, 2, \ldots$. Finally, we refer to the boundary $\partial\Omega$ as Lipschitz if $\varphi^{(x_0)}$ are Lipschitz continuous functions.

Near a point x_0, the boundary $\partial\Omega$ can be "flattened out." We define new coordinates:

$$\begin{cases} z_i = y_i, \quad i = 1, 2, \ldots, n-1, \\ z_n = y_n - \varphi^{(x_0)}(y_1, y_2, \ldots, y_{n-1}), \end{cases}$$

and write $z = \mathbf{\Phi}(y)$. Similarly, we set

$$\begin{cases} y_i = z_i, \quad i = 1, 2, \ldots, n-1, \\ y_n = z_n + \varphi^{(x_0)}(z_1, z_2, \ldots, z_{n-1}), \end{cases}$$

and write $y = \mathbf{\Psi}(z)$. Then $\mathbf{\Phi} = \mathbf{\Psi}^{-1}$, and the mapping $y \longmapsto \mathbf{\Phi}(y) = z$ flattens out the boundary near the point x_0. It is also worth noting that $\det(\nabla\mathbf{\Phi}) = \det(\nabla\mathbf{\Psi}) = 1$, where $\det(\nabla\mathbf{\Phi})$ is the Jacobian of $\mathbf{\Phi}$.

2.3.3 Sobolev Spaces

2.3.3.1 Definition of Sobolev Spaces

Let Ω be a domain in $\mathbb{R}^n$. We denote by $C^\infty(\Omega)$ the set of all infinitely differentiable functions in Ω and by $C_0^\infty(\Omega)$ the set of all functions from $C^\infty(\Omega)$ having compact support in Ω.

Assume $1 \le q \le \infty$, and define the norm of a function v in $L^q(\Omega)$ as follows:

$$\|v\|_{L^q(\Omega)} = \begin{cases} \left(\int\limits_\Omega |v(x)|^q \, dx \right)^{1/q}, & \text{if } 1 \le q < \infty, \\ \text{ess sup}_{x\in\Omega} |v(x)|, & \text{if } q = \infty. \end{cases}$$

$L^q(\Omega)$ represents the space of Lebesgue measurable functions in $\Omega \subset \mathbb{R}^n$ for which the norm $\|v\|_{L^q(\Omega)}$ is finite. For $1 \le q < \infty$ this space is a separable Banach space if we identify functions that coincide almost everywhere. For $1 < q < \infty$, spaces $L^q(\Omega)$ are reflexive. If $q = 2$, the space $L^2(\Omega)$ is a Hilbert space with the inner product $(u, v)_{L^2(\Omega)} = \int_\Omega u(x)v(x)\, dx$.

Suppose u and v are locally summable functions in Ω and α is a multi-index. We say that v is an αth-weak (generalized) partial derivative of u, written $D^\alpha u = v$, provided

$$\int_\Omega u\, D^\alpha \eta\, dx = (-1)^{|\alpha|} \int_\Omega v\eta\, dx$$

for all test functions $\eta \in C_0^\infty(\Omega)$.

Assume $1 \le q \le \infty$, and let k be a nonnegative integer. The Sobolev space $W^{k,q}(\Omega)$ consists of all locally summable functions v such that for each multi-index α with $|\alpha| \le k$, the derivatives $D^\alpha u$ exist in the weak sense and belong to $L^q(\Omega)$. The norm in $W^{k,q}(\Omega)$ is defined by

$$\|v\|_{W^{k,q}(\Omega)} = \begin{cases} \left(\sum_{|\alpha|\le k} \int_\Omega |D^\alpha v(x)|^q\, dx \right)^{1/q}, & \text{if } 1 \le q < \infty, \\ \sum_{|\alpha|\le k} \operatorname{ess\,sup}_{x\in\Omega} |D^\alpha v(x)|, & \text{if } q = \infty. \end{cases}$$

Obviously, $W^{0,q}(\Omega) = L^q(\Omega)$.

Sobolev spaces $W^{k,q}(\Omega)$ for $1 < q < \infty$ are separable, reflexive Banach spaces. The set $C^\infty(\Omega)$ is dense in these spaces $W^{k,q}(\Omega)$.

We write $v \in W^{k,q}_{\text{loc}}(\Omega)$ if $v \in W^{k,q}(\Omega')$ for any bounded subdomain Ω' with $\bar{\Omega}' \subset \Omega$.

We denote by $\overset{\circ}{W}{}^{k,q}(\Omega)$ the subspace of $W^{k,q}(\Omega)$ obtained as the closure of $C_0^\infty(\Omega)$ in $W^{k,q}(\Omega)$-norm.

If $q = 2$, then $W^{k,2}(\Omega)$ and $\overset{\circ}{W}{}^{k,2}(\Omega)$ are Hilbert spaces with the inner product

$$(u, v)_{W^{k,2}(\Omega)} = \sum_{|\alpha|\le k} \int_\Omega D_x^\alpha u(x) D_x^\alpha v(x)\, dx.$$

The Sobolev spaces can be defined for non-integer values of k. Let $\kappa \in (0, 1)$. $W^{\kappa,p}(\Omega)$ is an intermediary Banach space between $L^p(\Omega)$ and $W^{1,p}(\Omega)$, endowed with the norm

$$\|u\|_{W^{\kappa,p}(\Omega)} = \left(\int_\Omega |u(x)|^p dx + \int_\Omega \int_\Omega \frac{|u(x) - u(y)|}{|x - y|^{n+\kappa p}} dxdy \right)^{1/p}.$$

Remark 2.3.1 It is worth mentioning that other definitions of the space $W^{k,q}(\Omega)$ are available. However, for general domains, they may not be equivalent. For the results presented here, we assume that $\partial\Omega$ is Lipschitz, in which case $W^{k,q}(\Omega)$ is the closure of $C^\infty(\bar{\Omega})$ in the norm $\|\cdot\|_{W^{k,q}(\Omega)}$.

Remark 2.3.2 The space $W^{k,2}(\Omega)$ is often denoted as $H^k(\Omega)$, and $\overset{\circ}{W}{}^{k,2}(\Omega)$ as $H_0^k(\Omega)$. However, in this context, we avoid using these notations to preserve the letter H as the standard symbol for spaces of divergence-free functions (see Sect. 2.5).

Finally, we shall provide a characterization of bounded linear functionals in the space $\mathring{W}^{1,2}(\Omega)$. We denote by $W^{-1,2}(\Omega)$ the dual space to $\mathring{W}^{1,2}(\Omega)$. If $f \in W^{-1,2}(\Omega)$, then its norm is defined by

$$\|f\|_{W^{-1,2}(\Omega)} = \sup\left\{\langle f, u\rangle : u \in \mathring{W}^{1,2}(\Omega),\ \|u\|_{\mathring{W}^{1,2}(\Omega)} \leq 1\right\}.$$

Theorem 2.3.1 (Characterization of $W^{-1,2}(\Omega)$)

(i) Assume that $f \in W^{-1,2}(\Omega)$. Then there exist functions $f^{(0)}, f^{(1)}, \ldots, f^{(n)} \in L^2(\Omega)$ such that

$$\langle f, v\rangle = \int_\Omega f^{(0)} v\, dx + \sum_{j=1}^{n} f^{(j)} \frac{\partial v}{\partial x_i}\, dx \quad \text{for all } v \in \mathring{W}^{1,2}(\Omega). \tag{2.3.1}$$

(ii) Furthermore,

$$\|f\|_{W^{-1,2}(\Omega)} = \inf\left\{\left(\int_\Omega \sum_{j=0}^{n} |f^{(j)}|^2\, dx\right)^{1/2}\right\},$$

where the infimum is taken over all f satisfying (2.3.1) for $f^{(0)}, f^{(1)}, \ldots, f^{(n)} \in L^2(\Omega)$.

2.3.3.2 Traces

Let Ω be a bounded domain in $\mathbb{R}^n$, and let $u \in C(\bar{\Omega})$. It is evident that u possesses a value on the boundary of Ω (also known as the trace on $\partial\Omega$) in the conventional sense. However, a typical function $u \in W^{1,q}(\Omega)$ is not necessarily continuous; it is only defined almost everywhere in Ω. As $\partial\Omega$ has an n-dimensional Lebesgue measure equal to zero, the expression "u restricted on $\partial\Omega$" lacks a direct interpretation. The subsequent theorem resolves this issue.

Theorem 2.3.2 *Assume $1 \leq q < \infty$, $\Omega \subset \mathbb{R}^n$ is a bounded domain with Lipschitz boundary $\partial\Omega$. Then, there exists a bounded linear operator $T : W^{1,q}(\Omega) \to L^q(\partial\Omega)$ such that:*

$$Tu = u\big|_{\partial\Omega}, \quad \text{if } u \in W^{1,q}(\Omega) \cap C(\bar{\Omega}),$$

and

$$\|Tu\|_{L^q(\partial\Omega)} \leq c\|u\|_{W^{1,q}(\Omega)},$$

for each $u \in W^{1,q}(\Omega)$. The constant c depends solely on q and Ω.

The operator T is known as the trace operator, and Tu is referred to as the trace of $u \in W^{1,q}(\Omega)$ on $\partial\Omega$. Since the set $C^\infty(\bar{\Omega}) \subset C(\bar{\Omega})$ is dense in $W^{1,q}(\Omega)$, this definition is correct.

Functions $u \in \overset{\circ}{W}{}^{1,q}(\Omega)$ have zero traces on $\partial\Omega$. To be more precise, the following theorem holds:

Theorem 2.3.3 *Assume* $1 \le q < \infty$, Ω *is a bounded domain with Lipschitz boundary* $\partial\Omega$. *Suppose* $u \in W^{1,q}(\Omega)$. *Then* $u \in \overset{\circ}{W}{}^{1,q}(\Omega)$ *if and only if* $Tu = 0$ *on* $\partial\Omega$.

Let $k \ge 1$ be an integer. Then the Sobolev space $W^{k,q}(\partial\Omega)$ can be defined in the same way as in the domain Ω. Specifically, we say $w \in W^{k,q}(\partial\Omega)$ if, for each $|\alpha| \le k$, there exists a weak derivative $D^\alpha w \in L^q(\partial\Omega)$, and the norm is given by:

$$\|w\|_{W^{k,q}(\partial\Omega)} = \sum_{|\alpha|\le k} \|D^\alpha w\|_{L^q(\partial\Omega)},$$

where the norm is finite.

To extend the definition of the space $W^{k,q}(\partial\Omega)$ to non-integer $k \ge 1$, we introduce, for $\gamma \in (0, 1)$ and $1 \le q < \infty$, a semi-norm:

$$\langle w\rangle_{W^{\gamma,q}(\partial\Omega)} = \Big(\int\limits_{\partial\Omega}\int\limits_{\partial\Omega} \frac{|w(x) - w(y)|^q}{|x - y|^{n-1+q\gamma}}\, dS_x dS_y\Big)^{1/q}.$$

Let $[k]$ denote the integer part of k. We say $w \in W^{k,q}(\partial\Omega)$ if $w \in W^{[k],q}(\partial\Omega)$ and the norm

$$\|w\|_{W^{k,q}(\partial\Omega)} = \Big(\sum_{|\alpha|\le[k]} \|D^\alpha w\|^q_{L^q(\partial\Omega)} + \sum_{|\alpha|=[k]} \langle D^\alpha w\rangle^q_{W^{l-[l],q}(\partial\Omega)}\Big)^{1/q}$$

is finite.

Assume $1 \le q < \infty$, $k \ge 1$ is an integer, Ω is a bounded domain, and $\partial\Omega$ is Lipschitz. Then, for each function $u \in W^{k,q}(\Omega)$, there exists a trace w on the boundary $\partial\Omega$. Denote by $\mathcal{M}$ the set of traces of functions $u \in W^{k,q}(\Omega)$. In $\mathcal{M}$, we can introduce the natural norm:

$$|||w|||_{(k,q)} = \inf\big\{\|u\|_{W^{k,q}(\Omega)} : u\big|_{\partial\Omega} = w\big\}.$$

The set $\mathcal{M}$ with the norm $|||\cdot|||_{(k,q)}$ forms a Banach space. The following theorem demonstrates that the space $\mathcal{M}$ is equivalent to a Sobolev space with a fractional index.

Theorem 2.3.4 (Trace Theorem) *Assume* $1 \le q < \infty$, $k \ge 1$ *is an integer,* Ω *is a bounded domain, and* $\partial\Omega$ *is Lipschitz.*

(i) *If* $u \in W^{k,q}(\Omega)$, *then there exists a trace* $D^\alpha u|_{\partial\Omega} = w_\alpha$, $|\alpha| < k$, $w_\alpha \in W^{k-|\alpha|-1/q,q}(\partial\Omega)$, *and*

$$\|w_\alpha\|_{W^{k-|\alpha|-1/q,q}(\partial\Omega)} \le c\|u\|_{W^{k,q}(\Omega)}. \tag{2.3.2}$$

(ii) *For each function* $w \in W^{k-1/q,q}(\partial\Omega)$, *there exists an extension* $u \in W^{k,q}(\Omega)$ *such that* $u\big|_{\partial\Omega} = w$, *and*

$$\|u\|_{W^{k,q}(\Omega)} \le c\|w\|_{W^{k-1/q,q}(\partial\Omega)}. \tag{2.3.3}$$

(iii) *Norms* $|||\cdot|||_{(k,q)}$ *and* $\|\cdot\|_{W^{k-1/q,q}(\partial\Omega)}$ *are equivalent.*

The constants in inequalities (2.3.2) *and* (2.3.3) *are independent of* u *and* w.

Results related to the properties of Sobolev spaces are classical and can be found in books by R. Adams, J. Fournier [2], L.C. Evans [61], V.G. Maz'ya [120], and many others.

2.3.3.3 Inequalities

The following inequalities and facts are well-known. The proofs of these results can be found in books on Sobolev spaces (e.g., [2, 27, 106, 107, 120]).

Lemma 2.3.1 (Poincaré Inequality) *Let* Ω *be a bounded domain in* $\mathbb{R}^n$ *with Lipschitz boundary. For any* $u \in W^{1,2}(\Omega)$ *such that* $u\big|_S = 0$, *where* $S \subset \partial\Omega$, $|S| = mes(S) > 0$, *the inequality*

$$\int_\Omega |u|^2\,dx \le c\int_\Omega |\nabla u|^2\,dx \tag{2.3.4}$$

holds.

If $u \in \mathring{W}^{1,2}(\Omega)$, *then*

$$\int_\Omega |u|^2\,dx \le \frac{1}{\mu_1}\int_\Omega |\nabla u|^2\,dx, \tag{2.3.5}$$

where μ_1 *is the smallest eigenvalue of the Laplace operator:*

$$\begin{cases} -\Delta u(x) = \mu u(x), & x \in \Omega, \\ u(x) = 0, & x \in \partial\Omega. \end{cases}$$

Moreover, the following inequality

$$\int_{\Omega} |u - u_{\Omega}|^q \, dx \le c \int_{\Omega} |\nabla u|^q \, dx \tag{2.3.6}$$

holds, where $u_{\Omega} = \frac{1}{|\Omega|} \int_{\Omega} u(x)dx$, *and* $|\Omega|$ *is the Lebesgue measure of* Ω. *The constant* c *in* (2.3.4)–(2.3.6) *depends only on* q *and* Ω. *In particular, if* Ω *is a ball of radius* r, *then* $c = C(n, q)r^q$, *where* $C(n, q)$ *depends only on the dimension* n *and* q.

Lemma 2.3.2 (Ladyzhenskaya Inequalities [102]) *Let* $u \in \mathring{W}^{1,2}(\Omega)$.

(i) If $\Omega \subset \mathbb{R}^2$, *then*

$$\|u\|_{L^4(\Omega)}^4 \le 2\|u\|_{L^2(\Omega)}^2 \|\nabla u\|_{L^2(\Omega)}^2. \tag{2.3.7}$$

(ii) If $\Omega \subset \mathbb{R}^3$, *then*

$$\|u\|_{L^4(\Omega)}^4 \le \left(4/3\right)^{3/2} \|u\|_{L^2(\Omega)} \|\nabla u\|_{L^2(\Omega)}^3. \tag{2.3.8}$$

(iii) If $\Omega \subset \mathbb{R}^3$, *then*

$$\|u\|_{L^6(\Omega)} \le 48^{1/6} \|\nabla u\|_{L^2(\Omega)}. \tag{2.3.9}$$

Theorem 2.3.5 (Sobolev Embedding Theorem, e.g., [2, 61]) *Let* $\Omega \subset \mathbb{R}^n$ *be a bounded domain with Lipschitz boundary, and let* $u \in W^{l,q}(\Omega)$, *where* $l \ge 1$ *is an integer, and* $q \ge 1$.

(i) If $n > ql$, $r \le \dfrac{qn}{n - ql}$, *then* $u \in L^r(\Omega)$ *and*

$$\|u\|_{L^r(\Omega)} \le c\|u\|_{W^{l,q}(\Omega)}. \tag{2.3.10}$$

For $r < \dfrac{qn}{n - ql}$, *the embedding* $W^{l,q}(\Omega) \hookrightarrow L^r(\Omega)$ *is compact.*

(ii) If $n < ql$, *then* $u \in C^{h,\delta}(\bar{\Omega})$ *with* $h + \delta \le \dfrac{(ql - n)}{q}$, *and*

$$\|u\|_{C^{h,\delta}(\bar{\Omega})} \le c\|u\|_{W^{l,q}(\Omega)}. \tag{2.3.11}$$

For $h + \delta < \dfrac{(ql - n)}{q}$, *the embedding* $W^{l,q}(\Omega) \hookrightarrow C^{h,\delta}(\bar{\Omega})$ *is compact.*

(iii) If $ql < n$, $n - ql < s \le n$ *and* $q \le p \le q^* = \dfrac{sq}{n - ql}$, *then* $u \in L^p(\Omega_s)$, *where* $\Omega_s = \Omega \cap \mathbb{R}^s$, *and*

$$\|u\|_{L^p(\Omega_s)} \le c\|u\|_{W^{l,q}(\Omega)}. \tag{2.3.12}$$

For $1 \le p < \dfrac{qs}{n-ql}$, *the embedding* $W^{l,q}(\Omega) \hookrightarrow L^p(\Omega_s)$ *is compact.*

(iv) If $ql = n$, $1 \le s \le n$ *and* $q \le p < \infty$, *then* $u \in L^p(\Omega_s)$ *and*

$$\|u\|_{L^p(\Omega_s)} \le c\|u\|_{W^{l,q}(\Omega)}. \tag{2.3.13}$$

For $1 \le p < \infty$, *the embedding* $W^{l,q}(\Omega) \hookrightarrow L^p(\Omega_s)$ *is compact.*

(v) If $\partial\Omega$ *is of class* C^1, *and either a)* $1 < ql < n$ *and* $q \le p \le \dfrac{(n-1)q}{n-ql}$ *or b)* $ql = n$ *and* $q \le p < \infty$, *then* $u \in L^p(\partial\Omega)$ *and*

$$\|u\|_{L^p(\partial\Omega)} \le c\|u\|_{W^{l,q}(\Omega)}. \tag{2.3.14}$$

In case b), the embedding $W^{l,q}(\Omega) \hookrightarrow L^p(\partial\Omega)$ *is compact; in case a), it is compact if* $p < \dfrac{(n-1)q}{n-ql}$.

The constants in inequalities (2.3.10)–(2.3.14) *depend only on* Ω.

Remark 2.3.3 Compactness of the embedding $V \hookrightarrow U$ means that any weakly convergent sequence in V is strongly convergent in U. $\mathbb{R}^s$ for $s < n$ is a s-dimensional hyperplane.

Lemma 2.3.3 *Let* Ω *be a bounded domain in* $\mathbb{R}^n$ *with* C^2*-smooth boundary, and let* $u \in W^{2,2}(\Omega) \cap \mathring{W}^{1,2}(\Omega)$. *Then the following multiplicative inequality*

$$\|\nabla u\|^2_{L^2(\partial\Omega)} \le c\Big(\|\nabla u\|^2_{L^2(\Omega)} + \|\nabla u\|_{L^2(\Omega)}\|\Delta u\|_{L^2(\Omega)}\Big) \tag{2.3.15}$$

holds.

This inequality follows from the standard interpolation inequality (see [27]) and the second main inequality for elliptic equations [103].

Let us consider functions defined on the infinite interval $(0, +\infty)$. Below we prove some elementary properties of such functions, including a version of the Poincaré inequality for exponentially vanishing functions (see Lemma 2.3.5).

Lemma 2.3.4 *Let* $H \in W^{1,2}(0, \hat{T})$. *For any* $\gamma > 0$, *the following inequality holds:*

$$\int_0^{\hat{T}} H^2(\tau)\exp\{2\gamma\tau\}d\tau \le \frac{1}{\gamma^2}\int_0^{\hat{T}} (H'(\tau))^2\exp\{2\gamma\tau\}d\tau + \frac{1}{\gamma}H^2(\hat{T})\exp\{2\gamma\hat{T}\}. \tag{2.3.16}$$

Proof

By integrating by parts and applying the Young inequality, we obtain:

$$\int_0^{\hat{T}} H^2(\tau)\exp\{2\gamma\tau\}d\tau =$$

$$= \frac{1}{2\gamma}(H^2(\hat{T})\exp\{2\gamma\hat{T}\} - H^2(0)) - 2\int_0^{\hat{T}} H(\tau)H'(\tau)\frac{\exp\{2\gamma\tau\}}{2\gamma}d\tau$$

$$\leq 2\int_0^{\hat{T}} \{\frac{\gamma}{2}H^2(\tau) + \frac{1}{2\gamma}(H'(\tau))^2\}\frac{\exp\{2\gamma\tau\}}{2\gamma}d\tau + \frac{1}{2\gamma}H^2(\hat{T})\exp\{2\gamma\hat{T}\}.$$

Hence,

$$\int_0^{\hat{T}} H^2(\tau)\exp\{2\gamma\tau\}d\tau \leq \frac{1}{2}\int_0^{\hat{T}} H^2(\tau)\exp\{2\gamma\tau\}d\tau +$$

$$+\frac{1}{2\gamma^2}\int_0^{\hat{T}}(H'(\tau))^2\exp\{2\gamma\tau\}d\tau + \frac{1}{2\gamma}H^2(\hat{T})\exp\{2\gamma\hat{T}\},$$

and the lemma is proved. □

Lemma 2.3.5 *Suppose* $H \in W^{1,2}(0,\infty)$ *and* $\int_0^\infty |H'(\tau)|^2\exp\{2\gamma\tau\}d\tau < \infty$ *for some* $\gamma > 0$*. Then the following weighted Poincaré's inequality holds:*

$$\int_0^\infty H^2(\tau)\exp\{2\gamma\tau\}d\tau \leq \frac{1}{\gamma^2}\int_0^\infty |H'(\tau)|^2\exp\{2\gamma\tau\}d\tau. \tag{2.3.17}$$

Proof

As $H \in W^{1,2}(0,+\infty)$, there exists a sequence $\{\tau_k\} \to +\infty$ such that $\{H(\tau_k)\} \to 0$. Let us evaluate the variation of the function H. For any $\hat{\tau} > \tau_k$ we get:

$$|H(\tau_k) - H(\hat{\tau})|^2 = |\int_{\tau_k}^{\hat{\tau}} H'(t)dt|^2 = |\int_{\tau_k}^{\hat{\tau}} e^{-\gamma t}e^{\gamma t}H'(t)dt|^2$$

$$\leq \int_{\tau_k}^{\hat{\tau}} e^{-2\gamma t}dt \int_{\tau_k}^{\hat{\tau}} e^{2\gamma t}|H'(t)|^2dt$$

$$\leq \frac{1}{2\gamma}(e^{-2\gamma\tau_k} - e^{-2\gamma\hat{\tau}}) \int\limits_{\tau_k}^{+\infty} e^{2\gamma t}|H'(t)|^2 dt$$

$$\leq \frac{1}{2\gamma}e^{-2\gamma\tau_k}C_k\ , \quad C_k = \int\limits_{\tau_k}^{+\infty} e^{2\gamma t}|H'(t)|^2 dt\ .$$

Here the sequence $\{C_k\}$ converges to zero (due to the convergence of the integral $\int\limits_0^\infty |H'(\tau)|^2 \exp\{2\gamma\tau\}d\tau$). So,

$$\sup_{t\in[\tau_k,+\infty)} |H(t) - H(\tau_k)|^2 \leq \frac{1}{2\gamma}e^{-2\gamma\tau_k}C_k$$

and

$$\forall t \in [\tau_k, +\infty)\ |H(\tau_k) - H(t)|^2 \leq \frac{1}{2\gamma}e^{-2\gamma\tau_k}C_k\ .$$

Letting $t \to +\infty$, we get

$$|H(\tau_k)|^2 \leq \frac{1}{2\gamma}e^{-2\gamma\tau_k}C_k.$$

Thus,

$$e^{2\gamma\tau_k}H^2(\tau_k) \leq \frac{1}{2\gamma}C_k \to 0(k \to +\infty)\ .$$

Applying now Lemma 2.3.5 with $\hat{T} = \hat{\tau}_k$ and passing to the limit as $k \to \infty$, we finalize the proof. □

Remark 2.3.4 Condition $H \in W^{1,2}(0, +\infty)$ can be replaced by $H \in W^{1,2}_{\text{loc}}(0, +\infty)$ and $\lim\limits_{t\to+\infty} H(t) = 0$. The proof remains the same.

Lemma 2.3.6 *Let $f \in L^2_{loc}(0, +\infty)$. Suppose there exist constants $C > 0$ and α such that for any $\hat{T} > 1$,*

$$\int\limits_0^{\hat{T}} f^2(\tau)d\tau \leq C\hat{T}^\alpha$$

with $\alpha > 0$. Then, for any $\gamma > 0$:

$$\int_0^{+\infty} f^2(\tau)\exp\{-2\gamma\tau\}d\tau < \infty.$$

Proof
For any integer $N > 0$,

$$\int_0^N f^2(\tau)\exp\{-2\gamma\tau\}d\tau \le \sum_{j=0}^{N-1}\int_j^{j+1} f^2(\tau)d\tau\exp\{-2\gamma j\}$$
$$\le C\sum_{j=0}^{N-1}(j+1)^{\alpha}\exp\{-2\gamma j\}.$$

Evidently, this sum is uniformly bounded with respect to N. □

We shall also require the mean value theorem for Lebesgue integrals. This particular theorem was originally established by R. López Pouso [114]. Given the straightforward nature of the proof, we present it here for the reader's convenience.

Theorem 2.3.6 *If* $f : [a, b] \longmapsto \mathbb{R}$ *is Lebesgue integrable function, then*

$$\mathrm{mes}\Big(\Big\{c \in (a, b) : \; f(c) \le \frac{1}{a-b}\int_a^b f(x)dx\Big\}\Big) > 0, \tag{2.3.18}$$

and

$$\mathrm{mes}\Big(\Big\{c \in (a, b) : \; f(c) \ge \frac{1}{a-b}\int_a^b f(x)dx\Big\}\Big) > 0. \tag{2.3.19}$$

Here $\mathrm{mes}(A)$ *is the Lebesgue measure in* $\mathbb{R}$ *of the set* A.

Proof
The set

$$A = \Big\{c \in (a, b) : \; f(c) \le \frac{1}{a-b}\int_a^b f(x)dx\Big\}$$

$$= f^{-1}\Big([-\infty, \frac{1}{a-b}\int_a^b f(x)dx]\Big)$$

is Lebesgue measurable. If $\mathrm{mes}(A) = 0$, i.e., if for almost all $c \in [a, b]$, we have

$$f(c) > \frac{1}{a-b} \int_a^b f(x)dx.$$

Then, integrating both sides of the above inequality with respect to c, we obtain a contradiction:

$$\int_a^b f(x)dx > \int_a^b f(x)dx.$$

The proof of (2.3.19) is analogous. □

2.3.4 *Spaces of Functions Depending on x and t*

In this section, we define function spaces that depend on both the spacial variable x and the time variable t. Let $\Omega \subset \mathbb{R}^n$ be a bounded domain, and we denote $\Omega^T = \Omega \times (0, T)$. We introduce the space $W^{(2m,m),2}(\Omega^T)$, where $m \geq 1$ is an integer. This space consists of functions belonging to $L^2(\Omega^T)$ and possessing all generalized derivatives of the form $\partial_t^k \partial_x^\alpha$ with arbitrary k and α that satisfy the inequality $2k + |\alpha| \leq 2m$. The norm in $W^{(2m,m),2}(\Omega^T)$ is defined as follows:

$$\|u\|_{W^{(2m,m),2}(\Omega^T)} = \Big(\sum_{j=0}^{2m} \sum_{2k+|\alpha|=j} \int_0^T \int_\Omega |\partial_t^k \partial_x^\alpha u(x,t)|^2 \, dx \, dt\Big)^{1/2}.$$

Here, the summation in the second term is over all integers k and $|\alpha|$ satisfying the condition $2k + |\alpha| = j$. We also assume that $\partial\Omega$ is Lipschitz.

Furthermore, we introduce the spaces $W^{(1,1),2}(\Omega^T)$ and $W^{(1,0),2}(\Omega^T)$, which consist of functions with finite norms:

$$\|u\|^2_{W^{(1,1),2}(\Omega^T)} = \|u\|^2_{L^2(\Omega^T)} + \|\nabla_x u\|^2_{L^2(\Omega^T)} + \|\partial_t u\|^2_{L^2(\Omega^T)},$$

$$\|u\|^2_{W^{(1,0),2}(\Omega^T)} = \|u\|^2_{L^2(\Omega^T)} + \|\nabla_x u\|^2_{L^2(\Omega^T)}.$$

For all the spaces $W^{(\dots),2}(\Omega^T)$, we define $\mathring{W}^{(\dots),2}_{\partial\Omega}(\Omega^T)$ as the subspace of functions satisfying the condition $u|_{\partial\Omega} = 0$.

Next, we present the embedding and trace inequalities that are utilized in the book:

Theorem 2.3.7 *Let Ω be a Lipschitz bounded domain in $\mathbb{R}^n$.*

(i) If $u \in W^{(2m,m),2}(\Omega^T)$ and

$$p \geq 2, \quad 2m - 2r - s - \left(\frac{1}{2} - \frac{1}{p}\right)(n+2) \geq 0, \tag{2.3.20}$$

then $\partial_t^r \partial_x^s u \in L^p(\Omega^T)$ *and*

$$\|\partial_t^r \partial_x^s u\|_{L^p(\Omega^T)} \leq c\|u\|_{W^{(2m,m),2}(\Omega^T)}, \tag{2.3.21}$$

where the constant c *is independent of* T.

(ii) *If* $u \in W^{(2m,m),2}(\Omega^T)$ *and* $2r + s \leq 2m - 1$, *then the trace* $\partial_t^r \partial_x^s u\big|_{t=t_*} \in W^{2m-2r-s-1,2}(\Omega)$ *and*

$$\|\partial_t^r \partial_x^s u\big|_{t=t_*}\|_{W^{2m-2r-s-1,2}(\Omega)} \leq c\|u\|_{W^{(2m,m),2}(\Omega^T)} \tag{2.3.22}$$

with the constant c *independent of* $t_* \in [0, T]$ *and* T.

In particular, if $u \in W^{(2,1),2}(\Omega^T)$, *then*

$$\|u\big|_{t=t_*}\|_{W^{1,2}(\Omega)} \leq c\|u\|_{W^{(2,1),2}(\Omega^T)}. \tag{2.3.23}$$

For further information on the embedding for spaces of functions depending on x and t, one can refer to [107] and [27].

We also need to consider spaces of functions with values in Banach spaces, see [210, Sect. 1.2.6]. Let V be a Banach space. The space $L^2(0, T; V)$ consists of functions u such that $u(\cdot, t) \in V$ for almost all $t \in [0, T]$, and the norm of u in this space is given by:

$$\|u\|_{L^2(0,T;V)} = \Big(\int_0^T \|u(\cdot, t)\|_V^2 \, dt\Big)^{1/2},$$

where $\|u(\cdot, t)\|_V$ represents the norm of the function $u(\cdot, t)$ in the Banach space V.

Theorem 2.3.8 *Let* $H_1 \subset H_2$ *be separable Hilbert spaces having a common orthogonal base* $(\varphi_l, \ l = 1, 2, ...)$. *Let* $\mathcal{M}_1$ *be a subset of functions* u *from* $L^2(0, T; H_1)$ *having derivatives* $\partial_t u(t) \in H_2$, *such that for* $t, t + \tau \in [0, T]$,

$$\|\partial_t^m u(t + \tau) - \partial_t^m u(t)\|_{H_{m+1}} \to 0, \ m = 0, 1,$$

as $\tau \to 0$. *Let* H *be the space* $\{u \in L^2(0, T; H_1) : u_t \in L^2(0, T; H_2)\}$ *with the norm* $\|\cdot\|_H = \sqrt{\int_0^T \|u(t)\|_{H_1}^2 + \|u_t(t)\|_{H_2}^2}$. *Assume that there exists a constant* $C \geq 0$ *such that for any function* u *from the set* $\mathcal{M}_1$ *the following inequality holds:*

$$\|u\|_H \leq C \max_{t \in [0,T], m \in \{0,1\}} \|\partial_t^m u(t)\|_{H_{m+1}}.$$

Let $\mathcal{M}_2$ be a subset of functions u from $\mathcal{M}_1$ having the form $u(t) = \sum_{l=1}^{N} d_l(t)\varphi_l$ with $d_l \in C^1([0, T])$, and some $N \in \mathbb{N}$. If $\mathcal{M}_1$ is dense in H then $\mathcal{M}_2$ is dense in H.

Proof
Let u be a function from the space $\mathcal{M}_1$. Let us prove that $\mathcal{M}_2$ is dense in $\mathcal{M}_1$. Without loss of the generality, assume that the base $(\varphi_l, l = 1, 2, \ldots)$ is orthonormalized in H_2. Assume that the function u belongs to $\mathcal{M}_1$. Let ε be an arbitrary positive number. Let us divide the interval $[0, T]$ in K subintervals $[t_k, t_{k+1}]$, $k = 0, \ldots, K-1$, in such a way that

$$\|\partial_t^m u(t') - \partial_t^m u(t'')\|_{H_{m+1}} < \varepsilon/3,$$
$$m = 0, 1, \quad \forall t', t'' \in [t_k, t_{k+1}],\ k = 0, \ldots, K-1.$$

Then, we choose N so large that for all points t_k, $k = 0, \ldots, K$,

$$\|\partial_t^m u(t_k) - \sum_{l=1}^{N} d_l^{(m)}(t_k)\varphi_l\|_{H_{m+1}} < \varepsilon/3, \quad m = 0, 1,$$

where $d_l(t_k) = (u(t_k), \varphi_l)_{H_2}$ are inner products, i.e., $d_l(t_k)$ are the Fourier coefficients of the function $u(t_k)$ and $d_l^{(m)}(t_k) = (\partial_t^m u(t_k), \varphi_l)_{H_2}$ are the Fourier coefficients of $\partial_t^m u(t_k)$, $m = 0, 1$.

Then for any $t \in [t_k, t_{k+1}]$, we have

$$\|\partial_t^m u(t) - \sum_{l=1}^{N} d_l^{(m)}(t)\varphi_l\|_{H_{m+1}} \le \|\partial_t^m u(t) - \partial_t^m u(t_k)\|_{H_{m+1}}$$

$$+ \|\sum_{l=1}^{N} d_l^{(m)}(t_k)\varphi_l - \sum_{l=1}^{N} d_l^{(m)}(t)\varphi_l\|_{H_{m+1}} + \|\partial_t^m u(t_k) - \sum_{l=1}^{N} d_l^{(m)}(t_k)\varphi_l\|_{H_{m+1}}.$$

Since

$$\|\sum_{l=1}^{N} d_l^{(m)}(t_k)\varphi_l - \sum_{l=1}^{N} d_l^{(m)}(t)\varphi_l\|_{H_{m+1}} \le \|\partial_t^m u(t) - \partial_t^m u(t_k)\|_{H_{m+1}},$$

the right-hand side of the previous inequality can be estimated by

$$2\|\partial_t^m u(t) - \partial_t^m u(t_k)\|_{H_{m+1}} + \|\partial_t^m u(t_k) - \sum_{l=1}^{N} d_l^{(m)}(t_k)\varphi_l\|_{H_{m+1}} < \varepsilon,$$

and we deduce

$$\|\partial_t^m u(t) - \sum_{l=1}^{N} d_l^{(m)}(t)\varphi_l\|_{H_{m+1}} < \varepsilon.$$

Thus, for any function u from $\mathcal{M}_1$ and any $\varepsilon > 0$, there exists a function w from $\mathcal{M}_2$, such that

$$\max_{t\in[0,T],m\in\{0,1,\}} \|\partial_t^m u(t) - \partial_t^m w(t)\|_{H_{m+1}} \le \varepsilon,$$

and so,

$$\|u - w\|_H \le C\varepsilon,$$

which yields the density of $\mathcal{M}_2$ in $\mathcal{M}_1$. Hence, $\mathcal{M}_2$ is dense in H. □

Remark 2.3.5 Let $\theta(t) = \exp\{\frac{t^2}{t^2-1}\}$ for $|t| < 1$, $\theta(t) = 0$ for $|t| \ge 1$, $\kappa = \int_{-1}^{1} \theta(t)dt$. Introduce $u_h = \frac{1}{\kappa h} \int_{-\infty}^{+\infty} u(\cdot, \tau)\theta(\frac{t-\tau}{h})d\tau$. For Sobolev spaces $H_1 = W^{1,2}$, $H_2 = L^2$, it can be proved that there exists an extension of functions $u \in H$ to $(-\varepsilon, T + \varepsilon)$, $\varepsilon > 0$, such that the sequence of the averages u_h converges to u in the norm of H as $h \to 0$. In this case, $\mathcal{M}_1$ is dense in H.

2.3.5 *Function Spaces in Domains with Cylindrical Outlets to Infinity*

2.3.5.1 Definition of Domains with Cylindrical Outlets to Infinity

Consider the domain $\Omega \subset \mathbb{R}^n$, where $n = 2, 3$, with J cylindrical outlets to infinity: $\Omega = \Omega_0 \cup \left(\bigcup_{j=1}^{J} \Omega_j\right)$. Here, Ω_0 is a bounded domain, and $\Omega_0 \cap \Omega_j = \emptyset$ for $j \in \{1, \dots, J\}$, as well as $\Omega_j \cap \Omega_l = \emptyset$ for $j \ne l$, where $j, l \in \{1, \dots, J\}$. The outlets to infinity Ω_j in some coordinate systems $x^{(j)} = (x^{(j)\prime}, x_n^{(j)})$ with the origins within the boundary of domain Ω_0 are defined by

$$\Omega_j = \{x^{(j)} \in \mathbb{R}^n,\ x^{(j)\prime} \in \sigma_j,\ x_n^{(j)} \ge 0\},$$

where σ_j are bounded domains in $\mathbb{R}^{n-1}$, representing cross-sections of the cylinders (see Fig. 2.1).

Assume that for every $k \in \{1, \dots, J\}$ there exists a $\delta_j > 0$ such that the cylinder

$$\{x^{(j)} \in \mathbb{R}^n,\ x^{(j)\prime} \in \sigma_j,\ -\delta_j < x_n^{(j)} < 0\} \subset \Omega_0.$$

Fig. 2.1 Domain Ω

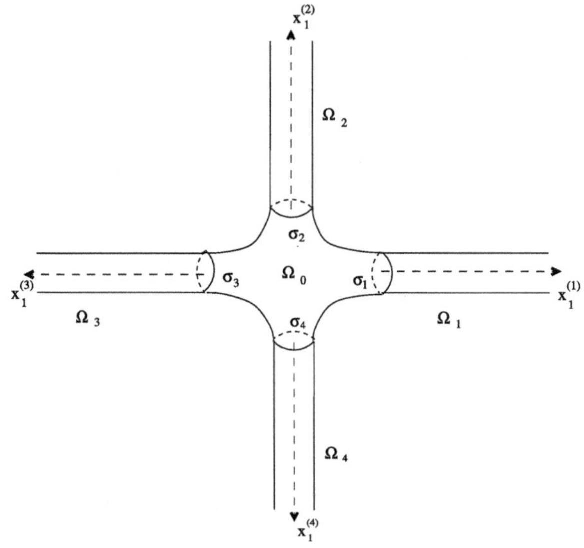

Let d_σ denote the maximal diameter of the cross-sections σ_j. We make the additional assumption that the intersection $\partial\Omega \cap \partial\Omega_0$ is not empty and has a positive measure. In particular, Ω can represent a semi-infinite cylinder: $\Omega = \{x \in \mathbb{R}^n,\ x' \in \sigma \subset \mathbb{R}^{n-1},\ x_n > 0\}$.

Clearly, there exists a positive real number $R > d_\sigma$ such that the ball $B_R = \{x \in \mathbb{R}^n,\ |x| < R\}$ contains Ω_0.

We introduce the following notation:

$$\Omega_{jk} = \{x \in \Omega_j : x_n^{(j)} < k\}, \quad \omega_{jk} = \Omega_{jk+1} \setminus \Omega_{jk}, \quad j = 1, \ldots, J,$$

$$\widehat{\omega}_{jk} = \omega_{jk-1} \cup \omega_{jk} \cup \omega_{jk+1}, \quad j = 1, \ldots, J, \tag{2.3.24}$$

$$\Omega^{(k)} = \Omega_0 \bigcup \Big(\bigcup_{j=0}^{J} \Omega_{jk} \Big),$$

where $k \geq 0$ is an integer. The domain with one outlet will be called a semi-cylinder.

Below all considered domains with cylindrical outlets to infinity are supposed to have at least Lipschitz boundary.

2.3.6 Definition of Weighted Function Spaces in Domains with Outlets to Infinity

In this section, we introduce the concept of weighted function spaces in domains with outlets to infinity. Let $\Omega \subset \mathbb{R}^n$ be a Lipschitz domain with J outlets to infinity. We define weighted function spaces within Ω. We use the notation $\boldsymbol{\beta} = (\beta, \dots, \beta)$ and introduce an infinitely smooth function $\phi_{\boldsymbol{\beta}}(x)$ that depends solely on $x_n^{(j)}$ in Ω_j, where $j = 1, 2, \dots, J$, and is defined as follows:

$$\phi_{\boldsymbol{\beta}}(x) = \begin{cases} 0, & x \in \Omega_0, \\ \beta x_n^{(j)}, & x \in \Omega_j, \quad x_n^{(j)} > 2, \quad j = 1, \dots, J, \end{cases} \tag{2.3.25}$$

We then define a function $E_{\boldsymbol{\beta}}(x)$ as follows:

$$E_{\boldsymbol{\beta}}(x) = \exp\{2\phi_{\boldsymbol{\beta}}(x)\}. \tag{2.3.26}$$

Next, we define the weighted function space $\mathcal{W}_{\boldsymbol{\beta}}^{l,q}(\Omega)$, where $l \geq 0$ and $q > 1$. It is the space of functions obtained by taking the closure of $C_0^\infty(\Omega)$ in the following norm:

$$\|u\|_{\mathcal{W}_{\boldsymbol{\beta}}^{l,q}(\Omega)} = \left(\sum_{|\alpha|=0}^{l} \int_\Omega E_{\boldsymbol{\beta}}^{q/2}(x) |D^\alpha u(x)|^q \, dx \right)^{1/q}.$$

Clearly, $\mathcal{W}_{\boldsymbol{\beta}}^{0,q}(\Omega) = \mathcal{L}_{\boldsymbol{\beta}}^q(\Omega)$. For $\beta > 0$, the functions in the space $\mathcal{W}_{\boldsymbol{\beta}}^{l,q}(\Omega)$ decay exponentially as $|x| \to \infty$.

Furthermore, it is worth noting that the following inclusions hold:

$$\mathcal{W}_{\boldsymbol{\beta}}^{l,q}(\Omega) \subset W^{l,q}(\Omega) \subset \mathcal{W}_{-\boldsymbol{\beta}}^{l,q}(\Omega) \quad \text{for} \quad \beta \geq 0.$$

Similarly, we define function spaces $\mathcal{W}_{\boldsymbol{\beta}}^{(2l,l),2}(\Omega^T)$, $\mathcal{W}_{\boldsymbol{\beta}}^{(1,1),2}(\Omega^T)$, and $\mathcal{W}_{\boldsymbol{\beta}}^{(1,0),2}(\Omega^T)$, which involve partial derivatives in both x and t directions. They are spaces of functions obtained as closures of infinitely many times differentiable functions with respect to x and t, which are equal to zero for large $|x|$. The norms for these function spaces are given by:

$$\|u\|_{\mathcal{W}_{\boldsymbol{\beta}}^{(2l,l),2}(\Omega^T)} = \left(\sum_{j=0}^{2l} \sum_{2r+|\alpha|=j} \int_0^T \int_\Omega E_{\boldsymbol{\beta}}(x) |\partial_t^r \partial_x^\alpha u(x,t)|^2 \, dx dt \right)^{1/2},$$

$$\|u\|_{\mathcal{W}_{\beta}^{(1,1),2}(\Omega^T)} = \left(\int_0^T \int_\Omega E_\beta(x) \left(|\partial_t u(x,t)|^2 + |u(x,t)|^2 + |\nabla u(x,t)|^2 \right) dx dt \right)^{1/2},$$

$$\|u\|_{\mathcal{W}_{\beta}^{(1,0),2}(\Omega^T)} = \left(\int_0^T \int_\Omega E_\beta(x) \left(|u(x,t)|^2 + |\nabla u(x,t)|^2 \right) dx dt \right)^{1/2}.$$

Finally, $\mathcal{L}^2_\beta(\Omega^T)$ is the space of functions with finite norm:

$$\|u\|_{\mathcal{L}^2_\beta(\Omega^T)} = \left(\int_0^T \int_\Omega E_\beta(x) |u(x,t)|^2 \, dx dt \right)^{1/2}.$$

These weighted function spaces allow us to study the decay properties of functions in domains with outlets to infinity. The following sections present embedding inequalities and theorems related to these function spaces.

2.3.7 Embedding Inequalities in Domains with Cylindrical Outlets

Lemma 2.3.7 (Poincaré Inequality) *For any function $u \in \mathcal{W}_\beta^{1,2}(\Omega)$ that is equal to zero on $\partial\Omega$, the following weighted Poincaré inequality holds:*

$$\|u\|^2_{\mathcal{L}^2_\beta(\Omega)} \le C \|\nabla u\|^2_{\mathcal{L}^2_\beta(\Omega)} \tag{2.3.27}$$

where the constant C is independent of β.

Proof

Since $E_\beta^{(k)}(x)$ is a smooth function and depends only on $x_n^{(j)}$ in $\{x \in \Omega_j : x_n^j > 2\}$ and is equal to 1 in Ω_0, we get (2.3.27) by applying Poincaré inequality (2.3.4) in the domains Ω_0, $\omega_{j1} \cup \omega_{j2}$, and on the cross-sections σ_j:

$$\int_\Omega E_\beta(x)|u(x)|^2 \, dx = \int_{\Omega_0} |u(x)|^2 \, dx + \sum_{j=1}^J \int_{\Omega_j} \exp\{2\phi_\beta(x)\}|u(x)|^2 \, dx$$

$$= \int_{\Omega_0} |u(x)|^2 \, dx + c \sum_{j=1}^J \int_{\omega_{j1} \cup \omega_{j2}} |u(x)|^2 \, dx$$

$$
\begin{aligned}
&+\sum_{j=1}^{J}\int_{2}^{\infty}\exp\{2\beta x_n^{(j)}\}\int_{\sigma_j}|u(x)|^2\,dx^{(j)\prime}dx_n^{(j)}\\
&\leq c\int_{\Omega^{(2)}}|\nabla u(x)|^2\,dx\\
&+c\sum_{j=1}^{J}\int_{2}^{\infty}\exp\{2\beta x_n^{(j)}\}\int_{\sigma_j}|\nabla' u(x)|^2\,dx^{(j)\prime}dx_n^{(j)}\\
&\leq c\int_{\Omega}E_{\boldsymbol{\beta}}(x)|\nabla u(x)|^2\,dx.
\end{aligned}
$$

□

Theorem 2.3.9 *Let $\Omega\subset\mathbb{R}^n$ be a domain with J cylindrical outlets to infinity. Let $u\in\mathcal{W}^{l,q}_{\boldsymbol{\beta}}(\Omega)$, where $l\geq 1$ is an integer, $q>1$, and $\beta>0$.*

(i) If $n>ql$ and $r\leq\frac{qn}{n-ql}$, then $u\in\mathcal{L}^r_{\boldsymbol{\beta}}(\Omega)$ and

$$\|u\|_{\mathcal{L}^r_{\boldsymbol{\beta}}(\Omega)}\leq c\|u\|_{\mathcal{W}^{l,q}_{\boldsymbol{\beta}}(\Omega)}. \tag{2.3.28}$$

(ii) If $n<ql$, then $u\in C^{h,\delta}(\bar{\Omega})$ with $h+\delta\leq\frac{ql-n}{q}$ and

$$\|E^{1/2}_{\boldsymbol{\beta}}u\|_{C^{h,\delta}(\bar{\Omega})}\leq c\|u\|_{\mathcal{W}^{l,q}_{\boldsymbol{\beta}}(\Omega)}. \tag{2.3.29}$$

Proof

(i) Let us represent the domain Ω as a union of bounded domains:

$$\Omega=\Omega_0\bigcup\Big(\bigcup_{j=1}^{J}\bigcup_{k=1}^{\infty}\omega_{jk}\Big).$$

By Theorem 2.3.5, in every ω_{jk}, we have the inequality

$$\|u\|_{L^r(\omega_{jk})}\leq c\|u\|_{W^{l,q}(\omega_{jk})}, \tag{2.3.30}$$

where the constant c is independent of k. Suppose that $r\geq q$. Multiplying inequalities (2.3.30) by $e^{\beta k}$ and having in mind that for large values of k, $e^{\beta k}\sim e^{\beta x_1^{(j)}}$ in ω_{jk}, we obtain

$$\|e^{\beta x_1^{(j)}} u\|^r_{L^r(\omega_{jk})} \le c\|e^{\beta x_1^{(j)}} u\|^r_{W^{l,q}(\omega_{jk})} \le c\|u\|^{r-q}_{\mathcal{W}^{l,q}_\beta(\Omega)}\|e^{\beta x_1^{(j)}} u\|^q_{W^{l,q}(\omega_{jk})}.$$

Here, the constants depend on β only. Summing these inequalities over k and j and adding the inequality

$$\|u\|^r_{L^r(\Omega^{(2)})} \le c\|u\|^{r-q}_{\mathcal{W}^{l,q}_\beta(\Omega)}\|u\|^q_{W^{1,2}(\Omega^{(2)})},$$

we obtain (2.3.28). The proof for $r < q$ is obvious due to the Hölder inequality.

(ii) Similarly, using the inequalities

$$\|u\|_{C^{h,\delta}(\bar\omega_{jk})} \le c\|u\|_{W^{l,q}(\omega_{jk})}, \quad \|u\|_{C^{h,\delta}(\bar\Omega^{(2)})} \le c\|u\|_{W^{l,q}(\Omega^{(2)})},$$

we get (2.3.29).

□

In particular, from the above theorem follows:

Lemma 2.3.8

(*i*) *For any function $u \in \mathcal{W}^{1,2}_\beta(\Omega)$,*

$$\|E_\beta^{1/2} u\|_{L^4(\Omega)} \le C\|u\|_{\mathcal{W}^{1,2}_\beta(\Omega)}. \tag{2.3.31}$$

(*ii*) *For any function $u \in \mathcal{W}^{2,2}_\beta(\Omega)$,*

$$\|E_\beta^{1/2} u\|_{L^\infty(\Omega)} \le C\|u\|_{\mathcal{W}^{2,2}_\beta(\Omega)}. \tag{2.3.32}$$

Theorem 2.3.10 *Let $\Omega \subset \mathbb{R}^n$ be a domain with J cylindrical outlets to infinity.*

(*i*) *If $u \in \mathcal{W}^{(2m,m),2}_\beta(\Omega^T)$ and*

$$p \ge 2, \quad 2m - 2r - s - \left(\frac{1}{2} - \frac{1}{p}\right)(n+2) \ge 0, \tag{2.3.33}$$

then $\partial_t^r \partial_x^s u \in \mathcal{L}^p_\beta(\Omega^T)$ and

$$\|\partial_t^r \partial_x^s u\|_{\mathcal{L}^p_\beta(\Omega^T)} \le c\|u\|_{\mathcal{W}^{(2m,m),2}_\beta(\Omega^T)}, \tag{2.3.34}$$

where the constant c is independent of T.

(ii) If $u \in \mathcal{W}_{\boldsymbol{\beta}}^{(2m,m),2}(\Omega^T)$ and $2r + s \le 2m - 1$, then the trace $\partial_t^r \partial_x^s u\big|_{t=t_*} \in \mathcal{W}_{\boldsymbol{\beta}}^{2m-2r-s-1,2}(\Omega)$ and

$$\|\partial_t^r \partial_x^s u\big|_{t=t_*}\|_{\mathcal{W}_{\boldsymbol{\beta}}^{2m-2r-s-1,2}(\Omega)} \le c\|u\|_{\mathcal{W}_{\boldsymbol{\beta}}^{(2m,m),2}(\Omega^T)}, \tag{2.3.35}$$

with the constant c independent of $t_* \in [0, T]$ and T.

The proof of this theorem can be conducted using the same ideas as in Theorem 2.3.9.

2.3.8 Multiplicative Inequalities in Tube Structures

In this section, we establish some multiplicative inequalities in the tube structure B_ε. We start by constructing two coverings of the domain B_ε.

Let us consider domains $A_{\varepsilon,k}^{(e_j)} = \{x \in \Pi_\varepsilon^{(e_j)} : x_n^{(e_j)} \in \varepsilon(k-2, k+2)\}$, where $j = 1, \dots, M$ and $k = 2, \dots, L_\varepsilon^j$, with $L_\varepsilon^j \sim |e|\varepsilon^{-1}$. Additionally, we define $A_{\varepsilon,k}^{(j)} = \omega_\varepsilon^j \cup \{x \in \Pi_\varepsilon^{(e_j)} : x_n^{(e_j)} \in (0, 2\varepsilon)\}$ when O^j are vertices and $A_{\varepsilon,k}^{(j)} = \omega_\varepsilon^j \cup \bigcup_{k_j}\{x \in \Pi_\varepsilon^{(e_{k_j})} : x_n^{(e_{k_j})} \in (0, 2\varepsilon)\}$ when O^j are nodes. Here, the union over k_j is taken over all edges of the bundle $\mathcal{B}^{(j)}$ associated with the node O^j.

In parallel with the covering

$$\mathfrak{A}_\varepsilon = \left(\bigcup_{j=1}^{M}\bigcup_{k=2}^{L_\varepsilon^j} A_{\varepsilon,k}^{(e_j)}\right) \bigcup \left(\bigcup_{j=1}^{N} A_{\varepsilon,k}^{(j)}\right),$$

we take the covering

$$\tilde{\mathfrak{A}}_\varepsilon = \left(\bigcup_{j=1}^{M}\bigcup_{k=2}^{\tilde{L}_\varepsilon^j} \tilde{A}_{\varepsilon,k}^{(e_j)}\right) \bigcup \left(\bigcup_{j=1}^{N} \tilde{A}_{\varepsilon,k}^{(j)}\right),$$

of B_ε containing larger domains: $\tilde{A}_{\varepsilon,k}^{(e_j)} = \{x \in \Pi_\varepsilon^{(e_j)} : x_n^{(e_j)} \in \varepsilon(k-3, k+3)\}$, where $j = 1, \dots, M$ and $k = 3, \dots, \tilde{L}_\varepsilon^j$, with $\tilde{L}_\varepsilon^j \sim |e|\varepsilon^{-1}$. Then we define $\tilde{A}_{\varepsilon,k}^{(j)} = \omega_\varepsilon^j \cup \{x \in \Pi_\varepsilon^{(e_j)} : x_n^{(e_j)} \in (0, 3\varepsilon)\}$ when O^j are vertices and $\tilde{A}_{\varepsilon,k}^{(j)} = \omega_\varepsilon^j \cup \bigcup_{k_j}\{x \in \Pi_\varepsilon^{(e_{k_j})} : x_n^{(e_{k_j})} \in (0, 3\varepsilon)\}$ when O^j are nodes.

Clearly, we have the inclusions:

$$A_{\varepsilon,k}^{(e_j)} \subset \tilde{A}_{\varepsilon,k}^{(e_j)}, \quad A_{\varepsilon,k}^{(j)} \subset \tilde{A}_{\varepsilon,k}^{(j)}.$$

Lemma 2.3.9 (Poincaré Inequality) *The following inequality holds for any $u \in \mathring{W}^{1,2}(B_\varepsilon)$:*

$$\|u\|_{L^2(B_\varepsilon)} \leq c\varepsilon \|\nabla u\|_{L^2(B_\varepsilon)}, \tag{2.3.36}$$

where c is a constant independent of ε.

The proof of the lemma follows immediately from Lemma 2.3.1 after the scaling $y = \varepsilon x$.

From Lemmas 2.3.2 and 2.3.9, we obtain the following inequalities:

Lemma 2.3.10 *For $B_\varepsilon \subset \mathbb{R}^2$, the following inequalities hold:*

$$\|u\|_{L^4(B_\varepsilon)} \leq c\varepsilon^{1/2} \|\nabla u\|_{L^2(B_\varepsilon)}, \quad \forall u \in \mathring{W}^{1,2}(B_\varepsilon), \tag{2.3.37}$$

and for $B_\varepsilon \subset \mathbb{R}^3$, the following inequalities hold:

$$\|u\|_{L^4(B_\varepsilon)} \leq c\varepsilon^{1/4} \|\nabla u\|_{L^2(B_\varepsilon)}, \quad \forall u \in \mathring{W}^{1,2}(B_\varepsilon). \tag{2.3.38}$$

The constant c in (2.3.37) and (2.3.38) is independent of ε.

Lemma 2.3.11 *Let $B_\varepsilon \subset \mathbb{R}^n$, $n = 2, 3$, and $u \in W^{1,2}(B_\varepsilon)$. Then we have the following inequalities:*

$$\|u\|_{L^4(B_\varepsilon)}^4 \leq c\varepsilon^{-2} \|u\|_{L^2(B_\varepsilon)}^2 \left(\|u\|_{L^2(B_\varepsilon)}^2 + \varepsilon^2 \|\nabla u\|_{L^2(B_\varepsilon)}^2 \right), \quad n = 2, \tag{2.3.39}$$

$$\|u\|_{L^4(B_\varepsilon)}^4 \leq c\varepsilon^{-3} \|u\|_{L^2(B_\varepsilon)} \left(\|u\|_{L^2(B_\varepsilon)}^2 + \varepsilon^2 \|\nabla u\|_{L^2(B_\varepsilon)}^2 \right)^{3/2}, \quad n = 3, \tag{2.3.40}$$

where c is a constant independent of ε.

Proof

Let $n = 2$. In any bounded Lipschitz domain Ω, the interpolation inequality (see [27, 106]) holds:

$$\|u\|_{L^4(\Omega)}^4 \leq c(\Omega) \|u\|_{L^2(\Omega)}^2 \left(\|u\|_{L^2(\Omega)}^2 + \|\nabla u\|_{L^2(\Omega)}^2 \right),$$

and, by Young's inequality,

$$\|u\|_{L^4(\Omega)}^4 \leq \delta^{-2} \|u\|_{W^{1,2}(\Omega)}^4 + c\delta^2 \|u\|_{L^2(\Omega)}^4, \quad \delta > 0.$$

By scaling, it is easy to see that for any domain A_ε in the covering $\mathfrak{A}_\varepsilon$, we get the estimate:

$$\|u\|^4_{L^4(A_\varepsilon)} \le \delta^{-2}\varepsilon^{-2}\left(\|u\|^2_{L^2(A_\varepsilon)} + \varepsilon^2\|\nabla u\|^2_{L^2(A_\varepsilon)}\right)^2 + c\delta^2\varepsilon^{-2}\|u\|^4_{L^2(A_\varepsilon)}, \quad \delta > 0.$$

Summing the above inequalities over all domains A_ε from the covering $\mathfrak{A}_\varepsilon$, we get

$$\|u\|^4_{L^4(B_\varepsilon)} \le \delta^{-2}\varepsilon^{-2}\left(\|u\|^2_{L^2(B_\varepsilon)} + \varepsilon^2\|\nabla u\|^2_{L^2(B_\varepsilon)}\right)^2 + c\delta^2\varepsilon^{-2}\|u\|^4_{L^2(B_\varepsilon)}, \quad \delta > 0.$$

Putting now $\delta^2 = \left(\|u\|^2_{L^2(B_\varepsilon)} + \varepsilon^2\|\nabla u\|^2_{L^2(B_\varepsilon)}\right)\|u\|^{-2}_{L^2(B_\varepsilon)}$ yields (2.3.39).

The case $n = 3$ can be proved by the same token. □

Lemma 2.3.12 *Let $B_\varepsilon \subset \mathbb{R}^n$, $u \in \mathring{W}^{1,2}(B_\varepsilon) \cap W^{2,2}(B_\varepsilon)$. Then we have the inequality:*

$$\|\nabla u\|^2_{L^2(B_\varepsilon)} \le c\|u\|_{L^2(B_\varepsilon)}\|\nabla^2 u\|_{L^2(B_\varepsilon)}, \tag{2.3.41}$$

and in particular,

$$\|\nabla u\|^2_{L^2(B_\varepsilon)} \le c\varepsilon^2\|\nabla^2 u\|^2_{L^2(B_\varepsilon)}, \tag{2.3.42}$$

where c is a constant independent of ε.

Proof
In any bounded Lipschitz domain Ω, the interpolation inequality (see [106]) holds:

$$\|\nabla u\|^2_{L^2(\Omega)} \le c\left(\delta\|\nabla^2 u\|^2_{L^2(\Omega)} + \delta^{-1}\|u\|^2_{L^2(\Omega)}\right), \quad \delta > 0.$$

By scaling in the ε-domain A_ε (i.e., A_ε in any direction is of "size" ε), we derive:

$$\varepsilon^2\|\nabla u\|^2_{L^2(A_\varepsilon)} \le c\left(\delta\varepsilon^4\|\nabla^2 u\|^2_{L^2(A_\varepsilon)} + \delta^{-1}\|u\|^2_{L^2(A_\varepsilon)}\right), \quad \delta > 0.$$

Summing the above inequalities over all A_ε from the covering $\mathfrak{A}_\varepsilon$, we get:

$$\varepsilon^2\|\nabla u\|^2_{L^2(B_\varepsilon)} \le c\left(\delta\varepsilon^4\|\nabla^2 u\|^2_{L^2(B_\varepsilon)} + \delta^{-1}\|u\|^2_{L^2(B_\varepsilon)}\right), \quad \delta > 0.$$

Putting $\delta = \varepsilon^{-2}\|u\|_{L^2(B_\varepsilon)}\|\nabla^2 u\|^{-1}_{L^2(B_\varepsilon)}$ implies (2.3.41). □

Lemma 2.3.13 *Let $B_\varepsilon \subset \mathbb{R}^3$, $u \in W^{1,2}(B_\varepsilon)$. Then we have the inequality:*

$$\|u\|^3_{L^3(B_\varepsilon)} \le c\varepsilon^{-3/2}\|u\|^{3/2}_{L^2(B_\varepsilon)}\left(\|u\|^2_{L^2(B_\varepsilon)} + \varepsilon^2\|\nabla u\|^2_{L^2(B_\varepsilon)}\right)^{3/4}, \tag{2.3.43}$$

where c is a constant independent of ε.

Proof
By the multiplicative inequality in a bounded Lipschitz domain Ω (e.g., [27, 106]):

$$\|u\|^3_{L^3(\Omega)} \le c\|u\|^{3/2}_{L^2(\Omega)} \left(\|u\|^2_{L^2(\Omega)} + \|\nabla u\|^2_{L^2(\Omega)}\right)^{3/4},$$

and, by the Young inequality:

$$\|u\|^3_{L^3(\Omega)} \le \delta^2\|u\|^2_{W^{1,2}(\Omega)} + c\delta^{-6}\|u\|^6_{L^2(\Omega)} \quad \text{for all } \delta > 0.$$

Then, by scaling, we get for any domains A_ε from the covering $\mathfrak{A}_\varepsilon$ the estimates:

$$\|u\|^3_{L^3(A_\varepsilon)} \le \delta^2\left(\|u\|^2_{L^2(A_\varepsilon)} + \varepsilon^2\|\nabla u\|^2_{L^2(A_\varepsilon)}\right) + c\delta^{-6}\varepsilon^{-6}\|u\|^6_{L^2(A_\varepsilon)},$$

and summing them over all A_ε from $\mathfrak{A}_\varepsilon$, we derive:

$$\|u\|^3_{L^3(B_\varepsilon)} \le \delta^2\left(\|u\|^2_{L^2(B_\varepsilon)} + \varepsilon^2\|\nabla u\|^2_{L^2(B_\varepsilon)}\right) + c\delta^{-6}\varepsilon^{-6}\|u\|^6_{L^2(B_\varepsilon)}.$$

Taking in the last inequality

$$\delta = \varepsilon^{-3/4}\|u\|^{3/4}_{L^2(B_\varepsilon)}\left(\|u\|^2_{L^2(B_\varepsilon)} + \varepsilon^2\|\nabla u\|^2_{L^2(B_\varepsilon)}\right)^{-1/8},$$

gives (2.3.43). □

Lemma 2.3.14 (Agmon's Inequality, $n = 3$**)** *Let* $B_\varepsilon \subset \mathbb{R}^3$, $u \in \mathring{W}^{1,2}(B_\varepsilon) \cap W^{2,2}(B_\varepsilon)$. *Then we have the inequality:*

$$\|u\|^4_{L^\infty(B_\varepsilon)} \le c\varepsilon^{-6}\|u\|_{L^2(B_\varepsilon)}\left(\|u\|^2_{L^2(B_\varepsilon)} + \varepsilon^2\|\nabla u\|^2_{L^2(B_\varepsilon)} + \varepsilon^4\|\nabla^2 u\|^2_{L^2(B_\varepsilon)}\right)^{3/2}, \tag{2.3.44}$$

where c is a constant independent of ε. *In particular,*

$$\|u\|_{L^\infty(B_\varepsilon)} \le c\varepsilon^{1/4}\|\nabla^2 u\|_{L^2(B_\varepsilon)}. \tag{2.3.45}$$

Proof
The proof of the first inequality (2.3.44) goes in the same way as the proof of Lemma 2.3.13 and is done by scaling. In any bounded Lipschitz domains Ω, we have the multiplicative inequality (see [5])

$$\|u\|^4_{L^\infty(\Omega)} \le c\|u\|_{L^2(\Omega)}\|u\|^3_{W^{2,2}(\Omega)}.$$

By scaling it is easy to see that in A_ε holds:

$$\|u\|^4_{L^\infty(A_\varepsilon)} \leq c\Big(\varepsilon^{-3/2}\|u\|_{L^2(A_\varepsilon)}\Big)\Big(\varepsilon^{-3}\|u\|^2_{L^2(A_\varepsilon)} + \varepsilon^{-1}\|\nabla u\|^2_{L^2(A_\varepsilon)}$$

$$+\varepsilon\|\nabla^2 u\|^2_{L^2(A_\varepsilon)}\Big)^{3/2} \leq c\varepsilon^{-6}\|u\|_{L^2(A_\varepsilon)}\Big(\|u\|^2_{L^2(A_\varepsilon)}$$

$$+\varepsilon^2\|\nabla u\|^2_{L^2(A_\varepsilon)} + \varepsilon^4\|\nabla^2 u\|^2_{L^2(A_\varepsilon)}\Big)^{3/2}$$

$$\leq c\varepsilon^{-6}\|u\|_{L^2(B_\varepsilon)}\Big(\|u\|^2_{L^2(B_\varepsilon)} + \varepsilon^2\|\nabla u\|^2_{L^2(B_\varepsilon)} + \varepsilon^4\|\nabla^2 u\|^2_{L^2(B_\varepsilon)}\Big)^{3/2}.$$

Taking the supremum over all $A_\varepsilon \in \mathfrak{A}_\varepsilon$, we obtain (2.3.44). The inequality (2.3.45) follows from (2.3.44), (2.3.36), and (2.3.42). □

Lemma 2.3.15 (Agmon's Inequality, $n = 2$) *Let $B_\varepsilon \subset \mathbb{R}^2$, and $u \in \mathring{W}^{1,2}(B_\varepsilon) \cap W^{2,2}(B_\varepsilon)$. Then, the following inequality holds:*

$$\|u\|^4_{L^\infty(B_\varepsilon)} \leq c\varepsilon^{-4}\|u\|^2_{L^2(B_\varepsilon)}\left(\|u\|^2_{L^2(B_\varepsilon)} + \varepsilon^2\|\nabla u\|^2_{L^2(B_\varepsilon)} + \varepsilon^4\|\nabla^2 u\|^2_{L^2(B_\varepsilon)}\right), \tag{2.3.46}$$

where c is a constant independent of ε. Consequently, we have:

$$\|u\|_{L^\infty(B_\varepsilon)} \leq c\varepsilon^{1/2}\|\nabla^2 u\|_{L^2(B_\varepsilon)}. \tag{2.3.47}$$

Proof
In any bounded Lipschitz two-dimensional domain Ω, the interpolation inequality (see [5]) holds:

$$\|u\|^4_{L^\infty(\Omega)} \leq c\|u\|^2_{L^2(\Omega)}\|u\|^2_{W^{2,2}(\Omega)}.$$

By scaling, we obtain:

$$\|u\|^4_{L^\infty(A_\varepsilon)} \leq c\varepsilon^{-4}\|u\|^2_{L^2(A_\varepsilon)}\left(\|u\|^2_{L^2(A_\varepsilon)} + \varepsilon^2\|\nabla u\|^2_{L^2(A_\varepsilon)} + \varepsilon^4\|\nabla^2 u\|^2_{L^2(A_\varepsilon)}\right),$$

where A_ε is any domain from the covering $\mathfrak{A}_\varepsilon$ of B_ε. Taking the supremum over all $A_\varepsilon \in \mathfrak{A}_\varepsilon$, we obtain (2.3.46). □

Lemma 2.3.16 *Let $B_\varepsilon \subset \mathbb{R}^n$, where $n = 2, 3$, and $u \in W^{1,2}(B_\varepsilon)$. Then the following inequalities hold:*

$$\|u\|^2_{L^2(\gamma^j_\varepsilon)} \leq c\varepsilon^{-1}\left(\|u\|^2_{L^2(B_\varepsilon)} + \varepsilon^2\|\nabla u\|^2_{L^2(B_\varepsilon)}\right) \leq c\varepsilon\|\nabla u\|^2_{L^2(B_\varepsilon)}, \tag{2.3.48}$$

where c is a constant independent of ε.

Proof
The inequality (2.3.48) follows immediately from the embedding estimate (see Theorem 2.3.5 (v)):

$$\|v\|_{L^2(\partial\Omega)} \leq c\|v\|_{W^{1,2}(\Omega)},$$

and a scaling argument. □

Let $\mathcal{G}$ be a domain in $\mathbb{R}^n$. In the Sobolev space $W^{l,2}(\mathcal{G})$, we introduce the equivalent (for the fixed ε) norm:

$$|||u|||^2_{l,\alpha,\mathcal{G}} = \sum_{k=0}^{l} \varepsilon^{-2(l-k)+2\alpha} \|\nabla^k u\|^2_{L^2(\mathcal{G})}.$$

Obviously,

$$|||u|||^2_{l,\alpha,\mathcal{G}} = \varepsilon^{2\alpha} |||u|||^2_{l,0,\mathcal{G}}.$$

Lemma 2.3.17 *Let $B_\varepsilon \subset \mathbb{R}^n$, where $n = 2, 3$, and $u \in W^{1,2}(B_\varepsilon)$. Then the following inequalities hold:*

$$\|u\|^4_{L^4(A_\varepsilon)} \leq c\varepsilon^{-n+4} |||u|||^4_{1,0,A_\varepsilon}, \tag{2.3.49}$$

where A_ε is any domain from the covering $\mathfrak{A}_\varepsilon$, and

$$\|u\|^4_{L^4(B_\varepsilon)} \leq c\varepsilon^{-n+4} |||u|||^4_{1,0,B_\varepsilon}. \tag{2.3.50}$$

Proof
By the embedding theorem, in any bounded Lipschitz domain Ω, the inequality

$$\|u\|^4_{L^4(\Omega)} \leq c(\Omega) \left(\|u\|^2_{L^2(\Omega)} + \|\nabla u\|^2_{L^2(\Omega)}\right)^2$$

holds, where $c(\Omega)$ is a constant depending on Ω. By scaling, it is easy to see that in any A_ε, the estimate

$$\|u\|^4_{L^4(A_\varepsilon)} \leq c\varepsilon^{-n+4} \left(\varepsilon^{-2}\|u\|^2_{L^2(A_\varepsilon)} + \|\nabla u\|^2_{L^2(A_\varepsilon)}\right)^2 = c\varepsilon^{-n+4} |||u|||^4_{1,0,A_\varepsilon}$$

holds with the constant c independent of ε. From this, it follows that

$$\|u\|^4_{L^4(A_\varepsilon)} \leq c\varepsilon^{-n+4} |||u|||^2_{1,0,B_\varepsilon} |||u|||^2_{1,0,A_\varepsilon}.$$

Summing the last inequalities over all domains A_ε from the covering $\mathfrak{A}_\varepsilon$, we get (2.3.50). □

Similarly, using the continuous embedding of $W^{2,2}(\Omega)$ into $L^\infty(\Omega)$, we obtain the following lemma:

Lemma 2.3.18 *Let $B_\varepsilon \subset \mathbb{R}^n$, and $u \in W^{2,2}(B_\varepsilon)$. Then the following inequalities hold:*

$$\begin{aligned}\|u\|_{L^\infty(A_\varepsilon)} &\le c\varepsilon^{-(n-4)/2}|||u|||_{2,0,A_\varepsilon},\\ \|u\|_{L^\infty(B_\varepsilon)} &\le c\varepsilon^{-(n-4)/2}|||u|||_{2,0,B_\varepsilon},\end{aligned} \tag{2.3.51}$$

where c is a constant independent of ε.

Proof
In a bounded Lipschitz domain Ω, the inequality

$$\|u\|^2_{L^\infty(\Omega)} \le c(\Omega)\left(\|u\|^2_{L^2(\Omega)} + \|\nabla u\|^2_{L^2(\Omega)} + \|\nabla^2 u\|^2_{L^2(\Omega)}\right)$$

holds (see Theorem 2.3.5). Then, by scaling, we obtain the estimate

$$\|u\|^2_{L^\infty(A_\varepsilon)} \le c\varepsilon^{-n+4}|||u|||^2_{2,0,A_\varepsilon}.$$

with the constant c independent of ε. From this, it follows that

$$\|u\|_{L^\infty(B_\varepsilon)} \le \sup_{A_\varepsilon\in\mathfrak{A}_\varepsilon} \|u\|_{L^\infty(A_\varepsilon)} \le c\varepsilon^{-(n-4)/2}|||u|||_{2,0,B_\varepsilon}.$$

□

From Lemmas 2.3.17 and 2.3.18, we deduce the following result:

Lemma 2.3.19 *Let $u \in W^{3,2}(B_\varepsilon)$. Then the following inequalities hold:*

$$\|\nabla u\|_{L^\infty(B_\varepsilon)} \le c\varepsilon^{-(n-4)/2}|||u|||_{3,0,B_\varepsilon}, \tag{2.3.52}$$

$$\|\nabla u\|_{L^4(B_\varepsilon)} \le c\varepsilon^{-(n-8)/4}|||u|||_{3,0,B_\varepsilon}, \tag{2.3.53}$$

$$\|\nabla^2 u\|_{L^4(B_\varepsilon)} \le c\varepsilon^{-(n-4)/4}|||u|||_{3,0,B_\varepsilon}. \tag{2.3.54}$$

The same inequalities hold for functions $u \in W^{3,2}(A_\varepsilon)$.

2.4 Some Problems of Vector Analysis

2.4.1 The Divergence Equation

This subsection addresses various problems associated with the vector analysis operators curl and div, which are of great significance in the study of divergence-free vector fields and in the general theory of the Navier–Stokes equations.

Let Ω be a bounded domain in $\mathbb{R}^n$, where $n = 2, 3$. We denote

$$\widehat{L}^2(\Omega) = \{h \in L^2(\Omega) : \int\limits_\Omega h(x)dx = 0\}.$$

We consider the following problem.

Problem I Given $h \in \widehat{L}^2(\Omega)$, find a vector field $\mathbf{w} \in \overset{\circ}{W}{}^{1,2}(\Omega)$ satisfying the equation

$$\operatorname{div}\mathbf{w} = h, \tag{2.4.1}$$

and the estimate

$$\|\nabla \mathbf{w}\|_{L^2(\Omega)} \le c\|h\|_{L^2(\Omega)} \tag{2.4.2}$$

where c is a constant dependent solely on Ω.

Note that the condition $\int_\Omega h(x)\,dx = 0$ is necessary for the solvability of the above problem. This requirement follows from the Stokes formula:

$$\int\limits_\Omega h\,dx = \int\limits_\Omega \operatorname{div}\mathbf{w}\,dx = \int\limits_{\partial\Omega} \mathbf{w}\cdot\mathbf{n}\,dS = 0,$$

where $\mathbf{n}$ is the outward unit normal to the boundary of Ω.

The following two lemmas were proven in [104].

Lemma 2.4.1 *Problem I admits at least one solution* $\mathbf{w} \in \overset{\circ}{W}{}^{1,2}(\Omega)$ *in any bounded domain* $\Omega \subset \mathbb{R}^n$, $n = 2, 3$, *with Lipschitz boundary* $\partial\Omega$.

The next result describes the representation of bounded linear functionals that vanish on divergence-free functions from $\overset{\circ}{W}{}^{1,2}(\Omega)$.

Lemma 2.4.2 *Let* Ω *be a bounded domain with Lipschitz boundary. Then, any bounded linear functional* $L(\mathbf{u})$ *defined on vector fields* $\mathbf{u} \in \overset{\circ}{W}{}^{1,2}(\Omega)$ *and vanishing on the set* $\{\mathbf{v} \in \overset{\circ}{W}{}^{1,2}(\Omega) : \operatorname{div}\mathbf{v} = 0\}$ *can be uniquely represented in the form*

$$L(\mathbf{u}) = \int\limits_\Omega p \operatorname{div}\mathbf{u}\,dx \quad \forall \mathbf{u} \in \overset{\circ}{W}{}^{1,2}(\Omega), \tag{2.4.3}$$

where $p \in \widehat{L}^2(\Omega)$, *and*

$$\|L\|_{(\overset{\circ}{W}{}^{1,2}(\Omega))^*} = \|p\|_{L^2(\Omega)}. \tag{2.4.4}$$

Now, let us consider a similar problem with a non-homogeneous boundary condition.

Problem I$'$ Given $h \in L^2(\Omega)$ and $\boldsymbol{\alpha} \in W^{1/2,2}(\partial\Omega)$ satisfying the compatibility condition

$$\int_\Omega h(x)\,dx = \int_{\partial\Omega} \boldsymbol{\alpha}(x)\cdot\mathbf{n}(x)\,dS, \tag{2.4.5}$$

find a vector field $\mathbf{w} \in W^{1,2}(\Omega)$ satisfying the boundary condition $\mathbf{w}|_{\partial\Omega} = \boldsymbol{\alpha}$, the equation

$$\mathrm{div}\mathbf{w}(x) = h(x), \tag{2.4.6}$$

and the estimate

$$\|\nabla\mathbf{w}\|_{L^2(\Omega)} \le c\Big(\|h\|_{L^2(\Omega)} + \|\boldsymbol{\alpha}\|_{W^{1/2,2}(\partial\Omega)}\Big), \tag{2.4.7}$$

where c is a constant dependent only on Ω.

Lemma 2.4.3 *Let $\Omega \subset \mathbb{R}^n$ be a bounded domain with Lipschitz boundary. Then, for any given $h \in L^2(\Omega)$ and $\boldsymbol{\alpha} \in W^{1/2,2}(\partial\Omega)$ satisfying the compatibility condition* (2.4.5)*, there exists a vector field $\mathbf{w} \in W^{1,2}(\Omega)$ such that*

$$\mathrm{div}\,\mathbf{w}(x) = h(x), \quad \mathbf{w}(x)|_{\partial\Omega} = \boldsymbol{\alpha}(x) \tag{2.4.8}$$

and the estimate (2.4.7) *is valid.*

Proof
By the Trace theorem 2.3.4, there exists an extension $\mathbf{A} \in W^{1,2}(\Omega)$ of $\boldsymbol{\alpha}$ into the domain Ω such that

$$\|\mathbf{A}\|_{W^{1,2}(\Omega)} \le c\|\boldsymbol{\alpha}\|_{W^{1/2,2}(\partial\Omega)}. \tag{2.4.9}$$

Next, we take $\mathbf{u} \in \overset{\circ}{W}{}^{1,2}(\Omega)$ as the solution of Problem I with the right-hand side $h_1 = h - \mathrm{div}\,\mathbf{A}$. Since

$$\int_\Omega \mathrm{div}\,\mathbf{A}(x)\,dx = \int_{\partial\Omega} \mathbf{A}\cdot\mathbf{n}\,dS = \int_{\partial\Omega} \boldsymbol{\alpha}\cdot\mathbf{n}\,dS,$$

by (2.4.5), we have

$$\int_\Omega h_1(x)\,dx = \int_\Omega h(x)\,dx - \int_\Omega \mathrm{div}\,\mathbf{A}(x)\,dx = \int_\Omega h(x)\,dx - \int_{\partial\Omega} \boldsymbol{\alpha}\cdot\mathbf{n}\,dS = 0.$$

Therefore, there exists $\mathbf{u} \in \overset{\circ}{W}{}^{1,2}(\Omega)$ with $\mathrm{div}\,\mathbf{u} = h_1$ and

$$\|\nabla \mathbf{u}\|_{L^2(\Omega)} \leq c\|h_1\|_{L^2(\Omega)} \leq c\Big(\|h\|_{L^2(\Omega)} + \|\nabla \mathbf{A}\|_{L^2(\Omega)}\Big). \tag{2.4.10}$$

Setting $\mathbf{w}(x) = \mathbf{u}(x) + \mathbf{A}(x)$, we obtain the vector field satisfying (2.4.8). The estimate (2.4.7) for $\mathbf{w}$ follows from (2.4.9) and (2.4.10). □

Remark 2.4.1 The analogous results in the L^q-setting were proved in [170, 172], and [31].

The following generalization of Lemma 2.4.1 was proved in [31].

Lemma 2.4.4 *Let $\Omega \subset \mathbb{R}^n$ be a bounded domain. For any given $h \in \widehat{L}^2(\Omega) \cap \mathring{W}^{k,2}(\Omega)$, there exists a vector field $\mathbf{w} \in \mathring{W}^{k+1,2}(\Omega)$ such that*

$$\operatorname{div} \mathbf{w}(x) = h(x) \tag{2.4.11}$$

and the estimate holds

$$\|\mathbf{w}\|_{W^{k+1,2}(\Omega)} \leq c\|h\|_{W^{k,2}(\Omega)}, \tag{2.4.12}$$

where c is a constant depending only on Ω.

This result, established in [31], is obtained using an explicit representation formula for the solution, adapting for the divergence operator a well-known Sobolev's representation formula.

If the boundary $\partial\Omega$ is more regular and $h(x)$ does not necessarily vanish on the boundary, it is possible to find a solution $\mathbf{w}(x)$ that also possesses more regularity.

Lemma 2.4.5 *Let $\Omega \subset \mathbb{R}^n$ be a bounded domain with $\partial\Omega \in C^{l+2}$ for $l \geq 0$. Then, for every given $h \in W^{l,2}(\Omega) \cap \widehat{L}^2(\Omega)$, there exists a vector field $\mathbf{w} \in W^{l+1,2}(\Omega) \cap \mathring{W}^{1,2}(\Omega)$ such that* $\operatorname{div}\mathbf{w}(x) = h(x)$ *and the following estimate*

$$\|\mathbf{w}\|_{W^{l+1,2}(\Omega)} \leq c\|h\|_{W^{l,2}(\Omega)} \tag{2.4.13}$$

holds. Moreover, if $\Omega_1 \subset \Omega_2$ are two subdomains of Ω and $\Gamma_i = \partial\Omega_i \setminus \partial\Omega$ for $i = 1, 2$ are finite unions of C^{l+2}-smooth surfaces, $\operatorname{dist}(\Gamma_1, \Gamma_2) > 0$, *and* $\operatorname{supp} h(x) \subset \overline{\Omega_1}$, *then there exists a solution $\mathbf{w}$ with*

$$\operatorname{supp}\, \mathbf{w}(x) \subset \overline{\Omega_2}.$$

The proof of Lemma 2.4.5 can be found in [85]. It is worth noting that, in contrast to Lemma 2.4.4, here we do not require $h(x)$ to be equal to zero in the neighborhood of the entire boundary $\partial\Omega$, which means that boundaries $\partial\Omega_1$, $\partial\Omega_2$, and $\partial\Omega$ may have nonzero intersections.

Remark 2.4.2 Let $\Omega \subset \mathbb{R}^n$ be a bounded domain with the boundary $\partial\Omega$ of Ω of class C^{l+2}, with $l \geq 0$. We consider functions $h \in W^{l,2}(\Omega)$ and $\boldsymbol{\alpha} \in W^{l+1/2,2}(\partial\Omega)$. Additionally, assume that the following equality holds:

$$\int\limits_{\Omega} h(x)dx = \int\limits_{\partial\Omega} \boldsymbol{\alpha} \cdot \mathbf{n} dS.$$

Consider the following problem

$$\begin{cases} \operatorname{div} \mathbf{w}(x) = h(x), \ x \in \Omega, \\ \quad \mathbf{w}(x) = \boldsymbol{\alpha}(x), \ x \in \partial\Omega. \end{cases} \tag{2.4.14}$$

This problem can be reduced to the case discussed in Lemma 2.4.5. Specifically, we can construct an extension $\widetilde{\mathbf{w}} \in W^{l+1,2}(\Omega)$ of $\boldsymbol{\alpha}(x)$ into Ω satisfying the inequality:

$$\|\widetilde{\mathbf{w}}\|_{W^{l+1,2}(\Omega)} \leq c\|\boldsymbol{\alpha}\|_{W^{l+1/2,2}(\partial\Omega)}.$$

We seek the solution $\mathbf{w}(x)$ for Eq. (2.4.15) in the form $\mathbf{w}(x) = \widehat{\mathbf{w}}(x) + \widetilde{\mathbf{w}}(x)$. Then, for the vector-function $\widehat{\mathbf{w}}(x)$, we obtain the problem:

$$\begin{cases} \operatorname{div} \widehat{\mathbf{w}}(x) = \widehat{h}(x), \ x \in \Omega, \\ \quad \widehat{\mathbf{w}}(x) = 0, \qquad x \in \partial\Omega, \end{cases} \tag{2.4.15}$$

where $\widehat{h}(x) = h(x) - \operatorname{div}\widetilde{\mathbf{w}}(x)$. It can be observed that $\widehat{h} \in \widehat{L}^2(\Omega) \cap W^{l,2}(\Omega)$. By applying Lemma 2.4.5, we find the existence of $\widehat{\mathbf{w}}$ that satisfies the following estimate:

$$\|\widehat{\mathbf{w}}\|_{W^{l+1,2}(\Omega)} \leq c\|\widehat{h}\|_{W^{l,2}(\Omega)}. \tag{2.4.16}$$

The sum $\mathbf{w} = \widehat{\mathbf{w}} + \widetilde{\mathbf{w}}$ solves (2.4.15), and we have the following estimate:

$$\|\mathbf{w}\|_{W^{l+1,2}(\Omega)} \leq c\left(\|h\|_{W^{l,2}(\Omega)} + \|\boldsymbol{\alpha}\|_{W^{l+1/2,2}(\partial\Omega)}\right). \tag{2.4.17}$$

Now, consider two subdomains $\Omega_1 \subset \Omega_2$ of Ω, which satisfy the conditions of Lemma 2.4.5. If $\operatorname{supp} h(x) \subset \overline{\Omega}_1$ and $\operatorname{supp} \boldsymbol{\alpha}(x) \subset \overline{\Omega}_1 \cap \partial\Omega$, then we can select the extension $\widetilde{\mathbf{w}}$ in such a way that $\operatorname{supp} \widetilde{\mathbf{w}}(x)$ is contained in a neighborhood of $\overline{\Omega}_1 \cap \partial\Omega$. Using Lemma 2.4.5, we deduce that

$$\operatorname{supp}\ \mathbf{w}(x) \subset \overline{\Omega}_2.$$

Now, we consider a bounded domain Ω in $\mathbb{R}^n$, where $n = 2, 3$, with a multiply connected boundary $\partial\Omega$ consisting of $J + 1$ disjoint components Γ_j:

$$\Omega = \Omega_0 \setminus \bigcup_{i=1}^{I} \Omega_i, \tag{2.4.18}$$

where Ω_i are bounded domains with connected boundaries Γ_i, $\bar{\Omega}_i \subset \Omega_0$, $i = 1, \dots, I$, and $\partial\Omega = \Gamma = \Gamma_0 \cup \Gamma_1 \cup \dots \cup \Gamma_J$ with $\Gamma_i \cap \Gamma_j = \emptyset$ for $i \neq j$.

Suppose $\boldsymbol{\alpha} \in W^{k-1/2,2}(\partial\Omega)$ satisfies the following condition:

$$\int\limits_{\Gamma_j} \boldsymbol{\alpha} \cdot \mathbf{n}\, dS = 0, \quad j = 0, 1, \dots, J. \tag{2.4.19}$$

Lemma 2.4.6 *Assume that* $\Gamma \in C^{k+1}$. *For a given* $\boldsymbol{\alpha} \in W^{k-1/2,2}(\Gamma)$ *with* $k \geq 1$ *satisfying* (2.4.19), *there exists a divergence-free vector field* $\mathbf{A} \in W^{k,2}(\Omega)$ *such that*

$$\mathbf{A}(x)|_\Gamma = \boldsymbol{\alpha}(x)|_\Gamma, \quad \mathbf{A}(x) = \operatorname{curl} \mathbf{B}(x) \tag{2.4.20}$$

where $\mathbf{B} \in W^{k+1,2}(\Omega)$, *and the following estimate is valid:*

$$\|\mathbf{B}\|_{W^{k+1,2}(\Omega)} \leq c\|\boldsymbol{\alpha}\|_{W^{k-1/2,2}(\partial\Omega)}. \tag{2.4.21}$$

Moreover, the vector field $\mathbf{A}$ *can be constructed in such a way that* $\operatorname{supp} \mathbf{A}$ *is contained in an arbitrarily small neighborhood of* $\operatorname{supp} \boldsymbol{\alpha} \subset \Gamma$.

In the two-dimensional case, the operator curl *is defined as* $\operatorname{curl} B(x) = \nabla^\perp B(x) = \left(\frac{\partial B(x)}{\partial x_2}, -\frac{\partial B(x)}{\partial x_1}\right)$, *where* B *is a scalar function.*

The above lemma is proven [85]. Note that condition (2.4.19) is necessary. If $\boldsymbol{\alpha}$ satisfies a more general condition

$$\int\limits_{\Gamma} \boldsymbol{\alpha} \cdot \mathbf{n}\, dS = \sum_{i=0}^{I} \int\limits_{\Gamma_j} \boldsymbol{\alpha} \cdot \mathbf{n}\, dS = 0,$$

where the integrals $\int_{\Gamma_j} \boldsymbol{\alpha} \cdot \mathbf{n}\, dS$ are not equal to zero for some j, then it is not possible to construct the extension in the form of curl.

Let us now consider a domain Ω with J cylindrical outlets to infinity and define the weight-function $E_{\boldsymbol{\beta}}(x)$ as given in (2.3.26). Additionally, we introduce the "step" weight-function $E_{\boldsymbol{\beta}}^{(k)}(x)$ as follows:

$$E_{\boldsymbol{\beta}}^{(k)}(x) = \begin{cases} 1, & x \in \Omega_0, \\ E_\beta(x_n^{(j)}), & x \in \Omega_{jk}, \quad j = 1, \dots, J, \\ E_\beta(k), & x \in \Omega_j \setminus \Omega_{jk}, \quad j = 1, \dots, J. \end{cases} \tag{2.4.22}$$

It is evident that $E_{\boldsymbol{\beta}}^{(k)}(x) = E_{\boldsymbol{\beta}}(x)$ for $x \in \Omega^{(k)}$. Furthermore, we have the inequality

$$|\nabla E_{\boldsymbol{\beta}}^{(k)}(x)| \le b_0|\beta| E_{\boldsymbol{\beta}}^{(k)}(x), \tag{2.4.23}$$

where b_0 is a positive constant. If $\beta \ge 0$, then

$$E_{\beta}^{(k)}(x) \le E_{\beta}(x) \quad \forall x \in \Omega.$$

Moreover, by definition,

$$\frac{\partial}{\partial x_n^{(j)}} E_{\boldsymbol{\beta}}^{(k)}(x) = \frac{\partial}{\partial x_n^{(j)}} E_{\boldsymbol{\beta}}(x) \text{ for } x \in \Omega_{jk}, \quad j = 1, \dots, J,$$

$$\frac{\partial}{\partial x_l^{(j)}} E_{\boldsymbol{\beta}}^{(k)}(x) = 0 \text{ for } l \ne n,$$

and

$$\nabla E_{\boldsymbol{\beta}}^{(k)}(x) = 0 \text{ for } x \in \Omega \setminus \Big(\bigcup_{j=1}^{J} \Omega_{jk}\Big).$$

Thus,

$$\operatorname{supp} \nabla E_{\boldsymbol{\beta}}^{(k)} \subset \bigcup_{j=1}^{J} \overline{\Omega}_{jk}. \tag{2.4.24}$$

We will need the following auxiliary result for the study of decay properties of the solutions to the Stokes and Navier–Stokes problems:

Lemma 2.4.7 *Let* $\mathbf{u} \in \mathring{W}^{1,2}(\Omega)$, $\operatorname{div} \mathbf{u}(x) = 0$. *Assume* $\mathbf{u}$ *satisfies the conditions:*

$$\int_{\sigma_j} \mathbf{u}(x) \cdot \mathbf{n}(x)\, ds = 0, \qquad j = 1, \dots, J. \tag{2.4.25}$$

Then there exists a vector-function $\mathbf{W}^{(k)} \in \mathring{W}^{1,2}(\Omega)$ *such that*

$$\operatorname{supp} \mathbf{W}^{(k)}(x) \subset \bigcup_{j=1}^{J} \Omega_{jk} \tag{2.4.26}$$

and

$$\operatorname{div} \mathbf{W}^{(k)}(x) = -\operatorname{div} \big(E_{\boldsymbol{\beta}}^{(k)}(x)\mathbf{u}(x)\big), \qquad x \in \Omega. \tag{2.4.27}$$

Moreover, the following estimate

$$\begin{aligned}\int\limits_{\Omega} E_{-\boldsymbol{\beta}}^{(k)}(x)|\nabla \mathbf{W}^{(k)}(x)|^2\,dx &\le c\beta^2 \int\limits_{\Omega} E_{\boldsymbol{\beta}}^{(k)}(x)|\mathbf{u}(x)|^2\,dx \\ &\le c\beta^2 \int\limits_{\Omega} E_{\boldsymbol{\beta}}^{(k)}(x)|\nabla \mathbf{u}(x)|^2\,dx\end{aligned} \tag{2.4.28}$$

holds. The constant c in (2.4.28) *is independent of k, β, and* $\mathbf{u}$.

Proof
We begin by noting that the support of the function $\operatorname{div}\big(E_{\boldsymbol{\beta}}^{(k)}(x)\mathbf{u}(x)\big) = \nabla E_{\boldsymbol{\beta}}^{(k)}(x)\cdot \mathbf{u}(x)$ is contained within the bounded domain $\bigcup\limits_{j=1}^{J} \overline{\Omega}_{jk}$ (as defined in Eq. (2.4.26)). Each Ω_{jk} can be represented as a union $\Omega_{jk} = \bigcup\limits_{s=0}^{k-1} \omega_{js}$. Additionally, due to condition (2.4.25), we have:

$$\int\limits_{\omega_{js}} \nabla E_{\boldsymbol{\beta}}^{(k)}(x)\cdot \mathbf{u}(x)\,dx = \int\limits_{s}^{s+1} \frac{\partial}{\partial x_n^{(j)}} E_{\beta}^{(k)}(x_n^{(j)}) \int\limits_{\sigma_j} u_n(x^{(j)\prime})\,dx^{(j)\prime} = 0.$$

Therefore, by Lemma 2.4.1, we know that there exist vector fields $\mathbf{w}_{js} \in \overset{\circ}{W}{}^{1,2}(\omega_{js})$ such that:

$$\operatorname{div} \mathbf{w}_{js}(x) = -\frac{\partial}{\partial x_n^{(j)}} E_{\beta}^{(k)}(x_n^{(j)})\, u_n(x), \quad x \in \omega_{js},$$

and using inequalities (2.4.23) and (2.4.2), we obtain the estimates:

$$\|\nabla \mathbf{w}_{js}\|^2_{L^2(\omega_{js})} \le c\beta^2 \|E_{\beta}^{(k)}\mathbf{u}\|^2_{L^2(\omega_{js})}. \tag{2.4.29}$$

The domains ω_{js} can be transformed into the fixed domain ω_{j0} by translation, allowing us to choose the constant in (2.4.29) to be independent of s. Multiplying (2.4.29) by $E_{-\beta}^{(k)}(s)$ and taking into account properties of the function $E_{\beta}^{(k)}(x)$, we derive:

$$\int\limits_{\omega_{js}} E_{-\beta}^{(k)}(x)|\nabla \mathbf{w}_{js}(x)|^2\,dx \le c\beta^2 \int\limits_{\omega_{js}} E_{\beta}^{(k)}(x)|\mathbf{u}(x)|^2\,dx. \tag{2.4.30}$$

Next, we extend the functions $\mathbf{w}_{js}(x)$ by zero into $\Omega \setminus \omega_{js}$ and define:

$$\mathbf{W}^{(k)}(x) = \sum_{j=1}^{J} \sum_{s=0}^{k-1} \mathbf{w}_{js}(x).$$

It is evident that $\mathbf{W}^{(k)}(x)$ satisfies equation (2.4.27) and $\operatorname{supp} \mathbf{W}^{(k)}(x) \subset \bigcup_{j=1}^{J} \overline{\Omega}_{jk}$. The estimate (2.4.28) follows from (2.4.30), the definition (2.4.22) of the step weight-function $E_{\boldsymbol{\beta}}^{(k)}(x)$, and the weighted Poincaré inequality (2.3.27) (the Poincaré inequality (2.3.27) remains valid if we replace in it the weight function $E_{\boldsymbol{\beta}}(x)$ by the step weight-function $E_{\boldsymbol{\beta}}^{(k)}(x)$. The constant in obtained inequality is independent of k). □

2.4.2 Divergence Equation with Time-Dependent Right-Hand Side

Let Ω be a bounded domain in $\mathbb{R}^n$. Consider Problem I within Ω with the right-hand side $h(x, t)$, which depends on t. In other words, we are considering the problem:

$$\begin{cases} \operatorname{div} \mathbf{w}(x, t) = h(x, t), \ x \in \Omega, \ t \in [0, \ T], \\ \quad \mathbf{w}(x, t) = 0, \qquad x \in \partial\Omega, \ t \in [0, \ T], \end{cases} \tag{2.4.31}$$

under the assumption that:

$$\int_{\Omega} h(x, t) dx = 0 \quad \forall t \in [0, T]. \tag{2.4.32}$$

Lemma 2.4.8 *Let $\Omega \subset \mathbb{R}^n$ be a bounded domain with a Lipschitz boundary $\partial\Omega$. If both $h(\cdot, t)$ and $\partial_t h(\cdot, t)$ belong to $L^2(\Omega)$ and the condition* (2.4.32) *holds for all $t \in [0, T]$, then the Eq.* (2.4.31) *can be "differentiated" with respect to t. This means that the solution $\mathbf{w}(x, t)$ obtained in Lemma 2.4.1 satisfies the following properties for all $t \in [0, T]$: $\mathbf{w}(\cdot, t)$ and $\partial_t \mathbf{w}(\cdot, t)$ belong to $\overset{\circ}{W}{}^{1,2}(\Omega)$. The derivative $\partial_t \mathbf{w}(\cdot, t)$ satisfies the relation:*

$$\operatorname{div} \partial_t \mathbf{w}(x, t) = \partial_t h(x, t). \tag{2.4.33}$$

Additionally, the following estimates hold, with the constant c being independent of h and $t \in [0, T]$:

$$
\begin{aligned}
\|\nabla \mathbf{w}(\cdot,t)\|_{L^2(\Omega)} &\le c\|h(\cdot,t)\|_{L^2(\Omega)}, \\
\|\nabla \partial_t \mathbf{w}(\cdot,t)\|_{L^2(\Omega)} &\le c\|\partial_t h(\cdot,t)\|_{L^2(\Omega)}.
\end{aligned} \tag{2.4.34}
$$

Moreover, if both h and $\partial_t h$ are in $L^2(\Omega^T)$, then the estimates hold for the space–time domain Ω^T as well:

$$
\begin{aligned}
\|\nabla \mathbf{w}\|_{L^2(\Omega^T)} &\le c\|h\|_{L^2(\Omega^T)}, \\
\|\nabla \partial_t \mathbf{w}\|_{L^2(\Omega^T)} &\le c\|\partial_t h\|_{L^2(\Omega^T)}.
\end{aligned} \tag{2.4.35}
$$

Lemma 2.4.9 *Let $\Omega \subset \mathbb{R}^n$ be a bounded domain with a Lipschitz boundary $\partial\Omega$. Consider the problem* (2.4.31) *with $h(\cdot,t) \in \mathring{W}^{1,2}(\Omega)$ and $\partial_t h(\cdot,t) \in L^2(\Omega)$, where the condition* (2.4.32) *holds for all $t \in [0,T]$. Then, the problem* (2.4.31) *has at least one solution $\mathbf{w}(\cdot,t) \in \mathring{W}^{2,2}(\Omega)$ with $\partial_t \mathbf{w}(\cdot,t) \in \mathring{W}^{1,2}(\Omega)$ that satisfies equation* (2.4.33)*. There hold estimates* (2.4.34) *and the estimate*

$$
\|\mathbf{w}(\cdot,t)\|_{W^{2,2}(\Omega)} \le c\|h(\cdot,t)\|_{W^{1,2}(\Omega)} \tag{2.4.36}
$$

where the constant c is independent of $h(x,t)$ and $t \in [0,T]$.

Furthermore, if $h \in \mathring{W}^{(1,0),2}_{\partial\Omega}(\Omega^T)$ and $\partial_t h \in L^2(\Omega^T)$, then the estimates (2.4.35) *hold, and we have the inequality:*

$$
\int_0^T \|\mathbf{w}(\cdot,t)\|^2_{W^{2,2}(\Omega)} dt \le c \int_0^T \|h(\cdot,t)\|^2_{W^{1,2}(\Omega)} dt. \tag{2.4.37}
$$

Lemmas 2.4.8 and 2.4.9 can be proved using an explicit representation formula proposed in [31]. This formula allows differentiation with respect to t, which enables us to establish Lemmas 2.4.8 and 2.4.9.

Using Lemma 2.4.5, we can easily demonstrate the following statement.

Lemma 2.4.10 *Let $\Omega \subset \mathbb{R}^n$ be a bounded domain with $\partial\Omega \in C^{2l+3}$ for $l \ge 0$. If $h \in W^{(2l+2,l+1),2}(\Omega^T)$ and*

$$
\int_\Omega h(x,t)dx = 0 \quad \forall t \in [0,T],
$$

then there exists a vector field $\mathbf{w} \in W^{(2l+2,l+1),2}(\Omega^T) \cap \mathring{W}^{1,2}(\Omega)$ with $\dfrac{\partial^{l+1}}{\partial t^{l+1}} \nabla \mathbf{w} \in L^2(\Omega^T)$ such that $\operatorname{div} \mathbf{w}(x,t) = h(x,t)$*. The following estimate holds:*

$$
\|\mathbf{w}\|^2_{W^{(2l+2,l+1),2}(\Omega^T)} + \|\frac{\partial^{l+1}}{\partial t^{l+1}} \nabla \mathbf{w}\|^2_{L^2(\Omega^T)} \le c\,\|h\|^2_{W^{(2l+2,l+1),2}(\Omega^T)}. \tag{2.4.38}
$$

Moreover, if $\Omega_1 \subset \Omega_2$ *are two subdomains of* Ω*, and* $\Gamma_i = \partial\Omega_i \setminus \partial\Omega$ *for* $i = 1, 2$*, are finite unions of* C^{l+2}*-smooth surfaces, and* $\operatorname{supp} h(x,t) \subset \overline{\Omega_1}$ $\forall t \in [0,T]$*, then there exists a solution* $\mathbf{w}(x,t)$ *with*

$$\operatorname{supp} \mathbf{w}(x,t) \subset \overline{\Omega_2} \quad \forall t \in [0,T].$$

Proof

From the inclusion $h \in W^{(2l+2,l+1),2}(\Omega^T)$, it follows that $\frac{\partial^s}{\partial t^s} h(\cdot,t) \in W^{2l+1-2s,2}(\Omega)$, $s = 0, 1, \ldots, l$, for almost all $t \in [0,T]$ and, by the embedding theorem for traces (Theorem 2.3.7),

$$\sum_{s=0}^{l} \|\frac{\partial^s}{\partial t^s} h(\cdot,t)\|_{W^{2l+1-2s,2}(\Omega)} \le c\|h\|_{W^{(2l+2,l+1),2}(\Omega^T)}. \tag{2.4.39}$$

By Lemma 2.4.5, for almost all $t \in [0,T]$, there exists a vector field $\mathbf{w}(\cdot,t) \in W^{2l+2,2}(\Omega) \cap \mathring{W}^{1,2}(\Omega)$ such that

$$\operatorname{div} \mathbf{w}(x,t) = h(x,t), \quad \operatorname{supp}_x \mathbf{w}(\cdot,t) \subset \overline{\Omega_2},$$

and

$$\|\mathbf{w}(\cdot,t)\|^2_{W^{2l+2,2}(\Omega)} \le c\|h(\cdot,t)\|^2_{W^{2l+1,2}(\Omega)}. \tag{2.4.40}$$

The constructions of Lemma 2.4.5 use only linear operations, which admit differentiation with respect to t (see [85]). Therefore,

$$\operatorname{div} \frac{\partial^s}{\partial t^s} \mathbf{w}(x,t) = \frac{\partial^s}{\partial t^s} h(x,t), \quad x \in \Omega, \qquad \frac{\partial^s}{\partial t^s} \mathbf{w}(x,t)\big|_{\partial\Omega} = 0, \quad s = 1, \ldots, l,$$

and

$$\|\frac{\partial^s}{\partial t^s} \mathbf{w}(\cdot,t)\|^2_{W^{2l+2-2s,2}(\Omega)} \le c\|\frac{\partial^s}{\partial t^s} h(\cdot,t)\|^2_{W^{2l+1-2s,2}(\Omega)}, \tag{2.4.41}$$

$s = 1, \ldots, l$. On the other hand, $\frac{\partial^{l+1}}{\partial t^{l+1}} h \in L^2(\Omega^T)$. Hence, $\frac{\partial^{l+1}}{\partial t^{l+1}} h(\cdot,t) \in L^2(\Omega)$ for almost all $t \in [0,T]$. Moreover,

$$\int_\Omega \frac{\partial^{l+1}}{\partial t^{l+1}} h(x,t)dx = 0 \quad \text{for almost all } t \in [0,T].$$

Therefore,

$$\operatorname{div} \frac{\partial^{l+1}}{\partial t^{l+1}} \mathbf{w}(x,t) = \frac{\partial^{l+1}}{\partial t^{l+1}} h(x,t), \qquad \frac{\partial^{l+1}}{\partial t^{l+1}} \mathbf{w}(x,t)\big|_{\partial\Omega} = 0,$$

and

$$\|\frac{\partial^{l+1}}{\partial t^{l+1}}\mathbf{w}(\cdot,t)\|^2_{W^{1,2}(\Omega)} \le c\|\frac{\partial^{l+1}}{\partial t^{l+1}}h(\cdot,t)\|^2_{L^2(\Omega)}. \tag{2.4.42}$$

Summing inequalities (2.4.40)–(2.4.42) and integrating them over t, we obtain the estimate

$$\begin{aligned}\sum_{s=0}^{l}\int_0^T \|\frac{\partial^s}{\partial t^s}\mathbf{w}(\cdot,t)\|^2_{W^{2l+2-2s,2}(\Omega)}\,dt + \int_0^T \|\frac{\partial^{l+1}}{\partial t^{l+1}}\mathbf{w}(\cdot,t)\|^2_{L^2(\Omega)}\,dt \\ + \int_0^T \|\frac{\partial^{l+1}}{\partial t^{l+1}}\nabla\mathbf{w}(\cdot,t)\|^2_{L^2(\Omega)}\,dt \le c\Big(\sum_{s=0}^{l}\int_0^T \|\frac{\partial^s}{\partial t^s}h(\cdot,t)\|^2_{W^{2l+1-2s,2}(\Omega)}\,dt \\ + \int_0^T \|\frac{\partial^{l+1}}{\partial t^{l+1}}h(\cdot,t)\|^2_{L^2(\Omega)}\,dt\Big) \le c\,\|h\|^2_{W^{(2l+2,l+1),2}(\Omega^T)}\end{aligned}$$

which is equivalent to (2.4.38). □

Let Ω be a domain with cylindrical outlets to infinity. From Lemma 2.4.8, just in the same way as in Lemma 2.4.7, it follows that

Lemma 2.4.11 *Let* $\mathbf{u}(\cdot,t) \in \mathring{W}^{1,2}(\Omega)$, $\partial_t\mathbf{u}(\cdot,t) \in L^2(\Omega)$, *and* $\mathrm{div}\mathbf{u}(x,t) = 0$ *for all* $t \in [0,T]$. *If*

$$\int_{\sigma_j} \mathbf{u}(x,t)\cdot\mathbf{n}(x)\,dS = 0, \qquad j = 1,\dots,J; \tag{2.4.43}$$

then there exists a vector field $\mathbf{W}^{(k)}(\cdot,t) \in \mathring{W}^{1,2}(\Omega)$ *with* $\partial_t\mathbf{W}^{(k)}(\cdot,t) \in \mathring{W}^{1,2}(\Omega)$ *such that*

$$\mathrm{div}\,\mathbf{W}^{(k)}(x,t) = -\mathrm{div}\,\big(E^{(k)}_{\boldsymbol{\beta}}(x)\mathbf{u}(x,t)\big). \tag{2.4.44}$$

Moreover, the following estimates hold:

$$\begin{aligned}\int_\Omega E^{(k)}_{-\boldsymbol{\beta}}(x)|\nabla\mathbf{W}^{(k)}(x,t)|^2\,dx &\le c\beta^2\int_\Omega E^{(k)}_{\boldsymbol{\beta}}(x)|\mathbf{u}(x,t)|^2\,dx \\ &\le c\beta^2\int_\Omega E^{(k)}_{\boldsymbol{\beta}}(x)|\nabla\mathbf{u}(x,t)|^2\,dx,\end{aligned} \tag{2.4.45}$$

$$\int_\Omega E^{(k)}_{-\boldsymbol{\beta}}(x)|\nabla\partial_t\mathbf{W}^{(k)}(x,t)|^2\,dx \le c\beta^2\int_\Omega E^{(k)}_{\boldsymbol{\beta}}(x)|\partial_t\mathbf{u}(x,t)|^2\,dx. \tag{2.4.46}$$

The constants in (2.4.45) *and* (2.4.46) *are independent of* k *and* $t \in [0,T]$.

Proof
As in Lemma 2.4.7, the solution $\mathbf{W}^{(k)}(x,t)$ of Eq. (2.4.44) is defined as a sum

$$\mathbf{W}^{(k)}(x,t) = \sum_{j=1}^{J}\sum_{s=0}^{k-1} \mathbf{w}_{js}(x,t),$$

where $\mathbf{w}_{js}(\cdot,t) \in \mathring{W}^{1,2}(\omega_{js})$ are solutions of the equation

$$\operatorname{div} \mathbf{w}_{js}(x,t) = -\frac{\partial}{\partial x_n^{(j)}} E_\beta^{(k)}(x_n^{(j)})\, u_n(x,t), \qquad x \in \omega_{js},$$

By Lemma 2.4.8, in addition to (2.4.29), $\mathbf{w}_{js}(x,t)$ satisfies the estimate

$$\|\nabla \mathbf{w}_{js,t}(\cdot,t)\|^2_{L^2(\omega_{js})} \le c\beta^2 \|E_{\beta_j}^{(k)} \mathbf{u}_t(\cdot,t)\|^2_{L^2(\omega_{js})}.$$

Therefore, we get (2.4.45), and

$$\int_\Omega E_{-\boldsymbol{\beta}}^{(k)}(x)|\nabla\partial_t \mathbf{W}^{(k)}(x,t)|^2\, dx \le c\beta^2 \int_\Omega E_{\boldsymbol{\beta}}^{(k)}(x)|\partial_t \mathbf{u}(x,t)|^2\, dx.$$

□

2.4.3 Divergence Equation in a Tube Structure

Let us consider Problem I (see (2.4.1)) in the tube structure domain, given by:

$$B_\varepsilon = \Big(\bigcup_{j=1}^{N} \Pi_\varepsilon^{(e_j)}\Big) \bigcup \Big(\bigcup_{k=1}^{N_1} \omega_\varepsilon^k\Big). \tag{2.4.47}$$

Assume that $h(\cdot,t) \in \mathring{W}^{1,2}(B_\varepsilon)$, $\partial_t h(\cdot,t) \in L^2(B_\varepsilon)$ and the condition holds:

$$\int_{B_\varepsilon} h(x,t)dx = 0 \quad \forall t \in [0,T]. \tag{2.4.48}$$

The existence of the solution $\mathbf{w}$ of problem (2.4.1) in B_ε follows from previous lemmas. However, the constants in the above estimate can depend on the geometry of the domain Ω. In other words, in the tube structure B_ε, these constants depend on ε. We aim to demonstrate the existence of a solution $\mathbf{w}$ that satisfies the following estimates:

$$\|\mathbf{w}(\cdot,t)\|_{W^{1,2}(B_\varepsilon)} \leq \frac{c}{\sqrt{\varepsilon^3}}\|h(\cdot,t)\|_{L^2(B_\varepsilon)},$$

$$\|\nabla\partial_t\mathbf{w}(\cdot,t)\|_{L^2(B_\varepsilon)} \leq \frac{c}{\sqrt{\varepsilon^3}}\|\partial_t h(\cdot,t)\|_{L^2(B_\varepsilon)},$$

$$\|\nabla^2\mathbf{w}(\cdot,t)\|_{L^2(B_\varepsilon)} \leq c\Big(\frac{1}{\sqrt{\varepsilon^5}}\|h(\cdot,t)\|_{L^2(B_\varepsilon)} + \frac{1}{\sqrt{\varepsilon^3}}\|\nabla h(\cdot,t)\|_{L^2(B_\varepsilon)}\Big),$$

where the constant c is independent of h, $t \in [0,\ T]$, and ε.

To achieve this, we will first investigate problem (2.4.1) in the nodal domains ω_ε^k and the cylinders $\Pi_\varepsilon^{(e_j)}$. The results for these cases will be used later to show the desired estimates for problem (2.4.1) in B_ε.

Lemma 2.4.12 *Let $h(\cdot,t) \in \mathring{W}^{1,2}(\omega_\varepsilon^k)$, $h_t(\cdot,t) \in L^2(\omega_\varepsilon^k)$, and*

$$\int\limits_{\omega_\varepsilon^k} h(x,t)\,dx = 0 \quad \forall t \in [0,T]. \tag{2.4.49}$$

Then, problem (2.4.1) *admits a solution $\mathbf{w}^{(k)}(\cdot,t) \in \mathring{W}^{2,2}(\omega_\varepsilon^k)$ with $\partial_t\mathbf{w}^{(k)}(\cdot,t) \in \mathring{W}^{1,2}(\omega_\varepsilon^k)$, satisfying the estimates:*

$$\varepsilon^{-1}\|\mathbf{w}^{(k)}(\cdot,t)\|_{L^2(\omega_\varepsilon^k)} + \|\nabla\mathbf{w}^{(k)}(\cdot,t)\|_{L^2(\omega_\varepsilon^k)} \leq c\|h(\cdot,t)\|_{L^2(\omega_\varepsilon^k)}, \tag{2.4.50}$$

$$\varepsilon^{-1}\|\partial_t\mathbf{w}^{(k)}(\cdot,t)\|_{L^2(\omega_\varepsilon^k)} + \|\nabla\partial_t\mathbf{w}^{(k)}(\cdot,t)\|_{L^2(\omega_\varepsilon^k)} \leq c\|\partial_t h(\cdot,t)\|_{L^2(\omega_\varepsilon^k)}, \tag{2.4.51}$$

$$\varepsilon\|\nabla^2\mathbf{w}^{(k)}(\cdot,t)\|_{L^2(\omega_\varepsilon^k)} \leq c\Big(\|h(\cdot,t)\|_{L^2(\omega_\varepsilon^k)} + \varepsilon\|\nabla h(\cdot,t)\|_{L^2(\omega_\varepsilon^k)}\Big), \tag{2.4.52}$$

where the constant c is independent of h, $t \in [0,\ T]$, and ε.

Proof
Denote $\widehat{h}(y,t) = h(x,t)|_{x=\varepsilon y+O_k}$. We consider the solution $\widehat{\mathbf{u}}^{(k)}$ of problem (2.4.1) satisfying assertions of Lemmas 2.4.8 and 2.4.9 in the domain ω^k:

$$\mathrm{div}_y\widehat{\mathbf{u}}^{(k)}(y,t) = \widehat{h}(y,t), \quad \widehat{\mathbf{u}}^{(k)}(y,t)|_{\partial\omega^k} = 0,$$

and the estimates (2.4.33)–(2.4.37) are valid. Obviously, the constants in these estimates are independent of ε. Put $\mathbf{w}^{(k)}(x,t) = \varepsilon\widehat{\mathbf{u}}^{(k)}(y,t)\big|_{y=\frac{x-O_k}{\varepsilon}}$. Then $\mathbf{w}^{(k)}$ is the required solution of problem (2.4.1) in the domain ω_ε^k. For this solution estimates (2.4.50)–(2.4.52) follow immediately by scaling from estimates (2.4.33)–(2.4.37) and the Poincaré inequality (2.3.9). □

Lemma 2.4.13 *Let* $h(\cdot,t) \in \mathring{W}^{1,2}(\Pi_\varepsilon^{(e_j)})$, $h_t(\cdot,t) \in L^2(\Pi_\varepsilon^{(e_j)})$, *and*

$$\int\limits_{\Pi_\varepsilon^{(e_j)}} h(x,t)\,dx = 0 \quad \forall t \in [0,T]. \tag{2.4.53}$$

Then, problem (2.4.1) *admits a solution* $\mathbf{w}^{(e_j)}(\cdot,t) \in \mathring{W}^{2,2}(\Pi_\varepsilon^{(e_j)})$ *with* $\partial_t \mathbf{w}^{(e_j)}(\cdot,t) \in \mathring{W}^{1,2}(\Pi_\varepsilon^{(e_j)})$, *satisfying the estimates:*

$$\begin{aligned}
&\varepsilon^{-1}\Big(\|\mathbf{w}^{(e_j)\prime}(\cdot,t)\|_{L^2(\Pi_\varepsilon^{(e_j)})} + \|\partial_n \mathbf{w}^{(e_j)\prime}(\cdot,t)\|_{L^2(\Pi_\varepsilon^{(e_j)})}\Big)\\
&+\|w_n^{(e_j)}(\cdot,t)\|_{L^2(\Pi_\varepsilon^{(e_j)})} + \|\nabla' \mathbf{w}^{(e_j)\prime}(\cdot,t)\|_{L^2(\Pi_\varepsilon^{(e_j)})}\\
&+\|\partial_n w_n^{(e_j)}(\cdot,t)\|_{L^2(\Pi_\varepsilon^{(e_j)})} + \varepsilon\|\nabla' w_n^{(e_j)}(\cdot,t)\|_{L^2(\Pi_\varepsilon^{(e_j)})}\\
&\le c\|h(\cdot,t)\|_{L^2(\Pi_\varepsilon^{(e_j)})},
\end{aligned} \tag{2.4.54}$$

$$\begin{aligned}
&\varepsilon^{-1}\Big(\|\partial_t \mathbf{w}^{(e_j)\prime}(\cdot,t)\|_{L^2(\Pi_\varepsilon^{(e_j)})} + \|\partial_n \partial_t \mathbf{w}^{(e_j)\prime}(\cdot,t)\|_{L^2(\Pi_\varepsilon^{(e_j)})}\Big)\\
&+\|\partial_t w_n^{(e_j)}(\cdot,t)\|_{L^2(\Pi_\varepsilon^{(e_j)})} + \|\nabla' \partial_t \mathbf{w}^{(e_j)\prime}(\cdot,t)\|_{L^2(\Pi_\varepsilon^{(e_j)})}\\
&+\|\partial_n \partial_t w_n^{(e_j)}(\cdot,t)\|_{L^2(\Pi_\varepsilon^{(e_j)})} + \varepsilon\|\nabla' \partial_t w_n^{(e_j)}(\cdot,t)\|_{L^2(\Pi_\varepsilon^{(e_j)})}\\
&\le c\|\partial_t h(\cdot,t)\|_{L^2(\Pi_\varepsilon^{(e_j)})},
\end{aligned} \tag{2.4.55}$$

$$\begin{aligned}
&\varepsilon^{-1}\|\partial_n^2 \mathbf{w}^{(e_j)\prime}(\cdot,t)\|_{L^2(\Pi_\varepsilon^{(e_j)})} + \|\nabla' \partial_n \mathbf{w}^{(e_j)\prime}(\cdot,t)\|_{L^2(\Pi_\varepsilon^{(e_j)})}\\
&+\varepsilon\|\nabla'^2 \mathbf{w}^{(e_j)\prime}(\cdot,t)\|_{L^2(\Pi_\varepsilon^{(e_j)})} + \|\partial_n^2 w_n^{(e_j)}(\cdot,t)\|_{L^2(\Pi_\varepsilon^{(e_j)})}\\
&+\varepsilon\|\nabla' \partial_n w_n^{(e_j)}(\cdot,t)\|_{L^2(\Pi_\varepsilon^{(e_j)})} + \varepsilon^2\|\nabla'^2 w_n^{(e_j)}(\cdot,t)\|_{L^2(\Pi_\varepsilon^{(e_j)})}\\
&\le c\Big(\|h(\cdot,t)\|_{L^2(\Pi_\varepsilon^{(e_j)})} + \|\partial_n h(\cdot,t)\|_{L^2(\Pi_\varepsilon^{(e_j)})}\\
&+\varepsilon\|\nabla' h(\cdot,t)\|_{L^2(\Pi_\varepsilon^{(e_j)})}\Big),
\end{aligned} \tag{2.4.56}$$

where the constant c *is independent of* h, $t \in [0,\ T]$, *and* ε.

Proof

For simplicity we will write $x^{(e_j)} = x$ and define $y' = \dfrac{x'}{\varepsilon}$, $y_n = x_n$, $\widehat{h}(y,t) = h(x,t)|_{x'=\varepsilon y', x_n=y_n}$. Consider in the domain $\widehat{\Pi}^{(e_j)} = \{y \in \mathbb{R}^n : y_n \in (0,|e_j|), y' \in \sigma^{(e_j)}\}$, which is independent of ε, the solution $\widehat{\mathbf{u}}^{(e_j)}$ of problem (2.4.1), i.e.,

$$\mathrm{div}_y \widehat{\mathbf{u}}^{(e_j)}(y,t) = \widehat{h}(y,t), \quad \widehat{\mathbf{u}}^{(e_j)}(y,t)|_{\partial\widehat{\Pi}^{(e_j)}} = 0.$$

Due to Lemmas 2.4.8 and 2.4.9, $\widehat{\mathbf{u}}^{(e_j)}$ satisfies the estimates (2.4.33)–(2.4.37). The required solution $\mathbf{w}^{(e_j)}$ of problem (2.4.1) in $\Pi_\varepsilon^{(e_j)}$ is defined by:

$$\mathbf{w}^{(e_j)}(x,t) = \left(\varepsilon\widehat{\mathbf{u}}^{(e_j)\prime}(y,t), \widehat{u}_n^{(e_j)}(y,t)\right)\Big|_{y'=\frac{x'}{\varepsilon}, y_n=x_n}.$$

Estimates (2.4.54)–(2.4.56) follow from (2.4.33)–(2.4.37) after passing to coordinates x. □

Lemma 2.4.14 *Let $h(\cdot,t) \in \overset{\circ}{W}{}^{1,2}(B_\varepsilon)$, $\partial_t h(\cdot,t) \in L^2(B_\varepsilon)$, and the condition* (2.4.48) *holds. Then, problem* (2.4.1) *admits a solution $\mathbf{w}(\cdot,t) \in \overset{\circ}{W}{}^{2,2}(B_\varepsilon)$ with $\partial_t \mathbf{w}(\cdot,t) \in \overset{\circ}{W}{}^{1,2}(B_\varepsilon)$, satisfying the estimates:*

$$\|\mathbf{w}(\cdot,t)\|_{W^{1,2}(B_\varepsilon)} \le \frac{c}{\sqrt{\varepsilon^3}}\|h(\cdot,t)\|_{L^2(B_\varepsilon)}, \tag{2.4.57}$$

$$\|\nabla\partial_t \mathbf{w}(\cdot,t)\|_{L^2(B_\varepsilon)} \le \frac{c}{\sqrt{\varepsilon^3}}\|\partial_t h(\cdot,t)\|_{L^2(B_\varepsilon)}, \tag{2.4.58}$$

$$\|\nabla^2 \mathbf{w}(\cdot,t)\|_{L^2(B_\varepsilon)} \le c\Big(\frac{1}{\sqrt{\varepsilon^5}}\|h(\cdot,t)\|_{L^2(B_\varepsilon)} + \frac{1}{\sqrt{\varepsilon^3}}\|\nabla h(\cdot,t)\|_{L^2(B_\varepsilon)}\Big). \tag{2.4.59}$$

The constant c is independent of h, $t \in [0, T]$, and ε.

Proof

We consider a partition of unity $\{\varphi_i(x)\}_{i=1}^{M+N_1}$ subordinated to the covering (2.4.47) of B_ε. This means that $\mathrm{supp}\,\varphi_i \subset \bar{\Pi}_\varepsilon^{(e_i)}$ for $i = 1,\ldots,M$, and $\mathrm{supp}\,\varphi_{M+k} \subset \omega_\varepsilon^k$ for $k = 1,\ldots,N_1$. These functions satisfy $|\nabla\varphi_i(x)| \le c\varepsilon^{-1}$ for all i. Using this partition of unity, we can represent the function h as follows:

$$h(x,t) = \sum_{i=1}^{M+N_1} \varphi_i(x)h(x,t) \quad \forall x \in B_\varepsilon$$

Let us denote $\mu_i(t) = \int\limits_{B_\varepsilon} \varphi_i(x)h(x,t)\,dx$. Evidently

$$|\mu_i(t)| \le c\varepsilon^{(n-1)/2}\|h(\cdot,t)\|_{L^2(B_\varepsilon)}.$$

Since

$$\int\limits_{B_\varepsilon} h(x,t)\,dx = 0,$$

we have

$$\sum_{i=1}^{M+N_1} \mu_i(t) = 0. \tag{2.4.60}$$

Put $\gamma_i = \Pi_\varepsilon^{(e_i)}$, $i = 1, \ldots, M$, $\gamma_{M+k} = \omega_\varepsilon^k$, $k = 1, \ldots, N_1$, and $\gamma_{il} = \gamma_i \cap \gamma_l$. Notice that meas$(\gamma_{il})$ is of order $\varepsilon^n, n = 2, 3$. For each pair of intersecting domains γ_i and γ_l, we introduce a C^2–regular function $\eta_{il}(x)$ defined on B_ε such that supp $\eta_{il} \subset \gamma_{il}$, $|\eta_{il}(x)| \leq \frac{c}{\varepsilon^n}$, $|\nabla \eta_{il}(x)| \leq \frac{c}{\varepsilon^{n+1}}$, and

$$\int_{\gamma_{il}} \eta_{il}(x)\, dx = 1. \tag{2.4.61}$$

Evidently, $\eta_{il}(x) = \eta_{li}(x)$.

Furthermore, we introduce a skew-symmetric matrix $\{K_{il}(t)\}_{i,l=1,\ldots,M+N_1}$ such that $K_{il}(t) = 0$ if γ_{il} is empty (and if $i = l$), and

$$\sum_{l=1}^{M+N_1} c_{il} K_{il}(t) = \mu_i(t),\ i = 1, \ldots, M + N_1, \tag{2.4.62}$$

where $\{c_{il}\}_{i,l=1,\ldots,M+N_1}$ is the adjacency matrix of the graph of the covering $\{\gamma_i\}_{i=1}^{M+N_1}$ of the tube structure B_ε. In other words, $c_{il} = 1$ if γ_{il} is nonempty and $i \neq l$, and $c_{il} = 0$ otherwise.

The proof of the existence of the matrix $\{K_{il}(t)\}_{i,l=1,\ldots,M+N_1}$ uses the induction on the number $M + N_1$. For the base case when $M + N_1 = 2$, the values of c_{12} and c_{21} are 1, and, according to Eq. (2.4.60), we establish $K_{12} = -K_{21} = \mu_1$ and $K_{11} = K_{22} = 0$.

For the induction step, we assume the assertion is true for $M + N_1 = s - 1$ and aim to prove it for $M + N_1 = s$. We assign $K_{il} = 0$ for all i and l where $c_{il} = 0$. According to Harary [78], one can always remove one vertex from a connected graph with more than two vertices, leaving the remaining graph connected. This implies the existence of an index λ where $1 \leq \lambda \leq s$, so that the matrix $\{c_{il}\}$, for $i \neq \lambda$ and $l \neq \lambda$, is the adjacency matrix of a connected graph with $(s - 1)$ vertices. We can assume $\lambda = s$ without loss of generality. Consider the s-th equation from (2.4.62):

$$\mu_s(t) = \sum_{l=1}^{s-1} c_{sl} K_{sl}(t). \tag{2.4.63}$$

Given that the initial graph is connected, an index l_1 exists such that $c_{sl_1} = 1$. We then establish $K_{sl_1}(t) = -K_{l_1 s}(t) = \mu_s(t)$ and $K_{sl}(t) = -K_{ls}(t) = 0$ for $l \neq l_1$.

The remaining equations from (2.4.62) can be expressed in the form:

$$\tilde{\mu}_i(t) = \sum_{l=1}^{s-1} c_{il} K_{il}(t), \qquad i = 1, \dots, s-1, \tag{2.4.64}$$

where $\tilde{\mu}_i(t) = \mu_i(t) - c_{is} K_{is}(t)$.

By virtue of Eqs. (2.4.60) and (2.4.63), we have

$$\sum_{i=1}^{s-1} \tilde{\mu}_i = 0.$$

Indeed,

$$\sum_{i=1}^{s-1} \tilde{\mu}_i = \sum_{i=1}^{s-1} \mu_i - \sum_{i=1}^{s-1} c_{is} K_{is} = \sum_{i=1}^{s-1} \mu_i + \sum_{i=1}^{s-1} c_{si} K_{si} = \sum_{i=1}^{s} \mu_i = 0.$$

In addition, as stated earlier, the matrix $\{c_{il}\}_{i,l=1,\dots,s-1}$ satisfies the induction hypothesis for $M + N_1 = s - 1$. Consequently, the system (2.4.64) has a solution with the required properties.

Note that the matrix $\{c_{il}\}_{i.l=1,\dots,M+N_1}$ is independent of ε, and it can be proven by induction (as previously shown) that

$$|K_{il}(t)| \le c \max\left\{|\mu_1(t)|, \dots, |\mu_{M+N_1}(t)|\right\},$$

where c is a constant independent of ε. Therefore, we have

$$|K_{il}(t)| \le c\varepsilon^{(n-1)/2} \|h(\cdot, t)\|_{L^2(B_\varepsilon)}.$$

Now, let us define the function $\eta_i(x, t)$ on B_ε as follows:

$$\eta_i(x, t) = \sum_{l=1}^{M+N_1} K_{il}(t) \eta_{il}(x), \qquad i = 1, \dots, M + N_1. \tag{2.4.65}$$

Using the skew-symmetry of $\{K_{il}(t)\}_{i,l=1,\dots,M+N_1}$ and the property $\operatorname{supp} \eta_{il} \subset \operatorname{supp} \eta_i \subset \gamma_i$, it is easy to observe that

$$\sum_{i=1}^{M+N_1} \eta_i(x, t) = 0 \qquad \forall x \in B_\varepsilon. \tag{2.4.66}$$

Let us define the functions $h^{(i)}(x, t)$ on B_ε as follows:

$$h^{(i)}(x, t) = \varphi_i(x) h(x, t) - \eta_i(x, t) \qquad \text{for} \quad x \in B_\varepsilon. \tag{2.4.67}$$

The functions $h^{(i)}$ possess the following properties:

(i) $h^{(i)}$ have the same regularity as h.
(ii) $\operatorname{supp} h^{(i)}(\cdot, t) \subset \gamma_i$.
(iii) $\sum\limits_{i=1}^{M+N_1} h^{(i)}(x,t) = h(x) \ \ \forall x \in B_\varepsilon$.
(iv) $\int\limits_{\gamma_i} h^{(i)}(x,t)\,dx = 0 \ \ \forall t \in [0,T],\ i = 1, \dots M, M+1, \dots, M+N_1$.

The property (iii) follows from (2.4.66), and the property (iv) results from the following equality:

$$\int\limits_{\gamma_i} h^{(i)}(x,t)\,dx = \int\limits_{\gamma_i} \varphi_i(x) h^{(i)}(x,t)\,dx$$

$$-\int\limits_{\gamma_i} \sum_{l=1}^{M+N_1} K_{il}(t)\eta_{il}(x)\,dx = \mu_i(t) - \sum_{l=1}^{M+N_1} K_{il}(t)c_{il} = 0.$$

Let $\mathbf{w}^{(e_i)}(x,t), i = 1, \dots, M$, be solutions in $\Pi_\varepsilon^{(e_i)}$ of problem (2.4.1) with the right-hand sides $h^{(i)}(x,t)$ and $\mathbf{w}^{(k)}(x,t), k = N+1, \dots, M+N_1$, solutions of (2.4.1) in ω_ε^k with the right-hand sides $h^{(k)}(x,t)$. For $\mathbf{w}^{(e_i)}$ and $\mathbf{w}^{(k)}$, estimates from Lemmas 2.4.13 and 2.4.12 hold, respectively. Extending the functions $\mathbf{w}^{(e_i)}$ and $\mathbf{w}^{(k)}$ by zero into B_ε, we define

$$\mathbf{w}(x,t) = \sum_{i=1}^{M} \mathbf{w}^{(e_i)}(x,t) + \sum_{k=M+1}^{M+N_1} \mathbf{w}^{(k)}(x,t).$$

Then,

$$\operatorname{div}\mathbf{w}(x,t) = \sum_{i=1}^{M+N_1} h^{(i)}(x,t) = h(x,t).$$

According to (2.4.65), we have

$$\|\eta_i(\cdot,t)\|_{L^2(B_\varepsilon)} \le c\varepsilon^{(n-1)/2} \|h(\cdot,t)\|_{L^2(B_\varepsilon)} \sum_{l=1}^{M+N_1} \|\eta_{il}\|_{L^2(\gamma_{il})} \le c\varepsilon^{-1/2} \|h(\cdot,t)\|_{L^2(B_\varepsilon)}.$$

Thus,

$$\|h^{(i)}(\cdot,t)\|_{L^2(B_\varepsilon)} \le \frac{c}{\sqrt{\varepsilon}} \|h(\cdot,t)\|_{L^2(B_\varepsilon)},$$

$$\|\nabla h^{(i)}(\cdot,t)\|_{L^2(B_\varepsilon)} \le \frac{c}{\sqrt{\varepsilon}}\Big(\frac{1}{\varepsilon}\|h(\cdot,t)\|_{L^2(B_\varepsilon)} + \|\nabla h(\cdot,t)\|_{L^2(B_\varepsilon)}\Big).$$

Therefore, by roughening estimates (2.4.50)–(2.4.52) and (2.4.54)–(2.4.56) for $\mathbf{w}^{(k)}$ and $\mathbf{w}^{(e_i)}$, we obtain the required estimates for $\mathbf{w}$ as stated in (2.4.57)–(2.4.59).

□

Remark 2.4.3 Please note that the estimates (2.4.57) and (2.4.58) of above Lemma are still applicable when B_ε satisfies the conditions defined in Remark 1.4.1 of Chap. 1.

Now, we assume $\partial B_\varepsilon \in C^4$ and introduce the following notation: $\Upsilon_\varepsilon^{(e_j)} = \Pi_\varepsilon^{(e_j)} \cap \{x^{(e_j)} : x_1^{(e_j)} \in (\frac{1}{6}|e_j|, \frac{1}{3}|e_j|)\}$ and $\widehat{\Upsilon}_\varepsilon^{(e_j)} = \Pi_\varepsilon^{(e_j)} \cap \{x^{(e_j)} : x_1^{(e_j)} \in (\frac{1}{8}|e_j|, \frac{1}{2}|e_j|)\}$.

Lemma 2.4.15 *Assume that $h(\cdot,t) \in W^{2,2}(\widehat{\Upsilon}_\varepsilon^{(e_j)})$, with $\operatorname{supp} h(\cdot,t) \subset \Upsilon_\varepsilon^{(e_j)}$ and $\int_{\widehat{\Upsilon}_\varepsilon^{(e_j)}} h(x,t)dx = 0$ for all $t \in [0,T]$. Then, there exists a solution $\mathbf{w}(\cdot,t) \in \mathring{W}^{1,2}(\widehat{\Upsilon}_\varepsilon^{(e_j)}) \cap W^{3,2}(\widehat{\Upsilon}_\varepsilon^{(e_j)})$ of the divergence equation (2.4.1). The following estimate holds:*

$$\|\mathbf{w}(\cdot,t)\|_{W^{3,2}(\widehat{\Upsilon}_j)} \le c\,\varepsilon^{-3}\|h(\cdot,t)\|_{W^{2,2}(\widehat{\Upsilon}_j)} \qquad \textit{for all } t \in [0,T], \tag{2.4.68}$$

where the constant c is independent of ε.

Moreover, the support of $\mathbf{w}(\cdot,t)$ is contained within $\overline{\{x : x_n \in (\frac{1}{7}|e_j|, \frac{2}{5}|e_j|), x' \in \sigma^j\}}$.

Proof
First, we consider the divergence equation in the domain $\widehat{\Xi}_j = \{y : y_n \in (\frac{1}{8}|e_j|, \frac{1}{2}|e_j|), y' \in \sigma^j\}$, assuming that $h(\cdot,t) \in W^{2,2}(\widehat{\Xi}_j)$, $\operatorname{supp} h(\cdot,t) \subset \Xi_j = \{y : y_n \in (\frac{1}{6}|e_j|, \frac{1}{3}|e_j|), y' \in \sigma^j\}$, and $\int_{\widehat{\Xi}_j} h(x,t)dx = 0$.

The lateral boundary of $\widehat{\Xi}_j$ is C^4-regular, $\operatorname{supp} h(\cdot,t) \subset \overline{\Xi}_j$, $\overline{\Xi}_j \subset \overline{\widehat{\Xi}}_j$, and $\overline{\Xi}_j \neq \overline{\widehat{\Xi}}_j$. By using Lemma 2.4.5, we can find a vector field $\widetilde{\mathbf{w}}(\cdot,t) \in \mathring{W}^{1,2}(\widehat{\Xi}_j) \cap W^{3,2}(\widehat{\Xi}_j)$ such that $\operatorname{div}_y \widetilde{\mathbf{w}}(y,t) = \widetilde{h}(y,t)$, and the following estimate holds:

$$\|\widetilde{\mathbf{w}}(\cdot,t)\|_{W^{3,2}(\widehat{\Xi}_j)} \le c\,\|\widetilde{h}(\cdot,t)\|_{W^{2,2}(\widehat{\Xi}_j)}. \tag{2.4.69}$$

Additionally, $\widetilde{\mathbf{w}}(y,t)$ can be constructed such that

$$\operatorname{supp} \widetilde{\mathbf{w}}(\cdot,t) \subset \overline{\{y : y_n \in (\frac{1}{7}|e_j|, \frac{2}{5}|e_j|), y' \in \sigma^j\}}.$$

Here, $\widetilde{h}(y,t) = h(x,t)|_{x=\mathcal{X}(y)}$. Consider the change of variables $x = \mathcal{X}(y)$ defined by $x_n = y_n$, $x' = \varepsilon y'$.

Next, we define the vector field $\mathbf{w}$ with components $w_n(x,t) = \widetilde{w}_n(y,t)|_{\mathcal{X}^{-1}(x)}$, $w_i(x,t) = \varepsilon\widetilde{w}_i(y,t)|_{\mathcal{X}^{-1}(x)}$, $i = 1,\ldots,n-1$. It can be shown that $\operatorname{div}_x \mathbf{w}(x,t) =$

$h(x,t)$ and $\mathbf{w}(x,t)|_{\widehat{\Upsilon}_\varepsilon^{(e_j)}} = 0$. By passing (2.4.69) to coordinates x, we obtain the estimate:

$$\int\limits_{\widehat{\Upsilon}_\varepsilon^{(e_j)}} \Big(\sum_{l=0}^{3} \varepsilon^{2l} |\nabla_x^{\prime l} \widetilde{\mathbf{w}}(x,t)|^2 + \sum_{l=0}^{3} \Big|\frac{\partial^l \widetilde{\mathbf{w}}(x,t)}{\partial x_n^l}\Big|^2\Big) dx$$
$$\le c \int\limits_{\widehat{\Upsilon}_\varepsilon^{(e_j)}} \Big(\sum_{l=0}^{2} \varepsilon^{2l} |\nabla_x^{\prime l} h(x,t)|^2 + \sum_{l=0}^{2} \Big|\frac{\partial^l h(x,t)}{\partial x_n^l}\Big|^2\Big) dx.$$

This implies the following inequality:

$$\sum_{l=0}^{3} \Big(\varepsilon^{l-1} \|\nabla^{\prime l} \mathbf{w}'\|_{L^2(\widehat{\Upsilon}_\varepsilon^{(e_j)})} + \varepsilon^{l} \|\nabla^{\prime l} w_n\|_{L^2(\widehat{\Upsilon}_\varepsilon^{(e_j)})} + \varepsilon^{-1} \|\frac{\partial^l \mathbf{w}'}{\partial x_n^l}\|_{L^2(\widehat{\Upsilon}_\varepsilon^{(e_j)})}$$
$$+ \|\frac{\partial^l w_n}{\partial x_n^l}\|_{L^2(\widehat{\Upsilon}_\varepsilon^{(e_j)})}\Big) \le c \sum_{l=0}^{2} \Big(\varepsilon^{l} \|\nabla^{\prime l} h\|_{L^2(\widehat{\Upsilon}_\varepsilon^{(e_j)})} + \|\frac{\partial^l h}{\partial x_n^l}\|_{L^2(\widehat{\Upsilon}_\varepsilon^{(e_j)})}\Big).$$

Thus, we obtain (2.4.68). □

Furthermore, it is clear that this lemma holds true if the numbers $\frac{1}{8}, \frac{1}{7}, \frac{1}{6}, \frac{1}{3}, \frac{2}{5}, \frac{1}{2}$ are replaced by any α_i, $i = 1, \dots, 6$, such that $0 < \alpha_1 < \alpha_2 < \alpha_3 < \alpha_4 < \alpha_5 < \alpha_6 < 1$.

The results mentioned in this subsection were obtained in [150].

2.5 Spaces of Divergence-Free Vector Fields

2.5.1 Definition and Properties of Spaces of Divergence-Free Vector Functions

In this section, we introduce and study spaces of divergence-free vector functions in a domain Ω in $\mathbb{R}^n$. We denote by $\overset{\circ}{D}{}^{1,2}(\Omega)$ the space of vector fields $\mathbf{u} = \big(u_1, \dots, u_n\big)$ obtained as the closure of the vector set $C_0^\infty(\Omega)$ in the Dirichlet norm:

$$\|\nabla \mathbf{u}\|_{L^2(\Omega)} = \Big(\int\limits_{\Omega} |\nabla \mathbf{u}(x)|^2\, dx\Big)^{1/2},$$

where $|\nabla \mathbf{u}(x)|^2 = \sum\limits_{j=1}^{n} \sum\limits_{k=1}^{n} \Big|\frac{\partial u_j(x)}{\partial x_k}\Big|^2$.

It is important to note that $\overset{\circ}{D}{}^{1,2}(\Omega)$ is a Hilbert space equipped with the inner product

$$[\mathbf{u}, \mathbf{v}]_{\overset{\circ}{D}^{1,2}(\Omega)} = \int_{\Omega} \nabla \mathbf{u} \cdot \nabla \mathbf{v}\, dx.$$

For bounded domains Ω, the norms $\|\cdot\|_{\overset{\circ}{W}^{1,2}(\Omega)}$ and $\|\cdot\|_{\overset{\circ}{D}^{1,2}(\Omega)}$ are equivalent due to Poincaré inequality, and consequently, $\overset{\circ}{W}^{1,2}(\Omega) = \overset{\circ}{D}^{1,2}(\Omega)$. However, for unbounded domains, the Poincaré inequality is not necessarily valid, leading to distinct spaces.

Now, we define the spaces of divergence-free vector fields. Let $J_0^{\infty}(\Omega)$ be the subset of the vector set $C_0^{\infty}(\Omega)$ consisting of all divergence-free vector fields, i.e.,

$$J_0^{\infty}(\Omega) = \{\mathbf{v} \in C_0^{\infty}(\Omega) : \ \mathrm{div}\mathbf{v} = 0\}.$$

We define $\overset{\circ}{H}^{1,2}(\Omega)$ as the closure of $J_0^{\infty}(\Omega)$ in the $\overset{\circ}{D}^{1,2}(\Omega)$-norm and $\overset{\circ}{J}^{1,2}(\Omega)$ as the closure of $J_0^{\infty}(\Omega)$ in the $W^{1,2}(\Omega)$-norm. Additionally, we introduce the spaces

$$\widehat{\overset{\circ}{H}}^{1,2}(\Omega) = \{\mathbf{u} \in \overset{\circ}{D}^{1,2}(\Omega) : \ \mathrm{div}\mathbf{u} = 0\}$$

and

$$\widehat{\overset{\circ}{J}}^{1,2}(\Omega) = \{\mathbf{u} \in \overset{\circ}{W}^{1,2}(\Omega) : \ \mathrm{div}\mathbf{u} = 0\}.$$

It is evident that

$$\overset{\circ}{H}^{1,2}(\Omega) \subset \widehat{\overset{\circ}{H}}^{1,2}(\Omega) \ \text{ and } \ \overset{\circ}{J}^{1,2}(\Omega) \subset \widehat{\overset{\circ}{J}}^{1,2}(\Omega).$$

It is worth mentioning that the spaces of solenoidal (divergence-free) vector fields may not coincide for general unbounded domains. This was demonstrated by J. Heywood in his work [79]. This result was extended for large class of domains with outlets to infinity in [104, 171, 199, 200]. However, it is known that these spaces coincide for bounded and exterior domains with Lipschitz boundaries (see [104]).

Theorem 2.5.1 *For a bounded domain $\Omega \subset \mathbb{R}^n$ with Lipschitz boundary, we have*

$$\overset{\circ}{H}^{1,2}(\Omega) = \widehat{\overset{\circ}{H}}^{1,2}(\Omega).$$

For bounded domains, the spaces $\widehat{\overset{\circ}{H}}^{1,2}(\Omega)$ and $\widehat{\overset{\circ}{J}}^{1,2}(\Omega)$ coincide. However, for unbounded domains, these spaces are generally distinct, since the norm $\|\mathbf{u}\|_{L^2(\Omega)}$ is not always dominated by $\|\nabla \mathbf{u}\|_{L^2(\Omega)}$. As a result, we need to consider both the pair of spaces $\overset{\circ}{H}^{1,2}(\Omega)$ and $\widehat{\overset{\circ}{H}}^{1,2}(\Omega)$, as well as the spaces $\overset{\circ}{J}^{1,2}(\Omega)$ and $\widehat{\overset{\circ}{J}}^{1,2}(\Omega)$.

Theorem 2.5.2 *For an exterior domain Ω with Lipschitz boundary, we have*

$$\overset{\circ}{H}{}^{1,2}(\Omega) = \widehat{\overset{\circ}{H}}{}^{1,2}(\Omega) \text{ and } \overset{\circ}{J}{}^{1,2}(\Omega) = \widehat{\overset{\circ}{J}}{}^{1,2}(\Omega).$$

For unbounded domains with cylindrical outlets to infinity, we have the following result (see [200]).

Theorem 2.5.3 *For a domain $\Omega \subset \mathbb{R}^n$ with cylindrical outlets to infinity and Lipschitz boundary, we have*

$$\overset{\circ}{H}{}^{1,2}(\Omega) = \widehat{\overset{\circ}{H}}{}^{1,2}(\Omega) \text{ and } \overset{\circ}{J}{}^{1,2}(\Omega) = \widehat{\overset{\circ}{J}}{}^{1,2}(\Omega).$$

Additionally, for any vector function $\mathbf{u} \in \overset{\circ}{H}{}^{1,2}(\Omega)$, the fluxes of $\mathbf{u}$ over all cross-sections σ_j are equal to zero:

$$\int_{\sigma_j} \mathbf{u} \cdot \mathbf{n} dS = 0, \quad j = 1, 2, \ldots, N. \tag{2.5.1}$$

Remark 2.5.1 Finally, we note that the spaces of divergence-free vector fields in L^q-setting have been studied in [170, 172].

Consider the space $J(\Omega)$, which is defined as the closure of the set $J_0^\infty(\Omega)$ in the $L^2(\Omega)$-norm, and the space $\widehat{J}(\Omega) = \{\boldsymbol{\eta} \in L^2(\Omega); \ \operatorname{div}\boldsymbol{\eta} = 0\}$.

Theorem 2.5.4 *Let Ω be a bounded domain with Lipschitz boundary. Then $J(\Omega) = \widehat{J}(\Omega)$.*

Proof
Clearly, $J(\Omega) \subset \widehat{J}(\Omega)$. We need to prove the opposite inclusion $\widehat{J}(\Omega) \subset J(\Omega)$. Let $\mathbf{u} \in \widehat{J}(\Omega) \subset L^2(\Omega)$. Then, there exists a sequence $\mathbf{U}_j \in C_0^\infty(\Omega)$ such that

$$\|\mathbf{u} - \mathbf{U}_j\|_{L^2(\Omega)} \to 0.$$

Our goal is to demonstrate that $\|\operatorname{div} \mathbf{U}_j\|_{L^2(\Omega)} \to 0$. We have the relation

$$\int_\Omega \mathbf{u} \cdot \nabla\varphi dx = 0 \text{ for all } \varphi \in C_0^\infty(\Omega).$$

Since $\mathbf{U}_j$ converges to $\mathbf{u}$ in L^2-norm, we have

$$\lim_{j\to\infty} \int_\Omega \mathbf{U}_j \cdot \nabla\varphi dx = 0 \text{ for all } \varphi \in C_0^\infty(\Omega).$$

This leads to

$$\lim_{j\to\infty}\int_{\Omega} \operatorname{div}\mathbf{U}_j\varphi dx = 0 \text{ for all } \varphi \in C_0^\infty(\Omega).$$

Since $\mathbf{U}_j \in C_0^\infty(\Omega)$, we have $\operatorname{div}\mathbf{U}_j \in C_0^\infty(\Omega)$, and by taking $\varphi = \operatorname{div}\mathbf{U}_j$, we obtain

$$\lim_{j\to\infty}\int_{\Omega} |\operatorname{div}\mathbf{U}_j|^2 dx = 0.$$

Consider the equation

$$\begin{cases} \operatorname{div}\mathbf{w}_j = -\operatorname{div}\mathbf{U}_j, \\ \quad \mathbf{w}_j|_{\partial\Omega} = 0. \end{cases} \tag{2.5.2}$$

Evidently,

$$\int_{\Omega} \operatorname{div}\mathbf{U}_j dx = \int_{\partial\Omega} \mathbf{U}_j \cdot \mathbf{n} dS = 0,$$

which guarantees the solvability of problem (2.5.2) in the space $\overset{\circ}{W}{}^{1,2}(\Omega)$ (see Lemma 2.4.1). Moreover, the following estimate holds

$$\|\nabla\mathbf{w}_j\|_{L^2(\Omega)} \le c\|\operatorname{div}\mathbf{U}_j\|_{L^2(\Omega)}.$$

Now, for any j, let $\mathbf{V}_j = \mathbf{w}_j + \mathbf{U}_j \in \overset{\circ}{W}{}^{1,2}(\Omega)$. It follows that $\operatorname{div}\mathbf{V}_j = 0$. Hence, by Theorem 2.5.1, we have $\mathbf{V}_j \in \overset{\circ}{H}{}^{1,2}(\Omega)$, and consequently, there exists a sequence $\mathbf{V}_{jk} \in J_0^\infty(\Omega)$ such that

$$\|\mathbf{V}_j - \mathbf{V}_{lk}\|_{W^{1,2}(\Omega)} \to 0 \text{ as } k \to \infty.$$

Thus, we can write

$$\begin{aligned} \|\mathbf{u} - \mathbf{V}_{jk}\|_{L^2(\Omega)} &\le \|\mathbf{u} - \mathbf{U}_j\|_{L^2(\Omega)} + \|\mathbf{V}_j - \mathbf{V}_{jk}\|_{L^2(\Omega)} + \|\mathbf{w}_j\|_{L^2(\Omega)}, \\ &\le \|\mathbf{u} - \mathbf{U}_j\|_{L^2(\Omega)} + \|\mathbf{V}_j - \mathbf{V}_{jk}\|_{L^2(\Omega)} + c_1\|\operatorname{div}\mathbf{U}_j\|_{L^2(\Omega)}. \end{aligned}$$

which further leads to $\|\mathbf{u} - \mathbf{V}_{jk}\|_{L^2(\Omega)} \le \varepsilon$ for any arbitrarily small $\varepsilon > 0$.

Thus, $\mathbf{u}$ can be approximated in L^2-norm by functions $\mathbf{V}_{jk} \in J_0^\infty(\Omega)$, which implies $J(\Omega) = \widehat{J}(\Omega)$. □

2.5.2 Helmholtz–Weyl Decomposition

The Helmholtz–Weyl decomposition is a fundamental result in vector analysis, pertaining to the vector space $L^2(\Omega)$, where $\Omega \subset \mathbb{R}^n$ is a bounded Lipschitz domain. Let us define the subspace $G(\Omega)$ as follows:

$$G(\Omega) = \{\mathbf{v} \in L^2(\Omega) : \ \mathbf{v} = \nabla p \ \text{ for } \ \text{ some } \ p \in W^{1,2}(\Omega)\},$$

Theorem 2.5.5 (Ladyzhenskaya [102], see also [194]) *For any bounded domain Ω with Lipschitz boundary $\partial\Omega$, the following decomposition holds:*

$$L^2(\Omega) = J(\Omega) \oplus G(\Omega). \tag{2.5.3}$$

Chapter 3
Poiseuille Flows

The motion of a Navier–Stokes fluid in an infinitely long straight pipe $\Pi = \{x = (x', x_n) \in \mathbb{R}^n : \ x' \in \sigma \subset \mathbb{R}^{n-1}, \ -\infty < x_n < \infty\}$ of constant cross-section σ is one of the fundamental and extensively studied problems in fluid dynamics (see, e.g., references [13] and [109]). Among all practical flow patterns, the *Poiseuille flow* [181] assumes a central role. This flow is characterized by a velocity field having only one nonzero component (normal velocity), $u(x', t)$, directed along the axis of the pipe Π and depending solely on the transversal (tangential) coordinates x' of points on σ and possibly on time t. The associated pressure field does not depend on x' and has a (nonzero) axial gradient (pressure slope) $-\alpha(t)$, which is a function of time only.

The asymptotic analysis of flows in tube structures reveals that the Poiseuille flow serves as an asymptotic approximation of the solution inside each tube, away from the tube ends. This chapter focuses on the main properties of Poiseuille flows, which are crucial for constructing and justifying asymptotic expansions of solutions to the Navier–Stokes equations in tube structures.

Generally, two distinct methods are used to determine a Poiseuille flow: either by specifying the pressure slope $-\alpha(t)$ or the flow rate (flux) $F(t) := \int_\sigma u(x', t)\, dx'$. In both cases, the velocity component of the Poiseuille flow will be denoted as u_{P_α}.

In the first case, the problem reduces to solving an initial-boundary value problem for the *heat* equation for $u_{P_\alpha} = u_{P_\alpha}(x', t)$ with a prescribed time-dependent forcing term $\alpha(t)$. The unique solvability of this problem is standard and well-established (see, for instance, [103]).

However, in the second case, when the flux $F = F(t)$ is given, we must find $u_{P_\alpha}(x', t)$ and $\alpha(t)$. This problem can be seen as an inverse problem where we prescribe the time-dependent flux $F(t)$ and seek to determine the evolution of an unknown constant (in space) pressure slope $\alpha(t)$. The relation between $\alpha(t)$ and $F(t)$ in this case is non-local and depends on the solution of the aforementioned inverse problem.

G. Panasenko, K. Pileckas, *Multiscale Analysis of Viscous Flows in Thin Tube Structures*, Advances in Mathematical Fluid Mechanics,
https://doi.org/10.1007/978-3-031-54630-3_3

Aside from its unique solvability, another crucial aspect of this problem is determining the relationship between $F(t)$ and $-\alpha(t)$. This relationship is fundamental in any fluid dynamical situation involving pipes, as it allows us to control the pressure drop in the flow by adjusting the corresponding flow rate, and vice versa. In the steady case, both the pressure slope α and the flux F are constants, leading to the Poisson equation for $u_{P_\alpha} = u_{P_\alpha}(x')$ with the constant forcing α. In the steady scenario, the relationship between F and α is trivial, as they are proportional through a constant that depends solely on σ. However, in the time-dependent case, this relationship is not immediately apparent.

This chapter focuses on the existence and properties of Poiseuille flows in both the steady and time-dependent cases. Classical and well-known results concerning steady Poiseuille flows are presented in Sect. 3.1 (see [13] and [109]). Results regarding the existence of time-dependent Poiseuille flows (Sect. 3.2) lead to inverse problems, which have been studied by various authors, starting with the works of J.R. Cannon [40] and [41] and then by others (e.g., [89, 148, 173, 175, 176, 178, 182, 202], etc.).

This book mainly presents results based on the author's works [68, 175], and [176] (Sect. 3.2). In Sect. 3.3, time-periodic Poiseuille flows (Womersley flow [208]) are studied. The existence of such flows was first proved by H. Beirão da Veiga [16]. However, the proofs presented in Sect. 3.3 belong to G.P. Galdi and A.M. Robertson [69], as they are simpler and provide an elementary relationship between the pressure drop $\alpha(t)$ and the flux $F(t)$. Section 3.4 explores the existence of Poiseuille flows under minimal regularity of the flow rate, based on the author's papers [177] and [86]. Lastly, Sect. 3.5 is based on the author's paper [148], where the asymptotics of time-dependent Poiseuille flows in thin cylinders was studied.

3.1 Steady Poiseuille Flow

Consider an infinite pipe $\Pi = \{x = (x', x_n) \in \mathbb{R}^n : x' \in \sigma \subset \mathbb{R}^{n-1}, -\infty < x_n < \infty\}$, where $n = 2, 3$ and σ is a bounded domain.[1] The steady Navier–Stokes system is given by

$$\begin{cases} -\nu\Delta\mathbf{u}(x) + (\mathbf{u}(x)\cdot\nabla)\mathbf{u}(x) + \nabla p(x) = 0, \ x \in \Pi, \\ \operatorname{div} \mathbf{u}(x) = 0, \ x \in \Pi, \\ \mathbf{u}(x) = 0, \ x \in \partial\Pi, \end{cases} \tag{3.1.1}$$

where ν is the kinematic viscosity, $\mathbf{u}(x)$ is the velocity vector, and $p(x)$ is the pressure field (more exactly, p is the pressure divided by the constant density of the fluid)

[1] If $n = 2$, Π represents an infinite strip in $\mathbb{R}^2$, and $\sigma = (0, h_0)$ is a bounded interval.

The steady Poiseuille flow with a prescribed pressure slope $-\alpha$ is an exact solution of problem (3.1.1) with a simple form (e.g., [13, 109]):

$$\mathbf{u}_{P_\alpha}(x) = (0, \ldots, 0, \ \alpha u_0(x')), \quad \mathcal{P}_{P_\alpha}(x) = -\alpha x_n + p_0,$$

where p_0 is an arbitrary constant and $u_0(x')$ is the solution of the Poisson equation on the cross-section σ:

$$\begin{cases} -\nu \Delta' u_0(x') = 1, \ x \in \sigma, \\ \quad u_0(x') = 0, \ x \in \partial\sigma. \end{cases} \tag{3.1.2}$$

In (3.1.2), Δ' is the Laplace operator with respect to variables x'. It is well-known (see, e.g., [103]) that problem (3.1.2) has a unique weak solution $u_0 \in \overset{\circ}{W}{}^{1,2}(\sigma)$, and its regularity depends on the regularity of the boundary $\partial\sigma$. The following theorem holds:

Theorem 3.1.1 *Let σ be a Lipschitz domain. Problem* (3.1.2) *has a unique weak solution $u_0 \in \overset{\circ}{W}{}^{1,2}(\sigma)$. There holds the estimate*

$$\|u_0\|_{\overset{\circ}{W}{}^{1,2}(\sigma)} \le c_0, \tag{3.1.3}$$

where the constant c_0 depends only on σ.

If the boundary $\partial\sigma$ is $C^{l+2+\delta}$-regular, then $u_0 \in C^{l+2+\delta}(\overline{\sigma})$.

Multiplying (3.1.2) by u_0 and integrating by parts, we obtain

$$\int_\sigma u_0(x')\, dx' = \nu \int_\sigma |\nabla' u_0(x')|^2\, dx' := \nu\kappa_0 > 0. \tag{3.1.4}$$

Here, κ_0 is a geometrical characteristic of the domain σ.

The flux F of the Poiseuille flow $u_{P_\alpha}(x') = \alpha u_0(x')$ is equal to

$$F = \alpha \int_\sigma u_0(x')\, dx',$$

i.e.,

$$F = \alpha\kappa_0\nu. \tag{3.1.5}$$

In the two-dimensional case, we have the following formulas:

$$u_0(x_1) = \frac{1}{2\nu} x_1 \big(h_0 - x_1\big), \quad F = \frac{\alpha h_0^3}{12\nu}. \tag{3.1.6}$$

In the three-dimensional case, for an axisymmetric pipe of radius r_0, we have

$$u_0(r) = \frac{1}{4\nu}\big(r_0^2 - r^2\big), \quad F = \frac{\alpha\pi r_0^4}{8\nu}, \tag{3.1.7}$$

where $r = \sqrt{x_1^2 + x_2^2}$.

3.2 Time-Dependent Poiseuille Flow

3.2.1 Definition of a Time-Dependent Poiseuille Flow

Let us consider the initial-boundary value problem for the Navier–Stokes equations in $\Pi^T = \Pi \times (0, T)$:

$$\begin{cases} \partial_t \mathbf{u}(x,t) - \nu\Delta\mathbf{u}(x,t) + (\mathbf{u}(x,t)\cdot\nabla)\mathbf{u}(x,t) + \nabla p(x,t) = 0, & (x,t)\in\Pi^T, \\ \operatorname{div} \mathbf{u}(x,t) = 0, & (x,t)\in\Pi^T, \\ \mathbf{u}(x,t) = 0, & x\in\partial\Pi, \\ \mathbf{u}(x,0) = \mathbf{u}_0(x), x\in\Pi, \end{cases} \tag{3.2.1}$$

We assume that the initial data $\mathbf{u}_0(x)$ takes the form:

$$\mathbf{u}_0(x) = \big(0, \dots, 0, u_{0n}(x')\big).$$

We seek a solution $(\mathbf{u}, p)$ of problem (3.2.1) in the form of a time-dependent Poiseuille flow:

$$\mathbf{u}_{P_\alpha}(x,t) = (0, \dots, 0, u_{P_\alpha}(x',t)), \quad \mathcal{P}_{P_\alpha}(x,t) = -\alpha(t)x_n + p_0(t), \tag{3.2.2}$$

where $p_0(t)$ is an arbitrary function. Substituting (3.2.2) into (3.2.1), we obtain the following initial-boundary value problem on the cross-section σ for $u_{P_\alpha}(x',t)$ and $\alpha(t)$:

$$\begin{cases} \partial_t u_{P_\alpha}(x',t) - \nu\Delta' u_{P_\alpha}(x',t) = \alpha(t), & (x',t)\in\sigma^T, \\ u_{P_\alpha}(x',t) = 0, & x'\in\partial\sigma, t\in(0,T), \\ u_{P_\alpha}(x',0) = u_{0n}(x'), \; x'\in\sigma. \end{cases} \tag{3.2.3}$$

Here $\sigma^T = \sigma \times (0, T)$ (see Sect. 2.3.4). Thus, to find the Poiseuille solution with the given pressure slope $-\alpha(t)$, we need to solve the heat equation (3.2.3) with the given heat forcing $\alpha(t)$. The well-known properties of the heat equation (see, for instance, [103]) ensure the existence and uniqueness of solutions for (3.2.3). The following theorem holds:

Theorem 3.2.1 *Let σ be a Lipschitz domain. If $\alpha \in L^2(0,T)$ and $u_{0n} \in \overset{\circ}{W}{}^{1,2}(\sigma)$, then problem* (3.2.3) *admits a unique weak solution $u_{P_\alpha} \in \overset{\circ}{W}{}^{(1,1),2}_{\partial\sigma}(\sigma^T)$ satisfying the estimate:*

$$\|u_{P_\alpha}\|_{W^{(1,1),2}(\sigma^T)} \leq c\Big(\|\alpha\|_{L^2(0,T)} + \|u_{0n}\|_{W^{1,2}(\sigma)}\Big), \tag{3.2.4}$$

where the constant c is independent of T. If the boundary $\partial\sigma$ is C^2, then $u_{P_\alpha} \in \overset{\circ}{W}{}^{(2,1),2}_{\partial\sigma}(\sigma^T)$ and:

$$\|u_{P_\alpha}\|_{W^{(2,1),2}(\sigma^T)} \leq c\Big(\|\alpha\|_{L^2(0,T)} + \|u_{0n}\|_{W^{1,2}(\sigma)}\Big). \tag{3.2.5}$$

However, if we prescribe the flux $F(t)$:

$$\int_\sigma u_{P_\alpha}(x',t)\,dx' = F(t), \tag{3.2.6}$$

the function $\alpha(t)$ in (3.2.3) becomes unknown and has to be determined to satisfy condition (3.2.6). Thus, Eq. (3.2.3) transforms into an inverse problem:

For a given u_{0n} and $F(t)$, the objective is to determine a pair $(u_{P_\alpha}(x',t), \alpha(t))$ that solves the initial-boundary value problem (3.2.3) *and fulfills the flux condition* (3.2.6).

In contrast to the steady case, where the constant α defining the pressure slope is directly proportional to the flux F (see (3.1.5)), in the time-dependent scenario, the function $\alpha(t)$ needs to be determined as a solution of the inverse problem. Consequently, in the time-dependent case, the formulation of the problem with the prescribed pressure slope $-\alpha(t)$ is not equivalent to the formulation with the prescribed flux $F(t)$. When prescribing the pressure slope $-\alpha(t)$, one must solve the direct parabolic problem (3.2.3) with known $\alpha(t)$. On the other hand, prescribing the flux condition (3.2.6) necessitates solving the inverse problem. Hence, the relation between $\alpha(t)$ and $F(t)$ depends on the solution $u_{P_\alpha}(x',t)$.

In the following sections, we investigate the inverse problem given by:

$$\begin{cases} \partial_t u_{P_\alpha}(x',t) - \nu\Delta' u_{P_\alpha}(x',t) = \alpha(t), & (x',t) \in \sigma^T, \\ u_{P_\alpha}(x',t) = 0, & x' \in \partial\sigma, t \in (0,T) \\ u_{P_\alpha}(x',0) = u_{0n}(x'), \; x' \in \sigma, & \\ \int_\sigma u_{P_\alpha}(x',t)\,dx' = F(t), & t \in (0,T), \end{cases} \tag{3.2.7}$$

where $u_{0n}(x')$ and $F(t)$ are known, while $u_{P_\alpha}(x',t)$ and $\alpha(t)$ are unknown. The data $u_{0n}(x')$ and $F(t)$ must satisfy the compatibility condition:

$$\int\limits_\sigma u_{0n}(x')\,dx' = F(0). \tag{3.2.8}$$

3.2.2 Construction of Approximate Solutions to Problem (2.7)

Consider the inverse problem (3.2.7) in the case where $u_{0n}(x) = 0$, $F \in W^{1,2}(0,T)$, and $F(0) = 0$.

Let $u_k(x') \in \overset{\circ}{W}{}^{1,2}(\sigma)$ and λ_k be eigenfunctions and eigenvalues of the Laplace operator:

$$\begin{cases} -\nu\Delta u_k(x') = \lambda_k u_k(x'), & x' \in \sigma, \\ u_k(x') = 0, & x' \in \partial\sigma. \end{cases} \tag{3.2.9}$$

It is well-known that $\lambda_k > 0$, and as $k \to \infty$, the eigenvalues $\{\lambda_k\}$ tend to infinity. The eigenfunctions $u_k(x')$ are orthogonal in $L^2(\sigma)$, and for simplicity, we assume that they are normalized in $L^2(\sigma)$. Consequently, the following relationships hold:

$$\nu\int\limits_\sigma |\nabla' u_k(x')|^2\,dx' = \lambda_k, \qquad \int\limits_\sigma \nabla' u_k(x')\cdot\nabla' u_l(x')\,dx' = 0, \quad k \neq l.$$

Moreover, the set $\{u_k(x')\}$ forms a basis in $L^2(\sigma)$ and $\overset{\circ}{W}{}^{1,2}(\sigma)$ (see, e.g., [103]).

We seek an approximate solution $\big(u^{(N)}_{P_\alpha}(x',t), \alpha^{(N)}(t)\big)$ to problem (3.2.7) in the form:

$$u^{(N)}_{P_\alpha}(x',t) = \sum_{k=1}^N w_k^{(N)}(t) u_k(x'), \tag{3.2.10}$$

Then, we choose the function $\alpha^{(N)}(t)$ to satisfy the flux condition:

$$\int\limits_\sigma u^{(N)}_{P_\alpha}(x',t)\,dx' = F(t) \quad \forall t \in [0,T]. \tag{3.2.11}$$

To find the coefficients $w_k^{(N)}$, we use the following equations:

$$\int_{\sigma} \partial_t u_{P_\alpha}^{(N)}(x',t)u_k(x')dx' + \nu \int_{\sigma} \nabla' u_{P_\alpha}^{(N)}(x',t) \cdot \nabla' u_k(x')dx' \\ = \alpha^{(N)}(t) \int_{\sigma} u_k(x')dx'. \tag{3.2.12}$$

These equations are equivalent to the following ordinary differential equations (ODEs):

$$\frac{dw_k^{(N)}(t)}{dt} + \lambda_k w_k^{(N)}(t) = \beta_k \alpha^{(N)}(t), \quad w_k^{(N)}(0) = 0, \quad k = 1, 2, \ldots,$$

where $\beta_k = \int_\sigma u_k(x')dx'$, $k = 1, 2, \ldots$, are the Fourier coefficients of the constant function 1. Hence, the coefficients $w_k^{(N)}(t)$ can be expressed as follows:

$$w_k^{(N)}(t) = \beta_k \int_0^t \exp(-\lambda_k(t-\tau))\alpha^{(N)}(\tau)\,d\tau.$$

Consequently, the approximate solution $u_{P_\alpha}^{(N)}(x',t)$ takes the form:

$$u_{P_\alpha}^{(N)}(x',t) = \sum_{k=1}^{N} \beta_k \Big(\int_0^t \exp(-\lambda_k(t-\tau))\alpha^{(N)}(\tau)\,d\tau \Big) u_k(x'). \tag{3.2.13}$$

By substituting the sum (3.2.13) into the flux condition (3.2.11), we obtain the relation:

$$\int_{\sigma} u_{P_\alpha}^{(N)}(x',t)\,dx' = \sum_{k=1}^{N} \beta_k \int_0^t \exp(-\lambda_k(t-\tau))\alpha^{(N)}(\tau)\,d\tau \int_{\sigma} u_k(x')\,dx' = F(t).$$

This relation is equivalent to a first-kind Volterra integral equation for the function $\alpha^{(N)}(t)$:

$$\sum_{k=1}^{N} \beta_k^2 \int_0^t \exp(-\lambda_k(t-\tau))\alpha^{(N)}(\tau)\,d\tau = F(t). \tag{3.2.14}$$

It is well-known that such equations, in general, are ill-posed in L^2-spaces. However, if $F'(t)$ exists, Eq. (3.2.14) can be differentiated, leading to the second-kind Volterra equation:

$$\alpha^{(N)}(t) - \frac{1}{\kappa_N}\sum_{k=1}^{N}\beta_k^2\lambda_k\int_0^t \exp(-\lambda_k(t-\tau))\alpha^{(N)}(\tau)\,d\tau = \varphi^{(N)}(t), \tag{3.2.15}$$

i.e.,

$$\alpha^{(N)}(t) - \int_0^t K^{(N)}(t,\tau)\alpha^{(N)}(\tau)\,d\tau = \varphi^{(N)}(t),$$

where the kernel is defined as:

$$K^{(N)}(t,\tau) = \kappa_N^{-1}\sum_{k=1}^{N}\beta_k^2\lambda_k \exp(-\lambda_k(t-\tau)).$$

Here, $\kappa_N = \sum\limits_{k=1}^{N}\beta_k^2$ and $\varphi^{(N)}(t) = F'(t)/\kappa_N$.

The theory of integral equations of the second kind is well-known. Consider an integral equation

$$q(t) - \int_0^t K(t,\tau)q(\tau)\,d\tau = \varphi(t). \tag{3.2.16}$$

The following theorem is proved (see, e.g., [94, 206]).

Theorem 3.2.2 *Let $\varphi \in L^2(0,T)$, $K \in L^2(Q_T)$, $Q_T = (0,T)\times(0,T)$. Then there exists a unique solution $q \in L^2(0,T)$ of Eq.* (3.2.16). *The following estimate*

$$\|q\|_{L^2(0,T)} \le c\|\varphi\|_{L^2(0,T)} \tag{3.2.17}$$

holds.

Since $\varphi^{(N)}(t) = F'(t)/\kappa_N \in L^2(0,T)$ and $K^{(N)} \in L^2(Q^T)$, the unique solution $\alpha^{(N)} \in L^2(0,T)$ exists for any finite N. However, the norm $\|K^{(N)}\|_{L^2(Q^T)}$ of the kernel $K^{(N)}(t,\tau)$ depends on N and tends to infinity as $N \to +\infty$. As a result, the constant $c(N)$ in the estimate

$$\|\alpha^{(N)}\|_{L^2(0,T)} \le c(N)\|F'\|_{L^2(0,T)}$$

diverges as $N \to +\infty$.

Moreover, from Eq. (3.2.15), it is evident that if $\varphi^{(N)}(0) = F'(0) = 0$, then $\alpha^{(N)}(0) = 0$.

3.2.3 A Priori Estimates

In this section, we establish a priori estimates for the approximate solution $\big(u^{(N)}_{P_\alpha}(x',t), \alpha^{(N)}(t)\big)$.

Lemma 3.2.1 *Let $F \in W^{1,2}(0,T)$ with $F(0) = 0$, where $T \in (0,\infty]$. Then, the following estimates hold:*

$$\|u^{(N)}_{P_\alpha}(\,\cdot\,,t)\|^2_{W^{1,2}(\sigma)} + \|u^{(N)}_{P_\alpha}\|^2_{W^{(1,1),2}(\sigma^t)} + \|\alpha^{(N)}\|^2_{L^2(0,t)} \le c\,\|F\|^2_{W^{1,2}(0,t)} \quad \textit{for all} \quad t \in (0,T]. \tag{3.2.18}$$

If $\partial\sigma \in C^2$, then additionally holds the estimate:

$$\|u^{(N)}_{P_\alpha}\|^2_{W^{(2,1),2}(\sigma^t)} + \|\alpha^{(N)}\|^2_{L^2(0,t)} \le c\,\|F\|^2_{W^{1,2}(0,t)} \quad \textit{for all} \quad t \in (0,T]. \tag{3.2.19}$$

The constants in (3.2.18) *and* (3.2.19) *are independent of t, T, and N.*

Proof
We start by multiplying equations (3.2.12) by $w^{(N)}(t)$, summing them over k, and integrating over the interval $(0,t)$ to obtain:

$$\begin{aligned}
&\frac{1}{2}\int_\sigma |u^{(N)}_{P_\alpha}(x',t)|^2\,dx' + \nu\int_0^t\int_\sigma |\nabla' u^{(N)}_{P_\alpha}(x',\tau)|^2\,dx'd\tau \\
&= \int_0^t \alpha^{(N)}(\tau)\int_\sigma u^{(N)}_{P_\alpha}(x',\tau)\,dx'd\tau = \int_0^t \alpha^{(N)}(\tau)F(\tau)\,d\tau \\
&\le \varepsilon\int_0^t |\alpha^{(N)}(\tau)|^2\,d\tau + \frac{1}{4\varepsilon}\int_0^t |F(\tau)|^2\,d\tau
\end{aligned} \tag{3.2.20}$$

where $\varepsilon > 0$.

Next, we multiply equations (3.2.12) by $\dfrac{dw^{(N)}(t)}{dt}$, sum them over k, and integrate over the interval $(0,t)$ to obtain:

$$\begin{aligned}
&\int_0^t\int_\sigma |\partial_\tau u^{(N)}_{P_\alpha}(x',\tau)|^2\,dx'd\tau + \frac{\nu}{2}\int_\sigma |\nabla' u^{(N)}_{P_\alpha}(x',t)|^2\,dx' \\
&= \int_0^t \alpha^{(N)}(\tau)\int_\sigma \partial_\tau u^{(N)}_{P_\alpha}(x',\tau)\,dx'd\tau = \int_0^t \alpha^{(N)}(\tau)F'(\tau)d\tau \\
&\le \varepsilon\int_0^t |\alpha^{(N)}(\tau)|^2 d\tau + \frac{1}{4\varepsilon}\int_0^t |F'(\tau)|^2 d\tau.
\end{aligned} \tag{3.2.21}$$

Let $u_0(x')$ be the solution of the Poisson problem (3.1.2), i.e.:

$$\begin{cases} -\nu\Delta' u_0(x') = 1, \; x' \in \sigma, \\ \qquad u_0(x') = 0, \; x' \in \partial\sigma. \end{cases}$$

The function u_0 belongs to $\overset{\circ}{W}{}^{1,2}(\sigma)$ and can be expanded as a series $u_0(x') = \sum\limits_{l=1}^{\infty} b_l u_l(x')$, where $b_l = \int\limits_{\sigma} u_0(x') u_l(x') dx'$. Denote $u_0^{(N)}(x') = \sum\limits_{l=1}^{N} b_l u_l(x')$. It is clear that $u_0^{(N)}$ converges to u_0 in the norm of the space $W^{1,2}(\sigma)$.

Multiplying equalities (3.2.12) by b_k and summing from 1 to N, we get:

$$\int\limits_{\sigma} \partial_t u_{P_\alpha}^{(N)}(x',t) u_0^{(N)}(x')\, dx' - \nu \int\limits_{\sigma} \Delta' u_{P_\alpha}^{(N)}(x',t) u_0^{(N)}(x')\, dx'$$

$$= \alpha^{(N)}(t) \int\limits_{\sigma} u_0^{(N)}(x')\, dx'.$$

Integrating twice by parts in the term $-\nu \int_\sigma \Delta' u_{P_\alpha,t}^{(N)}(x') u_0^{(N)}(x')\, dx'$ and using the equation $-\nu\Delta' u_0 = 1$, we see that

$$-\nu \int\limits_{\sigma} \Delta' u_{P_\alpha}^{(N)}(x',t) u_0^{(N)}(x')\, dx' = -\nu \int\limits_{\sigma} u_{P_\alpha}^{(N)}(x',t) \Delta' u_0^{(N)}(x')\, dx'$$

$$= -\nu \int\limits_{\sigma} u_{P_\alpha}^{(N)}(x',t) \Delta' u_0(x')\, dx' - \nu \int\limits_{\sigma} u_{P_\alpha}^{(N)}(x',t) \Delta'(u_0^{(N)}(x') - u_0(x'))\, dx'$$

$$= \int\limits_{\sigma} u_{P_\alpha}^{(N)}(x',t) dx' + \nu \int\limits_{\sigma} \nabla' u_{P_\alpha}^{(N)}(x',t) \cdot \nabla'(u_0^{(N)}(x') - u_0(x'))\, dx'$$

$$= F(t) + \nu \int\limits_{\sigma} \nabla' u_{P_\alpha}^{(N)}(x',t) \cdot \nabla'(u_0^{(N)}(x') - u_0(x'))\, dx'.$$

Therefore, the previous relation yields:

$$\begin{aligned} \kappa_0 \alpha^{(N)}(t) = \int\limits_{\sigma} \partial_t u_{P_\alpha}^{(N)}(x',t) u_0(x')\, dx' + F(t) \\ + \alpha^{(N)}(t) \int\limits_{\sigma} \big(u_0(x') - u_0^{(N)}(x')\big) dx' \\ + \nu \int\limits_{\sigma} \nabla' u_{P_\alpha}^{(N)}(x',t) \cdot \nabla'(u_0^{(N)}(x') - u_0(x'))\, dx'. \end{aligned} \tag{3.2.22}$$

where $\kappa_0 = \nu \int_\sigma |\nabla' u_0(x')|^2 \, dx' = \int_\sigma u_0(x') \, dx'$.

From (3.2.22), using the Cauchy-Schwarz-Bunyakovsky and Poincaré inequalities, we derive the estimate:

$$
\begin{aligned}
\kappa_0^2 \int_0^t |\alpha^{(N)}(\tau)|^2 \, d\tau &\le c \int_0^t |\alpha^{(N)}(\tau)|^2 \Big(\int_\sigma |u_0(x') - u_0^{(N)}(x')|^2 \, dx' \Big) \, d\tau \\
&+ c \Big[\int_0^t \Big(\int_\sigma |\partial_\tau u_{P_\alpha}^{(N)}(x', \tau)|^2 \, dx' \Big) \Big(\int_\sigma |u_0(x')|^2 \, dx' \Big) \, d\tau + \int_0^t |F(\tau)|^2 \, d\tau \\
&+ \int_0^t \Big(\int_\sigma |\nabla' u_{P_\alpha}^{(N)}(x', \tau)|^2 \, dx' \Big) \Big(\int_\sigma |\nabla' \big(u_0(x') - u_0^{(N)}(x') \big)|^2 \, dx' \Big) \, d\tau \Big] \\
&\le c_1 \int_\sigma |\nabla' u_0(x') - \nabla' u_0^{(N)}(x')|^2 \, dx' \int_0^t |\alpha^{(N)}(\tau)|^2 \, d\tau \\
&+ C_* \Big[\int_0^t \int_\sigma |\partial_\tau u_{P_\alpha}^{(N)}(x', \tau)|^2 \, dx' d\tau + \int_0^t \int_\sigma |\nabla' u_{P_\alpha}^{(N)}(x', \tau)|^2 \, dx' d\tau \Big] \\
&+ c_2 \int_0^t |F(\tau)|^2 \, d\tau.
\end{aligned} \tag{3.2.23}
$$

Here, C_* depends on the $W^{1,2}$-norm of u_0. Since $\int_\sigma |\nabla' u_0(x') - \nabla' u_0^{(N)}(x')|^2 \, dx' \to 0$ as $N \to +\infty$, for sufficiently large N, we have:

$$
c_1 \int_\sigma |\nabla' u_0(x') - \nabla' u_0^{(N)}(x')|^2 \, dx' \int_0^t |\alpha^{(N)}(\tau)|^2 \, d\tau \le \frac{\kappa_0^2}{2} \int_0^t |\alpha^{(N)}(\tau)|^2 \, d\tau
$$

and then (3.2.23) yields:

$$
\begin{aligned}
\int_0^t |\alpha^{(N)}(\tau)|^2 \, d\tau &\le c \Big[\int_0^t \int_\sigma |\partial_\tau u_{P_\alpha}^{(N)}(x', \tau)|^2 \, dx' d\tau \\
&+ \int_0^t \int_\sigma |\nabla' u_{P_\alpha}^{(N)}(x', \tau)|^2 \, dx' d\tau \Big] + c \int_0^t |F(\tau)|^2 \, d\tau.
\end{aligned} \tag{3.2.24}
$$

From estimates (3.2.20), (3.2.21), and (3.2.24), it follows that:

$$\begin{aligned}&\int\limits_{\sigma}\left(|u_{P_\alpha}^{(N)}(x',t)|^2+|\nabla' u_{P_\alpha}^{(N)}(x',t)|^2\right)dx'\\&+\int\limits_0^t\int\limits_{\sigma}\left(|\nabla' u_{P_\alpha}^{(N)}(x',\tau)|^2+|\partial_\tau u_{P_\alpha}^{(N)}(x',\tau)|^2\right)dx'd\tau\\&\le c_1\varepsilon\int\limits_0^t\int\limits_{\sigma}\left(|\nabla' u_{P_\alpha}^{(N)}(x',\tau)|^2+|\partial_\tau u_{P_\alpha}^{(N)}(x',\tau)|^2\right)dx'd\tau\\&\qquad+c(\varepsilon)\int\limits_0^t\left(|F(\tau)|^2+|F'(\tau)|^2\right)d\tau.\end{aligned}\tag{3.2.25}$$

Taking $\varepsilon=\frac{1}{2c_1}$ yields:

$$\begin{aligned}&\int\limits_{\sigma}\left(|u_{P_\alpha}^{(N)}(x',t)|^2+|\nabla' u_{P_\alpha}^{(N)}(x',t)|^2\right)dx'\\&+\int\limits_0^t\int\limits_{\sigma}\left(|\nabla' u_{P_\alpha}^{(N)}(x',\tau)|^2+|\partial_\tau u_{P_\alpha}^{(N)}(x',\tau)|^2\right)dx'd\tau\\&\qquad\le c\int\limits_0^t\left(|F(\tau)|^2+|F'(\tau)|^2\right)d\tau.\end{aligned}\tag{3.2.26}$$

Furthermore, from (3.2.24), we obtain:

$$\int\limits_0^t|\alpha^{(N)}(\tau)|^2\,d\tau\le c\int\limits_0^t\left(|F(\tau)|^2+|F'(\tau)|^2\right)d\tau.\tag{3.2.27}$$

The estimates (3.2.26) and (3.2.27) imply (3.2.18).

Now, let $\partial\sigma\in C^2$. Consider $u_{P_\alpha}(x',\cdot)$ as a solution of the equation:

$$\begin{cases}-\nu\Delta' u_{P_\alpha}^{(N)}(x',t)=\alpha^{(N)}(t)-\partial_t u_{P_\alpha}^{(N)}(x',t), & x'\in\sigma,\\ \quad u_{P_\alpha}^{(N)}(x',t)=0, & x'\in\partial\sigma.\end{cases}\tag{3.2.28}$$

The right-hand side of (3.2.28) belongs to $L^2(\sigma)$ for almost all $t\in[0,T]$. Therefore, the solution $u_{P_\alpha}^{(N)}(\cdot,t)\in W^{2,2}(\sigma)$, and there holds the estimate:

$$\|u_{P_\alpha}^{(N)}(\cdot,t)\|^2_{W_2^2(\sigma)}\le c\Big(|\alpha^{(N)}(t)|^2+\|\partial_t u_{P_\alpha}^{(N)}(x',t)\|^2_{L^2(\sigma)}\Big).$$

Integrating the later inequality with respect to t, we obtain, in virtue of (3.2.18):

$$\int_0^t \|u^{(N)}_{P_\alpha}(\cdot,\tau)\|^2_{W^2_2(\sigma)} d\tau \le c\Big(\int_0^t \big(|F(\tau)|^2 + |F'(\tau)|^2\big)\, d\tau\Big).$$

The last estimate together with (3.2.18) implies (3.2.19). □

3.2.4 Existence of a Weak Solution

In this section, we define the concept of a weak solution for problem (3.2.7).

Definition 3.2.1 A weak solution is represented by a pair of functions $\big(u_{P_\alpha}, \alpha\big) \in \mathring{W}^{(1,1),2}_{\partial\sigma}(\sigma^T) \times L^2(0,T)$ that satisfy the initial condition $u_{P_\alpha}(x',0) = u_{0n}(x')$, the flux condition

$$\int_\sigma u_{P_\alpha}(x',t)dx' = F(t),$$

and the integral identity

$$\begin{aligned} &\int_0^t\int_\sigma \partial_\tau u_{P_\alpha}(x',\tau)\eta(x',\tau)dx'd\tau \\ &\quad + \nu \int_0^t\int_\sigma \nabla' u_{P_\alpha}(x',\tau)\cdot\nabla'\eta(x',\tau)dx'd\tau \\ &= \int_0^t \alpha(\tau)\int_\sigma \eta(x',\tau)dx'd\tau \quad \forall\, \eta \in \mathring{W}^{(1,0),2}_{\partial\sigma}(\sigma^T), \quad \forall t\in(0,T]. \end{aligned} \tag{3.2.29}$$

Next, we consider problem (3.2.7) with the assumption that $u_{0n}(x') = 0$.

Theorem 3.2.3 *Let $F \in W^{1,2}(0,T)$, $F(0) = 0$, and $u_{0n}(x') = 0$. Then, for arbitrary $T \in (0,\infty]$, problem (3.2.7) admits a unique weak solution $\big(u_{P_\alpha}(x',t), \alpha(t)\big)$. Furthermore, the following estimate holds:*

$$\sup_{t\in[0,T]} \|u_{P_\alpha}(\cdot,t)\|_{W^{1,2}(\sigma)} + \|u_{P_\alpha}\|_{W^{(1,1),2}(\sigma^T)} + \|\alpha\|_{L_2(0,T)} \le c\|F\|_{W^{1,2}(0,T)}, \tag{3.2.30}$$

where c is a positive constant.

Moreover, if $\partial\sigma \in C^2$, then $\big(u_{P_\alpha}, \alpha\big) \in W^{(2,1),2}(\sigma^T) \times L^2(0, T)$ and

$$\|u_{P_\alpha}\|_{W^{(2,1),2}(\sigma^T)} \le c\|F\|_{W^{1,2}(0,T)}. \tag{3.2.31}$$

Proof

To begin the proof, we consider the sequence $\{(u_{P_\alpha}^{(N)}(x',t),\alpha^{(N)}(t))\}$ of approximate solutions constructed in Sect. 3.2.2. These solutions satisfy the identities (3.2.12). Let $d_k \in C([0,T])$, $k = 1,\dots,K$, where $K \in \mathbb{N}$. By multiplying (3.2.12) by $d_k(t)$ and summing over k from 1 to K, we obtain the identity

$$\begin{aligned}\int\limits_\sigma \partial_t u_{P_\alpha}^{(N)}(x',t)\eta(x',t)dx' + \nu\int\limits_\sigma \nabla' u_{P_\alpha}^{(N)}(x',t)\cdot\nabla'\eta(x',t)dx' \\ = \alpha^{(N)}(t)\int\limits_\sigma \eta(x',t)dx',\end{aligned} \tag{3.2.32}$$

which holds for any $\eta(x',t) = \sum\limits_{k=1}^{K} d_k(t)u_k(x')$ with arbitrary $K \geq 1$.

By using Lemma 3.2.1, the iterations $\{(u_{P_\alpha}^{(N)},\alpha^{(N)})\}$ satisfy the estimate (3.2.18) with a constant independent of N. Thus, there exists a subsequence $\{(u_{P_\alpha}^{(N_l)},\ \alpha^{(N_l)})\}$ such that $\{u_{P_\alpha}^{(N_l)}\}$, $\{\nabla' u_{P_\alpha}^{(N_l)}\}$, and $\{\partial_t u_{P_\alpha}^{(N_l)}\}$ are weakly convergent in $L^2(\sigma^T)$. The limit $u_{P_\alpha} \in \mathring{W}_{\partial\sigma}^{(1,1),2}(\sigma^T)$. Moreover, $\{\alpha^{(N_l)}\}$ weakly converges in $L^2(0,T)$ to $\alpha \in L^2(0,T)$.

Passing in (3.2.32) to the limit over the sequence $\{N_l\}$, we derive the identity

$$\begin{aligned}\int\limits_0^T\int\limits_\sigma \partial_t u_{P_\alpha}(x',t)\eta(x',t)dx'dt + \nu\int\limits_0^T\int\limits_\sigma \nabla' u_{P_\alpha}(x',t)\cdot\nabla'\eta(x',t)dx'dt \\ = \int\limits_0^T \alpha(t)\int\limits_\sigma \eta(x',t)dx'dt.\end{aligned}$$

The above identity is proved for test functions η having the form $\eta(x',t) = \sum\limits_{k=1}^{K} d_k(t)u_k(x')$. However, such linear combinations are dense in the space $\mathring{W}_{\partial\sigma}^{(1,0),2}(\sigma^T)$ (see Theorem 2.3.8 and Remark 2.3.5), and thus, it remains valid for any $\eta \in \mathring{W}_{\partial\sigma}^{(1,0),2}(\sigma^T)$.

Since the iterations $u_{P_\alpha}^{(N_l)}$ satisfy the flux condition (3.2.11), we can conclude that

$$\int\limits_\sigma u_{P_\alpha}(x',t)dx' = F(t). \tag{3.2.33}$$

Moreover, the approximate solutions $u_{P_\alpha}^{(N_l)}(x',t)$ are continuous in time. By Sobolev's embedding theorem, the embeddings $W^{(1,1),2}(\sigma^T) = W^{1,2}(\sigma^T) \hookrightarrow$

$L^2(\sigma^T)$ and $W^{(1,1),2}(\sigma^T) \hookrightarrow L^2(\sigma)$ are compact (see Theorem 2.3.5). Thus, the sequence $\{u_{P_\alpha}^{(N_l)}\}$ converges strongly in $L^2(\sigma^T)$, and $\{u_{P_\alpha}^{(N_l)}(\cdot,t)\}$ converges strongly in $L^2(\sigma)$ for $\forall t \in [0,T]$. By the triangle inequality, we have

$$\|u_{P_\alpha}(\cdot,t)\|_{L^2(\sigma)} \le \|u_{P_\alpha}(\cdot,t) - u_{P_\alpha}^{(N_l)}(\cdot,t)\|_{L^2(\sigma)} + \|u_{P_\alpha}^{(N_l)}(\cdot,t)\|_{L^2(\sigma)}. \tag{3.2.34}$$

Next, we will show that the first term on the right-hand side of (3.2.34) uniformly tends to zero. We begin by obtaining an inequality for the time derivative of the L^2-norm of the difference between $u_{P_\alpha}(\cdot,t)$ and $u_{P_\alpha}^{(N_l)}(\cdot,t)$. Evidently, we have

$$\frac{d}{dt}\|u_{P_\alpha}(\cdot,t) - u_{P_\alpha}^{(N_l)}(\cdot,t)\|^2_{L^2(\sigma)}$$
$$\le 2\int_\sigma |\partial_t u_{P_\alpha}(x',t) - \partial_t u_{P_\alpha}^{(N_l)}(x',t)|\,|u_{P_\alpha}(x',t) - u_{P_\alpha}^{(N_l)}(x',t)|dx'.$$

Integrating this inequality with respect to t and using estimate (3.2.18), we get

$$\|u_{P_\alpha}(\cdot,t) - u_{P_\alpha}^{(N_l)}(\cdot,t)\|^2_{L^2(\sigma)} \le \|u_{P_\alpha}(\cdot,0) - u_{P_\alpha}^{(N_l)}(\cdot,0)\|^2_{L^2(\sigma)}$$
$$+2\int_0^t\int_\sigma |\partial_\tau u_{P_\alpha}(x',\tau) - \partial_\tau u_{P_\alpha}^{(N_l)}(x',\tau)|\,|u_{P_\alpha}(x',\tau) - u_{P_\alpha}^{(N_l)}(x',\tau)|dx'd\tau$$
$$\le \|u_{P_\alpha}(\cdot,0) - u_{P_\alpha}^{(N_l)}(\cdot,0)\|^2_{L^2(\sigma)}$$
$$+c\Big(\int_0^T\int_\sigma \big(|\partial_t u_{P_\alpha}(x',t)|^2 + |\partial_t u_{P_\alpha}^{(N_l)}(x',t)|^2\big)dx'dt\Big)^{1/2}$$
$$\times\Big(\int_0^T\int_\sigma |u_{P_\alpha}(x',t) - u_{P_\alpha}^{(N_l)}(x',t)|^2dx'dt\Big)^{1/2}$$
$$\le \|u_{P_\alpha}(\cdot,0) - u_{P_\alpha}^{(N_l)}(\cdot,0)\|^2_{L^2(\sigma)} + c\|F\|_{W^{1,2}(0,T)}$$
$$\times\Big(\int_0^T\int_\sigma |u_{P_\alpha}(x',t) - u_{P_\alpha}^{(N_l)}(x',t)|^2dx'dt\Big)^{1/2}.$$

The right-hand side of this inequality is independent of t and tends to zero as $N_l \to \infty$. Thus, $\|u_{P_\alpha}(\cdot,t) - u_{P_\alpha}^{(N_l)}(\cdot,t)\|_{L^2(\sigma)}$ tends to zero uniformly with respect to t.

Hence, for arbitrary small $\varepsilon > 0$, we can find N_* such that

$$\|u_{P_\alpha}(\cdot, t) - u_{P_\alpha}^{(N_l)}(\cdot, t)\|_{L^2(\sigma)} \leq \varepsilon/2,$$

and then find $\delta > 0$ such that for $t \in (0, \delta)$,

$$\|u_{P_\alpha}^{(N_l)}(\cdot, t)\|_{L^2(\sigma)} \leq \varepsilon/2.$$

Using inequality (3.2.34), we obtain

$$\|u_{P_\alpha}(\cdot, t)\|_{L^2(\sigma)} \leq \varepsilon,$$

and therefore, $u_{P_\alpha}(x', 0) = 0$ for almost all $x' \in \sigma$.

Next, we prove the uniqueness of the weak solution $\big(u_{P_\alpha}(x', t),\ \alpha(t)\big)$. Assume that $F(t) = 0$. Taking in (3.2.29), $\eta(x', t) = u_{P_\alpha}(x', t)$, we obtain

$$\frac{1}{2}\int\limits_\sigma |u_{P_\alpha}(x', t)|^2\, dx' + \nu \int\limits_0^t \int\limits_\sigma |\nabla' u_{P_\alpha}(x', \tau)|^2\, dx' d\tau = \int\limits_0^t \alpha(\tau) \int\limits_\sigma u_{P_\alpha}(x', \tau)\, dx' d\tau.$$

Since

$$\int\limits_\sigma u_{P_\alpha}(x', \tau)\, dx' = F(t) = 0,$$

we have

$$\frac{1}{2}\int\limits_\sigma |u_{P_\alpha}(x', t)|^2\, dx' + \nu \int\limits_0^t \int\limits_\sigma |\nabla' u_{P_\alpha}(x', \tau)|^2\, dx' d\tau = 0,$$

and therefore, $u_{P_\alpha}(x', t) = 0$.

Now, from the integral identity (3.2.29), it follows that

$$\int\limits_0^T \alpha(t) \int\limits_\sigma \eta(x', t) dx' dt = 0 \quad \forall \eta \in \overset{\circ}{W}{}_{\partial\sigma}^{(1,0),2}(\sigma^T). \tag{3.2.35}$$

Taking $\eta(x', t) = b(t)\xi(x')$, where $b \in L^2(0, T)$, $\xi \in \overset{\circ}{W}{}^{1,2}(\sigma)$, and $\int_\sigma \xi(x')dx' = 1$, the last equality transforms to the form

$$\int\limits_0^T \alpha(t) b(t) dt = 0 \quad \forall b \in L^2(0, T).$$

Hence, $\alpha(t) = 0$.

Finally, if $\partial\sigma \in C^2$, $F \in W^{1,2}(0, T)$, and $F(0) = 0$, then by (3.2.19), $u_{P_\alpha}^{(N_l)} \in W^{(2,1),2}(\sigma^T)$, and without loss of generality, we can assume that $\{u_{P_\alpha}^{(N_l)}\}$ is weakly convergent in $W^{(2,1),2}(\sigma^T)$. Then the limit function u_{P_α} also belongs to the space $W^{(2,1),2}(\sigma^T)$. Obviously, for u_{P_α} and α, estimates (3.2.18) and (3.2.19) still hold. This completes the proof. □

We consider now the general case of nonzero initial data $u_{0n}(x')$.

Theorem 3.2.4 *Let $u_{0n} \in \overset{\circ}{W}{}^{1,2}(\sigma)$, and let the compatibility condition*

$$F(0) = \int\limits_{\sigma} u_{0n}(x')\,dx \tag{3.2.36}$$

holds. Then, for arbitrary $T \in (0, \infty]$, there exists a unique solution $\big(u_{P_\alpha}(x', t), \alpha(t)\big)$ of problem (3.2.7) *such that*

$$\begin{aligned} &\sup_{t\in[0,T]} \|u_{P_\alpha}(\cdot, t)\|_{W^{1,2}(\sigma)} + \|u_{P_\alpha}\|_{W^{(1,1),2}(\sigma^T)} + \|\alpha\|_{L^2(0,T)} \\ &\leq c\Big(\|F\|_{W^{1,2}(0,T)} + \|u_{0n}\|_{W^{1,2}(\sigma)}\Big). \end{aligned} \tag{3.2.37}$$

Moreover, if $\partial\sigma \in C^2$, then $u_{P_\alpha} \in \overset{\circ}{W}{}^{(2,1),2}_{\partial\sigma}(\sigma^T)$, and the following estimate holds:

$$\|u_{P_\alpha}\|_{W^{(2,1),2}(\sigma^T)} \leq c\left(\|F\|^2_{W^{1,2}(0,T)} + \|u_{0n}\|_{W^{1,2}(\sigma)}\right). \tag{3.2.38}$$

Proof
We look for the solution $\big(u_{P_\alpha}(x', t), \alpha(t)\big)$ of problem (3.2.7) in the form:

$$\big(u_{P_\alpha}(x', t), \alpha(t)\big) = \big(u_h^{(1)}(x', t), 0\big) + \big(u_{P_\alpha}^{(2)}(x', t), \alpha(t)\big),$$

where $u_h^{(1)}(x', t)$ is the solution of the heat equation:

$$\begin{cases} \partial_t u_h^{(1)}(x', t) - \nu\Delta' u_h^{(1)}(x', t) = 0, & (x', t) \in \sigma^T, \\ u_h^{(1)}(x', t) = 0, & x' \in \partial\sigma, t \in (0, T), \\ u_h^{(1)}(x', 0) = u_{0n}(x'),\ x' \in \sigma, & \end{cases} \tag{3.2.39}$$

and $\big(u_{P_\alpha}^{(2)}(x', t), \alpha(t)\big)$ is the solution of the inverse problem:

$$\begin{cases} \partial_t u^{(2)}_{P_\alpha}(x',t) - \nu\Delta' u^{(2)}_{P_\alpha}(x',t) = \alpha(t), \ (x',t) \in \sigma^T, \\ u^{(2)}_{P_\alpha}(x',t) = 0, \quad x' \in \partial\sigma, t \in (0,T), \\ u^{(2)}_{P_\alpha}(x',0) = 0, \quad x' \in \sigma, \\ \int\limits_\sigma u^{(2)}_{P_\alpha}(x',t)\,dx' = \widetilde{F}(t), \ t \in (0,T), \end{cases} \tag{3.2.40}$$

with $\widetilde{F}(t) = F(t) - \int_\sigma u^{(1)}_{P_\alpha}(x',t)\,dx'$.

By Theorem 3.2.1, problem (3.2.39) admits a unique solution $u^{(1)}_h(x',t)$, which satisfies the estimate:

$$\sup_{t\in[0,T]} \|u^{(1)}_h(\cdot,t)\|^2_{W^{1,2}(\sigma)} + \|u^{(1)}_h\|^2_{W^{(1,1),2}(\sigma^T)} \le c\|u_{0n}\|^2_{W^{1,2}(\sigma)}. \tag{3.2.41}$$

Moreover, if $\partial\sigma \in C^2$, then $u^{(1)}_h \in \mathring{W}^{(2,1),2}(\sigma^T)$, and the following estimate holds:

$$\|u^{(1)}_h\|^2_{W^{(2,1),2}(\sigma^T)} \le c\|u_{0n}\|^2_{W^{1,2}(\sigma)}. \tag{3.2.42}$$

The flux $\widetilde{F}(t)$ in $(3.2.40)_4$ satisfies the estimate:

$$\begin{aligned} \|\widetilde{F}\|^2_{W^{1,2}(0,T)} &\le c\Big(\|F\|^2_{W^{1,2}(0,T)} + \|u^{(1)}_h\|^2_{L^2(\sigma)} + \|\partial_t u^{(1)}_h\|^2_{L^2(\sigma)}\Big) \\ &\le c\Big(\|F\|^2_{W^{1,2}(0,T)} + \|u_{0n}\|^2_{W^{1,2}(\sigma)}\Big). \end{aligned} \tag{3.2.43}$$

Moreover, $\widetilde{F}(0) = F(0) - \int_\sigma u_{0n}(x')\,dx' = 0$. Therefore, by Theorem 3.2.3, problem (3.2.40) admits a unique solution $\big(u^{(2)}_{P_\alpha}(x',t), \alpha(t)\big)$ that satisfies estimate (3.2.30). If $\partial\sigma \in C^2$, then estimate (3.2.31) also holds.

Estimates (3.2.37) and (3.2.38) follow from inequalities (3.2.41), (3.2.42), (3.2.43), and (3.2.30), (3.2.31), respectively. □

3.2.5 An Integral Equation Relating the Flux and the Pressure Slope

The objective of this section is twofold. On the one hand, we prove that the flow rate $F(t)$ and the pressure slope $-\alpha(t)$ are related by a linear Volterra integral equation of the second kind with corresponding kernel $K(t)$, and we obtain a one-to-one correspondence between $F(t)$ and $\alpha(t)$. On the other hand, we furnish results of (global) existence and uniqueness of solutions to (3.2.7) in general Sobolev spaces.

Consider problem (3.2.7) with $u_{0n} = 0$. Assume that $\partial\sigma \in C^2$. Let $u_0(x')$ be the solution of Poisson problem (3.1.2) and $\Theta(x', t)$ be the solution to the following initial boundary value problem for the heat equation

$$\begin{cases} \partial_t \Theta(x', t) - \nu \Delta' \Theta(x', t) = 0, & (x', t) \in \sigma^T, \\ \Theta(x', t) = 0, & x' \in \partial\sigma, t \in (0, T), \\ \Theta(x', 0) = u_0(x'), \ x' \in \sigma. \end{cases} \tag{3.2.44}$$

Under the above assumptions, the solution u_0 of problem (3.1.2) belongs to the space $W^{2,2}(\sigma) \cap \mathring{W}^{1,2}(\sigma)$ and satisfies the estimate

$$\|u_0\|_{W^{2,2}(\sigma)} \le c_0, \tag{3.2.45}$$

where c_0 depends only on σ. Consequently, we conclude that (3.2.44) has a unique solution $\Theta \in \mathring{W}^{(2,1),2}_{\partial\sigma}(\sigma^T)$ such that

$$\|\Theta\|_{W^{(2,1),2}(\sigma^T)} \le c\|u_0\|_{W^{1,2}(\sigma)}.$$

see, e.g., [103]. Since $u_0 \in W^{2,2}(\sigma)$, we deduce also $\nabla' \partial_t \Theta \in L^2(\sigma^T)$ and that

$$\sup_{t\in[0,T]} \|\partial_t \Theta(\cdot, t)\|_{L^2(\sigma)} + \|\nabla' \partial_t \Theta(\cdot, t)\|_{L^2(\sigma^T)} \le c\|u_0\|_{W^{2,2}(\sigma)}\,; \tag{3.2.46}$$

where c is independent of T; see [103]. Finally, for arbitrary $\delta > 0$, we further find $\frac{\partial^2}{\partial t^2}\Theta \in L^2(\sigma \times (\delta, T))$ and $\frac{\partial}{\partial t}\Delta'\Theta \in L^2(\sigma \times (\delta, T))$.[2]

Now, let $u_{P_\alpha}(x', t)$ be a solution to (3.2.7). Multiplying both sides of Eq. (3.2.7) by $\Theta(x', \tau - t)$ and integrating by parts over σ, we find

$$\int_\sigma \partial_t u_{P_\alpha}(x', t)\Theta(x', \tau - t)\, dx'$$

$$= \nu \int_\sigma \Delta' u_{P_\alpha}(x', t)\Theta(x', \tau - t)\, dx' + \alpha(t) \int_\sigma \Theta(x', \tau - t)\, dx',$$

[2] Note that it is not possible to take $\delta = 0$, because the initial datum $u_0(x)$ does not satisfy the compatibility condition, i.e., $\Delta' u_0(x)\,|_{\partial\sigma} \neq 0$.

that is

$$\frac{d}{dt}\int\limits_{\sigma} u_{P_\alpha}(x',t)\Theta(x',\tau-t)\,dx' - \int\limits_{\sigma} u_{P_\alpha}(x',t)\partial_t\Theta(x',\tau-t)\,dx'$$

$$= \nu\int\limits_{\sigma} u_{P_\alpha}(x',t)\Delta'\Theta(x',\tau-t)\,dx' + \alpha(t)\int\limits_{\sigma}\Theta(x',\tau-t)\,dx'.$$

Integrating this last relation with respect to t from 0 to τ and taking into account that

$$\nu\Delta'\Theta(x',\tau-t) = \partial_\tau\Theta(x',\tau-t) = -\partial_t\Theta(x',\tau-t),$$

we get

$$\int\limits_{\sigma} u_{P_\alpha}(x',t)\Theta(x',\tau-t)dx' - \int\limits_{\sigma} u_{P_\alpha}(x',0)\Theta(x',\tau)dx' = \alpha(t)\int\limits_{\sigma}\Theta(x',\tau-t)dx'.$$

As

$$u_{P_\alpha}(x',0) = 0, \quad \Theta(x',\tau-t)\Big|_{t=\tau} = \Theta(x',0) = u_0(x'),$$

we derive

$$\int\limits_{\sigma} u_{P_\alpha}(x',\tau)u_0(x')\,dx' = \int\limits_{0}^{\tau} H(\tau-t)\alpha(t)\,dt\,, \tag{3.2.47}$$

with

$$H(s) = \int\limits_{\sigma}\Theta(x',s)\,dx'\,. \tag{3.2.48}$$

We next differentiate both sides of (3.2.47) with respect to τ to get

$$\int\limits_{\sigma}\partial_\tau u_{P_\alpha}(x',\tau)u_0(x')\,dx' = H(0)\alpha(\tau) + \int\limits_{0}^{\tau} H'(\tau-t)\alpha(t)\,dt.$$

Using Eqs. (3.2.7) and (3.2.44) and integrating twice by parts, we deduce

$$\int\limits_{\sigma}\partial_\tau u_{P_\alpha}(x',\tau)u_0(x')\,dx' = \nu\int\limits_{\sigma}\Delta' u_{P_\alpha}(x,\tau)u_0(x')\,dx' + \alpha(\tau)\int\limits_{\sigma} u_0(x')\,dx'$$

$$= \nu \int_\sigma u_{P_\alpha}(x',\tau)\Delta' u_0(x')\,dx' + \alpha(\tau)\int_\sigma u_0(x')\,dx' = -\int_\sigma u_{P_\alpha}(x',\tau)\,dx'$$

$$+\alpha(\tau)\int_\sigma u_0(x')\,dx' = -F(\tau) + \alpha(\tau)\int_\sigma u_0(x')\,dx',$$

where $F(\tau) = \int_\sigma u_{P_\alpha}(x',\tau)\,dx'$ is the flux. Since $H(0) = \int_\sigma \Theta(x',0)\,dx' = \int_\sigma u_0(x')\,dx' = \kappa_0$, from the two latter displayed equalities, we find that $F(t)$ and $\alpha(t)$ are related by the following Volterra integral equation of the first kind

$$-F(\tau) = \int_0^\tau H'(\tau - t)\alpha(t)\,dt. \tag{3.2.49}$$

If we differentiate (3.2.49) with respect to τ, we obtain

$$-F'(\tau) = H'(0)\alpha(\tau) + \int_0^\tau H''(\tau - t)\alpha(t)\,dt\,, \tag{3.2.50}$$

and since, by (3.1.2), (3.2.44), we have

$$H'(0) = \int_\sigma \frac{\partial}{\partial\tau}\Theta(x',\tau)\Big|_{\tau=0} dx' = \nu\int_\sigma \Delta u_0(x')\,dx' = -|\sigma|,$$

we conclude that (3.2.50) takes the following form

$$\alpha(\tau) = \int_0^\tau K(\tau - t)\alpha(t)\,dt + \varphi(\tau)\,, \tag{3.2.51}$$

where $\varphi(\tau) := \dfrac{F'(\tau)}{|\sigma|}$ and

$$K(s) := \frac{1}{|\sigma|}\frac{d^2}{ds^2}\int_\sigma \Theta(x',s)\,dx'\,. \tag{3.2.52}$$

Notice that (3.2.51) is a Volterra integral equation of the second type with respect to $\alpha(\tau)$.

Lemma 3.2.2 *The kernel $K(s)$, defined by* (3.2.52)*, belongs to the space $L^1(0, \infty)$.*

Proof
We begin to show integrability of K in $(0, T)$, for all $T > 0$. From (3.2.44), (3.1.2) we get

$$K(s) = \frac{\nu}{|\sigma|} \int_{\sigma} \Delta' \frac{\partial}{\partial s} \Theta(x', s)\, dx' = \frac{\nu}{|\sigma|} \int_{\partial\sigma} \frac{\partial}{\partial \mathbf{n}} \frac{\partial}{\partial s} \Theta(x', s)\, dS_{x'} \quad s \in (0, T).$$

By Cauchy-Schwarz-Bunyakovsky inequality and by well-known trace inequality (see Lemma 2.3.3), we have

$$\begin{aligned} |K(s)|^2 &\leq \frac{\nu|\partial\sigma|}{|\sigma|} \int_{\partial\sigma} \Big| \frac{\partial}{\partial \mathbf{n}} \frac{\partial}{\partial s} \Theta(x', s) \Big|^2 dS_{x'} \\ &\leq c\Big(\|\nabla' \frac{\partial}{\partial s} \Theta(\cdot, s)\|^2_{L^2(\sigma)} + \|\nabla' \frac{\partial}{\partial s} \Theta(\cdot, s)\|_{L^2(\sigma)} \|\Delta' \frac{\partial}{\partial s} \Theta(\cdot, s)\|_{L^2(\sigma)} \Big). \end{aligned} \tag{3.2.53}$$

Here we have also used the inequality

$$\sum_{|\alpha|=2} \|D^{\alpha}_{x'} w\|_{L^2(\sigma)} \leq c\|\Delta' w\|_{L^2(\sigma)}$$

which is valid for functions w vanishing on $\partial\sigma$ and bounded domains σ with C^2 boundary (see, e.g., [103]). Let us now show that

$$\|\Delta' \frac{\partial}{\partial t} \Theta(\cdot, t)\|_{L^2(\sigma)} \leq c\, t^{-1} \qquad t \in (0, T)\,. \tag{3.2.54}$$

Actually, setting $W(x', t) = \dfrac{\partial}{\partial t} \Theta(x', t)$, we find

$$\frac{\partial}{\partial t} W(x', t) = \nu \Delta' W(x', t).$$

Therefore, recalling that $u_0 \in W^{2,2}(\sigma)$, by classical estimates of the heat semigroup (e.g., [65]), we have

$$\|\nu \Delta' W(\cdot, t)\|_{L^2(\sigma)} = \|\frac{\partial}{\partial t} W(\cdot, t)\|_{L^2(\sigma)}$$

$$\leq c\, t^{-1} \|W(\cdot, 0)\|_{L^2(\sigma)} = c\, t^{-1} \|\Delta' u_0\|_{L^2(\sigma)} = \tilde{c}\, t^{-1},$$

which proves (3.2.54). Thus, from (3.2.53), (3.2.54), by the Hölder inequality with $q = 4$ and $q' = 4/3$, and by (3.2.46), it follows that

$$
\begin{aligned}
&\int_0^t |K(s)|\, ds \leq \\
&\leq c\Big(\int_0^t \|\nabla \frac{\partial}{\partial s}\Theta(\cdot,s)\|_{L^2(\sigma)}\, ds + \int_0^t \|\nabla \frac{\partial}{\partial s}\Theta(\cdot,s)\|^{1/2}_{L^2(\sigma)} \|\Delta \frac{\partial}{\partial s}\Theta(\cdot,s)\|^{1/2}_{L^2(\sigma)}\, ds\Big) \\
&\leq c\Big(t^{1/2}\|\nabla \frac{\partial}{\partial s}\Theta\|_{L^2(\sigma\times(0,t))} + \|\nabla \frac{\partial}{\partial s}\Theta\|^{1/4}_{L^2(\sigma\times(0,t))}\Big(\int_0^t \|\Delta \frac{\partial}{\partial s}\Theta(\cdot,s)\|^{2/3}_{L^2(\sigma)}\, ds\Big)^{3/4}\Big) \\
&\leq c(T)\Big(1 + \Big(\int_0^t s^{-2/3}\, ds\Big)^{3/4}\Big) \leq C(T),
\end{aligned}
\tag{3.2.55}
$$

which proves that $K \in L^1(0, T)$, for any finite $T > 0$. From (3.2.52) we also have

$$
|K(t)| \leq c\|\frac{\partial^2}{\partial t^2}\Theta(\cdot,t)\|_{L^2(\sigma)}\,. \tag{3.2.56}
$$

However, for any $t > t_0$ and any $t_0 > 0$, again by classical estimates for the heat semigroup [65], we get

$$
\begin{aligned}
\|\frac{\partial^2}{\partial t^2}\Theta(\cdot,t)\|_{L^2(\sigma)} &\leq c_1\,(t-t_0)^{-1}\|\frac{\partial}{\partial t}\Theta(\cdot,t_0)\|_{L^2(\sigma)} \\
&\leq c_2\,(t-t_0)^{-1}t_0^{-1}\|\Theta(\cdot,0)\|_{L^2(\sigma)}\,.
\end{aligned}
$$

By taking $t_0 = t/2$, from this relation and (3.2.56), it follows that $K \in L^1(a, \infty)$ for all $a > 0$ and the lemma is proved. From this and (3.2.55), we conclude $K \in L^1(0, \infty)$. □

Our next objective is to show that the kernel $K(t)$ admits a useful representation in terms of a suitable infinite series. The following result holds.

Lemma 3.2.3 *The kernel $K(t)$ admits the representation*

$$
K(t) = \frac{1}{|\sigma|}\sum_{k=1}^{\infty} \beta_k^2 \lambda_k \exp(-\lambda_k t)\,, \tag{3.2.57}
$$

where $\beta_k = \int_\sigma u_k(x')dx'$, λ_k and u_k, $k = 1, 2, \ldots$ are eigenvalues and eigenfunction of the Laplace equation (3.2.9)*. Thus, $K(t) > 0$ for all $t > 0$ and, moreover,*

$$\int_0^\infty K(t)\,dt = 1. \tag{3.2.58}$$

Proof

Since the constant function 1 belongs to $L^2(\sigma)$, we may write it as a Fourier series:

$$1 = \sum_{k=1}^{\infty} \beta_k u_k(x'),$$

where $\sum\limits_{k=1}^{\infty} \beta_k^2 = |\sigma|$. It is easy to see that the solution $u_0(x')$ to problem (3.1.2) can be represented as

$$u_0(x') = \sum_{k=1}^{\infty} \frac{\beta_k}{\lambda_k} u_k(x') \tag{3.2.59}$$

and that the series (3.2.59) converges in $W^{2,2}(\sigma)$. Now, solving problem (2.3) by the classical Fourier method, we find that $\Theta(x', t)$ has the representation

$$\Theta(x', t) = \sum_{k=1}^{\infty} \frac{\beta_k}{\lambda_k} \exp(-\lambda_k t) u_k(x').$$

Hence,

$$K(t) = \frac{1}{|\sigma|} \frac{d^2}{dt^2} \int_\sigma \Theta(x', t)\,dx' = \frac{1}{|\sigma|} \sum_{k=1}^{\infty} \beta_k \lambda_k \exp(-\lambda_k t) \Big(\int_\sigma u_k(x')\,dx' \Big)$$

$$= \frac{1}{|\sigma|} \sum_{k=1}^{\infty} \beta_k^2 \lambda_k \exp(-\lambda_k t),$$

and (3.2.57) follows. It remains to prove (3.2.58). Actually,

$$\int_0^\infty K(t)\,dt = \frac{1}{|\sigma|} \sum_{k=1}^{\infty} \beta_k^2 \lambda_k \int_0^\infty \exp(-\lambda_k t)\,dt = \frac{1}{|\sigma|} \sum_{k=1}^{\infty} \beta_k^2 = 1.$$

The lemma is therefore proved. □

3.2.6 Existence in General Sobolev Spaces

Once the summability property of the kernel K is proved, it is easy to establish existence of solutions to the integral equation (3.2.51) and the validity of corresponding estimates. In fact, the following result holds.

Lemma 3.2.4 *Let* $\varphi \in L^r(0,T)$, $T < \infty$, $1 \le r < \infty$. *Then Eq.* (3.2.51) *has one and only one solution* $\alpha \in L^r(0,T)$, *and the following inequality holds*

$$\|\alpha\|_{L^r(0,T)} \le \frac{c(T)}{|\sigma|} \|F'\|_{L^r(0,T)}. \tag{3.2.60}$$

Furthermore, $\alpha(t) \ge 0$ *(respectively,* $\alpha(t) \le 0$ *) in* $[0,T]$, *if* $F'(t) \ge 0$ *(respectively,* $F'(t) \le 0$ *) in* $[0,T]$.

Proof
For $r = 1$ the lemma is proved in [123]. Moreover, it is shown in [206] that the unique solution α is represented by the formula

$$\alpha(t) = \int_0^t R(t-s)\varphi(s)\,ds + \varphi(t), \quad t \in [0,T], \tag{3.2.61}$$

where $R(t)$ is the resolvent associated with a given kernel $K(t)$. Recall that $R(t)$ is defined as the unique L^1-solution of the equation

$$R(t) = \int_0^t K(t-s)R(s)\,ds + K(t),$$

and that, moreover, it satisfies the following estimate [123]

$$\|R\|_{L^1(0,T)} \le C(T)\|K\|_{L^1(0,T)}\,. \tag{3.2.62}$$

Applying Young's inequality for convolutions (see, e.g., [2]), from (3.2.61) and (3.2.62) we conclude $\alpha \in L^r(0,T)$ and that

$$\|\alpha\|_{L^r(0,T)} \le \|R\|_{L^1(0,T)}\|\varphi\|_{L^r(0,T)} + \|\varphi\|_{L^r(0,T)}$$

$$\le C\,\|K\|_{L^1(0,T)}\|\varphi\|_{L^r(0,T)} + \|\varphi\|_{L^r(0,T)} \le C_1\,\|\varphi\|_{L^r(0,T)} = \frac{C_1}{|\sigma|}\|F'\|_{L^r(0,T)}\,,$$

where $C_1 = C_1(T)$.

The final part of the lemma follows by classical results on Volterra's equations with non-negative kernels [123], using the fact that, by Lemma 3.2.3, $K(t) > 0$ for all $t > 0$ and recalling that $\varphi(t) = F'(t)/|\sigma|$. □

Let us now go back to the integral equation (3.2.49). From (3.2.46) it follows that $H' \in L^\infty(0,T)$. So, if $\alpha \in L^1(0,T)$, we conclude from (3.2.49) the necessary compatibility condition:

$$F(0) = 0\,. \tag{3.2.63}$$

Assume that $\varphi'(t) := |\sigma|^{-1}F''(t)$ is integrable in $(0,T)$. Then, from [123, Theorem 1] we obtain that $\alpha' \in L^1(0,T)$ and that, moreover, differentiating (3.2.51), $\alpha'(t)$ satisfies the following equation

$$\alpha'(t) = K(t)\alpha(0) + \int_0^t K(t-s)\alpha'(s)\,ds + \varphi'(t)\,. \tag{3.2.64}$$

Thus, if

$$\alpha(0) = 0\,, \tag{3.2.65}$$

we find that

$$\alpha'(t) = \int_0^t K(t-s)\alpha'(s)\,ds + \varphi'(t)\,. \tag{3.2.66}$$

Note that, by (3.2.51), condition (3.2.65) requires

$$F'(0) = |\sigma|\varphi(0) = 0\,. \tag{3.2.67}$$

Now, (3.2.66) is formally identical to (3.2.51), so that we can apply the results of Lemma 3.2.4, provided $F(t)$ obeys the compatibility condition (3.2.67). Obviously, this argument can be iterated.

From the above considerations and from Lemma 3.2.4, we thus deduce the following result.

Lemma 3.2.5 *Let $F \in W^{l+1,r}(0,T)$, $1 \le r < \infty$, l non-negative integer, $T < +\infty$, and assume the compatibility conditions*

$$F(0) = 0,\ \frac{dF}{dt}(0) = 0,\ \ldots,\ \frac{d^lF}{dt^l}(0) = 0\,. \tag{3.2.68}$$

Then integral equation (3.2.51) *admits a unique solution $\alpha \in W^{l,r}(0,T)$,*

$$\|\alpha\|_{W^{l,r}(0,T)} \le c(T)\|F\|_{W^{l+1,r}(0,T)}. \tag{3.2.69}$$

Moreover, for $l \ge 1$,

$$\alpha(0) = 0, \quad \frac{d\alpha}{dt}(0) = 0, \quad \dots, \quad \frac{d^{l-1}\alpha}{dt^{l-1}}(0) = 0. \tag{3.2.70}$$

We are now in a position to prove the unique solvability of (3.2.7) in general Sobolev spaces. Consider first the case $u_{0n}(x) = 0$.

Theorem 3.2.5 *Let $\partial\sigma \in C^{2l+2}$, $F \in W^{l+1,r}(0, T)$, $l \geq 0, 1 < r < \infty, r \neq 3/2, T < \infty$, $u_0(x) \equiv 0$, and let $F(t)$ satisfies the compatibility conditions* (3.2.68). *Then there exists a unique solution $\left(u_{P_\alpha}, \alpha\right) \in \overset{\circ}{W}{}_{\partial\sigma}^{(2l+2,l+1),r}(\sigma^T) \times W^{l,r}(0, T)$ of problem* (3.2.7), *and the following estimate holds*

$$\|u_{P_\alpha}\|_{W^{(2l+2,l+1),r}(\sigma^T)} + \|\alpha\|_{W^{l,r}(0,T)} \leq c\,\|F\|_{W_r^{l+1}(0,T)}\,. \tag{3.2.71}$$

Proof
By Lemma 3.2.5, the integral equation (3.2.51) has a unique solution $\alpha \in W^{l,r}(0, T)$, which satisfies estimate (3.2.69) and conditions (3.2.70). Consider now problem (3.2.3) with $u_{0n} = 0$. According to well-known results [107, 197], this problem has a unique solution $u_{P_\alpha} \in W^{(2l+2,l+2),r}(\sigma^T)$, $r \neq 3/2$, and

$$\|u_{P_\alpha}\|_{W^{(2l+2,l+1),r}(\sigma^T)} \leq c\|\alpha\|_{W^{(2l,l),r}(\sigma^T)} \leq c\|\alpha\|_{W^{l,r}(0,\,T)}\,. \tag{3.2.72}$$

Obviously, $u_{P_\alpha}(x', t)$ satisfies the flux condition (3.2.6). Estimate (3.2.71) follows from (3.2.72) and (3.2.69). □

For the general case of problem (3.2.7), we have the following result.

Theorem 3.2.6 *Let $\partial\sigma \in C^{2l+2}$, $F \in W^{l+1,r}(0, T)$, $u_{0n} \in W^{2l+2-2/r,r}(\sigma)$, $l \geq 0$, $1 < r < \infty$, $r \neq 3/2$, $T < \infty$. Suppose that the following compatibility conditions*

$$\begin{aligned} &\frac{d^s F}{dt^s}(0) = \int\limits_\sigma u_{0n}^{(s)}(x')\,dx, \quad s = 0, 1, \dots, l, \\ &u_{0n}^{(s)}(x')|_{\partial\sigma} = 0, \quad s = 0, 1, \dots, [l + 1 - 3/2r] \end{aligned} \tag{3.2.73}$$

hold, where

$$u_{0n}^{(0)}(x') := u_{0n}(x), \quad u_{0n}^{(s)}(x') := \nu\Delta' u_{0n}^{(s-1)}(x'), \; s = 1, 2, \dots, l. \tag{3.2.74}$$

Then problem (3.2.7) *admits a unique solution $\left(u_{P_\alpha}, \alpha\right) \in \overset{\circ}{W}{}_{\partial\sigma}^{(2l+2,l+1),r}(\sigma^T) \times W^{l,r}(0, T)$, and the following estimate*

$$\begin{aligned} &\|u_{P_\alpha}\|_{W^{(2l+2,l+1),r}(\sigma^T)} + \|\alpha\|_{W^{l,r}(0,T)} \\ &\leq c\left(\|F\|_{W^{l+1,r}(0,T)} + \|u_{0n}\|_{W^{2l+2-2/r,r}(\sigma)}\right) \end{aligned} \tag{3.2.75}$$

holds.

Proof
We look for the solution $\big(u_{P_\alpha}(x',t),\alpha(t)\big)$ of problem (3.2.7) in the form

$$\big(u_{P_\alpha}(x',t),\alpha(t)\big)=\big(u_h^{(1)}(x',t),0\big)+\big(u_{P_\alpha}^{(2)}(x',t),\alpha(t)\big),$$

where $u_h^{(1)}(x',t)$ is the solution of the heat equation (3.2.39) and $(u_{P_\alpha}^{(2)}(x',t),q(t))$ is the solution of the inverse problem (3.2.40) with the flux

$$\widetilde{F}(t)=F(t)-\int\limits_{\sigma}u_h^{(1)}(x',t)\,dx.$$

The solvability of problem (3.2.39) in the appropriate spaces is well-known; see, e.g., [107, 197]). Specifically, since $\partial\sigma\in C^{2l+2}$ and $u_0^{(s)}(x')|_{\partial\sigma}=0$, $s=0,1,\dots,[l+1-3/2r]$, there exists a unique solution $u_h^{(1)}\in W^{(2l+2,l+1),r}(\sigma^T)$. This solution satisfies

$$\|u_h^{(1)}\|_{W^{(2l+2,l+1),r}(\sigma^T)}\le c\|u_{0n}\|_{W^{2l+2-2/r,r}(\sigma)}. \tag{3.2.76}$$

The flux function $\widetilde{F}(t)$ satisfies the estimate

$$\begin{aligned}\|\widetilde{F}\|_{W^{l+1,r}(0,T)}&\le c\Big(\|F\|_{W^{l+1,r}(0,T)}+\sum_{j=0}^{l+1}\|\frac{\partial^j}{\partial t^j}u_h^{(1)}\|_{L^r(\sigma^T)}\Big)\\&\le c\left(\|F\|_{W^{l+1,r}(0,T)}+\|u_{0n}\|_{W^{2l+2-2/r,r}(\sigma)}\right).\end{aligned} \tag{3.2.77}$$

Moreover, from $(3.2.73)_1$, we conclude that

$$\frac{d^s\tilde{F}}{dt^s}(0)=0,\qquad s=0,1,\dots,l.$$

Therefore, by Theorem 3.2.5, the inverse problem with the flux $\widetilde{F}$ admits a unique solution $\big(u_{P_\alpha}^{(2)},\alpha\big)\in W^{(2l+2,l+1),r}(\sigma^T)\times W^{l,r}(0,T)$, and estimate (3.2.71) holds. Estimate (3.2.75) follows now from inequalities (3.2.76), (3.2.77), and (3.2.71). □

Remark 3.2.1 Note that results of this subsection are true for arbitrary finite $T>0$. In Sect. 3.2.4 the solvability of problem (3.2.7) is proved, by using Galerkin approximations, for the infinite time interval $[0,\infty]$ in spaces $W^{(2,1),2}(\sigma\times(0,\infty))\times L^2(0,\infty)$. In order to get a priori estimates for Galerkin approximations on the infinite time interval $[0,\infty]$, the whole system (3.2.7) was used. The regularity of this solution in L^2-setting was studied in [176].

3.2.7 *Behavior of Poiseuille Flow as* $t\to+\infty$

In this section, we examine the behavior of the Poiseuille flow as t approaches infinity.

Let $\mathbb{W}^{l,2}_{\mu}(0,\infty)$ be the space of functions with a finite norm given by

$$\|F\|_{\mathbb{W}^{l,2}_{\mu}(0,\infty)} = \|\exp(\mu t)F(t)\|_{W^{l,2}(0,\infty)}, \quad \mathbb{L}^2_{\mu}(0,\infty) = \mathbb{W}^{0,2}_{\mu}(0,\infty).$$

Theorem 3.2.7 *Let $\partial\sigma \in C^2$, $u_{0n}(x') = 0$, and $F \in \mathbb{W}^{1,2}_{\mu}(0,\infty)$ with $\mu > 0$. Assume that the compatibility condition* (3.2.8) *holds. Then the solution $\big(u_{P_\alpha}(x',t), \alpha(t)\big)$ of problem* (3.2.7) *satisfies the estimate*

$$\begin{aligned} &\exp(\gamma_* t)\Big(\int\limits_{\sigma} |u_{P_\alpha}(x',t)|^2\,dx' + \nu\int\limits_{\sigma} |\nabla' u_{P_\alpha}(x',t)|^2\,dx'\Big) \\ &+ \int\limits_0^t \exp(\gamma_*\tau)\int\limits_{\sigma} |\partial_\tau u_{P_\alpha}(x',\tau)|^2\,dx'd\tau + \int\limits_0^t \exp(\gamma_*\tau)|\alpha(\tau)|^2\,d\tau \\ &\qquad \le c\|F\|^2_{\mathbb{W}^{l,2}_{\mu}(0,\infty)} \end{aligned} \tag{3.2.78}$$

where $\gamma_* = \min\{\lambda_1,\ 1,\ 2\mu\}$ and λ_1 is the first eigenvalue of the Dirichlet Laplace operator (3.2.9).

Proof
By Theorem 3.2.4, the solution $\big(u_{P_\alpha}, \alpha\big) \in \overset{\circ}{W}{}^{(2,1),2}_{\partial\sigma}(\sigma\times(0,\infty))\times L^2(0,\infty)$ satisfies equations (3.2.7) for almost all (x',t). Following the approach used in the proof of Lemma 3.2.1, we obtain the estimates for the solutions $\big(u_{P_\alpha}, \alpha\big)$ as follows:

$$\begin{aligned} &\frac{1}{2}\frac{d}{dt}\int\limits_{\sigma} |u_{P_\alpha}(x',t)|^2\,dx' + \nu\int\limits_{\sigma} |\nabla' u_{P_\alpha}(x',t)|^2\,dx' \\ &\qquad \le \varepsilon|\alpha(t)|^2 + c_\varepsilon|F(t)|^2, \end{aligned} \tag{3.2.79}$$

$$\begin{aligned} &\frac{\nu}{2}\frac{d}{dt}\int\limits_{\sigma} |\nabla' u_{P_\alpha}(x',t)|^2\,dx' + \int\limits_{\sigma} |\partial_t u_{P_\alpha}(x',t)|^2\,dx' \\ &\qquad \le \varepsilon|\alpha(t)|^2 + c_\varepsilon|F'(t)|^2, \end{aligned} \tag{3.2.80}$$

$$|\alpha(t)|^2 \le c\Big(\int\limits_{\sigma} |\partial_t u_{P_\alpha}(x',t)|^2\,dx' + |F(t)|^2\Big), \tag{3.2.81}$$

where ε is an arbitrary positive number. For sufficiently small ε, inequalities (3.2.79)–(3.2.81) lead to

$$\begin{aligned} &\frac{1}{2}\frac{d}{dt}\int\limits_{\sigma} \Big(|u_{P_\alpha}(x',t)|^2 + \nu|\nabla' u_{P_\alpha}(x',t)|^2\Big)\,dx' \\ &+\nu\int\limits_{\sigma} |\nabla' u_{P_\alpha}(x',t)|^2\,dx' + \frac{1}{2}\int\limits_{\sigma} |\partial_t u_{P_\alpha}(x',t)|^2\,dx' \\ &\qquad \le c\Big(|F(t)|^2 + |F'(t)|^2\Big). \end{aligned} \tag{3.2.82}$$

Using the Poincaré inequality in the domain σ, we have

$$\int\limits_{\sigma} |u_{P_\alpha}(x',t)|^2\,dx' \le \frac{\nu}{\lambda_1}\int\limits_{\sigma} |\nabla' u_{P_\alpha}(x',t)|^2\,dx',$$

where λ_1 is the first eigenvalue of problem (3.2.9). Thus, from (3.2.82), we obtain

$$\begin{aligned}&\frac{d}{dt}\int\limits_{\sigma}\Big(|u_{P_\alpha}(x',t)|^2+\nu|\nabla' u_{P_\alpha}(x',t)|^2\Big)\,dx'\\&+\gamma_*\Big(\int\limits_{\sigma}|u_{P_\alpha}(x',t)|^2\,dx'+\nu\int\limits_{\sigma}|\nabla' u_{P_\alpha}(x',t)|^2\,dx'\Big)\\&+\int\limits_{\sigma}|\partial_t u_{P_\alpha}(x',t)|^2\,dx'\le c\Big(|F(t)|^2+|F'(t)|^2\Big)\end{aligned}\tag{3.2.83}$$

with $\gamma_* = \min\{\lambda_1,\ 1,\ 2\mu\}$. Multiplying (3.2.83) by $\exp(\gamma_* t)$ and integrating over t, we receive

$$\begin{aligned}&\exp(\gamma_* t)\Big(\int\limits_{\sigma}|u_{P_\alpha}(x',t)|^2dx'+\nu\int\limits_{\sigma}|\nabla' u_{P_\alpha}(x',t)|^2dx'\Big)\\&\quad+\int\limits_0^t\exp(\gamma_*\tau)\int\limits_{\sigma}|\partial_\tau u_{P_\alpha}(x',\tau)|^2dx'd\tau\\&\quad\le c\int\limits_0^t\exp(\gamma_*\tau)\Big(|F(\tau)|^2+|F'(\tau)|^2\Big)\,d\tau\\&\quad= c\|F\|^2_{\mathbb{W}^{1,2}_{\gamma_*}(0,\infty)}\le c\|F\|^2_{\mathbb{W}^{1,2}_{\mu}(0,\infty)}.\end{aligned}\tag{3.2.84}$$

The relations (3.2.84) and (3.2.81) yield

$$\int\limits_0^t\exp(\gamma_*\tau)|\alpha(\tau)|^2\,d\tau\le c\|F\|^2_{\mathbb{W}^{1,2}_{\mu}(0,\infty)}.\tag{3.2.85}$$

Since the constants in inequalities (3.2.84) and (3.2.85) are independent of t, it follows from these estimates that (3.2.78) holds. □

3.3 Time-Periodic Poiseuille Flow

In this section, we investigate a time-periodic Poiseuille flow. Specifically, we focus on the case where the external force is zero. To describe this, we consider the

Navier–Stokes equations within a pipe Π, subject to the following conditions:

$$\begin{cases} \partial_t \mathbf{u} - \nu\Delta\mathbf{u} + (\mathbf{u}\cdot\nabla)\mathbf{u} + \nabla p = 0, & x \in \Pi,\ t \in (0, 2\pi), \\ \operatorname{div} \mathbf{u} = 0, & x \in \Pi,\ t \in (0, 2\pi), \\ \mathbf{u}(x, t) = 0, & x \in \partial\Pi,\ t \in (0, 2\pi), \\ \mathbf{u}(x, 0) = \mathbf{u}(x, 2\pi),\ x \in \Pi. \end{cases} \tag{3.3.1}$$

For simplicity, we take the time period to be $T = 2\pi$. The solution $(\mathbf{u}(x,t), p(x,t))$ to problem (3.3.1) can be expressed in the following form:

$$\begin{aligned} \mathbf{u}(x, t) &= (0, \dots, 0, u_{P_\alpha}(x', t)), \\ p(x, t) &= -\alpha(t)x_n + p_0(t), \end{aligned} \tag{3.3.2}$$

where $u_{P_\alpha}(x', t)$ is the Poiseuille flow velocity profile and $\alpha(t)$ is the time-periodic pressure slope.

The Poiseuille solution can be obtained by prescribing either the time-periodic pressure slope $\alpha(t)$, $\alpha(-\pi) = \alpha(\pi)$ or the time-periodic flux $F(t)$, where the condition for the flux is given by:

$$\int\limits_{\sigma} u_n(x, t)\, dx' = F(t), \qquad F(0) = F(2\pi), \tag{3.3.3}$$

where σ represents the cross-section of the pipe and $u_n(x, t)$ is the n-th component of the velocity vector $\mathbf{u}(x, t)$.

Depending on whether the pressure slope $\alpha(t)$ or the flux $F(t)$ is prescribed, we obtain different partial differential equations for $u_{P_\alpha}(x', t)$. In the case of the prescribed pressure slope $\alpha(t)$, $u_{P_\alpha}(x', t)$ satisfies the heat equation on the cross-section σ with a given right-hand side $\alpha(t)$:

$$\begin{cases} \partial_t u_{P_\alpha}(x', t) - \nu\Delta' u_{P_\alpha}(x', t) = \alpha(t), & x' \in \sigma,\ t \in (0, 2\pi), \\ u_{P_\alpha}(x', t) = 0, & x' \in \partial\sigma,\ t \in (0, 2\pi), \\ u_{P_\alpha}(x', 0) = u_{P_\alpha}(x', 2\pi),\ x' \in \sigma. \end{cases} \tag{3.3.4}$$

It is well-known that problem (3.3.4) has a unique 2π-periodic solution satisfying certain regularity properties (see, for instance, [207]).

Theorem 3.3.1 *Let $\partial\sigma \in C^2$ and let $\alpha(t)$ be a 2π-periodic function with $\alpha \in L^2(0, 2\pi)$. Then problem* (3.3.4) *has a unique 2π-periodic solution such that*

$$\sup_{t\in[0,2\pi]} \|u_{P_\alpha}(\cdot, t)\|^2_{W^{1,2}(\sigma)} + \|u_{P_\alpha}\|^2_{W^{(2,1),2}(\sigma\times(0,2\pi))} \le c\|\alpha\|^2_{L^2(0,2\pi)}. \tag{3.3.5}$$

Moreover, if $\partial\sigma \in C^{l+2}$, $l \geq 0$, *and* $\alpha \in W^{l,2}(0, 2\pi)$, *then* $u_{P_\alpha} \in W^{(2l+2,l+1),2}(\sigma \times (0, 2\pi))$ *and*

$$\|u_{P_\alpha}\|^2_{W^{(2l+2,l+1),2}(\sigma\times(0,2\pi))} \leq c\|\alpha\|^2_{W^{l,2}(0,2\pi)}. \tag{3.3.6}$$

Similarly, when the flux $F(t)$ is prescribed, $u_{P_\alpha}(x', t)$ and $\alpha(t)$ are subject to the inverse time-periodic problem on the cross-section σ:

$$\begin{cases} \partial_t u_{P_\alpha}(x', t) - \nu\Delta' u_{P_\alpha}(x', t) = \alpha(t), & x' \in \sigma,\ t \in (0, 2\pi), \\ u_{P_\alpha}(x', t) = 0, & x' \in \partial\sigma,\ t \in (0, 2\pi), \\ u_{P_\alpha}(x', 0) = u_{P_\alpha}(x', 2\pi),\ x' \in \sigma, & \\ \int\limits_\sigma u_{P_\alpha}(x', t)\, dx' = F(t), & t \in (0, 2\pi). \end{cases} \tag{3.3.7}$$

To establish the existence of a unique time-periodic solution to problem (3.3.7) and determine the relationship between the pressure slope $-\alpha(t)$ and the flux $F(t)$, we consider the following sequences of problems:

$$\begin{cases} \partial_t U_n^{(c)}(x', t) - \nu\Delta' U_n^{(c)}(x', t) = \cos(nt), & x' \in \sigma,\ t \in (0, 2\pi), \\ U_n^{(c)}(x', t) = 0, & x' \in \partial\sigma, t \in (0, 2\pi), \\ U_n^{(c)}(x', 0) = U_n^{(c)}(x', 2\pi),\ x \in \sigma, & \end{cases} \tag{3.3.8}$$

where $n = 0, 1, \ldots$,

$$\begin{cases} \partial_t U_n^{(s)}(x', t) - \nu\Delta' U_n^{(s)}(x', t) = \sin(nt), & x' \in \sigma,\ t \in (0, 2\pi), \\ U_n^{(s)}(x', t) = 0, & x' \in \partial\sigma,\ t \in (0, 2\pi), \\ U_n^{(s)}(x', 0) = U_n^{(s)}(x', 2\pi),\ x' \in \sigma, & \end{cases} \tag{3.3.9}$$

where $n = 1, \ldots,$.

The solutions $U_n^{(c)}(x', t)$ and $U_n^{(s)}(x', t)$ to problems (3.3.8) and (3.3.9), respectively, have the following form:

$$\begin{aligned} U_n^{(c)}(x', t) &= \varphi_n(x')\cos(nt) - \psi_n(x')\sin(nt), \\ U_n^{(s)}(x', t) &= \psi_n(x')\cos(nt) + \varphi_n(x')\sin(nt), \end{aligned} \tag{3.3.10}$$

where the pair $\big(\varphi_n(x'), \psi_n(x')\big)$ is a solution to the following elliptic problem:

$$\begin{cases} -n\psi_n(x') = \nu\Delta'\varphi_n(x') + 1, \ x' \in \sigma, \\ n\varphi_n(x') = \nu\Delta'\psi_n(x'), \qquad x' \in \sigma, \\ \varphi_n(x') = \psi_n(x') = 0, \qquad x' \in \partial\sigma. \end{cases} \tag{3.3.11}$$

It can be proven through standard methods that for each non-negative integer n, problem (3.3.11) possesses a unique solution $(\varphi_n, \psi_n) \in W^{2,2}(\sigma) \cap \overset{\circ}{W}{}^{1,2}(\sigma)$. Specifically, (3.3.11) is a strongly elliptic system that can be reduced to an operator equation with a compact operator in the space $\overset{\circ}{W}{}^{1,2}(\sigma) \times \overset{\circ}{W}{}^{1,2}(\sigma)$. By applying the Fredholm theorem, it follows that (3.3.11) is solvable if and only if the corresponding homogeneous problem has only the trivial solution. The homogeneous problem is as follows:

$$\begin{cases} -n\psi_n(x') = \nu\Delta'\varphi_n(x'), \ x' \in \sigma, \\ n\varphi_n(x') = \nu\Delta'\psi_n(x'), \ x' \in \sigma, \\ \varphi_n(x') = \psi_n(x') = 0, \ x' \in \partial\sigma. \end{cases}$$

By multiplying the first equation by φ_n and the second by ψ_n, integrating by parts over σ, and summing the obtained equalities, it follows that:

$$\nu \int_\sigma |\nabla'\varphi_n(x')|^2 dx' = 0, \quad \nu \int_\sigma |\nabla'\psi_n(x')|^2 dx' = 0.$$

Therefore, $\varphi_n(x') = 0$, $\psi_n(x') = 0$, and the non-homogeneous problem (3.3.11) has a unique solution in $\overset{\circ}{W}{}^{1,2}(\sigma) \times \overset{\circ}{W}{}^{1,2}(\sigma)$. The regularity of the solution follows from existing results about the regularity of weak solutions to the Laplace equation (see, for instance, [103]).

Let us define the quantities a_n and b_n as follows:

$$a_n = \int_\sigma \varphi_n(x')\,dx', \quad b_n = -\int_\sigma \psi_n(x')\,dx', \quad n = 0, 1, 2, \ldots. \tag{3.3.12}$$

Now, we prove the following lemma:

Lemma 3.3.1 *Let* $(\varphi_n, \psi_n) \in W^{2,2}(\sigma) \cap \overset{\circ}{W}{}^{1,2}(\sigma)$ *be a solution to* (3.3.11). *Then, the following inequality holds:*

$$\nu\|\Delta'\varphi_n\|^2_{L^2(\sigma)} + \nu\|\Delta'\psi_n\|^2_{L^2(\sigma)} \le \frac{1}{\nu}|\sigma| \ \forall n = 0, 1, 2, \ldots \tag{3.3.13}$$

Moreover, the numbers a_n *and* b_n *have the following properties:*

$$
\begin{array}{lll}
(a) & a_n > 0 \ \forall n = 0, 1, 2, \ldots; \ b_0 = 0, \ b_n > 0 \ \forall n = 1, 2, \ldots; & \\
(b) & a_n \le \dfrac{|\sigma|}{n}, \qquad\qquad b_n \le \dfrac{|\sigma|}{n} \ \forall n = 1, 2, \ldots; & (3.3.14) \\
(c) & \lim\limits_{n\to\infty} (nb_n) = |\sigma|. &
\end{array}
$$

Proof

If we multiply $(3.3.11)_1$ by $\Delta'\varphi_n(x')$ and $(3.3.11)_2$ by $\Delta'\psi_n(x')$, add both resulting equalities, and integrate by parts over σ, we find the relation:

$$
\nu\|\Delta'\varphi_n\|^2_{L^2(\sigma)} + \nu\|\Delta'\psi_n\|^2_{L^2(\sigma)} = -\int\limits_\sigma \Delta'\varphi_n(x')\,dx'
$$

$$
\le \frac{\nu}{2}\|\Delta'\varphi_n\|^2_{L^2(\sigma)} + \frac{1}{2\nu}|\sigma|.
$$

From this, it immediately follows (3.3.13).

To show (a), we multiply $(3.3.11)_2$ by $\varphi_n(x')$ and $(3.3.11)_1$ by $\psi_n(x')$ and integrate by parts over σ. This delivers:

$$
-n\|\psi_n\|^2_{L^2(\sigma)} = -\nu\int\limits_\sigma \nabla'\varphi_n(x')\cdot\nabla'\psi_n(x')\,dx' - b_n
$$

and

$$
-n\|\varphi_n\|^2_{L_2(\sigma)} = \nu\int\limits_\sigma \nabla'\varphi_n(x')\cdot\nabla'\psi_n(x')\,dx'. \tag{3.3.15}
$$

Summing these two last displayed equations gives:

$$
b_n = n\Big(\|\varphi_n\|^2_{L^2(\sigma)} + \|\psi_n\|^2_{L^2(\sigma)}\Big). \tag{3.3.16}
$$

We now multiply $(3.3.11)_1$ by $\varphi_n(x')$ and $(3.3.11)_2$ by $\psi_n(x')$, integrate by parts, and add two resulting relations. We thus deduce:

$$
a_n = \nu\Big(\|\nabla'\varphi_n\|^2_{L^2(\sigma)} + \|\nabla'\psi_n\|^2_{L_2(\sigma)}\Big). \tag{3.3.17}
$$

Since (3.3.11) does not allow an identically zero solution, the proof of (a) follows from (3.3.16), (3.3.17).

Let us show (b). By Cauchy-Schwarz-Bunyakovsky inequality from (3.3.12), we get:

$$
a_n \le |\sigma|^{1/2}\|\varphi_n\|_{L^2(\sigma)}, \quad b_n \le |\sigma|^{1/2}\|\psi_n\|_{L^2(\sigma)}, \ n = 0, 1, 2, \ldots. \tag{3.3.18}
$$

Using relations (3.3.15) and (3.3.17), we deduce the inequality:

$$\|\varphi_n\|^2_{L^2(\sigma)} \leq \frac{\nu}{n}\Big(\|\nabla\varphi_n\|^2_{L^2(\sigma)} + \|\nabla\psi_n\|^2_{L^2(\sigma)}\Big) = \frac{a_n}{n}.$$

This inequality and (3.3.18) yield

$$a_n \leq |\sigma|^{1/2}\sqrt{\frac{a_n}{n}}.$$

On the other hand, Eq. (3.3.16) implies

$$\|\psi_n\|^2_{L^2(\sigma)} \leq \frac{b_n}{n},$$

and so, from the second inequality in (3.3.18), it follows that

$$b_n \leq |\sigma|^{1/2}\sqrt{\frac{b_n}{n}}.$$

Thus, we have proved the inequalities:

$$a_n \leq \frac{|\sigma|}{n}, \quad b_n \leq \frac{|\sigma|}{n}, \quad n = 1, 2, \ldots. \tag{3.3.19}$$

It remains to show property (c). From the second relation in (3.3.19) and from (3.3.16), it follows that:

$$n\|\psi_n\|_{L^2(\sigma)}| \leq |\sigma|^{1/2}, \quad n = 0, 1, 2. \ldots. \tag{3.3.20}$$

Using the first relation in (3.3.19) together with (3.3.17), we obtain:

$$\lim_{n\to\infty} \|\nabla'\varphi_n\|_{L^2(\sigma)} = 0. \tag{3.3.21}$$

Let us multiply both sides of $(3.3.11)_1$ by arbitrary $\chi \in C_0^\infty(\sigma)$ and integrate by parts over σ:

$$-n\int_\sigma \psi_n(x')\chi(x')\,dx' = -\nu\int_\sigma \nabla'\varphi_n(x')\cdot\nabla'\chi(x')\,dx' + \int_\sigma \chi(x')\,dx'.$$

By passing to the limit $n \to \infty$ in this relation and with the help of (3.3.21), we have:

$$\lim_{n\to\infty}\Big(-n\int_\sigma \psi_n(x')\chi(x')\,dx'\Big) = \int_\sigma \chi(x')\,dx'.$$

The set $C_0^\infty(\sigma)$ is dense in $L^2(\sigma)$. Therefore, the last relation remains valid for any $\chi \in L^2(\sigma)$. Taking $\chi(x') \equiv 1$ in it, we get property (c). □

We are now in a position to establish the relationship between the functions $\alpha(t)$ and $F(t)$. To accomplish this, we shall express both quantities in terms of their Fourier series representations (see [73]):

$$\begin{aligned} \alpha(t) &= \frac{\alpha_0^{(c)}}{2} + \sum_{n=1}^{\infty} \left\{ \alpha_n^{(c)} \cos(nt) + \alpha_n^{(s)} \sin(nt) \right\}, \\ F(t) &= \frac{F_0^{(c)}}{2} + \sum_{n=1}^{\infty} \left\{ F_n^{(c)} \cos(nt) + F_n^{(s)} \sin(nt) \right\}. \end{aligned} \tag{3.3.22}$$

The next lemma establishes a relationship between the Fourier coefficients of the function $\alpha(t)$ and those of the function $F(t)$.

Lemma 3.3.2 *The Fourier coefficients $\left(\alpha_n^{(c)}, \alpha_n^{(s)}\right)$ of $\alpha(t)$ and those $\left(F_n^{(c)}, F_n^{(s)}\right)$ of $F(t)$ are related to each other by the following formulas*

$$F_n^{(c)} = a_n \alpha_n^{(c)} - b_n \alpha_n^{(s)}, \quad F_n^{(s)} = b_n \alpha_n^{(c)} + a_n \alpha_n^{(s)}, \tag{3.3.23}$$

or, equivalently, by their inverse

$$\alpha_n^{(c)} = \frac{a_n F_n^{(c)} + b_n F_n^{(s)}}{a_n^2 + b_n^2}, \quad \alpha_n^{(s)} = \frac{a_n F_n^{(s)} - b_n F_n^{(c)}}{a_n^2 + b_n^2}, \tag{3.3.24}$$

$n = 0, 1, 2, \ldots.$

Proof
To prove the lemma, we start by noting that it suffices to establish the validity of (3.3.23). The inverse relations (3.3.24) can be directly derived from (3.3.23) since it holds that $a_n^2 + b_n^2 > 0$ for all n. Let us define the functions

$$\widehat{U}_n^{(c)}(x', t) \equiv U_n^{(c)}(x', -t), \quad \widehat{U}_n^{(s)}(x', t) \equiv U_n^{(s)}(x', -t), \quad t \in [0, 2\pi]. \tag{3.3.25}$$

We can then write the following partial differential equations for $\widehat{U}_n^{(s)}$ and $\widehat{U}_n^{(c)}$:

$$\begin{cases} \partial_t \widehat{U}_n^{(c)}(x', t) + \nu \Delta' \widehat{U}_n^{(c)}(x', t) = -\cos(nt), & x' \in \sigma, \ t \in (0, 2\pi), \\ \widehat{U}_n^{(c)}(x', t) = 0, & x' \in \partial\sigma, \ t \in (0, 2\pi), \\ \widehat{U}_n^{(c)}(x', 0) = \widehat{U}_n^{(c)}(x', 2\pi), \ x' \in \sigma, \end{cases} \tag{3.3.26}$$

and

$$\begin{cases} \partial_t \widehat{U}_n^{(s)}(x',t) + \nu\Delta'\widehat{U}_n^{(s)}(x',t) = -\sin(nt), & x' \in \sigma,\ t \in (0,2\pi), \\ \widehat{U}_n^{(s)}(x',t)\big|_{\partial\sigma} = 0, & x' \in \partial\sigma,\ t \in (0,2\pi), \\ \widehat{U}_n^{(s)}(x',0) = \widehat{U}_n^{(s)}(x',2\pi),\ x' \in \sigma. \end{cases} \tag{3.3.27}$$

Next, we multiply both sides of (3.3.7) by $\widehat{U}_n^{(c)}(x',t)$ and integrate it by parts over σ:

$$\frac{d}{dt}\int\limits_{\sigma} u_{P_\alpha}(x',t)\widehat{U}_n^{(c)}(x',t)\,dx' - \int\limits_{\sigma} u_{P_\alpha}(x',t)\left[\frac{\partial}{\partial t}\widehat{U}_n^{(c)}(x',t) + \nu\Delta'\widehat{U}_n^{(c)}(x',t)\right]dx'$$

$$= \int\limits_{\sigma} \alpha(t)\widehat{U}_n^{(c)}(x',t)\,dx'.$$

Furthermore, we integrate this relation over $t \in [0,2\pi]$. Taking into account (3.3.10), (3.3.12), (3.3.22), (3.3.26), and the fact that $u_{P_\alpha}(x',0) = u_{P_\alpha}(x',2\pi)$, we arrive at the following equality:

$$\int\limits_0^{2\pi} F(t)\cos(nt)\,dt = \int\limits_0^{2\pi} \alpha(t)\,(a_n\cos(nt) - b_n\sin(nt))\,dt,$$

which establishes the first relation in (3.3.23). Similarly, the second relation in (3.3.23) can be obtained using the same procedure, but replacing $\widehat{U}_n^{(c)}(x',t)$ with $\widehat{U}_n^{(s)}(x',t)$ in the previous argument. □

Our next objective is to investigate the convergence of the series defined by (3.3.22), as well as the relationship between the norms of the functions $\alpha(t)$ and $F(t)$. We begin by presenting the following lemma:

Lemma 3.3.3 *Suppose that the sequences* $\big(F_n^{(c)}, F_n^{(s)}\big)$ *and* $\big(\alpha_n^{(c)}, \alpha_n^{(s)}\big)$, *where* $n = 0, 1, 2, \ldots$, *satisfy equations* (3.3.23) (*or equivalently, Eqs.* (3.3.24)). *If the Fourier series* $(3.3.22)_1$ *converges to some function* $\alpha \in L^2(0,2\pi)$, *then the Fourier series* $(3.3.22)_2$ *also converges to some function* $F \in W^{1,2}(0,2\pi)$, *and we have the inequality:*

$$\|F\|_{W^{1,2}(0,2\pi)} \le c_1\|\alpha\|_{L^2(0,2\pi)}. \tag{3.3.28}$$

Conversely, if the Fourier series $(3.3.22)_2$ *converges to some function* $F \in W^{1,2}(0,2\pi)$, *then the Fourier series* $(3.3.22)_1$ *also converges to some function* $\alpha \in L^2(0,2\pi)$, *and we have the inequality:*

$$\|\alpha\|_{L^2(0,2\pi)} \le c_2\|F\|_{W^{1,2}(0,2\pi)}. \tag{3.3.29}$$

Moreover, if $F \in W^{l+1,2}(0, 2\pi)$, then $\alpha \in W^{l,2}(0, 2\pi)$, and the following inequality holds:

$$\|\alpha\|_{W^{l,2}(0,2\pi)} \leq c_2 \|F\|_{W^{l+1,2}(0,2\pi)}. \tag{3.3.30}$$

Proof
From Eq. (3.3.23) and Lemma 3.3.1 (a), we obtain the relations:

$$F_0^{(c)} = a_0 \alpha_0^{(c)}, \ |F_n^{(c)}|^2 + |F_n^{(s)}|^2 = \left(a_n^2 + b_n^2\right)\left(|\alpha_n^{(c)}|^2 + |\alpha_n^{(s)}|^2\right), \ n \geq 1. \tag{3.3.31}$$

Therefore, by applying Lemma 3.3.1 (b), we arrive at the inequality:

$$\frac{|F_0^{(c)}|^2}{2} + \sum_{n=1}^{\infty} \{|F_n^{(c)}|^2 + |F_n^{(s)}|^2\} \leq a_0^2 \frac{|\alpha_0^{(c)}|^2}{2} + 2|\sigma|^2 \sum_{n=1}^{\infty} \frac{1}{n^2} \{|\alpha_n^{(c)}|^2 + |\alpha_n^{(s)}|^2\}$$

which shows the convergence of the series $(3.3.22)_2$ in the $W^{1,2}(0, 2\pi)$ space if $\alpha \in L^2(0, 2\pi)$. Furthermore, inequality (3.3.28) follows from Parseval's equality.

Conversely, we derive from Eq. (3.3.31) the inequality:

$$|\alpha_n^{(c)}|^2 + |\alpha_n^{(s)}|^2 \leq \frac{1}{b_n^2}\left(|F_n^{(c)}|^2 + |F_n^{(s)}|^2\right), \quad n \geq 1. \tag{3.3.32}$$

Using Lemma 3.3.1 (c), we find a positive integer N_0 such that

$$b_n \geq \frac{|\sigma|}{2n} \ \text{ for all } \ n \geq N_0.$$

Let $b_* = \min\{b_1, \ldots, b_{N_0}\}$. Based on Lemma 3.3.1 (a), we observe that $b_* > 0$. Utilizing equations (3.3.32), Lemma 3.3.1 (b), and the assumptions on $F(t)$, we deduce:

$$\frac{|\alpha_0^{(c)}|^2}{2} + \sum_{n=1}^{\infty} \{|\alpha_n^{(c)}|^2 + |\alpha_n^{(s)}|^2\} \leq \frac{1}{a_0} \frac{|F_0^{(c)}|^2}{2} + \frac{1}{b_*^2} \sum_{n=1}^{\infty} \{|F_n^{(c)}|^2 + |F_n^{(s)}|^2\}$$

$$+ \frac{4}{|\sigma|^2} \sum_{n=1}^{\infty} n^2 \{|F_n^{(c)}|^2 + |F_n^{(s)}|^2\},$$

which proves the convergence of the series $(3.3.22)_1$ in the $L^2(0, 2\pi)$ space. Finally, inequality (3.3.29) follows from Parseval's equality.

Assuming that $F \in W^{l+1,2}(0, 2\pi)$, we can similarly prove that $\alpha \in W^{l,2}(0, 2\pi)$, and the estimate (3.3.30) holds. □

An immediate consequence of the preceding results is the existence, uniqueness, and regularity of a periodic Poiseuille flow under a given periodic flux function $F(t)$.

Theorem 3.3.2 *Let $\partial\sigma \in C^2$, and let F be a 2π–periodic function with $F \in W^{1,2}(0, 2\pi)$. Then, the problem (3.3.7) has one and only one 2π–periodic solution $\big(u_{P_\alpha}(x', t), \alpha(t)\big)$ satisfying the following inequality:*

$$\sup_{t\in[-0,2\pi]} \|u_{P_\alpha}(\cdot, t)\|^2_{W^{1,2}(\sigma)} + \|u_{P_\alpha}\|^2_{W^{(2,1),2}(\sigma\times(0,2\pi))} \leq c\|F\|^2_{W^{1,2}(0,2\pi)}. \tag{3.3.33}$$

Furthermore, if $\partial\sigma \in C^{l+2}$, $l \geq 0$, and $F \in W^{l+1,2}(0, 2\pi)$, then u_{P_α} and α belong to the spaces $W^{(2l+2,l+1),2}(\sigma \times (0, 2\pi))$ and $W^{l,2}(0, 2\pi)$, respectively, and are subject to the estimate:

$$\|u_{P_\alpha}\|^2_{W^{(2l+2,l+1),2}(\sigma\times(0,2\pi))} + \|\alpha\|^2_{W^{l,2}(0,2\pi)} \leq c\|F\|^2_{W^{l+1,2}(0,2\pi)}. \tag{3.3.34}$$

Proof

Let $F_n^{(c)}$, $F_n^{(s)}$ be the Fourier coefficients of $F(t)$, and consider the series $(3.3.22)_1$ with coefficients $\alpha_n^{(c)}$, $\alpha_n^{(s)}$ given in (3.3.24). By applying Lemma 3.3.2, we establish that the series $(3.3.22)_1$ converges to some $\alpha \in L^2(0, 2\pi)$, and inequality (3.3.29) holds. Next, we solve problem (3.3.4) with the given $\alpha(t)$. The existence part, along with the validity of estimate (3.3.33), follows from Theorem 3.3.1, Lemma 3.3.2, and inequality (3.3.29).

Similarly, if $\partial\sigma \in C^{l+2}$, $l \geq 0$, and $F \in W^{l+1,2}(0, 2\pi)$, we employ Theorem 3.3.1 to conclude that $u_{P_\alpha} \in W^{(2l+2,l+1),2}(\sigma \times (0, 2\pi))$, $\alpha \in W^{l,2}(0, 2\pi)$, and estimate (3.3.34) follows from (3.3.6).

To demonstrate uniqueness, we assume that $F(t) = 0$ and multiply both sides of (3.3.7) by $u_{P_\alpha}(x', t)$. We then integrate by parts over σ and integrate by t over the interval $[0, 2\pi]$. Utilizing the fact that $F(t) = 0$ and the periodicity of u_{P_α} with respect to t, we obtain $\int\limits_0^{2\pi}\int\limits_\sigma |\nabla' u_{P_\alpha}(x', t|^2 dx' dt = 0$, which implies $u_{P_\alpha}(x', t) = 0$. Returning to Eqs. (3.3.7), we deduce $\alpha(t) = 0$. □

Remark 3.3.1 A solution of problem (3.3.7) has the following simple representation:

$$\begin{aligned} u_{P_\alpha}(x', t) = \frac{\alpha_0^{(c)}}{2}\varphi_0(x') + \sum_{n=1}^{\infty} \Big\{ &\big(\alpha_n^{(c)}\varphi_n(x') + \alpha_n^{(s)}\psi_n(x')\big)\cos(nt) \\ &+ \big(\alpha_n^{(s)}\varphi_n(x') - \alpha_n^{(c)}\psi_n(x')\big)\sin(nt)\Big\}, \end{aligned}$$

where, for $n = 0, 1, 2, \ldots$, the functions $\varphi_n(x')$, $\psi_n(x')$ satisfy (3.3.11), while the numbers $\alpha_n^{(c)}$, $\alpha_n^{(s)}$ are given in (3.3.24). This representation easily follows from (3.3.8)–(3.3.11) along with inequalities (3.3.13) and (3.3.16), (3.3.17).

Remark 3.3.2 In the special case when the flux F is constant, then u_{P_α} does not depend on time, and we find $u_{P_\alpha}(x') = \frac{F}{\kappa_0 \nu} u_0(x')$.

3.4 Nonstationary Poiseuille Solutions Under Minimal Regularity Assumptions on $F(t)$

3.4.1 Definition of a Weak Solution in the Case $F \in L^2(0, T)$. Formulation of the Main Result

In the preceding sections, we examined problem (3.2.7) with the assumption that the flow rate $F(t)$ and the initial velocity u_{0n} have sufficient smoothness, such as $F \in W^{1,2}(0, T)$ and $u_{0n} \in \overset{\circ}{W}{}^{1,2}(\sigma)$. However, in practical applications and numerical computations, we often lack such regularity in the data, and we can only assert that $F \in L^2(0, T)$. Hence, we will now investigate problem (3.2.7) under weaker regularity assumptions on the data.

Since (3.2.7) is a linear problem, the solution $\left(u_{P_\alpha}, \alpha\right)$ can be expressed as the sum $u_{P_\alpha} = u_h^{(1)} + u_{P_\alpha}^{(2)}$, where $u_h^{(1)}$ is the solution of the following initial-boundary value problem for the heat equation:

$$\begin{cases} \partial_t u_h^{(1)}(x', t) - \nu \Delta' u_h^{(1)}(x', t) = 0, & (x', t) \in \sigma^T, \\ u_h^{(1)}(x', t) = 0, & x' \in \partial\sigma,\ t \in (0, T), \\ u_h^{(1)}(x', 0) = u_{0n}(x'),\ x' \in \sigma. \end{cases} \tag{3.4.1}$$

Meanwhile, $\left(u_{P_\alpha}^{(2)}, \alpha\right)$ is the solution of the inverse problem with the homogeneous initial condition:

$$\begin{cases} \partial_t u_{P_\alpha}^{(2)}(x', t) - \nu \Delta' u_{P_\alpha}^{(2)}(x', t) = \alpha(t), & (x', t) \in \sigma^T, \\ u_{P_\alpha}^{(2)}(x', t) = 0, & x' \in \partial\sigma,\ t \in (0, T), \\ u_{P_\alpha}^{(2)}(x', 0) = 0, & x' \in \sigma, \\ \int\limits_\sigma u_{P_\alpha}^{(2)}(x', t)\, dx' = F(t) - \int\limits_\sigma u_h^{(1)}(x', t)\, dx' = \widehat{F}(t),\ t \in (0, T). \end{cases} \tag{3.4.2}$$

If $u_{0n} \in L^2(\sigma)$, then it is well-known (e.g., [103]) that problem (3.4.1) has a unique solution $u_h^{(1)}$ in the class

$$\mathcal{M} = \Big\{ U : \max_{t \in [0,T]} \|U(\cdot, t)\|^2_{L^2(\sigma)} + \int\limits_0^T \|\nabla' U(\cdot, t)\|^2_{L^2(\sigma)}\, dt < +\infty \Big\}.$$

Thus, we can assume without loss of generality that $u_{0n} = 0$ and study the following problem:

$$\begin{cases} \partial_t u_{P_\alpha}(x', t) - \nu \Delta' u_{P_\alpha}(x', t) = \alpha(t), \ (x', t) \in \sigma^T, \\ u_{P_\alpha}(x', t) = 0, \qquad x' \in \partial\sigma, t \in (0, T), \\ u_{P_\alpha}(x', 0) = 0, \qquad x' \in \sigma, \\ \int\limits_\sigma u_{P_\alpha}(x', t)\, dx' = F(t), \ t \in (0, T), \ F(0) = 0. \end{cases} \tag{3.4.3}$$

Let us revisit the definition of a weak solution in the case where the function F belongs to the Sobolev space $W^{1,2}(0, T)$. A weak solution is characterized as a pair of functions, namely, $(u_{P_\alpha}, \alpha) \in \overset{\circ}{W}{}^{(1,1),2}_{\partial\sigma}(\sigma^T) \times L^2(0, T)$, which satisfy the initial condition $u_{P_\alpha}(x', 0) = 0$, the flux condition

$$\int\limits_\sigma u_{P_\alpha}(x', t)dx' = F(t), \quad F(0) = 0,$$

and the integral identity

$$\begin{aligned} \int\limits_0^T \int\limits_\sigma \partial_t u_{P_\alpha}(x', \tau)\eta(x', \tau)dx'd\tau + \nu \int\limits_0^T \int\limits_\sigma \nabla' u_{P_\alpha}(x', \tau) \cdot \nabla'\eta(x', \tau)dx'd\tau \\ = \int\limits_0^T \alpha(\tau) \int\limits_\sigma \eta(x', \tau)dx'd\tau \quad \forall\, \eta \in \overset{\circ}{W}{}^{(1,0),2}_{\partial\sigma}(\sigma^T). \end{aligned} \tag{3.4.4}$$

To extend the concept of weak solutions to the case where the data, specifically the flow rate $F(t)$, belongs only to $L^2(0, T)$, we will need to introduce some additional notation and function spaces.

For any L^2-function h, denote by $S_h(t)$ its primitive:

$$S_h(t) = \int\limits_0^t h(\tau)d\tau, \quad S_h(0) = 0.$$

We denote the linear set of all functions S_h, where $h \in C_0^\infty(0, T)$, as $\mathcal{L}(0, T)$. Clearly, $C_0^\infty(0, T)$ is a subset of $\mathcal{L}(0, T)$. Consequently, the closure of $\mathcal{L}(0, T)$ with respect to the L^2-norm gives the entire space $L^2(0, T)$. It should be noted that the L^2-limit of a sequence of primitives $\{S_{h_n}\}$ is not necessarily a primitive of some function from $L^2(0, T)$.

Let $W_T^{1,2}(0, T)$ be the closure of the set $C_T^\infty(0, T) = \{\varphi \in C^\infty(0, T) : \varphi(T) = 0\}$ with respect to the $W^{1,2}$-norm, and let $W_T^{-1,2}(0, T)$ be its dual, i.e.,

$W_T^{-1,2}(0,T) = \left(W_T^{1,2}(0,T)\right)^*$. In the next lemma, we prove that an element $h \in W_T^{-1,2}(0,T)$ possesses a primitive in the distributional sense.

Lemma 3.4.1 *Any functional $h \in W_T^{-1,2}(0,T)$ can be represented in the form:*

$$\langle h, \eta \rangle = \int_0^T H(t)\eta'(t)dt \quad \forall \eta \in W_T^{1,2}(0,T) \tag{3.4.5}$$

where $H \in L^2(0,T)$ is uniquely determined.

Proof
It is evident that the functional given by formula (3.4.5) satisfies the estimate:

$$\left|\langle h, \eta \rangle\right| \leq \|H\|_{L^2(0,T)} \|\eta\|_{W_T^{1,2}(0,T)}.$$

Hence, $h \in W_T^{-1,2}(0,T)$.

Now, consider an arbitrary functional $h \in W_T^{-1,2}(0,T)$ and show that it can be represented in the form (3.4.5). This assertion follows as a corollary of the Riesz theorem for the Hilbert space $W_T^{1,2}(0,T)$ equipped with the inner product:

$$(\varphi, \eta)_{W_T^{1,2}(0,T)} = \int_0^T \varphi'(t)\eta'(t)dt.$$

The theorem states that there exists a unique function $\varphi_0 \in W_T^{1,2}(0,T)$ such that:

$$\langle h, \eta \rangle = (\varphi_0, \eta)_{W_T^{1,2}(0,T)} \quad \forall \eta \in W_T^{1,2}(0,T),$$

which implies $H(t) = \varphi_0'(t) \in L^2(0,T)$. □

Remark 3.4.1 Note that if the functional h can be represented by $L^2(0,T)$-function, i.e., $\langle h, \eta \rangle = \int_0^T h(t)\eta(t)dt$, then $H(t) = -S_h(t)$. Indeed,

$$\langle h, \eta \rangle = \int_0^T h(t)\eta(t)dt = \int_0^T S_h'(t)\eta(t)dt = -\int_0^T S_h(t)\eta'(t)dt.$$

Therefore, even in the case of a general functional h, we will use the notation $H(t) = -S_h(t)$.

Let us take in integral identity (3.4.4) a test function η such that $\eta \in \overset{\circ}{W}{}^{(1,1),2}_{\partial\sigma}(\sigma^T)$ and $\eta(x', T) = 0$. Integrating by parts with respect to t, we can rewrite identity (3.4.4) in the following form:

$$\int_0^T \int_\sigma u_{P_\alpha}(x',t)\partial_t \eta(x',t)dx'dt + \nu \int_0^T \int_\sigma \nabla' S_{u_{P_\alpha}}(x',t) \cdot \nabla \partial_t \eta(x',t)dx'dt$$
$$= \int_0^T S_\alpha(t) \int_\sigma \partial_t \eta(x',t)dx'dt. \tag{3.4.6}$$

This equality intuitively indicates how a weak solution can be in this case where $F(t)$ belongs only to $L^2(0,T)$.

Definition 3.4.1 Let $F \in L^2(0,T)$, $F(0) = 0$. A *weak solution* of problem (3.4.3) is defined as a pair (u_{P_α}, α) such that $u_{P_\alpha} \in L^2(\sigma^T)$, $S_{u_{P_\alpha}} \in \overset{\circ}{W}{}^{(1,0),2}_{\partial\sigma}(\sigma^T)$,[3] $\alpha \in W_T^{-1,2}(0,T)$, and $u_{P_\alpha}(x',t)$ satisfies the initial condition:

$$u_{P_\alpha}(x',0) = 0,$$

the flux condition:

$$\int_\sigma u_{P_\alpha}(x',t)dx' = F(t), \tag{3.4.7}$$

and the pair (u_{P_α}, α) satisfies the integral identity (3.4.6) for any test function $\eta, \eta_t \in L^2(0,T; \overset{\circ}{W}{}^{1,2}(\sigma))$, $\eta(\cdot, T) = 0$.

The boundary condition $u_{P_\alpha}\big|_{\partial\sigma} = 0$ is understood in the usual sense of traces (see Chap. 2 or [2, 61]). Indeed, if $S_{u_{P_\alpha}} \in \overset{\circ}{W}{}^{(1,0),2}_{\partial\sigma}(\sigma^T)$, then $S_{u_{P_\alpha}} = \int_0^t u_{P_\alpha}(\cdot,\tau)d\tau \in \overset{\circ}{W}{}^{1,2}(\sigma)$ for almost all $t \in (0,T)$, and $\int_0^t u_{P_\alpha}(x',t)d\tau\big|_{\partial\sigma} = 0$ in the sense of traces. Then $u_{P_\alpha}(x',t)\big|_{\partial\sigma} = 0$ in the sense of traces for almost all $t \in (0,T)$.

However, if $F \in L^2(0,T)$, the compatibility condition

$$F(0) = 0 \tag{3.4.8}$$

[3] Since $u_{P_\alpha} \in L^2(\sigma^T)$, $S_{u_{P_\alpha}}$ is a primitive of u_{P_α}, i.e., $\partial_t S_{u_{P_\alpha}} = u_{P_\alpha}$, and we have the following inclusion $\partial_t S_{u_{P_\alpha}} \in L^2(\sigma^T)$.

in (3.4.3), may not have a well-defined meaning. To address this concern, we interpret this condition by assuming that $t = 0$ is a Lebesgue point of $|F(t)|^2$, and the following condition holds:

$$\lim_{t\to 0^+} \frac{1}{t}\int_0^t |F(\tau)|^2 d\tau = 0.$$

Similarly, we adopt a similar understanding for the initial condition. Specifically, we state that $u_{P_\alpha}(x',0) = 0$ if $t = 0$ is a Lebesgue point of $\|u_{P_\alpha}(\cdot,t)\|^2_{L^2(\sigma)}$, and the condition below is satisfied:

$$\lim_{t\to 0^+} \frac{1}{t}\int_0^t \|u_{P_\alpha}(\cdot,\tau)\|^2_{L^2(\sigma)} d\tau = 0. \tag{3.4.9}$$

The main result of this section is the following theorem:

Theorem 3.4.1 *Let $F \in L^2(0,T)$, $F(0) = 0$. Then for arbitrary $T < \infty$, problem* (3.4.3) *admits a unique weak solution $\big(u_{P_\alpha}(x',t), \alpha(t)\big)$. The following estimates hold:*

$$\int_0^T\int_\sigma |u_{P_\alpha}(x',t)|^2 dx'dt + \sup_{t\in[0,T]}\int_\sigma \big|\nabla' S_{u_{P_\alpha}}(x',t)\big|^2 dx' + \int_0^T |S_\alpha(t)|^2\,dt$$
$$\le c\Big(\int_0^T |S_F(t)|^2 dt + \int_0^T |F(t)|^2 dt\Big), \tag{3.4.10}$$

$$\int_0^T\int_\sigma \big|\nabla' S_{u_{P_\alpha}}(x',t)\big|^2 dx'dt \le cT(1+T^2)\int_0^T |F(t)|^2 dt \tag{3.4.11}$$

where the constant c in (3.4.10) *and* (3.4.11) *is independent of T.*

Remark 3.4.2 By uniqueness, for $F \in W^{1,2}(0,T)$, the solution $\big(u_{P_\alpha}(x',t), \alpha(t)\big)$ of Theorem 3.4.1 coincides with the regular solution found in Sect. 3.2. Thus, the proposed definition of the solution is a proper extension of the concept of weak solutions.

Remark 3.4.3 Later, we will construct a counterexample to show that in the case when F belongs only to $L^2(0,T)$, the regularity of the weak solution cannot be better than that stated in Theorem 3.4.1.

3.4.2 Construction of Approximate Solutions

As discussed in Sect. 3.2, we consider the base in the function spaces $L^2(\sigma)$ and $\overset{\circ}{W}{}^{1,2}(\sigma)$, which is composed of eigenfunctions $u_k(x')$ of the Laplace operator (3.2.9). We seek for an approximate solution $\big(u^{(N)}_{P_\alpha}(x',t), \alpha^{(N)}(t)\big)$ to problem (3.4.3) in the following form:

$$u^{(N)}_{P_\alpha}(x',t) = \sum_{k=1}^{N} w^{(N)}_k(t) u_k(x'), \tag{3.4.12}$$

where the coefficients are given by:

$$w^{(N)}_k(t) = \beta_k \int_0^t \exp(-\lambda_k(t-\tau))\alpha^{(N)}(\tau)\,d\tau,$$

with $\beta_k = \int_\sigma u_k(x')dx'$ and λ_k denoting the eigenvalues of problem (3.2.9) (see Sect. 3.2).

Computing the flux of $u^{(N)}_{P_\alpha}(x',t)$, we obtain the equality:

$$F(t) = \int_\sigma u^{(N)}_{P_\alpha}(x',t)\,dx' = \sum_{k=1}^{N} \beta_k \int_\sigma u_k(x')\,dx' \int_0^t \exp(-\lambda_k(t-\tau))\alpha^{(N)}(\tau)\,d\tau.$$

Thus, $\alpha^{(N)}$ can be obtained as the solution of the first-kind Volterra integral equation:

$$\int_0^t \sum_{k=1}^{N} \beta_k^2 \exp(-\lambda_k(t-\tau))\alpha^{(N)}(\tau)\,d\tau = F(t). \tag{3.4.13}$$

However, it is known (e.g., [108, 193]) that such equations, in general, suffer from ill-posedness in L^2-spaces. To mitigate this issue, we introduce a regularization of this equation considering the following second-kind Volterra equation for $\alpha^{(N,\varepsilon)}$:

$$\varepsilon\alpha^{(N,\varepsilon)}(t) + \int_0^t \sum_{k=1}^{N} \beta_k^2 \exp(-\lambda_k(t-\tau))\alpha^{(N,\varepsilon)}(\tau)\,d\tau = F(t), \tag{3.4.14}$$

where ε will later tend to zero.

Hence, we seek approximations $\big(u^{(N,\varepsilon)}_{P_\alpha}(x',t), \alpha^{(N,\varepsilon)}(t)\big)$ satisfying the following system of equations:

$$\begin{cases} \int\limits_{\sigma} \partial_t u_{P_\alpha}^{(N,\varepsilon)}(x',t)u_k(x')dx' + \nu \int\limits_{\sigma} \nabla' u_{P_\alpha}^{(N,\varepsilon)}(x',t) \cdot \nabla' u_k(x')dx' \\ \qquad = \alpha^{(N,\varepsilon)}(\tau) \int\limits_{\sigma} u_k(x')dx', \quad k = 1, 2, \ldots, N, \\ \varepsilon\alpha^{(N,\varepsilon)} + \int\limits_0^t \sum\limits_{k=1}^N \beta_k^2 \exp(-\lambda_k(t-\tau))\alpha^{(N,\varepsilon)}(\tau)\, d\tau = F(t), \end{cases} \tag{3.4.15}$$

where

$$u_{P_\alpha}^{(N,\varepsilon)}(x',t) = \sum_{k=1}^N w_k^{(N,\varepsilon)}(t)u_k(x'),$$

$$w_k^{(N,\varepsilon)}(t) = \beta_k \int\limits_0^t \exp(-\lambda_k(t-\tau))\alpha^{(N,\varepsilon)}(\tau)\, d\tau.$$

Note that the second-kind Volterra equation is uniquely solvable in $L^2(0,T)$ (see Theorem 3.2.2).

In order to establish the convergence, as $\varepsilon \to 0$ and $N \to +\infty$, of the approximate solutions $\big(u_{P_\alpha}^{(N,\varepsilon)}(x',t), \alpha^{(N,\varepsilon)}(t)\big)$ to the solution $\big(u_{P_\alpha}(x',t),\ \alpha(t)\big)$ of problem (3.4.3), we need to demonstrate uniform estimates with respect to N and ε for $\big(u_{P_\alpha}^{(N,\varepsilon)}(x',t),\ \alpha^{(N,\varepsilon)}(t)\big)$.

Remark 3.4.4 If $F \in W^{1,2}(0,T)$, it is possible to differentiate integral equation (3.4.13) directly, resulting in a second-kind Volterra equation for its solution. In this case, regularization is unnecessary.

3.4.3 Estimates of Approximate Solutions

Lemma 3.4.2 *Let $F \in L^2(0,T)$, $T < +\infty$. If $\varepsilon > 0$, then the following estimate holds:*

$$\int\limits_0^T \int\limits_{\sigma} |u_{P_\alpha}^{(N,\varepsilon)}(x',t)|^2 dx'dt + \nu \sup_{t\in[0,T]} \int\limits_{\sigma} \left|\nabla' S_{u_{P_\alpha}^{(N,\varepsilon)}}(x',t)\right|^2 dx'$$
$$+ \int\limits_0^T |S_{\alpha^{(N,\varepsilon)}}(t)|^2\, dt \le c \int\limits_0^T \left(|S_F(\tau)|^2 + |F(t)|^2\right) d\tau. \tag{3.4.16}$$

The constant c in (3.4.16) *is independent of ε, N, and T.*

Proof

Choose $s \in (0, T]$ and define $\gamma_k^{(N,\varepsilon)}(t,s) = \int\limits_t^s w_k^{(N,\varepsilon)}(\tau)d\tau$ for $t \leq s$ and $\gamma_k^{(N,\varepsilon)}(t,s) = 0$ for $t \geq s$. Multiply equation $(3.4.15)_1$ by $\gamma_k^{(N,\varepsilon)}(t,s)$, integrate them with respect to t from 0 to s, and sum the obtained relations from $k = 1$ to $k = N$:

$$\begin{aligned}
&\int\limits_0^s \int\limits_\sigma \Big(\partial_t u_{P_\alpha}^{(N,\varepsilon)}(x',t) \int\limits_t^s u_{P_\alpha}^{(N,\varepsilon)}(x',\tau)d\tau\Big)dx'dt \\
&+\nu \int\limits_0^s \int\limits_\sigma \Big(\nabla' u_{P_\alpha}^{(N,\varepsilon)}(x',t) \int\limits_t^s \nabla' u_{P_\alpha}^{(N,\varepsilon)}(x',\tau)d\tau\Big)dx'dt \qquad (3.4.17)\\
&= \int\limits_0^s \alpha^{(N,\varepsilon)}(t) \int\limits_t^s \int\limits_\sigma u_{P_\alpha}^{(N,\varepsilon)}(x',\tau)d\tau dx'dt.
\end{aligned}$$

Recall that $S_{\alpha^{(N,\varepsilon)}}(t) = \int\limits_0^t \alpha^{(N,\varepsilon)}(\tau)d\tau$, so $\dfrac{dS_{\alpha^{(N,\varepsilon)}}(t)}{dt} = \alpha^{(N,\varepsilon)}(t)$. From $(3.4.15)_2$ it follows that

$$\begin{aligned}
&\int\limits_0^s \alpha^{(N,\varepsilon)}(t) \int\limits_t^s \int\limits_\sigma u_{P_\alpha}^{(N,\varepsilon)}(x',\tau)d\tau dx'dt \\
&= \int\limits_0^s \frac{dS_{\alpha^{(N,\varepsilon)}}(t)}{dt} \int\limits_t^s \big(F(\tau) - \varepsilon\alpha^{(N,\varepsilon)}(\tau)\big)d\tau dt \\
&= \int\limits_0^s S_{\alpha^{(N,\varepsilon)}}(t)F(t)dt - \varepsilon \int\limits_0^s S_{\alpha^{(N,\varepsilon)}}(t)\frac{dS_{\alpha^{(N,\varepsilon)}}(t)}{dt}dt \qquad (3.4.18)\\
&= \int\limits_0^s S_{\alpha^{(N,\varepsilon)}}(t)F(t)dt - \frac{\varepsilon}{2}|S_{\alpha^{(N,\varepsilon)}}(s)|^2 \quad \forall s \in (0,T].
\end{aligned}$$

Integrating by parts with respect to t, we reduce equality (3.4.17) to the form

$$\begin{aligned}
&\int\limits_0^s \int\limits_\sigma |u_{P_\alpha}^{(N,\varepsilon)}(x',t)|^2 dx'dt + \frac{\nu}{2}\int\limits_\sigma \Big|\nabla' \int\limits_0^s u_{P_\alpha}^{(N,\varepsilon)}(x',t)dt\Big|^2 dx \\
&+ \frac{\varepsilon}{2}|S_{\alpha^{(N,\varepsilon)}}(s)|^2 = \int\limits_0^s S_{\alpha^{(N,\varepsilon)}}(t)F(t)dt. \qquad (3.4.19)
\end{aligned}$$

Applying Young's and Cauchy-Schwarz-Bunyakovsky inequalities, we get

$$\begin{aligned}&\int_0^s\int_\sigma |u_{P_\alpha}^{(N,\varepsilon)}(x',t)|^2 dt + \frac{\nu}{2}\int_\sigma \Big|\nabla'\int_0^s u_{P_\alpha}^{(N,\varepsilon)}(x',t)dt\Big|^2 dx\\ &\le \delta\int_0^s |S_{\alpha^{(N,\varepsilon)}}(t)|^2\,dt + \frac{1}{2\delta}\int_0^s |F(\tau)|^2\,d\tau \quad \forall s\in(0,T].\end{aligned}\tag{3.4.20}$$

Let $u_0 \in W^{2,2}(\sigma)\cap \mathring{W}^{1,2}(\sigma)$ be the solution of the Poisson problem (3.1.2). Recall that $\kappa_0 = \int_\sigma u_0(x')\,dx' = \nu\int_\sigma |\nabla' u_0(x')|^2\,dx' > 0$. The solution u_0 can be represented as a Fourier series in the space $\mathring{W}^{1,2}(\sigma)$:

$$u_0(x') = \sum_{k=1}^{\infty} a_k u_k(x'), \quad a_k \in \mathbb{R}.$$

Multiplying equations $(3.4.15)_1$ by a_k and summing over k give

$$\begin{aligned}&\int_\sigma \partial_t u_{P_\alpha}^{(N,\varepsilon)}(x',t)u_0(x')\,dx' + \nu\int_\sigma \nabla' u_{P_\alpha}^{(N,\varepsilon)}(x',t)\cdot\nabla' u_0(x')\,dx'\\ &= \alpha^{(N,\varepsilon)}(t)\int_\sigma u_0(x')\,dx' = \kappa_0\alpha^{(N,\varepsilon)}(t).\end{aligned}\tag{3.4.21}$$

On the other hand, multiplying equation (3.1.2) by $u_{P_\alpha}^{(N,\varepsilon)}(x',t)$ and integrating by parts over σ yield

$$\nu\int_\sigma \nabla' u_0(x')\cdot\nabla' u_{P_\alpha}^{(N,\varepsilon)}(x',t)\,dx' = \int_\sigma u_{P_\alpha}^{(N,\varepsilon)}(x',t)dx' = F(t) - \varepsilon\alpha^{(N,\varepsilon)}(t).$$

Therefore, from (3.4.21), it follows that

$$(\kappa_0+\varepsilon)\alpha^{(N,\varepsilon)}(t) = \frac{d}{dt}\int_\sigma u_{P_\alpha}^{(N,\varepsilon)}(x',t)u_0(x')\,dx' + F(t).\tag{3.4.22}$$

Integrating (3.4.22) by t over the interval $(0,\tau)$ gives

$$(\kappa_0+\varepsilon)S_{\alpha^{(N,\varepsilon)}}(\tau) = \int_\sigma u_{P_\alpha}^{(N,\varepsilon)}(x',\tau)u_0(x')\,dx' + \int_0^\tau F(t)dt.\tag{3.4.23}$$

Hence, applying the Cauchy-Schwarz-Bunyakovsky and Poincaré inequalities, we obtain

$$\begin{aligned}
&\kappa_0^2 \int_0^t |S_{\alpha^{(N,\varepsilon)}}(\tau)|^2 \, d\tau \le (\kappa_0+\varepsilon)^2 \int_0^t |S_{\alpha^{(N,\varepsilon)}}(\tau)|^2 \, d\tau \\
&\le 2\int_0^t \Big(\int_\sigma u_{P_\alpha}^{(N,\varepsilon)}(x',\tau) u_0(x')\, dx' \Big)^2 d\tau + 2\int_0^t \Big(\int_0^\tau F(s) ds \Big)^2 d\tau \\
&\le 2\int_0^t \Big(\int_\sigma |u_{P_\alpha}^{(N,\varepsilon)}(x',\tau)|^2 \, dx' \Big) \Big(\int_\sigma |u_0(x')|^2 \, dx' \Big) d\tau + 2\int_0^t |S_F(\tau)|^2 \, d\tau \\
&\le c\kappa_0 \int_0^t \int_\sigma |u_{P_\alpha}^{(N,\varepsilon)}(x',\tau)|^2 \, dx' d\tau + 2\int_0^t |S_F(\tau)|^2 \, d\tau.
\end{aligned} \tag{3.4.24}$$

Substituting (3.4.24) into (3.4.20) yields

$$\begin{aligned}
&\int_0^t \int_\sigma |u_{P_\alpha}^{(N,\varepsilon)}(x',\tau)|^2 d\tau + \frac{\nu}{2} \int_\sigma \Big|\nabla' \int_0^t u_{P_\alpha}^{(N,\varepsilon)}(x',\tau) d\tau\Big|^2 dx \\
&\le c_1\delta \int_0^t \int_\sigma |u_{P_\alpha}^{(N,\varepsilon)}(x',\tau)|^2 \, dx' d\tau + c_2(\delta) \int_0^t \big(|S_F(\tau)|^2 + |F(\tau)|^2 \big) \, d\tau.
\end{aligned} \tag{3.4.25}$$

Fixing $\delta = \dfrac{1}{2c_1}$ in (3.4.25), we obtain the inequality

$$\begin{aligned}
&\int_0^t \int_\sigma |u_{P_\alpha}^{(N,\varepsilon)}(x',\tau)|^2 d\tau + \nu \sup_{t\in[0,t]} \int_\sigma \Big|\nabla' S_{u_{P_\alpha}^{(N,\varepsilon)}}(x',\tau) d\tau\Big|^2 dx \\
&\le c_3 \int_0^t \big(|S_F(\tau)|^2 + |F(\tau)|^2 \big) \, d\tau \quad \forall t \in (0,T].
\end{aligned} \tag{3.4.26}$$

Finally, from (3.4.24) it follows

$$\int_0^t |S_{\alpha^{(N,\varepsilon)}}(\tau)|^2 \, d\tau \le c_4 \int_0^t \big(|S_F(\tau)|^2 + |F(t)|^2 \big) \, d\tau. \tag{3.4.27}$$

Therefore, we can conclude that the target inequality (3.4.16) is true. The constants c_1, c_2, c_3, c_4 are all independent of N, ε, and T. □

Remark 3.4.5 It is evident that the following inequality holds:

$$\int_0^T |S_F(\tau)|^2\,d\tau \le c_5 T^2 \int_0^t |F(\tau)|^2\,d\tau, \tag{3.4.28}$$

where the constant c_5 is independent of T. Combining (3.4.16) and (3.4.28), we can deduce the subsequent inequality:

$$\begin{aligned}&\int_0^T\int_\sigma |u_{P_\alpha}^{(N,\varepsilon)}(x',t)|^2 dt + \nu \sup_{t\in[0,T]}\int_\sigma \Big|\nabla' S_{u_{P_\alpha}^{(N,\varepsilon)}}(x',t)dt\Big|^2 dx'\\ &\quad + \int_0^T |S_{\alpha^{(N,\varepsilon)}}(t)|^2\,dt \le c(1+T^2)\int_0^T |F(t)|^2\,dt.\end{aligned} \tag{3.4.29}$$

Furthermore, the inequality given by

$$\int_0^T\int_\sigma \Big|\nabla' S_{u_{P_\alpha}^{(N,\varepsilon)}}(x',t)\Big|^2 dx'dt \le cT(1+T^2)\int_0^T |F(t)|^2\,dt \tag{3.4.30}$$

also holds.

3.4.4 Existence and Uniqueness of a Weak Solution

Proof of Existence We begin by demonstrating the existence of the weak solution for the problem at hand. The approximate solutions $\big(u_{P_\alpha}^{(N,\varepsilon)}(x',t), \alpha^{(N,\varepsilon)}(t)\big)$ constructed earlier satisfy equation $(3.4.15)_1$. Multiplying these equations by arbitrary functions $d_k \in C^2(0,T)$ such that $d_k(T) = 0$, summing over k from $k = 1$ to $k = M$, where $M \le N$, integrating with respect to t, and then integrating by parts, we obtain the following integral identity:

$$\begin{aligned}&\int_0^T\int_\sigma u_{P_\alpha}^{(N,\varepsilon)}(x',t)\partial_t\eta(x',t)dx'dt + \nu\int_0^T\int_\sigma \nabla' S_{u_{P_\alpha}^{(N,\varepsilon)}}(x',t)\cdot\nabla'\partial_t\eta(x',t)dx'dt\\ &\quad = \int_0^T S_{\alpha^{(N,\varepsilon)}}(t)\int_\sigma \partial_t\eta(x',t)dx'dt,\end{aligned} \tag{3.4.31}$$

where $\eta(x', t)$ is a test function of the form $\eta(x', t) = \sum_{k=1}^{M} d_k(t)u_k(x')$.

Recall that $\big(u^{(N,\varepsilon)}_{P_\alpha}(x', t), \alpha^{(N,\varepsilon)}(t)\big)$ satisfies estimates (3.4.16) and (3.4.29) with constants independent of ε and N. We fix N and choose a subsequence $\{(u^{(N,\varepsilon_l)}_{P_\alpha}, \alpha^{(N,\varepsilon_l)})\}$ such that $\{u^{(N,\varepsilon_l)}_{P_\alpha}\}$ and $\{\nabla' S_{u^{(N,\varepsilon_l)}_{P_\alpha}}\}$ converge weakly in $L^2(\sigma^T)$ as $\varepsilon_l \to 0$, to some $u^{(N)}_{P_\alpha}$,[4] while $\{\alpha^{(N,\varepsilon_l)}\}$ converges weakly in $W_T^{-1,2}(0, T)$ to $\alpha^{(N)}$ (i.e., $\{S_{\alpha^{(N,\varepsilon_l)}}\}$ converges weakly in $L^2(0, T)$). Clearly, for the limit functions $u^{(N)}_{P_\alpha}$ and $\alpha^{(N)}$, the estimates (3.4.16) and (3.4.29) continue to hold with constants independent of N. Taking $\varepsilon = \varepsilon_l$ in (3.4.31) and passing to the limit as $\varepsilon_l \to 0$, we obtain:

$$\begin{aligned}\int_0^T\int_\sigma u^{(N)}_{P_\alpha}(x', t)\partial_t\eta(x', t)dx'dt + \nu\int_0^T\int_\sigma \nabla S_{u^{(N)}_{P_\alpha}}(x', t)\cdot\nabla'\partial_t\eta(x', t)dx'dt \\ = \int_0^T S_{\alpha^{(N)}}(\tau)\int_\sigma \partial_t\eta(x', t)dx'dt.\end{aligned} \tag{3.4.32}$$

Next, we show that $u^{(N)}_{P_\alpha}(x', t)$ satisfies the flux condition:

$$\int_\sigma u^{(N)}_{P_\alpha}(x', t)dx' = F(t).$$

Equation $(3.4.15)_2$ for $\varepsilon = \varepsilon_l$ can be rewritten in the following equivalent form:

$$\varepsilon_l\alpha^{(N,\varepsilon_l)}(t) + \int_\sigma u^{(N,\varepsilon_l)}_{P_\alpha}(x', t)dx' = F(t).$$

Integrating this equation from 0 to t yields:

$$\varepsilon_l S_{\alpha^{(N,\varepsilon_l)}}(t) + \int_0^t\int_\sigma u^{(N,\varepsilon_l)}_{P_\alpha}(x', \tau)dx'd\tau = S_F(t). \tag{3.4.33}$$

[4] Through the weak convergence in the space $L^2(\sigma^T)$ of the sequence $\{\nabla' S_{u^{(N,\varepsilon_l)}_{P_\alpha}}\}$, it can be deduced, by means of the embedding theorem, that the sequence $\{S_{u^{(N,\varepsilon_l)}_{P_\alpha}}\}$ converges strongly. Additionally, as the sequence $\{u^{(N,\varepsilon_l)}_{P_\alpha}\}$ also converges weakly in the space $L^2(\sigma^T)$, we can establish that $\{\nabla' S_{u^{(N,\varepsilon_l)}_{P_\alpha}}\}$ converges (weakly) to $\nabla' S_{u^{(N)}_{P_\alpha}}$.

The sequence $\left\{\int\limits_{\sigma} u_{P_\alpha}^{(N,\varepsilon_l)}(x',\tau)dx'\right\}$ is bounded in $L^2(0,T)$, and we may assume, without loss of generality, that $\left\{\varphi_l^{(N)}(\tau)=\int\limits_{\sigma} u_{P_\alpha}^{(N,\varepsilon_l)}(x',\tau)dx'\right\}$ converges weakly to $\varphi^{(N)}$ in $L^2(0,T)$. Then, the sequence of primitives $S_{\varphi_l^{(N)}}(t)=\int\limits_0^t \varphi_l^{(N)}(\tau)d\tau \to S_{\varphi^{(N)}}(t)$ for all $t \in [0,T]$, and hence $\|S_{\varphi_l^{(N)}} - S_{\varphi^{(N)}}\|_{L^2(0,T)} \to 0$ as $l\to\infty$ ($\varepsilon_l \to 0$).

From (3.4.33), we have:

$$\|S_{\varphi_l^{(N)}} - S_F\|_{L^2(0,T)} = \varepsilon_l \|S_{\alpha^{(N,\varepsilon_l)}}\|_{L^2(0,T)} \le c\varepsilon_l \to 0 \quad \text{as } l\to\infty.$$

Therefore,

$$\int\limits_0^t \int\limits_{\sigma} u_{P_\alpha}^{(N)}(x',\tau)dx'd\tau = \int\limits_0^t F(\tau)d\tau \quad \text{for almost all } t\in[0,T],$$

and differentiating this equality with respect to t, we get

$$\int\limits_{\sigma} u_{P_\alpha}^{(N)}(x',t)dx'd\tau = F(t).$$

Since the pair $\left(u_{P_\alpha}^{(N)}(x',t), \alpha^{(N)}(t)\right)$ satisfies estimates (3.4.16) and (3.4.29) with constants independent of N, there exists a subsequence $\{(u_{P_\alpha}^{(N_l)}, \alpha^{(N_l)})\}$ such that $\{u_{P_\alpha}^{(N_l)}\}$ and $\{\nabla' S_{u_{P_\alpha}^{(N_l)}}\}$ converge weakly in $L^2(\sigma^T)$ to u_{P_α} and $\nabla' S_{u_{P_\alpha}}$, respectively, while $\{S_{\alpha^{(N_l)}}\}$ converges weakly in $L^2(0,T)$ to S_α. Passing in identity (3.4.32) to the limit as $N_l \to +\infty$, we obtain:

$$\begin{aligned}\int\limits_0^T \int\limits_{\sigma} u_{P_\alpha}(x',t)\partial_t\eta(x',t)dx'dt + \nu \int\limits_0^T \int\limits_{\sigma} \nabla S_{u_{P_\alpha}}(x',t)\cdot\nabla\partial_t\eta(x',t)dx'dt \\ = \int\limits_0^T S_\alpha(t)\int\limits_{\sigma} \partial_t\eta(x',t)dx'dt.\end{aligned} \tag{3.4.34}$$

Notice that, so far, the validity of identity (3.4.34) has been established only for test functions η that can be represented as the sums $\eta(x',t)=\sum\limits_{k=1}^M d_k(t)u_k(x')$, where

$d_k \in C^2(0,T)$ and $d_k(T) = 0$. It is worth noting that, after passing to the limit as $N_l \to +\infty$, the subscript M in the sum can be an arbitrarily large natural number. However, as demonstrated in Sect. 3.2, it is possible to show that such sums are dense in the space of functions $\eta \in \{\xi, \xi_t \in L^2(0,T; \overset{\circ}{W}{}^{1,2}(\sigma)), \xi(T) = 0\}$ (see Theorem 2.3.8). Consequently, identity (3.4.34) holds true for all such η. Furthermore, estimates (3.4.10) and (3.4.11) for $\big(u_{P_\alpha}(x',t), \alpha(t)\big)$ can be derived from the estimates (3.4.16) and (3.4.29) obtained for approximate solutions. □

Let us now show that the limit function $u_{P_\alpha}(x',t)$ satisfies the initial condition. From estimate (3.4.26), after passing to the limit as $\varepsilon_l \to 0$ and then as $N_k \to \infty$, we derive the inequality:

$$\int_0^t \int_\sigma |u_{P_\alpha}(x',s)|^2 dx' ds \le c\Big(\int_0^t |S_F(s)|^2 ds + \int_0^t |F(s)|^2 ds\Big)$$
$$\le c(t^2+1)\int_0^t |F(s)|^2 ds \le c\int_0^t |F(s)|^2 ds$$

with a constant c independent of $t \in [0,1]$. This implies:

$$\frac{1}{t}\int_0^t \int_\sigma |u_{P_\alpha}(x',\tau)|^2 dx' d\tau \le c\frac{1}{t}\int_0^t |F(\tau)|^2 d\tau. \tag{3.4.35}$$

By the assumptions on the flux $F(t)$, we have $\lim\limits_{t\to 0+} \frac{1}{t}\int_0^t |F(\tau)|^2 d\tau = F^2(0) = 0$. Therefore, (3.4.35) yields:

$$\lim_{t\to 0+} \frac{1}{t}\int_0^t \int_\sigma |u_{P_\alpha}(x',\tau)|^2 dx' d\tau \le c \lim_{t\to 0+} \frac{1}{t}\int_0^t |F(\tau)|^2 d\tau = 0.$$

By the Lebesgue differentiation theorem, $t = 0$ is the Lebesgue point for the function $H(t) = \int_\sigma |u_{P_\alpha}(x',t)|^2 dx$ and $H(0) = 0$, i.e., $\|u_{P_\alpha}(\cdot,0)\|_{L^2(\sigma)} = 0$.[5]

Thus, we have proved that $\big(u_{P_\alpha}(x',t),\ \alpha(t)\big)$ is a weak solution of problem (3.4.3).

Proof of Uniqueness Assuming that $F(t) = 0$, we consider $\eta(x',t) = \int_t^T \int_0^s u_{P_\alpha}(x',\tau) d\tau ds = \int_t^T S_{u_{P_\alpha}}(s) ds$. It is evident that such η is a suitable test

---

[5] These considerations do not establish the continuity of $\|u_{P_\alpha}(\cdot,t)\|_{L^2(\sigma)}$ at $t = 0$. To ensure this property, we require F to be an element of $W^{\varepsilon,2}(0,T)$ with arbitrary $\varepsilon > 0$ (see Remark 3.4.6).

function. By substituting it into identity (3.4.34), we obtain:

$$\int_0^T \int_\sigma u_{P_\alpha}(x',t) S_{u_{P_\alpha}}(x',t)\, dx'dt + \nu \int_0^T \int_\sigma |\nabla S_{u_{P_\alpha}}(x',t)|^2\, dx'dt \\ = \int_0^T S_\alpha(t) \int_\sigma S_{u_{P_\alpha}}(x',t)\, dx'dt. \qquad (3.4.36)$$

Since $\int_\sigma u_{P_\alpha}(x',t)\, dx' = 0$, according to Fubini's theorem, we have:

$$\int_\sigma S_{u_{P_\alpha}}(x',t)\, dx' = \int_\sigma \Big(\int_0^t u_{P_\alpha}(x',\tau)\, d\tau\Big) dx' = \int_0^t \Big(\int_\sigma u_{P_\alpha}(x',\tau)\, dx'\Big) d\tau = 0.$$

Thus, from (3.4.36), we obtain:

$$\frac{1}{2}\int_0^T \frac{d}{dt}\int_\sigma |S_{u_{P_\alpha}}(x',t)|^2 dx'dt + \nu \int_0^T \int_\sigma |\nabla S_{u_{P_\alpha}}(x',t)|^2\, dx'dt = 0$$

and

$$\frac{1}{2}\int_\sigma |S_{u_{P_\alpha}}(x',T)|^2\, dx' + \nu \int_0^T \int_\sigma |\nabla S_{u_{P_\alpha}}(x',\tau)|^2\, dx'd\tau = 0.$$

As a result, $\nabla S_{u_{P_\alpha}}(x',t) = 0$, and since $S_{u_{P_\alpha}}(x',t)|_{\partial\sigma} = 0$, we obtain $S_{u_{P_\alpha}}(x',t) = 0$ for almost all x' and t, i.e., $\int_0^t u_{P_\alpha}(x',\tau)d\tau = 0$ for almost all x' and t. This implies $u_{P_\alpha}(x',t) = 0$.

From identity (3.4.36), it now follows that:

$$\int_0^T S_\alpha(t) \int_\sigma \partial_t \eta(x',t) dx'dt = 0 \quad \forall \eta \in \overset{\circ}{W}{}^{(1,1),2}_{\partial\sigma}(\sigma^T), \quad \eta(x',T) = 0. \qquad (3.4.37)$$

Take $\eta(x,t) = \Big(\int_t^T b(\tau)d\tau\Big) u_0(x')$ in (3.4.37), where $b \in C^\infty([0,T])$ and $u_0(x')$ is the solution of problem (3.1.2). Recall that $\int_\sigma u_0(x')dx' = \kappa_0 \neq 0$. Then (3.4.37) takes the form:

$$\int_0^T S_\alpha(t) \int_\sigma \eta_t(x', t)dx'dt = -\kappa_0 \int_0^T S_\alpha(t)b(t)dt = 0 \quad \forall b \in C^\infty([0, T]).$$

As a result, $S_\alpha(t) = 0$ for almost all t, and, therefore, the functional $\alpha \equiv 0$. □

Remark 3.4.6 Let us make an additional assumption that the integral $\int_0^t \tau^{-\varepsilon}|F(\tau)|^2 d\tau$ is finite for any arbitrarily small positive ε and $t = 0$ is the Lebesgue point for the function $t^{-\varepsilon}|F(t)|^2$, which is equal to zero at $t = 0$. Under these conditions, inequality (3.4.35) implies the following result:

$$\frac{1}{t}\int_0^t \int_\sigma |u_{P_\alpha}(x', \tau)|^2 dx'd\tau \le c\frac{1}{t}\int_0^t |F(\tau)|^2 d\tau \le c\frac{t^{2\varepsilon}}{t}\int_0^t \tau^{-2\varepsilon}|F(\tau)|^2 d\tau.$$

Consequently, we obtain:

$$\lim_{t\to 0}\frac{1}{t^{1+2\varepsilon}}\int_0^t \int_\sigma |u_{P_\alpha}(x', \tau)|^2 dx'd\tau \le c \lim_{t\to 0}\frac{1}{t}\int_0^t \tau^{-2\varepsilon}|F(\tau)|^2 d\tau = 0.$$

Using l'Hôpital's rule, we find:

$$\begin{aligned}
0 &= \lim_{t\to 0}\frac{\frac{1}{t}\int_0^t \int_\sigma |u_{P_\alpha}(x', \tau)|^2 dx'd\tau}{t^{2\varepsilon}}\\
&= \lim_{t\to 0}\frac{-\frac{1}{t^2}\int_0^t \int_\sigma |u_{P_\alpha}(x', \tau)|^2 dx'd\tau + \frac{1}{t}\int_\sigma |u_{P_\alpha}(x', t)|^2 dx'}{2\varepsilon t^{2\varepsilon-1}}\\
&= -(2\varepsilon)^{-1}\lim_{t\to 0}\frac{1}{t^{1+2\varepsilon}}\int_0^t \int_\sigma |u_{P_\alpha}(x', \tau)|^2 dx'd\tau + (2\varepsilon)^{-1}\lim_{t\to 0}\frac{\int_\sigma |u_{P_\alpha}(x', t)|^2 dx'}{t^{2\varepsilon}}\\
&= 0 + (2\varepsilon)^{-1}\lim_{t\to 0}\frac{\int_\sigma |u_{P_\alpha}(x', t)|^2 dx'}{t^{2\varepsilon}}.
\end{aligned}$$

Thus, we conclude that $\lim\limits_{t\to 0}\dfrac{\int_\sigma |u_{P_\alpha}(x', t)|^2 dx'}{t^{2\varepsilon}} = 0$, which further implies:

$$\lim_{t\to 0}\|u_{P_\alpha}(\cdot, t)\|_{L^2(\sigma)} \le c \lim_{t\to 0} t^{2\varepsilon} = 0.$$

Note that if $\varphi \in W^{\varepsilon,2}(0,T)$ with $\varphi(0) = 0$, the inequality

$$\|t^{-\varepsilon}\varphi\|_{L^2(0,T)} \le c\|\varphi\|_{W^{\varepsilon,2}(0,T)}$$

holds (see [113], Section I.11.2). Thus, the continuity of $\|u_{P_\alpha}(\cdot,t)\|_{L^2(\sigma)}$ is ensured by the "minimal" regularity of the flux F: $F \in W^{\varepsilon,2}(0,T)$ for any small $\varepsilon > 0$. □

3.4.5 The Time-Periodic Case

In this section, we consider the time-periodic inverse problem (3.3.7), assuming that the periodic flux $F(t)$ belongs only to $L^2(0,2\pi)$.[6] Initially, we reduce problem (3.3.7) to the case where all involved functions have zero mean values. Let us denote the mean value of a function H as $\overline{H} = \frac{1}{2\pi}\int\limits_0^{2\pi} H(t)dt$. The pair $(\overline{u}_{P_\alpha}, \overline{\alpha})$ is a solution of the following problem on σ (the stationary Poiseuille solution corresponding to the flux $\overline{F}$):

$$\begin{cases} -\nu\Delta'\overline{u_{P_\alpha}}(x') = \overline{\alpha}, \ x' \in \sigma, \\ \overline{u_{P_\alpha}}(x') = 0, \ x' \in \partial\sigma, \\ \int\limits_\sigma \overline{u_{P_\alpha}}(x')dx' = \overline{F}. \end{cases} \tag{3.4.38}$$

The function $\overline{u_{P_\alpha}}(x')$ can be represented as $\overline{u_{P_\alpha}}(x') = \dfrac{\overline{F}}{\kappa_0}u_0(x')$, where $u_0(x')$ is the solution of problem (3.1.2), and

$$\overline{\alpha} = \frac{\overline{F}}{\kappa_0}, \quad \kappa_0 = \int\limits_\sigma u_0(x')dx' = \nu\int\limits_\sigma |\nabla' u_0(x')|^2 dx' > 0. \tag{3.4.39}$$

We then represent the solution (u_{P_α}, α) in the form:

$$u_{P_\alpha}(x',t) = v_{P_\alpha}(x',t) + \overline{u_{P_\alpha}}(x'), \quad \alpha(t) = s(t) + \overline{\alpha}. \tag{3.4.40}$$

It is clear that $\overline{v_{P_\alpha}} = 0$, $\overline{s} = 0$, and (v_{P_α}, s) satisfies the problem:

[6] For convenience, we assume the period is equal to 2π.

$$\begin{cases} \partial_t v_{P_\alpha}(x',t) - \nu\Delta' v_{P_\alpha}(x',t) = s(t), & (x',t) \in \sigma^{2\pi}, \\ v_{P_\alpha}(x',t) = 0, & x' \in \partial\sigma, , t \in (0,2\pi), \\ v_{P_\alpha}(x',0) = v_{P_\alpha}(x',2\pi), \ x' \in \sigma, & \\ \int\limits_\sigma v_{P_\alpha}(x',t)dx' = \widetilde{F}(t), & t \in (0,2\pi), \end{cases} \tag{3.4.41}$$

where $\widetilde{F}(t) = F(t) - \overline{F}$, $\overline{\widetilde{F}} = 0$. Thus, without loss of generality, we assume that $F(t) = \widetilde{F}(t)$, i.e., $\overline{F} = 0$.

Let us consider the space of smooth periodic functions $C^\infty_\wp(0,2\pi) = \{h \in C^\infty(\mathbb{R}) : h(t) = h(t+2\pi) \ \forall t \in \mathbb{R}\}$, and let $L^2(0,2\pi)$ be the Lebesgue space on the interval $(0,2\pi)$. We extend functions from $L^2(0,2\pi)$ to the whole set $\mathbb{R}$ by putting $h(t+2\pi) = h(t)$ for any t. To emphasize that functions are periodically extended to $\mathbb{R}$, we use the notation $L^2_\wp(0,2\pi)$.

Let $L^2_\sharp(0,2\pi) = \{h \in L^2_\wp(0,2\pi) : \int\limits_0^{2\pi} h(t)dt = 0\}$. Clearly, $L^2_\sharp(0,2\pi)$ is the closure of $C^\infty_\sharp(0,2\pi) = \{h \in C^\infty_\wp(0,2\pi) : \int\limits_0^{2\pi} h(t)dt = 0\}$ in $L^2(0,2\pi)$-norm, and it is a proper subspace of $L^2_\wp(0,2\pi)$.

Let $W^{1,2}_\wp(0,2\pi)$ be the closure of the set $C^\infty_\wp(0,2\pi)$ in the $W^{1,2}$-norm. Since a function h from $W^{1,2}_\wp(0,2\pi)$ coincides with a continuous function on a set whose complement has measure zero, we may assume that $h(0) = h(2\pi)$. We also denote $W^{1,2}_{\wp,0}(0,2\pi)$ as the subspace of functions h from $W^{1,2}_\wp(0,2\pi)$ satisfying the condition $h(2\pi) = 0$. Let $W^{-1,2}_\wp(0,2\pi)$ be the dual of $W^{1,2}_{\wp,0}(0,2\pi)$, i.e., $W^{-1,2}_\wp(0,2\pi) = \big(W^{1,2}_{\wp,0}(0,2\pi)\big)^*$.

For any function $h \in L^2_\wp(0,2\pi)$, we denote by $S_h(t)$ its primitive as follows:

$$S_h(t) = -\int\limits_t^{t_0+2\pi} h(\tau)d\tau, \quad \text{where } t_0 \in [0,2\pi), \ t \in [t_0, 2\pi + t_0]. \tag{3.4.42}$$

It is clear that $S_h(t_0+2\pi) = 0$. If $h \in L^2_\sharp(0,2\pi)$, then $S_h(t_0) = -\int\limits_{t_0}^{2\pi+t_0} h(\tau)d\tau = -\int\limits_0^{2\pi} h(t)dt = 0$. Moreover,

$$\int\limits_0^{2\pi} |S_h(t)|^2 dt \le 2\pi \int\limits_0^{2\pi}\int\limits_t^{t_0+2\pi} |h(\tau)|^2 d\tau dt \le 4\pi^2 \int\limits_{t_0}^{t_0+2\pi} |h(\tau)|^2 d\tau = 4\pi^2 \int\limits_0^{2\pi} |h(\tau)|^2 d\tau,$$

and $S_h(t)$ is a periodic function:

$$S_h(t+2\pi) = -\int\limits_{t+2\pi}^{t_0+2\pi} h(\tau)d\tau = -\int\limits_{t}^{t_0+2\pi} h(\tau)d\tau + \int\limits_{t}^{t+2\pi} h(\tau)d\tau$$
$$= S_h(t) - 2\pi\bar{h} = S_h(t).$$

Thus, $S_h \in L^2_{\wp}(0, 2\pi)$. Note that functions $S_h(t)$ defined by (3.4.42) with various t_0 differ from each other by a constant.

In the next lemma, we prove that any element $h \in W^{-1,2}_{\wp}(0, 2\pi)$ possesses a primitive in the distributional sense.

Lemma 3.4.3 *Any functional* $h \in W^{-1,2}_{\wp}(0, 2\pi)$ *can be represented in the form:*

$$\langle h, \eta\rangle = \int\limits_0^{2\pi} H(t)\eta'(t)dt \quad \forall \eta \in W^{1,2}_{\wp,0}(0, 2\pi) \tag{3.4.43}$$

with a uniquely determined $H \in L^2_{\sharp}(0, 2\pi)$.

The proof of this lemma laterally repeats the proof of Lemma 3.4.1.

Remark 3.4.7 Note that if the functional h can be represented by an $L^2_{\sharp}(0, 2\pi)$ function, that is, $\langle h, \eta\rangle = \int\limits_0^{2\pi} h(t)\eta(t)dt$ with $h \in L^2_{\sharp}(0, 2\pi)$ and arbitrary $\eta \in W^{1,2}_{\wp}(0, 2\pi)$, then $H(t) = S_h(t) = -\int\limits_t^{2\pi} h(\tau)d\tau$.

Now, we are in a position to define a weak solution.

Definition 3.4.2 Let $F \in L^2_{\sharp}(0, 2\pi)$. A *weak solution* of problem (3.4.41) is defined as a pair $\big(v_{P_\alpha}, s\big)$ such that $v_{P_\alpha} \in L^2_{\sharp}(0, 2\pi; L^2(\sigma))$, $S_{v_{P_\alpha}} \in L^2_{\wp}(0, 2\pi; \mathring{W}^{1,2}(\sigma))$; $v_{P_\alpha}(x', t)$ satisfies the flux condition:

$$\int\limits_\sigma v_{P_\alpha}(x', t)dx' = F(t) \tag{3.4.44}$$

and the pair $\big(v_{P_\alpha}, s\big)$ satisfies the integral identity:

$$\int_0^{2\pi}\int_\sigma v_{P_\alpha}(x',t)\partial_t\eta(x',t)dx'dt + \nu\int_0^{2\pi}\int_\sigma \nabla' S_{v_{P_\alpha}}(x',t)\cdot\nabla'\partial_t\eta(x',t)dx'dt$$
$$= \int_0^{2\pi} S_s(t)\int_\sigma \partial_t\eta(x',t)dx'dt \tag{3.4.45}$$

for any test function $\eta \in L^2_{\wp}(0,2\pi;\mathring{W}^{1,2}(\sigma))$ such that $\partial_t\eta \in L^2_{\sharp}(0,2\pi;W^{1,2}(\sigma))$.

Theorem 3.4.2 *Let $F \in L^2_{\sharp}(0,2\pi)$. Then problem* (3.4.41) *has a unique weak solution $\big(v_{P_\alpha}, s\big)$. Moreover, the following estimate holds:*

$$\int_0^{2\pi}\int_\sigma \left|v_{P_\alpha}(x',t)\right|^2 dx'dt + \int_0^{2\pi}\int_\sigma \left|\nabla' S_{v_{P_\alpha}}(x',t)\right|^2 dx'dt + \int_0^{2\pi}\left|S_s(\tau)\right|^2 d\tau$$
$$\le c\int_0^{2\pi}\left|F(\tau)\right|^2 d\tau, \tag{3.4.46}$$

where c is a constant depending solely on σ.

Remark 3.4.8 Since $\int\limits_0^{2\pi}\eta'(t)dt = 0$ for $\eta \in W^{1,2}_{\wp}(0,2\pi)$, and all primitives of the function $v_{P_\alpha}(x',t)$ differ from each other by a function independent of t, integral identity (3.4.45) remains valid for any primitive function $S_{v_{P_\alpha}}$. Hence, without loss of generality, we can assume $S_{v_{P_\alpha}}$ to be zero at the point $t = 2\pi$.

3.4.5.1 Construction of Galerkin Approximations

The proof of Theorem 3.4.2 employs a version of Galerkin approximations, similar to the approach used in the initial boundary value problem. To obtain estimates of the approximate solutions $\big(v^{(N,\varepsilon)}_{P_\alpha}(x',t), s^{(N,\varepsilon)}(t)\big)$, we introduce primitive functions defined by the integrals $\int\limits_t^{t_*+2\pi}\ldots$, where $t_* = t_*(\varepsilon, N) \in [0,2\pi)$ and $t \in (t_*, t_*+2\pi)$. Consequently, estimate (3.4.46) is valid only for the primitive function $S_{v_{P_\alpha}}$, obtained as the limit of the sequence $\big\{S_{v^{(N,\varepsilon)}_{P_\alpha}}\big\}$.

Let $u_k(x') \in \mathring{W}^{1,2}(\sigma)$ and λ_k be the eigenfunctions and eigenvalues of the Laplace operator (3.2.9), respectively. We seek approximate solutions of problem (3.4.41) in the form:

$$v_{P_\alpha}^{(N)}(x',t)=\sum_{k=1}^{N} w_k^{(N)}(t)u_k(x'). \tag{3.4.47}$$

The coefficients $w_k^{(N)}(t)$ and the function $s^{(N)}(t)$ are determined by solving the following linear problems:

$$\begin{aligned}
&\int\limits_\sigma \partial_t v_{P_\alpha}^{(N)}(x',t)u_k(x')dx' + \nu\int\limits_\sigma \nabla' v_{P_\alpha}^{(N)}(x',t)\cdot\nabla' u_k(x')dx'\\
&\qquad = s^{(N)}(t)\int\limits_\sigma u_k(x')dx', \quad k=1,2,\dots,N,\\
&\qquad w_k^{(N)}(0)=w_k^{(N)}(2\pi), \quad k=1,\dots,N,\\
&\qquad \int\limits_\sigma v_{P_\alpha}^{(N)}(x',t)\,dx' = F(t)
\end{aligned} \tag{3.4.48}$$

which, due to the orthonormality of functions $u_k(x')$, are equivalent to ordinary differential equations for the functions $w_k^{(N)}(t)$:

$$\begin{cases} (w_k^{(N)}(t))' + \lambda_k w_k^{(N)}(t) = \beta_k s^{(N)}(t), \ t\in(0,2\pi),\\ \qquad w_k^{(N)}(0)=w_k^{(N)}(2\pi), \end{cases} \tag{3.4.49}$$

where $\beta_k=\int_\sigma u_k(x')dx'$. The solution of problem (3.4.49) takes the form:

$$w_k^{(N)}(t)=\beta_k\int\limits_0^{2\pi} G_k(t,\tau)s^{(N)}(\tau)d\tau, \tag{3.4.50}$$

where the Green function $G_k(t,\tau)$ is given by (see [134]):

$$G_k(t,\tau)=\begin{cases} \dfrac{e^{-\lambda_k(t-\tau)}}{1-e^{-2\pi\lambda_k}}, & 0\le\tau\le t\le 2\pi,\\[2ex] \dfrac{e^{-\lambda_k(t-\tau+2\pi)}}{1-e^{-2\pi\lambda_k}}, & 0\le t<\tau\le 2\pi. \end{cases}$$

It can be observed that $G_k(0,\tau)=G_k(2\pi,\tau)$, and thus, $w_k^{(N)}(0)=w_k^{(N)}(2\pi)$. Substituting expression (3.4.50) into (3.4.47), we obtain:

$$v_{P_\alpha}^{(N)}(x',t)=\sum_{k=1}^{N}\beta_k\int\limits_0^{2\pi} G_k(t,\tau)s^{(N)}(\tau)d\tau\, u_k(x').$$

The flux condition now yields:

$$F(t) = \int_\sigma v_{P_\alpha}^{(N)}(x', t)dx' = \sum_{k=1}^N \beta_k \int_0^{2\pi} G_k(t, \tau)s^{(N)}(\tau)d\tau \int_\sigma u_k(x')dx'$$
$$= \sum_{k=1}^N \beta_k^2 \int_0^{2\pi} G_k(t, \tau)s^{(N)}(\tau)d\tau.$$

Thus, for the function $s^{(N)}$, we have derived a Fredholm integral equation of the first kind:

$$\int_0^{2\pi} \sum_{k=1}^N \beta_k^2 G_k(t, \tau)s^{(N)}(\tau)d\tau = F(t). \tag{3.4.51}$$

It is well-known (see, e.g., [108, 193]) that such equations, in general, are ill-posed in L^2-setting. To regularize equation (3.4.51), we consider the following Fredholm integral equation of the second kind:

$$\varepsilon s^{(N,\varepsilon)}(t) + \int_0^{2\pi} \sum_{k=1}^N \beta_k^2 G_k(t, \tau)s^{(N,\varepsilon)}(\tau)d\tau = F(t), \tag{3.4.52}$$

where ε will later tend to 0. Instead of problem (3.4.48), (3.4.51), we study the regularized problem:

$$\begin{aligned}
&\int_\sigma \partial_t v_{P_\alpha}^{(N,\varepsilon)}(x', t)u_k(x')dx' + \nu \int_\sigma \nabla' v_{P_\alpha}^{(N,\varepsilon)}(x', t) \cdot \nabla' u_k(x')dx' \\
&\qquad = s^{(N,\varepsilon)}(t) \int_\sigma u_k(x')dx', \quad k = 1, 2, \ldots, N, \\
&\qquad v_{P_\alpha}^{(N,\varepsilon)}(x', 0) = v_{P_\alpha}^{(N,\varepsilon)}(x', 2\pi), \\
&\varepsilon s^{(N,\varepsilon)}(t) + \int_0^{2\pi} \sum_{k=1}^N \beta_k^2 G_k(t, \tau)s^{(N,\varepsilon)}(\tau)d\tau = F(t),
\end{aligned} \tag{3.4.53}$$

where

$$v_{P_\alpha}^{(N,\varepsilon)}(x', t) = \sum_{k=1}^N \beta_k \int_0^{2\pi} G_k(t, \tau)s^{(N,\varepsilon)}(\tau)d\tau\, u_k(x').$$

Lemma 3.4.4 *Let $F \in L^2_\sharp(0, 2\pi)$. Then Eq. (3.4.52) has a unique solution $s^{(N,\varepsilon)} \in L^2_\sharp(0, 2\pi)$.*

Proof
First, we demonstrate the well-definedness of Eq. (3.4.52) in the space $L^2_\sharp(0, 2\pi)$. Clearly, if F is periodic and $s^{(N,\varepsilon)}$ is the solution of (3.4.52), then $s^{(N,\varepsilon)}$ is also a periodic function. Let us assume that the mean value $\overline{F} = 0$. Consequently, we have

$$\varepsilon\overline{s}^{(N,\varepsilon)} + \frac{1}{2\pi}\int_0^{2\pi}\sum_{k=1}^{N}\beta_k^2 s^{(N,\varepsilon)}(\tau)\int_0^{2\pi} G_k(t,\tau)dtd\tau = 0. \tag{3.4.54}$$

Given that

$$\int_0^{2\pi} G_k(t,\tau)dt = \frac{1}{1-e^{-2\pi\lambda_k}}\Big(\int_\tau^{2\pi} e^{-\lambda_k(t-\tau)}dt + \int_0^{\tau} e^{-\lambda_k(t-\tau+2\pi)}dt\Big) = \frac{1}{\lambda_k},$$

the second term in (3.4.54) can be expressed as $\sum_{k=1}^{N}\frac{\beta_k^2}{\lambda_k}\overline{s}^{(N,\varepsilon)}$, leading to

$$\Big(\varepsilon + \sum_{k=1}^{N}\frac{\beta_k^2}{\lambda_k}\Big)\overline{s}^{(N,\varepsilon)} = 0,$$

implying that $\overline{s}^{(N,\varepsilon)} = 0$. Consequently, the mean value $\overline{v_{P_\alpha}^{(N,\varepsilon)}}(x')$ of $v_{P_\alpha}^{(N,\varepsilon)}(x', t)$ also vanishes: $\overline{v_{P_\alpha}^{(N,\varepsilon)}}(x') = 0$.

It is well-known that Fredholm integral equations of the second kind (e.g., [94, 206]) satisfy the Fredholm alternative (see Theorem 2.2.8). Therefore, it suffices to prove the uniqueness of the solution to (3.4.52). Let $F(t) = 0$. By construction, we have

$$\int_0^{2\pi}\sum_{k=1}^{N}\beta_k^2 G_k(t,\tau)s_\alpha^{(N,\varepsilon)}(\tau)d\tau = \int_\sigma v_{P_\alpha}^{(N,\varepsilon)}(x', t)dx',$$

and the homogeneous equation (3.4.52) yields

$$\int_\sigma v_{P_\alpha}^{(N,\varepsilon)}(x', t)dx' = -\varepsilon s^{(N,\varepsilon)}(t).$$

Multiplying $(3.4.53)_1$ by $w_k^{(N,\varepsilon)}(t)$, summing by k from 1 to N, and integrating over the interval $(0, 2\pi)$ yield

$$\int_0^{2\pi}\int_\sigma \partial_t v_{P_\alpha}^{(N,\varepsilon)}(x',t) v_{P_\alpha}^{(N,\varepsilon)}(x',t)dx'dt + \nu \int_0^{2\pi}\int_\sigma |\nabla' v_{P_\alpha}^{(N,\varepsilon)}(x',t)|^2 dx'dt$$
$$= \int_0^{2\pi} s^{(N,\varepsilon)}(t) \int_\sigma v_{P_\alpha}^{(N,\varepsilon)}(x',\tau)dx'd\tau dt = -\varepsilon \int_0^{2\pi} |s_\alpha^{(N,\varepsilon)}(t)|^2 dt. \tag{3.4.55}$$

After integrating by parts with respect to t and taking into account the time-periodicity of $v_\alpha^{(N)}(x',t)$, we obtain

$$\nu \int_0^{2\pi}\int_\sigma |\nabla' v_{P_\alpha}^{(N,\varepsilon)}(x',t)|^2 dx'dt + \varepsilon \int_0^{2\pi} |s^{(N,\varepsilon)}(t)|^2 dt = 0.$$

Hence, $s^{(N,\varepsilon)}(t) = 0$ for almost all $t \in [0, 2\pi]$ and the lemma is proved. □

3.4.5.2 A Priori Estimates of the Galerkin Approximations

Consider the pair $(v_{P_\alpha}^{(N,\varepsilon)}(x',t), s^{(N,\varepsilon)}(t))$, which represents the solution of problem (3.4.53), and let $u_0(x')$ be the solution of problem (3.1.2). Consider the integral $\int_\sigma v_{P_\alpha}^{(N,\varepsilon)}(x',t)u_0(x')dx'$. Since the mean value $\overline{v_{P_\alpha}^{(N,\varepsilon)}}(x') = 0$, we have:

$$\int_0^{2\pi}\int_\sigma v_{P_\alpha}^{(N,\varepsilon)}(x',t)u_0(x')dx'dt = \int_\sigma u_0(x')\left(\int_0^{2\pi} v_{P_\alpha}^{(N,\varepsilon)}(x',t)dt\right)dx' = 0.$$

Thus, there exists $t_* = t_*(\varepsilon, N) \in [0, 2\pi)$ such that $\int_\sigma v_{P_\alpha}^{(N,\varepsilon)}(x',t_*)u_0(x')dx' = 0.$[7] Due to periodicity, we also have $\int_\sigma v_{P_\alpha}^{(N,\varepsilon)}(x',t_* + 2\pi)u_0(x')dx' = 0$.

Now, let $h \in L^2_\sharp(-\pi,\pi)$. For $t \in [t_*, t_* + 2\pi]$, we define the notation: $S_h^*(t) = \int_t^{t_*+2\pi} h(\tau)d\tau$. As the mean value of h vanishes, we have $S_h^*(t_*+2\pi) = S_h(t_*) = 0$. Moreover, $\dfrac{dS_h^*(t)}{dt} = -h(t)$.

[7] The value of $t_*(\varepsilon, N)$ depends on ε and N, but in this section, we denote it simply as t_* to simplify the notation.

Lemma 3.4.5 *The following estimate holds:*

$$\int_{t_*}^{t_*+2\pi}\int_{\sigma}|v_{P_\alpha}^{(N,\varepsilon)}(x',t)|^2dx'dt+\int_{t_*}^{t_*+2\pi}\int_{\sigma}\left|\nabla' S^*_{v_{P_\alpha}^{(N,\varepsilon)}}(x',t)\right|^2dx'dt$$
$$+\int_{t_*}^{t_*+2\pi}|S^*_{s^{(N,\varepsilon)}}(\tau)|^2d\tau\le c\int_{t_*}^{t_*+2\pi}|F(\tau)|^2d\tau, \tag{3.4.56}$$

where c is a constant independent of ε and N.

Proof
Let us define $\gamma_k^{(N,\varepsilon)}(t)=\int_t^{t_*+2\pi}w_k^{(N,\varepsilon)}(\tau)d\tau$ for $t_*\le t\le t_*+2\pi$. By multiplying $(3.4.53)_1$ by $\gamma_k^{(N,\varepsilon)}(t)$, summing the resulting equations over $k=1$ to $k=N$, and integrating them over the interval $(t_*,t_*+2\pi)$, we obtain:

$$\int_{t_*}^{t_*+2\pi}\int_{\sigma}\partial_t v_{P_\alpha}^{(N,\varepsilon)}(x',t)\int_{t}^{t_*+2\pi}v_{P_\alpha}^{(N,\varepsilon)}(x',\tau)d\tau dx'dt$$
$$+\nu\int_{t_*}^{t_*+2\pi}\int_{\sigma}\nabla' v_{P_\alpha}^{(N,\varepsilon)}(x',t)\int_{t}^{t_*+2\pi}\nabla' v_{P_\alpha}^{(N,\varepsilon)}(x',\tau)d\tau dx'dt \tag{3.4.57}$$
$$=\int_{t_*}^{t_*+2\pi}s^{(N,\varepsilon)}(t)\int_{t}^{t_*+2\pi}\int_{\sigma}v_{P_\alpha}^{(N,\varepsilon)}(x',\tau)dx'd\tau dt.$$

Using the relation:

$$\int_{\sigma}v_{P_\alpha}^{(N,\varepsilon)}(x',t)dx'=\sum_{k=1}^{N}\beta_k^2\int_0^{2\pi}G_k(t,\tau)s_\alpha^{(N,\varepsilon)}(\tau)d\tau$$

and $(3.4.53)_2$, we derive:

$$\int_{\sigma}v_{P_\alpha}^{(N,\varepsilon)}(x',t)dx'=F(t)-\varepsilon s_\alpha^{(N,\varepsilon)}(t).$$

Therefore, Eq. (3.4.57) can be expressed as follows:

$$\int_{t_*}^{t_*+2\pi}\int_{\sigma} \partial_t v_{P_\alpha}^{(N,\varepsilon)}(x',t) \int_{t}^{t_*+2\pi} v_{P_\alpha}^{(N,\varepsilon)}(x',\tau)d\tau dx'dt$$

$$+\nu \int_{t_*}^{t_*+2\pi}\int_{\sigma} \nabla' v_{P_\alpha}^{(N,\varepsilon)}(x',t) \int_{t}^{t_*+2\pi} \nabla' v_{P_\alpha}^{(N,\varepsilon)}(x',\tau)d\tau dx'dt \tag{3.4.58}$$

$$= \int_{t_*}^{t_*+2\pi} s^{(N,\varepsilon)}(t) \int_{t}^{t_*+2\pi} \Big(F(\tau) - \varepsilon s^{(N,\varepsilon)}(\tau)\Big) d\tau dt.$$

It is easy to calculate that

$$\int_{t_*}^{t_*+2\pi} s^{(N,\varepsilon)}(t) \int_{t}^{t_*+2\pi} \Big(F(\tau) - \varepsilon s^{(N,\varepsilon)}(\tau)\Big) d\tau dt$$

$$= - \int_{t_*}^{t_*+2\pi} \frac{dS^*_{s^{(N,\varepsilon)}}(t)}{dt} \int_{t}^{t_*+2\pi} \Big(F(\tau) - \varepsilon s^{(N,\varepsilon)}(\tau)\Big) d\tau dt$$

$$= - \int_{t_*}^{t_*+2\pi} S^*_{s^{(N,\varepsilon)}}(t)F(t)dt + \varepsilon \int_{t_*}^{t_*+2\pi} S^*_{s^{(N,\varepsilon)}}(t)s^{(N,\varepsilon)}(t)dt$$

$$= - \int_{t_*}^{t_*+2\pi} S^*_{s^{(N,\varepsilon)}}(t)F(t)dt - \varepsilon \int_{t_*}^{t_*+2\pi} S^*_{s^{(N,\varepsilon)}}(t)\frac{dS^*_{s^{(N,\varepsilon)}}(t)}{dt}dt$$

$$= - \int_{t_*}^{t_*+2\pi} S^*_{s^{(N,\varepsilon)}}(t)F(t)dt.$$

By a similar calculation for the two integrals on the left-hand side of Eq. (3.4.58), we derive:

$$\int_{t_*}^{t_*+2\pi}\int_{\sigma} \partial_t v_{P_\alpha}^{(N,\varepsilon)}(x',t) \int_{t}^{t_*+2\pi} v_{P_\alpha}^{(N,\varepsilon)}(x',\tau)d\tau dx'dt = \int_{t_*}^{t_*+2\pi}\int_{\sigma} \left|v_{P_\alpha}^{(N,\varepsilon)}(x',t)\right|^2 dx'dt$$

and

$$\nu \int_{t_*}^{t_*+2\pi} \int_{\sigma} \nabla' v_{P_\alpha}^{(N,\varepsilon)}(x',t) \int_{t}^{t_*+2\pi} \nabla' v_{P_\alpha}^{(N,\varepsilon)}(x',\tau)d\tau dx'dt = 0.$$

Hence, the relation (3.4.58) takes the form:

$$\int_{t_*}^{t_*+2\pi} \int_{\sigma} \left|v_{P_\alpha}^{(N,\varepsilon)}(x',t)\right|^2 dx'dt = - \int_{t_*}^{t_*+2\pi} S^*_{s^{(N,\varepsilon)}}(t)F(t)dt. \tag{3.4.59}$$

Now, let us multiply equation $(3.4.53)_1$ by $(t-t_*)\gamma_k^{(N,\varepsilon)}(t)$, sum from $k=1$ to $k=N$, and integrate over the interval $(t_*, t_*+2\pi)$:

$$\begin{aligned}
&\int_{t_*}^{t_*+2\pi} \int_{\sigma} (t-t_*)\partial_t v_{P_\alpha}^{(N,\varepsilon)}(x',t) \int_{t}^{t_*+2\pi} v_{P_\alpha}^{(N,\varepsilon)}(x',\tau)d\tau dx'dt \\
&+\nu \int_{t_*}^{t_*+2\pi} \int_{\sigma} (t-t_*)\nabla' v_{P_\alpha}^{(N,\varepsilon)}(x',t) \int_{t}^{t_*+2\pi} \nabla' v_{P_\alpha}^{(N,\varepsilon)}(x',\tau)d\tau dx'dt \\
&= \int_{t_*}^{t_*+2\pi} (t-t_*)s^{(N),\varepsilon}(t) \int_{t}^{t_*+2\pi} \big(F(\tau) - \varepsilon s^{(N,\varepsilon)}(\tau)\big)d\tau dt.
\end{aligned} \tag{3.4.60}$$

Integrating by parts with respect to t and considering that $\frac{d}{dt} \int_{t}^{t_*+2\pi} h(\tau)d\tau = -h(t)$, we calculate that:

$$\int_{t_*}^{t_*+2\pi} \int_{\sigma} (t-t_*)\partial_t v_{P_\alpha}^{(N,\varepsilon)}(x',t) \int_{t}^{t_*+2\pi} v_{P_\alpha}^{(N,\varepsilon)}(x',\tau)d\tau dx'dt$$

$$= - \int_{t_*}^{t_*+2\pi} \int_{\sigma} v_{P_\alpha}^{(N,\varepsilon)}(x',t) \int_{t}^{t_*+2\pi} v_{P_\alpha}^{(N,\varepsilon)}(x',\tau)d\tau dx'dt$$

$$+ \int_{t_*}^{t_*+2\pi} \int_{\sigma} (t-t_*)|v_{P_\alpha}^{(N,\varepsilon)}(x',t)|^2 dx'dt,$$

where the first integral in the right-hand side is equal to zero. Thus, Eq. (3.4.60) can be rewritten as follows:

$$\int\limits_{t_*}^{t_*+2\pi}\int\limits_{\sigma}(t-t_*)|v_{P_\alpha}^{(N,\varepsilon)}(x',t)|^2dx'dt+\frac{\nu}{2}\int\limits_{t_*}^{t_*+2\pi}\int\limits_{\sigma}|\nabla' S^*_{v_{P_\alpha}^{(N,\varepsilon)}}(x',t)|^2dx'dt$$
$$=-\int\limits_{t_*}^{t_*+2\pi}(t-t_*)S^*_{s(N,\varepsilon)}(t)F(t)dt-\frac{\varepsilon}{2}\int\limits_{t_*}^{t_*+2\pi}\left|S^*_{s(N,\varepsilon)}(t)\right|^2dt$$
$$+\int\limits_{t_*}^{t_*+2\pi}S^*_{s(N,\varepsilon)}(t)\int\limits_{t}^{t_*+2\pi}F(\tau)d\tau dt. \tag{3.4.61}$$

From (3.4.59) and (3.4.61), we obtain:

$$\int\limits_{t_*}^{t_*+2\pi}\int\limits_{\sigma}|v_{P_\alpha}^{(N,\varepsilon)}(x',t)|^2dx'dt\le\delta\int\limits_{t_*}^{t_*+2\pi}\left|S^*_{s(N,\varepsilon)}(t)\right|^2dt+\frac{1}{2\delta}\int\limits_{t_*}^{t_*+2\pi}|F(t)|^2dt, \tag{3.4.62}$$

and

$$\frac{\nu}{2}\int\limits_{t_*}^{t_*+2\pi}\int\limits_{\sigma}|\nabla' S^*_{v_{P_\alpha}^{(N,\varepsilon)}}(x',t)|^2dx'dt\le\Big|\int\limits_{t_*}^{t_*+2\pi}(t-t_*)S^*_{s(N,\varepsilon)}(t)F(t)dt\Big|$$
$$+\Big|\int\limits_{t_*}^{t_*+2\pi}S^*_{s(N,\varepsilon)}(t)\int\limits_{t}^{t_*+2\pi}F(\tau)d\tau dt\Big|\le\delta(4\pi^2+1)\int\limits_{t_*}^{t_*+2\pi}|S^*_{s(N,\varepsilon)}(t)|^2dt$$
$$+\frac{1}{2\delta}\int\limits_{t_*}^{t_*+2\pi}|F(t)|^2dt+\frac{1}{2\delta}\int\limits_{t_*}^{t_*+2\pi}\Big|\int\limits_{t}^{t_*+2\pi}F(\tau)d\tau\Big|^2dt \tag{3.4.63}$$
$$\le(4\pi^2+1)\Big(\delta\int\limits_{t_*}^{t_*+2\pi}|S^*_{s(N,\varepsilon)}(t)|^2dt+\frac{1}{2\delta}\int\limits_{t_*}^{t_*+2\pi}|F(t)|^2dt\Big).$$

Let us estimate the integral $\int\limits_{t_*}^{t_*+2\pi}\left|S^*_{s(N,\varepsilon)}(t)\right|^2dt$. Let $u_0\in\overset{\circ}{W}{}^{1,2}(\sigma)$ be a solution of problem (3.1.2). Then, u_0 can be expressed as a Fourier series in $\overset{\circ}{W}{}^{1,2}(\sigma)$. Let a_k be the Fourier coefficients of $u_0(x')$. Multiplying relations $(3.4.53)_1$ by a_k and summing them over k, we get:

$$\int_{\sigma} \partial_t v_{P_\alpha}^{(N,\varepsilon)}(x',t)u_0(x')dx' + \nu \int_{\sigma} \nabla' v_{P_\alpha}^{(N,\varepsilon)}(x',t) \cdot \nabla' u_0(x')dx' \\ = s^{(N,\varepsilon)}(t) \int_{\sigma} u_0(x')dx' = s^{(N,\varepsilon)}(t)\kappa_0. \tag{3.4.64}$$

On the other hand, multiplying the Poisson equation (3.1.2) by $v_{P_\alpha}^{(N,\varepsilon)}(x',t)$ and integrating by parts in σ, we obtain:

$$\nu \int_{\sigma} \nabla' u_0(x') \cdot \nabla' v_{P_\alpha}^{(N,\varepsilon)}(x',t)dx' = \int_{\sigma} v_{P_\alpha}^{(N,\varepsilon)}(x',t)dx' = F(t) - \varepsilon s^{(N,\varepsilon)}(t).$$

Substituting this into the previous equation, we get:

$$(\kappa_0 + \varepsilon)s_\alpha^{(N,\varepsilon)}(t) = \int_{\sigma} \partial_t v_{P_\alpha}^{(N,\varepsilon)}(x',t)u_0(x')dx' + F(t).$$

Integrating this with respect to t from τ to $t_* + 2\pi$, we obtain:

$$(\kappa_0 + \varepsilon) \int_{\tau}^{t_*+2\pi} s^{(N,\varepsilon)}(t)dt = (\kappa_0 + \varepsilon)S^*_{s^{(N,\varepsilon)}}(\tau) \\ = - \int_{\sigma} v_{P_\alpha}^{(N,\varepsilon)}(x',\tau)u_0(x')dx' + \int_{\tau}^{t_*+2\pi} F(t)dt. \tag{3.4.65}$$

Here we have used the choice of the point t_*, i.e.,

$$\int_{\sigma} v_{P_\alpha}^{(N,\varepsilon)}(x',t_*)u_0(x')dx' = \int_{\sigma} v_{P_\alpha}^{(N,\varepsilon)}(x',t_* + 2\pi)u_0(x')dx' = 0,$$

and hence,

$$\int_{\tau}^{t_*+2\pi} \int_{\sigma} \partial_t v_{P_\alpha}^{(N,\varepsilon)}(x',t)u_0(x')dx'dt = - \int_{\sigma} v_{P_\alpha}^{(N,\varepsilon)}(x',\tau)u_0(x')dx'.$$

From (3.4.65) it follows that

$$(\kappa_0+\varepsilon)^2\int\limits_{t_*}^{t_*+2\pi}\big|S^*_{s(N,\varepsilon)}(\tau)\big|^2d\tau\le 2\int\limits_{t_*}^{t_*+2\pi}\Big(\int\limits_{\sigma}v^{(N,\varepsilon)}_{P_\alpha}(x',\tau)u_0(x')dx'\Big)^2d\tau$$
$$+2\int\limits_{t_*}^{t_*+2\pi}\Big(\int\limits_{\tau}^{t_*+2\pi}F(t)dt\Big)^2d\tau\le 2\int\limits_{t_*}^{t_*+2\pi}\int\limits_{\sigma}\big|v^{(N,\varepsilon)}_{P_\alpha}(x',\tau)\big|^2dx'\int\limits_{\sigma}\big|u_0(x')\big|^2dx'd\tau$$
$$+2\int\limits_{t_*}^{t_*+2\pi}|S^*_F(\tau)|^2d\tau\le c\int\limits_{t_*}^{t_*+2\pi}\int\limits_{\sigma}\big|v^{(N,\varepsilon)}_{P_\alpha}(x',\tau)\big|^2dx'\int\limits_{\sigma}\big|\nabla' u_0(x')\big|^2dx'd\tau$$
$$+2\int\limits_{t_*}^{t_*+2\pi}|S^*_F(\tau)|^2d\tau\le c\Big(\int\limits_{t_*}^{t_*+2\pi}\int\limits_{\sigma}\big|v^{(N,\varepsilon)}_{P_\alpha}(x',\tau)\big|^2dx'd\tau+\int\limits_{t_*}^{t_*+2\pi}|F(\tau)|^2d\tau\Big).\tag{3.4.66}$$

Substituting (3.4.66) into (3.4.62) yields

$$\int\limits_{t_*}^{t_*+2\pi}\int\limits_{\sigma}\big|v^{(N,\varepsilon)}_{P_\alpha}(x',t)\big|^2dx'dt\le c\delta\int\limits_{t_*}^{t_*+2\pi}\int\limits_{\sigma}\big|v^{(N,\varepsilon)}_{P_\alpha}(x',t)\big|^2dx'dt+c_\delta\int\limits_{t_*}^{t_*+2\pi}\big|F(t)\big|^2dt$$

and, thus, choosing δ sufficiently small, we obtain:

$$\int\limits_{t_*}^{t_*+2\pi}\int\limits_{\sigma}\big|v^{(N,\varepsilon)}_{P_\alpha}(x',t)\big|^2dx'dt\le c\int\limits_{t_*}^{t_*+2\pi}\big|F(t)\big|^2dt.\tag{3.4.67}$$

Estimates (3.4.67) and (3.4.66) imply

$$\int\limits_{t_*}^{t_*+2\pi}\big|S^*_{s(N,\varepsilon)}(\tau)\big|^2d\tau\le c\int\limits_{t_*}^{t_*+2\pi}\big|F(\tau)\big|^2d\tau.\tag{3.4.68}$$

Finally, from (3.4.63) and (3.4.68), it follows that

$$\int\limits_{t_*}^{t_*+2\pi}\int\limits_{\sigma}\Big|\nabla' S^*_{v^{(N,\varepsilon)}_{P_\alpha}}(x',t)\Big|^2dx'dt\le c\int\limits_{t_*}^{t_*+2\pi}\big|F(t)\big|^2dt.\tag{3.4.69}$$

The constants in (3.4.67)–(3.4.69) are independent of ε and N. □

3.4.5.3 Convergence of Galerkin Approximations: Existence

The constructed approximate solutions $\left(v_{P_\alpha}^{(N,\varepsilon)}(x',t), s^{(N,\varepsilon)}(t)\right)$ satisfy the equalities $(3.4.53)_1$. We multiply these relations by arbitrary functions $d_k(t) \in L^2_{\wp}(0,2\pi)$ with $d_k'(t) \in L^2_\sharp(0,2\pi)$, sum over k from $k=1$ to $k=M$ with $M \le N$, integrate with respect to t, and then integrate by parts, resulting in the integral identity:

$$\int_0^{2\pi}\int_\sigma v_{P_\alpha}^{(N,\varepsilon)}(x',t)\partial_t\eta(x',t)dx'dt + \nu\int_0^{2\pi}\int_\sigma \nabla' S^*_{v_{P_\alpha}^{(N,\varepsilon)}}(x',t)\cdot\nabla'\partial_t\eta(x',t)dx'dt$$
$$= \int_0^{2\pi} S^*_{s^{(N,\varepsilon)}}(\tau)\int_\sigma \partial_t\eta(x',t)dx'dt \tag{3.4.70}$$

where the test functions η take the form $\eta(x',t) = \sum_{k=1}^{M} d_k(t)u_k(x')$.

Recall that $\left(v_{P_\alpha}^{(N,\varepsilon)}(x',t), s^{(N,\varepsilon)}(t)\right)$ satisfy the a priori estimate (3.4.56) with a constant c independent of ε and N. Since all functions in (3.4.56) are 2π-periodic, inequality (3.4.56) can be rewritten as:

$$\int_0^{2\pi}\int_\sigma |v_{P_\alpha}^{(N,\varepsilon)}(x',t)|^2 dx'dt + \int_0^{2\pi}\int_\sigma \left|\nabla' S^*_{v_{P_\alpha}^{(N,\varepsilon)}}(x',t)\right|^2 dx'dt + \int_0^{2\pi} |S^*_{s^{(N,\varepsilon)}}(\tau)|^2 d\tau$$
$$\le c\int_0^{2\pi} |F(\tau)|^2 d\tau. \tag{3.4.71}$$

Now, we select a fixed N and choose subsequences $\{\varepsilon_l\}$ and $\left\{\left(v_{P_\alpha}^{(N,\varepsilon_l)}(x',t), s^{(N,\varepsilon_l)}(t)\right)\right\}$ such that $\lim\limits_{l\to\infty}\varepsilon_l = 0$, and $\{v_{P_\alpha}^{(N,\varepsilon_l)}\}$ converges weakly in $L^2_\sharp(0,2\pi;L^2(\sigma))$ to some $v_{P_\alpha}^{(N)}$, $\{S^*_{v_{P_\alpha}^{(N,\varepsilon_l)}}\}$ converges weakly in $L^2_{\wp}(0,2\pi;\overset{\circ}{W}{}^{1,2}(\sigma))$ to $S_{v_{P_\alpha}^{(N)}}$, and $\{s^{(N,\varepsilon_l)}(t)\}$ converges weakly in $W^{-1,2}_{\wp}(0,2\pi)$ to $s^{(N)}(t)$. The last convergence means that

$$\lim_{l\to\infty}\int_0^{2\pi} S^*_{s^{(N,\varepsilon_l)}}(t)\eta'(t)dt = \int_0^{2\pi} S_{s^{(N)}}(t)\eta'(t)dt = \langle s^{(N)},\eta\rangle \quad \forall\eta\in W^{1,2}_{\wp}(0,2\pi),$$

where $s^{(N)} \in W^{-1,2}_{\wp}(0,2\pi)$ and $S_{s^{(N)}} \in L^2_\sharp(0,2\pi)$ is a primitive of $-s^{(N)}$ in the distributional sense.

It is evident that for the limit functions $v_{P_\alpha}^{(N)}$ and $S_{s^{(N)}}$, estimate (3.4.71) remains valid with a constant c independent of N. In equality (3.4.70), taking $\varepsilon = \varepsilon_l$ and passing to the limit as $\varepsilon_l \to 0$, we obtain:

$$\int_0^{2\pi}\int_\sigma v_{P_\alpha}^{(N)}(x',t)\partial_t\eta(x',t)dx'dt + \nu\int_0^{2\pi}\int_\sigma \nabla' S_{v_{P_\alpha}^{(N)}}(x',t)\cdot\nabla'\partial_t\eta(x',t)dx'dt$$
$$= \int_0^{2\pi} S_{s^{(N)}}(\tau)\int_\sigma \partial_t\eta(x',t)dx'dt. \tag{3.4.72}$$

Next, we show that $v_{P_\alpha}^{(N)}(x',t)$ satisfies the flux condition:

$$\int_\sigma v_{P_\alpha}^{(N)}(x',t)dx' = F(t). \tag{3.4.73}$$

Integrating equation (3.4.52) for $\varepsilon = \varepsilon_l$ from t to 2π, we have:

$$\varepsilon_l S_{s_{\varepsilon_l}^{(N)}}(t) + \int_t^{2\pi}\int_\sigma v_{P_\alpha}^{(N,\varepsilon_l)}(x',\tau)dx'd\tau = S_F(t). \tag{3.4.74}$$

Obviously, the sequence $\left\{\varphi_l^{(N)}(\tau) = \int\limits_\sigma v_{P_\alpha}^{(N,\varepsilon_l)}(x',\tau)dx'\right\}$ is bounded in $L^2(0,2\pi)$. So we may assume, without loss of generality, that $\{\varphi_l^{(N)}(\tau)\}$ is weakly convergent to $\varphi^{(N)}$ in $L^2(0,2\pi)$, and $S_{\varphi_l^{(N)}}(t) = \int\limits_t^{2\pi}\varphi_l^{(N)}(\tau)d\tau \to S_{\varphi^{(N)}}(t)$ for all $t\in[0,2\pi]$. Hence, $\|S_{\varphi_l^{(N)}} - S_{\varphi^{(N)}}\|_{L^2(0,2\pi)} \to 0$ as $l\to\infty$ ($\varepsilon_l \to 0$). From (3.4.74) we have:

$$\|S_{\varphi_l^{(N)}} - S_F\|_{L^2(0,2\pi)} = \varepsilon_l\|S_{s^{(N,\varepsilon)}}\|_{L^2(0,2\pi)} \le c\varepsilon_l \to 0 \quad \text{as } l\to\infty.$$

Therefore,

$$\int_t^{2\pi}\int_\sigma v_{P_\alpha}^{(N)}(x',\tau)dx'd\tau = \int_t^{2\pi} F(\tau)d\tau \quad \text{for} \quad \text{almost all} \quad t\in[0,2\pi],$$

and differentiating this equality with respect to t, we get (3.4.73).

Finally, we choose a subsequence $\left\{\left(v_{P_\alpha}^{(N_k)}(x',t),\ s^{(N_k)}(t)\right)\right\}$ such that $\{v_{P_\alpha}^{(N_k)}\}$ converges weakly in $L^2_\sharp(0,2\pi;L^2(\sigma))$ to some v_{P_α}, $\left\{S_{v_{P_\alpha}^{(N_k)}}\right\}$ converges weakly in $L^2_\wp(0,2\pi;\mathring{W}^{1,2}(\sigma))$ to $S_{v_{P_\alpha}}$, and $\{s^{(N_k)}\}$ converges weakly in $W_\wp^{-1,2}(0,2\pi)$ to s. In (3.4.72), passing to the limit over the subsequence $N_k \to +\infty$ yields:

$$\int_0^{2\pi}\int_\sigma v_{P_\alpha}(x',t)\eta_t(x',t)dx'dt + \nu\int_0^{2\pi}\int_\sigma \nabla' S_{v_{P_\alpha}}(x',t)\cdot\nabla'\eta_t(x',t)dx'dt$$
$$= \int_0^{2\pi} S_s(t)\int_\sigma \eta_t(x',t)dx'dt \tag{3.4.75}$$

By following similar arguments as above, it can be shown that $v_{P_\alpha}(x',t)$ satisfies the flux condition.

Integral identity (3.4.75) is proven for test functions η that can be represented as sums: $\eta(x',t) = \sum_{k=1}^{M} d_k(t)u_k(x')$, where $d_k(t) \in L^2_{\wp}(0, 2\pi)$ with the condition $d_k'(t) \in L^2_{\sharp}(0, 2\pi)$. After taking the limit in (3.4.72) as $N_k \to +\infty$, the subscript M in these sums can be an arbitrary large natural number. Such sums are dense in the space $\mathcal{V} = \{\eta : \eta \in L^2_{\wp}(0, 2\pi; \mathring{W}^{1,2}(\sigma)) : \partial_t\eta \in L^2_{\sharp}(0, 2\pi; \mathring{W}^{1,2}(\sigma))\}$. Hence, they are also dense in the subspace $\mathcal{V}_1 = \{\eta : \eta \in L^2_{\wp}(0, 2\pi; \mathring{W}^{1,2}(\sigma)) : \partial_t\eta \in L^2_{\sharp}(0, 2\pi; W^{1,2}(\sigma))\} \subset \mathcal{V}$, and therefore, (3.4.72) remains valid for $\forall\,\eta \in \mathcal{V}_1$. This proves that $\big(v_{P_\alpha}(x',t),\ s(t)\big)$ satisfies integral identity (3.4.45) and thus is a weak solution of problem (3.4.41). The estimate (3.4.46) for $\big(v_{P_\alpha}(x',t),\ s(t)\big)$ follows from estimate (3.4.71) for the approximate solutions.

Let us now prove the uniqueness of the solution. Assume that $F(t) = 0$. Let

$$\eta(x',t) = \int_t^{2\pi} \big(S_{v_{P_\alpha}}(x',\tau) - \overline{S_{v_{P_\alpha}}}(x')\big)d\tau$$
$$= \int_t^{2\pi}\Big(\int_\tau^{2\pi} v_{P_\alpha}(x',\mu)d\mu - \frac{1}{2\pi}\int_0^{2\pi} S_{v_{P_\alpha}}(x',t)dt\Big)d\tau.$$

It is evident that $\eta \in \mathcal{V}_1$. By substituting this η into identity (3.4.45), we obtain

$$\int_0^{2\pi}\int_\sigma v_{P_\alpha}(x',t)\big(S_{v_{P_\alpha}}(x',t) - \overline{S_{v_{P_\alpha}}}(x')\big)\,dx'dt$$
$$+\nu\int_0^{2\pi}\int_\sigma \nabla' S_{v_{P_\alpha}}(x',t)\cdot\nabla'\big(S_{v_{P_\alpha}}(x',t) - \overline{S_{v_{P_\alpha}}}(x')\big)\,dx'dt \tag{3.4.76}$$
$$= \int_0^{2\pi} S_s(t)\int_\sigma \big(S_{v_{P_\alpha}}(x',t) - \overline{S_{v_{P_\alpha}}}(x')\big)\,dx'dt.$$

Since

$$\int\limits_{\sigma} v_{P_\alpha}(x', t)\, dx' = F(t) = 0,$$

using Fubini's theorem and the time periodicity of the function $S_{v_{P_\alpha}}(x', t)$, we find

$$\int\limits_{\sigma} \big(S_{v_{P_\alpha}}(x', t) - \overline{S_{v_{P_\alpha}}}(x')\big)dx'$$

$$= \int\limits_{\sigma} \Big(\int\limits_{t}^{2\pi} v_{P_\alpha}(x', \tau)\, d\tau\Big)dx' - \frac{1}{2\pi}\int\limits_{\sigma} \Big(\int\limits_{0}^{2\pi}\int\limits_{t}^{2\pi} v_{P_\alpha}(x', \tau)\, d\tau dt\Big)dx'$$

$$= \int\limits_{t}^{2\pi} \Big(\int\limits_{\sigma} v_{P_\alpha}(x', \tau)\, dx'\Big)d\tau - \frac{1}{2\pi}\int\limits_{0}^{2\pi}\int\limits_{t}^{2\pi} \Big(\int\limits_{\sigma} v_{P_\alpha}(x', \tau)\, dx'\Big)d\tau dt = 0;$$

$$\int\limits_{0}^{2\pi}\int\limits_{\sigma} v_{P_\alpha}(x', t)\big(S_{v_{P_\alpha}}(x', t) - \overline{S_{v_{P_\alpha}}}(x')\big)\, dx'dt = -\frac{1}{2}\int\limits_{0}^{2\pi} \frac{d}{dt}\int\limits_{\sigma} |S_{v_{P_\alpha}}(x', t)|^2 dt$$

$$+\frac{1}{2}\int\limits_{0}^{2\pi} \frac{d}{dt}\int\limits_{\sigma} S_{v_{P_\alpha}}(x', t)\overline{S_{v_{P_\alpha}}}(x')dxdt = 0;$$

$$\int\limits_{0}^{2\pi}\int\limits_{\sigma} \nabla' S_{v_{P_\alpha}}(x', t) \cdot \nabla'\big(S_{v_{P_\alpha}}(x', t) - \overline{S_{v_{P_\alpha}}}(x')\big)\, dx'dt$$

$$= \int\limits_{0}^{2\pi}\int\limits_{\sigma} |\nabla'\big(S_{v_{P_\alpha}}(x', t) - \overline{S_{v_{P_\alpha}}}(x')\big)|^2\, dx'dt$$

$$+ \int\limits_{0}^{2\pi}\int\limits_{\sigma} \nabla'\overline{S_{v_{P_\alpha}}}(x') \cdot \nabla'\big(S_{v_{P_\alpha}}(x', t) - \overline{S_{v_{P_\alpha}}}(x')\big)dx'dt$$

$$= \int\limits_{0}^{2\pi}\int\limits_{\sigma} |\nabla'\big(S_{v_{P_\alpha}}(x', t) - \overline{S_{v_{P_\alpha}}}(x')\big)|^2 dx'dt$$

$$- \int\limits_{\sigma} \nabla'\overline{S_{v_{P_\alpha}}}(x') \cdot\nabla'\Big(\int\limits_{0}^{2\pi} \big(S_{v_{P_\alpha}}(x', t) - \overline{S_{v_{P_\alpha}}}(x')\big)\Big)dtdx'$$

$$= \int\limits_{0}^{2\pi}\int\limits_{\sigma} |\nabla'\big(S_{v_{P_\alpha}}(x', t) - \overline{S_{v_{P_\alpha}}}(x')\big)|^2\, dx'dt.$$

Thus, from (3.4.76), it follows that

$$\int_0^{2\pi}\int_\sigma |\nabla'\big(S_{v_{P_\alpha}}(x',t) - \overline{S_{v_{P_\alpha}}}(x')\big)|^2\,dx'dt = 0.$$

This implies $\nabla'\big(S_{v_{P_\alpha}}(x',t) - \overline{S_{v_{P_\alpha}}}(x')\big) = 0$, and hence, $S_{v_{P_\alpha}}(x',t) - \overline{S_{v_{P_\alpha}}}(x') = 0$, since $\big(S_{v_{P_\alpha}}(x',t) - \overline{S_{v_{P_\alpha}}}(x')\big) = 0$ is equal to zero on $\partial\sigma$. Thus, $S_{v_{P_\alpha}}(x',t) = \overline{S_{v_{P_\alpha}}}(x')$, and since $S_{v_{P_\alpha}}(x',0) = S_{v_{P_\alpha}}(x',2\pi) = 0$, we conclude that $S_{v_{P_\alpha}}(x',t) = 0$ for almost every $(x',t) \in \sigma \times [0,2\pi]$, which implies $\int_t^{2\pi} v_{P_\alpha}(x',\tau)d\tau = 0$ for almost every x' and t. Consequently, $v_{P_\alpha}(x',t) = 0$.

From identity (3.4.45), it now follows that

$$\int_0^{2\pi} S_s(t)\int_\sigma \partial_t\eta(x',t)dt = 0 \quad \forall\eta \in \mathcal{V}_1.$$

Taking $\eta = b(t)u_0(x')$, where $b \in W^{1,2}_{\wp}(0,2\pi)$ is arbitrary and $u_0(x')$ is the solution of problem (3.1.2), with $\int\limits_\sigma u_0(x')dx' = \kappa_0 \neq 0$, we obtain the expression

$$\int_0^{2\pi} S_s(t)\int_\sigma \partial_t\eta(x',t)dt = \int_0^{2\pi} S_s(t)b'(t)dt = 0 \quad \forall b \in W^{1,2}_{\wp}(0,2\pi).$$

Thus, it follows that the functional $s(t) = 0$. This concludes the proof. □

Remark 3.4.9 In general, if the flux function $F(t)$ belongs only to the space $L^2(0,2\pi)$, the regularity of the solution to problem (3.3.7) does not exceed the level specified in the given definition. To illustrate this point, consider the following example:

Let

$$F(t) = \sum_{n=2}^{\infty} \frac{1}{\sqrt{n}\ \ln^{\frac{1}{2}+\frac{1}{2}\varepsilon} n}\sin(nt), \tag{3.4.77}$$

where $\varepsilon > 0$ is an arbitrary constant. It can be shown that the L^2-norm of F on the interval $(0,2\pi)$ is finite:

$$\|F\|^2_{L^2(0,2\pi)} = \sum_{n=2}^{\infty} \frac{1}{n\ \ln^{1+\varepsilon} n} < \infty.$$

However, for any $\delta > 0$, the following series diverges:

$$\sum_{n=2}^{\infty} \frac{n^{2\delta}}{n\ \ln^{1+\varepsilon} n} = \infty.$$

As a result, F belongs to the space $L^2_\sharp(0, 2\pi)$, but it is not in $W^{\delta,2}(0, 2\pi)$ for any $\delta > 0$.

Now, let $\sigma = (0, \sqrt{2})$. By solving problem (3.3.7) using a Fourier series representation, we find that the solutions satisfy the following properties:

$$v_{P_\alpha} \in L^2(\sigma \times (0, 2\pi)), \quad \nabla S_{v_{P_\alpha}} \in L^2(\sigma \times (0, 2\pi)),$$

However, for $\delta > 0$, v_{P_α} does not have enough regularity to be in $W^{\delta,2}(\sigma \times (0, 2\pi))$.

3.5 Asymptotic Analysis of Poiseuille Flow in a Thin Pipe

In this section, we investigate the asymptotic behavior of Poiseuille flow in a thin pipe denoted by $\Pi_\varepsilon = \{x \in \mathbb{R}^3 : x' \in \sigma_\varepsilon = \varepsilon\sigma, \ x_3 \in \mathbb{R}\}$. Here, ε is a small positive parameter ($\varepsilon > 0$), and $\sigma \subset \mathbb{R}^2$ represents a bounded domain.

We consider an initial boundary value problem for the Navier–Stokes system within Π_ε:

$$\begin{cases} \partial_t \mathbf{u} - \nu\Delta\mathbf{u} + (\mathbf{u}\cdot\nabla)\mathbf{u} + \nabla p = \mathbf{f}, & (x,t) \in \Pi_\varepsilon^T, \\ \operatorname{div} \mathbf{u} = 0, & (x,t) \in \Pi_\varepsilon^T, \\ \mathbf{u}(x,t) = 0, & x \in \partial\Pi_\varepsilon,\ t \in (0,T), \\ \mathbf{u}(x,0) = \mathbf{u}_0(x),\ x \in \Pi_\varepsilon, & \end{cases} \tag{3.5.1}$$

with the prescribed flux condition:

$$\int_{\sigma_\varepsilon} u_3(x', x_3, t)\, dx' = F(t). \tag{3.5.2}$$

We assume that $\mathbf{u}_0 = (0, 0, u_{03})$ and $\mathbf{f} = (0, 0, f_3)$, where $u_{03} = u_{03}(x')$ and $f_3 = f_3(x', t)$ are independent of x_3. Finally, we assume that the necessary compatibility condition holds:

$$\int_{\sigma_\varepsilon} u_{03}(x')\, dx' = F(0). \tag{3.5.3}$$

We introduce the following scaling corresponding to the Reynolds number of order ε:

$$u_3 \sim 1; \quad f_3 \sim \varepsilon^{-2}; \quad F \sim \varepsilon^2.$$

Suppose that $f_3 \in L^2(\sigma_\varepsilon \times (0,\infty))$, $u_3 \in \mathring{W}^{1,2}(\sigma_\varepsilon)$, and $F \in W^{1,2}(0,\infty)$. The Poiseuille solution of problem (3.5.1), (3.5.2) has the same form as in Sect. 3.2. It is the unique solution $\big(u_{P_\alpha}(x',t), \alpha(t)\big) \in \mathring{W}^{(2,1),2}_{\partial\sigma_\varepsilon}(\sigma_\varepsilon \times (0,\infty)) \times L^2(0,\infty)$ of the inverse problem on the cross-section σ_ε (see Theorem 3.2.4):

$$\begin{cases} \partial_t u_{P_\alpha}(x',t) - \nu\Delta' u_{P_\alpha}(x',t) = \alpha(t) + f_3(x',t), \ (x',t) \in \sigma_\varepsilon^T, \\ u_{P_\alpha}(x',t) = 0, \quad x' \in \partial\sigma_\varepsilon,\ t \in (0,T), \\ u_{P_\alpha,t}(x',0) = u_{03}(x'), \quad x' \in \sigma_\varepsilon, \\ \int\limits_{\sigma_\varepsilon} u_{P_\alpha}(x',t)\,dx' = F(t), \quad t \in (0,T). \end{cases} \tag{3.5.4}$$

The solution $u_{P_\alpha}(x',t)$ can be represented as a sum:

$$u_{P_\alpha}(x',t) = u_h^{(1)}(x',t) + u_{P_\alpha}^{(2)}(x',t),$$

where $u_h^{(1)}(x',t)$ is the solution of the initial boundary value problem for the heat equation:

$$\begin{cases} \partial_t u_h^{(1)}(x',t) - \nu\Delta' u_h^{(1)}(x',t) = f_3(x',t), \ (x',t) \in \sigma_\varepsilon^T, \\ u_h^{(1)}(x',t) = 0, \quad x' \in \partial\sigma_\varepsilon,\ t \in (0,T), \\ u_h^{(1)}(x',0) = u_{03}(x'), \quad x' \in \sigma_\varepsilon, \end{cases} \tag{3.5.5}$$

and $\big(u_{P_\alpha}^{(2)}(x',t), \alpha(t)\big)$ is the solution of the inverse problem with the homogeneous initial condition:

$$\begin{cases} \partial_t u_{P_\alpha}^{(2)}(x',t) - \nu\Delta' u_{P_\alpha}^{(2)}(x',t) = \alpha(t), \quad (x',t) \in \sigma_\varepsilon^T, \\ u_{P_\alpha}^{(2)}(x',t) = 0, \quad x' \in \partial\sigma_\varepsilon,\ t \in (0,T), \\ u_{P_\alpha}^{(2)}(x',0) = 0, \quad x' \in \sigma_\varepsilon, \\ \int\limits_{\sigma_\varepsilon} u_{P_\alpha}^{(2)}(x',t)\,dx' = H(t),\ t \in (0,T). \end{cases} \tag{3.5.6}$$

Here, $H(t) = F(t) - \int_{\sigma_\varepsilon} u_h^{(1)}(x',t)\,dx'$, and from (3.5.3), we have

$$H(0) = 0. \tag{3.5.7}$$

Let the right-hand sides of problem (3.5.4) are defined by:

$$f_3(x',t) = \widetilde{f}(\frac{x'}{\varepsilon},t), \quad u_{03}(x') = \widetilde{a}(\frac{x'}{\varepsilon}), \tag{3.5.8}$$

where $y' = \frac{x'}{\varepsilon}$.

To proceed, we define the functions $\widetilde{f_3}(y',t)$, $\widetilde{a}(y')$, and $F(t)$ as power series expansions with respect to ε:

$$\begin{gathered} \widetilde{f_3}(y',t) = \sum_{j=0}^{J} \varepsilon^{j-2}\widetilde{f_j}(y',t), \quad \widetilde{a}(y') = \sum_{j=0}^{J} \varepsilon^{j}\widetilde{a}_j(y'), \\ F(t) = \sum_{j=0}^{J} \varepsilon^{j+2}F_j(t). \end{gathered} \tag{3.5.9}$$

Let J be a positive integer. Assume that the following regularity hypothesis holds:

$$\begin{gathered} \frac{\partial^s \widetilde{f_j}}{\partial t^s} \in L^2(\sigma \times (0,\infty)), \quad s = 0,1,\ldots,[\frac{J-j}{2}], \; j = 0,1,\ldots,J, \\ F_j \in W^{[\frac{J-j}{2}],2}(0,\infty), \quad \widetilde{a}_j \in \mathring{W}^{1,2}(\sigma), \quad j = 0,1,\ldots,J. \end{gathered} \tag{H_J}$$

Here the brackets $[\cdot]$ stand for the integer part of a number and functions $\widetilde{f_j}$, F_j and $\widetilde{a}_j$ are independent of ε.

3.5.1 Asymptotics of the Solution to the Heat Equation

This section discusses the asymptotics of the solution to the heat equation (3.5.5). After the change of variables $y' = \frac{x'}{\varepsilon}$, Eq. (3.5.5) takes the form

$$\begin{cases} \partial_t \widetilde{u}_h^{(1)}(y',t) - \frac{\nu}{\varepsilon^2}\Delta'_{y'}\widetilde{u}_h^{(1)}(y',t) = \widetilde{f}(y',t), \; (y',t) \in \sigma^T, \\ \qquad \widetilde{u}_h^{(1)}(y',t) = 0, \qquad y' \in \partial\sigma = 0, \; t \in (0,T) \\ \qquad \widetilde{u}_h^{(1)}(y',0) = \widetilde{u}_{03}(y'), \; y' \in \sigma, \end{cases} \tag{3.5.10}$$

where $\widetilde{u}_h^{(1)}(y',t) = u_h^{(1)}(\varepsilon y',t)$, and similar notation is used for other functions.

The goal is to find an asymptotic solution to problem (3.5.10) by employing a sum of two ansatzes. The first one is the regular part of the asymptotic expansion given as:

$$\widetilde{w}^{[J]}(y', t) = \sum_{j=0}^{J} \varepsilon^j \widetilde{w}_j(y', t), \tag{3.5.11}$$

The second ansatz introduces the boundary-layer-in-time term (we make a change of variables $\tau = \dfrac{t}{\varepsilon^2}$):

$$W^{[J]}(y', \tau) = \sum_{j=0}^{J} \varepsilon^j W_j(y', \tau). \tag{3.5.12}$$

It is observed that if the initial data vanish and the external force is zero in a neighborhood of $t = 0$, then the asymptotic expansion of the solution does not contain the boundary-layer-in-time correctors. This property also holds for the time-periodic problem.

Regular Part of the Asymptotic Expansion Rewriting problem (3.5.10) as:

$$\begin{cases} -\nu\Delta'_{y'}\widetilde{u}_h^{(1)}(y', t) = \varepsilon^2 \widetilde{f}(y', t) - \varepsilon^2 \partial_t \widetilde{u}_h^{(1)}(y', t), & (y', t) \in \sigma, \\ \widetilde{u}_h^{(1)}(y', t) = 0, & y' \in \partial\sigma, \\ \widetilde{u}_h^{(1)}(y', 0) = \widetilde{a}(y'), & y' \in \sigma, \end{cases} \tag{3.5.13}$$

we then considered it for every $t \in (0, T)$, with $T \leq +\infty$, as a stationary problem where the right-hand side depends on the parameter t. By substituting the sum (3.5.11) into Eq. (3.5.13), collecting coefficients for the same powers of ε, and applying appropriate boundary conditions, a recurrent sequence of problems for $\widetilde{w}_j(y', t)$ is derived. This sequence is represented as:

$$\begin{cases} -\nu\Delta'_{y'}\widetilde{w}_j(y', t) = \widetilde{f}_j(y', t), & y' \in \sigma, \\ \widetilde{w}_j(y', t) = 0, & y' \in \partial\sigma, \quad j = 0, 1, \end{cases} \tag{3.5.14}$$

and

$$\begin{cases} -\nu\Delta'_{y'}\widetilde{w}_j(y', t) = \widetilde{f}_j(y', t) - \partial_t \widetilde{w}_{j-2}(y', t), & y' \in \sigma, \\ \widetilde{w}_j(y', t) = 0, & y' \in \partial\sigma, \quad j = 2, 3, \dots, J. \end{cases} \tag{3.5.15}$$

The existence and uniqueness of solutions to problems (3.5.14) and (3.5.15) are established based on known results concerning the Poisson equation (e.g., [61, 103]).

Regularity Conditions To ensure regularity in $L^2(\sigma \times (0,\infty))$ with respect to t up to the order $l \geq 1$ for solutions $\widetilde{w}_{(j)}(y',t)$, specific regularity conditions are required. We assume that that there exist the following derivatives belonging to $L^2(\sigma \times (0,\infty))$:

$$\frac{\partial^l \widetilde{f}_j(y',t)}{\partial t^l}, \quad j = 0, 1,$$

$$\frac{\partial^l \widetilde{f}_j(y',t)}{\partial t^l}, \frac{\partial^{l+1} \widetilde{f}_{j-2}(y',t)}{\partial t^{l+1}}, \ldots, \frac{\partial^{l+s} \widetilde{f}_{j-2s}(y',t)}{\partial t^{l+s}}, \quad j = 2, 3, \ldots, J, \tag{3.5.16}$$

where $j-2s = 0$ or $j-2s = 1$. It means that due to the condition (H_J), there exist solutions $\widetilde{w}_j$ of problems(3.5.14), (3.5.15) for $j \leq J$.

Boundary Layer In cases where the initial data vanishes and all functions $\widetilde{f}_r(y',t)$, $r \leq J$, are zero in a neighborhood of $t = 0$, the solutions $\widetilde{w}_j(y',t)$ of problems (3.5.14) and (3.5.15) satisfy the initial conditions $\widetilde{w}_j(y',0) = 0$. However, for general cases, the initial condition is not satisfied by ansatz (3.5.11); a boundary layer needs to be constructed near the point $t = 0$. This is achieved by introducing the fast-time variable $\tau = \dfrac{t}{\varepsilon^2}$ in (3.5.10) and setting $\widetilde{f} = 0$, we considering the resulting problem:

$$\begin{cases} \dfrac{1}{\varepsilon^2}\partial_\tau W(y',\tau) - \dfrac{\nu}{\varepsilon^2}\Delta'_{y'} W(y',\tau) = 0, & y' \in \sigma,\ \tau \in (0, T/\varepsilon^2), \\ W(y',\tau) = 0, & y' \in \partial\sigma,\ \tau \in (0, T/\varepsilon^2), \\ W(y',0) = \widetilde{u}_{03}(y'),\ y' \in \sigma, & \end{cases} \tag{3.5.17}$$

where $W(y',\tau) = w(\varepsilon y', \varepsilon^2 \tau)$.

A boundary-layer-in-time asymptotic expansion for $W(y',\tau)$ is sought in the form of (3.5.12), where $W_j(y',\tau)$ are solutions to the following problems:

$$\begin{cases} \partial_\tau W_j(y',\tau) - \nu\Delta'_{y'} W_j(y',\tau) = 0, & y' \in \sigma,\ \tau \in (0, T/\varepsilon^2), \\ W_j(y',\tau) = 0, & y' \in \partial\sigma,\ \tau \in (0, T/\varepsilon^2), \\ W_j(y',0) = \widetilde{a}_j(y') - \widetilde{w}_j(y',0) \equiv \widetilde{b}_j(y'),\ y' \in \sigma. & \end{cases} \tag{3.5.18}$$

The existence and regularity of solutions to problems (3.5.18) are widely known and have been extensively studied in the literature (e.g., [61, 103]). The following theorem presents the corresponding results:

Theorem 3.5.1 *Let* $T = +\infty$*,* $\partial\sigma$ *is twice continuously differentiable, and* $\widetilde{b}_j$ *belongs to the space* $\overset{\circ}{W}{}^{1,2}(\sigma)$*. Under these conditions, the problem* (3.5.18) *has a unique solution* $W_j \in \overset{\circ}{W}{}^{(2,1),2}_{\partial\sigma}(\sigma \times (0,\infty))$*. Furthermore, the solution satisfies the following estimate:*

$$\|W_j\|_{W^{(2,1),2}(\sigma\times(0,\infty))} \leq c\|\widetilde{b}_j\|_{W^{1,2}(\sigma)}, \tag{3.5.19}$$

where c is a positive constant.

Moreover, as $\tau \to \infty$*,* $W_j(y', \tau)$ *exponentially tends to zero, as described by the estimate:*

$$\begin{aligned}&\max_{\tau\in[0,\infty)} \Big[\exp(\gamma_*\tau)\Big(\int_\sigma |W_j(y',\tau)|^2\,dy' + \nu\int_\sigma |\nabla' W_j(y',\tau)|^2\,dy'\Big)\Big]\\ &+\int_0^\infty \exp(\gamma_*\tau)\int_\sigma \Big|\frac{\partial W_j(y',\tau)}{\partial\tau}\Big|^2 dy'd\tau \leq c\|\widetilde{b}_j\|^2_{W^{1,2}(\sigma)}.\end{aligned} \tag{3.5.20}$$

Here, $\gamma_* = \min\{\lambda_1, 1\}$*, where* λ_1 *represents the first eigenvalue of the Dirichlet Laplacian in the domain* σ*.*

Remark 3.5.1 If the solution is time-periodic, i.e., $w(x', 0) = w(x', 2\pi)$, then it is not necessary to construct a boundary layer. In such cases, the solutions $\widetilde{w}_j(y', t)$ of problems (3.5.14) and (3.5.15) will be periodic in t, given that $\widetilde{f}_j(y', t)$ and its derivatives are periodic functions.

3.5.2 *Inverse Problem*

In this section, we address the inverse problem (3.5.6). The solution to problem (3.5.5) is represented as a series expansion in powers of ε. Therefore, we have the expression:

$$\begin{aligned}H(t) &= \sum_{j=0}^J \varepsilon^{j+2}F_j(t) - \sum_{j=0}^J \int_{\sigma_\varepsilon} \varepsilon^j \left(\widetilde{w}_j\left(\frac{x'}{\varepsilon}, t\right) + W_j\left(\frac{x'}{\varepsilon}, \frac{t}{\varepsilon^2}\right)\right) dx'\\ &= \sum_{j=0}^J \varepsilon^{j+2}F_j(t) - \sum_{j=0}^J \int_\sigma \varepsilon^{j+2}\left(\widetilde{w}_j(y', t) + W_j(y', \frac{t}{\varepsilon^2})\right) dy'\\ &= \sum_{j=0}^J \varepsilon^{j+2}\left(h_j(t) + H_j\left(\frac{t}{\varepsilon^2}\right)\right).\end{aligned}$$

Here,

$$h_j(t) = F_j(t) - \int\limits_{\sigma} \widetilde{w}_j(y',t)dy', \quad H_j(\tau) = -\int\limits_{\sigma} W_j(y',\tau)dy'.$$

In order to satisfy condition (3.5.7), we require the compatibility condition:

$$h_j(0) + H_j(0) = 0, \quad j = 0, 1, \ldots, J,$$

which is equivalent to the following expression:

$$F_j(0) = \int\limits_{\sigma} a_j(y')dy', \quad j = 0, 1, \ldots, J. \tag{3.5.21}$$

Performing the change of variables $y' = \frac{x'}{\varepsilon}$, we transform problem (3.5.6) into the following form:

$$\begin{cases} -\nu\Delta'_{y'}\widetilde{u}^{(2)}_{P_\alpha}(y',t) = \varepsilon^2\alpha(t) - \varepsilon^2\partial_t\widetilde{u}^{(2)}_{P_\alpha}(y',t), \; y' \in \sigma, & \\ \widetilde{u}^{(2)}_{P_\alpha}(y',t) = 0, & (y,t)' \in \partial\sigma^T, \\ \widetilde{u}^{(2)}_{P_\alpha}(y',0) = 0, & y' \in \sigma, \; t \in (0,T), \\ \int\limits_{\sigma} \widetilde{u}^{(2)}_{P_\alpha}(y',t)dy' = \varepsilon^{-2}H(t), & t \in (0,T). \end{cases} \tag{3.5.22}$$

Next, we seek the asymptotic expansion of the solution to problem (3.5.22) in the form of two ansatzes. The first ansatz, referred to as "the regular part" of the asymptotic decomposition, depends on (y', t) and is expressed as follows:

$$\widetilde{v}^{[J]}(y',t) = \sum_{j=0}^{J} \varepsilon^j \widetilde{v}_j(y',t), \quad s^{[J]}(t) = \sum_{j=0}^{J} \varepsilon^{j-2} s_j(t), \tag{3.5.23}$$

whereas the second ansatz corresponds to the boundary layer and depends on $(y', \tau) = (y', \frac{t}{\varepsilon^2})$:

$$V^{[J]}(y',\tau) = \sum_{j=0}^{J} \varepsilon^j V_j(y',\tau), \quad S^{[J]}(\tau) = \sum_{j=0}^{J} \varepsilon^{j-2} S_j(\tau). \tag{3.5.24}$$

Regular Part of the Asymptotic Expansion Neglecting the initial condition and terms depending on $\tau = \frac{t}{\varepsilon^2}$, we substitute series (3.5.23) into (3.5.22) and collect coefficients for the same powers of ε. This leads to the following equations:

$$\begin{cases} -\nu\Delta'_{y'}\widetilde{v}_j(y',t) = s_j(t), & y' \in \sigma, \\ \widetilde{v}_j(y',t) = 0, & y' \in \partial\sigma, \\ \int\limits_\sigma \widetilde{v}_j(y',t)dy' = h_j(t), \quad j = 0, 1; \end{cases} \tag{3.5.25}$$

$$\begin{cases} -\nu\Delta'_{y'}\widetilde{v}_j(y',t) = s_j(t) - \partial_t\widetilde{v}_{j-2}(y',t), & y' \in \sigma, \\ \widetilde{v}_j(y',t) = 0, & y' \in \partial\sigma, \\ \int\limits_\sigma \widetilde{v}_j(y',t)dy' = h_j(t), \quad j = 2, \dots, J. \end{cases} \tag{3.5.26}$$

For $j = 0$ and $j = 1$, the solutions $\widetilde{v}_j(y', t)$ can be expressed as

$$\widetilde{v}_j(y',t) = \kappa_0^{-1}h_j(t)u_0(y'),$$

where $u_0(y')$ is the solution of the Dirichlet problem for the Poisson equation (3.1.2). We choose $s_j(t)$ to satisfy the flux conditions:

$$s_j(t) = \kappa_0^{-1}h_j(t), \quad j = 0, 1, \tag{3.5.27}$$

where

$$\kappa_0 = \int\limits_\sigma u_0(y')dy' > 0$$

depends solely on σ. Thus, we have:

$$\widetilde{v}_j(y',t) = \kappa_0^{-1}h_j(t)u_0(y'), \quad j = 0, 1. \tag{3.5.28}$$

For $j = 2, \dots, J$, we use induction to find:

$$\widetilde{v}_{2k}(y',t) = s_{2k}(t)u_0(y') + \frac{ds_{2k-2}(t)}{dt}u_2(y') + \ldots + \frac{d^k s_0(t)}{dt^k}u_{2k}(y'), \tag{3.5.29}$$

$$\widetilde{v}_{2k+1}(y',t) = s_{2k+1}(t)u_0(y') + \frac{ds_{2k-1}(t)}{dt}u_2(y') + \ldots + \frac{d^k s_1(t)}{dt^k}u_{2k}(y'), \tag{3.5.30}$$

where $2k = j$ or $2k+1 = j$ and $k \geq 1$. The functions $u_{2r}(y')$ are defined recursively as the solutions to the problems:

$$\begin{cases} -\nu\Delta'_{y'}u_{2r}(y') = -u_{2r-2}(y'), & y' \in \sigma, \\ u_{2r}(y') = 0, & y' \in \partial\sigma, \quad r = 1, 2, \dots. \end{cases} \tag{3.5.31}$$

Substituting formulas (3.5.29) and (3.5.30) into the flux conditions, we find the coefficients $s_j(t)$ recursively as follows:

$$
\begin{aligned}
s_0(t) &= \kappa_0^{-1} h_0(t),\\
s_2(t) &= \kappa_0^{-1} h_2(t) - \kappa_0^{-2}\kappa_2 \tfrac{d}{dt} h_0(t),\\
s_4(t) &= \kappa_0^{-1} h_4(t) - \kappa_0^{-2}\kappa_2 \tfrac{d}{dt} h_2(t) + \kappa_0^{-3}\kappa_2^2 \tfrac{d}{dt} h_0(t) - \kappa_0^{-2}\kappa_4 \tfrac{d^2}{dt^2} h_0(t),\\
&\dots\\
s_{2k}(t) &= \kappa_0^{-1} h_{2k}(t) - \kappa_0^{-1}\kappa_2 \frac{ds_{2k-2}(t)}{dt} - \dots - \kappa_0^{-1}\kappa_{2k}\frac{d^k s_0(t)}{dt^k};
\end{aligned}
\tag{3.5.32}
$$

$$
\begin{aligned}
s_1(t) &= \kappa_0^{-1} h_1(t),\\
s_3(t) &= \kappa_0^{-1} h_3(t) - \kappa_0^{-2}\kappa_2 \tfrac{d}{dt} h_1(t),\\
&\dots\\
s_{2k+1}(t) &= \kappa_0^{-1} h_{2k+1}(t) - \kappa_0^{-1}\kappa_2 \frac{ds_{2k-1}(t)}{dt} - \dots - \kappa_0^{-1}\kappa_{2k}\frac{d^k s_1(t)}{dt^k},
\end{aligned}
\tag{3.5.33}
$$

where

$$\kappa_{2r} = \int_\sigma u_{2r}(y')dy'.$$

Let us demonstrate that $\kappa_{2r} \neq 0$. It can be deduced from (3.1.2) that

$$\kappa_0 = \nu \int_\sigma |\nabla' u_0|^2 dy' > 0.$$

Let us consider the case where $r = 1$. Utilizing (3.1.2) and performing two integrations by parts, we obtain

$$\kappa_2 = \int_\sigma u_2 dy' = -\nu \int_\sigma u_2 \Delta' u_0 dy' = -\nu \int_\sigma \Delta' u_2 u_0 dy' = -\int_\sigma |u_0|^2 dy' < 0.$$

Similarly, using (3.5.31) and (3.1.2), we arrive at

$$\kappa_{2r+2} = -\int_\sigma u_{2r} u_0 dy'.$$

For the case when $r = 2m$, we have

$$\kappa_{4m+2} = -\int\limits_{\sigma} u_{4m} u_0 dy' = -\nu \int\limits_{\sigma} u_{4m} \Delta' u_2 dy' = -\nu \int\limits_{\sigma} \Delta' u_{4m} u_2 dy'$$
$$= -\int\limits_{\sigma} u_{4m-2} u_2 dy' = \ldots = -\int\limits_{\sigma} |u_{2m}|^2 dy' < 0.$$

For the case when $r = 2m - 1$, we have

$$\kappa_{4m} = -\int\limits_{\sigma} u_{4m-2} u_0 dy' = \ldots = -\int\limits_{\sigma} u_{2m-2} u_{2m} dy' = -\nu \int\limits_{\sigma} \Delta' u_{2m} u_{2m} dy'$$
$$= \nu \int\limits_{\sigma} |\nabla' u_{2m}|^2 dy' > 0.$$

Thus, we can conclude that

$$\kappa_{4m} > 0, \quad \kappa_{4m+2} < 0, \quad m = 0, 1, 2, \ldots. \tag{3.5.34}$$

By induction, we can show that

$$s_{2k}(t) = a_0 h_{2k}(t) - \sum_{i=1}^{k} a_{2i} \frac{d^i}{dt^i} h_{2k-2i}(t), \; k = 0, 1, \ldots, \tag{3.5.35}$$

$$s_{2k+1}(t) = a_0 h_{2k+1}(t) - \sum_{i=1}^{k} a_{2i} \frac{d^i}{dt^i} h_{2k+1-2i}(t), \; k = 0, 1, \ldots, \tag{3.5.36}$$

where the coefficients are recursively computed as follows:

$$a_0 = \kappa_0^{-1}, \quad a_{2i} = \kappa_0^{-1}(-\sum_{j=1}^{i-1} \kappa_{2j} a_{2i-2j} + \kappa_{2i} a_0), \; i = 1, 2, \ldots, k. \tag{3.5.37}$$

Regularity Conditions Based on Eqs. (3.5.29), (3.5.30), (3.5.35), and (3.5.36), the functions $\widetilde{v}_j(y', t)$ and $s_j(t)$ possess l derivatives with respect to t, provided that the fluxes $h_j(t)$, where $j = 0, 1, 2, \ldots, J$, satisfy the following regularity conditions:

$$\begin{aligned} &\textit{For even } j = 2k: \; h_{2r} \textit{ have derivatives in } L^2(0,\infty) \textit{ up to} \\ &\qquad \textit{the orders } l + k - r, \; r = 0, 1, \ldots, k, \; k \geq 0; \\ &\textit{For odd } j = 2k+1: \; h_{2r+1}(t) \textit{ have derivatives in } L^2(0,\infty) \textit{ up} \\ &\qquad \textit{to the orders } l + k - r, \; r = 0, 1, \ldots, k, \; k \geq 0. \end{aligned} \tag{3.5.38}$$

Since

$$h_j(t) = F_j(t) - \int\limits_\sigma \widetilde{w}_j(y',t)dy',$$

the regularity conditions (3.5.38) require the regularity of $F_j(t)$ and $\widetilde{w}_j(y',t)$. We conclude from (3.5.14), (3.5.15) that F_j and $\widetilde{f}_j(y',t)$, $j = 0, 1, 2, \dots, J$ must satisfy the following regularity conditions:

$$\begin{gathered} \textit{For even } j = 2k: \ F_{2r}, \widetilde{f}_{2r} \textit{ must have derivatives with} \\ \textit{respect to } t \textit{ up to the orders } l + k - r, \ r = 0, 1, \dots, k, \ k \geq 0; \\ \textit{For odd } j = 2k+1: \ F_{2r+1}, \widetilde{f}_{2r+1} \textit{ must have derivatives with} \\ \textit{respect to } t \textit{ up to the orders } l + k - r, \ r = 0, 1, \dots, k, \ k \geq 0. \end{gathered} \tag{3.5.39}$$

These derivatives belong to $L^2(0,\infty)$ *and* $L^2(\sigma \times (0,\infty))$, *respectively.*

It is evident that the existence of a solution $(\widetilde{v}_j, s_j)$ for problems (3.5.14) and (3.5.15) is ensured by the fulfillment of condition (H_J) for all $j \leq J$.

Boundary Layer To construct the boundary layer, we perform a change of variables in (3.5.22) as $\tau = t/\varepsilon^2$ and $y' = x'/\varepsilon$, then substitute the series (3.5.24) into the obtained equation, and group terms with the same powers of ε. This process leads us to the following problems for the coefficients $\big(V_j(y',\tau), S_j(\tau)\big)$:

$$\begin{cases} \partial_\tau V_j(y',\tau) - \nu \Delta'_{y'} V_j(y',\tau) = S_j(\tau), & y' \in \sigma, \ \tau \in (0, T/\varepsilon^2) \\ V_j(y',\tau) = 0, & y' \in \partial\sigma, \ \tau \in (0, T/\varepsilon^2), \\ V_j(y',0) = -\widetilde{v}_j(y',0), & y' \in \sigma, \\ \int\limits_\sigma V_j(y',\tau)dy' = H_j(\tau), & \tau \in (0, T/\varepsilon^2). \end{cases} \tag{3.5.40}$$

It is evident that the following compatibility condition

$$H_j(0) = -\int\limits_\sigma \widetilde{v}_j(y',0)dy' \tag{3.5.41}$$

is necessary for the solvability of problem (3.5.40). Note that this condition coincides with the compatibility condition (3.5.21).

The solvability and decay properties of the solution to inverse problem (3.5.40) were studied in Sect. 3.2 (see Theorems 3.2.4 and 3.2.7).

From the construction we see that $\widetilde{v}_j(\cdot,0) \in \overset{\circ}{W}{}^{1,2}(\sigma)$. Then, from Theorem 3.5.1, it follows that $H_j \in \mathcal{W}^{1,2}_{\gamma_*/2}(0,\infty)$, where $\gamma_* = \min\{\lambda_1, 1\}$. Additionally, the compatibility condition (3.5.41) holds, which coincides with the condition (3.5.21). Consequently, Theorems 3.2.4 and 3.2.7 imply the existence of a unique solution $(V_j, S_j) \in \mathring{W}_{\partial\sigma}^{(1,1),2}(\sigma \times (0,\infty)) \times L^2(0,\infty)$ of the problem (3.5.40). Specificaly, the following theorem holds:

Theorem 3.5.2 *There exists a unique solution* $(V_j, S_j) \in \mathring{W}_{\partial\sigma}^{(1,1),2}(\sigma \times (0,\infty)) \times L^2(0,\infty)$ *of problem* (3.5.40) *such that*

$$\begin{aligned}\max_{\tau\in[0,\infty]} \|V_j(\cdot,\tau)\|_{W^{1,2}(\sigma)} + \|V_j\|_{W^{(1,1),2}(\sigma\times(0,\infty))} + \|S_j\|_{L^2(0,\infty)} \\ \leq c\Big(\|H_j\|_{W^{1,2}(0,\infty)} + \|\widetilde{v}_j(\cdot,0)\|_{W^{1,2}(\sigma)}\Big).\end{aligned} \tag{3.5.42}$$

If $\partial\sigma \in C^2$, *then* $V_j \in W^{(2,1),2}(\sigma \times (0,\infty))$ *and*

$$\|V_j\|_{W^{(2,1),2}(\sigma\times(0,\infty))} \leq c\Big(\|H_j\|_{W^{1,2}(0,\infty)} + \|\widetilde{v}(\cdot,0)\|_{W^{1,2}(\sigma)}\Big). \tag{3.5.43}$$

Furthermore,

$$\begin{aligned}&\max_{\tau\in[0,\infty]}\Big[\exp(\gamma_*\tau)\Big(\int\limits_\sigma |V_j(y',\tau)|^2\,dy' + \nu\int\limits_\sigma |\nabla' V_j(y',\tau)|^2\,dy'\Big)\Big] \\ &+\int\limits_0^\infty \exp(\gamma_*\tau)\int\limits_\sigma \Big|\frac{\partial V_j(y',\tau)}{\partial\tau}\Big|^2 dy'd\tau + \int\limits_0^\infty \exp(\gamma_*\tau)|S_j(\tau)|^2\,d\tau \\ &\leq c\Big(\|H_j\|^2_{\mathcal{W}^{1,2}_{\gamma_*/2}(0,\infty)} + \|\widetilde{v}_j(\cdot,0)\|^2_{W^{1,2}(\sigma)}\Big).\end{aligned} \tag{3.5.44}$$

3.5.3 Justification of the Asymptotic Expansion

By taking together the ansatzes (3.5.11), (3.5.12), (3.5.23), and (3.5.24), we define the following expansions:

$$\begin{aligned}\widehat{w}^{[J]}(x',t) &= \sum_{j=0}^J \varepsilon^j\left(\widetilde{w}_j\left(\frac{x'}{\varepsilon},t\right) + W_j\left(\frac{x'}{\varepsilon},\frac{t}{\varepsilon^2}\right)\right), \\ \widehat{v}^{[J]}(x',t) &= \sum_{j=0}^J \varepsilon^j\left(\widetilde{v}_j\left(\frac{x'}{\varepsilon},t\right) + V_j\left(\frac{x'}{\varepsilon},\frac{t}{\varepsilon^2}\right)\right), \\ \alpha^{[J]}(t) &= \sum_{j=0}^J \varepsilon^{j-2}\left(s_j(t) + S_j\left(\frac{t}{\varepsilon^2}\right)\right), \\ u^{[J]}_{P_\alpha}(x',t) &= \widehat{w}^{[J]}(x',t) + \widehat{v}^{[J]}(x',t),\end{aligned} \tag{3.5.45}$$

Next, we proceed to estimate the discrepancy and the remainder $(u_{P_\alpha} - u_{P_\alpha}^{[J]}, \alpha - \alpha^{[J]})$. Note that the pair $(u_{P_\alpha}^{[J]}, \alpha^{[J]})$ exactly satisfies the boundary and initial conditions, as well as the flux condition. Equations $(3.5.4)_1$ are satisfied with a residual term on the right-hand side given by

$$\varepsilon^{J-1}\left(\frac{\partial \widetilde{w}_{J-1}}{\partial t} + \frac{\partial \widetilde{v}_{J-1}}{\partial t}\right) + \varepsilon^{J}\left(\frac{\partial \widetilde{w}_{J}}{\partial t} + \frac{\partial \widetilde{v}_{J}}{\partial t}\right).$$

By applying estimate (3.2.37) (see Theorem 3.2.4), we obtain the following result:

Theorem 3.5.3 *Assume that the functions $\widetilde{f}_3$, $\widetilde{a}$, and F have the form* (3.5.8), (3.5.9) *and satisfy the condition (H_J) and the compatibility conditions* (3.5.21). *Then the following estimate holds:*

$$\max_{t\in[0,+\infty)} \|u_{P_\alpha}(\cdot,t) - u_{P_\alpha}^{[J]}(\cdot,t)\|_{W^{1,2}(\sigma_\varepsilon)} + \|u_{P_\alpha} - u_{P_\alpha}^{[J]}\|_{W^{(1,1),2}(\sigma_\varepsilon\times(0,\infty))}$$
$$+ \|\alpha - \alpha^{[J]}\|_{L^2(0,\infty)} \le C\varepsilon^{J-1}, \tag{3.5.46}$$

where the constant C is independent of ε.

Remark 3.5.2 The terms $W_j(\frac{x'}{\varepsilon}, \frac{t}{\varepsilon^2})$, $V_j(\frac{x'}{\varepsilon}, \frac{t}{\varepsilon^2})$, and $S_j(\frac{t}{\varepsilon^2})$ correspond to boundary layers with respect to the time variable, and they decay exponentially as $\frac{t}{\varepsilon^2} \to \infty$ with a rate of $O(e^{-c\frac{t}{\varepsilon^2}})$, where $c > 0$.

Remark 3.5.3 In the case where both the flux $F(t)$ and the right-hand side $f_3(x,t)$ are zero in the neighborhood of the point $t = 0$ (i.e., for $t \le t^*$, where $t^* > 0$), the asymptotic representation does not include the boundary layer part. Consequently, in this scenario, the asymptotic representation becomes a smooth function with respect to the x variables, and its regularity with respect to t is determined by the regularity of the flux $F(t)$ and the right-hand side $f_3(x,t)$ (see conditions (3.5.39)).

Remark 3.5.4 For the case of a time-periodic problem, all the results presented remain valid with the only difference being that the construction of boundary layers is unnecessary.

Remark 3.5.5 All the results obtained also hold (with appropriate adjustments in scaling) for the case of the viscous flow in a thin two-dimensional channel.

Chapter 4
Stokes Problem and Stokes Operator

4.1 Stationary Stokes Problem

Let Ω be a domain in $\mathbb{R}^n$, where $n = 2, 3$. We consider the homogeneous boundary value problem for the linear stationary Stokes system in Ω:

$$\begin{cases} -\nu\Delta\mathbf{u} + \nabla p = \mathbf{f}_0 + \sum\limits_{j=1}^{n} \dfrac{\partial \mathbf{f}_j}{\partial x_j}, & x \in \Omega, \\ \operatorname{div}\mathbf{u} = 0, & x \in \Omega, \\ \mathbf{u} = 0, & x \in \partial\Omega. \end{cases} \tag{4.1.1}$$

Definition 4.1.1 Let $\mathbf{f}_0 \in L^2(\Omega)$, $\mathbf{f}_j \in L^2(\Omega)$ for $j = 1, \ldots, n$. The function $\mathbf{u} \in \overset{\circ}{H}{}^{1,2}(\Omega)$ is called a weak solution of boundary value problem (4.1.1) if $\mathbf{u}$ satisfies the integral identity:

$$\nu \int\limits_{\Omega} \nabla\mathbf{u} \cdot \nabla\boldsymbol{\eta}\, dx = \int\limits_{\Omega} \mathbf{f}_0 \cdot \boldsymbol{\eta}\, dx - \int\limits_{\Omega} \sum_{j=1}^{n} \mathbf{f}_j \cdot \frac{\partial \boldsymbol{\eta}}{\partial x_j}\, dx \tag{4.1.2}$$

for every $\boldsymbol{\eta} \in \overset{\circ}{H}{}^{1,2}(\Omega)$.

The existence results that follow are well-known (see [66, 102]). For the reader's convenience, we present the proof.

Theorem 4.1.1 *Let Ω be a bounded domain with Lipschitz boundary. For any given functions $\mathbf{f}_0 \in L^2(\Omega)$, $\mathbf{f}_j \in L^2(\Omega)$ for $j = 1, \ldots, n$, problem* (4.1.1) *admits a unique weak solution $\mathbf{u} \in \overset{\circ}{H}{}^{1,2}(\Omega)$.*

G. Panasenko, K. Pileckas, *Multiscale Analysis of Viscous Flows in Thin Tube Structures*, Advances in Mathematical Fluid Mechanics,
https://doi.org/10.1007/978-3-031-54630-3_4

Moreover, there exists a unique pressure function $p \in \widehat{L}^2(\Omega)$ such that the pair $(\mathbf{u}, p)$ satisfies the following integral identity for all $\boldsymbol{\eta} \in \overset{\circ}{W}{}^{1,2}(\Omega)$:

$$\nu \int_\Omega \nabla \mathbf{u} \cdot \nabla \boldsymbol{\eta}\, dx - \int_\Omega p\, div \boldsymbol{\eta}\, dx = \int_\Omega \mathbf{f}_0 \cdot \boldsymbol{\eta}\, dx - \int_\Omega \sum_{j=1}^n \mathbf{f}_j \cdot \frac{\partial \boldsymbol{\eta}}{\partial x_j}\, dx. \tag{4.1.3}$$

Furthermore, the following estimate holds:

$$\|\mathbf{u}\|^2_{W^{1,2}(\Omega)} + \|p\|^2_{L^2(\Omega)} \le c \left(\|\mathbf{f}_0\|^2_{L^2(\Omega)} + \sum_{j=1}^n \|\mathbf{f}_j\|^2_{L^2(\Omega)} \right), \tag{4.1.4}$$

where c is a positive constant independent of the data.

Proof
By using the Cauchy-Schwarz-Bunyakovsky and Poincaré inequalities, we can estimate the right-hand side of (4.1.3) as follows:

$$\begin{aligned}
&\left| \int_\Omega \mathbf{f}_0 \cdot \boldsymbol{\eta}\, dx - \int_\Omega \sum_{j=1}^n \mathbf{f}_j \cdot \frac{\partial \boldsymbol{\eta}}{\partial x_j}\, dx \right| \\
&\quad \le c \left(\|\mathbf{f}_0\|_{L^2(\Omega)} \|\boldsymbol{\eta}\|_{L^2(\Omega)} + \sum_{j=1}^n \|\mathbf{f}_j\|_{L^2(\Omega)} \|\nabla \boldsymbol{\eta}\|_{L^2(\Omega)} \right) \\
&\quad \le c \left(\|\mathbf{f}_0\|_{L^2(\Omega)} + \sum_{j=1}^n \|\mathbf{f}_j\|_{L^2(\Omega)} \right) \|\nabla \boldsymbol{\eta}\|_{L^2(\Omega)},
\end{aligned}$$

where c is a positive constant. Hence, the expression $\int_\Omega \mathbf{f}_0 \cdot \boldsymbol{\eta}\, dx - \int_\Omega \sum_{j=1}^n \mathbf{f}_j \cdot \frac{\partial \boldsymbol{\eta}}{\partial x_j}\, dx$ defines a bounded linear functional on $\overset{\circ}{H}{}^{1,2}(\Omega)$. By the Riesz Representation Theorem, there exists a unique element $\mathbf{F} \in \overset{\circ}{H}{}^{1,2}(\Omega)$ such that

$$\int_\Omega \mathbf{f}_0 \cdot \boldsymbol{\eta}\, dx - \int_\Omega \sum_{j=1}^n \mathbf{f}_j \cdot \frac{\partial \boldsymbol{\eta}}{\partial x_j}\, dx = [\mathbf{F}, \boldsymbol{\eta}]_{\overset{\circ}{H}{}^{1,2}(\Omega)} \quad \forall \boldsymbol{\eta} \in \overset{\circ}{H}{}^{1,2}(\Omega),$$

and

$$\|\mathbf{F}\|_{\overset{\circ}{H}{}^{1,2}(\Omega)} \le c \left(\|\mathbf{f}_0\|_{L^2(\Omega)} + \sum_{j=1}^n \|\mathbf{f}_j\|_{L^2(\Omega)} \right). \tag{4.1.5}$$

Here, $[\mathbf{u}, \boldsymbol{\eta}]_{\mathring{H}^{1,2}(\Omega)} = \int_\Omega \nabla\mathbf{u} \cdot \nabla\boldsymbol{\eta}\, dx$ represents the inner product in the space $\mathring{H}^{1,2}(\Omega)$.

We rewrite the integral identity (4.1.2) as follows:

$$\nu[\mathbf{u}, \boldsymbol{\eta}]_{\mathring{H}^{1,2}(\Omega)} = [\mathbf{F}, \boldsymbol{\eta}]_{\mathring{H}^{1,2}(\Omega)} \quad \forall \boldsymbol{\eta} \in \mathring{H}^{1,2}(\Omega).$$

Thus, the weak solution $\mathbf{u}$ of problem (4.1.1) is given by

$$\mathbf{u} = \nu^{-1}\mathbf{F}. \tag{4.1.6}$$

Consider the linear functional

$$L(\boldsymbol{\eta}) = \nu \int_\Omega \nabla\mathbf{u} \cdot \nabla\boldsymbol{\eta}\, dx - \int_\Omega \mathbf{f}_0 \cdot \boldsymbol{\eta}\, dx + \int_\Omega \sum_{j=1}^{n} \mathbf{f}_j \cdot \frac{\partial \boldsymbol{\eta}}{\partial x_j}\, dx, \tag{4.1.7}$$

defined on $\mathring{W}^{1,2}(\Omega)$. Since $\mathbf{u}$ satisfies integral identity (4.1.2), we have

$$L(\boldsymbol{\eta}) = 0 \ \text{ for } \ \text{all } \ \boldsymbol{\eta} \in \mathring{H}^{1,2}(\Omega).$$

Therefore, we can apply Lemma 2.4.2 to find a unique $p \in \widehat{L^2}(\Omega)$ such that

$$L(\boldsymbol{\eta}) = \int_\Omega p \operatorname{div}\boldsymbol{\eta}\, dx \quad \forall \boldsymbol{\eta} \in \mathring{W}^{1,2}(\Omega), \tag{4.1.8}$$

with $\|p\|_{L^2(\Omega)} = \|L\|_{(\mathring{W}^{1,2}(\Omega))^*}$. By comparing (4.1.7) and (4.1.8), we obtain that $(\mathbf{u}, p)$ satisfies integral identity (4.1.3).

Using the Cauchy-Schwarz-Bunyakovsky inequality, we get

$$|L(\boldsymbol{\eta})| \le \|\nabla\mathbf{u}\|_{L^2(\Omega)} \|\nabla\boldsymbol{\eta}\|_{L^2(\Omega)} + \|\mathbf{F}\|_{\mathring{H}^{1,2}(\Omega)} \|\boldsymbol{\eta}\|_{\mathring{H}^{1,2}(\Omega)}.$$

Therefore, estimate (4.1.4) follows from (4.1.5) and (4.1.6).

Lastly, we prove the uniqueness of a weak solution to problem (4.1.1). Let $\mathbf{v} = \mathbf{u}^{(1)} - \mathbf{u}^{(2)}$ be the difference of two weak solutions $\mathbf{u}^{(1)}$ and $\mathbf{u}^{(2)}$. It satisfies the integral identity:

$$\nu \int_\Omega \nabla\mathbf{v} \cdot \nabla\boldsymbol{\eta}\, dx = 0 \quad \forall \boldsymbol{\eta} \in \mathring{H}^{1,2}(\Omega).$$

Taking $\boldsymbol{\eta} = \mathbf{v}$, we have

$$\nu \int_\Omega |\nabla\mathbf{v}|^2\, dx = 0,$$

and hence $\mathbf{v}(x) = \text{const}$ almost everywhere in Ω. Since the trace $\mathbf{v}|_{\partial\Omega} = 0$, we conclude that this constant is zero, i.e., $\mathbf{v}(x) = 0$ almost everywhere in Ω.

Now, let $q = p^{(1)} - p^{(2)}$. We get from (4.1.3) that

$$\int_\Omega q \,\mathrm{div}\boldsymbol{\eta}\, dx = 0 \quad \forall \boldsymbol{\eta} \in \mathring{W}^{1,2}(\Omega). \tag{4.1.9}$$

Since $q \in \widehat{L}^2(\Omega)$, i.e., $\int_\Omega q(x)\, dx = 0$, we find, according to Lemma 2.4.1, a function $\mathbf{w} \in \mathring{W}^{1,2}(\Omega)$ such that $\mathrm{div}\, \mathbf{w} = q$. Taking $\boldsymbol{\eta} = \mathbf{w}$ in (4.1.9) yields

$$\int_\Omega |q(x)|^2\, dx = 0,$$

which implies $q(x) = 0$ almost everywhere in Ω. This completes the proof. □

Remark 4.1.1 The first statement of the theorem remains valid if Ω is an arbitrary domain (the regularity and boundedness of Ω are not needed for the proof of the existence of the weak solution $\mathbf{u}$). However, the proof of the existence of a pressure p requires that Ω is a bounded Lipschitz domain.

It follows from Theorem 4.1.1 that the Definition 4.1.1 of the weak solution $\mathbf{u}$ of problem (4.1.1) is equivalent to the following one:

Definition 4.1.2 The pair $(\mathbf{u}, p) \in \mathring{H}^{1,2}(\Omega) \times \widehat{L}^2(\Omega)$ is called a weak solution of the boundary value problem (4.1.1) if $(\mathbf{u}, p)$ satisfies the integral identity (4.1.3) for every $\boldsymbol{\eta} \in \mathring{W}^{1,2}(\Omega)$.

Now, let us consider the Stokes problem with non-homogeneous boundary conditions:

$$\begin{cases} -\nu\Delta\mathbf{u} + \nabla p = \mathbf{f}_0 + \sum\limits_{j=1}^n \dfrac{\partial \mathbf{f}_j}{\partial x_j}, & x \in \Omega, \\ \mathrm{div}\, \mathbf{u} = 0, & x \in \Omega, \\ \mathbf{u} = \boldsymbol{\alpha}, & x \in \partial\Omega. \end{cases} \tag{4.1.10}$$

where $\boldsymbol{\alpha} \in W^{1/2,2}(\partial\Omega)$ and satisfies the compatibility condition:

$$\int_{\partial\Omega} \boldsymbol{\alpha} \cdot \mathbf{n} dS = 0. \tag{4.1.11}$$

Let Ω be a bounded Lipschitz domain.

Definition 4.1.3 By the weak solution of problem (4.1.10), we mean the pair $(\mathbf{u}, p) \in W^{1,2}(\Omega) \times \widehat{L}^2(\Omega)$ such that $\mathbf{u}|_{\partial\Omega} = \boldsymbol{\alpha}$, $\operatorname{div} \mathbf{u} = 0$ and it satisfies the following integral identity for any $\boldsymbol{\eta} \in \overset{\circ}{W}{}^{1,2}(\Omega)$:

$$\nu \int_\Omega \nabla \mathbf{u} \cdot \nabla \boldsymbol{\eta} \, dx - \int_\Omega p \operatorname{div} \boldsymbol{\eta} \, dx = \int_\Omega \mathbf{f}_0 \cdot \boldsymbol{\eta} \, dx - \int_\Omega \sum_{j=1}^n \mathbf{f}_j \cdot \frac{\partial \boldsymbol{\eta}}{\partial x_j} \, dx. \tag{4.1.12}$$

According to Lemma 2.4.3, there exists a divergence-free extension $\mathbf{A} \in W^{1,2}(\Omega)$ of $\boldsymbol{\alpha}$. Thus, the above definition is equivalent to the following one:

Definition 4.1.4 The weak solution of problem (4.1.10) is the pair $(\mathbf{u}, p) \in W^{1,2}(\Omega) \times \widehat{L}^2(\Omega)$ such that $\mathbf{u} - \mathbf{A} \in \overset{\circ}{H}{}^{1,2}(\Omega)$ and it satisfies the integral identity (4.1.12).

The following theorems state the existence, uniqueness, and regularity of the weak solution to the Stokes problem with non-homogeneous boundary conditions:

Theorem 4.1.2 *Let $\Omega \subset \mathbb{R}^n$ be a bounded domain with Lipschitz boundary. For any given functions $\mathbf{f}_0 \in L^2(\Omega)$, $\mathbf{f}_j \in L^2(\Omega)$ for $j = 1, \dots, n$, and $\boldsymbol{\alpha} \in W^{1/2,2}(\partial\Omega)$ satisfying* (4.1.11)*, problem* (4.1.10) *admits a unique weak solution* $(\mathbf{u}, p)$*. The following estimate holds:*

$$\|\mathbf{u}\|^2_{W^{1,2}(\Omega)} + \|p\|^2_{L^2(\Omega)} \leq c \left(\|\mathbf{f}_0\|^2_{L^2(\Omega)} + \sum_{j=1}^n \|\mathbf{f}_j\|^2_{L^2(\Omega)} + \|\boldsymbol{\alpha}\|^2_{W^{1/2,2}(\partial\Omega)} \right), \tag{4.1.13}$$

where c is a positive constant independent of the data.

The proof of this theorem follows from Theorem 4.1.1 and Lemma 2.4.3, and thus it will be omitted.

A solution of the Stokes problem (4.1.1) has the same regularity properties as the solution of the first boundary value problem for the Poisson equation. In particular, we have the following result (see [66, 102]).

Theorem 4.1.3 *Let Ω be a bounded domain in $\mathbb{R}^n$ with the boundary $\partial\Omega$ of class C^{k+2}, $k \geq 0$. Assume that $\mathbf{f} \in W^{k,2}(\Omega)$. Then the weak solution $(\mathbf{u}, p) \in \overset{\circ}{H}{}^{1,2}(\Omega) \times \widehat{L}^2(\Omega)$ of problem* (4.1.1) *has the following regularity $\mathbf{u} \in W^{k+2,2}(\Omega)$, $\nabla p \in W^k(\Omega)$. Moreover, the following estimate*

$$\|\mathbf{u}\|_{W^{k+2,2}(\Omega)} + \|\nabla p\|_{W^{k,2}(\Omega)} \leq c \|\mathbf{f}\|_{W^{k,2}(\Omega)} \tag{4.1.14}$$

holds.

4.2 Stokes Operator and Its Properties

In this section, we discuss the Stokes operator for the bounded domain $\Omega \subset \mathbb{R}^n$ with Lipschitz boundary.

Let us consider the Stokes problem (4.1.1). By virtue of Theorem 2.5.5, any given vector field $\mathbf{f} \in L^2(\Omega)$ can be uniquely represented in the form

$$\mathbf{f} = \widehat{\mathbf{f}} + \nabla q,$$

where $q \in W^{1,2}(\Omega)$ and $\widehat{\mathbf{f}} \in J(\Omega)$. Without loss of generality, we can assume that $\mathbf{f} \in J(\Omega)$ by adding the gradient part to the pressure.

We rewrite problem (4.1.1) in the operator form as follows:

$$-\widetilde{\Delta}\mathbf{u} = \mathbf{f}, \tag{4.2.1}$$

where $\widetilde{\Delta} = P\Delta$, $\widetilde{\Delta} : \overset{\circ}{H}{}^{1,2}(\Omega) \cap W^{2,2}(\Omega) \longrightarrow J(\Omega)$ is an unbounded operator with the domain $\mathcal{D}(\widetilde{\Delta}) = \overset{\circ}{H}{}^{1,2}(\Omega) \cap W^{2,2}(\Omega)$, and P is the projector from $L^2(\Omega)$ onto $J(\Omega)$, defined as $P\mathbf{f} = \widehat{\mathbf{f}}$ (also known as Leray's projector). Since the divergence of any vector in $J(\Omega)$ vanishes, we have

$$\int_\Omega \nabla p \cdot \mathbf{v}\, dx = 0 \quad \text{for} \quad \text{any} \quad \mathbf{v} \in J_0^\infty(\Omega)$$

and, thus, the operator $\widetilde{\Delta}\mathbf{w}$ is defined for any given $\mathbf{w} \in \overset{\circ}{H}{}^{1,2}(\Omega) \cap W^{2,2}(\Omega)$ as

$$\begin{aligned} -\int_\Omega \widetilde{\Delta}\mathbf{w} \cdot \mathbf{v}\, dx &= \int_\Omega (-\nu\Delta\mathbf{w} + \nabla p) \cdot \mathbf{v}\, dx \\ &= -\nu \int_\Omega \Delta\mathbf{w} \cdot \mathbf{v}\, dx = \nu \int_\Omega \nabla\mathbf{w} \cdot \nabla\mathbf{v}\, dx, \qquad \forall \mathbf{v} \in J_0^\infty(\Omega). \end{aligned} \tag{4.2.2}$$

The set $J_0^\infty(\Omega)$ is dense in $J(\Omega)$ and $\overset{\circ}{H}{}^{1,2}(\Omega)$. For any arbitrary function $\mathbf{v} \in J(\Omega)$, and a sequence $\{\mathbf{v}_k\} \subset J_0^\infty(\Omega)$ converging to $\mathbf{v}$ in the $L^2(\Omega)$-norm, we can write (4.2.2) for the test function $\mathbf{v}_k$ and pass to the limit. This gives us the equation

$$-\int_\Omega \widetilde{\Delta}\mathbf{w} \cdot \mathbf{v}\, dx = -\nu \int_\Omega \Delta\mathbf{w} \cdot \mathbf{v}\, dx \quad \forall \mathbf{v} \in J(\Omega). \tag{4.2.3}$$

From (4.2.3), we further deduce

$$-\int_\Omega \widetilde{\Delta}\mathbf{w} \cdot \mathbf{v}\, dx = \nu \int_\Omega \nabla\mathbf{w} \cdot \nabla\mathbf{v}\, dx \quad \forall \mathbf{v} \in \overset{\circ}{H}{}^{1,2}(\Omega), \tag{4.2.4}$$

and, consequently

$$\int_{\Omega} \widetilde{\Delta}\mathbf{w} \cdot \mathbf{v}\, dx = \int_{\Omega} \mathbf{w} \cdot \widetilde{\Delta}\mathbf{v}\, dx \quad \forall \mathbf{v} \in W^{2,2}(\Omega) \cap \overset{\circ}{H}{}^{1,2}(\Omega). \tag{4.2.5}$$

Lemma 4.2.1 *The operator $\widetilde{\Delta}$ is symmetric on $W^{2,2}(\Omega) \cap \overset{\circ}{H}{}^{1,2}(\Omega)$. The following estimates hold:*

$$\|\widetilde{\Delta}\mathbf{w}\|_{L^2(\Omega)} \leq \nu\|\Delta\mathbf{w}\|_{L^2(\Omega)}, \tag{4.2.6}$$

$$\|\widetilde{\Delta}\mathbf{w}\|_{(\overset{\circ}{H}{}^{1,2}(\Omega))^*} \leq \nu\|\nabla\mathbf{w}\|_{L^2(\Omega)} = \nu\|\mathbf{w}\|_{\overset{\circ}{H}{}^{1,2}(\Omega)}. \tag{4.2.7}$$

Here, $(\overset{\circ}{H}{}^{1,2}(\Omega))^*$ is the dual space of $\overset{\circ}{H}{}^{1,2}(\Omega)$.

If $\partial\Omega \in C^2$, then there exists an inverse operator $\widetilde{\Delta}^{-1} : J(\Omega) \to W^{2,2}(\Omega) \cap \overset{\circ}{H}{}^{1,2}(\Omega)$. Its restriction to the space $\overset{\circ}{H}{}^{1,2}(\Omega)$ is a compact operator from $\overset{\circ}{H}{}^{1,2}(\Omega)$ to $\overset{\circ}{H}{}^{1,2}(\Omega)$.

Proof

The symmetry of $\widetilde{\Delta}$ on $W^{2,2}(\Omega) \cap \overset{\circ}{H}{}^{1,2}(\Omega)$ follows immediately from (4.2.5). Taking $\mathbf{v} = \widetilde{\Delta}\mathbf{w}$ in (4.2.3), we obtain, using the Cauchy-Schwarz-Bunyakovsky inequality, the estimate (4.2.6). Furthermore, from (4.2.4) follows the estimate (4.2.7).

If $\partial\Omega \in C^2$, then, by Theorem 4.1.3, problem (4.1.1) admits a unique solution $(\mathbf{u}, p) \in W^{2,2}(\Omega) \times W^{1,2}(\Omega)$ and

$$\|\mathbf{u}\|_{W^{2,2}(\Omega)} + \|\nabla p\|_{L^2(\Omega)} \leq c\|\mathbf{f}\|_{L^2(\Omega)} = c\|\widetilde{\Delta}\mathbf{u}\|_{L^2(\Omega)}. \tag{4.2.8}$$

Therefore, the equation $\widetilde{\Delta}\mathbf{u} = \mathbf{f}$ is uniquely solvable for arbitrary $\mathbf{f} \in J(\Omega)$ and the inverse operator $\widetilde{\Delta}^{-1} : J(\Omega) \longrightarrow W^{2,2}(\Omega) \cap \overset{\circ}{H}{}^{1,2}(\Omega)$ exists. Moreover, because of (4.2.8), $\widetilde{\Delta}^{-1}$ is bounded and

$$\|\widetilde{\Delta}^{-1}\mathbf{f}\|_{W^{2,2}(\Omega)} \leq c\|\mathbf{f}\|_{L^2(\Omega)}. \tag{4.2.9}$$

Since, by Sobolev embedding theorem (Theorem 2.3.5), the embedding $\overset{\circ}{H}{}^{1,2}(\Omega) \hookrightarrow L^2(\Omega)$ is compact, we conclude that $\widetilde{\Delta}^{-1}$ is compact. □

Let us define an extension of the operator $\widetilde{\Delta}$ to the larger space $\overset{\circ}{H}{}^{1,2}(\Omega)$. The set $\mathcal{D}(\widetilde{\Delta}) = \overset{\circ}{H}{}^{1,2}(\Omega) \cap W^{2,2}(\Omega)$ is dense in $\overset{\circ}{H}{}^{1,2}(\Omega)$. For any function $\mathbf{u} \in \overset{\circ}{H}{}^{1,2}(\Omega) \setminus \mathcal{D}(\widetilde{\Delta})$, there exists a sequence $\{\mathbf{u}_k\} \subset \mathcal{D}(\widetilde{\Delta})$ that approximates $\mathbf{u}$ in the norm of $\overset{\circ}{H}{}^{1,2}(\Omega)$.

Using the expression for the Stokes operator (4.2.2), we find that the following equation holds for this sequence:

$$-\int_{\Omega} (\widetilde{\Delta}\mathbf{u}_k - \widetilde{\Delta}\mathbf{u}_l) \cdot \mathbf{v}\, dx = \nu \int_{\Omega} (\nabla\mathbf{u}_k - \nabla\mathbf{u}_l) \cdot \nabla\mathbf{v}\, dx \quad \forall \mathbf{v} \in \overset{\circ}{H}{}^{1,2}(\Omega).$$

From the estimate (4.2.7), we observe that the sequence $\{\widetilde{\Delta}\mathbf{u}_k\}$ converges in the norm of the space $(\mathring{H}^{1,2}(\Omega))^*$ to some functional $l \in (\mathring{H}^{1,2}(\Omega))^*$. We denote this limit by $\widetilde{\Delta}\mathbf{u} = l$. Consequently, the extended operator is defined on the entire space $\mathring{H}^{1,2}(\Omega)$, and we denote it again by $\widetilde{\Delta}$. This operator is commonly referred to as the *Stokes operator*.

It is evident that the Stokes operator satisfies the following equation:

$$-\int\limits_{\Omega} \widetilde{\Delta}\mathbf{w} \cdot \mathbf{v}\, dx = \nu \int\limits_{\Omega} \nabla\mathbf{w} \cdot \nabla\mathbf{v}\, dx \quad \forall \mathbf{w},\ \mathbf{v} \in \mathring{H}^{1,2}(\Omega). \tag{4.2.10}$$

The Stokes operator $\widetilde{\Delta}$ possesses properties similar to the Laplace operator under Dirichlet boundary conditions.

Theorem 4.2.1

(i) The Stokes operator $\widetilde{\Delta} : \mathring{H}^{1,2}(\Omega) \longrightarrow (\mathring{H}^{1,2}(\Omega))^*$ *is a bijection and is self-adjoint on* $\mathring{H}^{1,2}(\Omega)$. *The inverse operator* $\widetilde{\Delta}^{-1} : (\mathring{H}^{1,2}(\Omega))^* \longrightarrow \mathring{H}^{1,2}(\Omega)$ *is compact, and the following estimate holds:*

$$\|\widetilde{\Delta}^{-1}\mathbf{f}\|_{\mathring{H}^{1,2}(\Omega)} \le c\|\mathbf{f}\|_{(\mathring{H}^{1,2}(\Omega))^*}. \tag{4.2.11}$$

(ii) The operator $\widetilde{\Delta}$ *has a discrete spectrum:*

$$-\widetilde{\Delta}\mathbf{w} = \lambda\mathbf{w}, \quad \mathbf{w} \in \mathring{H}^{1,2}(\Omega), \quad \mathbf{w} \ne 0;$$

$0 < \lambda_1 < \lambda_2 < \ldots < \lambda_k < \ldots$ *and* $\lim_{k\to\infty} \lambda_k \to +\infty$. *The set* $\{\mathbf{w}_k\}$ *of eigenfunctions of* $\widetilde{\Delta}$ *forms an orthonormal basis in* $J(\Omega)$ *and an orthogonal basis in* $\mathring{H}^{1,2}(\Omega)$, *satisfying*

$$\int\limits_{\Omega} \nabla\mathbf{w}_k \cdot \nabla\mathbf{w}_l\, dx = \lambda_k\delta_{kl}, \quad \int\limits_{\Omega} \mathbf{w}_k \cdot \mathbf{w}_l\, dx = \delta_{kl}, \tag{4.2.12}$$

where δ_{kl} *is the Kronecker delta.*

(iii) If $\partial\Omega \in C^2$, *then there exist two constants* $c_1 > 0$ *and* $c_2 > 0$ *such that*

$$c_1\|\nabla^2\mathbf{w}\|_{L^2(\Omega)} \le \|\widetilde{\Delta}\mathbf{w}\|_{L^2(\Omega)} \le c_2\|\nabla^2\mathbf{w}\|_{L^2(\Omega)}. \tag{4.2.13}$$

Proof

(i) The bijectivity of the operator $\widetilde{\Delta}$ is evident from its definition. Additionally, from Eq. (4.2.10), we observe that $\widetilde{\Delta}$ is self-adjoint on the space $\mathring{H}^{1,2}(\Omega)$. The estimate (4.2.11) follows as a consequence of the estimate for the solution of the Stokes problem with the right-hand side in the dual space $(\mathring{H}^{1,2}(\Omega))^*$ (see Theorem 4.1.1). Moreover, the compactness of $\widetilde{\Delta}$ can be deduced from the compactness of the embedding $\mathring{H}^{1,2}(\Omega) \hookrightarrow L^2(\Omega)$, which is a subset of the dual space $(\mathring{H}^{1,2}(\Omega))^*$.

(ii) This statement can be derived from (i), Theorem 2.2.7, and the fact that the space $\mathring{H}^{1,2}(\Omega)$ is dense in $J(\Omega)$. The equalities (4.2.12) can be directly obtained from (4.2.10).

(iii) Assume that $\partial\Omega$ is of class C^2 and $\mathbf{f} \in J(\Omega)$. Then, Theorem 4.1.3 guarantees that the solution $\mathbf{u}$ of the Stokes problem (4.1.1) belongs to the space $W^{2,2}(\Omega) \cap \mathring{H}^{1,2}(\Omega)$. Furthermore, we have the inequality

$$\|\Delta \mathbf{w}\|_{L^2(\Omega)} \le c\|\nabla^2 \mathbf{w}\|_{L^2(\Omega)}. \tag{4.2.14}$$

Now, using estimate (4.2.6), we obtain

$$\|\widetilde{\Delta} \mathbf{w}\|_{L^2(\Omega)} \le c\|\Delta \mathbf{w}\|_{L^2(\Omega)} \le c\|\nabla^2 \mathbf{w}\|_{L^2(\Omega)}. \tag{4.2.15}$$

On the other hand, from the estimate (4.2.8), we have

$$\|\nabla^2 \mathbf{w}\|_{L^2(\Omega)} \le c\|\widetilde{\Delta} \mathbf{w}\|_{L^2(\Omega)},$$

which, combined with (4.2.15), yields

$$c_1\|\nabla^2 \mathbf{w}\|_{L^2(\Omega)} \le \|\widetilde{\Delta} \mathbf{w}\|^2_{L^2(\Omega)} \le c_2\|\nabla^2 \mathbf{w}\|_{L^2(\Omega)}. \tag{4.2.16}$$

Thus, we have established the desired results (i), (ii), and (iii).

□

4.3 The Stokes Operator in Tube Structures

In the tube structure B_ε, we consider the Dirichlet problem for the Stokes system:

$$\begin{cases} -\nu\Delta \mathbf{v} + \nabla p = \mathbf{f}, \ x \in B_\varepsilon, \\ \operatorname{div}\mathbf{v} = 0, \ x \in B_\varepsilon, \\ \mathbf{v} = 0, \ x \in \partial B_\varepsilon. \end{cases} \tag{4.3.1}$$

The weak solution $\mathbf{v} \in \mathring{H}^{1,2}(B_\varepsilon)$ to Eq. (4.3.1) satisfies the integral identity

$$\nu \int\limits_{B_\varepsilon} \nabla \mathbf{v} \cdot \nabla \eta dx = \int\limits_{B_\varepsilon} \mathbf{f} \cdot \eta dx \quad \forall \eta \in \mathring{H}^{1,2}(B_\varepsilon),$$

Hence, by Poincaré inequality (2.3.36), the following estimate

$$\|\nabla \mathbf{v}\|_{L^2(B_\varepsilon)} \leq c\varepsilon \|\mathbf{f}\|_{L^2(B_\varepsilon)} \tag{4.3.2}$$

holds.

Theorem 4.3.1 *Suppose $\partial B_\varepsilon \in C^{l+2}$ and $\mathbf{f} \in W^{l,2}(B_\varepsilon)$. Then $\mathbf{v} \in W^{l+2,2}(B_\varepsilon)$, $\nabla p \in W^{l,2}(B_\varepsilon)$, and the estimate*

$$|||\mathbf{v}|||^2_{l+2,\alpha,B_\varepsilon} + |||\nabla p|||^2_{l,\alpha,B_\varepsilon} \leq c|||\mathbf{f}|||^2_{l,\alpha,B_\varepsilon} \tag{4.3.3}$$

holds with the constant c being independent of ε.

Proof
Let $A_\varepsilon \subset \widetilde{A}_\varepsilon$ be the domains from the coverings $\mathfrak{A}_\varepsilon$ and $\widetilde{\mathfrak{A}}_\varepsilon$ of B_ε. We consider Eqs. (4.3.1) in $\widetilde{A}_\varepsilon$. By performing a change of variables $y = \varepsilon^{-1}x$, we transform A_ε and $\widetilde{A}_\varepsilon$ into fixed domains, independent of ε, denoted by A_0 and $\widetilde{A}_0$. The Stokes problem in coordinates y takes the form

$$\begin{cases} -\nu\Delta_y \mathbf{v} + \nabla_y(\varepsilon p) = \varepsilon^2 \mathbf{f}, \; y \in \widetilde{A}_0 \\ \mathrm{div}_y \mathbf{v} = 0, \quad y \in \widetilde{A}_0, \\ \mathbf{v} = 0, \quad y \in \partial B_\varepsilon \cap \partial \widetilde{A}_0. \end{cases} \tag{4.3.4}$$

Applying the ADN local estimates for elliptic problems (see [6, 198]), we get the inequality

$$\begin{aligned} &\sum_{m=0}^{l+2} \|\nabla^m \mathbf{v}\|^2_{L^2(A_0)} + \sum_{m=1}^{l+1} \|\nabla^m q\|^2_{L^2(A_0)} \\ &\leq c\left(\varepsilon^4 \sum_{m=0}^{l} \|\nabla^m \mathbf{f}\|^2_{L^2(\widetilde{A}_0)} + \|\mathbf{v}\|^2_{L^2(\widetilde{A}_0)} + \|q - \overline{q}\|^2_{L^2(\widetilde{A}_0)}\right), \end{aligned} \tag{4.3.5}$$

where $q(y) = \varepsilon p(y), \overline{q} = \frac{1}{|\widetilde{A}_0|}\int_{\widetilde{A}_0} q(y)dy$. Since $\int_{\widetilde{A}_0}(q(y) - \overline{q})dy = 0$, there exists a function $\mathbf{w} \in \overset{\circ}{W}{}^{1,2}(\widetilde{A}_0)$, such that $\mathrm{div}\mathbf{w} = q(y) - \overline{q}$ in $\widetilde{A}_0$, and the following estimate holds:

$$\|\nabla \mathbf{w}\|_{L^2(\widetilde{A}_0)} \leq c\|q - \overline{q}\|_{L^2(\widetilde{A}_0)},$$

(see Lemma 2.4.1). Multiplying (4.3.4) by $\mathbf{w}$ and integrating by parts, we obtain

$$\|q - \overline{q}\|^2_{L^2(\widetilde{A}_0)} = \int\limits_{\widetilde{A}_0} q(y)(q(y) - \overline{q})dy = \int\limits_{\widetilde{A}_0} q(y)\mathrm{div}\mathbf{w}dy$$

$$= \nu \int\limits_{\widetilde{A}_0} \nabla\mathbf{v} \cdot \nabla\mathbf{w}dy - \varepsilon^2 \int\limits_{\widetilde{A}_0} \mathbf{f} \cdot \mathbf{w}dy$$

$$\leq \|\nabla\mathbf{v}\|_{L^2(\widetilde{A}_0)} \|\nabla\mathbf{w}\|_{L^2(\widetilde{A}_0)} + \varepsilon^2 \|\mathbf{f}\|_{L^2(\widetilde{A}_0)} \|\mathbf{w}\|_{L^2(\widetilde{A}_0)}.$$

Using the estimates obtained earlier, we derive

$$\|q - \overline{q}\|_{L^2(\widetilde{A}_0)} \leq c \left(\|\nabla\mathbf{v}\|_{L^2(\widetilde{A}_0)} + \varepsilon^2 \|\mathbf{f}\|_{L^2(\widetilde{A}_0)} \right). \tag{4.3.6}$$

From (4.3.5), using (4.3.6) and the Poincaré inequality, we obtain the following estimate:

$$\begin{aligned} &\sum_{m=0}^{l+2} \|\nabla^m \mathbf{v}\|^2_{L^2(A_0)} + \sum_{m=1}^{l+1} \|\nabla^m q\|^2_{L^2(A_0)} \\ &\quad \leq c \left(\varepsilon^4 \sum_{m=0}^{l} \|\nabla^m \mathbf{f}\|^2_{L^2(\widetilde{A}_0)} + \|\nabla_y \mathbf{v}\|^2_{L^2(\widetilde{A}_0)} \right). \end{aligned} \tag{4.3.7}$$

Returning to the original coordinates, the inequality can be rewritten as

$$\begin{aligned} &\sum_{m=0}^{l+2} \varepsilon^{2m} \|\nabla^m_x \mathbf{v}\|^2_{L^2(A_\varepsilon)} + \varepsilon^2 \sum_{m=1}^{l+1} \varepsilon^{2m} \|\nabla^m_x p\|^2_{L^2(A_\varepsilon)} \\ &\quad \leq c \left(\varepsilon^4 \sum_{m=0}^{l} \varepsilon^{2m} \|\nabla^m \mathbf{f}\|^2_{L^2(\widetilde{A}_\varepsilon)} + \varepsilon^2 \|\nabla\mathbf{v}\|^2_{L^2(\widetilde{A}_\varepsilon)} \right). \end{aligned} \tag{4.3.8}$$

Summing up (4.3.8) over all domains $A_\varepsilon \subset \widetilde{A}_\varepsilon$ and estimating the last term on the right-hand side using (4.3.2) yields:

$$\begin{aligned} &\sum_{m=0}^{l+2} \varepsilon^{2m} \|\nabla^m_x \mathbf{v}\|^2_{L^2(B_\varepsilon)} + \varepsilon^2 \sum_{m=1}^{l+1} \varepsilon^{2m} \|\nabla^m_x p\|^2_{L^2(B_\varepsilon)} \\ &\quad \leq c\left(\varepsilon^4 \sum_{m=0}^{l} \varepsilon^{2m} \|\nabla^m \mathbf{f}\|^2_{L^2(B_\varepsilon)} + \varepsilon^2 \|\nabla_x \mathbf{v}\|^2_{L^2(B_\varepsilon)} \right) \\ &\quad \leq c\varepsilon^4 \sum_{m=0}^{l} \varepsilon^{2m} \|\nabla^m \mathbf{f}\|^2_{L^2(B_\varepsilon)}. \end{aligned} \tag{4.3.9}$$

By multiplying the last inequality by $\varepsilon^{-2(l+2)+2\alpha}$, we obtain (4.3.3). □

Remark 4.3.1 In particular, from Theorem 4.3.1, it follows that in the case where ∂B_ε is C^2, and $\mathbf{f} \in L^2(B_\varepsilon)$, the solution $\mathbf{v}$ possesses second derivatives, and the following inequality holds:

$$\|\nabla^2 \mathbf{v}\|_{L^2(B_\varepsilon)} \leq c\|\mathbf{f}\|_{L^2(B_\varepsilon)}, \tag{4.3.10}$$

where c is a constant independent of ε.

Now, we can apply the results from the previous section to tube structures. The problem (4.3.1) can be reformulated in the operator form (assuming, without loss of generality, $\mathbf{f} \in J(B_\varepsilon)$):

$$\widetilde{\Delta}\mathbf{v} = \mathbf{f}, \tag{4.3.11}$$

where $\widetilde{\Delta} = P\Delta : \overset{\circ}{H}{}^{1,2}(B_\varepsilon) \cap W^{2,2}(B_\varepsilon) \longmapsto J(B_\varepsilon)$ is an unbounded operator, with the domain $\overset{\circ}{H}{}^{1,2}(B_\varepsilon) \cap W^{2,2}(B_\varepsilon)$, and P represents the Leray projection onto divergence-free vector fields. We denote the Friedrichs' extension of this operator to the whole space $\overset{\circ}{H}{}^{1,2}(B_\varepsilon)$ by the same symbol $\widetilde{\Delta}$. It is referred to as the Stokes operator. Based on the results from Sect. 4.2 (also see [102] and [203]), the following properties hold:

(i) The Stokes operator has a discrete spectrum:

$$\widetilde{\Delta}\mathbf{w} = \lambda\mathbf{w}, \quad \mathbf{w} \in \overset{\circ}{H}{}^{1,2}(B_\varepsilon), \quad \mathbf{w} \neq 0;$$

where $\lambda_i > 0$, and $\lim\limits_{i\to\infty} \lambda_i \to +\infty$.

(ii) The eigenfunctions $\{\mathbf{w}_k\}_{k=1}^{\infty}$ of $\widetilde{\Delta}$ form an orthogonal basis in $J(B_\varepsilon)$ and $\overset{\circ}{H}{}^{1,2}(B_\varepsilon)$, with $\|\nabla \mathbf{w}_k\|_{L^2(B_\varepsilon)} = \sqrt{\lambda_k}$ and $\|\mathbf{w}_k\|_{L^2(B_\varepsilon)} = 1$. If $\partial B_\varepsilon \in C^2$, then $\mathbf{w}_k \in \overset{\circ}{H}{}^{1,2}(B_\varepsilon) \cap W^{2,2}(B_\varepsilon)$.

(iii) For any $\mathbf{w} \in \overset{\circ}{H}{}^{1,2}(B_\varepsilon) \cap W^{2,2}(B_\varepsilon)$, the following equality holds:

$$\int\limits_{B_\varepsilon} \widetilde{\Delta}\mathbf{w} \cdot \mathbf{v} dx = -\nu \int\limits_{B_\varepsilon} \Delta\mathbf{w} \cdot \mathbf{v} dx = \nu \int\limits_{B_\varepsilon} \nabla\mathbf{w} \cdot \nabla\mathbf{v} dx \quad \forall \mathbf{v} \in J_0^\infty(B_\varepsilon).$$

Furthermore, using density arguments, it can be shown that

$$\int\limits_{B_\varepsilon} |\widetilde{\Delta}\mathbf{w}|^2 dx = -\nu \int\limits_{B_\varepsilon} \Delta\mathbf{w} \cdot \widetilde{\Delta}\mathbf{w} dx.$$

(iv) The following estimates hold:

$$\|\widetilde{\Delta}\mathbf{w}\|_{L^2(B_\varepsilon)} \leq c\|\Delta\mathbf{w}\|_{L^2(B_\varepsilon)} \leq c\|\nabla^2\mathbf{w}\|_{L^2(B_\varepsilon)}, \tag{4.3.12}$$

$$\|\widetilde{\Delta}\mathbf{w}\|_{(\mathring{H}^{1,2}(B_\varepsilon))^*} \leq \nu\|\nabla\mathbf{w}\|_{L^2(B_\varepsilon)}$$

with constants c independent of ε.

From (4.3.10), we can deduce the following estimate:

$$\|\nabla^2\mathbf{w}\|_{L^2(B_\varepsilon)} \leq c\|\widetilde{\Delta}\mathbf{w}\|_{L^2(B_\varepsilon)},$$

which, together with (4.3.12), yields

$$c_1\|\nabla^2\mathbf{w}\|_{L^2(B_\varepsilon)} \leq \|\widetilde{\Delta}\mathbf{w}\|_{L^2(B_\varepsilon)} \leq c_2\|\nabla^2\mathbf{w}\|_{L^2(B_\varepsilon)}, \tag{4.3.13}$$

where c_1 and c_2 are constants independent of ε.

Lemma 4.3.1 *Let $B_\varepsilon \subset \mathbb{R}^n$, where $n = 2, 3$, and $\mathbf{w} \in \mathring{H}^{1,2}(B_\varepsilon) \cap W^{2,2}(B_\varepsilon)$. Then, the following inequalities hold:*

For $n = 3$:

$$\|\mathbf{w}\|_{L^\infty(B_\varepsilon)} \leq c\varepsilon^{1/4}\|\nabla^2\mathbf{w}\|_{L^2(B_\varepsilon)} \leq c\varepsilon^{1/4}\|\widetilde{\Delta}\mathbf{w}\|_{L^2(B_\varepsilon)}, \tag{4.3.14}$$

For $n = 2$:

$$\|\mathbf{w}\|_{L^\infty(B_\varepsilon)} \leq c\varepsilon^{1/2}\|\nabla^2\mathbf{w}\|_{L^2(B_\varepsilon)} \leq c\varepsilon^{1/2}\|\widetilde{\Delta}\mathbf{w}\|_{L^2(B_\varepsilon)}, \tag{4.3.15}$$

where the constant c is independent of ε.

Proof
The inequality (4.3.14) follows from (2.3.45) and (4.3.13), while the inequality (4.3.15) follows from (2.3.47) and (4.3.13). □

4.4 The Stationary Stokes Problem in Domains with Cylindrical Outlets

In this section, we address the stationary Stokes equations in domains with cylindrical outlets to infinity, similar to the domains discussed in the Introduction. Initially, we establish general theorems concerning the existence and uniqueness of the solution. Subsequently, we prove theorems concerning the exponential stabilization of the solution in the outlets at infinity. These types of theorems, pertaining to various partial differential equations, have been extensively studied in the literature, including elliptic equations [110, 111, 122, 132], elasticity equations [122, 132, 137, 138], the Stokes equations [105, 132, 174, 176], and non-Newtonian flow equations [117, 163, 166], etc. Such results are commonly referred to as Phrägmen–Lindelöf or Saint-Venant type theorems. We will use an energy estimate approach utilizing an appropriate exponential multiplier drawing from author's papers [174, 176].

Let $\Omega \subset \mathbb{R}^n$ be a domain with cylindrical outlets to infinity. We consider the stationary Stokes problem (4.1.1) in Ω. Notice that the first part of Theorem 4.1.1 remains applicable to domains with outlets to infinity. From now we assume that the right hand side of equation $(4.1.1)_1$ is a function $\mathbf{f} \in L^2(\Omega)$.

Theorem 4.4.1 *Let $\Omega \subset \mathbb{R}^n$ be a domain with cylindrical outlets to infinity. For any given function $\mathbf{f} \in L^2(\Omega)$, problem* (4.1.1) *admits a unique weak solution $\mathbf{u} \in \overset{\circ}{H}{}^{1,2}(\Omega)$ that satisfies the integral identity* (4.1.2). *Additionally, the following estimate holds:*

$$\|\mathbf{u}\|_{W^{1,2}(\Omega)} \leq c\|\mathbf{f}\|_{L^2(\Omega)}, \tag{4.4.1}$$

where c is a positive constant.

Next, we proceed to prove the decay properties of the weak solution $\mathbf{u} \in \overset{\circ}{H}{}^{1,2}(\Omega)$ of problem (4.1.1) as $|x| \to \infty$, with $x \in \Omega_j$, while assuming that the right-hand side $\mathbf{f}$ belongs to the weighted space $\mathcal{W}^{l,2}_{\boldsymbol{\beta}}(\Omega)$, where $l \geq 0$ and $\beta > 0$.

Theorem 4.4.2 *Let $\partial\Omega \in C^{l+2}$, and let $\mathbf{f} \in \mathcal{W}^{l,2}_{\boldsymbol{\beta}}(\Omega)$. There exists a number $\beta_* > 0$ such that if $\beta \in (0, \beta_*)$, then the weak solution $\mathbf{u}$ belongs to the space $\mathcal{W}^{l+2,2}_{\boldsymbol{\beta}}(\Omega)$, and there exists a pressure function p with $\nabla p \in \mathcal{W}^{l,2}_{\boldsymbol{\beta}}(\Omega)$. The pair $\big(\mathbf{u}(x),\ p(x)\big)$ satisfies the Stokes equations* (4.1.1) *almost everywhere in Ω. Furthermore, the following estimate holds:*

$$\|\mathbf{u}\|_{\mathcal{W}^{l+2,2}_{\boldsymbol{\beta}}(\Omega)} + \|\nabla p\|_{\mathcal{W}^{l,2}_{\boldsymbol{\beta}}(\Omega)} \leq c\, \|\mathbf{f}\|_{\mathcal{W}^{l,2}_{\boldsymbol{\beta}}(\Omega)}, \tag{4.4.2}$$

where c is a positive constant.

Proof
Since $\mathbf{u} \in \overset{\circ}{H}{}^{1,2}(\Omega)$, we have

$$\int\limits_{\sigma_j} \mathbf{u} \cdot \mathbf{n}\, dx^{(j)\prime} = 0, \quad j = 1, 2, \ldots, J,$$

(see Theorem 2.5.3). Thus, we can take in the integral identity (4.1.2) the function $\boldsymbol{\eta}(x) = E^{(k)}_{\boldsymbol{\beta}}(x)\mathbf{u}(x) + \mathbf{W}^{(k)}(x)$, where $E^{(k)}_{\boldsymbol{\beta}}(x)$ denotes the "step" weight-function (2.4.22), and $\mathbf{W}^{(k)}(x)$ is a vector field constructed in Lemma 2.4.7. Consequently, $\operatorname{div} \boldsymbol{\eta}(x) = 0$, and $\boldsymbol{\eta}(x)\big|_{\partial\Omega} = 0$. This leads to

$$\begin{aligned} \nu \int\limits_{\Omega} &\nabla \mathbf{u}(x) \cdot \nabla\big(E^{(k)}_{\boldsymbol{\beta}}(x)\mathbf{u}(x) + \mathbf{W}^{(k)}(x)\big)\, dx \\ &= \int\limits_{\Omega} \mathbf{f}(x) \cdot \big(E^{(k)}_{\boldsymbol{\beta}}(x)\mathbf{u}(x) + \mathbf{W}^{(k)}(x)\big)\, dx. \end{aligned} \tag{4.4.3}$$

Note that all integrals in (4.4.3) are finite since $E_{\beta}^{(k)}(x)$ is equal to a constant for large $|x|$ and $\mathbf{W}^{(k)}(x)$ has a compact support.

To estimate the right-hand side of (4.4.3), we represent the product $\mathbf{f}(x)\cdot\mathbf{W}^{(k)}(x)$ in the form $\big(E_{\beta}^{(k)}(x)\mathbf{f}(x)\big)\cdot\big(E_{-\beta}^{(k)}(x)\mathbf{W}^{(k)}(x)\big)$ and apply the Young inequality. Utilizing inequality (2.4.28) for $\mathbf{W}^{(k)}(x)$, we obtain

$$\begin{aligned}
&\Big|\int\limits_{\Omega}\mathbf{f}(x)\cdot\big(E_{\beta}^{(k)}(x)\mathbf{u}(x)+\mathbf{W}^{(k)}(x)\big)\,dx\Big|\le\frac{1}{4\varepsilon}\int\limits_{\Omega}E_{\beta}^{(k)}(x)|\mathbf{f}(x)|^2\,dx\\
&\quad+\varepsilon\Big(\int\limits_{\Omega}E_{\beta}^{(k)}(x)|\mathbf{u}(x)|^2\,dx+c\int\limits_{\Omega}E_{-\beta}^{(k)}(x)|\mathbf{W}^{(k)}(x)|^2\,dx\Big)\\
&\quad\le\frac{1}{4\varepsilon}\int\limits_{\Omega}E_{\beta}^{(k)}(x)|\mathbf{f}(x)|^2\,dx+c\varepsilon\Big(\int\limits_{\Omega}E_{\beta}^{(k)}(x)|\nabla\mathbf{u}(x)|^2\,dx\\
&\quad+\int\limits_{\Omega}E_{-\beta}^{(k)}(x)|\nabla\mathbf{W}^{(k)}(x)|^2\,dx\Big)\le\frac{1}{4\varepsilon}\int\limits_{\Omega}E_{\beta}^{(k)}(x)|\mathbf{f}(x)|^2\,dx\\
&\qquad+c_1\varepsilon\int\limits_{\Omega}E_{\beta}^{(k)}(x)|\nabla\mathbf{u}(x)|^2\,dx.
\end{aligned}\tag{4.4.4}$$

Next, we extract the term $\nu\int_{\Omega}E_{\beta}^{(k)}(x)|\nabla\mathbf{u}(x)|^2dx$ from the left-hand side of (4.4.3) and estimate the remaining terms. Using the same argument, together with the Cauchy-Schwarz-Bunyakovsky inequality and the weighted Poincaré inequality (2.3.27), and taking into account the estimate $|\nabla E_{\beta}^{(k)}(x)|\le c\beta E_{\beta}^{(k)}(x)$ (see (2.4.23)), we arrive at

$$\begin{aligned}
&\nu\Big|\int\limits_{\Omega}\nabla\mathbf{u}(x)\cdot\nabla E_{\beta}^{(k)}(x)\cdot\mathbf{u}(x)\,dx\Big|+\nu\Big|\int\limits_{\Omega}\nabla\mathbf{u}(x)\cdot\nabla\mathbf{W}^{(k)}(x)\,dx\Big|\\
&\le\nu\Big(\int\limits_{\Omega}E_{\beta}^{(k)}(x)|\nabla\mathbf{u}(x)|^2\,dx\Big)^{1/2}\Big[c\beta\Big(\int\limits_{\Omega}E_{\beta}^{(k)}(x)|\mathbf{u}(x)|^2\,dx\Big)^{1/2}\\
&+c\Big(\int\limits_{\Omega}E_{-\beta}^{(k)}(x)|\nabla\mathbf{W}^{(k)}(x)|^2\,dx\Big)^{1/2}\Big]\le c_2\beta\int\limits_{\Omega}E_{\beta}^{(k)}(x)|\nabla\mathbf{u}(x)|^2\,dx.
\end{aligned}\tag{4.4.5}$$

Combining (4.4.3)–(4.4.5), we obtain

$$\begin{aligned}
\nu\int\limits_{\Omega}E_{\beta}^{(k)}(x)|\nabla\mathbf{u}(x)|^2\,dx&\le\frac{1}{4\varepsilon}\int\limits_{\Omega}E_{\beta}^{(k)}(x)|\mathbf{f}(x)|^2\,dx\\
&+\big(c_1\varepsilon+c_2\beta\big)\int\limits_{\Omega}E_{\beta}^{(k)}(x)|\nabla\mathbf{u}(x)|^2\,dx.
\end{aligned}$$

Let $\beta \leq \dfrac{\nu}{4c_2}$. Choosing $\varepsilon = \dfrac{\nu}{4c_1}$, from the last inequality, one obtains the following estimate

$$\begin{aligned}\frac{\nu}{2}\int\limits_{\Omega^{(k)}} E_{\boldsymbol{\beta}}(x)|\nabla\mathbf{u}(x)|^2\,dx &\leq \frac{\nu}{2}\int\limits_{\Omega} E_{\boldsymbol{\beta}}^{(k)}(x)|\nabla\mathbf{u}(x)|^2\,dx \\ &\leq c\int\limits_{\Omega} E_{\boldsymbol{\beta}}^{(k)}(x)|\mathbf{f}(x)|^2\,dx \leq c\int\limits_{\Omega} E_{\boldsymbol{\beta}}(x)|\mathbf{f}(x)|^2\,dx.\end{aligned} \tag{4.4.6}$$

Since the right-hand side of (4.4.6) is independent of k, we can pass to the limit as $k \to \infty$ in (4.4.6). This gives

$$\frac{\nu}{2}\int\limits_{\Omega} E_{\boldsymbol{\beta}}(x)|\nabla\mathbf{u}(x)|^2\,dx \leq c\int\limits_{\Omega} E_{\boldsymbol{\beta}}(x)|\mathbf{f}(x)|^2\,dx. \tag{4.4.7}$$

This estimate proves the exponential decay at infinity of the velocity $\mathbf{u}$.

To obtain weighted estimates for the higher derivatives of $\mathbf{u}(x)$ and for the pressure $p(x)$, we use local estimates for ADN-elliptic problems from [6, 198]. It is well known that the Stokes problem is ADN-elliptic [198]. Therefore, $\mathbf{u} \in W^{l+2,2}_{\text{loc}}(\Omega)$, $p \in W^{l+1,2}_{\text{loc}}(\Omega)$, and the Stokes system is valid almost everywhere in Ω. Local estimates for ADN-elliptic problems yield

$$\begin{aligned}\|\mathbf{u}\|^2_{W^{l+2,2}(\omega_{jk})} + \|\nabla p\|^2_{W^{l,2}(\omega_{jk})} &\leq c\Big(\|\mathbf{f}\|^2_{W^{l,2}(\widehat{\omega}_{jk})} + \|\mathbf{u}\|^2_{L^2(\widehat{\omega}_{jk})} \\ &\quad + \|p - \bar{p}_{jk}\|^2_{L^2(\widehat{\omega}_{jk})}\Big), \\ j = 1, \ldots, J;&\quad k = 0, 1, 2, \ldots,\end{aligned} \tag{4.4.8}$$

where $\bar{p}_{jk} = |\widehat{\omega}_{jk}|^{-1}\int_{\widehat{\omega}_{jk}} p(x)\,dx$. Here, $\omega_{jk} \subset \widehat{\omega}_{jk}$ are subdomains of Ω, and c is a constant independent of k. Let us estimate the last norm at the right-hand side of (4.4.8). Since $\int_{\widehat{\omega}_{jk}} \big(p(x) - \bar{p}_{jk}\big)\,dx = 0$, by Lemma 2.4.1, there exist vector fields $\mathbf{w}^{(jk)} \in \overset{\circ}{W}{}^{1,2}(\widehat{\omega}_{jk})$ such that $\operatorname{div}\mathbf{w}^{(jk)}(x) = p(x) - \bar{p}_{jk}$ in $\widehat{\omega}_{jk}$. The following estimate holds:

$$\|\nabla\mathbf{w}^{(jk)}\|_{L_2(\widehat{\omega}_{jk})} \leq c\|p - \bar{p}_{jk}\|_{L_2(\widehat{\omega}_{jk})} \tag{4.4.9}$$

Then, using the Stokes system (4.1.1), integrating by parts in $\widehat{\omega}_{jk}$, and applying (4.4.9), we derive

$$\int\limits_{\widehat{\omega}_{jk}} |p(x) - \bar{p}_{jk}|^2\,dx = \int\limits_{\widehat{\omega}_{jk}} \big(p(x) - \bar{p}_{jk}\big)\operatorname{div}\mathbf{w}^{(jk)}(x)\,dx$$

$$= -\int\limits_{\widehat{\omega}_{jk}} \nabla p(x) \cdot \mathbf{w}^{(jk)}(x)\,dx = \nu \int\limits_{\widehat{\omega}_{jk}} \nabla \mathbf{u}(x) \cdot \nabla \mathbf{w}^{(jk)}(x)\,dx$$

$$- \int\limits_{\widehat{\omega}_{jk}} \mathbf{f}(x) \cdot \mathbf{w}^{(jk)}(x)\,dx \le c\Big(\|\nabla \mathbf{u}\|_{L^2(\widehat{\omega}_{jk})} \|\nabla \mathbf{w}^{(jk)}\|_{L^2(\widehat{\omega}_{jk})}$$

$$+ \|\mathbf{f}\|_{L^2(\widehat{\omega}_{jk})} \|\mathbf{w}^{(jk)}\|_{L^2(\widehat{\omega}_{jk})} \Big)$$

$$\le c\Big(\|\nabla \mathbf{u}\|_{L^2(\widehat{\omega}_{jk})} + \|\mathbf{f}\|_{L^2(\widehat{\omega}_{jk})} \Big) \|p - \bar{p}_{jk}\|_{L^2(\widehat{\omega}_{jk})}.$$

Thus, we have the estimate:

$$\|p - \bar{p}_{jk}\|^2_{L^2(\widehat{\omega}_{jk})} \le c\Big(\|\nabla \mathbf{u}\|^2_{L^2(\widehat{\omega}_{jk})} + \|\mathbf{f}\|^2_{L^2(\widehat{\omega}_{jk})} \Big). \tag{4.4.10}$$

Substituting (4.4.10) into (4.4.8), we find

$$\begin{aligned} \|\mathbf{u}\|^2_{W^{l+2,2}(\omega_{jk})} + \|\nabla p\|^2_{W^{l,2}(\omega_{jk})} &\le c\Big(\|\mathbf{f}\|^2_{W^{l,2}(\widehat{\omega}_{jk})} + \|\mathbf{u}\|^2_{L^2(\widehat{\omega}_{jk})} \\ &+ \|\nabla \mathbf{u}\|^2_{L^2(\widehat{\omega}_{jk})} \Big) \le c\Big(\|\mathbf{f}\|^2_{W^{l,2}(\widehat{\omega}_{jk})} + \|\mathbf{u}\|^2_{W^{l+2,2}(\widehat{\omega}_{jk})} \Big), \\ j = 1, \ldots, J;\quad k &= 0, 1, 2, \ldots. \end{aligned} \tag{4.4.11}$$

Similarly, we can establish the following estimate:

$$\|\mathbf{u}\|^2_{W^{l+2,2}(\Omega^{(1)})} + \|\nabla p\|^2_{W^{l,2}(\Omega^{(1)})} \le c\Big(\|\mathbf{f}\|^2_{W^{l,2}(\Omega^{(2)})} + \|\nabla \mathbf{u}\|^2_{L^2(\Omega^{(2)})} \Big) \tag{4.4.12}$$

Furthermore, by multiplying both sides of inequalities (4.4.11) by $E_\beta(k)$, using the properties of the weight-functions $E_\beta(x)$, and summing up the obtained relations, we obtain the following estimate:

$$\begin{aligned} \|\mathbf{u}\|^2_{\mathcal{W}^{l+2,2}_\beta(\Omega_j)} + \|\nabla p\|^2_{\mathcal{W}^{l,2}_\beta(\Omega_j)} &\le c\Big(\|\mathbf{f}\|^2_{\mathcal{W}^{l,2}_\beta(\Omega_j)} + \|\nabla \mathbf{u}\|^2_{\mathcal{L}^2_\beta(\Omega_j)} \Big), \\ j &= 1, 2, \ldots, J. \end{aligned} \tag{4.4.13}$$

Now, estimate (4.4.2) follows by combining (4.4.7), (4.4.12), and (4.4.13). □

Consider now the stationary Stokes problem with a nonhomogeneous divergence equation and boundary values:

$$\begin{cases} -\nu\Delta\mathbf{u} + \nabla p = \mathbf{f}, \ x \in \Omega, \\ \operatorname{div}\mathbf{u} = h, \ x \in \Omega, \\ \mathbf{u}|_{\partial\Omega} = \mathbf{a}, \ x \in \partial\Omega, \\ \int\limits_{\sigma_k} \mathbf{u}\cdot\mathbf{n}\, ds = 0, \ k = 1, \dots, J. \end{cases} \tag{4.4.14}$$

Suppose that the boundary value $\mathbf{a} \in W^{1/2,2}(\partial\Omega)$ and the right-hand side $h \in L^2(\Omega)$ of the divergence equation have compact supports: $\operatorname{supp} h \subset \overline{\Omega \cap B_{R_0}}$, $\operatorname{supp}\mathbf{a} \subset \partial\Omega \cap \overline{B}_{R_0}$, where $B_{R_0} = \{x \in \mathbb{R}^n : \ |x| < R_0\}$, and satisfy the compatibility condition:

$$\int\limits_{\Omega} h(x)\, dx = \int\limits_{\partial\Omega} \mathbf{a}\cdot\mathbf{n}\, dS. \tag{4.4.15}$$

Then there exists a vector field $\mathbf{G} \in W^{1,2}(\Omega)$ such that

$$\operatorname{div}\mathbf{G} = h, \quad \mathbf{G}|_{\partial\Omega} = \mathbf{a}, \quad \operatorname{supp}\mathbf{G} \subset \Omega \cap B_{2R_0},$$

and it satisfies the estimate:

$$\|\mathbf{G}\|_{W^{1,2}(\Omega)} \le c\Big(\|h\|_{L^2(\Omega)} + \|\mathbf{a}\|_{W^{1/2,2}(\partial\Omega)}\Big).$$

Moreover, if $\mathbf{a} \in W^{3/2,2}(\partial\Omega)$ and $h \in W^{1,2}(B_{R_0} \cap \Omega)$, then there exists $\mathbf{G} \in W^{2,2}(B_{R_0} \cap \Omega)$ such that

$$\|\mathbf{G}\|_{W^{2,2}(B_{R_0}\cap\Omega)} \le c\Big(\|h\|_{W^{1,2}(B_{R_0}\cap\Omega)} + \|\mathbf{a}\|_{W^{3/2,2}(\partial\Omega)}\Big)$$

(see Lemma 2.4.5 and Remark 2.4.2).

A weak solution of the nonhomogeneous problem (4.4.14) is a vector field $\mathbf{u} = \mathbf{v} + \mathbf{G} \in W^{1,2}(\Omega)$ with $\mathbf{v} \in \mathring{H}^{1,2}(\Omega)$ satisfying, for all $\boldsymbol{\eta} \in \mathring{H}^{1,2}(\Omega)$, the integral identity:

$$\nu\int\limits_{\Omega} \nabla\mathbf{v}\cdot\nabla\boldsymbol{\eta}\, dx = \int\limits_{\Omega} \mathbf{f}\cdot\boldsymbol{\eta}\, dx - \nu\int\limits_{\Omega} \nabla\mathbf{G}\cdot\nabla\boldsymbol{\eta}\, dx. \tag{4.4.16}$$

Since $\mathbf{G}$ has compact support, Theorem 4.4.2 implies the following:

Theorem 4.4.3 *Assume that* $\mathbf{a}$ *and* h *have compact supports and satisfy the compatibility condition* (4.4.15).

1. *If* $\mathbf{f} \in \mathcal{L}^2_{\boldsymbol{\beta}}(\Omega)$, $0 < \beta \le \beta_*$, $\mathbf{a} \in W^{1/2,2}(\partial\Omega)$, *and* $h \in L^2(\Omega)$, *then there exists a unique weak solution* $\mathbf{u}$ *of problem* (4.4.14), $\mathbf{u} = \mathbf{v} + \mathbf{G}$ *with* $\mathbf{v} \in \mathcal{W}^{1,2}_{\boldsymbol{\beta}}(\Omega) \cap \mathring{H}^{1,2}(\Omega)$ *and:*

$$\|\mathbf{u}\|_{\mathcal{W}^{1,2}_{\boldsymbol{\beta}}(\Omega)} \le c\Big(\|h\|_{L^2(\Omega)} + \|\mathbf{a}\|_{W^{1/2,2}(\partial\Omega)} + \|\mathbf{f}\|_{\mathcal{L}^2_{\boldsymbol{\beta}}(\Omega)}\Big). \tag{4.4.17}$$

2. *Moreover, if in addition,* $\partial\Omega \in C^2$, $\mathbf{a} \in W^{3/2,2}(\partial\Omega)$, *and* $h \in W^{1,2}(\Omega)$, *then* $\mathbf{u} \in \mathcal{W}^{2,2}_{\boldsymbol{\beta}}(\Omega)$ *and there exists a function* $p \in L^2_{\text{loc}}(\Omega)$ *with* $\nabla p \in \mathcal{L}^2_{\boldsymbol{\beta}}(\Omega)$ *such that the pair* $(\mathbf{u}, p)$ *satisfies equations* (4.4.14) *almost everywhere in* Ω. *The following estimate holds:*

$$\|\mathbf{u}\|_{\mathcal{W}^{2,2}_{\boldsymbol{\beta}}(\Omega)} + \|\nabla p\|_{\mathcal{L}^2_{\boldsymbol{\beta}}(\Omega)} \le c\Big(\|h\|_{W^{1,2}(\Omega)} + \|\mathbf{a}\|_{W^{3/2,2}(\partial\Omega)} + \|\mathbf{f}\|_{\mathcal{L}^2_{\boldsymbol{\beta}}(\Omega)}\Big). \tag{4.4.18}$$

In Theorem 4.4.2, it was shown that ∇p vanishes exponentially as $|x| \to +\infty$. This property allows us to establish the existence of constants p_k such that

$$\lim_{x\in\Omega_k, |x|\to\infty} p(x) = p_k, \quad k = 1, \dots, J. \tag{4.4.19}$$

Furthermore, we prove that the convergence in (4.4.19) is exponential. This result is based on the following lemma.

Lemma 4.4.1 *Consider a semi-cylinder* $\Pi^+ = \{x = (x', z) \in \mathbb{R}^n : x' \in \sigma,\ z \in (0, +\infty)\}$, *where* σ *is a bounded domain in* $\mathbb{R}^{n-1}$ *with Lipschitz boundary. Let* $p \in W^{1,2}_{loc}(\Pi^+)$ *and satisfy the condition*

$$\int\limits_{\Pi^+} |\nabla p(x)|^2 e^{2\beta z} dx' dz < +\infty, \quad \beta > 0.$$

Then there exists a constant p_0 *such that the following inequality holds:*

$$\int\limits_{\Pi^+} |p(z, x') - p_0|^2 e^{2\beta z} dx' dz \le \frac{c}{\beta^2} \int\limits_{\Pi^+} |\nabla p(x)|^2 e^{2\beta z} dx' dz, \tag{4.4.20}$$

where c is a positive constant.

Proof
First, we establish the boundedness of the average $\bar{p}(z) = \int_\sigma p(z, x')dx'$. Let $z > 0$. We can express $\bar{p}(z)$ as follows:

$$\bar{p}(z) = \bar{p}(0) + \int\limits_0^z \bar{p}'(r)dr,$$

and by using the Cauchy-Schwarz-Bunyakovsky inequality, we find

$$\Big(\int_0^z \bar{p}'(r)dr\Big)^2 \le \int_0^z |\bar{p}'(r)|^2 e^{2\beta r} dr \int_0^z e^{-2\beta r} dr.$$

Hence, we obtain the inequality:

$$\begin{aligned} |\bar{p}(z)| &\le |\bar{p}(0)| + c\Big(\int_0^{+\infty} |\bar{p}'(z)|^2 e^{2\beta z} dz\Big)^{1/2} \\ &\le |\bar{p}(0)| + c\Big(\int_0^{+\infty}\int_\sigma |\partial_z p(x',z)|^2 e^{2\beta z} dx'dz\Big)^{1/2} \le const. \end{aligned} \tag{4.4.21}$$

Let us show the existence of a constant p_0 such that

$$\lim_{z\to+\infty} \bar{p}(z) = p_0.$$

As $\bar{p}(z)$ is bounded, we can identify a sequence $\{z_k\}$ such that $\lim\limits_{k\to+\infty} \bar{p}(z_k) = p_0$ for some constant $p_0 < +\infty$. Consider

$$|\bar{p}(z) - \bar{p}(z_k)| = \Big|\int_z^{z_k} \bar{p}'(r)dr\Big| \le \Big(\int_z^{+\infty}\int_\sigma |\partial_z p(x',z)|^2 e^{2\beta z} dx'dz\Big)^{1/2}.$$

Taking the limit as $k \to +\infty$, we arrive at

$$|\bar{p}(z) - p_0| \le \Big(\int_z^{+\infty}\int_\sigma |\partial_z p(x',z)|^2 e^{2\beta z} dx'dz\Big)^{1/2} \to 0 \text{ as } z \to +\infty. \tag{4.4.22}$$

Next, using Poincaré's inequality (see (2.3.6)), we find

$$\begin{aligned} \Big|\frac{d}{dz}\int_\sigma |p(x',z) - \bar{p}(z)|^2 dx'\Big| &\le 2\int_\sigma |p(x',z) - \bar{p}(z)||\partial_z p(x',z)|dx' \\ &\quad +2\int_\sigma |p(x',z) - \bar{p}(z)||\bar{p}'(z)|dx' \\ \le 2\int_\sigma |p(x',z) - \bar{p}(z)|^2 dx' + c\int_\sigma &|\partial_z p(x',z)|^2 dx' \le c\int_\sigma |\nabla p(x',z)|^2 dx'. \end{aligned}$$

Let $z < \xi < r < 2z$. Integrating the last inequality from ξ to r gives us

$$\int_\sigma |p(x',\xi)-\bar{p}(\xi)|^2 dx' \le \int_\sigma |p(x',r)-\bar{p}(r)|^2 dx' + c\int_\xi^r \int_\sigma |\nabla p(x',z)|^2 dx' dz$$
$$\le c\int_\sigma |\nabla_{x'} p(x',r)|^2 dx' + c\int_\xi^r \int_\sigma |\nabla p(x',z)|^2 dx' dz.$$

Multiplying both sides of the last inequality by $e^{2\beta r}$ and integrating with respect to r from z to $2z$ yields

$$\frac{1}{2\beta}e^{2\beta z}(e^{2\beta z}-1)\int_\sigma |p(x',\xi)-\bar{p}(\xi)|^2 dx' \le c\int_z^{2z}\int_\sigma |\nabla_{x'} p(x',r)|^2 e^{2\beta r} dx' dr$$
$$+c\int_z^{2z} e^{2\beta r}\int_\xi^r \int_\sigma |\nabla p(x',y)|^2 dx' dy\, dr$$
$$\le c\int_z^{2z}\int_\sigma |\nabla_{x'} p(x',r)|^2 e^{2\beta r} dx' dr + \frac{c}{2\beta}e^{2\beta z}(e^{2\beta z}-1)$$
$$\times \int_z^{2z}\int_\sigma |\nabla_{x'} p(x',y)|^2 e^{2\beta y} dx' dy.$$

From this, we can deduce that

$$\int_\sigma |p(x',\xi)-\bar{p}(\xi)|^2 dx' \le c\frac{e^{-2\beta z}}{e^{2\beta z}-1}\int_z^{2z}\int_\sigma |\nabla_{x'} p(x',r)|^2 e^{2\beta r} dx' dr$$
$$+c\int_z^{2z}\int_\sigma |\nabla_{x'} p(x',y)|^2 e^{2\beta y} dx' dy \to 0 \ \text{ as } \ \xi\to+\infty.$$

Here, we used that $\xi\in[z,2z]$ and thus, as $\xi\to+\infty$, it implies $z\to+\infty$. By the triangle inequality, we conclude

$$\lim_{z\to+\infty}\int_\sigma |p(x',z)-p_0|^2 dx' = 0.$$

To finalize the proof, we use the inequality

$$\int_0^{+\infty} |f(t)|^2 e^{2\beta t} dt \le \frac{1}{\beta^2}\int_0^{+\infty} |f'(t)|^2 e^{2\beta t} dt, \tag{4.4.23}$$

which holds for any function f such that

$$\int_0^{+\infty} |f'(t)|^2 e^{2\beta t} dt < +\infty, \qquad \lim_{t\to+\infty} f(t) = 0$$

(see Lemma 2.3.5).

By applying (4.4.23) to $f(z) = \left(\int_\sigma |p(x',z) - p_0|^2 dx'\right)^{1/2}$, we obtain

$$\begin{aligned}
\int_0^{+\infty} e^{2\beta z} \int_\sigma |p(x',z)-p_0|^2 dx' dz &\le \frac{1}{\beta^2} \int_0^{+\infty} e^{2\beta z} \Big|\partial_z \Big(\int_\sigma |p(x',z)-p_0|^2 dx'\Big)^{1/2}\Big|^2 dz \\
\le \frac{1}{\beta^2} \int_0^{+\infty} e^{2\beta z} &\left| \frac{1}{\left(\int_\sigma |p(x',z) - p_0|^2 dx'\right)^{1/2}} \int_\sigma |p(x',z) - p_0||\partial_z p(x',z)| dx' \right|^2 dz \\
&\le \frac{1}{\beta^2} \int_0^{+\infty} e^{2\beta z} \int_\sigma |\partial_z p(x',z)|^2 dx' dz.
\end{aligned}$$

□

Remark 4.4.1 From the previous lemma, it can be deduced that when the following condition holds:

$$\int_\Omega E_{\boldsymbol{\beta}}(x) |\nabla p(x)|^2 dx < +\infty,$$

then there exist constants p_j, $j = 1, 2, \ldots, J$, such that

$$\begin{aligned}
\int_{\Omega_0} |p(x)|^2 dx + \sum_{j=1}^{J} \int_{\Omega_j} \exp\,(2\beta x_n^{(j)}) |p(x) - p_j|^2 dx \\
\le c \int_\Omega E_{\boldsymbol{\beta}}(x) |\nabla p(x)|^2 dx.
\end{aligned} \tag{4.4.24}$$

Lemma 4.4.2 *Assuming all conditions of Theorem* 4.4.3 *to be fulfilled, the following formulas hold:*

$$p_k = \int_\Omega \mathbf{f} \cdot \mathbf{U}_k dx + \int_{\partial\Omega} \mathbf{a} \cdot \mathbf{n} P_k dS - \nu \int_{\partial\Omega} \mathbf{a} \cdot \partial_n \mathbf{U}_k dS - \int_\Omega h P_k dx, \quad k = 2, \ldots, J. \tag{4.4.25}$$

Since the pressure p in the Stokes equations is defined up to an additive constant, we can assume, without loss of generality, that $p_1 = 0$. Here $\big(\mathbf{U}_k(x), P_k(x)\big)$ are the solutions of the following problems:

$$\begin{cases} -\nu\Delta\mathbf{U}_k + \nabla P_k = 0, & x \in \Omega, \\ \operatorname{div}\mathbf{U}_k = 0, & x \in \Omega, \\ \mathbf{U}_k(x) = 0, & x \in \partial\Omega, \\ \int\limits_{\sigma_1} \mathbf{U}_k \cdot \mathbf{n}\, dS = -1, \ \int\limits_{\sigma_k} \mathbf{U}_k \cdot \mathbf{n}\, dS = 1, \ k = 2, \dots, J, \\ \int\limits_{\sigma_m} \mathbf{U}_k \cdot \mathbf{n}\, dS = 0, & m \neq 1, m \neq k. \end{cases} \tag{4.4.26}$$

Proof

First, we proceed to construct solutions $\big(\mathbf{U}_k, P_k\big)$ for problems (4.4.26). Within the cylinders corresponding to the outlets Ω_1 and Ω_k, we consider stationary Poiseuille solutions denoted by $\big(\mathbf{u}^{(1)}_{P_{\alpha_1}}(x), \mathcal{P}^{(1)}_{P_{\alpha_1}}(x)\big)$ and $\big(\mathbf{u}^{(k)}_{P_{\alpha_k}}(x), \mathcal{P}^{(k)}_{P_{\alpha_k}}(x)\big)$, respectively. These solutions satisfy the following flux condition:

$$\int\limits_{\sigma_1} u^{(1)}_{P_{\alpha_1}}(x^{(1)\prime})dx^{(1)\prime} = -1, \qquad \int\limits_{\sigma_k} u^{(k)}_{P_{\alpha_k}}(x^{(k)\prime})dx^{(k)\prime} = 1.$$

Here, $u^{(j)}_{P_{\alpha_j}}(x^{(j)\prime})$ represents the last (normal) component of the velocity field of the Poiseuille flow in the cylinder corresponding to the outlet Ω_j, using local coordinates related to Ω_j. Importantly, this component is the only non-zero component of the Poiseuille flow and depends solely on the coordinates of the section σ_j. Additionally, note that the pressure component $\mathcal{P}^{(j)}_{P_{\alpha_j}}$ of the Poiseuille solution depends solely on the coordinate $x^{(j)}_n$.

We introduce the following expressions:

$$\mathbf{U}_k(x) = \zeta(x^{(1)}_n)\mathbf{u}^{(1)}_{P_{\alpha_1}}(x) + \zeta(x^{(k)}_n)\mathbf{u}^{(k)}_{P_{\alpha_k}}(x) + \widetilde{\mathbf{U}}_k(x),$$

$$P_k(x) = \zeta(x^{(1)}_n)\mathcal{P}^{(1)}_{P_{\alpha_1}}(x^{(1)}_n) + \zeta(x^{(k)}_n)\mathcal{P}^{(k)}_{P_{\alpha_k}}(x^{(k)}_n) + \widetilde{P}_k(x),$$

where ζ is a smooth cut-off function with the property that $\zeta(\tau) = 0$ for $\tau \leq 1$ and $\zeta(\tau) = 1$ for $\tau \geq 2$.

For the remaining part $\big(\widetilde{\mathbf{U}}_k(x), \widetilde{P}_k(x)\big)$, we solve the following problem:

$$\begin{cases} -\nu\Delta\widetilde{\mathbf{U}}_k + \nabla\widetilde{P}_k = \widetilde{\mathbf{f}}, \ x \in \Omega, \\ \operatorname{div}\widetilde{\mathbf{U}}_k = \widetilde{h}, \ x \in \Omega, \\ \widetilde{\mathbf{U}}_k(x) = 0, \ x \in \partial\Omega, \end{cases} \tag{4.4.27}$$

where the functions $\widetilde{\mathbf{f}}$ and $\widetilde{h}$ have compact supports and the condition $\int_\Omega \widetilde{h}(x)dx = 0$ holds.

Utilizing Theorem 4.4.3, we ascertain the existence of a unique solution $\big(\widetilde{\mathbf{U}}_k(x), \widetilde{P}_k(x)\big)$ to problem (4.4.27), satisfying the property that $\widetilde{\mathbf{U}}_k$ and $\nabla\widetilde{P}_k(x)$ exponentially vanish as $|x| \to \infty$. Furthermore, there exist constants p_{mk} such that

$$\lim_{|x|\to\infty,\ x\in\Omega_m} \widetilde{P}_k(x) = p_{mk}, \quad m = 1, \ldots, J.$$

This convergence is also exponential.

Let $\Omega(L) = \{x \in \Omega : y_n^{(m)} < L,\ m = 1, \ldots, J\}$, $\sigma_m(L) = \{x \in \Omega_m : y^{(m)} = L\}$, and $\sigma(L) = \bigcup_{m=1}^{J} \sigma_m(L)$. We multiply (4.4.14) by $\mathbf{U}_k$ and integrate by parts in the domain $\Omega(L)$, leading to the following relations:

$$\begin{aligned}
-\nu \int_{\Omega(L)} \Delta\mathbf{u}\cdot\mathbf{U}_k dx &= -\nu \int_{\Omega(L)} \mathbf{u}\cdot\Delta\mathbf{U}_k dx - \nu \int_{\partial\Omega(L)} \Big(\partial_n\mathbf{u}\cdot\mathbf{U}_k - \mathbf{u}\cdot\partial_n\mathbf{U}_k\Big) dS \\
&= -\int_{\Omega(L)} \mathbf{u}\cdot\nabla P_k dx + \nu \int_{\partial\Omega(L)\setminus\sigma(L)} \mathbf{u}\cdot\partial_n\mathbf{U}_k dS \\
-\nu \sum_{m=1}^{J} \int_{\sigma_m(L)} \Big(\partial_n\mathbf{u}\cdot\mathbf{U}_k - \mathbf{u}\cdot\partial_n\mathbf{U}_k\Big) dS &= \nu \int_{\partial\Omega(L)\setminus\sigma(L)} \mathbf{u}\cdot\partial_n\mathbf{U}_k dS \\
+ \int_{\Omega(L)} \operatorname{div}\mathbf{u} P_k dx - \int_{\partial\Omega(L)\setminus\sigma(L)} P_k\mathbf{u}\cdot\mathbf{n} dS &- \sum_{m=1}^{J} \int_{\sigma_m(L)} P_k\mathbf{u}\cdot\mathbf{n} dS \\
&- \nu \sum_{m=1}^{J} \int_{\sigma_m(L)} \Big(\partial_n\mathbf{u}\cdot\mathbf{U}_k - \mathbf{u}\cdot\partial_n\mathbf{U}_k\Big) dS \\
= -\int_{\partial\Omega(L)\setminus\sigma(L)} P_k\mathbf{u}\cdot\mathbf{n} dS &+ \nu \int_{\partial\Omega(L)\setminus\sigma(L)} \mathbf{u}\cdot\partial_n\mathbf{U}_k dS + \int_{\Omega(L)} \operatorname{div}\mathbf{u} P_k dx \\
&+ I_1(L) + I_2(L),
\end{aligned} \tag{4.4.28}$$

Here, $\partial_n = \nabla\cdot\mathbf{n}$ represents the normal derivative, and $I_1(L)$ and $I_2(L)$ are defined as follows:

$$I_1(L) = -\sum_{m=1}^{J} \int_{\sigma_m(L)} P_k \mathbf{u} \cdot \mathbf{n} \, dS, \quad I_2(L) = -\nu \sum_{m=1}^{J} \int_{\sigma_m(L)} \left(\partial_n \mathbf{u} \cdot \mathbf{U}_k - \mathbf{u} \cdot \partial_n \mathbf{U}_k\right) dS.$$

Since $\mathbf{u}$ exponentially vanishes as $|x| \to \infty$, it follows that $I_2(L) \to 0$ as $L \to \infty$. Further, we have

$$\lim_{L\to\infty} I_1(L) = \lim_{L\to\infty} \Big[-\zeta(L)\mathcal{P}^{(1)}_{P_{\alpha_1}}(L) \int_{\sigma_1(L)} \mathbf{u} \cdot \mathbf{n} \, dS - \zeta(L)\mathcal{P}^{(k)}_{P_{\alpha_k}}(L) \int_{\sigma_k(L)} \mathbf{u} \cdot \mathbf{n} \, dS$$

$$- \sum_{m\neq 1, m\neq k} \int_{\sigma_m(L)} \widetilde{P}_k \mathbf{u} \cdot \mathbf{n} \, dS \Big] - \sum_{m\neq 1, m\neq k} p_{mk} \int_{\sigma_m(L)} \mathbf{u} \cdot \mathbf{n} \, dS = 0.$$

Here, we have used the fact that $\int_{\sigma_m(L)} \mathbf{u} \cdot \mathbf{n} \, dS = 0$ for $m = 1, \ldots, J$ (since $\operatorname{div} \mathbf{u} = 0$ for sufficiently large $|x|$, and $\mathbf{u}$ tends to zero as x tends to infinity: $\mathbf{u} \in \mathcal{W}^{1,2}_{\boldsymbol{\beta}}(\Omega)$, $\beta > 0$).

On the other hand, we have

$$\int_{\Omega(L)} \nabla p \cdot \mathbf{U}_k dx = \sum_{m=1}^{J} \int_{\sigma_m(L)} p \mathbf{U}_k \cdot \mathbf{n} dS, \tag{4.4.29}$$

and

$$\lim_{L\to\infty} \sum_{m=1}^{J} \int_{\sigma_m(L)} p \mathbf{U}_k \cdot \mathbf{n} dS = \lim_{L\to\infty} \sum_{m=1}^{J} p_m \int_{\sigma_m(L)} \mathbf{U}_k \cdot \mathbf{n} dS = p_k. \tag{4.4.30}$$

Combining equations (4.4.28)–(4.4.30), we obtain the final expression for p_k:

$$p_k = \int_{\Omega} \mathbf{f} \cdot \mathbf{U}_k dx + \int_{\partial\Omega} \mathbf{a} \cdot \mathbf{n} P_k dS - \nu \int_{\partial\Omega} \mathbf{a} \cdot \partial_n \mathbf{U}_k dS - \int_{\Omega} h P_k dx, \quad k = 2, \ldots, J.$$

□

Remark 4.4.2 We shall demonstrate that by making a slight modification to the right-hand side $\mathbf{f}$ in the Stokes equations (4.4.14), we can obtain a solution $\big(\mathbf{u}(x), p_1(x)\big)$ with the same velocity field as before, while ensuring that the redefined pressure $p^{(1)}(x)$ vanishes as $|x| \to \infty$. Let us define the new right-hand side as follows:

$$\mathbf{f}_1(x) = \mathbf{f}(x) - \sum_{m=2}^{J} \nabla\zeta(x_n^{(m)}) p_m,$$

where $\zeta(x_n^{(m)})$ is a smooth cut-off function in Ω_m which equals one for large $x_n^{(m)}$ and equals zero in $\{x \in \Omega_m : x_n^{(m)} < 1\}$.

Using the expression (4.4.25) for p_k, we obtain

$$p_k^{(1)} = \int\limits_{\Omega} \mathbf{f}\cdot\mathbf{U}_k dx + \int\limits_{\partial\Omega} \mathbf{a}\cdot\mathbf{n} P_k dS - \sum_{m=2}^{J} p_m \int\limits_{\Omega} \nabla\zeta\cdot\mathbf{U}_k dx - \nu \int\limits_{\partial\Omega} \mathbf{a}\cdot\partial_n \mathbf{U}_k dS - \int\limits_{\Omega} h P_k dx$$

$$= p_k - \sum_{m=2}^{J} p_m \int\limits_{\sigma_m} \mathbf{U}_k \cdot \mathbf{n}\, dS = p_k - p_k = 0, \quad k = 2, \dots, J.$$

Thus, the redefined pressure $p^{(1)}(x)$ satisfies the relations

$$p_k^{(1)} = 0, \quad k = 1, 2, \dots, J.$$

This ensures that the modified solution retains the same velocity field, but the pressure distribution guarantees vanishing behavior as $|x| \to \infty$.

4.5 Time-Dependent Stokes Problem

4.5.1 Time-Dependent Stokes Problem in a Bounded Domain

Let $\Omega \subset \mathbb{R}^n$ be a bounded Lipschitz domain. Consider the time-dependent Stokes problem within Ω^T:

$$\begin{cases} \partial_t \mathbf{u}(x,t) - \nu\Delta\mathbf{u}(x,t) + \nabla p(x,t) = \mathbf{f}(x,t), & (x,t) \in \Omega^T, \\ \operatorname{div} \mathbf{u}(x,t) = 0, & (x,t) \in \Omega^T, \\ \mathbf{u}(x,t) = 0, & x \in \partial\Omega,\ t \in [0,T], \\ \mathbf{u}(x,0) = \mathbf{u}_0(x), & x \in \Omega. \end{cases} \tag{4.5.1}$$

Definition 4.5.1 Let $\mathbf{f} \in L^2(\Omega^T)$ and $\mathbf{u}_0 \in \overset{\circ}{H}{}^{1,2}(\Omega)$. A *weak solution* of problem (4.5.1) is a function $\mathbf{u} \in \overset{\circ}{W}{}^{(1,1),2}_{\partial\Omega}(\Omega^T)$ that satisfies the following conditions:

$$\mathbf{u}(x,0) = \mathbf{u}_0(x), \quad \operatorname{div}\mathbf{u}(x,t) = 0 \tag{4.5.2}$$

and for every divergence-free vector field $\boldsymbol{\eta} \in \overset{\circ}{W}{}^{(1,0),2}_{\partial\Omega}(\Omega^T)$, the integral identity holds:

$$\int_0^t \int_\Omega \partial_\tau \mathbf{u}(x,\tau)\cdot \boldsymbol{\eta}(x,\tau)\,dxd\tau + \nu \int_0^t \int_\Omega \nabla \mathbf{u}(x,\tau)\cdot \nabla \boldsymbol{\eta}(x,\tau)\,dxd\tau$$
$$= \int_0^t \int_\Omega \mathbf{f}(x,\tau)\cdot \boldsymbol{\eta}(x,\tau)\,dxd\tau \quad \forall t \in (0,\,T]. \tag{4.5.3}$$

Theorem 4.5.1 *Suppose* $\mathbf{f} \in L^2(\Omega^T)$ *and* $\mathbf{u}_0 \in \overset{\circ}{H}{}^{1,2}(\Omega)$. *Then problem* (4.5.1) *admits a unique weak solution* $\mathbf{u} \in \overset{\circ}{W}{}^{(1,1),2}_{\partial\Omega}(\Omega^T)$, *and the following estimate holds:*

$$\|\mathbf{u}\|_{W^{(1,1),2}(\Omega^T)} \le c\Big(\|\mathbf{f}\|_{L^2(\Omega^T)} + \|\mathbf{u}_0\|_{W^{1,2}(\Omega)}\Big), \tag{4.5.4}$$

where the constant c is independent of $\mathbf{f}$, $\mathbf{u}_0$, *and* T.

The proof of this theorem can be found in [102] and [203]. In particular, the Galerkin method can be employed.

Regarding more regular solutions of problem (4.5.1), it is well-known that besides the regularity of the boundary and the data of the problem, $\mathbf{f}(x,t)$ and $\mathbf{u}_0(x)$ need to satisfy certain compatibility conditions. Let us describe these conditions following the arguments in the papers [196] and [197].

Denote

$$g^{(k)}(x) = \frac{\partial^k}{\partial t^k} g(x,t)\big|_{t=0}.$$

For a smooth solution $\mathbf{u}(x,t)$ of (4.5.1), $\mathbf{u}^{(k)}(x)$ must satisfy the compatibility conditions at the edge $\partial\Omega \times \{t=0\}$ of the domain $\Omega^T = \Omega \times (0,\,T)$:

$$\mathbf{u}^{(k)}(x)\big|_{\partial\Omega} = 0, \tag{4.5.5}$$

which yields

$$\mathbf{u}^{(k)}(x)\cdot \mathbf{n}(x)\big|_{\partial\Omega} = 0, \tag{4.5.6}$$

where $\mathbf{n}(x)$ is the outward unit normal vector on $\partial\Omega$. We can calculate $\mathbf{u}^{(k)}(x)$ from system (4.5.1). By differentiating (4.5.1) $(k-1)$ times with respect to t and passing to the limit as $t \to 0$, we get

$$\mathbf{u}^{(k)}(x) = \nu\Delta \mathbf{u}^{(k-1)}(x) - \nabla p^{(k-1)}(x) + \mathbf{f}^{(k-1)}(x). \tag{4.5.7}$$

Obviously, from (4.5.3) we have $\operatorname{div} \mathbf{u}^{(k)}(x) = 0$, and therefore, from (4.5.7), it follows that

$$\Delta p^{(k-1)}(x) = \operatorname{div} \mathbf{f}^{(k-1)}(x) \tag{4.5.8}$$

and

$$\frac{\partial}{\partial \mathbf{n}} p^{(k-1)}(x)\big|_{\partial\Omega} = \left(\nu\Delta\mathbf{u}^{(k-1)}(x)\cdot\mathbf{n}(x) + \mathbf{f}^{(k-1)}(x)\cdot\mathbf{n}(x)\right)\big|_{\partial\Omega}. \tag{4.5.9}$$

Thus, $p^{(k-1)}(x)$ is defined over $\mathbf{u}^{(k-1)}(x)$ as a solution of the Neumann problem (4.5.8), (4.5.9). The necessary compatibility condition for the existence of a solution with the finite Dirichlet integral of problem (4.5.8), (4.5.9) is the following:

$$\int\limits_{\partial\Omega} \left(\nu\Delta\mathbf{u}^{(k-1)}(x)\cdot\mathbf{n}(x) + \mathbf{f}^{(k-1)}(x)\cdot\mathbf{n}(x)\right) dS = \int\limits_{\Omega} \operatorname{div}\mathbf{f}^{(k-1)}(x)\, dx. \tag{4.5.10}$$

By the Gauss–Ostrogradsky formula, we have

$$\int\limits_{\partial\Omega} \left(\nu\Delta\mathbf{u}^{(k-1)}(x)\cdot\mathbf{n}(x) + \mathbf{f}^{(k-1)}(x)\cdot\mathbf{n}(x)\right) dS = \int\limits_{\Omega} \operatorname{div}(\nu\Delta\mathbf{u}^{(k-1)}(x) + \mathbf{f}^{(k-1)}(x))dx.$$

Since $\operatorname{div}\mathbf{u}^{(k-1)} = 0$, the condition (4.5.10) follows from the later equality.

Thus, assuming that $\mathbf{u}^{(k)}(x)\cdot\mathbf{n}(x)|_{\partial\Omega} = 0$, we can define all $\mathbf{u}^{(k)}(x)$ by recurrent formulas (4.5.7)–(4.5.9), where $\mathbf{u}^{(0)}(x) = \mathbf{u}_0(x)$.

Definition 4.5.2 We say that the data $\mathbf{f}(x,t)$ and $\mathbf{u}_0(x)$ of problem (4.5.1) satisfy the compatibility conditions of order s if $\operatorname{div}\mathbf{u}_0 = 0$, and the functions $\mathbf{u}^{(k)}(x)$, $k = 0, 1, \ldots, s-1$, defined by recurrent formulas (4.5.7)–(4.5.9), satisfy condition (4.5.5) and $\operatorname{div}\mathbf{u}^{(k)} = 0$.

Remark 4.5.1 Note that the conditions $\operatorname{div}\mathbf{u}_0 = 0$ and (4.5.7)–(4.5.9) imply that $\operatorname{div}\mathbf{u}^{(k)} = 0$. Furthermore, the condition (4.5.9) yields (4.5.6). So, to check the compatibility conditions one needs to check (4.5.5) only for tangential entries of $\mathbf{u}^{(k)}$.

Theorem 4.5.2 *Assume that* $\partial\Omega \in C^{2l+2}$, *and let* $\mathbf{f} \in W^{(2l,l),2}(\Omega^T)$ *and* $\mathbf{u}_0 \in W^{2l+1}(\Omega) \cap \mathring{W}^{1,2}(\Omega)$. *If the compatibility conditions of order l hold, then problem (4.5.1) admits a unique solution* $\left(\mathbf{u}(x,t), p(x,t)\right)$ *with* $\mathbf{u} \in \mathring{W}_{\partial\Omega}^{(2l+2,l+1),2}(\Omega^T)$ *and* $\nabla p \in W^{(2l,l),2}(\Omega^T)$. *Furthermore, the following estimate is valid:*

$$\|\mathbf{u};\|_{W^{(2l+2,l+1),2}(\Omega^T)} + \|\nabla p\|_{W^{(2l,l),2}(\Omega^T)} \le c\Big(\|\mathbf{u}_0\|_{W^{2l+1,2}(\Omega)} + \|\mathbf{f}\|_{W^{(2l,l)}(\Omega^T)}\Big), \tag{4.5.11}$$

where c is a constant independent of $\mathbf{f}$, $\mathbf{u}_0$, *and T.*

The proof of this theorem can be found in [102, 196], and [203].

4.5.2 Time-Dependent Stokes Problem in Domains with Outlets to Infinity

Let $\Omega \subset \mathbb{R}^n$ be a domain with cylindrical outlets to infinity. Consider the time-dependent Stokes problem (4.5.1) in Ω^T. It should be noted that the definition of the weak solution and the existence theorem (Theorem 4.5.1) remain applicable to problem (4.5.1) in domains with cylindrical outlets. The proof of the theorem is standard. Initially, by invoking Theorem 4.5.1, we establish the existence of a weak solution $\mathbf{u}^{(k)}(x,t)$ in the bounded domain $\Omega^{(k)}$. Subsequently, we take the limit as $k \to \infty$ in the integral identity (4.5.3), resulting in the weak solution in the unbounded domain Ω (see, e.g., [102]). Additionally, it is worth noting that any divergence-free vector field $\mathbf{u}(x,t)$ that vanishes on $\partial\Omega$ and satisfies $\mathbf{u}(\cdot,t), \nabla\mathbf{u}(\cdot,t) \in L^2(\Omega)$ will possess zero fluxes over all cross-sections σ_j:

$$\int_{\sigma_j} \mathbf{u}(x,t)\cdot\mathbf{n}(x)\,dS = 0, \quad j = 1,\dots,J,$$

(see Theorem 2.5.3).

Next, let us demonstrate the spatial decay of the solution as $x_n^{(j)} \to \infty$ in the outlets to infinity Ω_j.

Theorem 4.5.3 *Let* $\mathbf{f} \in \mathcal{L}^2_{\boldsymbol{\beta}}(\Omega^T)$, $\mathbf{u}_0 \in \mathcal{W}^{1,2}_{\boldsymbol{\beta}}(\Omega) \cap \mathring{H}^{1,2}(\Omega)$, *with* $\beta \geq 0$. *There exists a constant* $\beta_0 > 0$ *such that if* $\beta \in (0, \beta_0]$, *then the weak solution* $\mathbf{u}$ *of problem* (4.5.1) *belongs to the space* $\mathcal{W}^{(1,1),2}_{\boldsymbol{\beta}}(\Omega^T)$, *and the following estimate holds:*

$$\sup_{t\in[0,T]} \|\mathbf{u}(\cdot,t)\|_{\mathcal{W}^{1,2}_{\boldsymbol{\beta}}(\Omega)} + \|\mathbf{u}\|_{\mathcal{W}^{(1,1),2}_{\boldsymbol{\beta}}(\Omega^T)} \leq c\Big(\|\mathbf{u}_0\|_{\mathcal{W}^{1,2}_{\boldsymbol{\beta}}(\Omega)} + \|\mathbf{f}\|_{\mathcal{L}^2_{\boldsymbol{\beta}}(\Omega^T)}\Big). \tag{4.5.12}$$

Moreover, if $\partial\Omega \in C^2$, *then* $\mathbf{u} \in \mathcal{W}^{(2,1),2}_{\boldsymbol{\beta}}(\Omega^T)$, *and there exists a function* p *with* $\nabla p \in \mathcal{L}^2_{\boldsymbol{\beta}}(\Omega^T)$ *such that the pair* $\big(\mathbf{u}(x,t),\ p(x,t)\big)$ *satisfies equations* (4.5.1) *almost everywhere in* Ω^T. *Additionally, the following estimate holds:*

$$\|\mathbf{u}\|_{\mathcal{W}^{(2,1),2}_{\boldsymbol{\beta}}(\Omega^T)} + \|\nabla p\|_{\mathcal{L}^2_{\boldsymbol{\beta}}(\Omega^T)} \leq c\Big(\|\mathbf{u}_0\|_{\mathcal{W}^{1,2}_{\boldsymbol{\beta}}(\Omega)} + \|\mathbf{f}\|_{\mathcal{L}^2_{\boldsymbol{\beta}}(\Omega^T)}\Big). \tag{4.5.13}$$

Proof

As in Theorem 4.4.2, we take in the integral identity (4.5.3) the test function $\boldsymbol{\eta}(x,t) = E^{(k)}_{\boldsymbol{\beta}}(x)\mathbf{u}(x,t) + \mathbf{W}^{(k)}(x,t)$, where $E^{(k)}_{\boldsymbol{\beta}}(x)$ is the "step" weight-function (2.4.22) and $\mathbf{W}^{(k)}(x,t)$ is a vector field constructed in Lemma 2.4.7. It is important to note that $\operatorname{div}\boldsymbol{\eta}(x,t) = 0$ and $\boldsymbol{\eta}(x,t)\big|_{\partial\Omega} = 0$. By proceeding as before, we obtain

$$\int_0^t\int_\Omega E_{\boldsymbol\beta}^{(k)}(x)\partial_\tau\mathbf{u}(x,\tau)\cdot\mathbf{u}\,dxd\tau+\int_0^t\int_\Omega\partial_\tau\mathbf{u}(x,\tau)\cdot\mathbf{W}^{(k)}(x,\tau)\,dxd\tau$$
$$+\nu\int_0^t\int_\Omega\nabla\mathbf{u}(x,\tau)\cdot\nabla\big(E_{\boldsymbol\beta}^{(k)}(x)\mathbf{u}(x,\tau)+\mathbf{W}^{(k)}(x,\tau)\big)\,dxd\tau \tag{4.5.14}$$
$$=\int_0^t\int_\Omega\mathbf{f}(x,\tau)\cdot\big(E_{\boldsymbol\beta}^{(k)}(x)\mathbf{u}(x,\tau)+\mathbf{W}^{(k)}(x,\tau)\big)\,dxd\tau.$$

Since

$$\int_0^t\int_\Omega E_{\boldsymbol\beta}^{(k)}(x)\partial_\tau\mathbf{u}(x,\tau)\cdot\mathbf{u}(x,\tau)\,dxd\tau=\frac{1}{2}\int_\Omega E_{\boldsymbol\beta}^{(k)}(x)|\mathbf{u}(x,t)|^2\,dx$$
$$-\frac{1}{2}\int_\Omega E_{\boldsymbol\beta}^{(k)}(x)|\mathbf{u}_0(x)|^2\,dx,$$

we rewrite relation (4.5.14) as

$$\frac{1}{2}\int_\Omega E_{\boldsymbol\beta}^{(k)}|\mathbf{u}(x,t)|^2\,dx+\nu\int_0^t\int_\Omega E_{\boldsymbol\beta}^{(k)}(x)|\nabla\mathbf{u}(x,\tau)|^2\,dxd\tau$$
$$=\frac{1}{2}\int_\Omega E_{\boldsymbol\beta}^{(k)}(x)|\mathbf{u}_0(x)|^2dx-\int_0^t\int_\Omega\partial_\tau\mathbf{u}(x,\tau)\cdot\mathbf{W}^{(k)}(x,\tau)dxd\tau$$
$$-\nu\int_0^t\int_\Omega\nabla\mathbf{u}(x,\tau)\cdot\big(\nabla E_{\boldsymbol\beta}^{(k)}(x)\mathbf{u}(x,\tau)+\nabla\mathbf{W}^{(k)}(x,\tau)\big)dxd\tau \tag{4.5.15}$$
$$+\int_0^t\int_\Omega\mathbf{f}(x,\tau)\cdot\big((E_{\boldsymbol\beta}^{(k)}(x)\mathbf{u}(x,\tau)+\mathbf{W}^{(k)}(x,\tau)\big)dxd\tau$$
$$=I_1(t)+I_2(t)+I_3(t)+I_4(t).$$

By following similar arguments as in the proof of Theorem 4.4.2, we obtain the estimates:

$$|I_3(t)|+|I_4(t)|\le\frac{1}{4\varepsilon}\int_0^t\int_\Omega E_{\boldsymbol\beta}^{(k)}(x)|\mathbf{f}(x,\tau)|^2\,dxd\tau$$

$$+ c(\varepsilon+\beta)\int_0^t\int_\Omega E_{\boldsymbol{\beta}}^{(k)}(x)|\nabla\mathbf{u}(x,\tau)|^2\,dxd\tau,$$

and

$$|I_2(t)| \le$$

$$\le c\Big(\int_0^t\int_\Omega E_{-\boldsymbol{\beta}}^{(k)}(x)|\mathbf{W}^{(k)}(x,\tau)|^2 dxd\tau\Big)^{1/2}\Big(\int_0^t\int_\Omega E_{\boldsymbol{\beta}}^{(k)}(x)|\partial_\tau\mathbf{u}(x,\tau)|^2 dxd\tau\Big)^{1/2}$$
$$\le c\Big(\int_0^t\int_\Omega E_{-\boldsymbol{\beta}}^{(k)}(x)|\nabla\mathbf{W}^{(k)}(x,\tau)|^2\,dxd\tau\Big)^{1/2}\Big(\int_0^t\int_\Omega E_{\boldsymbol{\beta}}^{(k)}(x)|\partial_\tau\mathbf{u}(x,\tau)|^2\,dxd\tau\Big)^{1/2}$$
$$\le c\beta\Big[\int_0^t\int_\Omega E_{\boldsymbol{\beta}}^{(k)}(x)|\nabla\mathbf{u}(x,\tau)|^2\,dxd\tau + \int_0^t\int_\Omega E_{\boldsymbol{\beta}}^{(k)}(x)|\partial_\tau\mathbf{u}(x,\tau)|^2\,dxd\tau\Big].$$

Here we utilized the weighted Poincaré inequality, properties of the weight-function $E_{\boldsymbol{\beta}}^{(k)}(x)$ and inequality (2.4.28) for $\mathbf{W}^{(k)}(x,t)$.

Combining these estimates with (4.5.15), we arrive at

$$\begin{aligned}
\frac{1}{2}\int_\Omega E_{\boldsymbol{\beta}}^{(k)}(x)|\mathbf{u}(x,t)|^2\,dx + \nu\int_0^t\int_\Omega E_{\boldsymbol{\beta}}^{(k)}(x)|\nabla\mathbf{u}(x,\tau)|^2\,dxd\tau \\
\le c(\varepsilon+\beta)\int_0^t\int_\Omega E_{\boldsymbol{\beta}}^{(k)}(x)|\nabla\mathbf{u}(x,\tau)|^2 dxd\tau \\
+c\beta\int_0^t\int_\Omega E_{\boldsymbol{\beta}}^{(k)}(x)|\partial_\tau\mathbf{u}(x,\tau)|^2 dxd\tau + \frac{1}{2}\int_\Omega E_{\boldsymbol{\beta}}^{(k)}(x)|\mathbf{u}_0(x)|^2\,dx \\
+\frac{1}{4\varepsilon}\int_0^t\int_\Omega E_{\boldsymbol{\beta}}^{(k)}(x)|\mathbf{f}(x,\tau)|^2\,dxd\tau.
\end{aligned} \tag{4.5.16}$$

Let us take now in (4.5.3),

$$\eta(x,t) = E_{\boldsymbol{\beta}}^{(k)}(x)\partial_t\mathbf{u}(x,t) + \partial_t\mathbf{W}^{(k)}(x,t).$$

This yields the following equation:

$$
\begin{aligned}
&\int_0^t\int_\Omega E^{(k)}_{\boldsymbol\beta}(x)|\partial_\tau\mathbf{u}(x,\tau)|^2\,dxd\tau+\frac{\nu}{2}\int_\Omega E^{(k)}_{\boldsymbol\beta}(x)|\nabla\mathbf{u}(x,t)|^2\,dxdt\\
&=\frac{\nu}{2}\int_\Omega E^{(k)}_{\boldsymbol\beta}(x)|\nabla\mathbf{u}_0(x)|^2\,dx-\int_0^t\int_\Omega \partial_\tau\mathbf{u}(x,\tau)\cdot\partial_\tau\mathbf{W}^{(k)}(x,\tau)\,dxd\tau\\
&-\nu\int_0^t\int_\Omega \nabla\mathbf{u}(x,\tau)\cdot\big(\nabla E^{(k)}_{\boldsymbol\beta}(x)\partial_\tau\mathbf{u}(x,\tau)+\nabla\partial_\tau\mathbf{W}^{(k)}(x,\tau)\big)\,dxd\tau\\
&+\int_0^t\int_\Omega \mathbf{f}(x,\tau)\cdot\big(E^{(k)}_{\boldsymbol\beta}(x)\partial_\tau\mathbf{u}(x,\tau)+\partial_\tau\mathbf{W}^{(k)}(x,\tau)\big)\,dxd\tau.
\end{aligned}
\tag{4.5.17}
$$

By estimating the right-hand side of (4.5.17), we deduce

$$
\begin{aligned}
&\int_0^t\int_\Omega E^{(k)}_{\boldsymbol\beta}(x)|\partial_\tau\mathbf{u}(x,\tau)|^2\,dxd\tau+\frac{\nu}{2}\int_\Omega E^{(k)}_{\boldsymbol\beta}(x)|\nabla\mathbf{u}(x,t)|^2\,dx\\
&\qquad\le c(\beta+\varepsilon)\int_0^t\int_\Omega E^{(k)}_{\boldsymbol\beta}(x)|\partial_\tau\mathbf{u}(x,\tau)|^2\,dxd\tau\\
&\qquad+c\beta\int_0^t\int_\Omega E^{(k)}_{\boldsymbol\beta}(x)|\nabla\mathbf{u}(x,\tau)|^2\,dxd\tau\\
&+\frac{\nu}{2}\int_\Omega E^{(k)}_{\boldsymbol\beta}(x)|\nabla\mathbf{u}_0(x)|^2\,dx+\frac{1}{4\varepsilon}\int_0^t\int_\Omega E^{(k)}_{\boldsymbol\beta}(x)|\mathbf{f}(x,\tau)|^2\,dxd\tau.
\end{aligned}
\tag{4.5.18}
$$

By summing up inequalities (4.5.16) and (4.5.18), we obtain

$$
\begin{aligned}
&\frac{1}{2}\int_\Omega E^{(k)}_{\boldsymbol\beta}(x)\big(|\mathbf{u}(x,t)|^2+\nu|\nabla\mathbf{u}(x,t)|^2\big)\,dx\\
&+\int_0^t\int_\Omega E^{(k)}_{\boldsymbol\beta}(x)\big(|\partial_\tau\mathbf{u}(x,\tau)|^2+\nu|\nabla\mathbf{u}(x,\tau)|^2\big)\,dxd\tau\\
&\le c_*\big(\beta+\varepsilon\big)\int_0^t\int_\Omega E^{(k)}_{\boldsymbol\beta}(x)\big(|\partial_\tau\mathbf{u}(x,\tau)|^2+|\nabla\mathbf{u}(x,\tau)|^2\big)\,dxd\tau\\
&\quad+\frac{1}{2}\int_\Omega E^{(k)}_{\boldsymbol\beta}(x)\big(|\mathbf{u}_0(x)|^2+\nu|\nabla\mathbf{u}_0(x)|^2\big)\,dx\\
&\quad+\frac{1}{2\varepsilon}\int_0^t\int_\Omega E^{(k)}_{\boldsymbol\beta}(x)|\mathbf{f}(x,\tau)|^2\,dxd\tau.
\end{aligned}
\tag{4.5.19}
$$

Next, let us fix $\varepsilon = \min\left\{\frac{\nu}{4c_*}, \frac{1}{4c_*}\right\}$. If the number β is sufficiently small, i.e., if

$$\beta \le \beta_* = \frac{1}{4c_*}\min\{\nu, 1\},$$

then (4.5.19) yields the inequality:

$$\begin{aligned}&\int\limits_{\Omega} E_{\boldsymbol{\beta}}^{(k)}(x)(|\mathbf{u}(x,t)|^2 + \nu|\nabla\mathbf{u}(x,t)|^2)\,dx \\ &+ \int\limits_0^t\int\limits_{\Omega} E_{\boldsymbol{\beta}}^{(k)}(x)\big(|\partial_\tau\mathbf{u}(x,\tau)|^2 + \nu|\nabla\mathbf{u}(x,\tau)|^2\big)\,dxd\tau \\ &\le c\int\limits_{\Omega} E_{\boldsymbol{\beta}}^{(k)}(x)(|\mathbf{u}_0(x)|^2 + \nu|\nabla\mathbf{u}_0(x)|^2)\,dx \\ &+c\int\limits_0^t\int\limits_{\Omega} E_{\boldsymbol{\beta}}^{(k)}(x)|\mathbf{f}(x,\tau)|^2\,dxd\tau \le c\Big(\|\mathbf{u}_0\|^2_{\mathcal{W}^{1,2}_{\boldsymbol{\beta}}(\Omega)} + \|\mathbf{f}\|^2_{\mathcal{L}^2_{\boldsymbol{\beta}}(\Omega^T)}\Big).\end{aligned} \tag{4.5.20}$$

The right-hand side of (4.5.20) does not depend on k. By taking the limit as $k \to \infty$ in (4.5.20), we obtain

$$\|\mathbf{u}(\cdot,t)\|^2_{\mathcal{W}^{1,2}_{\boldsymbol{\beta}}(\Omega)} + \|\mathbf{u}\|^2_{\mathcal{W}^{(1,1),2}_{\boldsymbol{\beta}}(\Omega^T)} \le c\Big(\|\mathbf{u}_0\|^2_{\mathcal{W}^{1,2}_{\boldsymbol{\beta}}(\Omega)} + \|\mathbf{f}\|^2_{\mathcal{L}^2_{\boldsymbol{\beta}}(\Omega^T)}\Big). \tag{4.5.21}$$

Now, let us assume that $\partial\Omega \in C^2$. From integral identity (4.5.3), it follows that for almost all $t \in [0, T]$, the identity

$$\nu\int\limits_{\Omega}\nabla\mathbf{u}(x,t)\cdot\nabla\boldsymbol{\eta}(x)\,dx = \int\limits_{\Omega}\big(\mathbf{f}(x,t) - \partial_t\mathbf{u}(x,t)\big)\cdot\boldsymbol{\eta}(x)\,dx$$

holds for any divergence-free test function $\boldsymbol{\eta}(x)$ belonging to the space $\mathring{W}^{1,2}(\Omega)$. Consequently, $\mathbf{u}(\cdot,t) \in \mathring{W}^{1,2}(\Omega)$ can be considered as a weak solution of the stationary Stokes problem with the right-hand side equal to $\mathbf{f}(x,t) - \partial_t\mathbf{u}(x,t)$:

$$\begin{cases} -\nu\Delta\mathbf{u}(x,t) + \nabla p(x,t) = \mathbf{f}(x,t) - \partial_t\mathbf{u}(x,t), & x \in \Omega, \\ \operatorname{div}\mathbf{u}(x,t) = 0, & x \in \Omega, \\ \mathbf{u}(x,t) = 0, & x \in \partial\Omega. \end{cases} \tag{4.5.22}$$

We have $\mathbf{f}(\cdot,t) - \partial_t\mathbf{u}(\cdot,t) \in \mathcal{L}^2_{\boldsymbol{\beta}}(\Omega)$ for almost all $t \in [0, T]$. If $\beta \in (0, \beta_*]$, then the weak solution $\mathbf{u}(x,t)$ of (4.5.22) belongs to the space $\mathcal{W}^{2,2}_{\boldsymbol{\beta}}(\Omega)$ (see Theorem 4.4.2). Moreover, there exists a function $p(x,t)$ with $\nabla p(\cdot,t) \in \mathcal{L}^2_{\boldsymbol{\beta}}(\Omega)$

such that the pair $(\mathbf{u}(x,t),\ p(x,t))$ satisfies equations (4.5.22) almost everywhere in Ω and

$$\begin{aligned}\|\mathbf{u}(\cdot,t)\|^2_{\mathcal{W}^{2,2}_\beta(\Omega)} + \|\nabla p(\cdot,t)\|^2_{\mathcal{L}^2_\beta(\Omega)} &\le c\|\big(\mathbf{f}(\cdot,t)-\mathbf{u}_t(\cdot,t)\big)\|^2_{\mathcal{L}^2_\beta(\Omega)}\\ &\le c\Big(\|\mathbf{f}(\cdot,t)\|^2_{\mathcal{L}^2_\beta(\Omega)} + \|\partial_t\mathbf{u}(\cdot,t)\|^2_{\mathcal{L}^2_\beta(\Omega)}\Big).\end{aligned} \tag{4.5.23}$$

Integrating inequality (4.5.23) with respect to t and using (4.5.21) for the estimate of $\|\partial_t\mathbf{u}(\cdot,t)\|^2_{\mathcal{L}^2_{2,\beta}(\Omega)}$, we derive

$$\int_0^t\int_\Omega E_\beta(x)\big(|\mathbf{u}(x,\tau)|^2 + |\nabla\mathbf{u}(x,\tau)|^2 + \sum_{|\alpha|=2}|D^\alpha_x\mathbf{u}(x,\tau)|^2\big)\,dxd\tau$$

$$+\int_0^t\int_\Omega E_\beta(x)|\nabla p(x,\tau)|^2\,dxd\tau \le c\Big(\|\mathbf{u}_0\|^2_{\mathcal{W}^{1,2}_\beta(\Omega)} + \|\mathbf{f}\|^2_{\mathcal{L}^2_\beta(\Omega^T)}\Big).$$

The last estimate together with (4.5.21) is equivalent to (4.5.13). □

Remark 4.5.2 In the proof of Theorem 4.5.3, we encountered a certain "inaccuracy." Specifically, for the function $\mathbf{u} \in W^{(1,1),2}(\Omega^T)$, the derivative $\partial_t\mathbf{u}(x,t)$ does not belong to the space $W^{(1,0),2}(\Omega^T)$. Consequently, taking $\boldsymbol{\eta}(x,t) = E^{(k)}_\beta(x)\partial_t\mathbf{u}(x,t) + \partial_t\mathbf{W}^{(k)}(x,t)$ in integral identity (4.5.3) is not formally permissible.

However, this inaccuracy can be addressed without compromising the final estimates. To achieve this, we first approximate the data $\mathbf{f} \in L^2(\Omega^T)$ and $\mathbf{u}_0 \in \mathring{W}^{1,2}(\Omega)$ using functions $\mathbf{f}_n$ with $\mathbf{f}_n,\ \partial_t\mathbf{f}_n \in L^2(\Omega^T)$ and $\mathbf{u}_{0n} \in W^{2,2}(\Omega)\cap\mathring{W}^{1,2}(\Omega)$. Each pair $\big(\mathbf{f}_n(x,t),\ \mathbf{u}_{0n}(x)\big)$ admits a unique solution $\mathbf{u}_n \in W^{(1,1),2}(\Omega^T)$ with $\nabla\partial_t\mathbf{u}_n \in L^2(\Omega^T)$ (see, e.g., the considerations in Chapter 6 of [102]). For such $\mathbf{u}_n(x,t)$, all the considerations of Theorem 4.5.3 remain valid and legitimate. Consequently, we obtain estimates (4.5.12), (4.5.13) with constants independent of n and subsequently take the limit as $n \to \infty$. The validity of the theorem's conclusions is thereby upheld, and the proof is complete.

Consider the regularity of the solution to problem (4.5.1) in domains with cylindrical outlets. Let $\mathbf{f}$ be an element of the space $\mathcal{W}^{(2l,l),2}_\beta(\Omega^T)$ and $\mathbf{u}_0$ an element of $\mathcal{W}^{2l+1,2}_\beta(\Omega)\cap\mathring{H}^{1,2}(\Omega)$. Suppose that $\mathbf{u}^{(k-1)}$ belongs to $\mathcal{W}^{2l+3-2k,2}_\beta(\Omega)$, where $k \le l$. Then, the following assertions hold:

$$\operatorname{div}\mathbf{f}^{(k-1)} \in \mathcal{W}^{2l-2k,2}_\beta(\Omega),\quad \big(\nu\Delta\mathbf{u}^{(k-1)}\cdot\mathbf{n} + \mathbf{f}^{(k-1)}\cdot\mathbf{n}\big)\big|_{\partial\Omega} \in \mathcal{W}^{2l-2k+1/2,2}_\beta(\partial\Omega).$$

Furthermore, for sufficiently small β, the solution $p^{(k-1)}(x)$ of problem (4.5.8), (4.5.9) satisfies the inclusion $\nabla p^{(k-1)} \in \mathcal{W}_{\beta}^{2l+1-2k,2}(\Omega)$, and there exists a constant c such that the estimate

$$\|\nabla p^{(k-1)}\|_{\mathcal{W}_{\beta}^{2l+1-2k,2}(\Omega)} \leq c\Big(\|\operatorname{div} \mathbf{f}^{(k-1)}\|_{\mathcal{W}_{\beta}^{2l-2k,2}(\Omega)} + \|\mathbf{u}^{(k-1)}\|_{\mathcal{W}_{\beta}^{2l+3-2k,2}(\Omega)}\Big)$$

$$\leq c\Big(\|\mathbf{f}\|_{\mathcal{W}_{\beta}^{(2l,l),2}(\Omega^T)} + \|\mathbf{u}^{(k-1)}\|_{\mathcal{W}_{\beta}^{2l+3-2k,2}(\Omega)}\Big)$$

holds. This estimate may be proved using the same method as in the proof of Theorem 4.4.2. (The proof of the above statements can be found also in [176], where results in weighted spaces with exponential weights are discussed).

From Eq. (4.5.7), it follows that $\mathbf{u}^{(k)} \in \mathcal{W}_{\beta}^{2l+1-2k,2}(\Omega)$, and we have the following estimate:

$$\|\mathbf{u}^{(k)}\|_{\mathcal{W}_{\beta}^{2l+1-2k,2}(\Omega)} \leq c\Big(\|\mathbf{f}\|_{\mathcal{W}_{\beta}^{(2l,l),2}(\Omega^T)} + \|\mathbf{u}^{(k-1)}\|_{\mathcal{W}_{\beta}^{2l+3-2k,2}(\Omega)}\Big).$$

Since $\mathbf{u}^{(0)} = \mathbf{u}_0 \in \mathcal{W}_{\beta}^{2l+1,2}(\Omega)$, the above inequality implies that

$$\|\mathbf{u}^{(k)}\|_{\mathcal{W}_{\beta}^{2l+1-2k,2}(\Omega)} \leq c\Big(\|\mathbf{f}\|_{\mathcal{W}_{\beta}^{(2l,l),2}(\Omega^T)} + \|\mathbf{u}_0\|_{\mathcal{W}_{\beta}^{2l+1,2}(\Omega)}\Big). \tag{4.5.24}$$

Finally, we observe that

$$\int\limits_{\sigma_j} \mathbf{u}^{(k)}(x) \cdot \mathbf{n}(x)\, ds = 0 \quad \forall j = 0, \ldots, J,$$

because divergence-free vector fields equal to zero on $\partial\Omega$ and having finite Dirichlet integrals have zero fluxes in domains with cylindrical outlets to infinity.

We can state the following theorem:

Theorem 4.5.4 *Let $\partial\Omega \in C^{2l+2}$. Assume that $\mathbf{f} \in \mathcal{W}_{\beta}^{(2l,l),2}(\Omega^T)$, $\mathbf{u}_0 \in \mathcal{W}_{\beta}^{2l+1,2}(\Omega)$, $\beta \geq 0$, and $l \geq 1$. Suppose that the compatibility conditions of order l are satisfied. There exists a number $\beta_\sharp > 0$ such that if $\beta \in (0, \beta_\sharp]$, then problem (4.5.1) admits a unique solution $\big(\mathbf{u}(x,t), p(x,t)\big)$ with $\mathbf{u} \in \mathcal{W}_{\beta}^{(2l+2,l+1),2}(\Omega^T)$ and $\nabla p \in \mathcal{W}_{\beta}^{(2l,l),2}(\Omega^T)$. Furthermore, the following estimate holds:*

$$\|\mathbf{u}\|_{\mathcal{W}_{\beta}^{(2l+2,l+1),2}(\Omega^T)} + \|\nabla p\|_{\mathcal{W}_{\beta}^{(2l,l),2}(\Omega^T)} \leq c\Big(\|\mathbf{u}_0\|_{\mathcal{W}_{\beta}^{2l+1,2}(\Omega)} + \|\mathbf{f}\|_{\mathcal{W}_{\beta}^{(2l,l),2}(\Omega^T)}\Big), \tag{4.5.25}$$

where c is a positive constant independent of β, T and the data of the problem.

Proof
Firstly, let us observe that if the compatibility conditions of the order l hold, then, as stated in Theorem 4.5.2, we know that $\mathbf{u} \in W^{(2l+2,l+1),2}(\Omega^T)$, $\nabla p \in W^{(2l,l),2}(\Omega^T)$, and the solution is unique. Therefore, our main objective is to demonstrate the validity of estimate (4.5.25).

For the case when $l = 0$, estimate (4.5.25) coincides with (4.5.13) from Theorem 4.5.3. Now, consider the case when $l \geq 1$. We assume that estimate (4.5.25) is true for $l \leq r-1$, and we aim to prove it for $l = r$. Let $\mathbf{f} \in \mathcal{W}_{\boldsymbol{\beta}}^{(2r,r),2}(\Omega^T)$ and $\mathbf{u}_0 \in \mathcal{W}_{\boldsymbol{\beta}}^{2r+1,2}(\Omega)$. According to Theorem 4.5.2 and our induction assumption, the solution $\mathbf{u} \in \mathcal{W}_{\boldsymbol{\beta}}^{(2r,r),2}(\Omega^T) \cap W^{(2r+2,r+1),2}(\Omega^T)$, $\nabla p \in \mathcal{W}_{\boldsymbol{\beta}}^{(2r-2,r-1),2}(\Omega^T) \cap W^{(2r,r),2}(\Omega^T)$ of problem (4.5.1) satisfies estimate (4.5.25) for $l = r-1$. Let us denote

$$\mathbf{u}^{\{s\}}(x,t) = \frac{\partial^s}{\partial t^s}\mathbf{u}(x,t), \quad p^{\{s\}}(x,t) = \frac{\partial^s}{\partial t^s}p(x,t),$$

$$\mathbf{f}^{\{s\}}(x,t) = \frac{\partial^s}{\partial t^s}\mathbf{f}(x,t), \quad s = 1,\dots,r.$$

By differentiating equation (4.5.1) s times with respect to t, we obtain the following problem:

$$\begin{cases} \partial_t \mathbf{u}^{\{s\}} - \nu\Delta\mathbf{u}^{\{s\}} + \nabla p^{\{s\}} = \mathbf{f}^{\{s\}}, & (x,t) \in \Omega^T, \\ \operatorname{div} \mathbf{u}^{\{s\}} = 0, & (x,t) \in \Omega^T \\ \mathbf{u}^{\{s\}} = 0, & x \in \partial\Omega, t \in [0,T], \\ \mathbf{u}^{\{s\}}(x,0) = \mathbf{u}^{(s)}(x), \ x \in \Omega. \end{cases} \tag{4.5.26}$$

It is evident that $\mathbf{f}^{\{s\}} \in \mathcal{W}_{\boldsymbol{\beta}}^{(2r-2s,r-s),2}(\Omega^T)$. Moreover, due to the compatibility conditions of the order r, we have $\mathbf{u}^{(s)} \in \mathcal{W}_{\boldsymbol{\beta}}^{2r+1-2s,2}(\Omega)$ and $\mathbf{u}^{(s)}$ satisfy the following conditions:

$$\operatorname{div} \mathbf{u}^{(s)}(x) = 0, \quad \mathbf{u}^{(s)}(x)\big|_{\partial\Omega} = 0, \quad \int\limits_{\sigma_j} \mathbf{u}^{(s)}(x) \cdot \mathbf{n}(x)\, ds = 0, \quad j = 1,\dots,J,$$

for $s = 1,\dots,r$. Therefore, based on estimate (4.5.25) at $l = r-s$ for $s = 1,\dots,r$, we can deduce the following inequalities:

$$\begin{aligned} &\|\mathbf{u}^{\{s\}}\|_{\mathcal{W}_{\boldsymbol{\beta}}^{(2r+2-2s,r+1-s),2}(\Omega^T)} + \|\nabla p^{\{s\}}\|_{\mathcal{W}_{\boldsymbol{\beta}}^{(2r-2s,r-s),2}(\Omega^T)} \\ &\leq c\Big(\|\mathbf{f}^{\{s\}}\|_{\mathcal{W}_{\boldsymbol{\beta}}^{(2r-2s,r-s),2}(\Omega^T)} + \|\mathbf{u}^{(s)}\|_{\mathcal{W}_{\boldsymbol{\beta}}^{2r+1-2s,2}(\Omega)}\Big). \end{aligned} \tag{4.5.27}$$

By estimating the right-hand side of (4.5.27) using (4.5.24) with $k = s$, we obtain

$$\begin{aligned}&\|\frac{\partial^s}{\partial t^s}\mathbf{u}\|_{\mathcal{W}_\beta^{(2r+2-2s,r+1-s),2}(\Omega^T)} + \|\frac{\partial^s}{\partial t^s}\nabla p\|_{\mathcal{W}_\beta^{(2r-2s,r-s),2}(\Omega^T)}\\ &\le c\Big(\|\mathbf{f}\|_{\mathcal{W}_\beta^{(2r,r),2}(\Omega^T)} + \|\mathbf{u}_0\|_{\mathcal{W}_\beta^{2r+1,2}(\Omega)}\Big), \quad s = 1, \dots, r.\end{aligned} \tag{4.5.28}$$

Next, we consider $\big(\mathbf{u}^{\{s\}}(x,t), p^{\{s\}}(x,t)\big)$ as a solution of the steady Stokes problem given by

$$\begin{cases} -\nu\Delta\mathbf{u}^{\{s\}} + \nabla p^{\{s\}} = \mathbf{f}^{\{s\}} - \partial_t\mathbf{u}^{\{s\}}, & x \in \Omega,\\ \operatorname{div}\mathbf{u}^{\{s\}} = 0, & x \in \Omega,\\ \mathbf{u}^{\{s\}} = 0, & x \in \partial\Omega. \end{cases} \tag{4.5.29}$$

According to Theorem 4.5.3, the following estimates hold:

$$\begin{aligned}&\|\mathbf{u}^{\{s\}}(\cdot,t)\|^2_{\mathcal{W}_\beta^{2r+2-2s,2}(\Omega)} + \|\nabla p^{\{s\}}(\cdot,t)\|^2_{\mathcal{W}_\beta^{2r-2s,2}(\Omega)}\\ &\le c\Big(\|\mathbf{f}^{\{s\}}(\cdot,t)\|^2_{\mathcal{W}_\beta^{2r-2s,2}(\Omega)} + \|\partial_t\mathbf{u}^{\{s\}}(\cdot,t)\|^2_{\mathcal{W}_\beta^{2r-2s,2}(\Omega)}\Big), \quad s = 1, \dots, r.\end{aligned}$$

By integrating these last inequalities from 0 to t and utilizing (4.5.28), we obtain

$$\begin{aligned}&\int_0^t \|\frac{\partial^s\mathbf{u}(\cdot,\tau)}{\partial\tau^s}\|^2_{\mathcal{W}_\beta^{2r+2-2s,2}(\Omega)}d\tau + \int_0^t \|\nabla\frac{\partial^s p(\cdot,\tau)}{\partial\tau^s}\|^2_{\mathcal{W}_\beta^{2r-2s,2}(\Omega)}d\tau\\ &\le c\Big(\|\mathbf{f}\|^2_{\mathcal{W}_\beta^{(2r,r),2}(\Omega^T)} + \|\mathbf{u}_0\|^2_{\mathcal{W}_\beta^{2r+1,2}(\Omega)}\Big) \quad \forall t \in [0,T],\end{aligned} \tag{4.5.30}$$

where $s = 1, \dots, r$. Finally, considering $\big(\mathbf{u}(x,t), p(x,t)\big)$ as a solution of the Stokes problem (4.5.29) for $s = 0$, we conclude that $\mathbf{u}(\cdot,t) \in \mathcal{W}_\beta^{2r+2,2}(\Omega)$. By using estimate (4.5.30), we obtain the estimate

$$\begin{aligned}&\int_0^t \|\mathbf{u}(\cdot,\tau)\|^2_{\mathcal{W}_\beta^{2r+2,2}(\Omega)}d\tau + \int_0^t \|\nabla p(\cdot,\tau)\|^2_{\mathcal{W}_\beta^{2r,2}(\Omega)}d\tau\\ &\le c\Big(\int_0^t \|\mathbf{f}(x,\tau)\|^2_{\mathcal{W}_\beta^{2r,2}(\Omega)}\,d\tau + \int_0^t \|\partial_\tau\mathbf{u}(x,\tau)\|^2_{\mathcal{W}_\beta^{2r,2}(\Omega)}\,d\tau\Big)\\ &\le c\Big(\|\mathbf{f}\|^2_{\mathcal{W}_\beta^{(2r,r),2}(\Omega^T)} + \|\mathbf{u}_0\|^2_{\mathcal{W}_\beta^{2r+1,2}(\Omega)}\Big).\end{aligned} \tag{4.5.31}$$

Estimate (4.5.25) for $l = r$ follows from inequalities (4.5.28), (4.5.30), (4.5.31) and the definition of the norm in the space $\mathcal{W}_\beta^{(2r,r),2}(\Omega^T)$. □

4.5.3 Behavior of a Solution to the Stokes Problem as $t \to +\infty$

In this section, we investigate the behavior of a solution to the Stokes problem (4.5.1) as t approaches infinity.

Let us first recall the notation used in Sect. 3.2.7 of Chap. 3. Let $\mathbb{W}^{l,2}_{\mu}(0,\infty)$ be the space of functions with a finite norm defined as follows:

$$\|F\|_{\mathbb{W}^{l,2}_{\mu}(0,\infty)} = \|\exp(\mu t)F(t)\|_{W^{l,2}(0,\infty)}, \quad \mathbb{L}^2_{\mu}(0,\infty) = \mathbb{W}^{0,2}_{\mu}(0,\infty).$$

Theorem 4.5.5 *Let $\Omega \subset \mathbb{R}^n$, $\partial\Omega \in C^2$, be a bounded domain. Assume that $\mathbf{f} \in \mathbb{L}^2_{\mu}(0,\infty; L^2(\Omega))$ with $\mu > 0$ and $\mathbf{u}_0 \in \overset{\circ}{H}{}^{1,2}(\Omega)$. There exists $\mu_* > 0$ such that for all $\mu \in (0,\mu_*)$, the weak solution $\mathbf{u}(x,t)$ of problem* (4.5.1) *satisfies the estimate*

$$\sup_{t\in[0,\infty]} \Big\{ \exp(\mu t)\|\mathbf{u}(\cdot,t)\|^2_{W^{1,2}(\Omega)} \Big\} + \|\mathbf{u}\|^2_{\mathbb{L}^2_{\mu}(0,\infty;\overset{\circ}{W}{}^{1,2}(\Omega))} + \|\partial_t\mathbf{u}\|^2_{\mathbb{L}^2_{\mu}(0,\infty;L^2(\Omega))} \\ \le c\Big(\|\mathbf{f}\|^2_{\mathbb{L}^2_{\mu}(0,\infty;L^2(\Omega))} + \|\mathbf{u}_0\|^2_{W^{1,2}(\Omega)} \Big), \tag{4.5.32}$$

Moreover, the pressure p with vanishing mean satisfies the following estimate:

$$\|p\|^2_{\mathbb{L}^2_{\mu}(0,\infty;L^2(\Omega))} \le c\Big(\|\mathbf{f}\|^2_{\mathbb{L}^2_{\mu}(0,\infty;L^2(\Omega))} + \|\mathbf{u}_0\|^2_{W^{1,2}(\Omega)} \Big). \tag{4.5.33}$$

Proof
Recall that $\Omega^\infty = \Omega \times (0,\infty)$. By Theorem 4.5.2, the solution $\mathbf{u} \in \overset{\circ}{W}{}^{(2,1),2}_{\partial\sigma}(\Omega^\infty)$. Moreover, there exists a pressure function p with $\nabla p \in L^2(\Omega^\infty)$ such that the pair $(\mathbf{u}, p)$ satisfies equations (4.5.1) for almost all (x', t). Multiplying equations (4.5.1) first by $\mathbf{u}$ and then by $\partial_t\mathbf{u}$ and integrating by parts in Ω, we obtain the estimates:

$$\frac{1}{2}\frac{d}{dt}\int_\Omega |\mathbf{u}(x,t)|^2\,dx + \nu\int_\Omega |\nabla\mathbf{u}(x,t)|^2\,dx \\ \le c_\varepsilon\int_\Omega |\mathbf{f}(x,t)|^2\,dx + \varepsilon\int_\Omega |\mathbf{u}(x,t)|^2\,dx \\ \le c_\varepsilon\int_\Omega |\mathbf{f}(x,t)|^2\,dx + \varepsilon c_*\int_\Omega |\nabla\mathbf{u}(x,t)|^2\,dx, \tag{4.5.34}$$

and

$$\frac{\nu}{2}\frac{d}{dt}\int_\Omega |\nabla\mathbf{u}(x,t)|^2\,dx + \int_\Omega |\partial_t\mathbf{u}(x,t)|^2\,dx \\ \le \frac{1}{2}\int_\Omega |\mathbf{f}(x,t)|^2\,dx + \frac{1}{2}\int_\Omega |\partial_t\mathbf{u}(x,t)|^2\,dx, \tag{4.5.35}$$

where ε is an arbitrary positive number and c_* is the constant in the Poincaré inequality for the domain Ω. For sufficiently small ε, inequalities (4.5.34)–(4.5.35) yield

$$\begin{aligned}&\frac{d}{dt}\int\limits_{\Omega}\Big(|\mathbf{u}(x,t)|^2+\nu|\nabla\mathbf{u}(x,t)|^2\Big)\,dx\\&+\nu\int\limits_{\Omega}|\nabla\mathbf{u}(x,t)|^2\,dx+\int\limits_{\Omega}|\partial_t\mathbf{u}(x,t)|^2\,dx\le c\int\limits_{\Omega}|\mathbf{f}(x,t)|^2\,dx.\end{aligned}\tag{4.5.36}$$

If

$$\mu_*\le\frac{\nu}{c_*+\nu},$$

then

$$\mu_*\int\limits_{\Omega}\Big(|\mathbf{u}(x,t)|^2+\nu|\nabla\mathbf{u}(x,t)|^2\Big)\,dx\le\nu\int\limits_{\Omega}|\nabla\mathbf{u}(x,t)|^2\,dx,$$

and from (4.5.36), it follows that

$$\begin{aligned}&\frac{d}{dt}\int\limits_{\Omega}\Big(|\mathbf{u}(x,t)|^2+\nu|\nabla\mathbf{u}(x,t)|^2\Big)\,dx\\&+\mu_*\int\limits_{\Omega}\Big(|\mathbf{u}(x,t)|^2+\nu|\nabla\mathbf{u}(x,t)|^2\Big)\,dx\\&+\int\limits_{\Omega}|\partial_t\mathbf{u}(x,t)|^2\,dx\le c\int\limits_{\Omega}|\mathbf{f}(x,t)|^2\,dx.\end{aligned}\tag{4.5.37}$$

Multiplying (4.5.37) by $\exp(\mu_* t)$ and integrating over t gives

$$\begin{aligned}&\exp(\mu_*t)\Big(\int\limits_{\Omega}|\mathbf{u}(x,t)|^2dx+\nu\int\limits_{\Omega}|\nabla\mathbf{u}(x,t))|^2dx\Big)\\&\quad+\int\limits_0^t\exp(\mu_*\tau)\int\limits_{\Omega}|\mathbf{u}(x,t)|^2dxd\tau\\&\le c\int\limits_0^t\exp(\mu_*\tau)\int\limits_{\Omega}|\mathbf{f}(x,t)|^2\,dxd\tau+\int\limits_{\Omega}\Big(|\mathbf{u}_0(x)|^2+\nu|\nabla\mathbf{u}_0(x))|^2\Big)dx\\&\le c\Big(\|\mathbf{f}\|^2_{\mathbb{L}^2_{\mu_*}(0,\infty;L^2(\Omega))}+\|\mathbf{u}_0\|^2_{W^{1,2}(\Omega)}\Big).\end{aligned}\tag{4.5.38}$$

Obviously, the constant c in inequality (4.5.38) is independent of t, and all arguments remain true for any $\mu \in (0, \mu_*]$. Therefore, from (4.5.38) follows (4.5.32).

Evidently, the linear functional

$$H(\boldsymbol{\eta}) = \nu \int\limits_{\Omega} \nabla \mathbf{u}(x,t) \cdot \nabla \boldsymbol{\eta}(x) dx - \int\limits_{\Omega} \big(\mathbf{f}(x,t) - \partial_t \mathbf{u}(x,t)\big) \cdot \boldsymbol{\eta}(x) dx$$

defined on $\boldsymbol{\eta} \in \overset{\circ}{W}{}^{1,2}(\Omega)$ is bounded and $H(\boldsymbol{\eta}) = 0$ for $\forall \boldsymbol{\eta} \in \overset{\circ}{H}{}^{1,2}(\Omega)$ and almost all $t \in [0, \infty)$. Therefore, by Lemma 2.4.2, there exists a function $p(\cdot, t) \in \widehat{L^2}(\Omega)$ such that

$$H(\boldsymbol{\eta}) = \int\limits_{\Omega} p(x,t) \operatorname{div} \boldsymbol{\eta}(x) \quad \forall \boldsymbol{\eta} \in \overset{\circ}{W}{}^{1,2}(\Omega).$$

Taking in this equality $p(x,t) = \operatorname{div} \mathbf{w}(x,t)$, where $\mathbf{w}(\cdot, t) \in \overset{\circ}{W}{}^{1,2}(\Omega)$ and

$$\|\mathbf{w}(\cdot,t)\|_{W^{1,2}(\Omega)} \leq c \|p(\cdot,t)\|_{L^2(\Omega)} \quad \text{for a.a. } t \in (0,\infty)$$

(see Lemmas 2.4.1 and 2.4.8), we obtain

$$\begin{aligned} \int\limits_{\Omega} |p(x,t)|^2 dx &= \nu \int\limits_{\Omega} \nabla \mathbf{u}(x,t) \cdot \nabla \mathbf{w}(x,t) dx \\ &- \int\limits_{\Omega} \big(\mathbf{f}(x,t) - \partial_t \mathbf{u}(x,t)\big) \cdot \mathbf{w}(x,t) dx \\ &\leq c \Big(\|\nabla \mathbf{u}\|_{L^2(\Omega)} + \|\partial_t \mathbf{u}\|_{L^2(\Omega)} + \|\mathbf{f}\|_{L^2(\Omega)} \Big) \|\mathbf{w}\|_{W^{1,2}(\Omega)} \\ &\leq c \Big(\|\nabla \mathbf{u}\|_{L^2(\Omega)} + \|\partial_t \mathbf{u}\|_{L^2(\Omega)} + \|\mathbf{f}\|_{L^2(\Omega)} \Big) \|p\|_{L^2(\Omega)}. \end{aligned} \tag{4.5.39}$$

From (4.5.39), it follows that

$$\|p(\cdot,t)\|^2_{L^2(\Omega)} \leq c \Big(\|\nabla \mathbf{u}(\cdot,t)\|^2_{L^2(\Omega)} + \|\partial_t \mathbf{u}(\cdot,t)\|^2_{L^2(\Omega)} + \|\mathbf{f}(\cdot,t)\|^2_{L^2(\Omega)} \Big).$$

Integrating the later inequality by t and utilizing (4.5.32), we recieve (4.5.33). □

4.6 Time-Periodic Stokes Problem

4.6.1 Time-Periodic Stokes Problem in Bounded Domain

In this section, we consider the time-periodic Stokes problem in a bounded Lipschitz domain $\Omega \subset \mathbb{R}^n$. We assume that the time period is equal to 2π. The problem is governed by the following partial differential equations:

$$\begin{cases} \partial_t \mathbf{u}(x,t) - \nu\Delta\mathbf{u}(x,t) + \nabla p(x,t) = \mathbf{f}(x,t), & x \in \Omega, t \in [0, 2\pi], \\ \operatorname{div}\mathbf{u}(x,t) = 0, & x \in \Omega, t \in [0, 2\pi], \\ \mathbf{u}(x,t) = 0, & x \in \partial\Omega, t \in [0, 2\pi], \\ \mathbf{u}(x,0) = \mathbf{u}(x,2\pi), \ x \in \Omega. \end{cases} \tag{4.6.1}$$

Definition 4.6.1 The weak solution of problem (4.6.1) is a time-periodic divergence-free vector-function $\mathbf{u} \in \mathring{W}^{(1,1),2}_{\partial\Omega}(\Omega \times (0, 2\pi))$ satisfying for every divergence-free time-periodic $\boldsymbol{\eta} \in \mathring{W}^{(1,0),2}_{\partial\Omega}(\Omega \times (0, 2\pi))$ the following integral identity:

$$\begin{aligned} &\int_0^{2\pi}\int_\Omega \partial_t \mathbf{u}(x,t)\cdot\boldsymbol{\eta}(x,t)dxdt + \int_0^{2\pi}\int_\Omega \nabla\mathbf{u}(x,t)\cdot\nabla\boldsymbol{\eta}(x,t)dxdt \\ &\quad = \int_0^{2\pi}\int_\Omega \mathbf{f}(x,t)\cdot\boldsymbol{\eta}(x,t)dxdt. \end{aligned} \tag{4.6.2}$$

In Sect. 3.4 we have introduced the Sobolev spaces for time-periodic functions, however from now we will use formulations applying the standard notations of functional spaces. Proceeding so we assume that all 2π-periodic in time functions f defined for $t \in [0, 2\pi]$ are extended periodically to the whole $\mathbb{R}$, and if f belongs to some function space W, then it is supposed that for any $\tau \in \mathbb{R}$ the "shifted" function $t \mapsto f(t+\tau)$ belongs to the same space W.

To construct the approximate solution $\mathbf{u}^{(N)}(x,t)$ of problem (4.6.1), we assume that the right-hand side $\mathbf{f}(x,t) \in L^2(\Omega^{2\pi})$ is a time-periodic function and can be represented as a Fourier series (see [73]):

$$\mathbf{f}(x,t) = \sum_{n=0}^{\infty}\Big(\mathbf{f}_{sn}(x)\sin(nt) + \mathbf{f}_{cn}(x)\cos(nt)\Big),$$

where $\mathbf{f}_{sn}(x)$ and $\mathbf{f}_{cn}(x)$ are square-integrable functions in Ω for $n = 0, 1, 2, \ldots$.

Now, we seek an approximate solution $\mathbf{u}^{(N)}(x,t)$ of the form

$$\mathbf{u}^{(N)}(x,t) = \sum_{n=0}^{N} \Big(\mathbf{a}_n(x)\sin(nt) + \mathbf{b_n}(x)\cos(nt)\Big), \tag{4.6.3}$$

where the coefficients $\big(\mathbf{a}_n,\, \mathbf{b}_n\big) \in \overset{\circ}{H}{}^{1,2}(\Omega) \times \overset{\circ}{H}{}^{1,2}(\Omega)$ are determined by solving the following system of integral identities:

$$\begin{cases} -n\displaystyle\int_\Omega \mathbf{b}_n(x)\cdot\boldsymbol{\eta}(x)dx + \nu\int_\Omega \nabla\mathbf{a}_n(x)\cdot\nabla\boldsymbol{\eta}(x)dx = \int_\Omega \mathbf{f}_{sn}(x)\cdot\boldsymbol{\eta}(x)dx, \\ n\displaystyle\int_\Omega \mathbf{a}_n(x)\cdot\boldsymbol{\xi}(x)dx + \nu\int_\Omega \nabla\mathbf{b}_n(x)\cdot\nabla\boldsymbol{\xi}(x)dx = \int_\Omega \mathbf{f}_{cn}(x)\cdot\boldsymbol{\xi}(x)dx, \end{cases} \tag{4.6.4}$$

where $\boldsymbol{\eta}(x)$ and $\boldsymbol{\xi}(x)$ are arbitrary functions from $\overset{\circ}{H}{}^{1,2}(\Omega)$. Thus, the pairs $(\mathbf{a}_n(x),\, \mathbf{b}_n(x))$, $n = 0, 1, 2, \ldots$, constitute weak solutions to the following stationary problems of "Stokes type":

$$\begin{cases} -n\mathbf{b}_n(x) - \nu\Delta\mathbf{a}_n(x) + \nabla r_n(x) = \mathbf{f}_{sn}(x), \; x \in \Omega, \\ n\mathbf{a}_n(x) - \nu\Delta\mathbf{b}_n(x) + \nabla s_n(x) = \mathbf{f}_{cn}(x), \; x \in \Omega, \\ \operatorname{div}\mathbf{a}_n(x) = 0, \quad \operatorname{div}\mathbf{b}_n(x) = 0, \qquad x \in \Omega, \\ \mathbf{a}_n(x) = 0, \quad \mathbf{b}_n(x) = 0, \qquad x \in \partial\Omega. \end{cases} \tag{4.6.5}$$

The uniqueness of the solution to problem (4.6.5) is straightforward to demonstrate. If we suppose $\mathbf{f}_{sn}(x) = \mathbf{f}_{cn}(x) = 0$ and substitude $\boldsymbol{\eta}(x) = \mathbf{a}_n(x)$ and $\boldsymbol{\xi}(x) = \mathbf{b}_n(x)$ into (4.6.4), the sum of the resulting inequalities leads us to the conclusion that

$$\nu\int_\Omega |\nabla\mathbf{a}_n(x)|dx + \nu\int_\Omega |\nabla\mathbf{b}_n(x)|dx = 0,$$

and thus, $\mathbf{a}_n(x) = \mathbf{b}_n(x) = 0$.

Lemma 4.6.1 *Assume that* $\mathbf{f}_{cn}, \mathbf{f}_{sn} \in L^2(\Omega)$. *Then the problems* (4.6.4) *admit unique weak solutions* $\big(\mathbf{a}_n, \mathbf{b}_n\big) \in \overset{\circ}{H}{}^{1,2}(\Omega) \times \overset{\circ}{H}{}^{1,2}(\Omega)$. *Furthermore, the following estimates are valid:*

$$\|\mathbf{a}_n\|^2_{L^2(\Omega)} + \|\mathbf{b}_n\|^2_{L^2(\Omega)} \le \frac{c}{n^2}\Big(\|\mathbf{f}_{sn}\|^2_{L^2(\Omega)} + \|\mathbf{f}_{cn}\|^2_{L^2(\Omega)}\Big) \;\; \forall n = 0, 1, \ldots; \tag{4.6.6}$$

$$\|\nabla\mathbf{a}_n\|^2_{L^2(\Omega)} + \|\nabla\mathbf{b}_n\|^2_{L^2(\Omega)} \le c\Big(\|\mathbf{f}_{sn}\|^2_{L^2(\Omega)} + \|\mathbf{f}_{cn}\|^2_{L^2(\Omega)}\Big) \;\; \forall n = 0, 1, \ldots \tag{4.6.7}$$

Moreover, if the boundary $\partial\Omega$ *is of class* C^{2l+2}, $l \geq 0$ *and* $\mathbf{f}_{sn}, \mathbf{f}_{cn} \in W^{2l,2}(\Omega)$, *then*

$$\|\nabla^{2k+1}\mathbf{a}_n\|^2_{L^2(\Omega)} + \|\nabla^{2k+1}\mathbf{b}_n\|^2_{L^2(\Omega)}$$

$$\leq c\Big[n^{2k}\|(\mathbf{f}_{sn}, \mathbf{f}_{cn})\|^2_{L^2(\Omega)} + n^{2k-2}\|(\nabla\mathbf{f}_{sn}, \nabla\mathbf{f}_{sn})\|^2_{L^2(\Omega)}$$

$$+n^{2k-4}\Big(\|\nabla^2\mathbf{f}_{sn}, \nabla^2\mathbf{f}_{cn}\|^2_{L^2(\Omega)} + \|(\nabla^3\mathbf{f}_{sn}, \nabla^3\mathbf{f}_{cn})\|^2_{L^2(\Omega)}\Big) + \dots \tag{4.6.8}$$

$$+n^2\Big(\|\nabla^{2k-4}\mathbf{f}_{sn}, \nabla^{2k-4}\mathbf{f}_{cn}\|^2_{L^2(\Omega)} + \|(\nabla^{2k-3}\mathbf{f}_{sn}, \nabla^{2k-3}\mathbf{f}_{cn})\|^2_{L^2(\Omega)}\Big)$$

$$+\Big(\|\nabla^{2k-2}\mathbf{f}_{sn}, \nabla^{2k-2}\mathbf{f}_{cn}\|^2_{L^2(\Omega)} + \|(\nabla^{2k-1}\mathbf{f}_{sn}, \nabla^{2k-1}\mathbf{f}_{cn})\|^2_{L^2(\Omega)}\Big)\Big];$$

$$\|\nabla^{2k+2}\mathbf{a}_n\|^2_{L^2(\Omega)} + \|\nabla^{2k+2}\mathbf{b}_n\|^2_{L^2(\Omega)} \leq c\Big[n^{2k}\|(\mathbf{f}_{sn}, \mathbf{f}_{cn})\|^2_{L^2(\Omega)}$$

$$+n^{2k-2}\Big(\|(\nabla\mathbf{f}_{sn}, \nabla\mathbf{f}_{sn})\|^2_{L^2(\Omega)} + \|(\nabla^2\mathbf{f}_{sn}, \nabla^2\mathbf{f}_{sn})\|^2_{L^2(\Omega)}\Big) + \dots \tag{4.6.9}$$

$$+n^2\Big(\|\nabla^{2k-3}\mathbf{f}_{sn}, \nabla^{2k-3}\mathbf{f}_{cn}\|^2_{L^2(\Omega)} + \|(\nabla^{2k-2}\mathbf{f}_{sn}, \nabla^{2k-2}\mathbf{f}_{cn})\|^2_{L^2(\Omega)}\Big)$$

$$+\Big(\|\nabla^{2k-1}\mathbf{f}_{sn}, \nabla^{2k-1}\mathbf{f}_{cn}\|^2_{L^2(\Omega)} + \|(\nabla^{2k}\mathbf{f}_{sn}, \nabla^{2k}\mathbf{f}_{cn})\|^2_{L^2(\Omega)}\Big)\Big],$$

where $k = 0, \dots, l$, $n = 0, 1, \dots$, *and* $\|h, g\|^2_{L^2(\Omega)} = \|h\|^2_{L^2(\Omega)} + \|g\|^2_{L^2(\Omega)}$. *The constant c in* (4.6.8), (4.6.9) *is independent of n.*

Proof
By reducing the integral identities (4.6.4) to operator equations with compact operators in the space $\overset{\circ}{H}{}^{1,2}(\Omega) \times \overset{\circ}{H}{}^{1,2}(\Omega)$ in a standard manner, we can rely on the Fredholm alternative (see Theorem 2.2.8) to deduce that the existence of the solution is a consequence of its uniqueness.

Estimates (4.6.6), (4.6.7) can be derived by taking in integral identities (4.6.4) first $\boldsymbol{\eta}(x) = -\mathbf{b}_n$, $\boldsymbol{\xi}(x) = \mathbf{a}_n$, and then $\boldsymbol{\eta}(x) = \mathbf{a}_n$, $\boldsymbol{\xi}(x) = \mathbf{b}_n$. Estimates (4.6.8), (4.6.9) can be proved by induction by moving the terms $n\mathbf{b}_n(x)$ and $n\mathbf{a}_n(x)$ in (4.6.5) to the-right hand side and applying the regularity results for Stokes system (4.1.1). □

Let us now demonstrate that the approximate solutions $\mathbf{u}^{(N)}(x, t)$ defined by (4.6.3) converge to a weak solution of problem (4.6.1).

Theorem 4.6.1 *Let* $\mathbf{f} \in L^2(\Omega \times (0, 2\pi))$ *be a time-periodic function. Then, the problem* (4.6.1) *admits a unique time-periodic weak solution* $\mathbf{u} \in \overset{\circ}{W}{}^{(1,1),2}_{\partial\Omega}(\Omega \times (0, 2\pi))$, *and the following estimate holds:*

$$\|\mathbf{u}\|_{W^{(1,1),2}(\Omega\times(0,2\pi))} \leq c\|\mathbf{f}\|_{L^2(\Omega\times(0,2\pi))}, \tag{4.6.10}$$

where c is a positive constant.

Moreover, if $\partial\Omega \in C^{2l+2}$ *and* $\mathbf{f} \in W^{(2l,l),2}(\Omega \times (0, 2\pi))$, *then* $\mathbf{u} \in W^{(2l+2,l+1),2}(\Omega \times (0, 2\pi))$, *and there exists a pressure* p *such that* $\nabla p \in W^{(2l,l),2}(\Omega \times (0, 2\pi))$. *The following estimate holds:*

$$\|\mathbf{u}\|_{W^{(2l+2,l+1),2}(\Omega\times(0,2\pi))} + c\|\nabla p\|_{W^{(2l,l),2}(\Omega\times(0,2\pi))} \le c\|\mathbf{f}\|_{W^{(2l,l),2}(\Omega\times(0,2\pi))}, \tag{4.6.11}$$

where c *is a positive constant.*

Proof

From (4.6.3) and (4.6.6), (4.6.7), it follows that

$$\|\mathbf{u}^{(N)}\|^2_{L^2(\Omega\times(0,2\pi))} = \sum_{n=0}^{N}\left(\|\mathbf{a}_n\|^2_{L^2(\Omega)} + \|\mathbf{b}_n\|^2_{L^2(\Omega)}\right)$$
$$\le c\sum_{n=0}^{N}\frac{1}{n^2}\left(\|\mathbf{f}_{sn}\|^2_{L^2(\Omega)} + \|\mathbf{f}_{cn}\|^2_{L^2(\Omega)}\right) \le c\|\mathbf{f}\|^2_{L^2(\Omega\times(0,2\pi))},$$

$$\|\nabla\mathbf{u}^{(N)}\|^2_{L^2(\Omega\times(0,2\pi))} = \sum_{n=0}^{N}\left(\|\nabla\mathbf{a}_n\|^2_{L^2(\Omega)} + \|\nabla\mathbf{b}_n\|^2_{L^2(\Omega)}\right)$$
$$\le c\sum_{n=0}^{N}\left(\|\mathbf{f}_{sn}\|^2_{L^2(\Omega)} + \|\mathbf{f}_{cn}\|^2_{L^2(\Omega)}\right) \le c\|\mathbf{f}\|^2_{L^2(\Omega\times(0,2\pi))},$$

$$\|\partial_t\mathbf{u}^{(N)}\|^2_{L^2(\Omega\times(0,2\pi))} = \sum_{n=0}^{N}n^2\left(\|\mathbf{a}_n\|^2_{L^2(\Omega)} + \|\mathbf{b}_n\|^2_{L^2(\Omega)}\right)$$
$$\le c\sum_{n=0}^{N}\left(\|\mathbf{f}_{sn}\|^2_{L^2(\Omega)} + \|\mathbf{f}_{cn}\|^2_{L^2(\Omega)}\right) \le c\|\mathbf{f}\|^2_{L^2(\Omega\times(0,2\pi))}.$$

Therefore, the sequence $\{\mathbf{u}^{(N)}\}$ converges in the norm $\|\cdot\|_{W^{(1,1),2}(\Omega\times(0,2\pi))}$, and the limit $\mathbf{u} \in \mathring{W}^{(1,1),2}_{\partial\Omega}(\Omega \times (0, 2\pi))$. Furthermore, we have the inequality

$$\|\mathbf{u}\|^2_{W^{(1,1),2}(\Omega\times(0,2\pi))} \le c\|\mathbf{f}\|^2_{L^2(\Omega\times(0,2\pi))} \tag{4.6.12}$$

From equations (4.6.3) and (4.6.4), we obtain the following integral identity for $\mathbf{u}^{(N)}(x, t)$:

$$\int_\Omega \mathbf{u}_t^{(N)}(x,t)\cdot\boldsymbol{\eta}(x)dx + \nu\int_\Omega \nabla\mathbf{u}^{(N)}(x,t)\cdot\nabla\boldsymbol{\eta}(x)dx$$
$$= \int_\Omega \mathbf{f}^{(N)}(x,t)\cdot\boldsymbol{\eta}(x)dx \quad \forall\boldsymbol{\eta} \in \mathring{H}^{1,2}(\Omega). \tag{4.6.13}$$

Here, $\mathbf{f}_n^{(N)}(x,t) = \sum\limits_{n=0}^{N} \Big(\mathbf{f}_{sn}(x)\sin(nt) + \mathbf{f}_{cn}(x)\cos(nt)\Big)$. To prove (4.6.13), we choose $\boldsymbol{\eta}(x) = \boldsymbol{\xi}(x)$ in (4.6.4), multiply $(4.6.4)_1$ by $\sin(nt)$, $(4.6.4)_2$ by $\cos(nt)$, and sum the obtained relations.

Next, let $\boldsymbol{\varphi}$ be a finite sum defined as

$$\boldsymbol{\varphi}(x,t) = \sum_{n=0}^{L} \Big(\boldsymbol{\varphi}_{sn}(x)\sin(nt) + \boldsymbol{\varphi}_{cn}(x)\cos(nt)\Big), \quad \boldsymbol{\varphi}_{sn}, \boldsymbol{\varphi}_{cn} \in \overset{\circ}{H}{}^{1,2}(\Omega). \tag{4.6.14}$$

From (4.6.13), we conclude that the integral identity also holds for any such test function $\boldsymbol{\varphi}(x,t)$:

$$\int_0^{2\pi}\int_\Omega \mathbf{u}_t^{(N)}(x,t)\cdot\boldsymbol{\varphi}(x,t)dxdt + \int_0^{2\pi}\int_\Omega \nabla\mathbf{u}^{(N)}(x,t)\cdot\nabla\boldsymbol{\varphi}(x,t)dxdt$$
$$= \int_0^{2\pi}\int_\Omega \mathbf{f}^{(N)}(x,t)\cdot\boldsymbol{\varphi}(x,t)dxdt.$$

By taking the limit as $N \to \infty$ in this identity, we obtain

$$\begin{aligned}\int_0^{2\pi}\int_\Omega \mathbf{u}_t(x,t)\cdot\boldsymbol{\varphi}(x,t)dxdt &+ \int_0^{2\pi}\int_\Omega \nabla\mathbf{u}(x,t)\cdot\nabla\boldsymbol{\varphi}(x,t)dxdt \\ &= \int_0^{2\pi}\int_\Omega \mathbf{f}(x,t)\cdot\boldsymbol{\varphi}(x,t)dxdt.\end{aligned} \tag{4.6.15}$$

The last identity is valid for test functions $\boldsymbol{\varphi}$ having the form (4.6.14). Moreover, since any time-periodic divergence-free function belonging to the space $\overset{\circ}{W}{}^{(1,0),2}_{\partial\Omega}(\Omega\times(0,2\pi))$ can be decomposed into a Fourier series with coefficients in $\overset{\circ}{W}{}^{1,2}(\Omega)$, the sums (4.6.14) are dense in $\overset{\circ}{W}{}^{(1,0),2}_{\partial\Omega}(\Omega\times(0,2\pi))$. Thus, integral identity (4.6.15) remains valid for any $\boldsymbol{\varphi} \in \overset{\circ}{W}{}^{(1,0),2}_{\partial\Omega}(\Omega\times(0,2\pi))$. As a result, $\mathbf{u}(x,t)$ is a weak solution of problem (4.6.1).

Next, let $\partial\Omega \in C^{2l+2}$ and $\mathbf{f} \in W^{(2l,l),2}(\Omega\times(0,2\pi))$. Using the definition of the space $W^{(2l,l),2}(\Omega\times(0,2\pi))$ and Parseval's identity, we find that for $2m+\mu \le 2l$, the following inequality holds:

$$\|\partial_t^m\nabla^\mu\mathbf{f}\|^2_{L^2(\Omega\times(0,2\pi))} = \sum_{n=0}^{\infty} n^{2m}\Big(\|\nabla^\mu\mathbf{f}_{sn}\|^2_{L^2(\Omega)} + \|\nabla^\mu\mathbf{f}_{cn}\|^2_{L^2(\Omega)}\Big) < \infty. \tag{4.6.16}$$

Then, by using the estimates (4.6.8) and (4.6.9), we find that

$$\|\partial_t^m \nabla^{2r+1}\mathbf{u}^{(N)}\|^2_{L^2(\Omega\times(0,2\pi))} = \sum_{n=0}^{N} n^{2m}\Big(\|\nabla^{2r+1}\mathbf{a}_n\|^2_{L^2(\Omega)} + \|\nabla^{2r+1}\mathbf{b}_n\|^2_{L^2(\Omega)}\Big)$$
$$\leq c\sum_{n=0}^{N} n^{2m}\Big(n^{2r}\|(\mathbf{f}_{sn},\mathbf{f}_{cn})\|^2_{L^2(\Omega)} + n^{2r-2}\|(\nabla\mathbf{f}_{sn},\nabla\mathbf{f}_{sn})\|^2_{L^2(\Omega)}\Big)$$
$$+c\sum_{n=0}^{N}\sum_{i=1}^{r-1} n^{2m+2r-2-2i}\Big(\|\nabla^{2i}\mathbf{f}_{sn},\nabla^{2i}\mathbf{f}_{cn})\|^2_{L^2(\Omega)} + \|(\nabla^{2i+1}\mathbf{f}_{sn},\nabla^{2i+1}\mathbf{f}_{cn})\|^2_{L^2(\Omega)}\Big), \tag{4.6.17}$$

and

$$\|\partial_t^m \nabla^{2r+2}\mathbf{u}^{(N)}\|^2_{L^2(\Omega\times(0,2\pi))} = \sum_{n=0}^{N} n^{2m}\Big(\|\nabla^{2r+2}\mathbf{a}_n\|^2_{L^2(\Omega)} + \|\nabla^{2r+2}\mathbf{b}_n\|^2_{L^2(\Omega)}\Big)$$
$$\leq c\sum_{n=0}^{N} n^{2m+2r}\|(\mathbf{f}_{sn},\mathbf{f}_{cn})\|^2_{L^2(\Omega)}$$
$$+c\sum_{n=0}^{N}\sum_{i=1}^{r-1} n^{2m+2r-2i}\Big(\|\nabla^{2i-1}\mathbf{f}_{sn},\nabla^{2i-1}\mathbf{f}_{cn})\|^2_{L^2(\Omega)} + \|(\nabla^{2i}\mathbf{f}_{sn},\nabla^{2i}\mathbf{f}_{cn})\|^2_{L^2(\Omega)}\Big). \tag{4.6.18}$$

From (4.6.16), it follows that the right-hand sides of (4.6.17) and (4.6.18) are bounded by $c\|\mathbf{f}\|^2_{W^{(2l,l),2}(\Omega\times(0,2\pi))}$. Therefore, the sequence $\{\mathbf{u}^{(N)}\}$ converges also in the space $W^{(2l,l),2}(\Omega\times(0,2\pi))$, and the limit function $\mathbf{u}$ satisfies the following estimate:

$$\|\mathbf{u}\|^2_{W^{(2l+2,l+1),2}(\Omega\times(0,2\pi))} \leq c\|\mathbf{f}\|^2_{W^{(2l,l),2}(\Omega\times(0,2\pi))}.$$

The gradient of the pressure ∇p can be estimated from system (4.6.1). Thus, we finally obtain the desired inequality (4.6.11). □

4.6.2 *Time-Periodic Stokes Problem in a Tube Structure*

We now study the time-periodic Stokes problem within a tube structure B_ε. To explore further applications, we consider a more general formulation of the problem. Specifically, we investigate the following case:

$$\begin{cases} \frac{1}{\varepsilon^\beta}\partial_t \mathbf{u}(x,t) - \nu\Delta\mathbf{u}(x,t) + \nabla p(x,t) = \mathbf{f}(x,t), & x \in \Omega,\ t \in [0,2\pi], \\ \mathrm{div}\mathbf{u}(x,t) = 0, & x \in \Omega,\ t \in [0,2\pi], \\ \mathbf{u}(x,t) = 0, & x \in \partial\Omega,\ t \in [0,2\pi], \\ \mathbf{u}(x,0) = \mathbf{u}(x,2\pi),\ x \in \Omega, \end{cases} \tag{4.6.19}$$

where $\beta = 0$[1] or $\beta = 2$.

Theorem 4.6.2 *Let* $\mathbf{f} \in L^2(B_\varepsilon \times (0,2\pi))$ *be a time-periodic function. Then, problem* (4.6.19) *admits a unique time-periodic solution* $(\mathbf{u}, p)$ *with the following properties:* $\mathbf{u} \in \mathring{W}^{(2,1),2}_{\partial\Omega}(B_\varepsilon \times (0,2\pi))$, $\nabla p \in L^2(B_\varepsilon \times (0,2\pi))$, *and the subsequent estimate holds:*

$$\begin{aligned} \frac{1}{\varepsilon^{2\beta}}\int_0^{2\pi} \|\partial_t \mathbf{u}(\cdot,t)\|^2_{L^2(B_\varepsilon)}dt + \int_0^{2\pi} |||\mathbf{u}(\cdot,t)|||^2_{2,0,B_\varepsilon}dt + \int_0^{2\pi} \|\nabla p(\cdot,t)\|^2_{L^2(B_\varepsilon)}dt \\ \le \frac{c}{\varepsilon^\beta}\int_0^{2\pi} \|\mathbf{f}(\cdot,t)\|^2_{L^2(B_\varepsilon)}dt \end{aligned} \tag{4.6.20}$$

where the constant c is independent of ε.

Proof
To begin, we multiply equations (4.6.19) by ε^β, thereby reducing the problem to the form (4.6.1) with a new viscosity coefficient $\widetilde{\nu} = \varepsilon^\beta\nu$, a new right-hand side $\widetilde{\mathbf{f}}(x,t) = \varepsilon^\beta\mathbf{f}(x,t)$, and a new pressure $\widetilde{p}(x,t) = \varepsilon^\beta p(x,t)$. Consequently, the existence and uniqueness of the solution to problem (4.6.19) follow from Theorem 4.6.1. The focus now shifts to proving the estimate (4.6.20).

By multiplying equations (4.6.19) by $\mathbf{u}(x,t)$, integrating by parts in B_ε, and integrating with respect to t over the interval $(0,2\pi)$, we obtain the following relation after using the periodicity of $\mathbf{u}(x,t)$:

$$\begin{aligned} \nu\int_0^{2\pi}\int_{B_\varepsilon} |\nabla\mathbf{u}(x,t)|^2 dxdt = \int_0^{2\pi}\int_{B_\varepsilon} \mathbf{f}(x,t)\cdot\mathbf{u}(x,t)dxdt \\ \le c\varepsilon^2\int_0^{2\pi}\int_{B_\varepsilon} |\mathbf{f}(x,t)|^2dxdt + \frac{\nu}{2}\int_0^{2\pi}\int_{B_\varepsilon} |\nabla\mathbf{u}(x,t)|^2dxdt. \end{aligned} \tag{4.6.21}$$

Next, we multiply (4.6.19) by $\partial_t\mathbf{u}(x,t)$ and integrate over B_ε and t:

[1] In this case, problem (4.6.19) coincides with problem (4.6.1).

$$\frac{1}{\varepsilon^\beta}\int_0^{2\pi}\int_{B_\varepsilon}|\partial_t\mathbf{u}(x,t)|^2dxdt = \int_0^{2\pi}\int_{B_\varepsilon}\mathbf{f}(x,t)\cdot\partial_t\mathbf{u}(x,t)dxdt$$
$$\leq \frac{1}{2}\int_0^{2\pi}\int_{B_\varepsilon}|\mathbf{f}(x,t)|^2dxdt + \frac{1}{2}\int_0^{2\pi}\int_{B_\varepsilon}|\partial_t\mathbf{u}(x,t)|^2dxdt. \tag{4.6.22}$$

Since $\frac{1}{\varepsilon^\beta} > 1$, inequalities (4.6.21) and (4.6.22) yield

$$\int_0^{2\pi}\int_{B_\varepsilon}|\nabla\mathbf{u}(x,t)|^2dxdt \leq c\varepsilon^2\int_0^{2\pi}\int_{B_\varepsilon}|\mathbf{f}(x,t)|^2dxdt,$$
$$\frac{1}{\varepsilon^\beta}\int_0^{2\pi}\int_{B_\varepsilon}|\partial_t\mathbf{u}(x,t)|^2dx \leq c\int_0^{2\pi}\int_{B_\varepsilon}|\mathbf{f}(x,t)|^2dxdt. \tag{4.6.23}$$

Now we can interpret $\mathbf{u}(\cdot,t)\in \overset{\circ}{H}{}^{1,2}(B_\varepsilon)$ as a weak solution to the following stationary Stokes problem:

$$\begin{cases} -\nu\Delta\mathbf{u}(x,t)+\nabla p(x,t) = \mathbf{f}(x,t) - \frac{1}{\varepsilon^\beta}\partial_t\mathbf{u}(x,t), & x\in\Omega,\\ \operatorname{div}\mathbf{u}(x,t) = 0, & x\in\Omega,\\ \mathbf{u}(x,t) = 0, & x\in\partial\Omega \end{cases} \tag{4.6.24}$$

By Theorem 4.3.1, we have the estimate (see (4.3.3)):

$$|||\mathbf{u}|||^2_{2,0,B_\varepsilon} + |||\nabla p|||^2_{0,0,B_\varepsilon} \leq c\Big(|||\mathbf{f}|||^2_{0,0,B_\varepsilon} + \frac{1}{\varepsilon^{2\beta}}|||\partial_t\mathbf{u}|||^2_{0,0,B_\varepsilon}\Big)$$
$$\leq c\Big(\|\mathbf{f}\|^2_{L^2(B_\varepsilon)} + \frac{1}{\varepsilon^{2\beta}}\|\partial_t\mathbf{u}\|^2_{L^2(B_\varepsilon)}\Big) \overset{(4.6.23)}{\leq} \frac{c}{\varepsilon^\beta}\|\mathbf{f}\|^2_{L^2(B_\varepsilon)}.$$

Integrating this inequality over t gives

$$\int_0^{2\pi}|||\mathbf{u}(\cdot,t)|||^2_{2,0,B_\varepsilon}dt + \int_0^{2\pi}\|\nabla p(\cdot,t)\|^2_{L^2(B_\varepsilon)}dt \leq \frac{c}{\varepsilon^\beta}\int_0^{2\pi}\|\mathbf{f}(\cdot,t)\|^2_{L^2(B_\varepsilon)}dt \tag{4.6.25}$$

From inequalities (4.6.25) and (4.6.23) follows the required estimate (4.6.20). □

4.6.3 Time-Periodic Stokes Problem in Domains with Outlets to Infinity

Let $\Omega \in \mathbb{R}^n$ be a domain with Lipschitz boundary with cylindrical outlets to infinity.

Theorem 4.6.3 *Let $\mathbf{f} \in \mathcal{L}^2_{\boldsymbol{\beta}}(\Omega \times (0, 2\pi))$, where $\beta > 0$, be a time-periodic function. There exists a value β_* such that if $\beta \in (0, \beta_*]$, the weak solution $\mathbf{u}(x, t)$ of the time-periodic Stokes problem* (4.6.1) *belongs to the space $\mathcal{W}^{(1,1),2}_{\boldsymbol{\beta}}(\Omega \times (0, 2\pi))$, and the following estimate holds:*

$$\|\mathbf{u}\|^2_{\mathcal{W}^{(1,1),2}_{\boldsymbol{\beta}}(\Omega\times(0,2\pi))} \le c\|\mathbf{f}\|^2_{\mathcal{L}^2_{\boldsymbol{\beta}}(\Omega\times(0,2\pi))}. \tag{4.6.26}$$

Moreover, if $\partial\Omega$ is of class C^2, then $\mathbf{u} \in \mathcal{W}^{(2,1),2}_{\boldsymbol{\beta}}(\Omega \times (0, 2\pi))$, $\nabla p \in \mathcal{L}^2_{\boldsymbol{\beta}}(\Omega \times (0, 2\pi))$, and the following estimate holds:

$$\|\mathbf{u}\|^2_{\mathcal{W}^{(2,1),2}_{\boldsymbol{\beta}}(\Omega\times(0,2\pi))} + \|\nabla p\|^2_{\mathcal{L}^2_{\boldsymbol{\beta}}(\Omega\times(0,2\pi))} \le c\|\mathbf{f}\|^2_{\mathcal{L}^2_{\boldsymbol{\beta}}(\Omega\times(0,2\pi))}. \tag{4.6.27}$$

Proof
Let us consider the integral identity (4.6.2) with $\boldsymbol{\eta}(x, t) = E^{(k)}_{\boldsymbol{\beta}}(x)\mathbf{u}(x, t) + \mathbf{W}^{(k)}(x, t)$, where $E^{(k)}_{\boldsymbol{\beta}}(x)$ is the "step" weight function, and $\mathbf{W}^{(k)}(x, t)$ is the vector field from Lemma 2.4.11. This yields

$$\begin{aligned}
&\int_0^{2\pi}\int_\Omega E^{(k)}_{\boldsymbol{\beta}}(x)\partial_t\mathbf{u}(x,t)\cdot\mathbf{u}(x,t)dxdt + \int_0^{2\pi}\int_\Omega \partial_t\mathbf{u}(x,t)\cdot\mathbf{W}^{(k)}(x,t)dxdt\\
&\qquad+\nu\int_0^{2\pi}\int_\Omega \nabla\mathbf{u}(x,t)\cdot\nabla\big(E^{(k)}_{\boldsymbol{\beta}}(x)\mathbf{u}(x,t)+\mathbf{W}^{(k)}(x,t)\big)dxdt\\
&\qquad=\int_0^{2\pi}\int_\Omega \mathbf{f}(x,t)\cdot\big(E^{(k)}_{\boldsymbol{\beta}}(x)\mathbf{u}(x,t)+\mathbf{W}^{(k)}(x,t)\big)dxdt.
\end{aligned} \tag{4.6.28}$$

Since

$$\int_0^{2\pi}\int_\Omega E^{(k)}_{\boldsymbol{\beta}}(x)\partial_t\mathbf{u}(x,t)\cdot\mathbf{u}(x,t)dxdt=\frac{1}{2}\int_\Omega E^{(k)}_{\boldsymbol{\beta}}(x)(|\mathbf{u}(x,2\pi)|^2-|\mathbf{u}(x,0)|^2)dx=0,$$

the relation (4.6.28) takes the form

$$\begin{aligned}
\nu\int_0^{2\pi}\int_\Omega E_{\boldsymbol\beta}^{(k)}(x)|\nabla\mathbf{u}(x,t)|^2dxdt &= -\int_0^{2\pi}\int_\Omega \partial_t\mathbf{u}(x,t)\cdot\mathbf{W}^{(k)}(x,t)dxdt\\
&-\nu\int_0^{2\pi}\int_\Omega \nabla\mathbf{u}(x,t)\cdot\big(\nabla E_{\boldsymbol\beta}^{(k)}(x)\mathbf{u}(x,t)+\nabla\mathbf{W}^{(k)}(x,t)\big)dxdt \qquad (4.6.29)\\
&+\int_0^{2\pi}\int_\Omega \mathbf{f}(x,t)\cdot(E_{\boldsymbol\beta}^{(k)}(x)\mathbf{u}(x,t)+\mathbf{W}^{(k)}(x,t)\big)dxdt = I_1+I_2+I_3.
\end{aligned}$$

We estimate the term I_1 using the Cauchy-Schwarz-Bunyakovsky inequality, the weighted Poincaré inequality, and properties of the function $\mathbf{W}^{(k)}(x,t)$ (see (2.4.45)):

$$\begin{aligned}
|I_1| &\le c\left(\int_0^{2\pi}\int_\Omega E_{-\boldsymbol\beta}^{(k)}(x)|\mathbf{W}^{(k)}|^2\,dx\,dt\right)^{1/2}\left(\int_0^{2\pi}\int_\Omega E_{\boldsymbol\beta}^{(k)}(x)|\partial_t\mathbf{u}|^2\,dx\,dt\right)^{1/2}\\
&\le c\left(\int_0^{2\pi}\int_\Omega E_{-\boldsymbol\beta}^{(k)}(x)|\nabla\mathbf{W}^{(k)}|^2\,dx\,dt\right)^{1/2}\left(\int_0^{2\pi}\int_\Omega E_{\boldsymbol\beta}^{(k)}(x)|\partial_t\mathbf{u}|^2\,dx\,dt\right)^{1/2}\\
&\le c\beta\left(\int_0^{2\pi}\int_\Omega E_{\boldsymbol\beta}^{(k)}(x)|\nabla\mathbf{u}|^2\,dx\,dt+\int_0^{2\pi}\int_\Omega E_{\boldsymbol\beta}^{(k)}(x)|\partial_t\mathbf{u}|^2\,dx\,dt\right).
\end{aligned} \qquad (4.6.30)$$

Similarly, for I_2, we have

$$\begin{aligned}
|I_2| &\le c\beta\left(\int_0^{2\pi}\int_\Omega E_{\boldsymbol\beta}^{(k)}(x)|\nabla\mathbf{u}|^2\,dx\,dt\right)^{1/2}\left(\int_0^{2\pi}\int_\Omega E_{\boldsymbol\beta}^{(k)}(x)|\mathbf{u}|^2\,dx\,dt\right)^{1/2}\\
&+c\left(\int_0^{2\pi}\int_\Omega E_{-\boldsymbol\beta}^{(k)}(x)|\nabla\mathbf{W}^{(k)}|^2\,dx\,dt\right)^{1/2}\left(\int_0^{2\pi}\int_\Omega E_{\boldsymbol\beta}^{(k)}(x)|\nabla\mathbf{u}|^2\,dx\,dt\right)^{1/2}\\
&\le c\beta\int_0^{2\pi}\int_\Omega E_{\boldsymbol\beta}^{(k)}(x)|\nabla\mathbf{u}|^2\,dx\,dt.
\end{aligned} \qquad (4.6.31)$$

Finally, for I_3, we obtain

$$
\begin{aligned}
|I_3| &\le c_\varepsilon \int_0^{2\pi}\int_\Omega E_{\boldsymbol\beta}^{(k)}(x)|\mathbf{f}(x,t)|^2\,dx\,dt + \varepsilon\left(\int_0^{2\pi}\int_\Omega E_{\boldsymbol\beta}^{(k)}(x)|\mathbf{u}(x,t)|^2\,dx\,dt\right.\\
&\quad\left. + \int_0^{2\pi}\int_\Omega E_{-\boldsymbol\beta}^{(k)}(x)|\mathbf{W}^{(k)}(x,t)|^2\,dx\,dt\right)\\
&\le c_\varepsilon \int_0^{2\pi}\int_\Omega E_{\boldsymbol\beta}^{(k)}(x)|\mathbf{f}(x,t)|^2\,dx\,dt + c\varepsilon\int_0^{2\pi}\int_\Omega E_{\boldsymbol\beta}^{(k)}(x)|\nabla\mathbf{u}(x,t)|^2\,dx\,dt.
\end{aligned}
\tag{4.6.32}
$$

By combining inequalities (4.6.29)–(4.6.32), we deduce that

$$
\begin{aligned}
\nu\int_0^{2\pi}\int_\Omega E_{\boldsymbol\beta}^{(k)}(x)|\nabla\mathbf{u}(x,t)|^2\,dx\,dt &\le c_1(\varepsilon+\beta)\int_0^{2\pi}\int_\Omega E_{\boldsymbol\beta}^{(k)}(x)|\nabla\mathbf{u}(x,t)|^2\,dx\,dt\\
&+c_2\beta\int_0^{2\pi}\int_\Omega E_{\boldsymbol\beta}^{(k)}(x)|\partial_t\mathbf{u}(x,t)|^2\,dx\,dt + c_\varepsilon\int_0^{2\pi}\int_\Omega E_{\boldsymbol\beta}^{(k)}(x)|\mathbf{f}(x,t)|^2\,dx\,dt.
\end{aligned}
\tag{4.6.33}
$$

Next, we substitute $\boldsymbol\eta(x,t) = E_{\boldsymbol\beta}^{(k)}(x)\partial_t\mathbf{u}(x,t) + \partial_t\mathbf{W}^{(k)}(x,t)$ into the integral identity (4.6.2). This gives

$$
\begin{aligned}
&\int_0^{2\pi}\int_\Omega E_{\boldsymbol\beta}^{(k)}(x)|\partial_t\mathbf{u}(x,t)|^2\,dx\,dt + \frac{\nu}{2}\int_\Omega E_{\boldsymbol\beta}^{(k)}(x)(|\nabla\mathbf{u}(x,2\pi)|^2 - |\nabla\mathbf{u}(x,0)|^2)\,dx\\
&\qquad = -\int_0^{2\pi}\int_\Omega \partial_t\mathbf{u}(x,t)\cdot\partial_t\mathbf{W}^{(k)}(x,t)\,dx\,dt\\
&\qquad -\nu\int_0^{2\pi}\int_\Omega \nabla\mathbf{u}(x,t)\cdot\big(\nabla E_{\boldsymbol\beta}^{(k)}(x)\partial_t\mathbf{u}(x,t) + \nabla\partial_t\mathbf{W}^{(k)}(x,t)\big)\,dx\,dt\\
&\qquad +\int_0^{2\pi}\int_\Omega \mathbf{f}(x,t)\cdot\big(\nabla E_{\boldsymbol\beta}^{(k)}(x)\partial_t\mathbf{u}(x,t) + \partial_t\mathbf{W}^{(k)}(x,t)\big)\,dx\,dt.
\end{aligned}
\tag{4.6.34}
$$

Since the problem is periodic, the second term on the left-hand side of (4.6.34) is equal to zero. By estimating the right-hand side as above and using inequality (2.4.46), we obtain

$$\int_0^{2\pi}\int_\Omega E_{\boldsymbol\beta}^{(k)}(x)|\partial_t\mathbf{u}(x,t)|^2\,dx\,dt$$
$$\leq c_3(\varepsilon+\beta)\int_0^{2\pi}\int_\Omega E_{\boldsymbol\beta}^{(k)}(x)|\partial_t\mathbf{u}(x,t)|^2\,dx\,dt + c_4\beta\int_0^{2\pi}\int_\Omega E_{\boldsymbol\beta}^{(k)}(x)|\nabla\mathbf{u}(x,t)|^2\,dx\,dt$$
$$+\,c_\varepsilon\int_0^{2\pi}\int_\Omega E_{\boldsymbol\beta}^{(k)}(x)|\mathbf{f}(x,t)|^2\,dx\,dt. \tag{4.6.35}$$

Combining inequalities (4.6.33) and (4.6.35), we deduce that

$$\int_0^{2\pi}\int_\Omega E_{\boldsymbol\beta}^{(k)}(x)\big(|\partial_t\mathbf{u}(x,t)|^2+|\nabla\mathbf{u}(x,t)|^2\big)\,dx\,dt$$
$$\leq c_5(\varepsilon+\beta)\int_0^{2\pi}\int_\Omega E_{\boldsymbol\beta}^{(k)}(x)(|\partial_t\mathbf{u}(x,t)|^2+|\nabla\mathbf{u}(x,t)|^2)\,dx\,dt \tag{4.6.36}$$
$$+\,c_\varepsilon\int_0^{2\pi}\int_\Omega E_{\boldsymbol\beta}^{(k)}(x)|\mathbf{f}(x,t)|^2\,dx\,dt.$$

Let us fix $\varepsilon=\frac{1}{4c_5}\min\{\nu,1\}$. If

$$\beta\leq\frac{1}{4c_5}\min\{\nu,1\},$$

then from (4.6.36), it follows that

$$\|\partial_t\mathbf{u}\|^2_{\mathcal{L}^2_{\boldsymbol\beta}(\Omega\times(0,2\pi))}+\|\nabla\mathbf{u}\|^2_{\mathcal{L}^2_{\boldsymbol\beta}(\Omega\times(0,2\pi))}\leq c\|\mathbf{f}\|^2_{\mathcal{L}^2_{\boldsymbol\beta}(\Omega\times(0,2\pi))}. \tag{4.6.37}$$

Let $\partial\Omega\in C^2$. From the integral identity (4.6.2), it follows that for almost all $t\in[0,2\pi]$ there holds the identity:

$$\nu\int_\Omega\nabla\mathbf{u}(x,t)\cdot\nabla\boldsymbol\eta(x)\,dx=\int_\Omega(\mathbf{f}(x,t)-\partial_t\mathbf{u}(x,t))\cdot\boldsymbol\eta(x)\,dx,$$

where $\boldsymbol\eta(x)$ is an arbitrary function belonging to the space $\overset{\circ}{H}{}^{1,2}(\Omega)$. Hence, $\mathbf{u}(\cdot,t)\in\overset{\circ}{H}{}^{1,2}(\Omega)$ can be considered, for almost all $t\in(0,2\pi)$, as a weak solution of the stationary Stokes problem with the right-hand side equal to $\mathbf{f}(x,t)-\partial_t\mathbf{u}(x,t)$:

$$\begin{cases} -\nu\Delta\mathbf{u}(x,t) + \nabla p(x,t) = \mathbf{f}(x,t) - \partial_t\mathbf{u}(x,t), & x \in \Omega, \\ \mathrm{div}\mathbf{u}(x,t) = 0, & x \in \Omega, \\ \mathbf{u}(x,t) = 0, & x \in \partial\Omega. \end{cases} \tag{4.6.38}$$

We have $\mathbf{f}(\cdot,t) - \partial_t\mathbf{u}(\cdot,t) \in \mathcal{L}^2_{\boldsymbol{\beta}}(\Omega)$. If β is sufficiently small, then the weak solution $\mathbf{u}(x,t)$ of (4.6.38) belongs to the space $\mathcal{W}^{2,2}_{\boldsymbol{\beta}}(\Omega)$. Moreover, there exists a function $p(x,t)$ with $\nabla p(\cdot,t) \in \mathcal{L}^2_{\boldsymbol{\beta}}(\Omega)$ such that $(\mathbf{u}(x,t), p(x,t))$ satisfy equations (4.6.38) almost everywhere in Ω and the following estimate

$$\begin{aligned} \|\mathbf{u}(\cdot,t)\|^2_{\mathcal{W}^{2,2}_{\boldsymbol{\beta}}(\Omega)} + \|\nabla p(\cdot,t)\|^2_{\mathcal{L}^2_{\boldsymbol{\beta}}(\Omega)} &\le c\|\mathbf{f}(\cdot,t) - \partial_t\mathbf{u}(\cdot,t)\|^2_{\mathcal{L}^2_{\boldsymbol{\beta}}(\Omega)} \\ &\le c\Big(\|\mathbf{f}(\cdot,t)\|^2_{\mathcal{L}^2_{\boldsymbol{\beta}}(\Omega)} + \|\partial_t\mathbf{u}(\cdot,t)\|^2_{\mathcal{L}^2_{\boldsymbol{\beta}}(\Omega)}\Big) \end{aligned} \tag{4.6.39}$$

holds (see Theorem 4.4.2). Integrating inequality (4.6.39) with respect to t and utilizing (4.6.37), we derive

$$\int_0^{2\pi} \|\mathbf{u}(\cdot,t)\|^2_{\mathcal{W}^{2,2}_{\boldsymbol{\beta}}(\Omega)}dt + \int_0^{2\pi} \|\nabla p(\cdot,t)\|^2_{\mathcal{L}^2_{\boldsymbol{\beta}}(\Omega)}dt \le c\int_0^{2\pi} \|\mathbf{f}(\cdot,t)\|^2_{\mathcal{L}^2_{\boldsymbol{\beta}}(\Omega)}dt.$$

The last estimate together with (4.6.37) is equivalent to (4.6.27). □

Consider the regularity of the solution to problem (4.6.1) in domains with cylindrical outlets.

Theorem 4.6.4 *Let $\partial\Omega \in C^{2l+2}$. Assume that $\mathbf{f} \in \mathcal{W}^{(2l,l),2}_{\boldsymbol{\beta}}(\Omega \times (0,2\pi))$, where $\beta \ge 0$ and $l \ge 1$ is an integer, and $\mathbf{f}$ is a time-periodic function. There exists a positive constant β_{**} such that if $\beta \in (0,\beta_{**}]$, then problem (4.6.1) admits a unique solution $\big(\mathbf{u}(x,t), p(x,t)\big)$ with $\mathbf{u} \in \mathcal{W}^{(2l+2,l+1),2}_{\boldsymbol{\beta}}(\Omega \times (0,2\pi))$ and $\nabla p \in \mathcal{W}^{(2l,l),2}_{\boldsymbol{\beta}}(\Omega \times (0,2\pi))$. The following estimate holds:*

$$\|\mathbf{u}\|_{\mathcal{W}^{(2l+2,l+1),2}_{\boldsymbol{\beta}}(\Omega\times(0,2\pi))} + \|\nabla p\|_{\mathcal{W}^{(2l,l),2}_{\boldsymbol{\beta}}(\Omega\times(0,2\pi))} \le c\|\mathbf{f}\|_{\mathcal{W}^{(2l,l),2}_{\boldsymbol{\beta}}(\Omega\times(0,2\pi))}, \tag{4.6.40}$$

where c is a positive constant.

Proof

Firstly, it is important to note that by classical regularity results for the linear time-periodic Stokes problem (4.6.1), the solution is unique and satisfies $\mathbf{u} \in W^{(2l+2,l+1),2}(\Omega \times (0,2\pi))$, $\nabla p \in W^{(2l,l),2}(\Omega \times (0,2\pi))$ (see Theorem 4.6.1). Hence, we need only prove the weighted estimate (4.6.40).

We shall proceed by induction. For the base case $l = 0$, estimate (4.6.40) coincides with (4.6.27) (see Theorem 4.6.3). Now, let $l \ge 1$, and assume that (4.6.40) is valid for $l \le r-1$, and we aim to prove it for $l = r$. Suppose

$\mathbf{f} \in \mathcal{W}_{\boldsymbol{\beta}}^{(2r,r),2}(\Omega \times (0, 2\pi))$. Then, by Theorem 4.6.1 and the induction assumption, the solution $(\mathbf{u}, p)$ of problem (4.6.1) satisfies the relations

$$\mathbf{u} \in \mathcal{W}_{\boldsymbol{\beta}}^{(2r,r),2}(\Omega \times (0, 2\pi)) \cap W^{(2r+2,r+1),2}(\Omega \times (0, 2\pi)),$$

$$\nabla p \in \mathcal{W}_{\boldsymbol{\beta}}^{(2r-2,r-1),2}(\Omega \times (0, 2\pi)) \cap W^{(2r,r),2}(\Omega \times (0, 2\pi)$$

and the estimate (4.6.40) when $l = r - 1$ holds.

Let us denote

$$\mathbf{u}^{\{s\}}(x,t) = \frac{\partial^s}{\partial t^s}\mathbf{u}(x,t), \quad p^{\{s\}}(x,t) = \frac{\partial^s}{\partial t^s}p(x,t), \quad \mathbf{f}^{\{s\}}(x,t) = \frac{\partial^s}{\partial t^s}\mathbf{f}(x,t),$$

where $s = 1, \ldots, r$. By differentiating equation (4.5.1) s times with respect to t, we obtain the problem

$$\begin{cases} \partial_t \mathbf{u}^{\{s\}} - \nu\Delta\mathbf{u}^{\{s\}} + \nabla p^{\{s\}} = \mathbf{f}^{\{s\}}, & x \in \Omega, t \in (0, 2\pi), \\ \operatorname{div} \mathbf{u}^{\{s\}} = 0, & x \in \Omega, t \in (0, 2\pi), \\ \mathbf{u}^{\{s\}} = 0, & x \in \partial\Omega, \ t \in (0, 2\pi), \\ \mathbf{u}^{\{s\}}(x,0) = \mathbf{u}^{\{s\}}(x, 2\pi), \ x \in \Omega. \end{cases} \tag{4.6.41}$$

Obviously, $\mathbf{f}^{\{s\}} \in \mathcal{W}_{\boldsymbol{\beta}}^{(2r-2s,r-s),2}(\Omega \times (0, 2\pi))$. Therefore, from estimate (4.6.40) for $l = r - s$, $s = 1, \ldots, r$, we obtain the inequalities

$$\|\mathbf{u}^{\{s\}}\|_{\mathcal{W}_{\boldsymbol{\beta}}^{(2r+2-2s,r+1-s),2}(\Omega\times(0,2\pi))} + \|\nabla p^{\{s\}}\|_{\mathcal{W}_{\boldsymbol{\beta}}^{(2r-2s,r-s),2}(\Omega\times(0,2\pi))}$$
$$\leq c\|\mathbf{f}^{\{s\}}\|_{\mathcal{W}_{\boldsymbol{\beta}}^{(2r-2s,r-s),2}(\Omega\times(0,2\pi))} \leq c\|\mathbf{f}\,|_{\mathcal{W}_{\boldsymbol{\beta}}^{(2r,r),2}(\Omega\times(0,2\pi))}, \quad s = 1, \ldots, r. \tag{4.6.42}$$

Now, consider $\big(\mathbf{u}^{\{s\}}(x,t), p^{\{s\}}(x,t)\big)$ as a solution of the steady Stokes problem:

$$\begin{cases} -\nu\Delta\mathbf{u}^{\{s\}} + \nabla p^{\{s\}} = \mathbf{f}^{\{s\}} - \partial_t \mathbf{u}^{\{s\}}, \ x \in \Omega, \\ \operatorname{div} \mathbf{u}^{\{s\}} = 0, \qquad x \in \Omega, \\ \mathbf{u}^{\{s\}} = 0, \qquad x \in \partial\Omega. \end{cases} \tag{4.6.43}$$

Using Theorem 4.4.2, we obtain the following estimates:

$$\|\mathbf{u}^{\{s\}}(\cdot,t)\|^2_{\mathcal{W}_{\boldsymbol{\beta}}^{2r+2-2s,2}(\Omega)} + \|\nabla p^{\{s\}}(\cdot,t)\|^2_{\mathcal{W}_{\boldsymbol{\beta}}^{2r-2s,2}(\Omega)}$$
$$\leq c\Big(\|\mathbf{f}^{\{s\}}(\cdot,t)\|^2_{\mathcal{W}_{\boldsymbol{\beta}}^{2r-2s,2}(\Omega)} + \|\partial_t \mathbf{u}^{\{s\}}(\cdot,t)\|^2_{\mathcal{W}_{\boldsymbol{\beta}}^{2r-2s,2}(\Omega)}\Big), \quad s = 1, \ldots, r,$$

where c is a positive constant. Integrating the last inequality from 0 to t and using (4.6.42), we deduce

$$\int_0^{2\pi} \|\frac{\partial^s \mathbf{u}(\cdot,\tau)}{\partial\tau^s}\|^2_{\mathcal{W}^{2r+2-2s,2}_{\boldsymbol{\beta}}(\Omega)} d\tau + \int_0^{2\pi} \|\nabla\frac{\partial^s p(\cdot,\tau)}{\partial\tau^s}\|^2_{\mathcal{W}^{2r-2s,2}_{\boldsymbol{\beta}}(\Omega)} d\tau \\ \le c\|\mathbf{f}\|^2_{\mathcal{W}^{(2r,r),2}_{\boldsymbol{\beta}}(\Omega\times(0,2\pi))}, \tag{4.6.44}$$

where $s = 1, \ldots, r$.

Finally, considering $\big(\mathbf{u}(x,t), p(x,t)\big)$ as the solution of the Stokes problem (4.6.43) for $s = 0$, we conclude that $\mathbf{u}(\cdot,t) \in \mathcal{W}^{2r+2,2}_{\boldsymbol{\beta}}(\Omega)$, and using (4.6.44) we obtain the estimate:

$$\int_0^{2\pi} \|\mathbf{u}(\cdot,\tau)\|^2_{\mathcal{W}^{2r+2,2}_{\boldsymbol{\beta}}(\Omega)} d\tau + \int_0^{2\pi} \|\nabla p(\cdot,\tau)\|^2_{\mathcal{W}^{2r,2}_{\boldsymbol{\beta}}(\Omega)} d\tau \\ \le c\Big(\int_0^{2\pi} \|\mathbf{f}(x,\tau)\|^2_{\mathcal{W}^{2r,2}_{\boldsymbol{\beta}}(\Omega)} dt + \int_0^{2\pi} \|\partial_\tau \mathbf{u}(x,\tau)\|^2_{\mathcal{W}^{2r,2}_{\boldsymbol{\beta}}(\Omega)} d\tau\Big) \\ \le c\|\mathbf{f}\|^2_{\mathcal{W}^{(2r,r),2}_{\boldsymbol{\beta}}(\Omega\times(0,2\pi))}. \tag{4.6.45}$$

Thus, estimate (4.6.40) for $l = r$ follows from inequalities (4.6.42), (4.6.44), (4.6.45) and from the definition of the norm in the space $\mathcal{W}^{(2r,r),2}_{\boldsymbol{\beta}}(\Omega \times (0, 2\pi))$.

□

Chapter 5
Steady-State Stokes and Navier–Stokes Equations in Tube Structures

In Chap. 1, it was previously mentioned that tube structures can serve as an abstract geometric model for networks of blood vessels, as well as for pipelines and analogous industrial configurations. In the present chapter, we delve into the examination of the stationary Stokes and Navier–Stokes equations within thin tube structures. The groundwork for asymptotic analysis of these equations was initially established in [28, 144], and in the books [146, 147]. The obtained asymptotic expansions of solutions were used to formulate hybrid-dimensional models using the Method of Asymptotic Partial Decomposition of the Domain (MAPDD). This method merges the one-dimensional depiction of viscous flow with small three-dimensional zoooms, as seen in [143, 145] concerning stationary problems. This strategic approach effectively reduces the computational burden on solvers dealing with flows within intricately shaped domains, thereby significantly expediting computations.

The first section revisits the formulation of the Stokes problem (1.4.3) from the Introduction, where velocities are prescribed at inlets and outlets, and with a no-slip boundary condition on the remaining boundary parts. An asymptotic expansion of the solution within the tube structure $B_\varepsilon \subset \mathbb{R}^n$, with $n = 2, 3$, is constructed. This expansion is justified first by an error estimate for the velocity in $W^{1,2}(B_\varepsilon)$ and, then for the pressure in $L^2(B_\varepsilon)$.

In Sect. 5.1, we justify the method of asymptotic partial decomposition of the domain introduced in Chap. 1 and prove an error estimate for the velocity in $W^{1,2}(B_\varepsilon)$. However, for the numerical realization of MAPDD, it is imperative to derive the weak formulation incorporating pressure. Section 5.2 introduces an extended adaptation of MAPDD to this case and furnishes proofs for velocity and pressure estimates in suitable norms, thus justifying this version. Section 5.3 navigate through the necessary modifications required to tackle the Navier–Stokes equations. Section 5.4 is dedicated to the study of Stokes equations within B_ε, with the given pressure at the inlets and outlets. The well-posedness of this problem is studied, and an asymptotic expansion of the solution is constructed. Error

G. Panasenko, K. Pileckas, *Multiscale Analysis of Viscous Flows in Thin Tube Structures*, Advances in Mathematical Fluid Mechanics,
https://doi.org/10.1007/978-3-031-54630-3_5

estimates are established for asymptotic approximations of velocity and pressure, alongside analogous estimates for MAPDD applied to this case. Lastly, Sect. 5.5 delves into scrutinizing the steady-state Navier–Stokes problem within a thin tube structure, encompassing boundary conditions that include Bernoulli pressure; see, e.g., [20–23, 51–53, 95, 100, 101, 136, 180, 192]. The setting of the Stokes and the Navier–Stokes equations with boundary conditions involving pressure is related to the following challenges. In computational scenarios pertaining to domains with cylindrical outlets arise concerns about the accuracy and correctness of prescribed boundary conditions on truncated surfaces. These artificial boundary conditions must fulfill dual criteria: (a) they should align with physical realism, highlighting pertinent phenomena and (b) resulting boundary value problems must exhibit mathematical well-posedness and suitability for numerical algorithms. A spectrum of potential choices for artificial boundary conditions has been explored by various researchers; see, e.g., [15, 17, 19, 29, 32, 34, 37, 64, 76, 80, 88, 99, 131, 133, 191, 201], and other works. This analysis underscores the significance of the proposed model: the inflow and outflow boundary conditions incorporate the "Bernoulli pressure" $p + \frac{1}{2}|\mathbf{u}|^2$, a pivotal parameter for high Reynolds numbers, as it remains constant along Euler flow streamlines. Furthermore, these boundary conditions are deemed more intuitive for numerical solutions.

5.1 Asymptotic Expansion of the Solution to the Stationary Stokes Equations

This section is dedicated to the study of a simplest model concerning blood flow within a network of blood vessels. We direct our focus on the Stokes equations in a tube structure B_ε. The precise definition of B_ε is provided in Introduction, and for our analysis, we adhere to the definition of B_ε with C^2-smooth boundary.

5.1.1 Problem Formulation

Consider the Stokes equations, previously introduced in Chaps. 1 and 4 (see problems (1.4.3) and (4.3.1)), within the the tube structure B_ε:

$$\begin{cases} -\nu\Delta\mathbf{v} + \nabla p = 0, \; x \in B_\varepsilon, \\ \operatorname{div} \mathbf{v} = 0, \; x \in B_\varepsilon, \\ \mathbf{v} = \mathbf{g}, \; x \in \partial B_\varepsilon. \end{cases} \tag{5.1.1}$$

Assume that the fluid velocity $\mathbf{g}$ at the boundary ∂B_ε has the following structure: $\mathbf{g} = 0$ on ∂B_ε except for small parts of the boundary denoted by

$\gamma_\varepsilon^{N_1+1}, \ldots, \gamma_\varepsilon^N$, each associated to vertices $O_{N_1+1}, \ldots, O_N$. In other words, $\gamma_\varepsilon^j = \partial\omega_\varepsilon^j \cap \partial B_\varepsilon$,

$$\begin{aligned} \mathbf{g}(x)\big|_{\gamma_\varepsilon^j} = \mathbf{g}^j\left(\frac{x-O_j}{\varepsilon}\right)\big|_{\gamma_\varepsilon^j}, \quad j = N_1+1, \ldots, N, \\ \mathbf{g}(x)\big|_{\partial B_\varepsilon \setminus \gamma_\varepsilon} = 0. \end{aligned} \tag{5.1.2}$$

Here, $\gamma_\varepsilon = \bigcup\limits_{j=N_1+1}^{N} \gamma_\varepsilon^j$, and $\mathbf{g}^j$ is a function of $C_0^2(\gamma_\varepsilon^j)$. Additionally, we assume that $\mathbf{g}$ satisfies the compatibility condition:

$$\int\limits_{\partial B_\varepsilon} \mathbf{g}\cdot\mathbf{n}\, dS = 0,$$

where $\mathbf{n}$ is the outward normal vector.

Suppose there exists a divergence-free extension of $\mathbf{g}$ into B_ε (denoted by the same symbol $\mathbf{g}$), such that $\mathbf{g} \in W^{2,2}(B_\varepsilon)$, and the following asymptotic estimates hold:

$$\begin{aligned} \sup_{x\in B_\varepsilon} |\mathbf{g}(x)| \le c, \quad \|\mathbf{g}\|_{L^2(B_\varepsilon)} \le c\varepsilon^{\frac{n-1}{2}}, \\ \|\nabla\mathbf{g}\|_{L^2(B_\varepsilon)} \le c\varepsilon^{\frac{n-3}{2}}, \quad \|\nabla^2\mathbf{g}\|_{L^2(B_\varepsilon)} \le c\varepsilon^{\frac{n-5}{2}}, \quad n = 2, 3, \end{aligned} \tag{5.1.3}$$

where c is a constant independent of ε. These conditions are satisfied by the asymptotic expansion of the solution, as constructed in the subsequent subsection.

With these considerations, the definition of the weak solution of problem (5.1.1) can be formulated as follows: The weak solution to problem (5.1.1) is a divergence-free vector-valued function $\mathbf{v} \in W^{1,2}(B_\varepsilon)$, such thus the difference $\mathbf{u} = \mathbf{v} - \mathbf{g}$ lies in $\mathring{H}^{1,2}(B_e)$ and satisfies the integral identity

$$\nu \int\limits_{B_\varepsilon} \nabla\mathbf{u}\cdot\nabla\eta\, dx = -\nu \int\limits_{B_\varepsilon} \nabla\mathbf{g}\cdot\nabla\eta\, dx \quad \text{for all } \eta \in \mathring{H}^{1,2}(B_\varepsilon). \tag{5.1.4}$$

The existence, uniqueness, regularity, and stability of a solution to the Stokes problem in an arbitrary domain are established in Chap. 4, as presented in Theorem 4.1.1.

5.1.2 *Asymptotic Expansion of the Solution*

In this section, we outline the procedure for constructing the asymptotic expansion of the solution. Let $\zeta \in C^2(\mathbb{R})$ denote an even function that is independent of ε, satisfying $\zeta(t) = 0$ when $|t| \le 1/3$, and $\zeta(t) = 1$ when $|t| \ge 2/3$. Denote $|e|_{\min} = \min\{|e_1|, \ldots, |e_M|\}$, where $|e_j|$ is the length of the edge e_i.

The asymptotic expansion of the velocity field is sought in the form

$$\begin{aligned}\mathbf{u}^{(J)}(x) = & \sum_{e=\overline{O_\alpha O_\beta};\alpha,\beta\le N} \zeta\left(\frac{x_n^{(e)}}{3r\varepsilon}\right)\zeta\left(\frac{|e|-x_n^{(e_j)}}{3r\varepsilon}\right)\mathbf{U}^{[e,J]}\left(\frac{x^{(e)'}}{\varepsilon}\right)\\ & +\sum_{l=1}^{N}\left(1-\zeta\left(\frac{|x-O_l|}{|e|_{\min}}\right)\right)\mathbf{U}^{[BLO_l,J]}\left(\frac{x-O_l}{\varepsilon}\right),\end{aligned} \tag{5.1.5}$$

where the first sum extends over all edges, each term associated with edge e being extended by zero outside the cylinder $\Pi_\varepsilon^{(e)}$, and each term in the second sum being extended by zero outside the bundle corresponding to the node O_l. The functions $\mathbf{U}^{[e,J]}$ represent velocities corresponding to the Poiseuille flows (see Chap. 3) within cylinders $\Pi_\varepsilon^{(e)}$, and they are expanded in powers of ε. Likewise, the functions $\mathbf{U}^{[BLO_l,J]}$ are of boundary layer type and exhibit exponential decay at infinity, also expanded in powers of ε:

$$\begin{cases} \mathbf{U}^{[e,J]}(y^{(e)\prime}) = \left(P^{(e)}\right)^*(0,\ldots,0,\widetilde{U}^{[e,J]}(y^{(e)\prime}))^*,\\ \mathbf{U}_j^{(e)}(y^{(e)\prime}) = \left(P^{(e)}\right)^*(0,\ldots,0,\widetilde{U}_j^{(e)}(y^{(e)\prime}))^*, \quad j=0,1,\ldots,J,\\ \widetilde{U}^{[e,J]}(y^{(e)\prime}) = \sum_{j=0}^{J}\varepsilon^j\widetilde{U}_j^{(e)}(y^{(e)\prime}),\\ \mathbf{U}^{[BLO_l,J]}(y) = \sum_{j=0}^{J}\varepsilon^j\mathbf{U}_j^{[BLO_l]}(y). \end{cases} \tag{5.1.6}$$

Thus, the expansion terms defined by (5.1.5) can be expressed as compositions of functions defined by (5.1.6) and linear mappings $y^{(e)\prime} = \frac{x^{(e)\prime}}{\varepsilon}$ and $y = \frac{x-O_l}{\varepsilon}$.

In the main part of domain B_ε, within the cylinders $\Pi_\varepsilon^{(e)}$ at some distance from the nodes and vertices, this expansion is close to the Poiseuille velocity (denoted $\mathbf{U}^{[e,J]}\left(\frac{x^{(e)\prime}}{\varepsilon}\right)$). This Poiseuille velocity is multiplied by the cut-off functions vanishing near the ends of the edge e: $\zeta\left(\frac{x_n^{(e)}}{3r\varepsilon}\right)$ and $\zeta\left(\frac{|e|-x_n^{(e_j)}}{3r\varepsilon}\right)$. In the neighborhood of the nodes and vertices O_l, the expansion $\mathbf{u}^{(J)}$ differs from the Poiseuille flow and is described by the sum of the truncated Poiseuille velocity and so-called boundary layer corrector $\mathbf{U}^{[BLO_l,J]}\left(\frac{x-O_l}{\varepsilon}\right)$. Note that the cut-off function $\zeta\left(\frac{|e|-x_n^{(e_j)}}{3r\varepsilon}\right)$ "truncates" the Poiseuille velocity only near the end of the edge e opposite to O_l, so that in the half-bundle HB_{O_l}, asymptotic approximation $\mathbf{u}^{(J)}(x)$ is given by

$$\mathbf{u}^{(J)}(x) = \sum_{e \subset \mathcal{B}_l} \zeta\left(\frac{x_n^{(e)}}{3r\varepsilon}\right) \mathbf{U}^{[e,J]}\left(\frac{x^{(e)\prime}}{\varepsilon}\right) + \left(1 - \zeta\left(\frac{|x - O_l|}{|e|_{\min}}\right)\right) \mathbf{U}^{[BLO_l,J]}\left(\frac{x - O_l}{\varepsilon}\right),$$

where the sum is taken over all edges e having O_l as an end point.

The asymptotic expansion of the pressure has a similar structure: linear function of $x_n^{(e)}$ (Poiseulle pressure increment with respect to its value at the point O_l, the end of the edge e_s) in the cylinders $\Pi_\varepsilon^{(e)}$ is multiplied by the same cut-off function $\zeta\left(\frac{x_n^{(e)}}{3r\varepsilon}\right)$, while the boundary layer corrector depends on $\frac{x-O_l}{\varepsilon}$, so that for each half-bundle HB_{O_l} with $l = 1, \ldots, N_1$ (where O_l is the origin of the local coordinate system), the asymptotic expansion of the pressure is given by

$$p^{(J)}(x) = \sum_{e \subset \mathcal{B}_l} \zeta\left(\frac{x_n^{(e)}}{3r\varepsilon}\right) \left(-s^{(e)} x_n^{(e)} + a^{(e)} - a^{(e_s)}\right) + a^{(e_s)} + \frac{1}{\varepsilon}\left(1 - \zeta\left(\frac{|x - O_l|}{|e|_{\min}}\right)\right) P^{[BLO_l,J]}\left(\frac{x - O_l}{\varepsilon}\right), \tag{5.1.7}$$

where

$$s^{(e)} = \frac{1}{\varepsilon^2}\sum_{j=0}^{J} \varepsilon^j s_j^{(e)}, \quad a^{(e)} = \frac{1}{\varepsilon^2}\sum_{j=0}^{J} \varepsilon^j a_j^{(e)}, \tag{5.1.8}$$

and

$$P^{[BLO_l,J]}(y) = \sum_{j=0}^{J} \varepsilon^j P_j^{[BLO_l,J]}(y). \tag{5.1.9}$$

The scaling order of the terms of this expansion is motivated by the order of the inlet/outlet velocities $\mathbf{g}^l$. Here and below, in every bundle of edges $\mathcal{B}^l$, we choose an arbitrary edge e_s and fix it. This edge will be called *the selected edge* e_s. The terms $s_j^{(e)}, a_j^{(e)}$ are constants associated with the edge e, independent of ε. Function $p^{(e)}$ defined on each edge e by the formula $p^{(e)}(x_n^{(e)}) = -s^{(e)} x_n^{(e)} + a^{(e)}$ will be called macroscopic pressure. Its value at the end of the edge e_s in the point O_l is equal to $a^{(e_s)}$.

Let us comment the terms of Poiseuille flow velocities $\widetilde{U}_j^{(e)}(y^{(e)\prime})$. We set $\widetilde{U}_j^{(e)}(y^{(e)\prime}) = u_{P_{s_j^{(e)}}}(y^{(e)\prime}) = s_j^{(e)} u_0(y^{(e)\prime})$. Here, u_0 represents the solution to the Poisson problem (3.1.2) in the dilated domain $\sigma = \sigma^{(e)}$. Let κ_e be the flux corresponding to this velocity: $\kappa_e = \int_{\sigma^{(e)}} u_0(y^{(e)\prime})dy^{(e)\prime}$. This constant characterizes the permeability of a tube with the cross-section $\sigma = \sigma^{(e)}$.

Along with the Poiseuille flows in the dilated cylinders with the cross-section $\sigma^{(e)}$ with the diameter of order one, we will consider the thin cylinders $\Pi_\varepsilon^{(e)}$ (parts of B_ε) where the cross-section $\sigma_\varepsilon^{(e)} = \{x^{(e)\prime} : x^{(e)\prime}/\varepsilon \in \sigma^{(e)}\}$ has diameter of the order ε. In this case the solution to problem (3.1.2) in $\sigma = \sigma_\varepsilon^{(e)}$ is given by formula $u_{0,\varepsilon}(x^{(e)\prime}) = \varepsilon^2 u_0(x^{(e)\prime}/\varepsilon)$, where u_0 is the solution of the same Poisson equation in $\sigma = \sigma^{(e)}$. The permeability of the thin tube $\Pi_\varepsilon^{(e)}$ is equal to

$$\kappa_\varepsilon^{(e)} = \int\limits_{\sigma_\varepsilon^{(e)}} \varepsilon^2 u_0(x^{(e)\prime}/\varepsilon)dx^{(e)\prime} = \varepsilon^{n+1}\kappa_e.$$

This formula establishes the relation between the permeabilities of the cylinders with dilated and non-dilated cross-sections.

The algorithm for iteratively determining the terms within the asymptotic expansions (5.1.5) and (5.1.7) is as follows.

We begin with the first-order approximation for the Poiseuille pressure $p^{(e)} = -s^{(e)}x_n^{(e)} + a^{(e)}$.

Base Case Solve the conductivity problem on the graph to find the function p_0:

$$\begin{cases} -\kappa_e \dfrac{\partial^2 p_0^{(e)}}{\partial x_n^{(e)2}}(x_n^{(e)}) = 0, & x_n^{(e)} \in (0, |e|),\ \ \forall e \subset \mathcal{B}^l, \\ -\displaystyle\sum_{e: O_l \in e} \kappa_e \dfrac{\partial p_0^{(e)}}{\partial x_n^{(e)}}(0) = 0, & l = 1, \ldots, N_1, \\ -\kappa_e \dfrac{\partial p_0^{(e)}}{\partial x_n^{(e)}}(0) = F_l, & l = N_1 + 1, \ldots, N, \\ p_0^{(e)}(0) = p_0^{(e_s)}(0),\ \forall e \subset \mathcal{B}^l, \end{cases} \tag{5.1.10}$$

where

$$F_l = -\int\limits_{\gamma^l} \mathbf{g}^l(y) \cdot \mathbf{n}\, dS,$$

Here $\mathbf{n}$ is an outward normal for Ω_l, so that

$$F_l = -\frac{1}{\varepsilon^{n-1}} \int\limits_{\gamma_\varepsilon^l} \mathbf{g}^l\left(\frac{x - O_l}{\varepsilon}\right) \cdot \mathbf{n} dS,$$

e_s is a selected (fixed) edge of the bundle $\mathcal{B}^l$.

It is worth noting that p_0 is the function defined on the graph. Its value on each edge e is denoted by $p_0^{(e)}$. The last equation of (5.1.10) means that the values of p_0 at the point $x_n^{(e)} = 0$ coincide for all edges e of the bundle $\mathcal{B}^l$. The first equation in (5.1.10) is written on every edge e and it means that $p_0^{(e)}$ is linear function. The second equation of (5.1.10) is written in every node; it is the Kirchhoff type junction condition and it means the the total flux at every node is equal to zero. The third equation is written in the vertices and it means that the fluxes at vertices are known. The local coordinates $x^{(e)}$ are defined so that all of them for the bundle $\mathcal{B}^l$ have the same origin O_l.

For the sake of reader's convenience, we provide herein the proof concerning the existence of a solution to the Eq. (5.1.10). The weak formulation of this problem employs the Sobolev space $W^{1,2}(\mathcal{B})$ of continuous functions defined on the graph $\mathcal{B}$ and belonging, on each edge e, to the space $W^{1,2}(e)$.

5.1.2.1 Problem on the Graph

We now consider a more general problem set on the graph.

Given real constants Ψ_l, where l ranges from 1 to N, positive constants κ_{e_i}, $i = 1, \dots, M$, and functions f belonging to the space $L^2(\mathcal{B})$, the objective is to determine a function p belonging to the Sobolev space $W^{1,2}(\mathcal{B})$ that satisfies the following equations:

$$\begin{cases} -\dfrac{\partial}{\partial x_n^{(e)}}\left(\kappa_e \dfrac{\partial p^{(e)}}{\partial x_n^{(e)}}(x_n^{(e)})\right) = f^{(e)}(x_n^{(e)}),\ x_n^{(e)} \in (0, |e|), \forall e \subset \mathcal{B}, \\ -\sum\limits_{e: O_l \in e} \left(\kappa_e \dfrac{\partial p^{(e)}}{\partial x_n^{(e)}}\right)(0) = \Psi_l, \qquad l = 1, \dots, N_1, \\ -\left(\kappa_e \dfrac{\partial p^{(e)}}{\partial x_n^{(e)}}\right)(0) = \Psi_l, \qquad l = N_1 + 1, \dots, N, \\ p^{(e)}(0) = p^{(e_s)}(0), \quad \forall e \subset \mathcal{B}^l, \end{cases} \tag{5.1.11}$$

where the equations hold for all edges $e = e_i$. Here $f^{(e)}$ is the restriction of f to the edge e.

We shall proceed by investigating this problem under the assumption that the following compatibility condition is fulfilled:

$$\sum_{i=1}^{M} \int_0^{|e_i|} f^{(e_i)} dx_n^{(e_i)} + \sum_{l=1}^{N} \Psi_l = 0. \tag{5.1.12}$$

Problem (5.1.11) admits a variational formulation. A weak solution of problem (5.1.11) is defined as a function p belonging to the Sobolev space $W^{1,2}(\mathcal{B})$, which satisfies the integral identity for any test function $q \in W^{1,2}(\mathcal{B})$:

$$\sum_{i=1}^{M}\int_{0}^{|e_i|}\kappa_{e_i}\frac{\partial p^{(e_i)}}{\partial x_n^{(e_i)}}\frac{\partial q^{(e_i)}}{\partial x_n^{(e_i)}}dx_n^{(e_i)}=\sum_{i=1}^{M}\int_{0}^{|e_i|}f^{(e_i)}q^{(e_i)}dx_n^{(e_i)}+\sum_{l=1}^{N}\Psi_l q(O_l), \tag{5.1.13}$$

where $q^{(e_i)}$ is the restriction of q to the edge e_i.

The equivalence between the weak formulation and the classical formulation can be demonstrated using standard techniques, such as the integration by parts formula and the proof of the equivalence between classical and weak formulations for second-order ordinary differential equations with constant coefficients.

Consider the subspace $W^{1,2}_{O_N}(\mathcal{B})$ of the space $W^{1,2}(\mathcal{B})$ consisting of functions that vanish at a single vertex, denoted as O_N.[1] By employing standard arguments, it is possible to establish the Poincaré inequality for functions $u \in W^{1,2}_{O_N}(\mathcal{B})$.

Denote $L = \sum\limits_{i=1}^{M} |e_i|$, $|e|_{max} = \max\limits_{i\in\{1,\dots,M\}} |e_i|$, and $C_{PF} = L|e|_{max}$.

Lemma 5.1.1 *Consider $u \in W^{1,2}_{O_N}(\mathcal{B})$. The following inequality holds:*

$$\sum_{i=1}^{M}\int_{0}^{|e_i|}|u^{(e_i)}|^2 dx_n^{(e_i)} \le C_{PF}\sum_{i=1}^{M}\int_{0}^{|e_i|}\left|\frac{\partial u^{(e_i)}}{\partial x_n^{(e_i)}}\right|^2 dx_n^{(e_i)} \tag{5.1.14}$$

Additionally, we have

$$\sup_{x\in\mathcal{B}}|u(x)| \le \sum_{i=1}^{M}\int_{0}^{|e_i|}\left|\frac{\partial u^{(e_i)}}{\partial x_n^{(e_i)}}\right| dx_n^{(e_i)} \le \left(|e|_{max}\sum_{i=1}^{M}\int_{0}^{|e_i|}\left|\frac{\partial u^{(e_i)}}{\partial x_n^{(e_i)}}\right|^2 dx_n^{(e_i)}\right)^{1/2}. \tag{5.1.15}$$

Proof

Consider an edge e of the graph $\mathcal{B}$. Due to the graph's connectedness, there exists a path (a connected chain of edges) that includes edge e and has O_N as an endpoint. Let x be a point on edge e. Denote by $\mathcal{B}_x$ the portion of this path between the points O_N and x. The difference $u(x) - u(O_N)$ (which equals $u(x)$ since $u(O_N) = 0$) can be represented as the integral over $\mathcal{B}_x$ of the derivative $\dfrac{\partial u^{(e_i)}}{\partial x_n^{(e_i)}}$. Evaluating the absolute value of this integral, we obtain

$$|u(x)| \le \int_{\mathcal{B}_x}\left|\frac{\partial u^{(e)}}{\partial x_n^{(e)}}\right| dx_n^{(e)}, \tag{5.1.16}$$

[1] It should be noted that functions in $W^{1,2}_{O_N}(\mathcal{B})$ are not assumed to be zero at the remaining vertices O_l, where $l = N_1 + 1, \dots, N - 1$.

where e represents the edges of the path. This leads to the following inequality:

$$\sup_{x\in\mathcal{B}} |u(x)| \leq \sum_{i=1}^{M} \int_{0}^{|e_i|} \Big|\frac{\partial u^{(e_i)}}{\partial x_n^{(e_i)}}\Big| dx_n^{(e_i)}, \tag{5.1.17}$$

and applying the Cauchy-Schwarz-Bunyakovsky inequality to the right-hand side, we obtain relation (5.1.15), and so

$$u^2(x) \leq |e|_{max} \sum_{i=1}^{M} \int_{0}^{|e_i|} \Big|\frac{\partial u^{(e_i)}}{\partial x_n^{(e_i)}}\Big|^2 dx_n^{(e_i)}, \tag{5.1.18}$$

which after integrating over $\mathcal{B}$ subsequently results in inequality (5.1.14). This concludes the proof of the lemma. □

It is evident that a solution to problem (5.1.11) (or equivalently (5.1.13)) is defined up to an arbitrary additive constant. Therefore, one can search for a solution of problem (5.1.13) that additionally satisfies the condition $p(O_N) = 0$, i.e, $p \in W^{1,2}_{O_N}(\mathcal{B})$.

Theorem 5.1.1 *Let $f \in L^2(\mathcal{B})$. Problem* (5.1.13) *has a unique solution $p \in W^{1,2}_{O_N}(\mathcal{B})$ if and only if the compatibility condition* (5.1.12) *holds. For the solution p, the following inequality holds:*

$$\|p\|_{W^{1,2}(\mathcal{B})} \leq C\Big(\|f\|_{L^2(\mathcal{B})} + \big(\sum_{l=1}^{N} \Psi_l^2\big)^{1/2}\Big) \tag{5.1.19}$$

Here

$$C = \frac{C_{PF}+1}{\kappa_{min}} \max\{1, \sqrt{|e|_{max} N}\}, \quad \kappa_{min} = \min_{i\in\{1,\ldots,M\}} \kappa_{e_i}, \tag{5.1.20}$$

$$\|f\|^2_{L_2(\mathcal{B})} = \sum_{i=1}^{M} \int_{0}^{|e_i|} |f^{(e_i)}|^2 dx_n^{(e_i)}.$$

Proof

Consider the function space $W^{1,2}_{O_N}(\mathcal{B})$ equipped with the inner product denoted as $[u, v]$, which is defined as follows:

$$[u, v] = \sum_{i=1}^{M} \int_{0}^{|e_i|} \kappa_{e_i} \frac{\partial u^{(e_i)}}{\partial x_n^{(e_i)}} \frac{\partial v^{(e_i)}}{\partial x_n^{(e)}} dx_n^{(e_i)}.$$

(Recall that the conventional inner product in $W^{1,2}(\mathcal{B})$ is given by

$$(u, v) = \sum_{i=1}^{M} \int_{0}^{|e_i|} \left(u^{(e_i)} v^{(e_i)} + \frac{\partial u^{(e_i)}}{\partial x_n^{(e_i)}} \frac{\partial v^{(e_i)}}{\partial x_n^{(e)}} \right) dx_n^{(e_i)}.$$

The Poincaré inequality establishes the equivalence of the norms corresponding to these two inner products.)

The right-hand side of integral identity (5.1.13) defines a linear functional on $W^{1,2}_{O_N}(\mathcal{B})$:

$$\Phi(q) = \sum_{i=1}^{M} \int_{0}^{|e_i|} f^{(e_i)} q^{(e_i)} \, dx_n^{(e_i)} + \sum_{l=1}^{N} \Psi_l q(O_l).$$

By virtue of (5.1.14) and (5.1.15), this functional is bounded. Therefore, applying the Riesz theorem (see Theorem 2.2.6), we deduce the existence of a unique function p in $W^{1,2}_{O_N}(\mathcal{B})$ such that

$$\Phi(q) = [p, q]$$

holds for all $q \in W^{1,2}_{O_N}(\mathcal{B})$. In other words, there exists a unique $p \in W^{1,2}_{O_N}(\mathcal{B})$ that satisfies identity (5.1.13) for all $q \in W^{1,2}_{O_N}(\mathcal{B})$.

Next, let us consider an arbitrary function $q \in W^{1,2}(\mathcal{B})$. We decompose it into a sum:

$$q(x) = q(O_N) + \tilde{q}(x),$$

where $\tilde{q} \in W^{1,2}_{O_N}(\mathcal{B})$. Let us check whether the integral identity (5.1.13) still holds. Clearly, the constant $q(O_N)$ does not affect the left-hand side. Furthermore, it does not influence the right-hand side either, owing to the compatibility condition (5.1.12). Consequently, the function p satisfies identity (5.1.13) for all $q \in W^{1,2}(\mathcal{B})$. Thus, the existence of a unique weak solution to (5.1.13) that vanishes at O_N has been established.

The necessity of condition (5.1.12) follows from (5.1.13) with $q = const$.

The estimate (5.1.19) is a consequence of the identity (5.1.13) with $q = p$, along with the estimates (5.1.14) and (5.1.15). Indeed, by the Poincaré inequality (5.1.14) and (5.1.15), we have

$$\kappa_{min} \|p\|^2_{W^{1,2}(\mathcal{B})} \leq \left(C_{PF} + 1 \right) \sum_{i=1}^{M} \kappa_{e_i} \int_{0}^{|e_i|} \left| \frac{\partial p^{(e_i)}}{\partial x_n^{(e_i)}} \right|^2 dx_n^{(e_i)}$$

$$= \Big(C_{PF}+1\Big)\Big(\sum_{i=1}^{M}\int_0^{|e_i|} f^{(e_i)}(x_n^{(e_i)})p^{(e_i)}(x_n^{(e_i)})\,dx_n^{(e_i)} + \sum_{l=1}^{N}\Psi_l p(O_l)\Big)$$

$$\le \Big(C_{PF}+1\Big)\Big(\|f\|_{L^2(\mathcal{B})}\|p\|_{W^{1,2}(\mathcal{B})} + \sqrt{N}\big(\sum_{l=1}^{N}\Psi_l^2\big)^{1/2}\max_{x\in\mathcal{B}}|p(x)|\Big)$$

$$\le \Big(C_{PF}+1\Big)\Big(\|f\|_{L^2(\mathcal{B})} + \sqrt{|e|_{max}N}\big(\sum_{l=1}^{N}\Psi_l^2\big)^{1/2}\Big)\|p\|_{W^{1,2}(\mathcal{B})}.$$

Thus, we have established inequality (5.1.19) with the constant C defined by (5.1.20). □

By solving the conductivity problem (5.1.10), we define for every edge e the constants $s_0^{(e)}$ and $a_0^{(e)}$ such that

$$p_0^{(e)}(x^{(e)}) = -s_0^{(e)}x_n^{(e)} + a_0^{(e)}$$

and define the velocity

$$\tilde{U}_{(0)}^{(e)}(y^{(e)\prime}) = s_{(0)}^{(e)}U_0^{(e)}(y^{(e)\prime}),\quad \mathbf{U}_0^{(e)} = \big(\mathcal{P}^{(e)}\big)^*(0,\ldots,0,\tilde{U}_{(0)}^{(e)})^*, \tag{5.1.21}$$

where $U_0^{(e)}(y^{(e)\prime})$ is determined as the solution of the Poisson problem:

$$\begin{cases} -\nu\Delta_{(y^{(e)\prime})}U_0^{(e)}(y^{(e)\prime}) = 1,\ y^{(e)\prime}\in\sigma^{(e)},\\ \qquad\qquad U_0^{(e)}(y^{(e)\prime}) = 0,\ y^{(e)\prime}\in\partial\sigma^{(e)}. \end{cases} \tag{5.1.22}$$

Here we used the notation $U_0^{(e)}(y^{(e)\prime})$ for the solution u_0 of the Poisson problem on $\sigma^{(e)}$.

5.1.2.2 The Boundary Layer Corrector

The boundary layer corrector compensates the residual generated by the multiplication of the functions $\tilde{U}_0^{(e)}\Big(\frac{x^{(e)}}{\varepsilon}\Big)$, $P_0^{(e)}(x^{(e)})$ by the cut-off function $\zeta\Big(\frac{x_n^{(e)}}{3r\varepsilon}\Big)$. The boundary layer problem for $l = 1,\ldots,N_1$ involves the unknowns $(\mathbf{U}_0^{[BLO_l]}(y), P_0^{[BLO_l]}(y))$, defined by the equations

$$\begin{cases} -\nu\Delta_y \mathbf{U}_0^{[BLO_l]}(y) + \nabla_y P_0^{[BLO_l]}(y) = \mathbf{f}_0^{[REGO_l]}(y), & y \in \Omega_l, \\ \operatorname{div}_y \mathbf{U}_0^{[BLO_l]}(y) = h_0^{[REGO_l]}(y), & y \in \Omega_l, \\ \mathbf{U}_0^{[BLO_l]}(y) = 0, & y \in \partial\Omega_l, \end{cases} \tag{5.1.23}$$

where

$$\mathbf{f}_0^{[REGO_l]}(y) = -\sum_{e:O_l\in e}\Big\{s_0^{(e)}\Big(-\nu\Delta_y\Big(\zeta\Big(\frac{y_n^{(e)}}{3r}\Big)(\mathcal{P}^{(e)})^*(0,\dots,0,U_0^{(e)}(y^{(e)\prime}))^*\Big) \\ -\nabla_y\Big(\zeta\Big(\frac{y_n^{(e)}}{3r}\Big)y_n^{(e)}\Big)\Big) + (a_1^{(e)} - a_1^{(e_s)})\,\nabla_y\Big(\zeta\Big(\frac{y_n^{(e)}}{3r}\Big)\Big)\Big\}, \tag{5.1.24}$$

$$h_0^{[REGO_l]}(y) = -\operatorname{div}_y \sum_{e:O_l\in e}\Big\{s_0^{(e)}\zeta\Big(\frac{y_n^{(e)}}{3r}\Big)(\mathcal{P}^{(e)})^*(0,\dots,0,U_0^{(e)}(y^{(e)\prime}))^*\Big\} \\ = -\sum_{e:O_l\in e}\Big\{s_0^{(e)}\nabla\zeta\Big(\frac{y_n^{(e)}}{3r}\Big)\cdot(\mathcal{P}^{(e)})^*(0,\dots,0,U_0^{(e)}(y^{(e)\prime}))^*\Big\}. \tag{5.1.25}$$

Here the sum $\sum_{e:O_l\in e}$ is taken over all edges e having ends in the node O_l.

The integral of the function $h_0^{[REGO_l]}$ over Ω_l is equal to zero due to condition $(5.1.10)_2$ of the problem on the graph. Indeed, the support of the function $h_0^{[REGO_l]}$ confined within a ball at O_l and having a radius of $3r$ (out of this ball, the gradient of the cut-off function vanishes). As per the Gauss-Ostrogradsky formula, the integral is equal to the sum of surface integrals of the Poiseuille velocities $\tilde{U}_0^{(e)}(y^{(e)\prime})$ across the cross-sections of all cylinders. These cumulative contributions align precisely with the left-hand side of condition $(5.1.10)_2$ i.e. the sum of fluxes at the node is zero. On account of this, problem (5.1.23) can be simplified to the Stokes problem with the divergence-free condition, and so, it admits a unique solution.

Of note, an indeterminate quantity emerges within the right-hand side of (5.1.24), specifically the constant $a_1^{(e)} - a_1^{(e_s)}$. Denote the solution to problem (5.1.23) without the last term $(a_1^{(e)} - a_1^{(e_s)})\nabla_y\Big(\zeta\Big(\frac{y_n^{(e)}}{3r}\Big)\Big)$ in the right-hand side $\mathbf{f}_0^{[REGO_l]}(y)$ by $(\mathbf{U}_0^{[BLO_l]}(y), \widehat{P}_0^{[BLO_l]}(y))$ (it should be noted that this term has a gradient form and permits incorporation into the pressure, causing solutions to differ only with respect to pressure components). The right-hand sides of system (5.1.23) have compact supports. Therefore, applying Theorem 4.4.3 and Lemma 4.4.1, we conclude that the pressure $\widehat{P}_0^{[BLO_l]}(y)$ attains exponential stabilization in each outlet (linked to the edge e) towards a constant value denoted as $\widehat{a}_0^{[BLO_l,e]}$. This convergence is valid in the integral sense

$$\lim_{k\to+\infty} \int\limits_{\{y_n^{(e)}\in(k,k+1)\}\cap\Omega_l} e^{2\beta y_n^{(e)}} |\widehat{P}_0^{[BLO_l]}(y) - \widehat{a}_0^{[BLO_l,e]}|^2 dy = 0 \qquad (5.1.26)$$

for some $\beta > 0$.

Additional observations can be made regarding the structure of the right-hand sides within the boundary layer problems associated with node O_l. Firstly, the truncated Poiseuille velocities $\zeta\Big(\frac{x_n^{(e)}}{3r\varepsilon}\Big)\zeta\Big(\frac{|e| - x_n^{(e_j)}}{3r\varepsilon}\Big)\mathbf{U}^{[e,J]}\Big(\frac{x^{(e)'}}{\varepsilon}\Big)$, as delineated by formula (5.1.5), are influenced solely by a single cut-off function $\zeta\Big(\frac{x_n^{(e)}}{3r\varepsilon}\Big)$, which corresponds to the node O_l. The second cut-off function $\zeta\Big(\frac{|e| - x_n^{(e_j)}}{3r\varepsilon}\Big)$ influences the right-hand side of the problem associated to the second end of the edge e. Second, out of the finite support of the right-hand side $h_0^{[REGO_l]}$ of the divergence equation $(5.1.23)_2$, the solution $\mathbf{U}_0^{[BLO_l]}(y)$ has vanishing flux because it tends to zero at infinity and is a solenoidal function.

For $l = N_1 + 1, \ldots, N$, the boundary layer term $(\mathbf{U}_0^{[BLO_l]}(y), P_0^{[BLO_l]}(y))$ is the solution of the following system of equations within the semi-cylinder Ω_l:

$$\begin{cases} -\nu\Delta_y \mathbf{U}_0^{[BLO_l]}(y) + \nabla_y P_0^{[BLO_l]}(y) = \mathbf{f}_0^{[REGO_l]}(y), & y \in \Omega_l, \\ \mathrm{div}_y \mathbf{U}_0^{[BLO_l]}(y) = h_0^{[REGO_l]}(y), & y \in \Omega_l, \\ \mathbf{U}_0^{[BLO_l]}(y) = 0, & y \in \partial\Omega_l\setminus\gamma^l, \\ \mathbf{U}_0^{[BLO_l]}(y) = \mathbf{g}^l(y), & y \in \gamma^l, \end{cases} \qquad (5.1.27)$$

where

$$\mathbf{f}_0^{[REGO_l]}(y) = -\sum_{e:O_l\in e} s_0^{(e)}\Big(-\nu\Delta_y\Big(\zeta\Big(\frac{y_n^{(e)}}{3r}\Big)(\mathcal{P}^{(e)})^*(0,\ldots,0,U_0^{(e)}(y^{(e)'}))^*\Big) - \nabla_y\Big(\zeta\Big(\frac{y_n^{(e)}}{3r}\Big)y_n^{(e)}\Big)\Big), \qquad (5.1.28)$$

and $h_0^{[REGO_l]}(y)$ is defined by formula (5.1.25). As before, the integral of the function $h_0^{[REGO_l]}$ over Ω is equal to zero. To show this, we mention that the function $h_0^{[REGO_l]}$ has a support contained in the ball with the center at O_l and the radius $3r$. So, due to the Gauss-Ostrogradsky formula, for $l = N_1 + 1, \ldots, N$, this integral is equal to the difference of the integral over the cross-section of the cylinder of the Poiseuille velocitiy $\widetilde{U}_0^{(e)}(y^{(e)'})$ and the surface integral of the normal velocity $\mathbf{g}^l$ in the right-hand side of condition $(5.1.27)_4$. According to the condition $(5.1.10)_3$, this difference is equal to zero. Thus, problem (5.1.27) can be reduced to the Stokes problem with the divergence-free condition and vanishing boundary conditions. So, it admits a unique solution. The pressure $P_0^{[BLO_l]}(y)$ can be normalized so, that it tends to zero as $|y| \to \infty$ (in the sense of (5.1.26)).

5.1.2.3 Asymptotic Expansion of the Solution: Higher Order Terms

Consider the conductivity problem of rank 1 on the graph for the function p_1:

$$\begin{cases} -\kappa_e \dfrac{\partial^2 p_1^{(e)}}{\partial x_n^{(e)2}}(x_n^{(e)}) = 0, & x_n^{(e)} \in (0, |e|), \ \ \forall e \subset \mathcal{B}^l, \\ -\sum\limits_{e:O_l \in e} \kappa_e \dfrac{\partial p_1^{(e)}}{\partial x_n^{(e)}}(0) = 0, & l = 1, \dots, N_1, \\ -\kappa_e \dfrac{\partial^2 p_1^{(e)}}{\partial x_n^{(e)2}}(0) = 0, & l = N_1 + 1, \dots, N, \\ p_1^{(e)}(0) - p_1^{(e_s)}(0) = \widehat{a}_0^{[BLO_l,e]}, \ \forall e \subset \mathcal{B}^l, \ e \neq e_s, & \end{cases} \tag{5.1.29}$$

where e_s is the selected edge of the bundle. In this graph-posed problem, the solution can exhibit discontinuities at the nodes. Specifically, at each node O_l, there are prescribed jumps of $p_1^{(e)}$ between any pair of edges e and e_s of the bundle. This problem also possesses a unique solution p_1 that vanishes at O_N.

Indeed, the problem (5.1.29) can be transformed into a problem for a continuous unknown function (without jumps), yet with a right-hand side in $L^2(\mathcal{B})$. By introducing the function $\left(1 - \zeta\left(\frac{x_n^{(e)}}{|e|}\right)\right)\widehat{a}_0^{[BLO_l,e]}$ on each edge e, the unknown function $p_1^{(e)}$ can be modified to

$$\tilde{p}_1^{(e)} = p_1^{(e)} - \left(1 - \zeta\left(\frac{x_n^{(e)}}{|e|}\right)\right)\widehat{a}_0^{[BLO_l,e]}$$

on every edge e. This leads to the new unknown function $\tilde{p}_1^{(e)}$ satisfying equations (5.1.11) with

$$f^{(e)}(x_n^{(e)}) = \kappa_e |e|^{-2} \zeta''\left(\frac{x_n^{(e)}}{|e|}\right)\widehat{a}_0^{[BLO_l,e]}, \quad \Psi_l = 0.$$

This problem has a weak formulation similar to (5.1.13) with the replacement of $f^{(e)}(x_n^{(e)})$ by the above expression.

Now, the constants $s_1^{(e)}$ and $a_1^{(e)}$ are known: $p_1^{(e)}(x_n^{(e)}) = -s_1^{(e)} x_n^{(e)} + a_1^{(e)}$, and the pressure in the boundary layer problem (5.1.23) can be completely determined as

$$P_0^{[BLO_l]}(y) = \widehat{P}_0^{[BLO_l]}(y) - \sum_{e:O_l \in e,\, e \neq e_s} \zeta\left(\frac{y_n^{(e)}}{3r}\right)\hat{a}_0^{[BLO_l,e]}.$$

Consider the scenario where all terms in the expansion (5.1.5)–(5.1.9) that correspond to ranks less than or equal to $j-1$ are already known. Additionally, let us assume that the macroscopic pressure on the graph p_j is also given and satisfies condition $(5.1.29)_3$. We will now outline the process of transitioning from rank $j-1$ to rank j.

Step 1 Given the known macroscopic pressure p_j on the graph, we define constants $s_j^{(e)}$ and $a_j^{(e)}$ for each edge e as follows:

$$p_j^{(e)}(x^{(e)}) = -s_j^{(e)} x_n^{(e)} + a_j^{(e)}.$$

Note that due to $(5.1.29)_3$, for all edges having vertex as an end point, $s_j^{(e)} = 0$ and $\mathbf{U}_j^{(e)} = 0$. Additionally, we define

$$\begin{aligned} \widetilde{U}_j^{(e)}(y^{(e)\prime}) &= s_j^{(e)} U_0^{(e)}(y^{(e)\prime}), \\ \mathbf{U}_j^{(e)} &= (\mathcal{P}^{(e)})^*(0, \ldots, 0, \widetilde{U}_j^{(e)})^*. \end{aligned} \tag{5.1.30}$$

Step 2 The boundary layer solution consists of a pair $(\mathbf{U}_j^{[BLO_l]}, P_j^{[BLO_l]})$ that solves the following Stokes system within each subdomain Ω_l, where $l = 1, \ldots, N_1$:

$$\begin{cases} -\nu\Delta_y \mathbf{U}_j^{[BLO_l]}(y) + \nabla_y P_j^{[BLO_l]}(y) = \mathbf{f}_j^{[REGO_l]}(y), & y \in \Omega_l, \\ \operatorname{div}_y \mathbf{U}_j^{[BLO_l]}(y) = h_j^{[REGO_l]}(y), & y \in \Omega_l, \\ \mathbf{U}_j^{[BLO_l]}(y) = 0, & y \in \partial\Omega_l, \quad j = 0, \ldots, J, \end{cases} \tag{5.1.31}$$

where

$$\begin{aligned} \mathbf{f}_j^{[REGO_l]}(y^{(e)}) = -\sum_{e:O_l\in e} \Big\{ &-\nu\Delta_y\Big[\zeta\Big(\frac{y_n^{(e)}}{3r}\Big)\mathbf{U}_j^{(e)}(y^{(e)\prime})\Big] \\ &+\nabla_y\Big[\zeta\Big(\frac{y_n^{(e)}}{3r}\Big)\Big(-s_j^{(e)} y_n^{(e)}\Big)\Big] + \widehat{a}_j^{(e)}\,\nabla_y\zeta\Big(\frac{y_n^{(e)}}{3r}\Big)\Big\} \end{aligned} \tag{5.1.32}$$

(for $j = J$ the coefficient $\widehat{a}_j^{(e)}$ is omitted),

$$h_j^{[REGO_l]}(y) = -\sum_{e:O_l\in e} \operatorname{div}_y\Big(\zeta\Big(\frac{y_n^{(e)}}{3r}\Big)\mathbf{U}_j^{(e)}(y^{(e)\prime})\Big). \tag{5.1.33}$$

Here the sum $\sum\limits_{e:O_l\in e}$ is taken over all edges e having ends in the node O_l, and by convention, the terms with the negative subscripts j are equal to zero.

Further, we determine the pair $(\mathbf{U}_j^{[BLO_l]}, \widehat{P}_j^{[BLO_l]})$ that satisfies the same problem (5.1.31) but without the last term in the definition of $\mathbf{f}_j^{[REGO_l]}$. It can be proved by induction, using Theorem 4.4.3 and Lemma 4.4.1, that the solution $\mathbf{U}_j^{[BLO_l]}$ asymptotically approaches zero as $|y| \to +\infty$, while the associated pressure function $\widehat{P}_j^{[BLO_l]}$ stabilizes towards certain constants $\widehat{a}_j^{[BLO_l,e]}$ in the sense of (5.1.26); these constants may be different for different outlets. Since the pressure function is defined up to an additive constant, we can fix the limit constant equal to zero for the outlet corresponding the the selected edge e_s.

Subsequently, we solve the problems in semi-cylinders Ω_l, $l = N_1 + 1, \ldots, N$:

$$\begin{cases} -\nu\Delta_y\mathbf{U}_j^{[BLO_l]}(y) + \nabla_y\widehat{P}_j^{[BLO_l]}(y) = \mathbf{f}_j^{[REGO_l]}(y), & y \in \Omega_l, \\ \mathrm{div}_y\mathbf{U}_j^{[BLO_l]}(y) = h_j^{[REGO_l]}(y), & y \in \Omega_l, \\ \mathbf{U}_j^{[BLO_l]}(y) = 0, & y \in \partial\Omega_l\backslash\gamma^l, \\ \mathbf{U}_j^{[BLO_l]}(y) = 0, & y \in \gamma^l, \end{cases} \tag{5.1.34}$$

where

$$\begin{aligned} \mathbf{f}_j^{[REGO_l]}(y^{(e)}) = -\sum_{e:O_l\in e}\Big\{ & -\nu\Delta_y\Big[\zeta\Big(\frac{y_n^{(e)}}{3r}\Big)\mathbf{U}_j^{(e)}(y^{(e)\prime})\Big] \\ & +\nabla_y\Big[\zeta\Big(\frac{y_n^{(e)}}{3r}\Big)\big(-s_j^{(e)}y_n^{(e)}\big)\Big]\Big\} \end{aligned} \tag{5.1.35}$$

and $h_j^{[REGO_l]}(y)$ is defined by formula (5.1.33), which has a form in this case:

$$h_j^{[REGO_l]}(y) = -\mathrm{div}_y\Big(\zeta\Big(\frac{y_n^{(e)}}{3r}\mathbf{U}_j^{(e)}(y^{(e)})\Big)\Big),$$

so that

$$\int\limits_{\Omega_l} h_j^{[REGO_l]}(y)dy = -\int\limits_{\sigma} \tilde{U}_j^{(e)}(y^{(e)\prime})dy^{(e)\prime} = 0,$$

where σ is the cross-section of the cylindric part of Ω_l.

Step 3 We solve the conductivity problem on the graph for the function $p_{j+1}^{(e)}$ ($j < J$):

$$\begin{cases} -\kappa_e \dfrac{\partial^2 p_{j+1}^{(e)}}{\partial x_n^{(e)2}}(x_n^{(e)}) = 0, & x_n^{(e)} \in (0, |e|), \\ -\sum\limits_{e: O_l \in e} \kappa_e \dfrac{\partial p_{j+1}^{(e)}}{\partial x_n^{(e)}}(0) = 0, & l = 1, \ldots, N_1, \\ \kappa_e \dfrac{\partial p_{j+1}^{(e^{O_l})}}{\partial x_n^{(e)}}(0) = 0, & l = N_1 + 1, \ldots, N, \\ p_{j+1}^{(e)}(0) - p_{j+1}^{(e_s)}(0) = \widehat{a}_j^{[BLO_l,e]}, \ \forall e \subset \mathcal{B}^l, e \neq e_s, \end{cases}$$

where the local coordinates $x^{(e)}$ are redefined so that all of them have the same origin O_l. Note, that $a_{j+1}^{(e)} - a_{j+1}^{(e_s)} = \widehat{a}_j^{[BLO_l,e]}$.

Step 4 Finally, we define the pressure $P_j^{[BLO_l]}(y)$ within the boundary layer problem (5.1.31), (5.1.32) for $l = 1, \ldots, N_1$. The expression for $P_j^{[BLO_l]}(y)$ is given by

$$P_j^{[BLO_l]}(y) = \widehat{P}_j^{[BLO_l]}(y) - \sum_{e: O_l \in e,\ e \neq e_s} \zeta\Big(\frac{y_n^{(e_{\alpha m})}}{3r}\Big) \widehat{a}_j^{[BLO_l,e]}. \tag{5.1.36}$$

This concludes the process of transitioning from rank j to rank $j+1$.
Here the pressure $P_j^{[BLO_l]}(y)$ tends to zero as $|y| \to \infty$.

5.1.2.4 Simplified Algorithm for Constructing Asymptotic Expansions for Tree-Type Graphs

The simplest category of graphs that correspond to tube structures is the tree type, which excludes cycles.[2] Let the graph $\mathcal{B}$ be a tree (Fig. 5.1). Assume its edges are organized as follows: we designate the "root vertex" as O_N and select the edge e for which it serves as an endpoint. This edge is relabeled as e_1 and referred to as the "first-level" edge. The other endpoint of e_1, denoted as O_{l_1}, serves as a common endpoint for second-level edges $e_{21}, e_{22}, \ldots, e_{2r_2}$; these second-level endpoints, labeled as $O_{l_{21}}, \ldots, O_{l_{2r_2}}$, are termed "second-level nodes/vertices." These, in turn, serve as common endpoints for third-level edges:

$$\begin{array}{lll} O_{l_{21}} & \text{for} & e_{311}, \ldots, e_{31\mu_{31}}, \\ O_{l_{22}} & \text{for} & e_{321}, \ldots, e_{32\mu_{32}}, \\ & \ldots & \ldots \quad \ldots \\ O_{l_{2r_2}} & \text{for} & e_{3r_21}, \ldots, e_{3r_2\mu_{3r_2}}, \end{array}$$

[2] A cycle is a finite sequence of distinct edges $\overline{O_\alpha O_{\alpha_1}}, \overline{O_{\alpha_1} O_{\alpha_2}}, \ldots, \overline{O_{\alpha_n} O_\alpha}$.

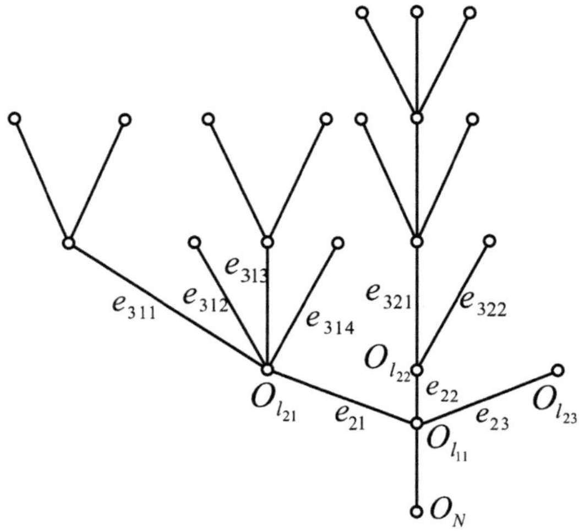

Fig. 5.1 Tree-type graph

and so forth. Notably, each second endpoint of a k-level edge, also known as a k-level "node/vertex," acts as a shared endpoint for several (≥ 0) edges of level $k+1$. The finite nature of the total number of levels should be highlighted.

It is worth recalling that each edge e permits the establishment of a local coordinate system in two distinct ways, contingent on choosing its origin at either first or second end of the edge. For the present purpose, the directionality for tree-type graphs is fixed from lower levels to higher levels; in other words, when the edge e is of level k, the direction of $x_n^{(e)}$ is chosen from the end of level $k-1$ to the end of level k.

In the context of tree-type graphs, for each edge e, the flow rate $F_\varepsilon^{(e)}$ can be determined. Commencing with edges e that end at vertices O_l, where $l = N_1+1, \ldots, N$, the flow rates $F_\varepsilon^{(e)}$ are computed as integrals $\int_{\gamma_\varepsilon^l} \mathbf{g}^l\left(\frac{x-O_l}{\varepsilon}\right) \cdot \mathbf{n}\, dS$, with $\mathbf{n}$ representing the outward normal vector. Notably, this can be expressed as $\varepsilon^{n-1}\int_{\gamma^l} \mathbf{g}^l(y) \cdot \mathbf{n}\, dy$, wherein the integral $F^{(e)} = \int_{\gamma^l} \mathbf{g}^l(y) \cdot \mathbf{n}\, dy$ remains independent of ε. Progressing from higher to lower levels and employing the flow rate balance at nodes (wherein the sum of flow rates at each node equals zero), the flow rates for edges at all levels are ascertained. Thus, the graph equation is partially solved through analytical means: the pressure slopes $-s^{(e)} = F_\varepsilon^{(e)}/\kappa_{e,\varepsilon} = \varepsilon^{-2}F^{(e)}/\kappa_e$ are analytically defined (see the previous section for the relation of the permeabilities κ_e and $\kappa_{e,\varepsilon}$ for the dilated and non-dilated tubes). Consequently, the regular part of the asymptotic expansion (in local coordinates) takes the form of the Poiseuille solution:

$$\mathbf{U}^{[e]}(y^{(e)\prime}) = (P^{(e)})^*(0, \ldots, 0, \tilde{U}^{[e]}(y^{(e)\prime}))^*,$$

wherein

$$\tilde{U}^{[e]}(y^{(e)\prime}) = s^{(e)} U_0^{(e)}(y^{(e)\prime}).$$

Here $y^{(e)} = x^{(e)}/\varepsilon$. Note that here $U^{[e]}$ replaces $U^{[e,J]}$ of the previous section but now it is independent of J.

In this scenario, the construction of the asymptotic expansion can be notably simplified, obviating the necessity for expansion in powers of ε. To clarify, the macroscopic pressure $p^{(e)}(x_n^{(e)}) = -s^{(e)} x_n^{(e)} + a^{(e)}$ and the boundary layer functions $\mathbf{U}^{[BLO_l]}$, $P^{[BLO_l]}$ can be derived with greater ease as they are lifted from lower levels to higher ones. Commencing from the edge with endpoint O_N, the macroscopic pressure is defined as $p^{(e)}(x_n^{(e)}) = -s^{(e)} x_n^{(e)}$ (with $a^{(e)}$ at this edge set to zero). Subsequently, an inductive approach is adopted: assuming we know the macroscopic pressure $p^{(e)}(x_n^{(e)}) = -s^{(e)} x_n^{(e)} + a^{(e)}$ for all edges at level k, it is extended as an affine function to edges at level $k+1$, preserving continuity at nodes and the known slopes for edges e at level $k+1$.

Next, for every edge e we multiply the Poiseuille velocity $\mathbf{U}^{[e]}$ and macroscopic pressure increment $p^{(e)}(x_n^{(e)}) - p^{(e)}(x_n^{(e_s)} = 0)$ at the point O_l (the end of the selected edge e_s) by the cut-off function:

$$\zeta\left(\frac{x_n^{(e)}}{3r\varepsilon}\right)\zeta\left(\frac{|e| - x_n^{(e)}}{3r\varepsilon}\right).$$

Here $p^{(e)}(x_n^{(e_s)} = 0) = a^{(e_s)}$ is the value of the pressure at the end of the edge e_s of the level k. This modification engenders a residual in the equation. To compensate it, boundary layer correctors are introduced. For every node O_l that connects an edge of level k to edges at level $k+1$, the bundle $\mathcal{B}_l$ is considered. The selected edge of this bundle at level k is denoted as e_s, while all other edges of the bundle at level $k+1$ are labeled as e (presuming O_l is a node rather than a vertex). In this context, the local coordinate system's origin is once again placed at O_l, encompassing the edge e_s as well.

Let us introduce new variables $y = \frac{x-O_l}{\varepsilon}$. The derivatives with respect to x and y are related by the evident rule: $\nabla_x = \varepsilon^{-1}\nabla_y$. Now we are in position to apply an appropriate scaling of the boundary layer correctors for the velocity and the pressure and to formulate the boundary layer problem. It will compensate the residual caused by the multiplication of the Poiseuille flow by the cut-off functions. The boundary layer solution is a pair $\left(\mathbf{U}^{[BLO_l]}, \varepsilon^{-1} P^{[BLO_l]}\right)$ solving the following Stokes system in $\Omega_l, l = 1, \ldots, N_1$:

$$\begin{cases} -\nu\Delta_y \mathbf{U}^{[BLO_l]}(y) + \nabla_y P^{[BLO_l]}(y) = \mathbf{f}^{[REGO_l]}(y), & y \in \Omega_l, \\ \operatorname{div}_y \mathbf{U}^{[BLO_l]} = h^{[REGO_l]}(y), & y \in \Omega_l, \\ \mathbf{U}^{[BLO_l]} = 0, & y \in \partial\Omega_l, \end{cases} \tag{5.1.37}$$

where

$$\mathbf{f}^{[REGO_l]}(y^{(e)}) = -\sum_{e:O_l\in e}\Big\{-\nu\Delta_y\Big[\zeta\Big(\frac{y_n^{(e)}}{3r}\Big)\mathbf{U}^{[e]}(y^{(e)\prime})\Big]$$
$$+\varepsilon\Big[\varepsilon\nabla_y\Big[\zeta\Big(\frac{y_n^{(e)}}{3r}\Big)\big(-s^{(e)}y_n^{(e)}\big)\Big]+(a^{(e)}-a^{(e_s)})\,\nabla_y\zeta\Big(\frac{y_n^{(e)}}{3r}\Big)\Big]\Big\} \tag{5.1.38}$$

$$h^{[REGO_l]}(y) = -\sum_{e:O_l\in e}\operatorname{div}_y\Big(\zeta\Big(\frac{y_n^{(e)}}{3r}\Big)\mathbf{U}^{[e]}(y^{(e)\prime})\Big). \tag{5.1.39}$$

Here the sum $\sum\limits_{e:O_l\in e}$ is taken over all edges e having ends in the node O_l.

However, at this stage, the constants $a^{(e)}$ are not yet defined for the edges of level $k+1$. Consequently, we must proceed as follows. Initially, we address an auxiliary boundary layer problem to determine the unknown pair $\big(\widehat{\mathbf{U}}^{[BLO_l]}, \widehat{P}^{[BLO_l]}\big)$.

$$\begin{cases} -\nu\Delta_y\widehat{\mathbf{U}}^{[BLO_l]}(y)+\nabla_y\widehat{P}^{[BLO_l]}(y) = \widehat{\mathbf{f}}^{[REGO_l]}(y), & y\in\Omega_l,\\ \operatorname{div}_y\widehat{\mathbf{U}}^{[BLO_l]} = h^{[REGO_l]}(y), & y\in\Omega_l,\\ \widehat{\mathbf{U}}^{[BLO_l]} = 0, & y\in\partial\Omega_l, \end{cases} \tag{5.1.40}$$

In this system, the right-hand side remains as previously stated, excluding the final term in $\mathbf{f}^{[REGO_l]}(y)$, which is indicated as

$$\widehat{\mathbf{f}}^{[REGO_l]}(y^{(e)}) = -\sum_{e:O_l\in e}\Big\{-\nu\Delta_y\Big[\zeta\Big(\frac{y_n^{(e)}}{3r}\Big)\mathbf{U}^{[e]}(y^{(e)\prime})\Big]$$
$$+\varepsilon^2\nabla_y\Big[\zeta\Big(\frac{y_n^{(e)}}{3r}\Big)\big(-s^{(e)}y_n^{(e)}\big)\Big]. \tag{5.1.41}$$

The function $h^{[REGO_l]}$ possesses compact support, and $\int_{\Omega_l} h^{[REGO_l]}(y)dy$ equals zero due to the same reasoning as presented earlier (see explanations after formula (5.1.25)). According to the theory of Stokes equations in domains with outlets to infinity (see Sect. 4.4 of Chap. 4), there exists a solution $\widehat{\mathbf{U}}^{[BLO_l]}\in W^{2,2}(\Omega_l)$ and $\widehat{P}^{[BLO_l]}\in W^{1,2}_{\mathrm{loc}}(\Omega_l)$. This solution has the property that $\widehat{P}^{[BLO_l]}(y)$ approaches a certain constant $\widehat{a}^{(e)}$ in each cylindrical outlet of Ω_l associated with the edge e of level $k+1$. Moreover, $\widehat{P}^{[BLO_l]}(y)$ tends to zero in the outlet corresponding to the unique edge e_s in the bundle of level k. Thus, by defining

$$a^{(e)} = \widehat{a}^{(e)} + a^{(e_s)},$$

and

$$\mathbf{U}^{[BLO_l]}(y) = \widehat{\mathbf{U}}^{[BLO_l]}(y),$$

$$P^{[BLO_l]}(y) = \widehat{P}^{[BLO_l]}(y) - \sum_{e:O_l\in e,\ e\neq e_s} \zeta\Big(\frac{y_n^{(e)}}{3r}\Big)\widehat{a}^{(e)},$$

we observe that $\big(\mathbf{U}^{[BLO_l]},\ P^{[BLO_l]}\big)$ satisfy equations (5.1.40).

If O_l is a vertex, the problem to be resolved becomes

$$\begin{cases} -\nu\Delta_y\mathbf{U}^{[BLO_l]}(y) + \nabla_y P^{[BLO_l]}(y) = \mathbf{f}^{[REGO_l]}(y), & y\in\Omega_l,\\ \operatorname{div}_y\mathbf{U}^{[BLO_l]} = h^{[REGO_l]}(y), & y\in\Omega_l,\\ \mathbf{U}^{[BLO_l]} = 0, & y\in\partial\Omega_l\backslash\gamma^l,\\ \mathbf{U}^{[BLO_l]}(y) = \mathbf{g}^l(y), & y\in\gamma^l. \end{cases} \tag{5.1.42}$$

In this setup, the modified right-hand side is given by

$$\begin{aligned}\mathbf{f}^{[REGO_l]}(y^{(e)}) &= -\Big\{ -\nu\Delta_y\Big[\zeta\Big(\frac{y_n^{(e)}}{3r}\Big)\mathbf{U}^{[e]}(y^{(e)\prime})\Big]\\ &+\varepsilon\Big[\varepsilon\nabla_y\Big[\zeta\Big(\frac{y_n^{(e)}}{3r}\Big)\big(-s^{(e)}y_n^{(e)}\big)\Big] + a^{(e)}\,\nabla_y\zeta\Big(\frac{y_n^{(e)}}{3r}\Big)\Big]\Big\}\end{aligned} \tag{5.1.43}$$

and the function $h^{[REGO_l]}$ is defined as

$$h^{[REGO_l]}(y) = -\operatorname{div}_y\Big(\zeta\Big(\frac{y_n^{(e)}}{3r}\Big)\mathbf{U}^{[e]}(y^{(e)\prime})\Big). \tag{5.1.44}$$

Here, e is a unique edge of level k, so that $a^{(e)}$ is a known constant. Thus, the boundary layer pressure $P^{[BLO_l]}$ can be defined in Ω_l so that $P^{[BLO_l]}\in W^{1,2}(\Omega_l)$. Indeed, boundary layer pressure is defined up to an additive constant, which we choose so that the pressure tends to zero at infinity.

One can verify that the linear expressions $p^{(e)}(x) = -s^{(e)}x_n^{(e)} + a^{(e)}$ satisfy on each edge the following problem on the graph (where the local coordinate system's origin is positioned at O_l):

$$
\begin{cases}
-\kappa_\varepsilon^{(e)} \dfrac{\partial^2 p^{(e)}}{\partial x_n^{(e)2}}(x_n^{(e)}) = 0, & x_n^{(e)} \in (0, |e|), \ \forall e \subset \mathcal{B}^l, \\
-\sum\limits_{e: O_l \in e} \kappa_\varepsilon^{(e)} \dfrac{\partial p^{(e)}}{\partial x_n^{(e)}}(0) = 0, & l = 1, \dots, N_1, \\
-\kappa_\varepsilon^{(e)} \dfrac{\partial p^{(e)}}{\partial x_n^{(e)}}(0) = F_\varepsilon^l, & l = N_1 + 1, \dots, N, \\
p^{(e)}(0) - p^{(e_s)}(0) = a^{(e)} - a^{(e_s)}, \ \forall e \subset \mathcal{B}^l, \ e \neq e_s, & \forall e \subset \mathcal{B}^l,
\end{cases}
\tag{5.1.45}
$$

where

$$
F_\varepsilon^l = \int\limits_{\gamma_\varepsilon^l} \mathbf{g}^l\left(\frac{x - O_l}{\varepsilon}\right) \cdot \mathbf{n} dS,
$$

$$
\kappa_\varepsilon^{(e)} = \int\limits_{\sigma_\varepsilon^{(e)}} u_{0,\varepsilon}(x^{(e)\prime}) dx^{(e)\prime} = \int\limits_{\sigma_\varepsilon^{(e)}} \varepsilon^2 u_0(x^{(e)\prime}/\varepsilon) dx^{(e)\prime} = \varepsilon^{n+1} \kappa_e.
$$

Finally, in each half-bundle HB_{O_l}, the asymptotic approximation for the velocity of order J can be defined as

$$
\mathbf{u}^{(J)}(x) = \sum_{e \subset \mathcal{B}_l} \zeta\left(\frac{x_n^{(e)}}{3r\varepsilon}\right) \mathbf{U}^{[e]}\left(\frac{x^{(e)\prime}}{\varepsilon}\right) + \left(1 - \zeta\left(\frac{|x - O_l|}{|e|_{\min}}\right)\right) \mathbf{U}^{[BLO_l]}\left(\frac{x - O_l}{\varepsilon}\right),
$$

where the sum is taken over all edges e having O_l as an end point. For the pressure, we get

$$
\begin{aligned}
p^{(J)}(x) = &\sum_{e \subset \mathcal{B}_l} \zeta\left(\frac{x_n^{(e)}}{3r\varepsilon}\right) \left(-s^{(e)} x_n^{(e)} + a^{(e)} - a^{(e_s)}\right) + a^{(e_s)} \\
&+ \frac{1}{\varepsilon}\left(1 - \zeta\left(\frac{|x - O_l|}{|e|_{\min}}\right)\right) p^{[BLO_l]}\left(\frac{x - O_l}{\varepsilon}\right),
\end{aligned}
\tag{5.1.46}
$$

This algorithm of constructing an asymptotic expansion is relatively straightforward and gives exponentially small residual, but it can only be employed for graph structures of the tree type.

5.1.3 Residual Analysis

Let us consider the asymptotic expansion $\left(\mathbf{u}^{(J)}, p^{(J)}\right)$ of order J (as defined in Eqs. (5.1.5) and (5.1.7)). By its construction, we have the following properties:

$$\mathbf{u}^{(J)} \in W^{2,2}(B_\varepsilon), \quad p^{(J)} \in W^{1,2}(B_\varepsilon). \tag{5.1.47}$$

Furthermore, it can be shown that $\|\mathbf{u}^{(J)}\|^4_{L^4(B_\varepsilon)} \leq c\varepsilon^{(n-1)}$. This inequality particularly holds for the Poiseuille part of $\mathbf{u}^{(J)}$. The $\|\cdot\|_{L^4(B_\varepsilon^{(i)})}$-norm of the boundary layer functions in each bundle $B_\varepsilon^{(i)}$ can be estimated by the L^4-norm in the unbounded dilated domain Ω_i, multiplied by $\varepsilon^{n/4}$. Considering the exponential decay of the boundary layers, this leads to the desired estimate.

Let $\mathcal{L}(\mathbf{u}, p) = (-\nu\Delta\mathbf{u} + \nabla p)$. We shall compute $\mathcal{L}(\mathbf{u}^{(J)}, p^{(J)})$ within the half-bundles HB_{O_l}, where $l = 1, \ldots, N_1$. This calculation yields

$$\mathcal{L}(\mathbf{u}^{(J)}, p^{(J)}) = \varepsilon^{J-2} \sum_{e:O_l\in e} \widehat{a}_J^{[BLO_l,e]} \nabla_y \zeta\Big(\frac{y_n^{(e)}}{3r}\Big)$$

$$- \mathcal{L}\Big(\zeta\Big(\frac{|x-O_l|}{|e|_{min}}\Big)\mathbf{U}^{[BLO_l,J]}(y), \zeta\Big(\frac{|x-O_l|}{|e|_{min}}\Big)P^{[BLO_l,J]}(y)\Big)\chi(x).$$

Here, $y = \dfrac{x-O_l}{\varepsilon}$, $y^{(e)} = \dfrac{x^{(e)}}{\varepsilon}$, and $\chi = \chi_{\text{supp }\zeta\left(\frac{|x-O_l|}{|e|_{min}}\right)}$ denotes the characteristic function of the set supp $\zeta\Big(\dfrac{|x-O_l|}{|e|_{min}}\Big)$. The first line arises from the pressure gradient term, which is the only term not compensated by the boundary layer-in-space problems. The last line represents the residual generated by the multiplication of the boundary layer correctors by the cut-off function $1-\zeta\Big(\dfrac{|x-O_l|}{|e|_{min}}\Big)$. Notably, the terms in this last line exponentially diminish because within the set supp $\zeta\Big(\dfrac{|x-O_l|}{|e|_{min}}\Big)$ (where $\chi \neq 0$), the order in L^2-norm of this term is $O(e^{-c_1/\varepsilon})$ with a positive constant c_1 (see Theorem 4.4.2).

Let $\mathbf{f}^{(J)} = -\mathcal{L}(\mathbf{u}^{(J)}, p^{(J)})$. From the derived formulas, it is evident that

$$\|\mathbf{f}^{(J)}\|_{L^2(B_\varepsilon)} = \|\mathcal{L}(\mathbf{u}^{(J)}, p^{(J)})\|_{L^2(B_\varepsilon)} = O(\varepsilon^{J-2}), \tag{5.1.48}$$

In the vertex-associated cylinders B_{O_l}, where $l = N_1+1, \ldots, N$, the residual is simpler and lacks the term $\nabla_y\zeta\Big(\dfrac{y_n^{(e)}}{3r}\Big)\widehat{a}_j^{[BLO_l,e]}$.

Let us compute the divergence of the vector field $\mathbf{u}^{(J)}$. We find

$$\operatorname{div}\mathbf{u}^{(J)} = -\sum_{l=1}^{N} \nabla\zeta\left(\frac{|x-O_l|}{|e|_{\min}}\right) \cdot \mathbf{U}^{[BLO_l,J]}(y) = h^{(J)}(y).$$

Clearly, the function $h^{(J)}$ belongs to the $W^{1,2}(B_\varepsilon)$ space. As the support of the function $\nabla\zeta\left(\frac{|x-O_l|}{|e|_{\min}}\right)$ is confined to the central third of each cylinder, the following relationship holds:

$$\|h^{(J)}\|_{W^{1,2}(B_\varepsilon)} = O(e^{-c_2/\varepsilon}) \tag{5.1.49}$$

for a positive constant c_2.

Finally, the fulfilment of the boundary conditions is evident.

It can be observed that the integral of $h^{(J)}$ over B_ε is zero:

$$\int\limits_{B_\varepsilon} h^{(J)}(y)\,dy = 0.$$

This assertion can be justified by noting that outside the finite support of the function $h_j^{[REGO_l]}(y)$ as given in (5.1.39), the solution $\mathbf{U}_j^{[BLO_l]}(y)$ of the boundary layer has zero flux since it approaches zero at infinity and is divergence-free. Thus, for each cylinder $\Pi_\varepsilon^{(e)}$, we have

$$\int\limits_{\Pi_\varepsilon^{(e)}} \nabla\zeta\left(\frac{|x-O_l|}{|e|_{\min}}\right)\cdot\mathbf{U}^{[BLO_l,J]}\left(\frac{x-O_l}{\varepsilon}\right)dx$$

$$= \int\limits_{\{x\in\Pi_\varepsilon^{(e)},x_n^{(e)}\in(|e|_{\min}/3,2|e|_{\min}/3\}} \nabla\cdot\left(\zeta\left(\frac{|x-O_l|}{|e|_{\min}}\right)\mathbf{U}^{[BLO_l,J]}\left(\frac{x-O_l}{\varepsilon}\right)\right)dx$$

$$= \int\limits_{\{x\in\Pi_\varepsilon^{(e)},\,x_n^{(e)}=|e|_{\min}/3\}} \mathbf{U}^{[BLO_l,J]}\left(\frac{x-O_l}{\varepsilon}\right)\cdot\mathbf{n}\,dS$$

$$= \varepsilon^{n-1}\int\limits_{\sigma^{(e)}} \mathbf{U}^{[BLO_l,J]}\left(y^{(e)\prime},\frac{|e|_{\min}}{3\varepsilon}\right)\cdot\mathbf{n}\,dy^{(e)\prime} = 0.$$

So, finally,

$$\int\limits_{\Pi_\varepsilon^{(e)}} \nabla\zeta\Big(\frac{|x-O_l|}{|e|_{min}}\Big)\cdot\mathbf{U}^{[BLO_l,J]}\Big(\frac{x-O_l}{\varepsilon}\Big)dx = 0. \tag{5.1.50}$$

As a result, based on Lemma 2.4.14, we can construct a vector field $\mathbf{w}^{(J)} \in \mathring{W}^{1,2}(B_\varepsilon)$ such that $\mathrm{div}\mathbf{w}^{(J)} = -h^{(J)}$. Additionally, the following estimate is valid:

$$\|\mathbf{w}^{(J)}\|_{W^{1,2}(B_\varepsilon)} \le \varepsilon^{-3/2}c\|h^{(J)}\|_{L^2(B_\varepsilon)} \tag{5.1.51}$$

It is noteworthy that the solution to the divergence equation in this case can be selected from the space $W^{2,2}(B_\varepsilon)$. For the compact support G_1 of the function $h^{(J)}$ within the cylinder $\Pi_\varepsilon^{(e)}$, there exists a subregion $\tilde{\Pi}_\varepsilon^{(e)}$ with a height of $|e|/3+2\varepsilon$ still containing G_1, and its bases are situated at a distance greater than ε from G_1. Then there exists a subdomain G_2 within the cylinder $\Pi_\varepsilon^{(e)}$ with C^2-smooth boundary, which contains the cylinder $\tilde{\Pi}_\varepsilon^{(e)}$. It is possible to construct a solution $\mathbf{w}^{(J,e)}$ to the divergence equation in G_2, such that $\mathbf{w}^{(J,e)} \in W^{2,2}(\Pi_\varepsilon^{(e)})$, and $\mathbf{w}^{(J,e)}$ equals zero on the shared boundary of G_2 and $\Pi_\varepsilon^{(e)}$, as well as outside the domain G_2. Moreover, the subsequent estimate is valid:

$$\|\mathbf{w}^{(J,e)}\|_{W^{2,2}(\Pi_\varepsilon^{(e)})} \leq c\varepsilon^{-2}\|h^{(J)}\|_{W^{1,2}(\Pi_\varepsilon^{(e)})}$$

(see Lemma 2.4.13). We extend all $\mathbf{w}^{(J,e)}$ by zero into $B_\varepsilon \setminus \Pi_\varepsilon^{(e)}$.

Constructing such solution for all $\Pi_\varepsilon^{(e)}$ (for all part of the support of $h^{(J)}$), we get the solution $\mathbf{w}^{(J)}$ in the whole B_ε satisfying the estimate

$$\|\mathbf{w}^{(J)}\|_{W^{2,2}(B_\varepsilon)} \leq c\varepsilon^{-2}\|h^{(J)}\|_{W^{1,2}(B_\varepsilon)}. \tag{5.1.52}$$

Let $\widetilde{\mathbf{u}}^{(J)} = \mathbf{u}^{(J)} + \mathbf{w}^{(J)}$. Consequently, $\operatorname{div}\widetilde{\mathbf{u}}^{(J)} = 0$, and $\widetilde{\mathbf{u}}^{(J)}$ satisfies the boundary conditions precisely. Furthermore, the pair $(\widetilde{\mathbf{u}}^{(J)}, p^{(J)})$ satisfies the Eqs. $(5.1.1)_1$ with the residual $\mathcal{L}(\widetilde{\mathbf{u}}^{(J)}, p^{(J)}) = -\mathbf{f}_1^{(J)}$ of order $O(\varepsilon^{J-2})$, i.e,

$$\|\mathbf{f}_1^{(J)}\|_{L^2(B_\varepsilon)} = O(\varepsilon^{J-2}). \tag{5.1.53}$$

5.1.4 Error Estimate

It is worth noting that the function $\widetilde{\mathbf{u}}^{(J)} \in W^{2,2}(B_\epsilon)$ satisfies the conditions specified in (5.1.3). Consequently, we can consider $\widetilde{\mathbf{u}}^{(J)}$ as an extension of the boundary value $\mathbf{g}$. Subsequently, the integral identity (5.1.4) takes on the following form:

$$\nu\int\limits_{B_\varepsilon} \nabla\mathbf{u}\cdot\nabla\boldsymbol{\eta}\,dx = -\nu\int\limits_{B_\varepsilon} \nabla\widetilde{\mathbf{u}}^{(J)}\cdot\nabla\boldsymbol{\eta}\,dx \quad \forall\boldsymbol{\eta} \in \mathring{H}^{1,2}(B_\epsilon), \tag{5.1.54}$$

where $\mathbf{u} = \mathbf{v} - \widetilde{\mathbf{u}}^{(J)}$. With $\operatorname{div}\boldsymbol{\eta} = 0$, application of integration by parts enables the transformation of the right-hand side of (5.1.54) as follows:

$$-\nu\int\limits_{B_\varepsilon} \nabla\widetilde{\mathbf{u}}^{(J)}\cdot\nabla\boldsymbol{\eta}\,dx = -\int\limits_{B_\varepsilon} \nu\Delta\widetilde{\mathbf{u}}^{(J)}\cdot\boldsymbol{\eta}\,dx + \int\limits_{B_\varepsilon} p^{(J)}\operatorname{div}\boldsymbol{\eta}\,dx$$

$$= -\int_{B_\varepsilon} \Big(-\nu\Delta\widetilde{\mathbf{u}}^{(J)} + \nabla p^{(J)}\Big) \cdot \eta \, dx = \int_{B_\varepsilon} \mathbf{f}_1^{(J)} \cdot \eta \, dx.$$

This leads us to the identity

$$\nu \int_{B_\varepsilon} \nabla \mathbf{u} \cdot \nabla \eta \, dx = \int_{B_\varepsilon} \mathbf{f}_1^{(J)} \cdot \eta \, dx \quad \forall \eta \in \mathring{H}^{1,2}(B_\epsilon). \tag{5.1.55}$$

By invoking Theorem 4.1.1 and making use of the estimates (5.1.53) and (4.3.2), we establish the following inequality:

$$\|\mathbf{v} - \widetilde{\mathbf{u}}^{(J)}\|_{W^{1,2}(B_\varepsilon)} = O(\varepsilon^{J-1}). \tag{5.1.56}$$

Moving forward, evaluating the norm of the difference $\widetilde{\mathbf{u}}^{(J)} - \widetilde{\mathbf{u}}^{(J+2)}$ (as defined in (5.1.5) and (5.1.6)), we acquire

$$\|\widetilde{\mathbf{u}}^{(J)} - \widetilde{\mathbf{u}}^{(J+2)}\|_{W^{1,2}(B_\varepsilon)} = O(\varepsilon^{J+(n-1)/2}). \tag{5.1.57}$$

Substituting J with $J+2$ in (5.1.56) results in

$$\|\mathbf{v} - \widetilde{\mathbf{u}}^{(J+2)}\|_{W^{1,2}(B_\varepsilon)} = O(\varepsilon^{J+1}).$$

As a result, employing the triangle inequality leads us to the subsequent theorem.

Theorem 5.1.2 *The following error estimate holds*

$$\|\mathbf{v} - \widetilde{\mathbf{u}}^{(J)}\|_{W^{1,2}(B_\varepsilon)} = O(\varepsilon^{J+(n-1)/2}). \tag{5.1.58}$$

5.1.5 *Justification of the Method of Asymptotic Partial Decomposition of the Domain (MAPDD) for Stationary Stokes Equations*

In Introduction the method of asymptotic partial decomposition of the domain (MAPDD) for the stationary Stokes equations was described. This method reduces the dimension on the main part of B_ε replacing in cylinders the solution by the Poiseuille velocities and keeping the original full dimension in small neighbourhoods of the nodes and vertices. The proposed approach effectively minimizes computational expenses and expedites conventional numerical strategies.

Let us recall notation and the definitions of some spaces introduced in Introduction. Let $W_\Gamma^{1,2}(B_\varepsilon)$ denote the space of vector-valued functions in $W^{1,2}(B_\varepsilon)$ that vanish for $x \in \Gamma = \partial B_\varepsilon \setminus (\bigcup_{j=N_1+1}^{N} \gamma_\varepsilon^j)$. Similarly, $H_\Gamma^{1,2}(B_\varepsilon)$ represents the space of divergence-free vector functions from $W_\Gamma^{1,2}(B_\varepsilon)$.

This section describes the algorithm of the MAPDD for the Stokes problem in a tube structure B_ε. Let δ be a small positive quantity significantly larger than ε (it will be chosen on the order of $O(\varepsilon|\ln\varepsilon|)$). We choose $\delta = C_J\varepsilon|\ln\varepsilon||e|_{min}$, where the constant C_J is determined to satisfy (5.1.60). For each edge $e = \overline{O_i O_j}$ within the graph, we introduce two orthogonal hyperplanes intersecting this edge at a distance δ from its ends. Denote the cross-sections of the cylinder $\Pi_\varepsilon^{(e)}$ by these two hyperplanes respectively, by $S_{i,j}$ (the cross-section at the distance δ from O_i) and $S_{j,i}$ (the cross-section at the distance δ from O_j), and denote the part of the cylinder $\Pi_\varepsilon^{(e)}$ between these two cross-sections by $B_{ij}^{\text{dec},\varepsilon}$ (see Fig. 5.2).

Let $B_i^{\varepsilon,\delta}$ denote the connected portion of B_ε, truncated by the cross-sections $S_{i,j}$ and containing the vertex or node O_i (see Fig. 5.3).

Define the subspace $W_{\text{dec}}(B_\varepsilon, \delta)$ of $W_\Gamma^{1,2}(B_\varepsilon)$ such that, on every truncated cylinder $B_{ij}^{\text{dec},\varepsilon}$, its elements (vector-valued functions) coincide with the corresponding Poiseuille velocities (see Chap. 3). We also consider $W_{0,\text{dec}}(B_\varepsilon, \delta)$, a subspace of $W_{\text{dec}}(B_\varepsilon, \delta)$, with elements vanishing on the entire boundary ∂B_ε. Additionally, $L^2_{\text{dec}}(B_\varepsilon, \delta)$ is a subspace of $L^2(B_\varepsilon)$, with elements being affine functions of $x_1^{(e)}$ on each truncated cylinder $B_{ij}^{\text{dec},\varepsilon}$. The subspace $H_{\text{dec}}(B_\varepsilon, \delta)$ consists of divergence-free vector functions within $W_{\text{dec}}(B_\varepsilon, \delta)$.

The weak formulation of MAPDD reads as follows:

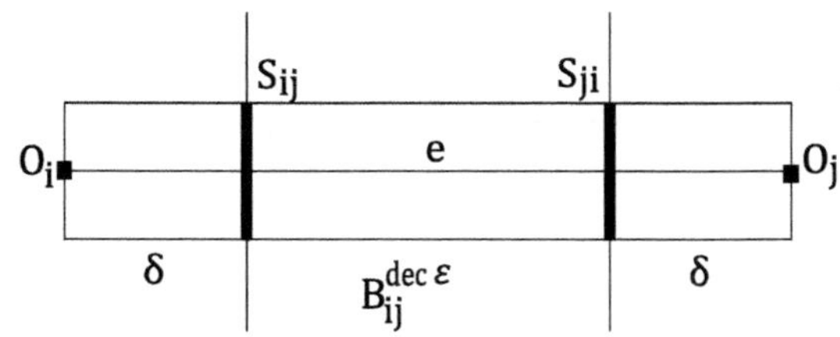

Fig. 5.2 Truncation of the cylinder $\Pi_\varepsilon^{(e)}$

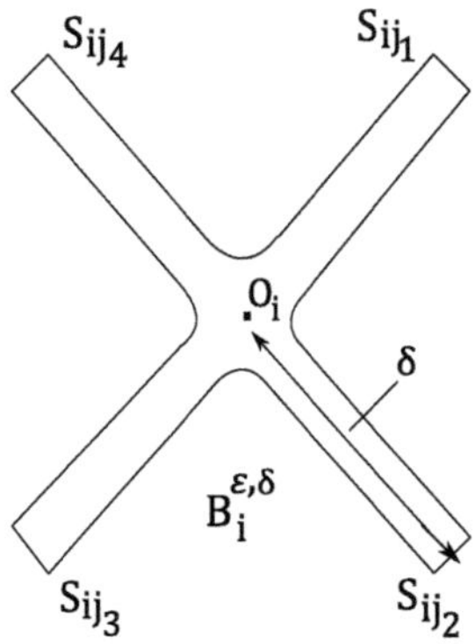

Fig. 5.3 Connected component $B_i^{\varepsilon,\delta}$

Definition 5.1.1 Find $\mathbf{v}_d \in H_{\text{dec}}(B_\varepsilon, \delta)$, where $\mathbf{v}_d|_{\partial B_\varepsilon} = \mathbf{g}$, such that for all test functions $\boldsymbol{\eta} \in H_{0,\text{dec}}(B_\varepsilon, \delta)$, the integral identity

$$\nu \int_{B_\varepsilon} \nabla \mathbf{v}_d \cdot \nabla \boldsymbol{\eta} dx = 0 \tag{5.1.59}$$

is satisfied. Here, $H_{0,\text{dec}}(B_\varepsilon, \delta)$ is the subspace of functions within $H_{\text{dec}}(B_\varepsilon, \delta)$ that vanish on the entire boundary.

The MAPDD reduces the domain of computations to some subdomains of the diameter of order δ (in the remaining parts the solution is Poiseuille type flows). So, the parameter δ is supposed to be chosen as small as possible (however large enough to insure the high accuracy of the solution to problem (5.1.59)). On the other hand, the justification of the method will be based on the comparison of the solution $\mathbf{v}_d$ with the asymptotic approximation $\widetilde{\mathbf{u}}^{(J+2)}$ in the space $H_{\text{dec}}(B_\varepsilon, \delta)$. Unfortunately, this asymptotic approximation does not belong to this space for small δ because the cut-off function $\zeta\left(\frac{x-O_l}{|e|_{\min}}\right)$ removes the "tails" of the boundary layer correctors too far from the nodes, only for $x_n^{(e)} > \frac{2}{3}|e|_{\min}$. In order to fix this problem we will change the constructed approximation modifying the cut-off factor and replacing it by $\zeta\left(\frac{x-O_l}{\delta}\right)$ with an appropate δ.

Remark 5.1.1 The asymptotic expansion (5.1.5)–(5.1.9) can be slightly modified without compromising accuracy. Specifically, the argument $\frac{|x-O_l|}{|e|_{\min}}$ of the cut-off function ζ can be substituted with $\frac{|x-O_l|}{\delta}$, where $\delta = C_J\varepsilon|\ln\varepsilon||e|_{\min}$, where the constant C_J is chosen such that

$$\begin{aligned} E_{\delta/\varepsilon} &= \|\mathbf{U}^{[BLO_l,J]}\|_{W^{2,2}(\Omega_{l,\delta/\varepsilon})} + \|P^{[BLO_l,J]}\|_{W^{1,2}(\Omega_{l,\delta/\varepsilon})} \\ &= O(\varepsilon^{J+(n-1)/2+2}). \end{aligned} \tag{5.1.60}$$

Here, $\Omega_{l,R} = \Omega_l \cap \{|y| > R\}$.

Indeed, $\mathbf{U}^{[BLO_l,J]}(y)$ and $P^{[BLO_l,J]}(y)$ tend to zero exponentially as $|y| \to \infty$, leading to the inequality:

$$E_R \leq c_1 e^{-c_2 R}, \quad c_1, c_2 > 0.$$

Consequently, the estimate $E_{\delta/\varepsilon} = O(\varepsilon^{J+(n-1)/2+2})$ is valid for δ such that:

$$c_2\delta/\varepsilon \geq (J + (n-1)/2 + 2)|\ln\varepsilon|,$$

implying $\delta = C_J\varepsilon|\ln\varepsilon|$.

Notably, the constant $C_J = c_3 J + c_4$, where c_3 and c_4 are independent of J. This adjustment along with (5.1.52), (5.1.60) and the change of variables $\nabla_x = \varepsilon^{-1}\nabla_y$

yields that $W^{2,2}(B_\varepsilon)$-norm of the difference between the constructed asymptotic expansion $\widetilde{\mathbf{u}}^{(J)}$ and the modified one is of order $O(\varepsilon^{J+(n-1)/2})$.

Remark 5.1.2 An observation of note is that the definitions of the aforementioned function spaces are contingent on the choice of δ, itself dependent on the approximation order J. Consequently, the MAPDD solution $\mathbf{v}_d$ implicitly hinges on J.

By invoking similar reasoning as presented in the proof of Theorem 4.1.1 (using the Riesz theorem), existence and uniqueness of the solution $\mathbf{v}_d$ can be established.

It is noteworthy that the refined asymptotic solution $\widetilde{\mathbf{u}}^{(J)}$, modified according to Remark 5.1.1, belongs to the space $H_{\text{dec}}(B_\varepsilon, \delta)$. It satisfies the weak formulation, retaining the same residual as before, namely, of order $O(\varepsilon^{J-2})$ within the $L^2(B_\varepsilon)$-norm (see (5.1.48), (5.1.49), and (5.1.53)):

$$\nu \int\limits_{B_\varepsilon} \nabla \widetilde{\mathbf{u}}^{(J)} \cdot \nabla \boldsymbol{\eta} dx = -\int\limits_{B_\varepsilon} \mathbf{f}_1^{(J)} \cdot \boldsymbol{\eta} dx \quad \forall\, \boldsymbol{\eta} \in \overset{\circ}{H}{}^{1,2}(B_\varepsilon),$$

where

$$\|\mathbf{f}_1^{(J)}\|_{L^2(B_\varepsilon)} = O(\varepsilon^{J-2}). \tag{5.1.61}$$

Respectively, $\widetilde{\mathbf{u}}^{(J+2)}$ satisfies

$$\nu \int\limits_{B_\varepsilon} \nabla \widetilde{\mathbf{u}}^{(J+2)} \cdot \nabla \boldsymbol{\eta} dx = -\int\limits_{B_\varepsilon} \mathbf{f}_1^{(J+2)} \cdot \boldsymbol{\eta} dx \quad \forall\, \boldsymbol{\eta} \in \overset{\circ}{H}{}^{1,2}(B_\varepsilon), \tag{5.1.62}$$

where

$$\|\mathbf{f}_1^{(J+2)}\|_{L^2(B_\varepsilon)} = O(\varepsilon^{J}). \tag{5.1.63}$$

The identity (5.1.62) evidently remains valid for $\boldsymbol{\eta} \in H_{0,\text{dec}}(B_\varepsilon, \delta)$, since $H_{0,\text{dec}}(B_\varepsilon, \delta)$ is a subspace of $\overset{\circ}{H}{}^{1,2}(B_\varepsilon)$. Clearly, the difference $\mathbf{v}_d - \widetilde{\mathbf{u}}^{(J+2)}$ lies within $H_{0,\text{dec}}(B_\varepsilon, \delta)$ and satisfies the integral identity

$$\nu \int\limits_{B_\varepsilon} \nabla(\mathbf{v}_d - \widetilde{\mathbf{u}}^{(J+2)}) \cdot \nabla \boldsymbol{\eta} dx = \int\limits_{B_\varepsilon} \mathbf{f}_1^{(J+2)} \cdot \boldsymbol{\eta} dx,$$

for every $\boldsymbol{\eta} \in H_{0,\text{dec}}(B_\varepsilon, \delta)$.

Selecting $\boldsymbol{\eta} = \mathbf{v}_d - \widetilde{\mathbf{u}}^{(J+2)}$ yields the estimate

$$\|\mathbf{v}_d - \widetilde{\mathbf{u}}^{(J+2)}\|_{W^{1,2}(B_\varepsilon)} = O(\varepsilon^{J+1}). \tag{5.1.64}$$

By utilizing estimates (5.1.64), (5.1.58), (5.1.57) and applying the triangle inequality, we obtain

$$\|\mathbf{v}-\mathbf{v}_d\|_{W^{1,2}(B_\varepsilon)} \le \|\mathbf{v}-\widetilde{\mathbf{u}}^{(J)}\|_{W^{1,2}(B_\varepsilon)} + \|\widetilde{\mathbf{u}}^{(J+2)}-\widetilde{\mathbf{u}}^{(J)}\|_{W^{1,2}(B_\varepsilon)}$$
$$+ \|\mathbf{v}_d-\widetilde{\mathbf{u}}^{(J+2)}\|_{W^{1,2}(B_\varepsilon)} = O(\varepsilon^{J+(n-1)/2}).$$

Hence, the following result can be deduced:

Theorem 5.1.3 *Let δ be $C_{J+2}\varepsilon|\ln\varepsilon|$, where the constant C_{J+2} satisfies* (5.1.60) *with J replaced by $J+2$. Then, the following error estimate holds:*

$$\|\mathbf{v}-\mathbf{v}_d\|_{W^{1,2}(B_\varepsilon)} = O(\varepsilon^{J+(n-1)/2}) \tag{5.1.65}$$

This estimate validates the application of MAPDD.

5.1.6 Pressure Reconstruction

The outcomes of the preceding sections pertain to velocity estimates. We shall now establish an estimate for the discrepancy between the exact pressure p and its asymptotic approximation $p^{(J)}$. Subsequently, we will present a variant of the Method of Asymptotic Partial Differential Equations (MAPDD) that incorporates the pressure component and justify the error estimate associated with this methodology.

Let us consider the difference between the exact solution $(\mathbf{v}, p)$ of the problem (5.1.1) and its asymptotic approximation $(\mathbf{u}^{(J)}, p^{(J)})$, wherein

$$\begin{aligned}\mathbf{u}^{(J)}(x) = &\sum_{O_l, l=N_1+1,\dots,N;\, e=\overline{O_l O_{i_l}}} \zeta\Big(\frac{|e|-x_n^{(e)}}{3r\varepsilon}\Big)\mathbf{U}^{[e,J]}\Big(\frac{x^{(e)\prime}}{\varepsilon}\Big)\\
+ &\sum_{e=\overline{O_\alpha O_\beta};\,\alpha,\beta\le N_1} \zeta\Big(\frac{x_n^{(e)}}{3r\varepsilon}\Big)\zeta\Big(\frac{|e|-x_n^{(e_j)}}{3r\varepsilon}\Big)\mathbf{U}^{[e,J]}\Big(\frac{x^{(e)\prime}}{\varepsilon}\Big)\\
+ &\sum_{l=1}^{N}\Big(1-\zeta\Big(\frac{|x-O_l|}{\delta}\Big)\Big)\mathbf{U}^{[BLO_l,J]}\Big(\frac{x-O_l}{\varepsilon}\Big),\end{aligned} \tag{5.1.66}$$

and the expression for $p^{(J)}(x)$ on each half-bundle HB_{O_l}, $l=1,\dots,N_1$ (with O_l denoting the origin of the local coordinate system) is provided by the formula

$$\begin{aligned}p^{(J)}(x) = &\sum_{e\subset\mathcal{B}_l} \zeta\Big(\frac{x_n^{(e_i)}}{3r\varepsilon}\Big)\Big(-s^{(e)}x_n^{(e)} + a^{(e)} - a^{(e_s)}\Big) + a^{(e_s)}\\
+ &\frac{1}{\varepsilon}\Big(1-\zeta\Big(\frac{|x-O_l|}{\delta}\Big)\Big)P^{[BLO_l,J]}\Big(\frac{x-O_l}{\varepsilon}\Big).\end{aligned} \tag{5.1.67}$$

It is worth noting that a pressure function $p \in \widehat{L}^2(B_\varepsilon)$ exists such that the pair $(\mathbf{v}, p)$ satisfies the following integral identity (as it is proved in Theorem 4.1.1):

$$\nu \int_{B_\varepsilon} \nabla \mathbf{v} \cdot \nabla \boldsymbol{\eta} dx - \int_{B_\varepsilon} p \,\mathrm{div} \boldsymbol{\eta} dx = 0 \tag{5.1.68}$$

for any test function $\boldsymbol{\eta} \in \overset{\circ}{W}{}^{1,2}(B_\varepsilon)$.

Now, consider the pair $(\widetilde{\mathbf{u}}^{(J+3)}, p^{(J+3)}) \in W^{1,2}(B_\varepsilon) \times L^2(B_\varepsilon)$. For every $\boldsymbol{\eta} \in \overset{\circ}{W}{}^{1,2}(B_\varepsilon)$, it satisfies the integral identity

$$\nu \int_{B_\varepsilon} \nabla \widetilde{\mathbf{u}}^{(J+3)} \cdot \nabla \boldsymbol{\eta} dx - \int_{B_\varepsilon} p^{(J+3)} \mathrm{div} \boldsymbol{\eta} dx = - \int_{B_\varepsilon} \mathbf{f}_1^{(J+3)} \cdot \boldsymbol{\eta} dx. \tag{5.1.69}$$

We introduce the differences $\mathbf{u} = \mathbf{v} - \widetilde{\mathbf{u}}^{(J+3)}$ and $q = p - p^{(J+3)}$. This yields $\mathbf{u} \in \overset{\circ}{W}{}^{1,2}(B_\varepsilon)$, $q \in L^2(B_\varepsilon)$ and the equation

$$\nu \int_{B_\varepsilon} \nabla \mathbf{u} \cdot \nabla \boldsymbol{\eta} dx - \int_{B_\varepsilon} q \,\mathrm{div} \boldsymbol{\eta} dx = \int_{B_\varepsilon} \mathbf{f}_1^{(J+3)} \cdot \boldsymbol{\eta} dx \quad \forall \boldsymbol{\eta} \in \overset{\circ}{W}{}^{1,2}(B_\varepsilon). \tag{5.1.70}$$

Denote $\langle \cdot \rangle$ the average operator, i.e., $\langle \cdot \rangle = (\mathrm{mes} B_\varepsilon)^{-1} \int_{B_\varepsilon} \cdot \, dx$. Let us assume $\langle q \rangle = 0$. We select $\boldsymbol{\eta} = \mathbf{w} \in \overset{\circ}{W}{}^{1,2}(B_\varepsilon)$, where $\mathbf{w}$ satisfies $\mathrm{div}\mathbf{w}(x) = q(x)$ in B_ε and

$$\|\mathbf{w}\|_{W^{1,2}(B_\varepsilon)} \le C\varepsilon^{-3/2} \|q\|_{L^2(B_\varepsilon)},$$

in line with Lemma 2.4.14. Substituting this choice into (5.1.70), we obtain

$$\|q\|^2_{L^2(B_\varepsilon)} \le c\varepsilon^{-3/2} \Big(\|\nabla \mathbf{u}\|_{L^2(B_\varepsilon)} + \varepsilon \|\mathbf{f}_1^{(J+3)}\|_{L^2(B_\varepsilon)} \Big) \|q\|_{L^2(B_\varepsilon)}.$$

By simplifying the term $\|q\|_{L^2(B_\varepsilon)}$ and taking into account the estimates (5.1.58) and (5.1.53) with J replaced by $J+3$, we deduce

$$\|q\|_{L^2(B_\varepsilon)} \le C\varepsilon^{J+1/2}.$$

Upon evaluating the difference

$$\|p^{(J+3)} - p^{(J)}\|_{L^2(B_\varepsilon)} \le C\varepsilon^{J-1+(n-1)/2} = O(\varepsilon^{J+(n-3)/2}),$$

and utilizing the triangle inequality, we arrive at the subsequent theorem:

Theorem 5.1.4 *Let s denote the normalized discrepancy $p - p^{(J)}$, specifically*

$$s = p - p^{(J)} - \langle p - p^{(J)} \rangle.$$

The ensuing error estimate holds

$$\|s\|_{L^2(B_\varepsilon)} = O(\varepsilon^{J+(n-3)/2}). \tag{5.1.71}$$

5.2 Generalized Version of the MAPDD

The variational formulation of the MAPDD using divergence-free test functions for the Stokes equations is not too suitable for numerical implementation. It motivates a weak formulation that incorporates pressure terms. Thus, we introduce an adapted version of the MAPDD, which entails substituting the function spaces $H_{\text{dec}}(B_\varepsilon, \delta)$ and $H_{0,\text{dec}}(B_\varepsilon, \delta)$ with elments exhibiting Poiseuille profiles on the cylinders $B_{ij}^{dec,\varepsilon}$, with broader spaces $\mathcal{H}_{\text{dec}}(B_\varepsilon, \delta)$ and $\mathcal{H}_{0,\text{dec}}(B_\varepsilon, \delta)$. The elements in these new spaces possess vanishing tangential velocity field entries and normal velocity field components independent of the normal variable on $B_{ij}^{dec,\varepsilon}$.

Let $\mathcal{W}_{\text{dec}}(B_\varepsilon, \delta)$ denote the space of vector-valued functions $\mathbf{w}$ belonging to $W^{1,2}(B_\varepsilon)$, and for all $e = \overline{O_i O_j}$ having the shape $\mathbf{w}|_{B_{ij}^{dec,\varepsilon}} = (\mathcal{P}^{(e)})^*\Big(0, \ldots, 0, \tilde{w}_n(x^{(e)\prime})\Big)$, where $\tilde{w}_n \in \mathring{W}^{1,2}(\sigma_\varepsilon^{(e)})$:

$$\begin{gathered}\mathcal{W}_{\text{dec}}(B_\varepsilon, \delta) = \Big\{\mathbf{w} \in W^{1,2}(B_\varepsilon),\ \mathbf{w}|_{\partial B_\varepsilon \setminus \gamma_\varepsilon} = 0, \\ \mathbf{w}|_{B_{ij}^{dec,\varepsilon}} = (\mathcal{P}^{(e)})^*\Big(0, \ldots, 0, \tilde{w}_n(x^{(e)\prime})\Big),\ \tilde{w}_n \in \mathring{W}^{1,2}(\sigma_\varepsilon^{(e)})\Big\}.\end{gathered} \tag{5.2.1}$$

Furthermore, let $\mathcal{H}_{\text{dec}}(B_\varepsilon, \delta)$ represent the subspace of $\mathcal{W}_{\text{dec}}(B_\varepsilon, \delta)$ consisting of divergence-free vector fields:

$$\mathcal{H}_{\text{dec}}(B_\varepsilon, \delta) = \Big\{\mathbf{w} \in \mathcal{W}_{\text{dec}}(B_\varepsilon, \delta),\ \text{div}\mathbf{w} = 0, \Big\}. \tag{5.2.2}$$

Additionally, we define

$$\mathcal{H}_{0,\text{dec}}(B_\varepsilon, \delta)=\mathcal{H}_{\text{dec}}(B_\varepsilon, \delta)\cap\mathring{W}^{1,2}(B_\varepsilon), \quad \mathcal{W}_{0,\text{dec}}(B_\varepsilon, \delta)=\mathcal{W}_{\text{dec}}(B_\varepsilon, \delta)\cap\mathring{W}^{1,2}(B_\varepsilon).$$

It is important to note that the spaces $\mathcal{H}_{\text{dec}}(B_\varepsilon, \delta)$ and $\mathcal{H}_{0,\text{dec}}(B_\varepsilon, \delta)$ differ from the original spaces $H_{\text{dec}}(B_\varepsilon, \delta)$ and $H_{0,\text{dec}}(B_\varepsilon, \delta)$ due to the relaxation of the requirement that the functions $\tilde{w}_n$ must satisfy the Dirichlet boundary value problem for the Poisson equation (5.1.22) on the cross-section $\sigma_\varepsilon^{(e)}$.

The revised (generalized) form of the weak formulation for the MAPDD is as follows:

Definition 5.2.1 Find $\mathbf{v}_d \in \mathcal{H}_{\text{dec}}(B_\varepsilon, \delta)$ such that $\mathbf{v}_d|_{\partial B_\varepsilon} = \mathbf{g}$, and for all $\boldsymbol{\eta} \in \mathcal{H}_{0,\text{dec}}(B_\varepsilon, \delta)$, the integral identity

$$\nu \int_{B_\varepsilon} \nabla \mathbf{v}_d \cdot \nabla \eta dx = 0 \tag{5.2.3}$$

is satisfied.

As for the formulation (5.1.59), one can prove the existence of a unique solution of the MAPDD and a corresponding estimate analogous to (5.1.65).

To distinguish this new type of flows in cylinders from the Poiseuille flows (Chap. 3), we will call them Womersley type flows.

Given the aforementioned challenges posed by formulations "without pressure" in terms of constructing suitable divergence-free bases for finite element methods, we propose an alternative equivalent formulation.

Let $\widetilde{W}^{1,2}\big(\bigcup\limits_{i=1}^{N} B_i^{\varepsilon,\delta}\big)$ denote the space of vector-valued functions $\mathbf{w}$ defined on $\bigcup\limits_{i=1}^{N} B_i^{\varepsilon,\delta}$ such that their restrictions on $B_i^{\varepsilon,\delta}$ belong to $W^{1,2}(B_i^{\varepsilon,\delta})$, $i = 1, \ldots, N$, and for each pair (i, j) where O_i and O_j are connected by the edge $\overline{O_i O_j} = e$, the traces $\mathbf{w}|_{S_{ij}}$ and $\mathbf{w}|_{S_{ji}}$ have the form

$$\mathbf{w}|_{S_{ij}} = \mathbf{w}|_{S_{ji}} = \left(\mathcal{P}^{(e)}\right)^* \left(0, \ldots, 0, w_n(x^{(e)\prime})\right)$$

with $w_n \in \overset{\circ}{W}{}^{1,2}(\sigma_\varepsilon^{(e)})$, while the trace of $\mathbf{w}$ is zero on $\partial B_i^{\varepsilon,\delta} \cap \partial B_\varepsilon$, $i = 1, \ldots, N_1$, as well as on $\left(\partial B_i^{\varepsilon,\delta} \cap \partial B_\varepsilon\right) \backslash \gamma_\varepsilon^i$, $i = N_1 + 1, \ldots, N$.

We aim to demonstrate the equivalence between the previously stated formulation and the subsequent one.

Definition 5.2.2 Find the divergence-free vector field $\mathbf{v}_d \in \widetilde{W}^{1,2}\big(\bigcup\limits_{i=1}^{N} B_i^{\varepsilon,\delta}\big)$ and the "MAPDD pressure" $p_d \in L^2(B_i^{\varepsilon,\delta})$, $i = 1, \ldots, N$, such that $\mathbf{v}_d = \mathbf{g}_\varepsilon$ on γ_ε and

$$\begin{aligned} \nu \sum_{i=1}^{N} \int_{B_i^{\varepsilon,\delta}} \nabla \mathbf{v}_d \cdot \nabla \eta dx + \nu \sum_{l=1}^{M} d_l \int_{\sigma_\varepsilon^{(e_l)}} \nabla_{x^{(e_l)\prime}} \mathbf{v}_d \cdot \nabla_{x^{(e_l)\prime}} \eta dx^{(e_l)\prime} \\ = \sum_{i=1}^{N} \int_{B_i^{\varepsilon,\delta}} p_d \mathrm{div} \eta dx \end{aligned} \tag{5.2.4}$$

for all $\eta \in \widetilde{W}^{1,2}\big(\bigcup\limits_{i=1}^{N} B_i^{\varepsilon,\delta}\big)$ vanishing on $\partial B_\varepsilon \cap \partial B_i^{\varepsilon,\delta}$.

Here $d_l = |e| - 2\delta$, the distance between the cross-sections S_{ij} and S_{ji} cutting the segment e. For numerical implementation of the MAPDD the condition $\mathrm{div}\mathbf{v}_d = 0$ can be taken into consideration within the integral identity. Namely, as in (1.4.9) the weak formulation reads:

Find the vector field $\mathbf{v}_d \in \tilde{W}^{1,2}(\bigcup_{i=1}^{N} B_i^{\varepsilon,\delta})$, and the MAPDD pressure $p_d \in L^2(B_i^{\varepsilon,\delta})$, $i = 1, ..., N$, such that $\mathbf{v}_d = \mathbf{g}_\varepsilon$ on γ_ε and

$$\nu \sum_{i=1}^{N} \int_{B_i^{\varepsilon,\delta}} \nabla \mathbf{v}_d \cdot \nabla \boldsymbol{\eta} dx + \nu \sum_{l=1}^{M} d_l \int_{\sigma_\varepsilon^{(e_l)}} \nabla_{x^{(e_l)\prime}} \mathbf{v}_d \cdot \nabla_{x^{(e_l)\prime}} \boldsymbol{\eta} dx^{(e_l)\prime}$$

$$= \sum_{i=1}^{N} \int_{B_i^{\varepsilon,\delta}} \big(p_d \mathrm{div} \boldsymbol{\eta} - q \mathrm{div} \mathbf{v}_d\big) dx$$

for all $\boldsymbol{\eta} \in \tilde{W}^{1,2}(\bigcup_{i=1}^{N} B_i^{\varepsilon,\delta})$ vanishing on $\partial B_\varepsilon \cap \partial B_i^{\varepsilon,\delta}$ and $q \in L^2(B_i^{\varepsilon,\delta})$, $i = 1, ..., N$.

To establish the equivalence between the formulations (5.2.3) and (5.2.4), we will first prove three auxiliary lemmas.

Denote $\langle \cdot \rangle = (\mathrm{mes} B_\varepsilon)^{-1} \int_{B_\varepsilon} \cdot \, dx$, $\langle \cdot \rangle_i = \big(\mathrm{mes} B_i^{\varepsilon,\delta}\big)^{-1} \int_{B_i^{\varepsilon,\delta}} \cdot \, dx$ and define

$$\mathcal{Q}_i(\mathbf{w}) = \int_{\partial B_i^{\varepsilon,\delta}} \mathbf{w} \cdot \mathbf{n} \, dS.$$

Lemma 5.2.1 *There exist $(N-1)$ vector-valued functions $\mathbf{U}_i \in \mathcal{W}_{0,dec}(B_\varepsilon, \delta)$, for $i = 1, \ldots, N-1$, such that $\mathcal{Q}_j(\mathbf{U}_i) = \delta_{ij}$ for $j = 1, \ldots, N-1$, and the following estimates hold:*

$$\|\mathbf{U}_i\|^2_{L^2(B_\varepsilon)} \le C_* \varepsilon^{-(n-1)}, \quad \|\mathbf{U}_i\|^2_{W^{1,2}(B_\varepsilon)} \le C_* \varepsilon^{-(n+1)} \tag{5.2.5}$$

with a constant C_* that is independent of ε and δ. Consequently, $\mathcal{Q}_N(\mathbf{U}_l) = -1$ for $l = 1, \ldots, N-1$.

Proof
Let us consider for any $i = 1, \ldots, N-1$ the following problem on the graph $\mathcal{B}$: Find a function $q_i \in W^{1,2}(\mathcal{B})$ that is linear on each edge e of the graph ($q_i(x) = -s_i^{(e)} x_1^{(e)} + a_i^{(e)}$) and satisfies the conditions $q_i(O_N) = 0$ and

$$-\sum_{e:O_j \in e} \kappa_\varepsilon^{(e)} s_i^{(e)} = \delta_{ij}, \tag{5.2.6}$$

for each node or vertex O_j, $j = 1, \ldots, N-1$, where the local coordinate system has its origin at O_j. Here, $u_{0,\varepsilon} \in \mathring{W}^{1,2}(\sigma_\varepsilon^{(e)})$ represents a solution to the problem

$$\begin{cases} -\nu\Delta'_{x^{(e)\prime}} u_{0,\varepsilon}(x^{(e)\prime}) = 1, \ x^{(e)\prime} \in \sigma_\varepsilon^{(e)}, \\ u_{0,\varepsilon}(x^{(e)\prime}) = 0, \ x^{(e)\prime} \in \partial\sigma_\varepsilon^{(e)}, \end{cases} \tag{5.2.7}$$

and $\kappa_\varepsilon^{(e)} = \int_{\sigma_\varepsilon^{(e)}} u_{0,\varepsilon}(x^{(e)\prime})\, dx^{(e)\prime}$.

It is evident that $\kappa_\varepsilon^{(e)} = \varepsilon^{n+1}\kappa^{(e)}$, where $\kappa^{(e)} = \int_{\sigma^{(e)}} u_0(y^{(e)\prime})\, dy^{(e)\prime}$ and u_0 is a unique solution of the problem

$$\begin{cases} -\nu\Delta'_{y^{(e)\prime}} u_0(y^{(e)\prime}) = 1, \ y^{(e)\prime} \in \sigma^{(e)}, \\ u_0(y^{(e)\prime})|_{\partial\sigma^{(e)}} = 0, \ y^{(e)\prime} \in \partial\sigma^{(e)}. \end{cases} \tag{5.2.8}$$

Importantly, $\kappa^{(e)}$ does not depend on ε.

By adding the condition $\kappa_\varepsilon^{(e)} s_i^{(e)} = -1$ at the vertex O_N, this problem becomes a specific instance of problems on the graph (5.1.11) and it possesses a unique solution. The relationship between $\kappa_\varepsilon^{(e)}$ and $\kappa^{(e)}$ yields: $s_i^{(e)} = \varepsilon^{-(n+1)}\tilde{s}_i^{(e)}$, where $\tilde{s}_i^{(e)}$ represents the scaled pressure slopes, independent of ε. Now, for each cylinder $B_{ij}^{dec,\varepsilon}$ corresponding to the edge e, we construct a function $\mathbf{U}_i$ as a Poiseuille velocity, which is equal, in the local variables $x^{(e)}$, to $s_i^{(e)}(\mathcal{P}^{(e)})^*\Big(0, \ldots, 0, u_{0,\varepsilon}(x^{(e)\prime})\Big)$. Notably, the cross-sectional area of the cylinder has an order of ε^{n-1}, and the magnitude of $u_{0,\varepsilon}(x^{(e)\prime})$ in the Poiseuille velocity is of order ε^2 (and accordingly, ε^4 for the square of the magnitude of $u_{0,\varepsilon}$), while its derivatives are of order ε. Thus, we have the following inequalities:

$$\|\mathbf{U}_i\|^2_{L^2(B_{ij}^\varepsilon)} \le c\varepsilon^{-2(n+1)+(n-1)+4} = c\varepsilon^{-n+1}, \quad \|\mathbf{U}_i\|^2_{W^{1,2}(B_{ij}^\varepsilon)} \le c\varepsilon^{-n-1}.$$

Subsequently, we extend $\mathbf{U}_i$ within the domains $B_k^{\varepsilon,\delta}$ as a function from $W^{1,2}(B_k^{\varepsilon,\delta})$ with the given boundary values. This extension can be accomplished through the multiplication of the Poiseuille velocities $\mathbf{U}_i(x^{(e)\prime})$ by cut-off functions $\eta_\varepsilon = 1 - \zeta$ that depend solely on the longitudinal variable $\dfrac{x_n^{(e)} - \delta}{\varepsilon}$. In other words, $\eta_\varepsilon(x_n^{(e)} - \delta) = 1 - \zeta(|\dfrac{x_n^{(e)} - \delta}{\varepsilon}|)$ (refer to Sect. 5.1). By doing so, we achieve the same estimates in $B_k^{\varepsilon,\delta}$:

$$\|\mathbf{U}_i\|^2_{L^2(B_k^{\varepsilon,\delta})} \le c\varepsilon^{-n+1}, \quad \|\mathbf{U}_i\|^2_{W^{1,2}(B_k^{\varepsilon,\delta})} \le c\varepsilon^{-(n+1)}.$$

While the last estimates include an additional factor of ε, it does not enhance the overall result. For example, let us calculate the norm:

$$\begin{aligned}\|\nabla(\eta_\varepsilon \mathbf{U}_i)\|^2_{L^2(B_k^{\varepsilon,\delta})} &= \|\eta'_\varepsilon \mathbf{U}_i\|^2_{L^2(B_k^{\varepsilon,\delta})} + \|\eta\nabla'\mathbf{U}_i)\|^2_{L^2(B_k^{\varepsilon,\delta})}\\ &= \|\eta'_\varepsilon \mathbf{U}_i\|^2_{L^2(B_k^{\varepsilon,\delta})} + \|\eta\nabla\mathbf{U}_i)\|^2_{L^2(B_k^{\varepsilon,\delta})}\\ &\le c\|\eta'_\varepsilon\|^2_{L^2((0,\varepsilon))}\|\mathbf{U}_i\|^2_{L^2(B_{ij}^{\varepsilon})} + \|\eta\|^2_{L^2((0,\varepsilon))}\|\nabla\mathbf{U}_i\|^2_{L^2(B_k^{\varepsilon,\delta})}\\ &\le c(\varepsilon^{-1}\varepsilon^{-n+1} + \varepsilon\varepsilon^{-(n+1)}) = c\varepsilon^{-n}.\end{aligned}$$

Here, we utilized the estimates $\eta_\varepsilon = O(1)$ and $\eta'_\varepsilon = O(\varepsilon^{-1})$, resulting in $\|\eta'_\varepsilon\|^2_{L^2((0,\varepsilon))} = O(\varepsilon^{-1})$ and $\|\eta_\varepsilon\|^2_{L^2((0,\varepsilon))} = O(\varepsilon)$.

By summing up these estimates for $\mathbf{U}_i$, we obtain the estimate (5.2.5).

Now, let us calculate $\mathcal{Q}_N(\mathbf{U}_l)$ for $l = 1, \dots, N-1$. Notice that $\sum\limits_{k=1}^{N} \mathcal{Q}_k(\mathbf{U}_l)$ represents the summation of fluxes through all cross sections S_{ij} and S_{ji}, as well as through the parts of the boundary ∂B_ε corresponding to γ_ε. According to the lemma's assumption, $\mathbf{U}_j = 0$ over the entire boundary ∂B_ε, while for each neighboring pair of cross sections S_{ij} and S_{ji}, the fluxes are of the same absolute value but with opposite signs. Consequently, it follows that $\sum\limits_{k=1}^{N} \mathcal{Q}_k(\mathbf{U}_l) = 0$. Given that, as asserted by the lemma, $\mathcal{Q}_k(\mathbf{U}_l) = \delta_{kl}$ for $k = 1, \dots, N-1$, we conclude that $\mathcal{Q}_N(\mathbf{U}_l) = -1$. The proof of the Lemma is completed. □

Lemma 5.2.2 *Let p be a function defined in the union $\bigcup\limits_{i=1}^{N} B_i^{\varepsilon,\delta}$, and belonging to the space $L^2(B_i^{\varepsilon,\delta})$ for each $i = 1, \dots, N$. Additionally, assume that p satisfies the condition*

$$\sum_{i=1}^{N} \int_{B_i^{\varepsilon,\delta}} p(x)dx = 0.$$

Then, there exists a vector-valued function $\mathbf{U} \in \mathcal{W}_{0,\mathrm{dec}}(B_\varepsilon, \delta)$ such that

$$\operatorname{div}\mathbf{U}(x) = \begin{cases} p(x), & x \in \bigcup\limits_{i=1}^{N} B_i^{\varepsilon,\delta},\\ 0, & x \in B_\varepsilon \setminus \bigcup\limits_{i=1}^{N} B_i^{\varepsilon,\delta}.\end{cases} \tag{5.2.9}$$

Furthermore, the following estimate holds:

$$\|\mathbf{U}\|^2_{W^{1,2}(B_\varepsilon)} \le C(\varepsilon^{-2n-3}\delta^2 + \varepsilon^{-2n}\delta^{-1}) \sum_{i=1}^{N} \|p\|^2_{L^2(B_i^{\varepsilon,\delta})}, \tag{5.2.10}$$

where C is a positive constant independent of ε and δ.

Proof
Let us consider the sum

$$\boldsymbol{\Psi} = \sum_{i=1}^{N-1} \langle p \rangle_i \mathbf{U}_i,$$

where $\mathbf{U}_i$ are functions constructed in the previous lemma. Using Gauss–Ostrogradsky formula, we obtain $\int_{B_k^{\varepsilon,\delta}} \mathrm{div}\mathbf{U}_k dx = \mathcal{Q}_j(\mathbf{U}_i) = \delta_{ij} = 1$. This leads to the conclusion that

$$\int\limits_{B_k^{\varepsilon,\delta}} \mathrm{div}\boldsymbol{\Psi} dx = \langle p \rangle_k \int\limits_{B_k^{\varepsilon,\delta}} \mathrm{div}\mathbf{U}_k dx = \langle p \rangle_k.$$

Hence, we have

$$\int\limits_{B_k^{\varepsilon,\delta}} (p - \mathrm{div}\boldsymbol{\Psi}) dx = 0.$$

We now need to solve the divergence equation within $B_k^{\varepsilon,\delta}$, i.e., to construct a function $\boldsymbol{\Theta} \in \mathring{W}^{1,2}(B_k^{\varepsilon,\delta})$ such that

$$\mathrm{div}\boldsymbol{\Theta} = -\mathrm{div}\boldsymbol{\Psi} + p. \tag{5.2.11}$$

The existence of a solution of (5.2.11) follows from general results (see Lemma 2.4.1). However, in order to obtain an appropriate estimate (with a constant independent of ε and δ), we need to transform this problem into the same problem in the δ^{-1}-times dilated domain

$$\delta^{-1} B_k^{\varepsilon,\delta} = \left\{x : \delta x + O_k \in B_k^{\varepsilon,\delta}\right\}.$$

This domain can be thought of as a thin tube structure with ε/δ as a small parameter, replacing the previous small parameter ε (recall that $\delta = c\varepsilon|\ln \varepsilon|$). Thus, we can apply Lemma 2.4.14 and construct a solution satisfying the estimate in the original variables:

$$\begin{aligned} \|\boldsymbol{\Theta}\|^2_{W^{1,2}(B_k^{\varepsilon,\delta})} &\le c(\varepsilon/\delta)^{-3} \| - \mathrm{div}\boldsymbol{\Psi} + p\|^2_{L^2(B_k^{\varepsilon,\delta})} \\ &\le c(\varepsilon/\delta)^{-3} \left(\|\nabla \boldsymbol{\Psi}\|^2_{L^2(B_k^{\varepsilon,\delta})} + \|p\|^2_{L^2(B_k^{\varepsilon,\delta})} \right). \end{aligned} \tag{5.2.12}$$

Here and below c is a generic constant independent of small parameters.

Continuing, we evaluate the norm $\|\Psi\|_{W^{1,2}(B_k^{\varepsilon,\delta})}$. It can be bounded by the sum:

$$\sum_{i=1}^{N-1} |\langle p\rangle_i| \|\mathbf{U}_i\|_{W^{1,2}(B_k^{\varepsilon,\delta})},$$

where

$$|\langle p\rangle_i| \leq c\varepsilon^{-(n-1)}\delta^{-1}\Big|\int\limits_{B_k^{\varepsilon,\delta}} p dx\Big| \leq \varepsilon^{-(n-1)/2}\delta^{-1/2}\|p\|_{L^2(B_k^{\varepsilon,\delta})},$$

and

$$\|\mathbf{U}_i\|_{W^{1,2}(B_k^{\varepsilon,\delta})} \leq \|\mathbf{U}_i\|_{W^{1,2}(B_\varepsilon)} \leq C\varepsilon^{-(n+1)/2}$$

(see the previous lemma). Consequently, we obtain the following inequalities:

$$\begin{aligned}
\|\Psi\|^2_{W^{1,2}(B_k^{\varepsilon,\delta})} &\leq C\varepsilon^{-2n}\delta^{-1}\|p\|^2_{L^2(B_k^{\varepsilon,\delta})},\\
\|\Theta\|^2_{W^{1,2}(B_k^{\varepsilon,\delta})} &\leq C(\varepsilon/\delta)^{-3}\big(\varepsilon^{-2n}\delta^{-1}+1\big)\|p\|^2_{L^2(B_k^{\varepsilon,\delta})}\\
\leq C\big(\varepsilon^{-2n-3}\delta^2+\varepsilon^{-3}\delta^3\big)\|p\|^2_{L^2(B_k^{\varepsilon,\delta})} &\leq C\varepsilon^{-2n-3}\delta^2\|p\|^2_{L^2(B_k^{\varepsilon,\delta})}.
\end{aligned} \tag{5.2.13}$$

Taking $\mathbf{U} = \Psi + \Theta$, and extending Θ by zero to the cylinders $B_{ij}^{\mathrm{dec},\varepsilon}$, we have $\mathbf{U} = \Psi$ in $B_{ij}^{\mathrm{dec},\varepsilon}$. Recall that $\mathrm{div}\Psi = 0$ for $x \in B_\varepsilon \setminus \bigcup_{i=1}^{N} B_i^{\varepsilon,\delta}$. Using estimates (5.2.13), we finalize the proof. □

Remark 5.2.1 In the case where $\delta = C_J\varepsilon|\ln\varepsilon|$, the estimate (5.2.10) can be roughened as follows:

$$\|\mathbf{U}\|^2_{W^{1,2}(B_\varepsilon)} \leq C\varepsilon^{-2n-3/2}\sum_{i=1}^{N}\|p\|^2_{L^2(B_i^{\varepsilon,\delta})}.$$

Now, we are prepared to prove the existence of the solution $(\mathbf{v}_d, p_d)$ to problem (5.2.4) where test functions from the space $\mathcal{W}_{0,\mathrm{dec}}(B_\varepsilon,\delta)$ are used.

Let us introduce the space $\widehat{L^2_{dec}}(B_\varepsilon)$ as the set of scalar functions $q \in L^2_{dec}(B_\varepsilon)$ with a mean value $\langle q\rangle = \dfrac{1}{|B_\varepsilon|}\int_{B_\varepsilon} q(x)dx = 0$, and such that $q(x) = 0$ on $B_\varepsilon \setminus \bigcup_{i=1}^{N} B_i^{\varepsilon,\delta}$, i.e., on all cylinders B_{ij}^{ε}.

Theorem 5.2.1 *There exists a unique* $p_d \in \widehat{L^2_{dec}}(B_\varepsilon)$ *satisfying the integral identity* (5.2.4).

Proof
Le us consider the linear bounded functional defined on $\mathcal{W}_{0,\mathrm{dec}}(B_\varepsilon, \delta)$:

$$f(\eta) = \nu \int_{B_\varepsilon} \nabla \mathbf{v}_d \cdot \nabla \eta dx. \tag{5.2.14}$$

This functional is nullified over the space $\mathcal{H}_{0,\mathrm{dec}}(B_\varepsilon, \delta)$ of divergence-free functions. Invoking Lemma 5.2.2, for every $q \in \widehat{L^2_{dec}}(B_\varepsilon)$, there exists a function $\mathbf{w} = \boldsymbol{\Psi} + \boldsymbol{\Theta}$ belonging to the space $\mathcal{W}_{0,\mathrm{dec}}(B_\varepsilon, \delta)$ and satisfying the relation (5.2.9) and the estimate (5.2.10). It can be verified that the formula $\mathbf{w} = \boldsymbol{\Psi} + \boldsymbol{\Theta}$ determines a bounded linear operator M^{-1} that maps from $\widehat{L^2_{dec}}(B_\varepsilon)$ to $\mathcal{W}_{0,\mathrm{dec}}(B_\varepsilon, \delta)$.

Let us introduce the concept of equivalence classes within $\mathcal{W}_{0,\mathrm{dec}}(B_\varepsilon, \delta)$. Functions $\mathbf{w}$ and $\mathbf{v}$ are deemed equivalent if they share the same divergence, i.e., $\mathrm{div}\mathbf{w} = \mathrm{div}\mathbf{v}$. We can formulate the sum of two classes W and V as the equivalence class encompassing $\mathbf{w} + \mathbf{v}$, where $\mathbf{w}$ and $\mathbf{v}$ are the representatives of W and V respectively. Similarly, the product αW is determined by the equivalence class of $\alpha\mathbf{w}$, given that $\mathbf{w}$ represents class W.

This framework establishes a vector space of equivalence classes, frequently termed the quotient space. For this space, we define the inner product as:

$$(W, V) = \int_{B_\varepsilon} \mathrm{div}\mathbf{w} \cdot \mathrm{div}\mathbf{v} dx. \tag{5.2.15}$$

This definition not only respects the axioms of the inner product (bilinearity, symmetry and positivity of the associated norm) but is also robust regarding the choice of representatives $\mathbf{w}$ and $\mathbf{v}$.

Given any $\mathbf{w} \in \mathcal{W}_{0,\mathrm{dec}}(B_\varepsilon, \delta)$, let W be its associated equivalence class corresponding to its divergence $q \in \widehat{L^2_{dec}}(B_\varepsilon)$, i.e., $\mathbf{w} \in W$ and $M^{-1}q \in W$. Then $f(\mathbf{w}) = f(M^{-1}q)$, and because $\mathbf{w} - M^{-1}q$ is divergence free, $f(\mathbf{w} - M^{-1}q) = 0$. So, $f(\mathbf{w})$ is uniquely defined for all functions of the class W, and one can consider f as a linear functional on the vector space of equivalence classes. As the functional f is bounded with respect to the norm of $\mathcal{W}_{0,\mathrm{dec}}(B_\varepsilon, \delta)$, and the operator M^{-1} is also bounded,

$$|f(M^{-1}q)| \le c_\varepsilon \|q\|_{L^2(B_\varepsilon)} = c_\varepsilon \|\mathrm{div}\mathbf{w}\|_{L^2(B_\varepsilon)}.$$

So, f is bounded on the space of equivalence classes with inner product (5.2.15) and, according to the Riesz theorem, there exists a unique equivalence class U such that for a representative $\mathbf{u}$ of U and for any $\eta \in \mathcal{W}_{0,\mathrm{dec}}(B_\varepsilon, \delta)$ the functional f can be represented in a form of the inner product

$$f(\eta) = \int_{B_\varepsilon} \mathrm{div}\mathbf{u} \cdot \mathrm{div}\eta dx. \tag{5.2.16}$$

Taking $p_d = \mathrm{div}\mathbf{u}$, we complete the proof of the existence of p_d. Its uniqueness follows from the uniqueness of the equivalence class U, since for all $\mathbf{u}$ from U the divergence $\mathrm{div}\mathbf{u}$ is the same. □

Remark 5.2.2 It is important to observe that the function $p_d(x)$ constructed in the theorem is equal to zero on all cylinders B^{ε}_{ij}.

Remark 5.2.3 From Theorem 5.2.1 follows the equivalence between the two definitions (5.2.3) and (5.2.4) concerning the weak MAPDD solution.

By employing the definition (5.2.3), we can establish the following statement, akin to the reasoning found in Theorem 5.1.3:

Theorem 5.2.2 *Let δ be $C_{J+2}\varepsilon|\ln\varepsilon|$ where the constant C_{J+2} satisfies* (5.1.60) *with J replaced by $J+2$. Then the solution $\mathbf{v}_d$ of the problem* (5.2.3) *satisfies the following error estimate*

$$\|\mathbf{v} - \mathbf{v}_d\|_{W^{1,2}(B_\varepsilon)} = O(\varepsilon^{J+(n-1)/2}). \tag{5.2.17}$$

Let us prove now the estimate for the MAPDD pressure p_d.

Theorem 5.2.3 *Under the assumption that δ is chosen as $C_{J+2n+1}\varepsilon|\ln\varepsilon|$ where the constant C_{J+2n+1} satisfies* (5.1.60) *with J replaced by $J+2n+1$, there exists a function p, the solution of* (5.1.1) *(p is defined up to an additive constant), such that the following estimate holds:*

$$\|p_d - p\|_{L^2(B_k^{\varepsilon,\delta})} = O(\varepsilon^{J+n/2+(n-3)/2}), \quad k = 1, \ldots N, \tag{5.2.18}$$

Proof
The asymptotic expansion $(\widetilde{\mathbf{u}}^{(J)}, p^{(J)})$ constructed in Sect. 5.1 satisfies the integral identity:

$$\nu \int_{B_\varepsilon} \nabla\widetilde{\mathbf{u}}^{(J)} \cdot \nabla\eta dx = \int_{B_\varepsilon} p^{(J)} \mathrm{div}\eta dx - \int_{B_\varepsilon} \mathbf{f}_1^{(J)} \cdot \eta dx \quad \forall \eta \in \mathring{W}^{1,2}(B_\varepsilon), \tag{5.2.19}$$

where $\mathbf{f}_1^{(J)} = -\mathcal{L}(\widetilde{\mathbf{u}}^{(J)}, p^{(J)})$. Recall that

$$\|\mathbf{f}_1^{(J)}\|_{L^2(B_\varepsilon)} = O(\varepsilon^{J-2}). \tag{5.2.20}$$

As $p^{(J)}$ is defined up to an additive constant we assume further that the mean value of $p^{(J)}$ coincides with the mean value of p. Next, we consider the approximations $(\widetilde{\mathbf{u}}^{(J+2n+1)}, p^{(J+2n+1)})$, modified according to Remark 5.1.1, and denote $J' = J+2n+1$. We choose $\delta = C_{J'}\varepsilon|\ln\varepsilon|$ with a constant $C_{J'}$ that ensures the estimate (5.2.17) and (5.1.71) with J' replacing J. Consider the difference $\widetilde{q}_d = p_d - p^{(J')}$ with the property that $\sum\limits_{i=1}^{N} \int_{B_i^{\varepsilon,\delta}} \widetilde{q}_d(x,t)dx = 0$. This property holds because both $p^{(J')}$ and p_d are defined up to arbitrary constants that can be chosen to satisfy this condition.

Applying Lemma 5.2.2, we can construct a function $\mathbf{U} \in \mathcal{W}_{0,\mathrm{dec}}(B_\varepsilon, \delta)$ such that

$$\mathrm{div}\mathbf{U}(x) = \begin{cases} \widetilde{q}_d(x), \; x \in \bigcup_{i=1}^{N} B_i^{\varepsilon,\delta}, \\ 0, \; x \in B_\varepsilon \setminus \bigcup_{i=1}^{N} B_i^{\varepsilon,\delta}, \end{cases} \tag{5.2.21}$$

and

$$\|\mathbf{U}\|^2_{W^{1,2}(B_\varepsilon)} \le C\varepsilon^{-2n-3/2} \sum_{i=1}^{N} \|\widetilde{q}_d\|^2_{L^2(B_i^{\varepsilon,\delta})} \tag{5.2.22}$$

with some constant C independent of ε and δ.

Using $\mathbf{U}$ as a test function in (5.2.19) (with $J+2n+1$ replacing J) and in (5.2.4), and considering the difference between these integral identities, we arrive at[3]

$$\nu \int_{B_\varepsilon} \nabla\widetilde{\mathbf{u}}_d \cdot \nabla\mathbf{U} dx = \sum_{i=1}^{N} \int_{B_i^{\varepsilon,\delta}} \widetilde{q}_d \mathrm{div}\mathbf{U} dx + \int_{B_\varepsilon} \mathbf{f}_1^{(J')} \cdot \mathbf{U} dx, \tag{5.2.23}$$

where $\widetilde{\mathbf{u}}_d = \mathbf{v}_d - \widetilde{\mathbf{u}}^{(J')}$, and thus,

$$\sum_{i=1}^{N} \int_{B_i^{\varepsilon,\delta}} |\widetilde{q}_d|^2 dx \le c\Big(\|\nabla\widetilde{\mathbf{u}}_d\|_{L^2(B_\varepsilon)} + \varepsilon\|\mathbf{f}_1^{(J')}\|_{L^2(B_\varepsilon)}\Big)\|\nabla\mathbf{U}\|_{L^2(B_\varepsilon)}.$$

This implies, by using estimates (5.2.20), (5.2.22), and (5.1.64) (which also holds for the generalized version of MAPDD), that

$$\begin{aligned} \Big(\sum_{i=1}^{N} \int_{B_i^{\varepsilon,\delta}} |\widetilde{q}_d|^2 dx\Big)^{\frac{1}{2}} &\le c_2\varepsilon^{-(n+3/4)}\Big(\|\nabla\widetilde{\mathbf{u}}_d\|_{L^2(B_\varepsilon)} + \varepsilon\|\mathbf{f}_1^{(J')}\|_{L^2(B_\varepsilon)}\Big) \\ &= \varepsilon^{-(n+3/4)}\big(O(\varepsilon^{J'+(n-1)/2}) + O(\varepsilon^{J'-1})\big) = O(\varepsilon^{J+n-3/4}). \end{aligned} \tag{5.2.24}$$

Thus, applying the triangle inequality and estimate (5.1.71) for $p - p^{(J')}$ with zero mean of the difference, we obtain

$$\|p_d - p\|_{L^2(B_k^{\varepsilon,\delta})} \le \|p_d - p^{(J')}\|_{L^2(B_k^{\varepsilon,\delta})} + \|p^{(J')} - p\|_{L^2(B_k^{\varepsilon,\delta})}$$

---

[3] It is crucial that $\mathrm{div}\mathbf{U} = 0$ in $B_\varepsilon \setminus \bigcup_{i=1}^{N} B_i^{\varepsilon,\delta}$, which means that the integration in the term $\int_{B_\varepsilon} p^{(J)} \mathrm{div}\mathbf{U} dx$ is only over the domain $\bigcup_{i=1}^{N} B_i^{\varepsilon,\delta}$.

$$\leq c\Big(\varepsilon^{J+n-3/4}+\varepsilon^{J+2n+1+(n-3)/2}\Big) \leq c\varepsilon^{J+n-3/4} \leq c\varepsilon^{J+n/2+(n-3)/2}, \ k=1,\dots,N.$$

The theorem is proved. □

Let us proceed to extend the function p_d to the cylinders B^{ε}_{ij} as an affine function, employing the subsequent expression:

$$p_d(x_1^{(e)}) = \langle p_d\rangle_{C_{ij}} + \frac{\langle p_d\rangle_{C_{ji}} - \langle p_d\rangle_{C_{ij}}}{|e| - 2\delta + \varepsilon}\left(x_1^{(e)} - \delta + \frac{\varepsilon}{2}\right), \tag{5.2.25}$$

Here, for any function $s \in L^2(B_\varepsilon)$, $\langle s\rangle_{C_{ij}}$ denotes the mean value of s within the cylinder $C_{ij} = \{x_1^{(e)} \in (\delta-\varepsilon, \delta)\}\times S_{ij}$, which has one of its bases located at the cross section S_{ij} and a height of the length ε. Notably, for the integer J' introduced in Theorem 5.2.3, the asymptotic approximation $p^{(J')}$ conforms to a relation analogous to (5.2.25):

$$p^{(J')}(x_1^{(e)}) = \langle p^{(J')}\rangle_{C_{ij}} + \frac{\langle p^{(J')}\rangle_{C_{ji}} - \langle p^{(J')}\rangle_{C_{ij}}}{|e| - 2\delta + \varepsilon}\left(x_1^{(e)} - \delta + \frac{\varepsilon}{2}\right). \tag{5.2.26}$$

This is due to its nature as an affine function of $x_1^{(e)}$ within the cylinder $\big\{x_1^{(e)} \in (\delta - \varepsilon, |e| - \delta + \varepsilon)\big\} \times \sigma_\varepsilon^{(e)}$. On a different note, the following inequality holds:

$$|\langle p_d\rangle_{C_{ij}} - \langle p^{(J')}\rangle_{C_{ij}}|^2 \leq c_5(\text{mes}C_{ij})^{-1}\|p_d - p^{(J')}\|^2_{L^2(B_i^{\varepsilon,\delta})}. \tag{5.2.27}$$

Applying estimate (5.2.24) from Theorem 5.2.3, we deduce

$$|\langle p_d\rangle_{C_{ij}} - \langle p^{(J')}\rangle_{C_{ij}}| = O(\varepsilon^{J+(n-3)/2}). \tag{5.2.28}$$

Consequently, it can be concluded that

$$\|p_d - p^{(J')}\|_{L^2(B_\varepsilon)} = O(\varepsilon^{J+(n-3)/2}). \tag{5.2.29}$$

Finally, the application of the triangle inequality, analogously to the proof presented in Theorem 5.2.3, yields the subsequent proposition.

Theorem 5.2.4 *Given a constant C_{J+2n+1} satisfying* (5.1.60) *with J replaced by $J + 2n + 1$, let δ be $C_{J+2n+1}\varepsilon|\ln \varepsilon|$. There exists a function p, a solution of* (5.1.1), *such that*

$$\|p_d - p\|_{L^2(B_\varepsilon)} = O(\varepsilon^{J+(n-3)/2}). \tag{5.2.30}$$

Remark 5.2.4 In contrast to the classical version of the MAPDD, as formulated in Sect. 5.1.5, the solution $\mathbf{v}_d$ of the generalized version may differ from the Poiseuille solution velocity within the truncated cylinders $B^{dec,\varepsilon}_{ij}$.

5.3 Stationary Navier–Stokes Equations

5.3.1 Formulation of the Problem. Existence and Uniqueness

In this section, we study the stationary Navier-Stokes problem within the context of a tube structure B_ε:

$$\begin{cases} -\nu\Delta\mathbf{v} + (\mathbf{v}\cdot\nabla)\mathbf{v} + \nabla p = 0,\ x \in B_\varepsilon, \\ \operatorname{div}\mathbf{v} = 0,\ x \in B_\varepsilon, \\ \mathbf{v} = \mathbf{g},\ x \in \partial B_\varepsilon, \end{cases} \tag{5.3.1}$$

where the velocity $\mathbf{g}$ on the boundary ∂B_ε satisfies conditions (5.1.2).

Definition 5.3.1 A weak solution of problem (5.3.1) is defined as a divergence-free vector field $\mathbf{v} \in W^{1,2}(B_\varepsilon)$ that adheres to the boundary condition $\mathbf{v}|_{\partial B_e} = \mathbf{g}$, and satisfies the integral identity for all test functions $\boldsymbol{\eta} \in \overset{\circ}{H}{}^{1,2}(B_\varepsilon)$:

$$\nu \int\limits_{B_\varepsilon} \nabla\mathbf{v}\cdot\nabla\boldsymbol{\eta}dx + \int\limits_{B_\varepsilon} (\mathbf{v}\cdot\nabla)\mathbf{v}\cdot\boldsymbol{\eta}dx = 0. \tag{5.3.2}$$

The above identity is derived from the Eqs. (5.3.1) through multiplication by $\boldsymbol{\eta} \in \overset{\circ}{H}{}^{1,2}(B_\varepsilon)$ and integration by parts within B_ε.

Let $\mathbf{g} \in W^{2,2}(B_e)$ be a divergence-free extension of $\mathbf{g}$ (denoted by the same symbol $\mathbf{g}$), subject to inequalities (5.1.3). Since $\mathbf{g}$ vanishes on the lateral boundary of B_ε, it satisfies the Poincaré inequality

$$\|\mathbf{g}\|_{L^2(B_\varepsilon)} \le c\varepsilon\|\nabla\mathbf{g}\|_{L^2(B_\varepsilon)}, \tag{5.3.3}$$

where the constant c is independent of ε.

The weak solution $\mathbf{v}$ of problem (5.3.1) is sought in the form $\mathbf{v} = \mathbf{u} + \mathbf{g}$. Subsequently, for $\mathbf{u}$, the following variational formulation emerges:

Definition 5.3.2 Find $\mathbf{u} \in \overset{\circ}{H}{}^{1,2}(B_\varepsilon)$ such that for all test functions $\boldsymbol{\eta} \in \overset{\circ}{H}{}^{1,2}(B_\varepsilon)$, the integral identity is satisfied:

$$\begin{aligned} &\nu \int\limits_{B_\varepsilon} \nabla\mathbf{u}\cdot\nabla\boldsymbol{\eta}dx - \int\limits_{B_\varepsilon} \big((\mathbf{u}+\mathbf{g})\cdot\nabla\big)\boldsymbol{\eta}\cdot\mathbf{u}dx - \int\limits_{B_\varepsilon} (\mathbf{u}\cdot\nabla)\boldsymbol{\eta}\cdot\mathbf{g}dx \\ &= -\int\limits_{B_\varepsilon} \Big(\nu\nabla\mathbf{g}\cdot\nabla\boldsymbol{\eta} + (\mathbf{g}\cdot\nabla)\mathbf{g}\cdot\boldsymbol{\eta}\Big)dx. \end{aligned} \tag{5.3.4}$$

Here we used the formula

$$\int_{B_\varepsilon} (\mathbf{u} \cdot \nabla)\mathbf{v} \cdot \mathbf{w} dx = - \int_{B_\varepsilon} (\mathbf{u} \cdot \nabla)\mathbf{w} \cdot \mathbf{v} dx$$

which holds if $\operatorname{div}\mathbf{u} = 0$ and one of the three functions vanishes on the boundary.

Theorem 5.3.1 *There exists a positive constant c_*, independent of ε, such that if the condition*

$$c_* \varepsilon^\alpha \|\nabla \mathbf{g}\|_{L^2(B_\varepsilon)} < \nu \tag{5.3.5}$$

holds, where $\alpha = 1$ for $n = 2$ and $\alpha = 1/2$ for $n = 3$, then the problem (5.3.1) *admits at least one weak solution $\mathbf{u} \in \overset{\circ}{H}{}^{1,2}(B_\varepsilon)$. Additionally, the estimate*

$$\|\nabla \mathbf{u}\|_{L^2(B_\varepsilon)} \leq c\Big(\|\nabla \mathbf{g}\|_{L^2(B_\varepsilon)} + \varepsilon^\alpha \|\nabla \mathbf{g}\|^2_{L^2(B_\varepsilon)}\Big) \tag{5.3.6}$$

holds, with the constant c independent of ε.

Proof
The proof closely follows the concepts presented in the book by O.A. Ladyzhenskaya [102]. In the Hilbert space $\overset{\circ}{H}{}^{1,2}(B_\varepsilon)$, we define the inner product

$$[\mathbf{u}, \boldsymbol{\eta}] = \int_{B_\varepsilon} \nabla \mathbf{u} \cdot \nabla \boldsymbol{\eta} dx,$$

which corresponds to the Dirichlet norm. By utilizing Hölder and Poincaré inequalities along with Lemma 2.3.11, we obtain the following estimates:

$$\begin{aligned}\Big|\int_{B_\varepsilon} (\mathbf{u} \cdot \nabla)\mathbf{u} \cdot \boldsymbol{\eta} dx\Big| &\leq \|\mathbf{u}\|_{L^4(B_\varepsilon)} \|\nabla \mathbf{u}\|_{L^2(B_\varepsilon)} \|\boldsymbol{\eta}\|_{L^4(B_\varepsilon)} \\ &\leq c\varepsilon^\alpha \|\nabla \mathbf{u}\|^2_{L^2(B_\varepsilon)} \|\nabla \boldsymbol{\eta}\|_{L^2(B_\varepsilon)},\end{aligned}$$

$$\Big|\int_{B_\varepsilon} (\mathbf{g} \cdot \nabla)\mathbf{u} \cdot \boldsymbol{\eta} dx\Big| + \Big|\int_{B_\varepsilon} (\mathbf{u} \cdot \nabla)\mathbf{g} \cdot \boldsymbol{\eta} dx\Big| \leq c\varepsilon^\alpha \|\nabla \mathbf{g}\|_{L^2(B_\varepsilon)} \|\nabla \mathbf{u}\|_{L^2(B_\varepsilon)} \|\nabla \boldsymbol{\eta}\|_{L^2(B_\varepsilon)}.$$

Moreover, estimating the integral in the right-hand side of (5.3.4), we get

$$\begin{aligned}&\Big|\int_{B_\varepsilon} \Big(\nu \nabla \mathbf{g} \cdot \nabla \boldsymbol{\eta} + (\mathbf{g} \cdot \nabla)\mathbf{g} \cdot \boldsymbol{\eta}\Big) dx\Big| \\ &\leq \nu \|\nabla \mathbf{g}\|_{L^2(B_\varepsilon)} \|\nabla \boldsymbol{\eta}\|_{L^2(B_\varepsilon)} + c\varepsilon^\alpha \|\nabla \mathbf{g}\|^2_{L^2(B_\varepsilon)} \|\nabla \boldsymbol{\eta}\|_{L^2(B_\varepsilon)}.\end{aligned} \tag{5.3.7}$$

We can deduce from the aforementioned estimates and the Riesz theorem that the integral identity (5.3.4) is equivalent to the operator equation in the space $\overset{\circ}{H}{}^{1,2}(B_\varepsilon)$:

$$\mathbf{u} = \mathcal{A}\mathbf{u}, \tag{5.3.8}$$

where the operator $\mathcal{A}$ is defined for all $\boldsymbol{\eta} \in \overset{\circ}{H}{}^{1,2}(B_\varepsilon)$ as

$$[\mathcal{A}\mathbf{u}, \boldsymbol{\eta}] = \nu^{-1} \int\limits_{B_\varepsilon} \Big((\mathbf{u} \cdot \nabla)\boldsymbol{\eta} \cdot \mathbf{u} + (\mathbf{g} \cdot \nabla)\boldsymbol{\eta} \cdot \mathbf{u} + (\mathbf{u} \cdot \nabla)\boldsymbol{\eta} \cdot \mathbf{g}$$
$$- \nu \nabla \mathbf{g} \cdot \nabla \boldsymbol{\eta} - (\mathbf{g} \cdot \nabla)\mathbf{g} \cdot \boldsymbol{\eta} \Big) dx.$$

By exploiting the compactness of the embedding $W^{1,2}(B_\varepsilon) \hookrightarrow L^4(B_\varepsilon)$, it becomes standard to demonstrate that the operator $\mathcal{A} : \overset{\circ}{H}{}^{1,2}(B_\varepsilon) \mapsto \overset{\circ}{H}{}^{1,2}(B_\varepsilon)$ is compact (as seen in, for instance, [102]). Consequently, the existence of at least one solution for (5.3.8) follows through the Leray–Schauder fixed point theorem (see Theorem 2.2.10) if we can show that all possible solutions $\mathbf{u}^{(\lambda)}$ of the equation

$$\mathbf{u}^{(\lambda)} = \lambda \mathcal{A}\mathbf{u}^{(\lambda)}, \quad \lambda \in [0, 1] \tag{5.3.9}$$

are uniformly bounded with respect to λ.

A solution $\mathbf{u}^{(\lambda)}$ of (5.3.9) satisfies the subsequent integral identity for all $\boldsymbol{\eta} \in \overset{\circ}{H}{}^{1,2}(B_\varepsilon)$:

$$\begin{aligned} &\nu \int\limits_{B_\varepsilon} \nabla \mathbf{u}^{(\lambda)} \cdot \nabla \boldsymbol{\eta} dx + \lambda \int\limits_{B_\varepsilon} (\mathbf{u}^{(\lambda)} \cdot \nabla)\mathbf{u}^{(\lambda)} \cdot \boldsymbol{\eta} dx \\ &+ \lambda \int\limits_{B_\varepsilon} (\mathbf{g} \cdot \nabla)\mathbf{u}^{(\lambda)} \cdot \boldsymbol{\eta} dx + \lambda \int\limits_{B_\varepsilon} (\mathbf{u}^{(\lambda)} \cdot \nabla)\mathbf{g} \cdot \boldsymbol{\eta} dx \\ &= -\lambda \int\limits_{B_\varepsilon} \Big(\nu \nabla \mathbf{g} \cdot \nabla \boldsymbol{\eta} + (\mathbf{g} \cdot \nabla)\mathbf{g} \cdot \boldsymbol{\eta} \Big) dx. \end{aligned} \tag{5.3.10}$$

Taking $\boldsymbol{\eta} = \mathbf{u}^{(\lambda)}$ in (5.3.10) yields

$$\nu \int\limits_{B_\varepsilon} |\nabla \mathbf{u}^{(\lambda)}|^2 dx = -\lambda \int\limits_{B_\varepsilon} (\mathbf{u}^{(\lambda)} \cdot \nabla)\mathbf{g} \cdot \mathbf{u}^{(\lambda)} dx$$
$$-\lambda \int\limits_{B_\varepsilon} \Big(\nu \nabla \mathbf{g} \cdot \nabla \mathbf{u}^{(\lambda)} + (\mathbf{g} \cdot \nabla)\mathbf{g} \cdot \mathbf{u}^{(\lambda)} \Big) dx.$$

By utilizing (5.3.7), we derive

$$\nu \|\nabla \mathbf{u}^{(\lambda)}\|^2_{L^2(B_\varepsilon)} \le c \|\nabla \mathbf{g}\|_{L^2(B_\varepsilon)} \|\mathbf{u}^{(\lambda)}\|^2_{L^4(B_\varepsilon)}$$
$$+ c\Big(\|\nabla \mathbf{g}\|_{L^2(B_\varepsilon)} + \varepsilon^\alpha \|\nabla \mathbf{g}\|^2_{L^2(B_\varepsilon)} \Big) \|\nabla \mathbf{u}^{(\lambda)}\|_{L^2(B_\varepsilon)}$$
$$\le c_* \varepsilon^\alpha \|\nabla \mathbf{g}\|_{L^2(B_\varepsilon)} \|\nabla \mathbf{u}^{(\lambda)}\|^2_{L^2(B_\varepsilon)}$$
$$+ c\Big(\|\nabla \mathbf{g}\|_{L^2(B_\varepsilon)} + \varepsilon^\alpha \|\nabla \mathbf{g}\|^2_{L^2(B_\varepsilon)} \Big) \|\nabla \mathbf{u}^{(\lambda)}\|_{L^2(B_\varepsilon)}.$$

Hence, provided condition (5.3.5) holds, we deduce the estimate

$$\|\nabla \mathbf{u}^{(\lambda)}\|_{L^2(B_\varepsilon)} \le c\Big(\|\nabla \mathbf{g}\|_{L^2(B_\varepsilon)} + \varepsilon^\alpha \|\nabla \mathbf{g}\|^2_{L^2(B_\varepsilon)}\Big).$$

The constant c in the latter inequality is independent of λ and ε, thus completing the theorem's proof. □

Remark 5.3.1 It should be noted that in both cases where $n = 2$ and $n = 3$, according to conditions (5.1.3), the inequality $\varepsilon^\alpha \|\nabla \mathbf{g}\|_{L^2(B_\varepsilon)} \le c\varepsilon^{1/2}$ holds. Thus, condition (5.3.5) is satisfied for sufficiently small values of ε.

Theorem 5.3.2 *There exists a positive constant c_0, independent of ε, such that if*

$$c_0\varepsilon^\alpha\Big(\|\nabla \mathbf{g}\|_{L^2(B_\varepsilon)} + \varepsilon^\alpha \|\nabla \mathbf{g}\|^2_{L^2(B_\varepsilon)}\Big) < \nu, \tag{5.3.11}$$

then the weak solution $\mathbf{v}$ *of problem* (5.3.1) *is unique.*

Proof
Suppose there are two weak solutions, $\mathbf{v}_1$ and $\mathbf{v}_2$, satisfying the identity (5.3.2). By subtracting the identity (5.3.2) for $\mathbf{v}_2$ from the one for $\mathbf{v}_1$, we obtain the equation

$$\nu \int\limits_{B_\varepsilon} \nabla \mathbf{w} \cdot \nabla \boldsymbol{\eta} dx + \int\limits_{B_\varepsilon} (\mathbf{w}\cdot\nabla)\mathbf{v}_2 \cdot \boldsymbol{\eta} + (\mathbf{v}_1\cdot\nabla)\mathbf{w}\cdot\boldsymbol{\eta} dx = 0, \tag{5.3.12}$$

where $\mathbf{w} = \mathbf{v}_1 - \mathbf{v}_2$. Since $\text{div}\mathbf{w} = 0$ and $\mathbf{w}|_{\partial B_\varepsilon} = 0$, it follows that $\mathbf{w} \in \overset{\circ}{H}{}^{1,2}(B_\varepsilon)$, and by choosing $\boldsymbol{\eta} = \mathbf{w}$ in Eq. (5.3.12), we deduce

$$\nu \int\limits_{B_\varepsilon} |\nabla \mathbf{w}|^2 dx = -\int\limits_{B_\varepsilon} (\mathbf{w}\cdot\nabla)\mathbf{v}_2 \cdot \mathbf{w} dx.$$

By applying Lemma 2.3.11 and using the estimates (5.3.6) for $\mathbf{u}_2$, we arrive at

$$\nu \int\limits_{B_\varepsilon} |\nabla \mathbf{w}|^2 dx \le c_0\varepsilon^\alpha\Big(\|\nabla \mathbf{g}\|_{L^2(B_\varepsilon)} + \varepsilon^\alpha \|\nabla \mathbf{g}\|^2_{L^2(B_\varepsilon)}\Big)\|\nabla \mathbf{w}\|^2_{L^2(B_\varepsilon)},$$

where the constant c_0 is independent of ε. If (5.3.11) holds true, then the above inequality yields $\int_{B_\varepsilon} |\nabla \mathbf{w}|^2 dx = 0$, which implies $\mathbf{w} = 0$. □

Remark 5.3.2 It is also noteworthy that the weak solution $\mathbf{v}$ of problem (5.3.1) belongs to the space $W^{2,2}(B_\varepsilon)$ whenever $\mathbf{f} \in L^2(B_\varepsilon)$, $\mathbf{g} \in W^{3/2,2}(\partial B_\varepsilon)$, and $\partial B_\varepsilon \in C^2$. Furthermore, the corresponding pressure belongs to $W^{1,2}(B_\varepsilon)$ (see [102]).

Let us proceed with the evaluation of the pressure p, which corresponds to the weak solution $\mathbf{v}$ of the problem (5.3.1). For this purpose, we introduce the linear functional $f(\boldsymbol{\eta})$ defined for $\boldsymbol{\eta}$ belonging to $\overset{\circ}{W}{}^{1,2}(B_\varepsilon)$ as follows:

$$f(\eta) = \nu \int_{B_\varepsilon} \nabla \mathbf{v} \cdot \nabla \eta \, dx + \int_{B_e} (\mathbf{v} \cdot \nabla)\mathbf{v} \cdot \eta \, dx. \tag{5.3.13}$$

It is evident that $f(\eta)$ is bounded, and according to (5.3.2), $f(\eta) = 0$ for all $\eta \in \mathring{H}^{1,2}(B_\varepsilon) \subset \mathring{W}^{1,2}(B_\varepsilon)$. Consequently, by virtue of Lemma 2.4.2, there exists a function p belonging to $\widehat{L}^2(B_\varepsilon)$ such that

$$f(\eta) = \int_{B_\varepsilon} p \operatorname{div}\eta \, dx \quad \forall \eta \in \mathring{W}^{1,2}(B_\varepsilon). \tag{5.3.14}$$

Therefore, the variational formulation (5.3.2) is equivalent to the following one:

Definition 5.3.3 Find a pair $(\mathbf{v}, p) \in W^{1,2} \times \widehat{L}^2(B_\varepsilon)$ such that $\mathbf{v}|_{\partial B_\varepsilon} = \mathbf{g}$, $\operatorname{div}\mathbf{v} = 0$, and for every $\eta \in \mathring{W}^{1,2}(B_\varepsilon)$, the integral identity holds

$$\nu \int_{B_\varepsilon} \nabla \mathbf{v} \cdot \nabla \eta \, dx + \int_{B_\varepsilon} (\mathbf{v} \cdot \nabla)\mathbf{v} \cdot \eta \, dx = \int_{B_\varepsilon} p\operatorname{div}\eta \, dx. \tag{5.3.15}$$

Let $\mathbf{w} \in \mathring{W}^{1,2}(B_\varepsilon)$ be such that $\operatorname{div}\mathbf{w} = p$ in B_ε and it satisfies

$$\|\nabla \mathbf{w}\|_{L^2(B_\varepsilon)} \le c\varepsilon^{-3/2} \|p\|_{L^2(B_\varepsilon)} \tag{5.3.16}$$

(see Lemma 2.4.14). By substituting $\eta = \mathbf{w}$ into (5.3.15), we obtain

$$\int_{B_\varepsilon} |p|^2 \, dx = \nu \int_{B_\varepsilon} \nabla \mathbf{v} \cdot \nabla \mathbf{w} \, dx + \int_{B_\varepsilon} (\mathbf{v} \cdot \nabla)\mathbf{v} \cdot \mathbf{w} \, dx. \tag{5.3.17}$$

By estimating the right-hand side of (5.3.17) using inequalities (2.3.37), (2.3.38), and (5.3.16), we arrive at

$$\begin{aligned} \|p\|^2_{L^2(B_\varepsilon)} &\le \nu \|\nabla \mathbf{v}\|_{L^2(B_\varepsilon)} \|\nabla \mathbf{w}\|_{L^2(B_\varepsilon)} + \|\mathbf{v}\|_{L^4(B_\varepsilon)} \|\nabla \mathbf{v}\|_{L^2(B_\varepsilon)} \|\mathbf{w}\|_{L^4(B_\varepsilon)} \\ &\le c\Big(\varepsilon^{-3/2} \|\nabla \mathbf{v}\|_{L^2(B_\varepsilon)} + \varepsilon^{\alpha-3/2} \|\nabla \mathbf{v}\|^2_{L^2(B_\varepsilon)}\Big) \|p\|_{L^2(B_\varepsilon)}. \end{aligned}$$

Applying estimate (5.3.6), the above inequality simplifies to

$$\|p\|_{L^2(B_\varepsilon)} \le c\Big(\varepsilon^{-3/2} \|\nabla \mathbf{g}\|_{L^2(B_\varepsilon)} + \varepsilon^{\alpha-3/2} \|\nabla \mathbf{g}\|^2_{L^2(B_\varepsilon)} + \varepsilon^{3\alpha-3/2} \|\nabla \mathbf{g}\|^4_{L^2(B_\varepsilon)}\Big). \tag{5.3.18}$$

Thus, we have established the following theorem:

Theorem 5.3.3 *The variational formulations* (5.3.2) *and* (5.3.15) *are equivalent. The pressure p satisfies the estimate* (5.3.18).

5.3.2 Remarks on Asymptotic Approximation of the Solution

In this section, we consider the stationary Navier–Stokes problem (5.3.1). The process of constructing and justifying the asymptotic expansion of the solution for (5.3.1) is quite similar to that of the stationary Stokes problem. However, there are additional terms in the right-hand sides of the boundary layer problems. We will briefly outline the necessary modifications. The sought-after asymptotic expansion is given by (5.1.5)–(5.1.9). We will begin with the base case ($j = 0$), which is the same as for the Stokes problem (equations on the graph (5.1.10), Poiseuille flow (5.1.21), and the boundary layer problem (5.1.23)). The induction from j to $j + 1$ is carried out in four steps, much like before. The first step of induction remains unchanged: determining the Poiseuille flow (5.1.30). However, the second step involves some modifications.

Step 2 The boundary layer solution consists of a pair $\left(\mathbf{U}_j^{[BLO_l]}, P_j^{[BLO_l]}\right)$ that satisfies the following Stokes system within Ω_l, where $l = 1, \ldots, N_1$:

$$\begin{cases} -\nu\Delta_y \mathbf{U}_j^{[BLO_l]} + \nabla_y P_j^{[BLO_l]} = \mathbf{f}_j^{[REGO_l]} + \mathbf{f}_j^{[BLO_l]}, & y^{(e)} \in \Omega_l, \\ \operatorname{div}_y \mathbf{U}_j^{[BLO_l]} = h_j^{[REGO_l]}, & y^{(e)} \in \Omega_l, \\ \mathbf{U}_j^{[BLO_l]} = 0, & y^{(e)} \in \partial\Omega_l, \end{cases} \tag{5.3.19}$$

where

$$\begin{aligned} \mathbf{f}_j^{[REGO_l]}(y^{(e)}) = &- \sum_{e:O_l\in e} \Big\{ -\nu\Delta_y\Big[\zeta\Big(\frac{y_n^{(e)}}{3r}\Big)\mathbf{U}_j^{[e]}(y^{(e)\prime})\Big] + \nabla_y\Big[\zeta\Big(\frac{y_n^{(e)}}{3r}\Big)\big(-s_j^{(e)} y_n^{(e)}\big)\Big] \\ &+ \sum_{p+r=j-1} \zeta\Big(\frac{y_n^{(e)}}{3r}\Big)\big(\mathbf{U}_p^{[e]}(y^{(e)\prime}) \cdot \nabla_y\big)\Big(\zeta\Big(\frac{y_n^{(e)}}{3r}\Big)\mathbf{U}_r^{[e]}(y^{(e)\prime})\Big) + \widehat{a}_j^{(e)}\, \nabla_y \zeta\Big(\frac{y_n^{(e)}}{3r}\Big)\Big\} \end{aligned} \tag{5.3.20}$$

(for $j = J$ the coefficient $\widehat{a}_j^{(e)}$ is omitted),

$$\begin{aligned} \mathbf{f}_j^{[BLO_l]}(y^{(e)}) = &- \sum_{e:O_l\in e} \Big\{ \sum_{p+r=j-1} \zeta\Big(\frac{y_n^{(e)}}{3r}\Big)\big(\mathbf{U}_p^{[e]}(y^{(e)\prime}) \cdot \nabla_y\big)\mathbf{U}_r^{[BLO_l]}(y) \\ &+ \sum_{p+r=j-1} \big(\mathbf{U}_p^{[BLO_l]}(y) \cdot \nabla_y\big)\Big(\zeta\Big(\frac{y_n^{(e)}}{3r}\Big)\mathbf{U}_r^{[e]}(y^{(e)\prime})\Big)\Big\} \\ &- \sum_{p+r=j-1} \big(\mathbf{U}_p^{[BLO_l]}(y) \cdot \nabla_y\big)\mathbf{U}_r^{[BLO_l]}(y), \end{aligned} \tag{5.3.21}$$

$$h_j^{[REGO_l]}(y) = -\sum_{e:O_l\in e} \mathrm{div}_y\Big(\zeta\Big(\frac{y_n^{(e)}}{3r}\Big)\mathbf{U}_j^{[e]}(y^{(e)\prime})\Big). \tag{5.3.22}$$

Here, as before, the sum $\sum\limits_{e:O_l\in e}$ is taken over all edges e having ends in the node O_l; the terms of the sum are extended by zero out of cylinders $\Pi_\varepsilon^{(e)}$; and by convention, the terms with the negative subscripts j are equal to zero.

The third and fourth steps remain unchanged from before.

As in the case of the Stokes equation, the constructed function $\mathbf{u}^{(J)}$ is not divergence-free because of the multiplication of the boundary layer terms $\mathbf{U}^{[BLO_l,J]}$ in each bundle of cylinders by the cut-off factor $\zeta(\frac{|x-O_l|}{|e|_{min}})$. Let us construct a corrector to the approximation $\mathbf{u}^{(J)}$. Namely, we have

$$\mathrm{div}\mathbf{u}^{(J)} = h^{(J)}(x),$$

where

$$h^{(J)}(x) = -\sum_{l=1}^{N} \nabla\zeta(\frac{|x-O_l|}{|e|_{min}})\cdot \mathbf{U}^{[BLO_l,J]}(\frac{x-O_l}{\varepsilon}).$$

Obviously, $h^{(J)} \in \mathring{W}^{1,2}(B_\varepsilon)$. Since the support of the function $\nabla\zeta(\dfrac{|x-O_l|}{|e|_{min}})$ belongs to the middle third of every cylinder, the following relation holds

$$\|h^{(J)}\|_{W^{1,2}(B_\varepsilon)} = O(e^{-c_1/\varepsilon}), \tag{5.3.23}$$

for some $c_1 > 0$.

As observed in the Stokes problem, we find that

$$\int_{B_\varepsilon} h^{(J)}(y)\,dy = 0.$$

Hence, as previously, a vector field $\mathbf{w}^{(J)} \in \mathring{W}^{1,2}(B_\varepsilon)$ exists such that $\mathrm{div}\mathbf{w}^{(J)} = -h^{(J)}$. Moreover, the following estimate holds:

$$\|\mathbf{w}^{(J)}\|_{W^{1,2}(B_\varepsilon)} \le \varepsilon^{-3/2}c\|h^{(J)}\|_{L^2(B_\varepsilon)}. \tag{5.3.24}$$

Additionally, akin to the Stokes problem, we can establish the estimate

$$\|\mathbf{w}^{(J)}\|_{W^{2,2}(B_\varepsilon)} \le c\varepsilon^{-2}\|h^{(J)}\|_{W^{1,2}(B_\varepsilon)}. \tag{5.3.25}$$

Let us define $\widetilde{\mathbf{u}}^{(J)} = \mathbf{u}^{(J)} + \mathbf{w}^{(J)}$. Consequently, $\mathrm{div}\widetilde{\mathbf{u}}^{(J)}$ equals zero, $\widetilde{\mathbf{u}}^{(J)}$ fulfills the boundary conditions.

Additionally, in the analysis of the residual, it should be noted that the Navier–Stokes operator now includes an additional convective (inertial) term, i.e., $\mathcal{L}(\mathbf{u}, p) = -\nu\Delta\mathbf{u} + (\mathbf{u}\cdot\nabla)\mathbf{u} + \nabla p$. Let us calculate $\mathcal{L}(\widetilde{\mathbf{u}}^{(J)}, p^{(J)})$ in a half-bundle $HB_{O_l}, l = 1, \dots, N$:

$$-\mathbf{f}_1^{(J)}(x) = \mathcal{L}(\tilde{\mathbf{u}}^{(J)}, p^{(J)})$$

$$= \sum_{J+1\le j\le 2J} \varepsilon^{j-2}\Bigg(\sum_{p+r=j-1=j-1} \Big(\mathbf{U}_p^{(e)}(y^{(e)\prime})\zeta\Big(\frac{y_n^{(e)}}{3r}\Big)\cdot\nabla_y\Big)\mathbf{U}_r^{(e)}(y^{(e)\prime})\zeta\Big(\frac{y_n^{(e)}}{3r}\Big)$$

$$+ \sum_{p+r=j-1} \zeta\Big(\frac{y_n^{(e)}}{3r}\Big)\big(\mathbf{U}_p^{(e)}(y^{(e)\prime})\cdot\nabla_y\big)\mathbf{U}_r^{[BLO_l]}(y)$$

$$+ \sum_{p+r=j-1} \big(\mathbf{U}_p^{[BLO_l]}(y)\cdot\nabla_y\big)\Big(\zeta\Big(\frac{y_n^{(e)}}{3r}\Big)\mathbf{U}_{(r)}^{(e)}(y^{(e)\prime})\Big)$$

$$+ \sum_{p+r=j-1} \big(\mathbf{U}_p^{[BLO_l]}(y)\cdot\nabla_y\big)\mathbf{U}_r^{[BLO_l]}(y)\Bigg)$$

$$+\nu\Delta\Big(\mathbf{U}^{[BLO_l,J]}(y)\zeta\big(\frac{|x-O_l|}{|e|_{min}}\big)\Big)\chi(x)$$

$$+\Big(\mathbf{U}^{[BLO_l,J]}(y)\Big(1-\zeta\big(\frac{|x-O_l|}{|e|_{min}}\big)\Big)\cdot\nabla\Big)\Big(\mathbf{U}^{[BLO_l,J]}(y)\big(1-\zeta\big(\frac{|x-O_l|}{|e|_{min}}\big)\big)\Big)$$

$$-\Big(\mathbf{U}^{[BLO_l,J]}(y)\zeta\big(\frac{|x-O_l|}{|e|_{\min}}\big)\cdot\nabla\Big)\mathbf{U}^{[e,J]}(y^{(e)\prime})$$

$$-\Big(\mathbf{U}^{[e,J]}(y^{(e)\prime})\cdot\nabla\Big)\Big(\mathbf{U}^{[BLO_l,J]}(y)\zeta\big(\frac{|x-O_l|}{|e|_{\min}}\big)\Big)$$

$$-\ (\mathbf{U}^{[BLO_l,J]}(y)\cdot\nabla)\mathbf{U}^{[BLO_l,J]}(y)\Big)\chi(x)$$

$$-\nu\Delta\mathbf{w}^{(J)} + (\mathbf{w}^{(J)}\cdot\nabla)\mathbf{w}^{(J)} + (\mathbf{u}^{(J)}\cdot\nabla)\mathbf{w}^{(J)} + (\mathbf{w}^{(J)}\cdot\nabla)\mathbf{u}^{(J)}$$

$$+\varepsilon^{J-2}\sum_{e:O_l\in e} \widehat{a}_J^{[BLO_l,e]}\nabla_y\zeta\Big(\frac{y_n^{(e)}}{3r}\Big)$$

$$-\nabla\Big(\zeta\big(\frac{|x-O_l|}{|e|_{min}}\big)p^{[BLO_l,J]}(y)\Big)\chi(x).$$

Here $y = \dfrac{x - O_l}{\varepsilon}$; $y^{(e)} = \dfrac{x^{(e)}}{\varepsilon}$; $\chi = \chi_{\mathrm{supp}\zeta\left(\frac{|x-O_l|}{|e|_{min}}\right)}$ is the characteristic function of the set $\mathrm{supp}\zeta\left(\frac{|x-O_l|}{|e|_{min}}\right)$. As before the terms of the sums $\sum\limits_{e:O_l\in e}$ are extended by zero out of cylinders $\Pi_\varepsilon^{(e)}$.

In the presented equations, the initial four lines originate from the inertial term, encompassing all possible combinations of $\mathbf{U}_\beta^{(e)}$ and $\mathbf{U}_\beta^{[BLO_l]}$ with orders higher than $J - 2$. The subsequent five terms represent the residual resulting from the multiplication of the boundary layer correctors by the cut-off function $1 - \zeta\left(\dfrac{|x - O_l|}{|e|_{min}}\right)$. Notably, these terms tend to vanish exponentially due to their exponential decay within the set $\mathrm{supp}\zeta\left(\dfrac{|x - O_l|}{|e|_{min}}\right)$ (where $\chi \neq 0$). Specifically, the order of this term in the L^2-norm diminishes as $O(e^{-c_1/\varepsilon})$, where c_1 is a positive constant. The next four terms are generated by the addition of the "divergence corrector" $\mathbf{w}^{(J)}$ to the expansion $\mathbf{u}^{(J)}$, and their support also belongs to $\mathrm{supp}\zeta\left(\frac{|x-O_l|}{|e|_{min}}\right)$. The last two terms come from the pressure gradient term: the first one remains uncompensated by the boundary layer-in-space problems, while the second one results from the multiplication of the boundary layer correctors by the cut-off function $\zeta\left(\dfrac{|x - O_l|}{|e|_{min}}\right)$ and so diminishes exponentially.

From the derived formulas, it becomes evident that

$$\|\mathbf{f}_1^{(J)}\|_{L^2(B_\varepsilon)} = \|\mathcal{L}(\tilde{\mathbf{u}}^{(J)}, p^{(J)})\|_{L^2(B_\varepsilon)} = O(\varepsilon^{J-2}). \tag{5.3.26}$$

Much like in the linear case, we establish the subsequent theorem.

Theorem 5.3.4 *For sufficiently small ε, the following error estimate holds*

$$\|\mathbf{v} - \widetilde{\mathbf{u}}^{(J)}\|_{W^{1,2}(B_\varepsilon)} = O(\varepsilon^{J+(n-1)/2}). \tag{5.3.27}$$

Regarding the normalized difference $q = p - p^{(J)}$, where $\langle q\rangle = \frac{1}{\mathrm{mes}B_\varepsilon}\int_{B_\varepsilon}(p - p^{(J)})dx = 0$, we have the subsequent error estimate

$$\|q\|_{L^2(B_\varepsilon)} = \|p - p^{(J)}\|_{L^2(B_\varepsilon)} = O(\varepsilon^{J+(n-3)/2}). \tag{5.3.28}$$

Proof

Notably, $\widetilde{\mathbf{u}}^{(J)} \in W^{2,2}(B_\varepsilon)$ complies with conditions (5.1.3). It is essential to observe that $\widetilde{\mathbf{u}}^{(J)}$ extends the boundary value $\mathbf{g}$. Consequently, integral identity (5.3.4) adopts the form

$$\nu\int\limits_{B_\varepsilon}\nabla\mathbf{u}\cdot\nabla\eta dx-\int\limits_{B_\varepsilon}((\mathbf{u}+\widetilde{\mathbf{u}}^{(J)})\cdot\nabla)\eta\cdot\mathbf{u}dx-\int\limits_{B_\varepsilon}(\mathbf{u}\cdot\nabla)\eta\cdot\widetilde{\mathbf{u}}^{(J)}dx$$
$$=-\int\limits_{B_\varepsilon}\Big(\nu\nabla\widetilde{\mathbf{u}}^{(J)}\cdot\nabla\eta+(\widetilde{\mathbf{u}}^{(J)}\cdot\nabla)\widetilde{\mathbf{u}}^{(J)}\cdot\eta\Big)dx\quad\forall\,\eta\in\mathring{H}^{1,2}(B_\varepsilon),\tag{5.3.29}$$

where $\mathbf{u}=\mathbf{v}-\widetilde{\mathbf{u}}^{(J)}$. By following the reasoning in Sect. 5.1.4, we convert the right-hand side of (5.3.29) to the form

$$-\int\limits_{B_\varepsilon}\Big(\nu\nabla\widetilde{\mathbf{u}}^{(J)}\cdot\nabla\eta+(\widetilde{\mathbf{u}}^{(J)}\cdot\nabla)\widetilde{\mathbf{u}}^{(J)}\cdot\eta\Big)dx$$

$$=-\int\limits_{B_\varepsilon}\Big(-\nu\Delta\widetilde{\mathbf{u}}^{(J)}+(\widetilde{\mathbf{u}}^{(J)}\cdot\nabla)\widetilde{\mathbf{u}}^{(J)}\Big)\cdot\eta\,dx+\int\limits_{B_\varepsilon}p^{(J)}\mathrm{div}\eta dx$$

$$=-\int\limits_{B_\varepsilon}\Big(-\nu\Delta\widetilde{\mathbf{u}}^{(J)}+(\widetilde{\mathbf{u}}^{(J)}\cdot\nabla)\widetilde{\mathbf{u}}^{(J)}+\nabla p^{(J)}\Big)\cdot\eta\,dx=\int\limits_{B_\varepsilon}\mathbf{f}_1^{(J)}\cdot\eta\,dx.$$

By setting $\eta=\mathbf{u}$ in (5.3.29) and utilizing the estimates

$$\Big|\int\limits_{B_\varepsilon}\mathbf{f}_1^{(J)}\cdot\eta\,dx\Big|\le c\varepsilon\|\mathbf{f}_1^{(J)}\|_{L^2(B_\varepsilon)}\|\nabla\mathbf{u}\|_{L^2(B_\varepsilon)}$$

and (5.3.26), we acquire

$$\nu\|\nabla\mathbf{u}\|^2_{L^2(B_\varepsilon)}\le c\varepsilon^\alpha\|\nabla\widetilde{\mathbf{u}}^{(J)}\|_{L^2(B_\varepsilon)}\|\nabla\mathbf{u}\|^2_{L^2(B_\varepsilon)}+c\varepsilon\|\mathbf{f}_1^{(J)}\|_{L^2(B_\varepsilon)}\|\nabla\mathbf{u}\|_{L^2(B_\varepsilon)}$$
$$\le O(\varepsilon^{1/2})\|\nabla\mathbf{u}\|^2_{L^2(B_\varepsilon)}+O(\varepsilon^{J-1})\|\nabla\mathbf{u}\|_{L^2(B_\varepsilon)}.$$

For sufficiently small ε, the last inequality gives

$$\|\mathbf{v}-\widetilde{\mathbf{u}}^{(J)}\|_{W^{1,2}(B_\varepsilon)}=\|\mathbf{u}\|_{W^{1,2}(B_\varepsilon)}=O(\varepsilon^{J-1}).\tag{5.3.30}$$

Evaluating the norm of the difference $\widetilde{\mathbf{u}}^{(J)}-\widetilde{\mathbf{u}}^{(J+2)}$ results in

$$\|\widetilde{\mathbf{u}}^{(J)}-\widetilde{\mathbf{u}}^{(J+2)}\|_{W^{1,2}(B_\varepsilon)}=O(\varepsilon^{J+(n-1)/2}).\tag{5.3.31}$$

By replacing J with $J+2$ in (5.3.30), we get

$$\|\mathbf{v}-\widetilde{\mathbf{u}}^{(J+2)}\|_{W^{1,2}(B_\varepsilon)}=O(\varepsilon^{J+1}).$$

Thus, applying the triangle inequality, we arrive at (5.3.27).

Let us compute the pressure that corresponds to the weak solution $\mathbf{v}$ of problem (5.3.1). As established in Sect. 5.3.1, there exists a pressure function $p \in \widehat{L}^2(B_\varepsilon)$ associated with the weak solution $\mathbf{v}$, such that the pair $(\mathbf{v}, p)$ satisfies for all $\boldsymbol{\eta} \in \overset{\circ}{W}{}^{1,2}(B_\varepsilon)$ the integral identity

$$\nu \int_{B_\varepsilon} \nabla \mathbf{v} \cdot \nabla \boldsymbol{\eta}\, dx + \int_{B_\varepsilon} (\mathbf{v} \cdot \nabla)\mathbf{v} \cdot \boldsymbol{\eta}\, dx = \int_{B_\varepsilon} p\mathrm{div}\boldsymbol{\eta}\, dx. \tag{5.3.32}$$

Consider the pair $(\widetilde{\mathbf{u}}^{(J)}, p^{(J)}) \in W^{1,2}(B_\varepsilon) \times L^2(B_\varepsilon)$. For all $\boldsymbol{\eta} \in \overset{\circ}{W}{}^{1,2}(B_\varepsilon)$, it satisfies the integral identity:

$$\nu \int_{B_\varepsilon} \nabla \widetilde{\mathbf{u}}^{(J)} \cdot \nabla \boldsymbol{\eta} dx + \int_{B_\varepsilon} (\widetilde{\mathbf{u}}^{(J)} \cdot \nabla)\widetilde{\mathbf{u}}^{(J)} \cdot \boldsymbol{\eta}\, dx - \int_{B_\varepsilon} p^{(J)}\mathrm{div}\boldsymbol{\eta} dx = -\int_{B_\varepsilon} \mathbf{f}_1^{(J)} \cdot \boldsymbol{\eta} dx. \tag{5.3.33}$$

Let $\mathbf{u} = \mathbf{v} - \widetilde{\mathbf{u}}^{(J)}$ and $q = p - p^{(J)}$. Then, $\mathbf{u} \in \overset{\circ}{W}{}^{1,2}(B_\varepsilon)$, $q \in L^2(B_\varepsilon)$, and the following integral identity holds

$$\begin{aligned} \nu \int_{B_\varepsilon} \nabla \mathbf{u} \cdot \nabla \boldsymbol{\eta} dx - \int_{B_\varepsilon} ((\mathbf{u} + \widetilde{\mathbf{u}}^{(J)}) \cdot \nabla)\boldsymbol{\eta} \cdot \mathbf{u} dx - \int_{B_\varepsilon} (\mathbf{u} \cdot \nabla)\boldsymbol{\eta} \cdot \widetilde{\mathbf{u}}^{(J)} dx \\ - \int_{B_\varepsilon} q\, \mathrm{div}\boldsymbol{\eta} dx = \int_{B_\varepsilon} \mathbf{f}_1^{(J)} \cdot \boldsymbol{\eta} dx \quad \forall \boldsymbol{\eta} \in \overset{\circ}{W}{}^{1,2}(B_\varepsilon). \end{aligned} \tag{5.3.34}$$

Given that $p^{(J)}$ is defined up to an additive constant, we can assume without loss of generality that $\int_{B_\varepsilon} p^{(J)}(x)dx = 0$, leading to $q \in \widehat{L}^2(B_\varepsilon)$. Let $\mathbf{w} \in \overset{\circ}{W}{}^{1,2}(B_\varepsilon)$ be such that $\mathrm{div}\mathbf{w} = q$ in B_ε, and satisfy

$$\|\nabla \mathbf{w}\|_{L^2(B_\varepsilon)} \le c\varepsilon^{-3/2}\|q\|_{L^2(B_\varepsilon)} \tag{5.3.35}$$

(see Lemma 2.4.14). By setting $\boldsymbol{\eta} = \mathbf{w}$ in (5.3.15), we obtain the inequality

$$\begin{aligned} \int_{B_\varepsilon} |q|^2\, dx &= \nu \int_{B_\varepsilon} \nabla \mathbf{u} \cdot \nabla \mathbf{w}\, dx - \int_{B_\varepsilon} (\mathbf{u} \cdot \nabla)\mathbf{w} \cdot \mathbf{u}\, dx \\ &\quad - \int_{B_\varepsilon} (\widetilde{\mathbf{u}}^{(J)} \cdot \nabla)\mathbf{w} \cdot \mathbf{u} dx - \int_{B_\varepsilon} (\mathbf{u} \cdot \nabla)\mathbf{w} \cdot \widetilde{\mathbf{u}}^{(J)} dx - \int_{B_\varepsilon} \mathbf{f}_1^{(J)} \cdot \mathbf{w} dx \\ &\le \Big(\nu \|\nabla \mathbf{u}\|_{L^2(B_\varepsilon)} + \|\mathbf{u}\|^2_{L^4(B_\varepsilon)} + \|\widetilde{\mathbf{u}}^{(J)}\|_{L^4(B_\varepsilon)}\|\mathbf{u}\|_{L^4(B_\varepsilon)} \\ &\quad + c\varepsilon\|\mathbf{f}_1^{(J)}\|_{L^2(B_\varepsilon)}\Big)\|\nabla \mathbf{w}\|_{L^2(B_\varepsilon)}. \end{aligned} \tag{5.3.36}$$

Thus, applying (5.3.35) and simplifying the factor $\|q\|_{L^2(B_\varepsilon)}$, we obtain

$$\|q\|_{L^2(B_\varepsilon)} \le c\Big(\varepsilon^{-3/2}\|\nabla \mathbf{u}\|_{L^2(B_\varepsilon)} + \varepsilon^{\alpha-3/2}\|\nabla \mathbf{u}\|^2_{L^2(B_\varepsilon)}$$

$$+ \varepsilon^{\alpha-3/2}\|\nabla \widetilde{\mathbf{u}}^{(J)}\|_{L^2(B_\varepsilon)}\|\nabla \mathbf{u}\|_{L^2(B_\varepsilon)} + \varepsilon^{-1/2}\|\mathbf{f}_1^{(J)}\|_{L^2(B_\varepsilon)}\Big).$$

Taking into account (5.3.26), (5.3.27), and $\|\nabla \widetilde{\mathbf{u}}^{(J)}\|_{L^2(B_\varepsilon)} \le c\varepsilon^{\frac{n-3}{2}}$, yields

$$\|q\|_{L^2(B_\varepsilon)} \le c\varepsilon^{J-5/2}.$$

The last inequality with J replaced by $J+3$ implies

$$\|p - p^{(J+3)}\|_{L^2(B_e)} \le c\varepsilon^{J+1/2}.$$

Evaluating the difference $p^{(J+3)} - p^{(J)}$, we find

$$\|p^{(J+3)} - p^{(J)}\|_{L^2(B_\varepsilon)} \le C\varepsilon^{J-1+(n-1)/2} = O(\varepsilon^{J+(n-3)/2}).$$

Finally, the last two estimates and the triangle inequality yield

$$\|q\|_{L^2(B_\varepsilon)} = \|p - p^{(J)}\| = O(\varepsilon^{J+(n-3)/2}).$$

□

5.3.3 MAPDD for the Navier–Stokes Equations

The refined version of the weak formulation the of MAPDD for the Navier–Stokes equations is presented in this subsection:

Definition 5.3.4 Find $\mathbf{v}_d \in \mathcal{H}_{\text{dec}}(B_\varepsilon, \delta)$, where $\mathbf{v}_d|_{\partial B_\varepsilon} = \mathbf{g}$, such that for all $\boldsymbol{\eta} \in \mathcal{H}_{0,\text{dec}}(B_\varepsilon, \delta)$, the following integral identity holds

$$\nu \int_{B_\varepsilon} \nabla \mathbf{v}_d \cdot \nabla \boldsymbol{\eta} dx + \int_{B_\varepsilon} (\mathbf{v}_d \cdot \nabla)\mathbf{v}_d \cdot \nabla \boldsymbol{\eta} dx = 0. \tag{5.3.37}$$

The function spaces are defined as in Sect. 5.2.

To enhance the numerical implementation, we consider the following equivalent formulation with pressure:

Definition 5.3.5 Find the vector field $\mathbf{v}_d \in \mathcal{H}_{\text{dec}}(B_\varepsilon, \delta)$ and the "MAPDD pressure" p_d defined in $L^2(B_i^{\varepsilon,\delta})$ for all $i = 1, \ldots, N$, satisfying $\mathbf{v}_d = \mathbf{g}$ on B_ε, and the following integral identity:

$$
\begin{aligned}
&\nu\sum_{i=1}^{N}\int\limits_{B_i^{\varepsilon,\delta}}\nabla\mathbf{v}_d\cdot\nabla\eta dx+\sum_{i=1}^{N}\int\limits_{B_i^{\varepsilon,\delta}}(\mathbf{v}_d\cdot\nabla)\mathbf{v}_d\cdot\eta dx\\
&+\nu\sum_{l=1}^{M}d_l\int\limits_{\sigma_\varepsilon^{(e_l)}}\nabla_{x^{(e_l)\prime}}\mathbf{v}_d\cdot\nabla_{x^{(e_l)\prime}}\eta dx^{(e_l)\prime}=\sum_{i=1}^{N}\int\limits_{B_i^{\varepsilon,\delta}}p_d\operatorname{div}\eta dx
\end{aligned}
\tag{5.3.38}
$$

for all $\eta\in\mathcal{W}_{0,\mathrm{dec}}(B_\varepsilon,\delta)$. Here, $d_l=|e|-2\delta$ represents the distance between the cross-sections S_{ij} and S_{ji} intersecting the segment e.

Theorem 5.3.5 *For δ chosen as $C_{J+2n+1}\varepsilon|\ln\varepsilon|$, where C_{J+2n+1} satisfies* (5.1.60) *with J replaced by $J+2n+1$, and sufficiently small ε, there exists a unique solution $(\mathbf{v}_d, p_d)$ of the problem* (5.3.1). *This solution satisfies the following error estimate:*

$$
\|\mathbf{v}-\mathbf{v}_d\|_{W^{1,2}(B_\varepsilon)}=O(\varepsilon^{J+(n-1)/2}).
\tag{5.3.39}
$$

Furthermore, there exists a pressure p, which is a solution of (5.3.1) *(p is defined up to an additive constant), such that the subsequent error estimate holds*

$$
\|p-p_d\|_{L^2(B_\varepsilon)}=O(\varepsilon^{J+(n-3)/2}).
\tag{5.3.40}
$$

Proof
We first prove the existence of $\mathbf{v}_d\in\mathcal{H}_{\mathrm{dec}}(B_\varepsilon,\delta)$ that satisfies identity (5.3.37). The proof is similar to the proof of Theorem 5.3.1. So, we only emphasize the pivotal aspects of the argument, focusing on crucial steps.

We seek the solution $\mathbf{v}_d$ in the form $\mathbf{v}_d=\widetilde{\mathbf{u}}_d+\widetilde{\mathbf{u}}^{(J)}$, where $\widetilde{\mathbf{u}}_d\in\mathcal{H}_{0,\mathrm{dec}}(B_\varepsilon,\delta)$. As was the case in the proof of Theorem 5.3.4, we reformulate the integral identity (5.3.37) as follows:

$$
\begin{aligned}
&\nu\sum_{i=1}^{N}\int\limits_{B_i^{\varepsilon,\delta}}\nabla\widetilde{\mathbf{u}}_d\cdot\nabla\eta dx+\nu\sum_{l=1}^{M}d_l\int\limits_{\sigma_\varepsilon^{(e_l)}}\nabla_{x^{(e_l)\prime}}\widetilde{\mathbf{u}}_d\cdot\nabla_{x^{(e_l)\prime}}\eta dx^{(e_l)\prime}\\
&+\sum_{i=1}^{N}\int\limits_{B_i^{\varepsilon,\delta}}((\widetilde{\mathbf{u}}_d+\widetilde{\mathbf{u}}^{(J)})\cdot\nabla)\widetilde{\mathbf{u}}_d\cdot\eta dx+\sum_{i=1}^{N}\int\limits_{B_i^{\varepsilon,\delta}}(\widetilde{\mathbf{u}}_d\cdot\nabla)\widetilde{\mathbf{u}}^{(J)}\cdot\eta dx\\
&=\int\limits_{B_\varepsilon}\mathbf{f}_1^{(J)}\cdot\eta dx\quad\forall\eta\in\mathcal{H}_{0,\mathrm{dec}}(B_\varepsilon,\delta).
\end{aligned}
\tag{5.3.41}
$$

Here, $\mathbf{f}_1^{(J)}=-\big(-\nu\Delta\widetilde{\mathbf{u}}^{(J)}+(\widetilde{\mathbf{u}}^{(J)}\cdot\nabla)\widetilde{\mathbf{u}}^{(J)}+\nabla p^{(J)}\big)$.

It is evident that the first two terms on the left-hand side define an inner product within the Hilbert space $\mathcal{H}_{0,\mathrm{dec}}(B_\varepsilon,\delta)$, while the terms

$$\sum_{i=1}^{N}\int\limits_{B_i^{\varepsilon,\delta}}((\widetilde{\mathbf{u}}_d+\widetilde{\mathbf{u}}^{(J)})\cdot\nabla)\widetilde{\mathbf{u}}_d\cdot\eta dx+\sum_{i=1}^{N}\int\limits_{B_i^{\varepsilon,\delta}}(\widetilde{\mathbf{u}}_d\cdot\nabla)\widetilde{\mathbf{u}}^{(J)}\cdot\eta dx-\int\limits_{B_\varepsilon}\mathbf{f}_1^{(J)}\cdot\eta dx$$

contribute to a bounded linear functional over $\mathcal{H}_{0,\text{dec}}(B_\varepsilon,\delta)$. Similar to Theorem 5.3.1, this identity can be recast as an operator equation in the space $\mathcal{H}_{0,\text{dec}}(B_\varepsilon,\delta)$:

$$\widetilde{\mathbf{u}}_d=\mathcal{A}\widetilde{\mathbf{u}}_d, \tag{5.3.42}$$

where $\mathcal{A}$ is a compact operator. Thus, the Leray–Schauder theorem assures the existence of a fixed point for (5.3.42), provided an a priori estimate of the solution is demonstrated.

Choosing $\eta=\widetilde{\mathbf{u}}_d$ in (5.3.41), and utilizing inequalities (5.3.26) and $\|\nabla\widetilde{\mathbf{u}}^{(J)}\|_{L^2(B_\varepsilon)}\le c\varepsilon^{\frac{n-3}{2}}$, yields

$$\begin{aligned}\nu\|\nabla\widetilde{\mathbf{u}}_d\|^2_{L^2(B_\varepsilon)}&=-\sum_{i=1}^{N}\int\limits_{B_i^{\varepsilon,\delta}}(\widetilde{\mathbf{u}}_d\cdot\nabla)\widetilde{\mathbf{u}}^{(J)}\cdot\widetilde{\mathbf{u}}_d dx+\int\limits_{B_\varepsilon}\mathbf{f}_1^{(J)}\cdot\widetilde{\mathbf{u}}_d dx\\&\le\|\widetilde{\mathbf{u}}_d\|^2_{L^4(B_\varepsilon)}\|\nabla\widetilde{\mathbf{u}}^{(J)}\|_{L^2(B_\varepsilon)}+\|\mathbf{f}_1^{(J)}\|_{L^2(B_\varepsilon)}\|\widetilde{\mathbf{u}}_d\|_{L^2(B_\varepsilon)}\\&\le c\varepsilon^{\alpha}\|\nabla\widetilde{\mathbf{u}}^{(J)}\|_{L^2(B_\varepsilon)}\|\nabla\widetilde{\mathbf{u}}_d\|^2_{L^2(B_\varepsilon)}+c\varepsilon\|\mathbf{f}_1^{(J)}\|_{L^2(B_\varepsilon)}\|\nabla\widetilde{\mathbf{u}}_d\|_{L^2(B_\varepsilon)}\\&\le c_*\varepsilon^{1/2}\|\nabla\widetilde{\mathbf{u}}_d\|^2_{L^2(B_\varepsilon)}+c\varepsilon^{J-1}\|\nabla\widetilde{\mathbf{u}}_d\|_{L^2(B_\varepsilon)}.\end{aligned} \tag{5.3.43}$$

For $\nu>c_*\varepsilon^{1/2}$, the last inequality leads to

$$\|\mathbf{v}_d-\widetilde{\mathbf{u}}^{(J)}\|_{\mathcal{H}_{0,\text{dec}}(B_\varepsilon,\delta)}=\|\widetilde{\mathbf{u}}_d\|_{\mathcal{H}_{0,\text{dec}}(B_\varepsilon,\delta)}=\|\nabla\widetilde{\mathbf{u}}_d\|_{L^2(B_\varepsilon)}\le c\varepsilon^{J-1}. \tag{5.3.44}$$

Consequently, the problem (5.3.41) possesses a solution $\widetilde{\mathbf{u}}_d\in\mathcal{H}_{0,\text{dec}}(B_\varepsilon,\delta)$, and this solution satisfies estimate (5.3.44).

We now adopt the approximations $(\widetilde{\mathbf{u}}^{(J+2n+1)},p^{(J+2n+1)})$ instead of $(\widetilde{\mathbf{u}}^{(J)},p^{(J)})$. Accordingly, $\|\mathbf{f}_1^{(J')}\|_{L^2(B_\varepsilon)}=O(\varepsilon^{J'-2})=O(\varepsilon^{J+2n-1})$, with $J'=J+2n+1$,

$$\|\mathbf{v}_d-\widetilde{\mathbf{u}}^{(J')}\|_{\mathcal{H}_{0,\text{dec}}(B_\varepsilon,\delta)}\le c\varepsilon^{J+2n}. \tag{5.3.45}$$

From (5.3.27) and the triangle inequality, we deduce that (5.3.39) holds.

We now consider the functional

$$f(\boldsymbol{\eta}) = \nu \int\limits_{B_\varepsilon} \nabla \mathbf{v}_d \cdot \nabla \boldsymbol{\eta} dx + \int\limits_{B_\varepsilon} (\mathbf{v}_d \cdot \nabla)\mathbf{v}_d \cdot \boldsymbol{\eta} dx,$$

defined on $\boldsymbol{\eta} \in \mathcal{W}_{0,\mathrm{dec}}(B_\varepsilon, \delta)$. Evidently, $f(\boldsymbol{\eta})$ is a linear bounded functional, and $f(\boldsymbol{\eta}) = 0$ on $\mathcal{H}_{0,\mathrm{dec}}(B_\varepsilon, \delta)$. As per Theorem 5.2.1, this functional can be expressed as

$$f(\boldsymbol{\eta}) = \int\limits_{B_\varepsilon} p_d \mathrm{div} \boldsymbol{\eta} dx,$$

where $p_d \in \widehat{L}^2_{dec}(B_\varepsilon)$ (recalling that p_d is zero on all cylinders B^ε_{ij}). This leads the pair $(\mathbf{v}_d,\ p_d)$ to satisfy integral identity (5.3.38). By taking $p_d = \widetilde{q}_d + p^{(J')}$ in (5.3.38), we rephrase it as

$$\begin{aligned}
&\nu \sum_{i=1}^{N} \int\limits_{B_i^{\varepsilon,\delta}} \nabla \widetilde{\mathbf{u}}_d \cdot \nabla \boldsymbol{\eta} dx + \nu \sum_{l=1}^{M} d_l \int\limits_{\sigma_\varepsilon^{(e_l)}} \nabla_{x^{(e_l)\prime}} \widetilde{\mathbf{u}}_d \cdot \nabla_{x^{(e_l)\prime}} \boldsymbol{\eta} dx^{(e_l)\prime} \\
&+ \sum_{i=1}^{N} \int\limits_{B_i^{\varepsilon,\delta}} ((\widetilde{\mathbf{u}}_d + \widetilde{\mathbf{u}}^{(J')}) \cdot \nabla)\widetilde{\mathbf{u}}_d \cdot \boldsymbol{\eta} dx + \sum_{i=1}^{N} \int\limits_{B_i^{\varepsilon,\delta}} (\widetilde{\mathbf{u}}_d \cdot \nabla)\widetilde{\mathbf{u}}^{(J')} \cdot \boldsymbol{\eta} dx \\
&= \sum_{i=1}^{N} \int\limits_{B_i^{\varepsilon,\delta}} \widetilde{q}_d \mathrm{div} \boldsymbol{\eta} dx + \int\limits_{B_\varepsilon} \mathbf{f}_1^{(J')} \cdot \boldsymbol{\eta} dx \quad \forall \boldsymbol{\eta} \in \mathcal{W}_{0,\mathrm{dec}}(B_\varepsilon, \delta).
\end{aligned} \tag{5.3.46}$$

Select the test function $\boldsymbol{\eta} = \mathbf{U} \in \mathcal{W}_{0,\mathrm{dec}}(B_\varepsilon, \delta)$, where

$$\mathrm{div}\mathbf{U}(x) = \begin{cases} \widetilde{q}_d(x),\ x \in \bigcup\limits_{i=1}^{N} B_i^{\varepsilon,\delta}, \\ 0,\ x \in B_\varepsilon \setminus \bigcup\limits_{i=1}^{N} B_i^{\varepsilon,\delta}, \end{cases} \tag{5.3.47}$$

and

$$\|\mathbf{U}\|^2_{W^{1,2}(B_\varepsilon)} \le C\varepsilon^{-2n-3/2} \sum_{i=1}^{N} \|\widetilde{q}_d\|^2_{L^2(B_i^{\varepsilon,\delta})} \tag{5.3.48}$$

with some constant C independent of ε and δ (see Lemma 5.2.2). This gives

$$
\begin{aligned}
\sum_{i=1}^{N}\int\limits_{B_i^{\varepsilon,\delta}} |\widetilde{q}_d|^2 dx &= -\nu\sum_{i=1}^{N}\int\limits_{B_i^{\varepsilon,\delta}} \nabla\widetilde{\mathbf{u}}_d\cdot\nabla\mathbf{U}dx \\
-\nu\sum_{l=1}^{M} d_l \int\limits_{\sigma_\varepsilon^{(e_l)}} \nabla_{x^{(e_l)\prime}}\widetilde{\mathbf{u}}_d\cdot\nabla_{x^{(e_l)\prime}}\mathbf{U}dx^{(e_l)\prime} &- \sum_{i=1}^{N}\int\limits_{B_i^{\varepsilon,\delta}} ((\widetilde{\mathbf{u}}_d+\widetilde{\mathbf{u}}^{(J')})\cdot\nabla)\widetilde{\mathbf{u}}_d\cdot\mathbf{U}dx \\
-\sum_{i=1}^{N}\int\limits_{B_i^{\varepsilon,\delta}} (\widetilde{\mathbf{u}}_d\cdot\nabla)\widetilde{\mathbf{u}}^{(J')}\cdot\mathbf{U}dx &- \int\limits_{B_\varepsilon}\mathbf{f}_1^{(J')}\cdot\mathbf{U}dx \\
\leq c\Big(\big(\|\nabla\widetilde{\mathbf{u}}_d\|_{L^2(B_\varepsilon)}+\varepsilon\|\mathbf{f}_1^{(J')}\|\big)\|\nabla\mathbf{U}\|_{L^2(B_\varepsilon)} &+ \big(\|\widetilde{\mathbf{u}}_d\|_{L^4(B_\varepsilon)}\|\nabla\widetilde{\mathbf{u}}^{(J')}\|_{L^2(B_\varepsilon)} \\
+\|\widetilde{\mathbf{u}}_d\|_{L^4(B_\varepsilon)}\|\nabla\widetilde{\mathbf{u}}_d\|_{L^2(B_\varepsilon)} &+ \|\widetilde{\mathbf{u}}^{(J')}\|_{L^4(B_\varepsilon)}\|\nabla\widetilde{\mathbf{u}}_d\|_{L^2(B_\varepsilon)}\big)\|\mathbf{U}\|_{L^4(B_\varepsilon)}\Big) \\
\leq c\Big(\varepsilon^{J'-1}+\varepsilon^{\alpha}\big(\|\nabla\widetilde{\mathbf{u}}_d\|^2_{L^2(B_\varepsilon)} &+ \|\nabla\widetilde{\mathbf{u}}^{(J')}\|_{L^2(B_\varepsilon)}\|\nabla\widetilde{\mathbf{u}}_d\|_{L^2(B_\varepsilon)}\big)\Big)\|\nabla\mathbf{U}\|_{L^2(B_\varepsilon)} \\
\leq c\big(\varepsilon^{J'-1}+\varepsilon^{\alpha}\big(\varepsilon^{2J'-2}+\varepsilon^{J'-1+(n-3)/2}\big)\big)\varepsilon^{-n-3/4}&\Big(\sum_{i=1}^{N}\|\widetilde{q}_d\|^2_{L^2(B_i^{\varepsilon,\delta})}\Big)^{1/2} \\
\leq c\varepsilon^{J'-n-7/4}&\Big(\sum_{i=1}^{N}\|\widetilde{q}_d\|^2_{L^2(B_i^{\varepsilon,\delta})}\Big)^{1/2}.
\end{aligned}
\tag{5.3.49}
$$

Recalling that $J' = J + 2n + 1$, from the last inequality we receive the estimate

$$
\Big(\sum_{i=1}^{N}\|p_d - p^{(J')}\|^2_{L^2(B_i^{\varepsilon,\delta})}\Big)^{1/2} \leq O(\varepsilon^{J+n-3/4}). \tag{5.3.50}
$$

Now (5.3.50), estimate (5.3.28) for $p - p^{(J')}$ and triangle inequality yield the estimate

$$
\Big(\sum_{i=1}^{N}\|p - p_d\|^2_{L^2(B_i^{\varepsilon,\delta})}\Big)^{1/2} \leq O(\varepsilon^{J+n-3/2}). \tag{5.3.51}
$$

Extending p_d as in the linear case (see the proof of Theorem 5.2.4), we receive for the extended MAPDD pressure p_d the required estimate (5.3.40). □

5.4 Stokes Equations with Given Pressure Inflow-Outflow Boundary Conditions

In this section, we focus on the study of the Stokes equations with pressure inflow-outflow boundary conditions in a tube structure B_ε. The mathematical setup for this investigation is delineated below.

Let us consider a thin tube structure B_ε, which was introduced in Remark 1.4.2. Within this structure, the boundaries γ_ε^j constitute the bases of the cylinders that encompass the vertices O_j. These boundaries are orthogonal to the lateral boundaries of the corresponding cylinders.

We define $\Gamma = \partial B_\varepsilon \setminus \bigcup_{j=N_1+1}^{N} \gamma_\varepsilon^j$ as the lateral surface of the domain B_ε. Within this tube structure, we introduce the function spaces $\widehat{W}^{1,2}(B_\varepsilon)$ and $\widehat{K}^{1,2}(B_\varepsilon)$ as follows:

$$\widehat{W}^{1,2}(B_\varepsilon) = \Big\{\eta \in W^{1,2}(B_\varepsilon) : \eta|_\Gamma = 0,\ \eta_\tau|_{\gamma_\varepsilon^j} = 0,\ j = N_1+1,\dots,N\Big\},$$
$$\widehat{K}^{1,2}(B_\varepsilon) = \Big\{\eta \in \widehat{W}^{1,2}(B_\varepsilon) : \mathrm{div}\eta = 0\Big\}, \tag{5.4.1}$$

where $\eta_\tau = \eta - (\eta\cdot\mathbf{n})\mathbf{n}$ represents the tangential component of the vector η and $\mathbf{n}$ signifies the unit normal vector to γ_ε^j.

In the following subsection, we explain the problem formulation and delve into the topics of existence, uniqueness, and stability of its solution.

5.4.1 *Problem Formulation: Existence, Uniqueness, and Stability of a Solution*

We now focus on the following boundary value problem for the stationary Stokes equations in the tube structure B_ε:

$$\begin{cases} -\nu\Delta\mathbf{v} + \nabla p = \mathbf{f}, & x \in B_\varepsilon, \\ \mathrm{div}\mathbf{v} = 0, & x \in B_\varepsilon, \\ \mathbf{v} = 0, & x \in \partial B_\varepsilon \setminus \bigcup_{j=N_1+1}^{N} \gamma_\varepsilon^j, \\ \mathbf{v}_\tau = 0, & x \in \gamma_\varepsilon^j, \\ -\nu\partial_n\mathbf{v}\cdot\mathbf{n} + p = c^j/\varepsilon^2, & x \in \gamma_\varepsilon^j,\ j = N_1+1,\dots,N, \end{cases} \tag{5.4.2}$$

Here, $\mathbf{n}$ is the unit normal vector to γ_ε^j, $\mathbf{v}_\tau = \mathbf{v} - (\mathbf{v}\cdot\mathbf{n})\mathbf{n}$ is the tangential component of the vector $\mathbf{v}$, $\partial_n g = \nabla g \cdot \mathbf{n}$ is the normal derivative of g, c^j are constants, and $\mathbf{f}$ is the given source term.

To analyze this problem, we first rewrite it in an equivalent form to facilitate the investigation. From the boundary condition $\mathbf{v}_\tau|_{\gamma_\varepsilon^j} = 0$ and the divergence equation $\operatorname{div}\mathbf{v} = 0$, it follows that $-\nu\partial_n\mathbf{v}\cdot\mathbf{n}|_{\gamma_\varepsilon^j} = 0$. Thus, we transform the equations into

$$\begin{cases} -\nu\Delta\mathbf{v} + \nabla p = \mathbf{f}, & x \in B_\varepsilon, \\ \operatorname{div}\mathbf{v} = 0, & x \in B_\varepsilon, \\ \mathbf{v} = 0, & x \in \partial B_\varepsilon \setminus \bigcup\limits_{j=N_1+1}^{N} \gamma_\varepsilon^j, \\ \mathbf{v}_\tau = 0, & x \in \gamma_\varepsilon^j, \\ p = p_j, \; x \in \gamma_\varepsilon^j, & j = N_1+1, \ldots, N, \end{cases} \tag{5.4.3}$$

where p_j corresponds to the constants c^j/ε^2.

To proceed with the analysis, we define a weak solution of problem (5.4.3).

Definition 5.4.1 The weak solution of problem (5.4.3) is a vector field $\mathbf{v}$ belonging to the function space $\widehat{K}^{1,2}(B_\varepsilon)$ and satisfying the integral identity:

$$\nu \int\limits_{B_\varepsilon} \nabla\mathbf{v}\cdot\nabla\boldsymbol{\eta}\,dx = -\sum_{j=N_1+1}^{N} p_j \int\limits_{\gamma_\varepsilon^j} \boldsymbol{\eta}\cdot\mathbf{n}\,dS + \int\limits_{B_\varepsilon} \mathbf{f}\cdot\boldsymbol{\eta}\,dx \tag{5.4.4}$$

for every $\boldsymbol{\eta} \in \widehat{K}^{1,2}(B_\varepsilon)$.

The main goal is to establish the existence, uniqueness, and stability of this solution.

Introduce $p_j^* = p_j - p_N$, $j = N_1, \ldots, N$, and consider an alternative weak formulation:

Definition 5.4.2 Find a vector field $\mathbf{v} \in \widehat{K}^{1,2}(B_\varepsilon)$ satisfying the integral identity:

$$\nu \int\limits_{B_\varepsilon} \nabla\mathbf{v}\cdot\nabla\boldsymbol{\eta}\,dx = -\sum_{j=N_1+1}^{N-1} p_j^* \int\limits_{\gamma_\varepsilon^j} \boldsymbol{\eta}\cdot\mathbf{n}\,dS + \int\limits_{B_\varepsilon} \mathbf{f}\cdot\boldsymbol{\eta}\,dx \tag{5.4.5}$$

for every $\boldsymbol{\eta} \in \widehat{K}^{1,2}(B_\varepsilon)$.

The equivalence between these formulations is established through the following relationship:

$$\sum_{j=N_1+1}^{N-1} p_j^* \int\limits_{\gamma_\varepsilon^j} \boldsymbol{\eta}\cdot\mathbf{n}\,dS = \sum_{j=N_1+1}^{N} p_j \int\limits_{\gamma_\varepsilon^j} \boldsymbol{\eta}\cdot\mathbf{n}\,dxS,$$

which can be derived as a consequence of the following relation:

$$\sum_{j=N_1+1}^{N} \int_{\gamma_\varepsilon^j} \boldsymbol{\eta} \cdot \mathbf{n} \, dS = 0,$$

valid for a solenoidal vector-valued function $\boldsymbol{\eta}$.

Below we explain this weak formulation heuristically; the rigorous analysis of the equivalence of the weak formulation and the classical one requires to study the regularity of the weak solution; see [59] for the methods.

The derivation of identity (5.4.4) follows from the multiplication of Eqs. (5.4.3) by $\boldsymbol{\eta} \in \widehat{K}^{1,2}(B_\varepsilon)$ and subsequent integration by parts over B_ε. Conversely, for a sufficiently smooth solution $\mathbf{u}$ that satisfies (5.4.4), a pressure field p can be identified such that the pair $(\mathbf{v}, p)$ almost everywhere satisfies equations $(5.4.3)_{1,2}$ within B_ε. The boundary conditions $(5.4.3)_{3,4,5}$ are adhered to in terms of traces (see the definition of the space $\widehat{K}^{1,2}(B_\varepsilon)$). Although function p is inherently defined up to an additive constant, this constant can be chosen to ensure satisfaction of $(5.4.3)_5$. Indeed, select a smooth solenoidal function $\boldsymbol{\eta}$ in (5.4.4) that meets the boundary conditions $\boldsymbol{\eta}|_\Gamma = 0$, $\boldsymbol{\eta}_{\boldsymbol{\tau}}|_{\gamma_\varepsilon^j} = 0$ for $j = N_1 + 1, \ldots, N$. Integrating by parts yield

$$\begin{aligned} \int_{B_\varepsilon} (-\nu \Delta \mathbf{v} - \mathbf{f}) \cdot \boldsymbol{\eta} dx &= -\nu \sum_{j=N_1+1}^{N} \int_{\gamma_\varepsilon^j} \partial_n \mathbf{v} \cdot \boldsymbol{\eta} \, dS - \sum_{j=N_1+1}^{N} p_j \int_{\gamma_\varepsilon^j} \boldsymbol{\eta} \cdot \mathbf{n} \, dS \\ &= - \sum_{j=N_1+1}^{N} p_j \int_{\gamma_\varepsilon^j} \boldsymbol{\eta} \cdot \mathbf{n} \, dS. \end{aligned} \tag{5.4.6}$$

Recall that $J_0^\infty(B_\varepsilon) = \left\{ C_0^\infty(B_\varepsilon) : \operatorname{div} \boldsymbol{\eta} = 0 \right\}$. It becomes evident from the previous equation that

$$\int_{B_\varepsilon} (-\nu \Delta \mathbf{v} - \mathbf{f}) \cdot \boldsymbol{\eta} \, dx = 0 \ \ \forall \boldsymbol{\eta} \in J_0^\infty(B_\varepsilon).$$

This leads to the conclusion that a pressure function p can be identified such that

$$-\nu \Delta \mathbf{v} + \nabla p = \mathbf{f} \ \text{ a.e. in } B_\varepsilon.$$

From here, by considering the equation,

$$\int\limits_{B_\varepsilon} (-\nu\Delta \mathbf{v} - \mathbf{f}) \cdot \boldsymbol{\eta}\, dx = -\int\limits_{B_\varepsilon} \nabla p \cdot \boldsymbol{\eta}\, dx = -\sum_{j=N_1+1}^{N} \int\limits_{\gamma_\varepsilon^j} p\,\boldsymbol{\eta} \cdot \mathbf{n}\, dS,$$

for every solenoidal $\boldsymbol{\eta} \in \widehat{K}^{1,2}(B_\varepsilon)$. Then, from (5.4.6), the relation

$$\sum_{j=N_1+1}^{N} \int\limits_{\gamma_\varepsilon^j} p\,\boldsymbol{\eta} \cdot \mathbf{n}\, dS = \sum_{j=N_1+1}^{N} p_j \int\limits_{\gamma_\varepsilon^j} \boldsymbol{\eta} \cdot \mathbf{n}\, dS,$$

can be deduced. Henceforth,

$$\sum_{j=N_1+1}^{N} \int\limits_{\gamma_\varepsilon^j} (p - p_j)\,\boldsymbol{\eta} \cdot \mathbf{n}\, dx^{(e_j)'} = 0 \text{ for all } \boldsymbol{\eta} \in \widehat{K}^{1,2}(B_\varepsilon). \tag{5.4.7}$$

Let us consider an arbitrary fixed index j from the set $\{N_1+1, \dots, N\}$. By selecting $\boldsymbol{\eta} \in \widehat{K}^{1,2}(B_\varepsilon)$ such that $\boldsymbol{\eta}|_{\gamma_\varepsilon^k} = 0$ for $k \neq j$, we obtain

$$\int\limits_{\gamma_\varepsilon^j} (p - p_j)\,\boldsymbol{\eta} \cdot \mathbf{n}\, dx^{(e_j)'} = 0. \tag{5.4.8}$$

The normal traces $(\boldsymbol{\eta} \cdot \mathbf{n})$ of functions $\boldsymbol{\eta}$ belonging to $\widehat{K}^{1,2}(B_\varepsilon)$ and satisfying the conditions $\boldsymbol{\eta}|_{\gamma_\varepsilon^k} = 0$ for $k \neq j$ form the entire space $W^{1/2,2}(\gamma_\varepsilon^j) \cap \widehat{L}^2(\gamma_\varepsilon^j)$. To prove this, consider the following boundary value problem:

$$\begin{cases} \operatorname{div}\mathbf{w} = 0, \ x \in B_\varepsilon, \\ \mathbf{w} = 0, \ x \in \partial B_\varepsilon \setminus \Big(\bigcup\limits_{k=N_1+1,\dots,N, k\neq j} \gamma_\varepsilon^k \Big), \\ \mathbf{w} \cdot \boldsymbol{\tau} = 0, \ x \in \gamma_\varepsilon^j, \\ \mathbf{w} \cdot \mathbf{n} = h, \ x \in \gamma_\varepsilon^j. \end{cases}$$

By virtue of Lemma 2.4.3, this problem has a solution $\mathbf{w} \in W^{1,2}(B_\varepsilon)$ for every $h \in W^{1/2,2}(\gamma_\varepsilon^j) \cap \widehat{L}^2(\gamma_\varepsilon^j)$. Moreover,

$$\mathbf{w}|_{\gamma_\varepsilon^k} = 0 \ \text{ for } \ k \neq j, \ \text{ for } \ k = N_1 + 1, \dots, N, \ \text{ and} \int\limits_{\gamma_\varepsilon^j} \mathbf{w} \cdot \mathbf{n}\, dx^{(e_l)'} = 0.$$

Consequently, $\mathbf{w} \in \widehat{K}^{1,2}(B_\varepsilon)$. On the other hand, any $\boldsymbol{\eta} \in \widehat{K}^{1,2}(B_\varepsilon)$ satisfying $\boldsymbol{\eta}|_{\gamma_\varepsilon^k} = 0$ for $k \neq j$ has its normal trace $\boldsymbol{\eta} \cdot \mathbf{n}|_{\gamma_\varepsilon^j}$ lying in $W^{1/2,2}(\gamma_\varepsilon^j) \cap \widehat{L}^2(\gamma_\varepsilon^j)$.

Hence, condition (5.4.8) holds for any $h \in W^{1/2,2}(\gamma_\varepsilon^j) \cap \widehat{L}^2(\gamma_\varepsilon^j)$:

$$\int\limits_{\gamma_\varepsilon^j} (p - p_j)\, h\, dx^{(e_j)\prime} = 0.$$

Since $C_0^\infty(\gamma_\varepsilon^j) \cap \widehat{L}^2(\gamma_\varepsilon^j)$ is a subspace of $W^{1/2,2}(\gamma_\varepsilon^j) \cap \widehat{L}^2(\gamma_\varepsilon^j)$ and is dense in $\widehat{L}^2(\gamma_\varepsilon^j)$, the last equality implies that

$$(p - p_j)|_{\gamma_\varepsilon^j} = c_j,$$

where c_j is a constant.

Utilizing these relationships, and by now selecting a test function $\boldsymbol{\eta} \in \widehat{K}^{1,2}(B_\varepsilon)$ satisfying $\int_{\gamma_\varepsilon^k} \boldsymbol{\eta} \cdot \mathbf{n}\, dx' = 0$ for $k \neq j$ and $k \neq N$, $\int_{\gamma_\varepsilon^j} \boldsymbol{\eta} \cdot \mathbf{n}\, dx' = 1$, and $\int_{\gamma_\varepsilon^N} \boldsymbol{\eta} \cdot \mathbf{n}\, dx' = -1$, we derive

$$\sum_{j=N_1+1}^{N} c_j \int\limits_{\gamma_\varepsilon^j} \boldsymbol{\eta} \cdot \mathbf{n}\, dx' = c_j - c_N \Rightarrow c_j = c_N.$$

Hence,

$$c_j = c_N \text{ for all } j = N_1 + 1, \ldots, N. \tag{5.4.9}$$

As the pressure p in the weak formulation is defined up to an additive constant, we can set $c_j = c_N = 0$ for $j = N_1 + 1, \ldots, N$. Then, according to (5.4.9), we obtain

$$p|_{\gamma_\varepsilon^j} = p_j \text{ for } j = N_1 + 1, \ldots, N.$$

These considerations serve to justify the definition of the weak solution.

Theorem 5.4.1 *For any given $\mathbf{f} \in L^2(B_\varepsilon)$ and $p_j^* \in \mathbb{R}$, where $j = N_1+1, \ldots, N-1$, the problem (5.4.3) has a unique weak solution $\mathbf{v} \in \widehat{K}^{1,2}(B_\varepsilon)$. Additionally, the following estimate holds:*

$$\|\nabla \mathbf{v}\|_{L^2(B_\varepsilon)} \leq c\Big(\varepsilon^{n/2} \sum_{j=N_1+1}^{N-1} |p_j^*| + \varepsilon \|\mathbf{f}\|_{L^2(B_\varepsilon)}\Big), \tag{5.4.10}$$

where the constant c is independent of ε.

Proof
Consider the inner product $[\mathbf{v}, \boldsymbol{\eta}] = \int_{B_\varepsilon} \nabla \mathbf{v} \cdot \nabla \boldsymbol{\eta}\, dx$ in the space $\widehat{K}^{1,2}(B_\varepsilon)$, corresponding to the Dirichlet norm. Utilizing the Poincaré inequality and the Lemma 2.3.16, we derive two inequalities:

$$\Big|\sum_{j=N_1+1}^{N-1} p_j^* \int\limits_{\gamma_\varepsilon^j} \boldsymbol{\eta}\cdot\mathbf{n}\,dx'\Big| \le \sum_{j=N_1+1}^{N-1} |p_j^*|\Big(\int\limits_{\gamma_\varepsilon^j} |\boldsymbol{\eta}|^2\,dx\Big)^{1/2}|\gamma_\varepsilon^j|^{1/2}$$
$$\le c\varepsilon^{n/2}\sum_{j=N_1+1}^{N-1}|p_j^*|\|\nabla\boldsymbol{\eta}\|_{L^2(B_\varepsilon)}, \tag{5.4.11}$$

$$\Big|\int\limits_{B_\varepsilon}\mathbf{f}\cdot\boldsymbol{\eta}\,dx\Big| \le \Big(\int\limits_{B_\varepsilon}|\mathbf{f}|^2\,dx\Big)^{1/2}\Big(\int\limits_{B_\varepsilon}|\boldsymbol{\eta}|^2\,dx\Big)^{1/2} \le c\varepsilon\|\mathbf{f}\|_{L^2(B_\varepsilon)}\|\nabla\boldsymbol{\eta}\|_{L^2(B_\varepsilon)}. \tag{5.4.12}$$

Consider the linear functional $\mathcal{N}:\widehat{K}^{1,2}(B_\varepsilon)\to\mathbb{R}$ defined as

$$\mathcal{N}(\boldsymbol{\eta}) = \frac{1}{\nu}\Big(-\sum_{j=N_1+1}^{N-1} p_j^*\int\limits_{\gamma_\varepsilon^j}\boldsymbol{\eta}\cdot\mathbf{n}\,dx' + \int\limits_{B_\varepsilon}\mathbf{f}\cdot\boldsymbol{\eta}\,dx\Big).$$

By applying estimates (5.4.11) and (5.4.12), it can be seen that $\mathcal{N}$ is bounded in the Hilbert space $\widehat{K}^{1,2}(B_\varepsilon)$. Therefore, by the Riesz theorem, a unique function $\mathbf{v}\in\widehat{K}^{1,2}(B_\varepsilon)$ exists such that for all $\boldsymbol{\eta}\in\widehat{K}^{1,2}(B_\varepsilon)$,

$$\mathcal{N}(\boldsymbol{\eta}) = [\mathbf{v},\boldsymbol{\eta}].$$

This function $\mathbf{v}$ is the unique solution to the weak formulation (5.4.3). Furthermore, due to estimates (5.4.11) and (5.4.12), it satisfies the estimate (5.4.10). □

Remark 5.4.1 It is important to note that the weak solution $\mathbf{u}$ of problem (5.4.3) belongs to the space $W^{2,2}(B_\varepsilon)$ when $\mathbf{f}\in L^2(B_\varepsilon)$. Similarly, the associated pressure p belongs to $W^{1,2}(B_\varepsilon)$. This can be proven by extending the solutions and the data through reflection over the sections γ_ε^j to a larger domain. Integrating by parts in (5.4.4) and varying $\boldsymbol{\eta}$, it is evident that $(\mathbf{u}, p)$ becomes the classical solution to problem (5.4.2).

Let us now recover and estimate the pressure p corresponding to the weak solution $\mathbf{v}\in\widehat{W}^{1,2}(B_\varepsilon)$ of problem (5.4.3). To achieve this, we need to demonstrate that a linear bounded functional f defined on the space $\widehat{W}^{1,2}(B_\varepsilon)$ and vanishing on the subspace $\widehat{K}^{1,2}(B_\varepsilon)$ can be expressed in the form $f(\boldsymbol{\eta}) = \int_{B_\varepsilon} p(x)\mathrm{div}\boldsymbol{\eta}(x)dx$, where $p\in L^2(B_\varepsilon)$. To do so, we first examine the divergence equation in the space $\widehat{W}^{1,2}(B_\varepsilon)$ with the right-hand side from $L^2(B_\varepsilon)$.

Let σ be a bounded domain with a Lipschitz boundary in $\mathbb{R}^{n-1}$. Consider a cylinder Π defined as $\{x : x'\in\sigma,\ 0<x_n<1\}$ in $\mathbb{R}^n$. We begin with the divergence equation within the cylinder Π.

Lemma 5.4.1 *Let h belong to* $L^2(\Pi)$. *Then, the divergence equation*

$$div\,\mathbf{w}(x) = h(x), \quad x \in \Pi, \tag{5.4.13}$$

admits at least one solution $\mathbf{w} \in W^{1,2}(\Pi)$ *that vanishes on the portion of the boundary* $\partial\Pi \setminus \{x : x_n = 0\}$ *and satisfies the condition* $\mathbf{w}_{\boldsymbol{\tau}}|_{x_n=0} = 0$. *The solution also adheres to the following estimate:*

$$\|\nabla \mathbf{w}\|_{L^2(\Pi)} \leq c\|h\|_{L^2(\Pi)}, \tag{5.4.14}$$

where c is a constant.

Proof
We extend the function h as an odd function to a larger cylinder $\widetilde{\Pi} = \{x : x' \in \sigma, -1 < x_n < 1\}$. In $\widetilde{\Pi}$, we consider the divergence equation:

$$\begin{cases} \operatorname{div} \mathbf{W} = \tilde{h}(x), \; x \in \widetilde{\Pi}, \\ \quad \mathbf{W} = 0, \qquad x \in \partial\widetilde{\Pi}, \end{cases} \tag{5.4.15}$$

where

$$\tilde{h}(x) = \begin{cases} h(x), & \text{if } x \in \Pi, \\ -h(x', -x_n), & \text{if } x \in \widetilde{\Pi} \setminus \Pi. \end{cases}$$

It is worth noting that $\int_{\widetilde{\Pi}} \tilde{h}(x)dx = 0$.

By Lemma 2.4.1, a solution $\mathbf{W} \in \overset{\circ}{W}{}^{1,2}(B_\varepsilon)$ exists, satisfying the estimate

$$\|\mathbf{W}\|_{W^{1,2}(\widetilde{\Pi})} \leq c\|\tilde{h}\|_{L^2(\widetilde{\Pi})} \leq c\|h\|_{L^2(\Pi)}.$$

We can assume that $\mathbf{W}$ comprises an odd component $\mathbf{W}'$ and an even component W_n with respect to x_n, without loss of generality. To ensure this, we introduce the function $\widetilde{\mathbf{W}}$ defined as follows:

$$\widetilde{W}_n(x', x_n) = \frac{1}{2}(W_n(x', x_n) + W_n(x', -x_n)),$$

$$\widetilde{\mathbf{W}}'(x', x_n) = \frac{1}{2}(\mathbf{W}'(x', x_n) - \mathbf{W}'(x', -x_n)).$$

$\widetilde{\mathbf{W}}$ still satisfies the divergence equation (5.4.15) and adheres to the aforementioned parity conditions with respect to x_n. Consequently, $\widetilde{\mathbf{W}}' = 0$ on the section $x_n = 0$, implying $\widetilde{\mathbf{W}}_{\boldsymbol{\tau}} = 0$ for $x_n = 0$. The restriction $\mathbf{w}$ of $\widetilde{\mathbf{W}}$ onto Π becomes a solution to problem (5.4.13). □

Lemma 5.4.2 *Suppose h is a function from* $L^2(B_\varepsilon)$. *Then, the divergence equation*

$$div\,\mathbf{w}(x) = h(x), \quad x \in B_\varepsilon, \tag{5.4.16}$$

has at least one solution $\mathbf{w} \in \widehat{W}^{1,2}(B_\varepsilon)$, *satisfying the following estimate:*

$$\|\nabla \mathbf{w}\|_{L^2(B_\varepsilon)} \leq c\varepsilon^{-3/2}\|h\|_{L^2(B_\varepsilon)}, \tag{5.4.17}$$

where c is a constant.

Proof
Consider a cylinder $\Pi_\varepsilon^{(e)}$ with the base γ_ε^N. Cut this cylinder at a distance $|e|/2$ from the base γ_ε^N, denoting the cut piece as C_ε. Define the mean value of h in B_ε as $\langle h\rangle_{B_\varepsilon} = \int_{B_\varepsilon} h(x)dx/\text{mes}(B_\varepsilon)$. Let χ_A represent the characteristic function of the set A. For the function $h_1 = h - \alpha\langle h\rangle_{B_\varepsilon}\chi_{C_\varepsilon}(x)$, $\alpha = \text{mes}B_\varepsilon/\text{mes}C_\varepsilon = O(1)$, we construct a solution of the divergence equation

$$\text{div}\mathbf{w}_1(x) = h_1(x), \quad x \in B_\varepsilon, \tag{5.4.18}$$

from the space $\mathring{W}^{1,2}(B_\varepsilon)$. By applying Lemma 2.4.14, it satisfies the estimate

$$\|\nabla \mathbf{w}_1\|_{L^2(B_\varepsilon)} \leq c\varepsilon^{-3/2}\|h_1\|_{L^2(B_\varepsilon)} \leq c\varepsilon^{-3/2}\|h\|_{L^2(B_\varepsilon)}, \tag{5.4.19}$$

since

$$\|\alpha\langle h\rangle_{B_\varepsilon}\chi_{C_\varepsilon}\|_{L^2(B_\varepsilon)} = \frac{1}{\text{mes}C_\varepsilon}\left|\int_{B_\varepsilon} hdx\right|\sqrt{\text{mes}C_\varepsilon} \leq c\|h\|_{L^2(B_\varepsilon)}. \tag{5.4.20}$$

Next, we construct a solution $\mathbf{w}_0 \in \widehat{W}^{1,2}(C_\varepsilon)$ within the cylinder C_ε for the divergence equation

$$\text{div}\,\mathbf{w}_0(x) = \alpha\langle h\rangle_{B_\varepsilon}\chi_{C_\varepsilon}(x), \quad x \in C_\varepsilon, \tag{5.4.21}$$

satisfying $\mathbf{w}_{0\tau}|_{\gamma_\varepsilon^N} = 0$ and the estimate

$$\|\nabla \mathbf{w}_0\|_{L^2(C_\varepsilon)} \leq c\varepsilon^{-1}\|\alpha\langle h\rangle_{B_\varepsilon}\chi_{C_\varepsilon}\|_{L^2(C_\varepsilon)} \leq c\varepsilon^{-1}\|h\|_{L^2(B_\varepsilon)}, \tag{5.4.22}$$

where the first inequality in (5.4.22) follows from Lemma 5.4.1 after the scaling transform $\Pi \to C_\varepsilon$ introduced in the proof of Lemma 2.4.13, and the second inequality follows from (5.4.20). As $\mathbf{w}_0$ is constructed based on Lemma 5.4.1, it vanishes on the second base of the cylinder C_ε and can be zero-extended into $B_\varepsilon\backslash C_\varepsilon$. Thus, $\mathbf{w}_0 \in \widehat{W}^{1,2}(B_\varepsilon)$. Finally, by choosing $\mathbf{w} = \mathbf{w}_1 + \mathbf{w}_0$, we conclude the proof.

□

Remark 5.4.2 Consider the linear operator $M : \widehat{W}^{1,2}(B_\varepsilon) \to L^2(B_\varepsilon)$, which maps a function $\mathbf{u} \in \widehat{W}^{1,2}(B_\varepsilon)$ to the corresponding divergence $\text{div}\,\mathbf{u} = q \in L^2(B_\varepsilon)$. It is evident that this operator is bounded. As shown in Lemma 5.2.2, there exists a bounded inverse operator M^{-1} that is defined on the entire space $L^2(B_\varepsilon)$.

Lemma 5.4.3 *Let f be a linear bounded functional defined on the space $\widehat{W}^{1,2}(B_\varepsilon)$ and vanishing on the subspace $\widehat{K}^{1,2}(B_\varepsilon)$. Then, there exists a unique function $p \in L^2(B_\varepsilon)$ such that the expression $f(\eta)$ can be represented in the form*

$$f(\eta) = \int_{B_\varepsilon} p(x) \operatorname{div} \eta(x)\, dx.$$

Proof
The Riesz theorem guarantees the existence of a unique vector function $\mathbf{u} \in \widehat{W}^{1,2}(B_\varepsilon)$ such that the functional f can be expressed as an inner product within the space $\widehat{W}^{1,2}(B_\varepsilon)$:

$$f(\eta) = [\mathbf{u}, \eta],$$

where $[\mathbf{u}, \eta] = \int_{B_\varepsilon} \nabla \mathbf{u} \cdot \nabla \eta\, dx$. Since $f(\eta) = 0$ for $\eta \in \widehat{K}^{1,2}(B_\varepsilon)$, we deduce that $f(\eta) = [\mathbf{u}, P\eta]$, where P is the projector onto the subspace $\widehat{W}^{1,2}(B_\varepsilon) \ominus \widehat{K}^{1,2}(B_\varepsilon)$. Making use of Remark 5.4.2, each $\eta \in \widehat{W}^{1,2}(B_\varepsilon) \ominus \widehat{K}^{1,2}(B_\varepsilon)$ can be represented as $\eta = M^{-1}h$ with $h \in L^2(B_\varepsilon)$. Consequently, the functional $f(\eta) = f(M^{-1}h) = [\mathbf{u}, PM^{-1}h]$ can be regarded as a linear bounded functional on the space $L^2(B_\varepsilon)$. By applying the Riesz theorem once more within the space $L^2(B_\varepsilon)$, we obtain a function $p \in L^2(B_\varepsilon)$, such that

$$f(\eta) = f(M^{-1}h) = \int_{B_\varepsilon} p\, h\, dx = \int_{B_\varepsilon} p \operatorname{div} \eta\, dx.$$

This concludes the proof. □

We can now introduce an alternative weak formulation for problem (5.4.3), referred to as the "with pressure" formulation.

Definition 5.4.3 Find $\mathbf{u} \in \widehat{W}^{1,2}(B_\varepsilon)$ and $p \in L^2(B_\varepsilon)$ that satisfy the following integral identity:

$$\nu \int_{B_\varepsilon} \nabla \mathbf{v} \cdot \nabla \eta\, dx - \int_{B_\varepsilon} p \operatorname{div} \eta\, dx = - \sum_{j=N_1+1}^{N} p_j \int_{\gamma_\varepsilon^j} \eta \cdot \mathbf{n}\, dS + \int_{B_\varepsilon} \mathbf{f} \cdot \eta\, dx, \tag{5.4.23}$$

for all $\eta \in \widehat{W}^{1,2}(B_\varepsilon)$.

Theorem 5.4.2 *For any $\mathbf{f} \in L^2(B_\varepsilon)$ and $p_j \in \mathbb{R}$, $j = N_1 + 1, \ldots, N$, the problem (5.4.3) possesses a unique weak solution $(\mathbf{v}, p)$, where $\mathbf{v} \in \widehat{W}^{1,2}(B_\varepsilon)$ and $p \in L^2(B_\varepsilon)$. Additionally, the following estimate holds:*

$$\varepsilon^{-3/2}\|\nabla \mathbf{v}\|_{L^2(B_\varepsilon)} + \|p\|_{L^2(B_\varepsilon)} \le c\varepsilon^{-3/2}\Big(\varepsilon^{n/2}\sum_{j=N_1+1}^{N}|p_j| + \varepsilon\|\mathbf{f}\|_{L^2(B_\varepsilon)}\Big), \tag{5.4.24}$$

where the constant c is independent of ε.

Proof
Applying Lemma 5.4.3 to the functional

$$f(\boldsymbol{\eta}) = \nu\int_{B_\varepsilon}\nabla \mathbf{v}\cdot\nabla\boldsymbol{\eta}\,dx + \sum_{j=N_1+1}^{N} p_j\int_{\gamma_\varepsilon^j}\boldsymbol{\eta}\cdot\mathbf{n}\,dS - \int_{B_\varepsilon}\mathbf{f}\cdot\boldsymbol{\eta}\,dx, \tag{5.4.25}$$

defined for $\boldsymbol{\eta}\in\widehat{W}^{1,2}(B_\varepsilon)$, we establish the existence and uniqueness of the pressure $p\in L^2(B_\varepsilon)$ such that

$$f(\boldsymbol{\eta}) = \int_{B_\varepsilon} p\mathrm{div}\,\boldsymbol{\eta}dx.$$

Consequently, the functions $\mathbf{v}$ and p fulfill the integral identity (5.4.23). Moreover, it is evident from the estimates (5.4.11) and (5.4.12) that

$$|f(\boldsymbol{\eta})| \le \Big(\|\nabla\mathbf{v}\|_{L^2(B_\varepsilon)} + \varepsilon^{n/2}\sum_{j=N_1+1}^{N}|p_j| + \varepsilon\|\mathbf{f}\|_{L^2(B_\varepsilon)}\Big)\|\nabla\boldsymbol{\eta}\|_{L^2(B_\varepsilon)}. \tag{5.4.26}$$

By Lemma 5.4.2, we can choose a test function $\boldsymbol{\eta}$ for which $\mathrm{div}\boldsymbol{\eta} = p(x)$ and

$$\|\nabla\boldsymbol{\eta}\|_{L^2(B_\varepsilon)} \le c\varepsilon^{-3/2}\|p\|_{L^2(B_\varepsilon)}.$$

Subsequently, the inequality (5.4.26) yields

$$\begin{aligned}|f(\boldsymbol{\eta})| &= \|p\|^2_{L^2(B_\varepsilon)}\\ &\le c\varepsilon^{-3/2}\Big(\|\nabla\mathbf{v}\|_{L^2(B_\varepsilon)} + \varepsilon^{n/2}\sum_{j=N_1+1}^{N}|p_j| + \varepsilon\|\mathbf{f}\|_{L^2(B_\varepsilon)}\Big)\|p\|_{L^2(B_\varepsilon)},\end{aligned} \tag{5.4.27}$$

and this, along with (5.4.10), in turn, implies (5.4.24). □

It should be noted that the vector field $\mathbf{v}$ remains the same across both formulations.

5.4.2 Asymptotic Expansion of the Solution

This section elucidates the procedure for constructing the asymptotic expansion. Consider a function $\zeta \in C^2(\mathbb{R})$ that is independent of ε and such that $\zeta(t) = 0$ when $|t| \le 1/3$ and $\zeta(t) = 1$ when $|t| \ge 2/3$. Let $e = e_{O_j}$ denotes the edge having the end point O_j, and let $x^{(e)}$ be the Cartesian coordinates corresponding to the origin O_j and the edge e. In other words, $x^{(e)} = \mathcal{P}^{(e)}(x - O_j)$, where $\mathcal{P}^{(e)}$ signifies the orthogonal matrix that links the global coordinates x with the local coordinates $x^{(e)}$.

We consider problem (5.4.2). Notably, the asymptotic expansion in this context deviates slightly from the solution expansion in (5.1.1) (as given by formula (5.1.5)). This is due to the absence of a need to construct boundary layers in the neighborhood of vertices, given that the Poiseuille velocity and the pressure satisfy exactly the boundary conditions on γ_ε^j.

The sought asymptotic expansion for the velocity field is expressed as follows:

$$\begin{aligned}\mathbf{u}^{(J)}(x) = & \sum_{O_l, l=N_1+1,\dots,N; e=\overline{O_l O_{i_l}}} \mathbf{U}^{[e,J]}\left(\frac{x^{(e)\prime}}{\varepsilon}\right)\zeta\left(\frac{|e| - x_n^{(e)}}{3r\varepsilon}\right) \\ & + \sum_{e=\overline{O_\alpha O_\beta};\alpha,\beta\le N_1} \zeta\left(\frac{x_n^{(e)}}{3r\varepsilon}\right)\zeta\left(\frac{|e| - x_n^{(e)}}{3r\varepsilon}\right)\mathbf{U}^{[e,J]}\left(\frac{x^{(e)\prime}}{\varepsilon}\right) \\ & + \sum_{l=1}^{N}\left(1 - \zeta\left(\frac{|x - O_l|}{|e|_{\min}}\right)\right)\mathbf{U}^{[BLO_l,J]}\left(\frac{x - O_l}{\varepsilon}\right),\end{aligned} \tag{5.4.28}$$

where the first sum is taken over all edges having a vertex as an end point (and with the origin of the local coordinate system at the vertex), the second sum is taken over all remaining edges, and all terms in these sums are extended by zero out of cylinders $\Pi_\varepsilon^{(e)}$; the terms of the third sum are extended by zero out of the corresponding bundles;

$$\begin{cases} \mathbf{U}^{[e,J]} = \left(P^{(e)}\right)^*(0,\dots,0,\widetilde{U}^{[e,J]})^*, \\ \mathbf{U}_{(j)}^{(e)} = \left(P^{(e)}\right)^*(0,\dots,0,\widetilde{U}_j^{(e)})^*, \quad j = 0,1,\dots,J, \\ \widetilde{U}^{[e,J]}(y^{(e)\prime}) = \sum_{j=0}^{J}\varepsilon^j \widetilde{U}_j^{(e)}(y^{(e)\prime}), \\ \mathbf{U}^{[BLO_l,J]}(y) = \sum_{j=0}^{J}\varepsilon^j \mathbf{U}_j^{[BLO_l]}(y). \end{cases} \tag{5.4.29}$$

The asymptotic expansion of the pressure for every half-cylinder $\Pi_\varepsilon^{(e)}$, $x_n < |e|/2$, corresponding to the edge $e = \overline{O_l O_{i_l}}$, $l = N_1 + 1, \dots, N$ (O_l is the origin of the local coordinate system) is sought in the form

$$p^{(J)}(x) = -s^{(e)} x_n^{(e)} + a^{(e)}, \tag{5.4.30}$$

and on every half-bundle $HB_{O_l}, l = 1, \dots, N_1$, we define

$$\begin{aligned} p^{(J)}(x) = &\sum_{e \subset \mathcal{B}_l} \zeta\Big(\frac{x_n^{(e)}}{3r\varepsilon}\Big)\Big(- s^{(e)} x_n^{(e)} + a^{(e)} - a^{(e_s)}\Big) + a^{(e_s)} \\ &+ \frac{1}{\varepsilon}\Big(1 - \zeta\Big(\frac{|x - O_l|}{|e|_{min}}\Big)\Big) P^{[BLO_l,J]}\Big(\frac{x - O_l}{\varepsilon}\Big), \end{aligned} \tag{5.4.31}$$

where the terms of the sum are extended by zero out of cylinders $\Pi_\varepsilon^{(e)}$,

$$s^{(e)} = \frac{1}{\varepsilon^2} \sum_{j=0}^{J} \varepsilon^j s_j^{(e)}, \quad a^{(e)} = \frac{1}{\varepsilon^2} \sum_{j=0}^{J} \varepsilon^j a_j^{(e)} \tag{5.4.32}$$

and

$$P^{[BLO_l,J]}(y) = \sum_{j=0}^{J} \varepsilon^j P_j^{[BLO_l,J]}(y). \tag{5.4.33}$$

Here e_s is the selected edge of the bundle (arbitrary chosen among edges of the bundle) and the local coordinates $x^{(e)}$ are redefined so that all of them have the same origin O_l.

The process of successively determining terms in asymptotic expansions (5.4.28) and (5.4.30) closely mirrors the algorithm outlined in Sect. 5.1.2. This iterative procedure shares several key aspects. Firstly, the construction of boundary layer correctors corresponding to vertices is unnecessary. Secondly, distinct boundary conditions apply to problems within the graph. Specifically, a prescribed pressure value at the vertices is considered, aligning with pressure-based boundary conditions on γ_ε^J for $j = N_1 + 1, \dots, N$. Consequently, the base case leads to a conductivity problem on the graph, for the function p_0 and is expressed as follows:

$$\begin{cases} -\kappa_e \dfrac{\partial^2 p_0^{(e)}}{\partial x_n^{(e)2}}(x_n^{(e)}) = 0, & x_n^{(e)} \in (0, |e|), \\ -\displaystyle\sum_{e: O_l \in e} \kappa_e \dfrac{\partial p_0^{(e)}}{\partial x_n^{(e)}}(0) = 0, & l = 1, \dots, N_1, \\ p_0^{(e)}(0) = c^l, & l = N_1 + 1, \dots, N, \\ p_0^{(e)}(0) = p_0^{(e_s)}(0), \ \forall e \subset \mathcal{B}_l. \end{cases} \tag{5.4.34}$$

Proof of the existence and uniqueness of the solution for (5.4.34) relies on the application of the Riesz theorem. The details are available in the proof of Theorem 5.1.1.

Consequently, for each edge e, constants $s_0^{(e)}$ and $a_0^{(e)}$ exist, yielding the expression

$$p_0^{(e)}(x^{(e)}) = -s_0^{(e)} x_n^{(e)} + a_0^{(e)}.$$

The corresponding velocity takes the form

$$\widetilde{U}_0^{(e)}(y^{(e)\prime}) = s_0^{(e)} U_0^{(e)}(y^{(e)\prime}), \mathbf{U}_0^{(e)}(y^{(e)\prime}) \;\; = (\mathcal{P}^{(e)})^*(0, \dots, 0, \widetilde{U}_{(0)}^{(e)})^*(y^{(e)\prime}), \tag{5.4.35}$$

where $U_0^{(e)}(y^{(e)\prime})$ represents the solution of the Dirichlet problem (5.1.22).

The boundary layer problem for $(\mathbf{U}_0^{[BLO_l]}(y), P_0^{[BLO_l]}(y))$ retains the same structure as (5.1.23), although its application is abolished in the cases where $l = 1, \dots, N_1$.

The conductivity problem of rank 1 on the graph for the function p_1 can be formulated as follows:

$$\begin{cases} -\kappa_e \dfrac{\partial^2 p_1^{(e)}}{\partial x_n^{(e)2}}(x_n^{(e)}) = 0, & x_n^{(e)} \in (0, |e|), \\ -\displaystyle\sum_{e:O_l \in e} \kappa_e \dfrac{\partial p_1^{(e)}}{\partial x_n^{(e)}}(0) = 0, & l = 1, \dots, N_1, \\ p_1^{(e)}(x_n^{(e)} = 0) = \widehat{a}_0^{[BLO_{l'},e]}, & l = N_1 + 1, \dots, N, \\ p_1^{(e)}(0) - p_1^{(e_s)}(0) = \widehat{a}_0^{[BLO_l,e]}, & \forall e \subset \mathcal{B}_l, e \neq e_s, \end{cases} \tag{5.4.36}$$

Here, e_s represents the selected edge of the bundle. Throughout, $O_{l'}$ denotes the node linked by an edge $e = O_{l'}O_l$ with the vertex O_l, and $\widehat{a}_0^{[BLO_{l'},e]}$ stands for the limit of $\widehat{P}_0^{[BLO_{l'},e]}(y)$ at the outlet corresponding to edge $e = O_{l'}O_l$. In this context, the solution on the graph may display discontinuities at nodes. Specificaly, at each node O_l there are prescribed jumps of $p_1^{(e)}$ between the edges e and e_s of the bundle. This problem also possesses a unique solution p_1.

Assume that all terms within expansion (5.4.28)–(5.4.33) corresponding to ranks less than or equal to $j-1$ are known and that the macroscopic pressure on the graph p_j is also established. The transition from rank $j-1$ to rank j follows a similar adjustment: we solve the conductivity problem on the graph for the function $p_{j+1}^{(e)}$ $(j < J)$:

$$\begin{cases} -\kappa_e \dfrac{\partial^2 p_{j+1}^{(e)}}{\partial x_n^{(e)2}}(x_n^{(e)}) = 0, & x_n^{(e)} \in (0, |e|), \\ -\displaystyle\sum_{e:O_l \in e} \kappa_e \dfrac{\partial p_{j+1}^{(e)}}{\partial x_n^{(e)}}(0) = 0, & l = 1, \dots, N_1, \\ p_{j+1}^{(e)}(0) = \widehat{a}_j^{[BLO_{l'},e]}, & l = N_1+1, \dots, N, \\ p_{j+1}^{(e)}(0) - p_{j+1}^{(e_s)}(0) = \widehat{a}_j^{[BLO_l,e]}, & \forall e \subset \mathcal{B}_l, e \neq e_s, \end{cases} \quad (5.4.37)$$

Here, the local coordinates $x^{(e)}$ are redefined so that they all share the same origin at O_l.

The remaining part of the induction step is just the same as before.

5.4.3 Residual

As in Sect. 5.1.2, the asymptotic expansion $\left(\mathbf{u}^{(J)}, p^{(J)}\right)$ of order J satisfies the following regularity conditions:

$$\mathbf{u}^{(J)} \in W^{2,2}(B_\varepsilon), \quad p^{(J)} \in W^{1,2}(B_\varepsilon).$$

It is worth noting that for the regularity of the boundary layer functions, an approach similar to that described in [59] can be employed, which involves extending solutions and data through reflection over the sections γ_ε^j to an enlarged domain.

The residual $\mathbf{f}^{(J)} = -\mathcal{L}(\mathbf{u}^{(J)}, p^{(J)})$, where $\mathcal{L}(\mathbf{u}, p) = -\nu\Delta\mathbf{u} + \nabla p$, remains consistent with Sect. 5.1.3. Specifically, we have

$$\|\mathbf{f}^{(J)}\|_{L^2(B_\varepsilon)} = \|\mathcal{L}(\mathbf{u}^{(J)}, p^{(J)})\|_{L^2(B_\varepsilon)} = O(\varepsilon^{J-2}). \quad (5.4.38)$$

The divergence of $\mathbf{u}^{(J)}$ can be expressed as

$$\mathrm{div}\mathbf{u}^{(J)} = -\sum_{l=1}^{N} \nabla\zeta\left(\frac{|x - O_l|}{|e|_{min}}\right) \cdot \mathbf{U}^{[BLO_l,J]}(y) = h^{(J)}(y),$$

where

$$\|h^{(J)}\|_{W^{1,2}(B_\varepsilon)} = O(e^{-c/\varepsilon}) \quad (5.4.39)$$

with a constant $c > 0$. It is important to note that the support of the right-hand side resides in the central portions (middle thirds) of the cylinders $\Pi_\varepsilon^{(e)}$ which compose the tube structure. Since the boundary layer velocities $\mathbf{U}^{[BLO_l,J]}(y)$ have zero fluxes due to their convergence to zero, the integral of $h^{(J)}$ over each middle portion of the cylinder is equal to zero. By invoking Lemma 2.4.14, a vector field $\mathbf{w}^{(J)}$ can be introduced such that $\mathrm{div}\mathbf{w}^{(J)} = -h^{(J)}$, and the following estimate holds:

$$\|\mathbf{w}^{(J)}\|_{W^{1,2}(B_\varepsilon)} \le \varepsilon^{-3/2} c \|h^{(J)}\|_{L^2(B_\varepsilon)}.$$

Moreover, $\mathbf{w}^{(J)} \in W^{2,2}(B_\varepsilon)$ and

$$\|\mathbf{w}^{(J)}\|_{W^{2,2}(B_\varepsilon)} \le \varepsilon^{-2} c \|h^{(J)}\|_{W^{1,2}(B_\varepsilon)}.$$

By defining $\widetilde{\mathbf{u}}^{(J)} = \mathbf{u}^{(J)} + \mathbf{w}^{(J)}$, it is ensured that $\mathrm{div}\widetilde{\mathbf{u}}^{(J)} = 0$, and further:

$$\|\mathbf{f}_1^{(J)}\|_{L^2(B_\varepsilon)} = O(\varepsilon^{J-2}), \tag{5.4.40}$$

where $\mathbf{f}_1^{(J)} = -\mathcal{L}(\widetilde{\mathbf{u}}^{(J)}, p^{(J)})$.

It is pertinent to observe that an additional residual arises in the boundary condition on γ_ε^l, specifically $\varepsilon^{J-1}\widehat{a}_J^{[BLO_{l'},e]}$. This discrepancy emerges due to the presence of uncompensated values $\widehat{a}_J^{[BLO_{l'},e]}, l = N_1 + 1, \dots, N$ (limits of $\widehat{P}_J^{[BLO_{l'},e]}$ at the outlets corresponding to edges $\overline{O_{l'}O_l}$, $l = N_1 + 1, \dots, N$).

5.4.4 Error Analysis

Theorem 5.4.3 *The following error estimates hold:*

$$\|\mathbf{v} - \widetilde{\mathbf{u}}^{(J)}\|_{W^{1,2}(B_\varepsilon)} = O(\varepsilon^{J+(n-1)/2}), \quad \|p - p^{(J)}\|_{L^2(B_\varepsilon)} = O(\varepsilon^{J+(n-3)/2}). \tag{5.4.41}$$

Proof

Let $\mathbf{u} = \mathbf{v} - \widetilde{\mathbf{u}}^{(J)}$ and $q = p - p^{(J)}$. Then the following integral identity

$$\nu \int_{B_\varepsilon} \nabla\mathbf{u} \cdot \nabla\eta \, dx - \int_{B_\varepsilon} q(x) \mathrm{div}\eta \, dx$$

$$= - \sum_{j=N_1+1}^{N} \varepsilon^{J-1} \widehat{a}_J^{[BLO_{l'},e]} \int_{\gamma_\varepsilon^j} \eta \cdot \mathbf{n} \, dS + \int_{B_\varepsilon} \mathbf{f}_1^{(J)} \cdot \eta \, dx.$$

is valid for every $\eta \in \widehat{W}^{1,2}(B_\varepsilon)$. Here, the norm of $\mathbf{f}_1^{(J)}$ is given by (5.4.40).

By applying Theorem 5.4.2 along with the estimate (5.4.24), we obtain

$$\|\mathbf{u}\|_{W^{1,2}(B_\varepsilon)} = \|\mathbf{v} - \widetilde{\mathbf{u}}^{(J)}\|_{W^{1,2}(B_\varepsilon)} = O(\varepsilon^{J-1}) \tag{5.4.42}$$

and

$$\|q\|_{L^2(B_\varepsilon)} = \|p - p^{(J)}\|_{L^2(B_\varepsilon)} = O(\varepsilon^{J-5/2}). \tag{5.4.43}$$

Evaluating the norm of the difference $\mathbf{u}^{(J)}$ and $\mathbf{u}^{(J+2)}$, we have

$$\|\widetilde{\mathbf{u}}^{(J)} - \widetilde{\mathbf{u}}^{(J+2)}\|_{W^{1,2}(B_\varepsilon)} = O(\varepsilon^{J+(n-1)/2}).$$

Replacing J with $J+2$ in (5.4.42) yields

$$\|\mathbf{u} - \widetilde{\mathbf{u}}^{(J+2)}\|_{W^{1,2}(B_\varepsilon)} = O(\varepsilon^{J+1}).$$

Consequently, the triangle inequality leads to the first estimate in (5.4.41). By applying the same argument to $p^{(J)}$ and $p^{(J+3)}$, the second estimate in (5.4.41) is established. □

5.4.5 *MAPDD for the Inflow–Outflow Boundary Conditions Involving Pressure*

For boundary conditions involving pressure, we need to introduce modifications to the previously described version of the MAPDD algorithm. Specifically, we introduce the "octopus-shaped" domains denoted as $B_l^{\varepsilon,\delta}$, which are defined for $l = 1, \dots, N_1$, i.e., corresponding only to the nodes O_l. Consequently, the cylinders $\Pi_\varepsilon^{(e)}$ containing a vertex O_l, where $l = N_1 + 1, \dots, N$, in one of its bases, do not undergo truncation from the side of this vertex at a distance δ from it. This adjustment ensures that the cylinders $B_{ij}^{dec,\varepsilon}$ intended for $j = N_1 + 1, \dots, N$ have γ_ε^j as one of their bases.

The subspace $\widehat{\mathcal{W}}_{\text{dec}}(B_\varepsilon, \delta)$ is defined in a similar manner as before. It is the subspace of vector functions from $\widehat{W}^{1,2}(B_\varepsilon)$ having on $B_{ij}^{dec,\varepsilon}$, $j = N_1 + 1, \dots, N$, zero tangential entries of the vector fields and independent of the normal variable normal components. So, on each base γ_ε^j the vector functions have vanishing tangential component. Likewise, $\widehat{\mathcal{H}}_{\text{dec}}(B_\varepsilon, \delta)$ represents a subspace of divergence-free functions within $\widehat{\mathcal{W}}_{\text{dec}}(B_\varepsilon, \delta)$.

The weak formulation for the MAPDD "without pressure" is expressed as follows:

Definition 5.4.4 Find $\mathbf{v}_d \in \widehat{\mathcal{H}}_{\text{dec}}(B_\varepsilon, \delta)$, satisfying for all $\boldsymbol{\eta} \in \widehat{\mathcal{H}}_{\text{dec}}(B_\varepsilon, \delta)$ the following integral identity:

$$\nu \int_{B_\varepsilon} \nabla \mathbf{v}_d \cdot \nabla \boldsymbol{\eta} dx = - \sum_{j=N_1+1}^{N} p_j \int_{\gamma_\varepsilon^j} \boldsymbol{\eta} \cdot \mathbf{n} dS. \tag{5.4.44}$$

Denote $\Gamma = \partial B_\varepsilon \setminus \gamma_\varepsilon$, where $\gamma_\varepsilon = \bigcup_{j=N_1+1,\dots,N} \gamma_\varepsilon^j$.

Applying the same approach as in the case of the Dirichlet boundary condition, we can prove the analogue of Lemmas 5.2.1 and 5.2.2 (see [154]).

Lemma 5.4.4 *There exists a set of N_1 vector-valued functions $\mathbf{U}_i \in \widehat{\mathcal{W}}_{\text{dec}}(B_\varepsilon, \delta)$, satisfying the conditions $\int_{\partial B_j^{\varepsilon,\delta}} \mathbf{U}_i \cdot \mathbf{n} dS = \delta_{ij}$ for $j = 1, \dots, N_1$. Furthermore, the following inequalities hold:*

$$\|\mathbf{U}_i\|^2_{L^2(B_\varepsilon)} \le C_* \varepsilon^{-(n-1)}, \quad \|\nabla \mathbf{U}_i\|^2_{L^2(B_\varepsilon)} \le C_* \varepsilon^{-(n+1)} \tag{5.4.45}$$

with a constant C_* independent of ε and δ.

Lemma 5.4.5 *Given a function p defined in $\bigcup_{i=1}^{N_1} B_i^{\varepsilon,\delta}$ and belonging to $L^2(B_i^{\varepsilon,\delta})$ for all $i = 1, \dots, N_1$, there exists a vector-valued function $\mathbf{U} \in \widehat{\mathcal{W}}_{\text{dec}}(B_\varepsilon, \delta)$ such that*

$$\operatorname{div}\mathbf{U}(x) = \begin{cases} p(x), \ x \in \bigcup_{i=1}^{N_1} B_i^{\varepsilon,\delta}, \\ 0, \ x \in B_\varepsilon \setminus \bigcup_{i=1}^{N_1} B_i^{\varepsilon,\delta}. \end{cases} \tag{5.4.46}$$

Furthermore, the following estimate holds:

$$\|\mathbf{U}\|^2_{W^{1,2}(B_\varepsilon)} \le C\left(\varepsilon^{-2n}\delta^{-1} + \varepsilon^{-2n-3}\delta^2\right) \sum_{i=1}^{N_1} \|p\|^2_{L^2(B_i^{\varepsilon,\delta})} \tag{5.4.47}$$

where C is a positive constant independent of ε and δ. In the case where $\delta = O(\varepsilon|\ln\varepsilon|)$, the estimate (5.4.47) *becomes*

$$\|\mathbf{U}\|^2_{W^{1,2}(B_\varepsilon)} \le C\varepsilon^{-2n-3/2} \sum_{i=1}^{N_1} \|p\|^2_{L^2(B_i^{\varepsilon,\delta})}. \tag{5.4.48}$$

Let us formulate the MAPDD "with pressure":

Definition 5.4.5 Find $\mathbf{v}_d \in \widehat{\mathcal{H}}_{\text{dec}}(B_\varepsilon, \delta)$ and $p_d \in L^2(B_i^{\varepsilon,\delta})$ for $i = 1, \dots, N_1$, such that the following integral identity holds for all $\boldsymbol{\eta} \in \widehat{\mathcal{W}}_{\text{dec}}(B_\varepsilon, \delta)$:

$$\nu \int_{B_\varepsilon} \nabla \mathbf{v}_d \cdot \nabla \boldsymbol{\eta} dx = \sum_{i=1}^{N_1} \int_{B_i^{\varepsilon,\delta}} p_d \operatorname{div}\boldsymbol{\eta} dx - \sum_{j=N_1+1}^{N} p_j \int_{\gamma_\varepsilon^j} \boldsymbol{\eta} \cdot \mathbf{n} dS. \tag{5.4.49}$$

This formulation is equivalent to the one employing functions solely defined on $\bigcup_{i=1}^{N} B_i^{\varepsilon,\delta}$.

Let $\widehat{\widehat{W}}^{1,2}(\bigcup_{i=1}^{N} B_i^{\varepsilon,\delta})$ be the space of vector-valued functions $\mathbf{w}$ defined on $\bigcup_{i=1}^{N} B_i^{\varepsilon,\delta}$, such that their restrictions on $B_i^{\varepsilon,\delta}$ belong to $W^{1,2}(B_i^{\varepsilon,\delta})$ for $i = 1, \ldots, N$. Additionally, for each pair (i, j) where O_i and O_j are connected by the edge $O_iO_j = e$, the traces $\mathbf{w}|_{S_{ij}}$ and $\mathbf{w}|_{S_{ji}}$ possess the form

$$\mathbf{w}|_{S_{ij}} = \mathbf{w}|_{S_{ji}} = \left(\mathcal{P}^{(e)}\right)^*\left(0, \ldots, 0, w_n(x^{(e)\prime})\right) \tag{5.4.50}$$

with $w_n \in \mathring{W}^{1,2}(\sigma_\varepsilon^{(e)})$, and the trace of $\mathbf{w}$ is zero on $\partial B_i^{\varepsilon,\delta} \cap \partial B_\varepsilon$ for $i = 1, \ldots, N_1$ and on $\left(\partial B_i^{\varepsilon,\delta} \cap \partial B_\varepsilon\right) \setminus \gamma_\varepsilon^i$ for $i = N_1 + 1, \ldots, N$. Moreover, $\mathbf{w}_\tau|_{\gamma_\varepsilon^i} = 0$.

The corresponding equivalent formulation is as follows:

Definition 5.4.6 Find the divergence-free vector-field $\mathbf{v}_d \in \widehat{\widehat{W}}^{1,2}(\bigcup_{i=1}^{N} B_i^{\varepsilon,\delta})$ and the "MAPDD pressure" $p_d \in L^2(\bigcup_{i=1}^{N} B_i^{\varepsilon,\delta})$, such that

$$\begin{aligned}
&\nu \sum_{i=1}^{N} \int_{B_i^{\varepsilon,\delta}} \nabla \mathbf{v}_d \cdot \nabla \boldsymbol{\eta} dx + \nu \sum_{l=1}^{M} d_l \int_{\sigma_\varepsilon^{(e_l)}} \nabla_{x^{(e_l)\prime}} \mathbf{v}_d \cdot \nabla_{x^{(e_l)\prime}} \boldsymbol{\eta} dx^{(e_l)\prime} \\
&= \sum_{i=1}^{N} \int_{B_i^{\varepsilon,\delta}} p_d \operatorname{div} \boldsymbol{\eta} dx - \sum_{j=N_1+1}^{N} p_j \int_{\gamma_\varepsilon^j} \boldsymbol{\eta} \cdot \mathbf{n} dS \quad \forall \boldsymbol{\eta} \in \widehat{\widehat{W}}^{1,2}\Big(\bigcup_{i=1}^{N} B_i^{\varepsilon,\delta}\Big).
\end{aligned} \tag{5.4.51}$$

Here $d_l = |e| - 2\delta$ is the distance between the cross-sections S_{ij} and S_{ji} cutting the segment e, for segments where both endpoints are nodes (e connects two nodes). If one of the endpoints is a vertex, i.e., e connects a node and a vertex, $d_l = |e| - \delta$. In the last case, in (5.4.50) with $j = N_1 + 1, \ldots, N$ or $l = N_1 + 1, \ldots, N$, by convention $S_{lj} = \gamma_\varepsilon^j$ (respectively, $S_{lj} = \gamma_\varepsilon^l$).

By utilizing Lemmas 5.4.4 and 5.4.5, we can establish, analogous to Sect. 5.2, the subsequent theorem:

Theorem 5.4.4 *There exists a unique pair of functions* $\mathbf{v}_d \in \widehat{\widehat{W}}^{1,2}(\bigcup_{i=1}^{N} B_i^{\varepsilon,\delta})$, $p_d \in L^2(B_i^{\varepsilon,\delta})$ *for all* $i = 1, \ldots, N_1$, *which satisfies integral identity* (5.4.51).

The justification of the MAPDD in this case is quite analogous to the case of the Dirichlet boundary conditions. The estimates are of the same order.

Theorem 5.4.5 *Consider* $\delta = C_{J+2}\varepsilon|\ln\varepsilon|$, *where the constant* C_{J+2} *satisfies condition* (5.1.60) *with* J *replaced by* $J+2$. *In this case, the following error estimate holds:*

$$\|\mathbf{v} - \mathbf{v}_d\|_{W^{1,2}(B_\varepsilon)} = O(\varepsilon^{J+(n-1)/2}). \tag{5.4.52}$$

Likewise, for $\delta = C_{J+2n+1}\varepsilon|\ln\varepsilon|$, *where* C_{J+2n+1} *satisfies* (5.1.60) *with* J *replaced by* $J+2n+1$, *the following relationship holds:*

$$\|p_d - p\|_{L^2(B_k^{\varepsilon,\delta})} = O(\varepsilon^{J+n/2+(n-3)/2}), \quad k = 1, \ldots N_1. \tag{5.4.53}$$

Now, let us extend p_d to the cylinders $B_{ij}^{dec,\varepsilon}$ following the same approach as previously described in Sect. 5.2, i.e.,

$$p_d(x_n^{(e)}) = \langle p_d\rangle_{S_{ij}} + \frac{\langle p_d\rangle_{S_{ji}} - \langle p_d\rangle_{S_{ij}}}{|e| - 2\delta + \varepsilon}\left(x_n^{(e)} - \delta + \frac{\varepsilon}{2}\right). \tag{5.4.54}$$

Subsequently, we are able to establish the following estimate for this extended function:

$$\|p_d - p\|_{L^2(B_\varepsilon)} = O(\varepsilon^{J+(n-3)/2}). \tag{5.4.55}$$

5.4.6 Comments on the Mixed Boundary Conditions and on the Navier–Stokes Equations

Consider the case where the pressure conditions (5.4.2) are prescribed on certain surfaces γ_ε^j, where $j = N_1 + 1, \ldots, N_2$, while on others γ_ε^j, $j = N_2 + 1, \ldots, N$, the inflow/outflow velocity is given. This case is treated in the same way as in previously considered cases of given velocity on all surfaces γ_ε^j or given pressure on all surfaces γ_ε^j. The procedure for constructing boundary layers aligns with the approach utilized under Dirichlet boundary conditions. Notably, in this context, there is no necessity for a compatibility condition concerning the integrals of the boundary value function $\mathbf{g} \cdot \mathbf{n}$ over γ_ε^j for $j = N_2 + 1, \ldots, N$, if $N_2 < N_1$. Consequently, the equations on the graph include boundary conditions for fluxes at the vertices O_j for $j = N_2 + 1, \ldots, N$ and boundary conditions for the unknown macroscopic pressure p at the vertices O_j for $j = N_1 + 1, \ldots, N_2$. The condition of pressure at vertex O_N no longer remains constrained to zero.

In the case of the Navier–Stokes equations with pressure involved boundary conditions, all considerations are quite similar to that of Sects. 5.1–5.4. Notably, the construction of the asymptotic expansion differs from one of this section only by the supplementary terms (as in the case of the Dirichlet boundary condition). Another point of distinction arises in the validation of the asymptotic expansion, particularly concerning the establishment of existence and uniqueness theorems for the Navier–Stokes equations subjected to inflow-outflow boundary conditions involving pressure. These theorems are slated for formulation and substantiation in the subsequent section, within the context of the more intricate Bernoulli pressure inflow-outflow boundary conditions.

5.5 Stationary Navier–Stokes Equations with Given Bernoulli Pressure Inflow-Outflow Boundary Conditions

5.5.1 Formulation of the Problem: Existence and Uniqueness of a Solution

In this section, we address the formulation, existence, and uniqueness of a solution to the following boundary value problem for the stationary Navier–Stokes equations in a tube structure B_ε:

$$\begin{cases} -\nu\Delta\mathbf{v} + (\mathbf{v}\cdot\nabla)\mathbf{v} + \nabla p = \mathbf{f}, & x \in B_\varepsilon, \\ \operatorname{div}\mathbf{v} = 0, & x \in B_\varepsilon, \\ \mathbf{v} = 0, & x \in \partial B_\varepsilon \setminus \bigcup\limits_{j=N_1+1}^{N} \gamma_\varepsilon^j, \\ \mathbf{v}_\tau = 0, & x \in \gamma_\varepsilon^j,\ j = N_1+1,\dots,N, \\ -\nu\partial_n(\mathbf{v}\cdot\mathbf{n}) + (p + \frac{1}{2}|\mathbf{v}|^2) = c^j/\varepsilon^2, & x \in \gamma_\varepsilon^j,\ j = N_1+1,\dots,N. \end{cases} \tag{5.5.1}$$

Here, ν is a positive constant, $\mathbf{n}$ represents the unit normal vector to γ_ε^j, $\mathbf{v}_\tau = \mathbf{v} - (\mathbf{v}\cdot\mathbf{n})\mathbf{n}$ is the tangential component of the vector $\mathbf{v}$, and c^j are constants.

From the boundary condition $\mathbf{v}_\tau|_{\gamma_\varepsilon^j} = 0$ and the divergence equation $\operatorname{div}\mathbf{v} = 0$, it follows that $\partial_n \mathbf{v}_\mathbf{n}|_{\gamma_\varepsilon^j} = 0$. Thus, by using the identity $\nabla(\frac{1}{2}\mathbf{v}^2) = \mathbf{v}\cdot(\nabla\mathbf{v})^t = \sum\limits_{i=1}^{n} v_i \nabla v_i$, where the upper index t means the transposed matrix, we can rewrite the problem (5.5.1) in the following form:

$$\begin{cases} -\nu\Delta\mathbf{v} + (\mathbf{v}\cdot\nabla)\mathbf{v} - \mathbf{v}\cdot(\nabla\mathbf{v})^t + \nabla\Phi = \mathbf{f}, & x \in B_\varepsilon, \\ \operatorname{div}\mathbf{v} = 0, & x \in B_\varepsilon, \\ \mathbf{v} = 0, & x \in \partial B_\varepsilon \setminus \bigcup\limits_{j=N_1+1}^{N} \gamma_\varepsilon^j, \\ \mathbf{v}_\tau = 0, & x \in \gamma_\varepsilon^j,\ j = N_1+1,\dots,N, \\ \Phi = p_j, & x \in \gamma_\varepsilon^j,\ j = N_1+1,\dots,N, \end{cases} \tag{5.5.2}$$

where $\Phi = (p + \frac{1}{2}|\mathbf{v}|^2)$ is the Bernoulli pressure, p_j stand for the constants c^j/ε^2.

Definition 5.5.1 We define a weak solution to problem (5.5.2) as a vector field $\mathbf{v} \in \widehat{K}^{1,2}(B_\varepsilon)$ that satisfies, for all test functions $\boldsymbol{\eta} \in \widehat{K}^{1,2}(B_\varepsilon)$, the integral identity:

$$\begin{aligned} \nu \int\limits_{B_\varepsilon} \nabla \mathbf{v} \cdot \nabla \boldsymbol{\eta} dx + \int\limits_{B_\varepsilon} (\mathbf{v} \cdot \nabla)\mathbf{v} \cdot \boldsymbol{\eta} dx - \int\limits_{B_\varepsilon} (\boldsymbol{\eta} \cdot \nabla)\mathbf{v} \cdot \mathbf{v} dx \\ = - \sum_{j=N_1+1}^{N} p_j \int\limits_{\gamma_\varepsilon^j} \boldsymbol{\eta} \cdot \mathbf{n} dS + \int\limits_{B_\varepsilon} \mathbf{f} \cdot \boldsymbol{\eta} dx. \end{aligned} \tag{5.5.3}$$

The identity (5.5.3) is derived from Eqs. (5.5.2) by multiplying them with $\boldsymbol{\eta} \in \widehat{K}^{1,2}(B_\varepsilon)$ and performing integration by parts within the domain B_ε. Conversely, as demonstrated in the preceding section, for a sufficiently regular solution $\mathbf{v}$ that satisfies condition (5.5.3), there exists a corresponding pressure field p. This pair $(\mathbf{v}, p)$ then fulfills equations $(5.5.2)_{1,2}$ almost everywhere in B_ε, while boundary conditions $(5.5.2)_{3,4}$ are satisfied in a trace sense. Boundary condition $(5.5.2)_5$ is fulfilled up to a constant. By selecting an appropriate constant for the pressure term, it is possible to assume that $p_N = 0$ and $p_j = p_j^*, \ j = N_1 + 1, \dots, N - 1$. Ultimately, the integral identity (5.5.3) can be reformulated as follows:

$$\begin{aligned} \nu \int\limits_{B_\varepsilon} \nabla \mathbf{v} \cdot \nabla \boldsymbol{\eta} dx + \int\limits_{B_\varepsilon} (\mathbf{v} \cdot \nabla)\mathbf{v} \cdot \boldsymbol{\eta} dx - \int\limits_{B_\varepsilon} (\boldsymbol{\eta} \cdot \nabla)\mathbf{v} \cdot \mathbf{v} dx \\ = - \sum_{j=N_1+1}^{N-1} p_j^* \int\limits_{\gamma_\varepsilon^j} \boldsymbol{\eta} \cdot \mathbf{n} dS + \int\limits_{B_\varepsilon} \mathbf{f} \cdot \boldsymbol{\eta} dx \quad \forall \boldsymbol{\eta} \in \widehat{K}^{1,2}(B_\varepsilon). \end{aligned} \tag{5.5.4}$$

Theorem 5.5.1 *For arbitrary* $\mathbf{f} \in L^2(B_\varepsilon)$, $p_j^* \in \mathbb{R}, \ j = N_1 + 1, \dots, N - 1$, *and sufficiently small* ε, *problem* (5.5.2) *admits at least one weak solution* $\mathbf{v} \in \widehat{K}^{1,2}(B_\varepsilon)$. *The following estimate holds:*

$$\|\nabla \mathbf{v}\|_{L^2(B_\varepsilon)} \le c\Big(\varepsilon^{n/2} \sum_{j=N_1+1}^{N-1} |p_j^*| + \varepsilon \|\mathbf{f}\|_{L^2(B_\varepsilon)}\Big) \tag{5.5.5}$$

Here, c is a constant that is independent of ε.

Proof
Define in $\widehat{K}^{1,2}(B_\varepsilon)$ the inner product $[\mathbf{u}, \boldsymbol{\eta}] = \int_{B_\varepsilon} \nabla \mathbf{u} \cdot \nabla \boldsymbol{\eta} dx$, which corresponds to the Dirichlet norm. As in Theorem 5.3.1, by employing Hölder and Poincaré inequalities along with Lemmas 2.3.11 and 2.3.16, we establish the estimates

$$\Big|\int\limits_{B_\varepsilon}(\boldsymbol{\eta}\cdot\nabla)\mathbf{v}\cdot\mathbf{v}dx\Big|+\Big|\int\limits_{B_\varepsilon}(\mathbf{v}\cdot\nabla)\mathbf{v}\cdot\boldsymbol{\eta}dx\Big|\le c\varepsilon^{\alpha}\|\nabla\mathbf{v}\|^2_{L^2(B_\varepsilon)}\|\nabla\boldsymbol{\eta}\|_{L^2(B_\varepsilon)}, \tag{5.5.6}$$

$$\Big|\sum_{j=N_1+1}^{N-1}p_j^*\int\limits_{\gamma_\varepsilon^j}\boldsymbol{\eta}\cdot\mathbf{n}dS\Big|\le c\varepsilon^{n/2}\sum_{j=N_1+1}^{N-1}|p_j^*|\|\nabla\boldsymbol{\eta}\|_{L^2(B_\varepsilon)}. \tag{5.5.7}$$

$$\Big|\int\limits_{B_\varepsilon}\mathbf{f}\cdot\boldsymbol{\eta}dx\Big|\le c\varepsilon\|\mathbf{f}\|_{L^2(B_\varepsilon)}\|\nabla\boldsymbol{\eta}\|_{L^2(B_\varepsilon)}, \tag{5.5.8}$$

where $\alpha=1$ for $n=2$ and $\alpha=1/2$ for $n=3$.

From these estimates and the Riesz theorem, it follows that the integral identity (5.5.4) is equivalent to the operator equation in the space $\widehat{K}^{1,2}(B_\varepsilon)$:

$$\mathbf{v}=\mathcal{A}\mathbf{v}, \tag{5.5.9}$$

where the compact operator $\mathcal{A}$ (see [102]) is defined by

$$[\mathcal{A}\mathbf{v},\boldsymbol{\eta}]=\int\limits_{B_\varepsilon}\nu^{-1}\Big[-(\mathbf{v}\cdot\nabla)\mathbf{v}\cdot\boldsymbol{\eta}+(\boldsymbol{\eta}\cdot\nabla)\mathbf{v}\cdot\mathbf{v}+\mathbf{f}\cdot\boldsymbol{\eta}\Big]dx$$
$$-\nu^{-1}\sum_{j=N_1+1}^{N-1}p_j^*\int\limits_{\gamma_\varepsilon^j}\boldsymbol{\eta}\cdot\mathbf{n}dS\qquad\forall\,\boldsymbol{\eta}\in\widehat{K}^{1,2}(B_\varepsilon).$$

The existence of at least one solution to the Eq. (5.5.9) can be established by appealing to the Leray–Schauder fixed point theorem (see Theorem 2.2.10). To apply this theorem, we need to show that all potential solutions $\mathbf{v}^{(\lambda)}$ of the equation

$$\mathbf{v}^{(\lambda)}=\lambda\mathcal{A}\mathbf{v}^{(\lambda)},\quad\lambda\in[0,1] \tag{5.5.10}$$

remain uniformly bounded with respect to the parameter λ.

A solution $\mathbf{v}^{(\lambda)}$ of (5.5.10) satisfies the integral identity

$$\nu\int\limits_{B_\varepsilon}\nabla\mathbf{v}^{(\lambda)}\cdot\nabla\boldsymbol{\eta}dx+\lambda\int\limits_{B_\varepsilon}(\mathbf{v}^{(\lambda)}\cdot\nabla)\mathbf{v}^{(\lambda)}\cdot\boldsymbol{\eta}dx-\lambda\int\limits_{B_\varepsilon}(\boldsymbol{\eta}\cdot\nabla)\mathbf{v}^{(\lambda)}\cdot\mathbf{v}^{(\lambda)}dx$$
$$=-\lambda\sum_{j=N_1+1}^{N-1}p_j^*\int\limits_{\gamma_\varepsilon^j}\boldsymbol{\eta}\cdot\mathbf{n}dS+\lambda\int\limits_{B_\varepsilon}\mathbf{f}\cdot\boldsymbol{\eta}dx\quad\forall\,\boldsymbol{\eta}\in\widehat{K}^{1,2}(B_\varepsilon). \tag{5.5.11}$$

Substituting $\boldsymbol{\eta} = \mathbf{v}^{(\lambda)}$ into (5.5.11) leads to the relation

$$\nu \int\limits_{B_\varepsilon} |\nabla \mathbf{v}^{(\lambda)}|^2 dx = -\lambda \sum_{j=N_1+1}^{N-1} p_j^* \int\limits_{\gamma_\varepsilon^j} \mathbf{v}^{(\lambda)} \cdot \mathbf{n} dS + \lambda \int\limits_{B_\varepsilon} \mathbf{f} \cdot \mathbf{v}^{(\lambda)} dx.$$

By utilizing (5.5.7) and (5.5.8), we derive the inequality

$$\|\nabla \mathbf{v}^{(\lambda)}\|^2_{L^2(B_\varepsilon)} \le c\Big(\varepsilon^{n/2} \sum_{j=N_1+1}^{N-1} |p_j^*| + \varepsilon \|\mathbf{f}\|_{L^2(B_\varepsilon)}\Big) \|\nabla \mathbf{v}^{(\lambda)}\|_{L^2(B_\varepsilon)}.$$

Consequently,

$$\|\nabla \mathbf{v}^{(\lambda)}\|_{L^2(B_\varepsilon)} \le c\Big(\varepsilon^{n/2} \sum_{j=N_1+1}^{N-1} |p_j^*| + \varepsilon \|\mathbf{f}\|_{L^2(B_\varepsilon)}\Big).$$

It is noteworthy that the constant c in the latter inequality is independent of λ and ε. This concludes the proof of the theorem. □

Theorem 5.5.2

1. *There exists a positive constant c_0, independent of ε such that if the condition*

$$c_0 \varepsilon^\alpha \Big(\varepsilon^{n/2} \sum_{j=N_1+1}^{N-1} |p_j^*| + \varepsilon \|\mathbf{f}\|_{L^2(B_\varepsilon)}\Big) < \nu \tag{5.5.12}$$

is satisfied, then the weak solution $\mathbf{v} \in \widehat{K}^{1,2}(B_\varepsilon)$ of problem (5.5.2) *is unique.*

2. *Let $\{p_{1j}^*\}$ and $\{p_{2j}^*\}$, where $j = N_1+1, \ldots, N$, be two sets of real constants, along with functions $\mathbf{f}_1$ and $\mathbf{f}_2$ that both belong to $L^2(B_\varepsilon)$. Suppose $\{p_{1j}^*\}$, $\mathbf{f}_1$ and $\{p_{2j}^*\}$, $\mathbf{f}_2$ satisfy condition* (5.5.12)*. Let $\mathbf{v}_1$ and $\mathbf{v}_2$ be weak solutions of problem* (5.5.2) *corresponding to $\{p_{1j}^*\}$, $\mathbf{f}_1$ and $\{p_{2j}^*\}$, $\mathbf{f}_2$ respectively. Then there exist a constant C, independent of ε, such that the following inequality holds:*

$$\|\nabla \mathbf{v}_1 - \nabla \mathbf{v}_2\|_{L^2(B_\varepsilon)} \le \varepsilon C \|\mathbf{f}_1 - \mathbf{f}_2\|_{L^2(B_\varepsilon)} + c\varepsilon^{n/2} \sum_{j=N_1+1}^{N-1} |p_{1j}^* - p_{2j}^*|. \tag{5.5.13}$$

Proof

1. Suppose there exist two weak solutions $\mathbf{v}_1$ and $\mathbf{v}_2$ that satisfy identity (5.5.4). By subtracting the identity corresponding to $\mathbf{v}_2$ from the identity corresponding to $\mathbf{v}_1$, we obtain the following expression:

$$\nu\int\limits_{B_\varepsilon}\nabla\mathbf{w}\cdot\nabla\boldsymbol{\eta}dx+\int\limits_{B_\varepsilon}\big((\mathbf{w}\cdot\nabla)\mathbf{v}_1\cdot\boldsymbol{\eta}+(\mathbf{v}_2\cdot\nabla)\mathbf{w}\cdot\boldsymbol{\eta}\big)dx$$
$$-\int\limits_{B_\varepsilon}\big((\boldsymbol{\eta}\cdot\nabla)\mathbf{v}_1\cdot\mathbf{w}+(\boldsymbol{\eta}\cdot\nabla)\mathbf{w}\cdot\mathbf{v}_2\big)dx=0, \tag{5.5.14}$$

where $\mathbf{w}=\mathbf{v}_1-\mathbf{v}_2$. Considering $\boldsymbol{\eta}=\mathbf{w}$ in equality (5.5.14), we deduce, in light of Lemma 2.3.11 and estimate (5.5.5) for $\mathbf{v}_2$, the following inequality:

$$\nu\int\limits_{B_\varepsilon}|\nabla\mathbf{w}|^2dx=\int\limits_{B_\varepsilon}(\mathbf{w}\cdot\nabla)\mathbf{w}\cdot\mathbf{v}_2dx-\int\limits_{B_\varepsilon}(\mathbf{v}_2\cdot\nabla)\mathbf{w}\cdot\mathbf{w}dx$$
$$\le\|\mathbf{w}\|_{L^4(B_\varepsilon)}\|\nabla\mathbf{w}\|_{L^2(B_\varepsilon)}\|\mathbf{v}_2\|_{L^4(B_\varepsilon)}\le c\varepsilon^\alpha\|\nabla\mathbf{w}\|^2_{L^2(B_\varepsilon)}\|\nabla\mathbf{v}_2\|_{L^2(B_\varepsilon)}$$
$$\le c_0\varepsilon^\alpha\Big(\varepsilon^{n/2}\sum_{j=N_1+1}^{N-1}|p_j^*|+\varepsilon\|\mathbf{f}\|_{L^2(B_\varepsilon)}\Big)\|\nabla\mathbf{w}\|^2_{L^2(B_\varepsilon)},$$

where the constant c_0 is independent of ε. If the condition (5.5.12) holds, the last inequality implies that

$$\int\limits_{B_\varepsilon}|\nabla\mathbf{w}|^2dx=0,$$

and consequently, we have $\mathbf{w}=0$.

2. By subtracting the identity corresponding to $\mathbf{v}_2$ from the one corresponding to $\mathbf{v}_1$, we derive the following expression:

$$\nu\int\limits_{B_\varepsilon}\nabla\mathbf{w}\cdot\nabla\boldsymbol{\eta}dx+\int\limits_{B_\varepsilon}(\mathbf{w}\cdot\nabla)\mathbf{v}_1\cdot\boldsymbol{\eta}+(\mathbf{v}_2\cdot\nabla)\mathbf{w}\cdot\boldsymbol{\eta}dx$$
$$-\int\limits_{B_\varepsilon}(\boldsymbol{\eta}\cdot\nabla)\mathbf{v}_1\cdot\mathbf{w}+(\boldsymbol{\eta}\cdot\nabla)\mathbf{w}\cdot\mathbf{v}_2dx \tag{5.5.15}$$
$$=\int\limits_{B_\varepsilon}(\mathbf{f}_1-\mathbf{f}_2)\cdot\boldsymbol{\eta}dx-\sum_{j=N_1+1}^{N-1}\big(p_{1j}^*-p_{2j}^*\big)\int\limits_{\gamma_\varepsilon^j}\boldsymbol{\eta}\cdot\mathbf{n}dS,$$

where $\mathbf{w}=\mathbf{v}_1-\mathbf{v}_2$. By setting $\boldsymbol{\eta}=\mathbf{w}$ in identity (5.5.15), and utilizing Lemma 2.3.11 as well as estimate (5.5.5) for $\mathbf{v}_2$, we obtain the following expression:

$$\nu\int\limits_{B_\varepsilon}|\nabla\mathbf{w}|^2dx=-\int\limits_{B_\varepsilon}(\mathbf{v}_2\cdot\nabla)\mathbf{w}\cdot\mathbf{w}dx+\int\limits_{B_\varepsilon}(\mathbf{w}\cdot\nabla)\mathbf{w}\cdot\mathbf{v}_2dx+\int\limits_{B_\varepsilon}(\mathbf{f}_1-\mathbf{f}_2)\cdot\mathbf{w}dx$$

$$-\sum_{j=N_1+1}^{N-1}\left(p^*_{1j}-p^*_{2j}\right)\int\limits_{\gamma^j_\varepsilon}\mathbf{w}\cdot\mathbf{n}dS\le\|\mathbf{w}\|^2_{L^4(B_\varepsilon)}\|\nabla\mathbf{v}_2\|_{L^2(B_\varepsilon)}$$

$$+\|\nabla\mathbf{w}\|_{L^2(B_\varepsilon)}\|\mathbf{v}_2\|_{L^4(B_\varepsilon)}\|\mathbf{w}\|_{L^4(B_\varepsilon)}$$

$$+\|\mathbf{f}_2-\mathbf{f}_1\|_{L^2(B_\varepsilon)}\|\mathbf{w}\|_{L^2(B_\varepsilon)}$$

$$+c\varepsilon^{n/2}\sum_{j=N_1+1}^{N-1}|p^*_{1j}-p^*_{2j}|\|\nabla\mathbf{w}\|^2_{L^2(B_\varepsilon)}$$

$$\le c_0\varepsilon^\alpha\Big(\varepsilon^{n/2}\sum_{j=N_1+1}^{N-1}|p^*_{2j}|+\varepsilon\|\mathbf{f}_2\|_{L^2(B_\varepsilon)}\Big)\|\nabla\mathbf{w}\|^2_{L^2(B_\varepsilon)}$$

$$+\varepsilon c_*\|\nabla\mathbf{w}\|_{L^2(B_\varepsilon)}\|\mathbf{f}_2-\mathbf{f}_1\|_{L^2(B_\varepsilon)}$$

$$+c\varepsilon^{n/2}\sum_{j=N_1+1}^{N-1}|p^*_{1j}-p^*_{2j}|\|\nabla\mathbf{w}\|^2_{L^2(B_\varepsilon)},$$

where εc_* is the constant from the Poincaré inequality (2.3.36) for the domain B_ε (c_* is independent of ε). If condition (5.5.12) holds, the last inequality yields (5.5.13).

□

5.5.2 Recovery of the Bernoulli Pressure

Let us consider the weak solution $\mathbf{v}\in\widehat{K}^{1,2}(B_\varepsilon)$ to problem (5.5.2), satisfying the integral identity (5.5.4). We define a linear functional on $\widehat{W}^{1,2}(B_\varepsilon)$ as follows:

$$\begin{aligned}f(\eta)=\nu\int\limits_{B_\varepsilon}\nabla\mathbf{v}\cdot\nabla\eta\,dx+\int\limits_{B_e}(\mathbf{v}\cdot\nabla)\mathbf{v}\cdot\eta\,dx\\-\int\limits_{B_\varepsilon}(\eta\cdot\nabla)\mathbf{v}\cdot\mathbf{v}\,dx+\sum_{j=N_1+1}^{N}p_j\int\limits_{\gamma^j_\varepsilon}\eta\cdot\mathbf{n}dS-\int\limits_{B_\varepsilon}\mathbf{f}\cdot\eta\,dx.\end{aligned}\tag{5.5.16}$$

It is evident that $f(\eta)$ is bounded and, as a consequence of (5.5.4), we have $f(\eta)=0$ for every $\eta\in\widehat{K}^{1,2}(B_\varepsilon)\subset\widehat{W}^{1,2}(B_\varepsilon)$. Thus, due to Lemma 5.4.3, there exists a function $\Phi\in L^2(B_\varepsilon)$ such that

$$f(\boldsymbol{\eta}) = \int_{B_\varepsilon} \Phi \,\mathrm{div}\boldsymbol{\eta}\, dx \quad \forall \boldsymbol{\eta} \in \widehat{W}^{1,2}(B_\varepsilon). \tag{5.5.17}$$

Hence, the variational formulation (5.5.4) is equivalently stated as:

Find a pair $(\mathbf{v}, \Phi) \in \widehat{K}^{1,2}(B_\varepsilon) \times L^2(B_\varepsilon)$, such that for every $\boldsymbol{\eta} \in \widehat{W}^{1,2}(B_\varepsilon)$ the following integral identity holds:

$$\nu \int_{B_\varepsilon} \nabla\mathbf{v} \cdot \nabla\boldsymbol{\eta}\, dx + \int_{B_\varepsilon} (\mathbf{v}\cdot\nabla)\mathbf{v}\cdot\boldsymbol{\eta}\, dx - \int_{B_\varepsilon} (\boldsymbol{\eta}\cdot\nabla)\mathbf{v}\cdot\mathbf{v}\, dx$$
$$= \int_{B_\varepsilon} \Phi\,\mathrm{div}\boldsymbol{\eta}\, dx - \sum_{j=N_1+1}^{N} p_j \int_{\gamma_\varepsilon^j} \boldsymbol{\eta}\cdot\mathbf{n}\, dS + \int_{B_\varepsilon} \mathbf{f}\cdot\boldsymbol{\eta}\, dx. \tag{5.5.18}$$

Let $\mathbf{w} \in \widehat{W}^{1,2}(B_\varepsilon)$ be such that $\mathrm{div}\mathbf{w} = \Phi$ in B_ε and satisfy the condition

$$\|\nabla\mathbf{w}\|_{L^2(B_\varepsilon)} \le c\varepsilon^{-3/2}\|\Phi\|_{L^2(B_\varepsilon)}. \tag{5.5.19}$$

Substituting $\boldsymbol{\eta} = \mathbf{w}$ into Eq. (5.5.18), we derive

$$\int_{B_\varepsilon} |\Phi|^2\, dx = \nu \int_{B_\varepsilon} \nabla\mathbf{v}\cdot\nabla\mathbf{w}\, dx + \int_{B_\varepsilon} (\mathbf{v}\cdot\nabla)\mathbf{v}\cdot\mathbf{w}\, dx$$
$$- \int_{B_\varepsilon} (\mathbf{w}\cdot\nabla)\mathbf{v}\cdot\mathbf{v}\, dx + \sum_{j=N_1+1}^{N} p_j \int_{\gamma_\varepsilon^j} \mathbf{w}\cdot\mathbf{n}\, dS - \int_{B_\varepsilon} \mathbf{f}\cdot\mathbf{w}\, dx. \tag{5.5.20}$$

By estimating the right-hand side of (5.5.20) using inequalities (2.3.37), (2.3.38), (2.3.16), and (5.5.19), we obtain

$$\|\Phi\|^2_{L^2(B_\varepsilon)} \le c\Big(\|\nabla\mathbf{v}\|_{L^2(B_\varepsilon)}\|\nabla\mathbf{w}\|_{L^2(B_\varepsilon)} + 2\|\mathbf{v}\|_{L^4(B_\varepsilon)}\|\nabla\mathbf{v}\|_{L^2(B_\varepsilon)}\|\mathbf{w}\|_{L^4(B_\varepsilon)}$$
$$+ c\varepsilon^{n/2-1} \sum_{j=N_1+1}^{N} |p_j|\|\mathbf{w}\|_{L^2(B_\varepsilon)} + \|\mathbf{f}\|_{L^2(B_\varepsilon)}\|\mathbf{w}\|_{L^2(B_\varepsilon)}\Big)$$

$$\le c\Big(\varepsilon^{-3/2}\|\nabla\mathbf{v}\|_{L^2(B_\varepsilon)} + \varepsilon^{\alpha-3/2}\|\nabla\mathbf{v}\|^2_{L^2(B_\varepsilon)}$$
$$+ \varepsilon^{-1/2}\|\mathbf{f}\|_{L^2(B_\varepsilon)} + \varepsilon^{n/2-3/2} \sum_{j=N_1+1}^{N} |p_j|\Big)\|\Phi\|_{L^2(B_\varepsilon)},$$

where $\alpha = 1$ if $n = 2$ and $\alpha = 1/2$ if $n = 3$. Applying estimate (5.5.5), from the last inequality we derive

$$\|\Phi\|_{L^2(B_\varepsilon)} \le c\Big(\varepsilon^{n/2-3/2} \sum_{j=N_1+1}^{N} |p_j| + \varepsilon^{-1/2}\|\mathbf{f}\|_{L^2(B_\varepsilon)} + \varepsilon^{\alpha-3/2}\big[\varepsilon^{n/2} \sum_{j=N_1+1}^{N-1} |p_j^*| + \varepsilon\|\mathbf{f}\|_{L^2(B_\varepsilon)}\big]^2\Big). \tag{5.5.21}$$

Taking into consideration that $\alpha > 0$, we have proved the following theorem:

Theorem 5.5.3 *The variational formulations* (5.5.4) *and* (5.5.18) *are equivalent. The Bernoulli pressure* Φ *satisfies the estimate* (5.5.22)

$$\|\Phi\|_{L^2(B_\varepsilon)} \le c\Big(\varepsilon^{n/2-3/2} \sum_{j=N_1+1}^{N} |p_j| + \varepsilon^{-1/2}\|\mathbf{f}\|_{L^2(B_\varepsilon)}\Big). \tag{5.5.22}$$

5.5.3 Asymptotic Expansion of the Solution

In this section, we outline the procedure for constructing the asymptotic expansion of the solution to problem (5.5.2) when $\mathbf{f} = 0$. This construction closely resembles the approach employed in Sects. 5.3 and 5.4. The primary distinction lies in the introduction of boundary layer correctors for the vertices. The corresponding Stokes problems are formulated in semi-infinite cylinders with Neumann-type boundary conditions on the bases of these cylinders.

The asymptotic expansion is sought in the form (5.1.5)–(5.1.9) with only one modification for the half-bundles HB_{O_l}, $l = N_1 + 1, \dots, N$, i.e., for each half-cylinder having the base γ_ε^l (containing the vertex O_l): there is no use to multiply the Poiseuille terms for the velocity and the pressure by the cut-off function $\zeta\left(\frac{x_n^{(e)}}{3r\varepsilon}\right)$. So, the asymptotic expansions in these cylinders are given by formulas

$$\begin{aligned} \mathbf{u}^{(J)}(x) &= \mathbf{U}^{[e,J]}\left(\tfrac{x^{(e)\prime}}{\varepsilon}\right) + \left(1 - \zeta\left(\tfrac{|x-O_l|}{|e|_{\min}}\right)\right) \mathbf{U}^{[BLO_l,J]}\left(\tfrac{x-O_l}{\varepsilon}\right), \\ p^{(J)}(x) &= -s^{(e)}x_n^{(e)} + a^{(e)} + \tfrac{1}{\varepsilon}\left(1 - \zeta\left(\tfrac{|x-O_l|}{|e|_{\min}}\right)\right) P^{[BLO_l,J]}\left(\tfrac{x-O_l}{\varepsilon}\right), \end{aligned} \tag{5.5.23}$$

where e is the edge associated to the cylinder with a base γ_ε^l.

The equations for p_0, $\mathbf{U}_0^{(e)}$, $\mathbf{U}_0^{[BLO_l]}$, and $P_0^{[BLO_l]}$ are the same as in Sect. 5.4: see (5.4.34), (5.1.21), and (5.1.23), respectively. For the cylinders in Ω_l associated to vertices, i.e., when $l = N_1 + 1, \dots, N$, $\mathbf{U}_0^{[BLO_l]} = 0$ and $P_0^{[BLO_l]} = 0$. Next, p_1 satisfies problem (5.4.36) as in Sect. 5.4 because the first approximation of the pressure does not include the contribution of the terms of $\frac{1}{2}|\mathbf{v}|^2$, as the leading term of the pressure is of order ε^{-2} while the velocity magnitude is of order 1. The further terms are constructed as always by induction.

Steps 1 and 2 For $j > 0$ the construction of terms $\mathbf{U}_j^{(e)}$, $\mathbf{U}_j^{[BLO_l]}$ and $P_j^{[BLO_l]}$, $l = 1, \dots, N_1$, concern formulas (5.1.30) and (5.3.19)–(5.3.22) respectively (solving first as before this problem for $\mathbf{U}_j^{[BLO_l]}$, $\widehat{P}_j^{[BLO_l]}$) without the last term in (5.3.20)). Only for the terms $\mathbf{U}_j^{[BLO_l]}$ and $P_j^{[BLO_l]}$, $l = N_1 + 1, \dots, N$, the construction differs from Sect. 5.4. In this case when O_l are vertices, $\mathbf{U}_j^{[BLO_l]}$ and $\widehat{P}_j^{[BLO_l]}$ are solutions of the boundary layer type problems in the half-infinite cylinders with Neumann's boundary condition on the base. Subsequently, we address the problems within semi-cylinders Ω_l, $l = N_1 + 1, \dots, N$:

$$\begin{cases} -\nu\Delta_y \mathbf{U}_j^{[BLO_l]} + \nabla_y \widehat{P}_j^{[BLO_l]} = \mathbf{f}_j^{[REGO_l]} + \mathbf{f}_j^{[BLO_l]}, \quad y \in \Omega_l, \\ \mathrm{div}_y \mathbf{U}_j^{[BLO_l]} = 0, \quad y \in \Omega_l, \\ \mathbf{U}_j^{[BLO_l]} = 0, \quad y \in \partial\Omega_l \setminus \{y_n = 0\}, \\ \mathbf{U}_{j\tau}^{[BLO_l]} = 0, \quad y_n = 0, \\ \left(-\nu \dfrac{\partial \mathbf{U}_j^{[BLO_l]} \cdot \mathbf{n}}{\partial n} + \widehat{P}_j^{[BLO_l]} \right) = -\dfrac{1}{2} \sum\limits_{p+r=j-1} (\mathbf{U}_p^{(e)}(y^{(e)\prime}) + \mathbf{U}_p^{[BLO_l]}) \times \\ \qquad \times (\mathbf{U}_r^{(e)}(y^{(e)\prime}) + \mathbf{U}_r^{[BLO_l]}), \quad y_n = 0, \end{cases} \tag{5.5.24}$$

where the term $\dfrac{\partial \mathbf{U}_j^{[BLO_l]} \cdot \mathbf{n}}{\partial n}$ is fictive (is equal to zero),

$$\mathbf{f}_j^{[REGO_l]}(y^{(e)}) = 0, \tag{5.5.25}$$

$$\mathbf{f}_j^{[BLO_l]}(y^{(e)}) = - \sum_{p+r=j-1} (\mathbf{U}_p^{(e)}(y^{(e)\prime}) \cdot \nabla_y) \mathbf{U}_r^{[BLO_l]}(y)$$
$$- \sum_{p+r=j-1} (\mathbf{U}_p^{[BLO_l]}(y) \cdot \nabla_y) \mathbf{U}_r^{(e)}(y^{(e)\prime}) - \sum_{p+r=j-1} (\mathbf{U}_p^{[BLO_l]}(y) \cdot \nabla_y) \mathbf{U}_r^{[BLO_l]}(y). \tag{5.5.26}$$

Here, the pressure $\widehat{P}_j^{[BLO_l]}$ converges to a constant $\widehat{a}_j^{[BLO_l,e]}$ as $|y| \to \infty$. When $j = J$, the right-hand side of the boundary condition is substituted by

$$-\frac{1}{2} \sum_{J-1 \le p+r \le 2J} \varepsilon^{p+r-J+1} \big(\mathbf{U}_p^{(e)}(y^{(e)\prime}) + \mathbf{U}_p^{[BLO_l]}|_{y_n^{(e)}=0}\big)\big(\mathbf{U}_r^{(e)}(y^{(e)\prime}) + \mathbf{U}_r^{[BLO_l]}|_{y_n^{(e)}=0}\big), \; j = 0, \dots, J. \tag{5.5.27}$$

Problem (5.5.24) admits a unique solution $\mathbf{U}_j^{[BLO_l]} \in W^{1,2}(\Omega_l)$, $\nabla P_j^{[BLO_l]} \in L^2(\Omega_l)$. It can be proved in a standard way via the Riesz theorem. Using the same technique as in the case of Dirichlet boundary condition, one can prove the theorem on the exponential stabilization of the solution at infinity.

Note that in Sect. 5.4 the boundary layer correctors $\mathbf{U}_j^{[BLO_l]}$ and $P_j^{[BLO_l]}$ were absent in these cylinders, because the pressure boundary condition was satisfied exactly by the Poiseuille velocity.

At the **Step 3**, for p_{j+1} we set problem on the graph similar to (5.4.37): solve the conductivity problem on the graph for the function $p_{j+1}^{(e)}$ $(j < J)$:

$$\begin{cases} -\kappa_e \dfrac{\partial^2 p_{j+1}^{(e)}}{\partial x_n^{(e)2}}(x_n^{(e)}) = 0, & x_n^{(e)} \in (0, |e|), \\ -\displaystyle\sum_{e: O_l \in e} \kappa_e \dfrac{\partial p_{j+1}^{(e)}}{\partial x_n^{(e)}}(0) = 0, & l = 1, \dots, N_1, \\ \quad p_{j+1}^{(e)}(0) = \widehat{a}_j^{[BLO_l,e]}, \; l = N_1 + 1, \dots, N, & \\ p_{j+1}^{(e)}(0) - p_{j+1}^{(e_s)}(0) = \widehat{a}_j^{[BLO_l,e]}, \; \forall e \subset \mathcal{B}^l, e \neq e_s, & \end{cases}$$

Step 4 Finally, as in previous settings, we determine the pressure $P_j^{[BLO_l]}(y)$ for the boundary layer problems for $l = 1, \dots, N_1$:

$$P_j^{[BLO_l]}(y) = \widehat{P}_j^{[BLO_l]}(y) - \sum_{e: O_l \in e,\; e \neq e_s} \zeta\Big(\frac{y_n^{(e_{\alpha m})}}{3r}\Big)\widehat{a}_j^{[BLO_l,e]}.$$

Similarly, in the boundary layer problem (5.5.24), (5.5.25) for $l = N_1+1, \dots, N$, the pressure $P_j^{[BLO_l]}(y)$ is given by the equation

$$P_j^{[BLO_l]}(y) = \widehat{P}_j^{[BLO_l]}(y) - \widehat{a}_j^{[BLO_l,e]}$$

This step concludes the transition from the iteration j to $j + 1$.

5.5.4 Residual

Consider the asymptotic expansion $\left(\mathbf{u}^{(J)}, p^{(J)}\right)$ of order J (as given in Eqs. (5.1.5), (5.1.9) and (5.5.23)). As in the previous sections, it follows from the construction that

$$\mathbf{u}^{(J)} \in W^{2,2}(B_\varepsilon), \quad p^{(J)} \in W^{1,2}(B_\varepsilon). \tag{5.5.28}$$

Additionally, the magnitude of $\mathbf{u}^{(J)}$ in the L^4-norm is bounded by $\|\mathbf{u}^{(J)}\|_{L^4(B_\varepsilon)} \leq c\varepsilon^{(n-1)/4}$.

As before we introduce the divergence corrector, function $\mathbf{w}^{(J)}$ such that the sum $\widetilde{\mathbf{u}}^{(J)} = \mathbf{u} + \mathbf{w}^{(J)}$ is divergence free. Here $\mathbf{w}^{(J)}$ is the solution to the divergence equation

$$\operatorname{div}\mathbf{w}^{(J)} = \sum_{l=1}^{N} \nabla\zeta\left(\frac{|x - O_l|}{|e|_{min}}\right) \cdot \mathbf{U}^{[BLO_l,J]}(y) = -h^{(J)}(y).$$

Compute the residual $\mathbf{f}_1^{(J)} = -\mathcal{L}(\widetilde{\mathbf{u}}^{(J)}, p^{(J)})$, which coincides with the residual in Sect. 5.3. However, the boundary condition now is not satisfied completely: $(\widetilde{\mathbf{u}}^{(J)}, p^{(J)})$ has the constant residual in the boundary condition on γ_ε^l equal to $-\varepsilon^{J-1}\widehat{a}_J^{[BLO_l,e]}$.

It follows that

$$\|\mathbf{f}_1^{(J)}\|_{L^2(B_\varepsilon)} = O(\varepsilon^{J-2}), \tag{5.5.29}$$

where $\mathbf{f}_1^{(J)} = -\mathcal{L}(\widetilde{\mathbf{u}}^{(J)}, p^{(J)})$.

5.5.5 Error Estimate

Introduce the approximate Bernoulli pressure $\Phi^{(J)} = p^{(J)} + \frac{1}{2}(\mathbf{u}^{(J)})^2$.

Theorem 5.5.4 *The following error estimates are valid:*

$$\|\mathbf{v} - \widetilde{\mathbf{u}}^{(J)}\|_{W^{1,2}(B_\varepsilon)} = O(\varepsilon^{J+(n-1)/2}), \quad \|\Phi - \Phi^{(J)}\|_{L^2(B_\varepsilon)} = O(\varepsilon^{J+(n-3)/2}) \tag{5.5.30}$$

Proof
We introduce $\mathbf{u} = \mathbf{v} - \widetilde{\mathbf{u}}^{(J)}$. Integral identity (5.5.3) then takes the form

$$\nu \int\limits_{B_\varepsilon} \nabla \mathbf{u} \cdot \nabla \boldsymbol{\eta} dx + \int\limits_{B_\varepsilon} (\mathbf{u} \cdot \nabla)\mathbf{u} \cdot \boldsymbol{\eta} dx + \int\limits_{B_\varepsilon} (\mathbf{u} \cdot \nabla)\widetilde{\mathbf{u}}^{(J)} \cdot \boldsymbol{\eta} dx$$
$$+ \int\limits_{B_\varepsilon} (\widetilde{\mathbf{u}}^{(J)} \cdot \nabla)\mathbf{u} \cdot \boldsymbol{\eta} dx - \int\limits_{B_\varepsilon} (\boldsymbol{\eta} \cdot \nabla)\mathbf{u} \cdot \mathbf{u} dx$$
$$- \int\limits_{B_\varepsilon} (\boldsymbol{\eta} \cdot \nabla)\mathbf{u} \cdot \widetilde{\mathbf{u}}^{(J)} dx - \int\limits_{B_\varepsilon} (\boldsymbol{\eta} \cdot \nabla)\widetilde{\mathbf{u}}^{(J)} \cdot \mathbf{u} dx$$
$$+\varepsilon^{J-1} \sum_{l=N_1+1}^{N} \widehat{a}_J^{[BLO_l,e]} \int\limits_{\gamma_\varepsilon^l} \boldsymbol{\eta} \cdot \mathbf{n} dS - \int\limits_{B_\varepsilon} \mathbf{f}_1^{(J)} \cdot \boldsymbol{\eta} dx = 0,$$

where $\boldsymbol{\eta} \in \widehat{J}^{1,2}(B_\varepsilon)$.

By setting $\boldsymbol{\eta} = \mathbf{u}$, we derive

$$\nu \int\limits_{B_\varepsilon} |\nabla \mathbf{u}|^2 dx = -\varepsilon^{J-1} \sum_{l=N_1+1}^{N} \widehat{a}_J^{[BLO_l,e]} \int\limits_{\gamma_\varepsilon^l} \mathbf{u} \cdot \mathbf{n} dS$$
$$+ \int\limits_{B_\varepsilon} (\mathbf{u} \cdot \nabla)\mathbf{u} \cdot \widetilde{\mathbf{u}}^{(J)} dx - \int\limits_{B_\varepsilon} (\widetilde{\mathbf{u}}^{(J)} \cdot \nabla)\mathbf{u} \cdot \mathbf{u} dx + \int\limits_{B_\varepsilon} \mathbf{f}_1^{(J)} \cdot \mathbf{u} dx.$$

Applying Theorem 2.3.4 and Lemma 2.3.16, we can bound the term involving the boundary integrals as follows:

$$\left| \varepsilon^{J-1} \sum_{l=N_1+1}^{N} \widehat{a}_J^{[BLO_l,e]} \int\limits_{\gamma_\varepsilon^l} \mathbf{u} \cdot \mathbf{n} dS \right| \leq c\varepsilon^{J-1+n/2} \sum_{l=N_1+1}^{N} |\widehat{a}_J^{[BLO_l,e]}| \|\nabla \mathbf{u}\|_{L^2(B_\varepsilon)}.$$

Using the Hölder inequality, (2.3.37), (2.3.38), and the estimate $\|\widetilde{\mathbf{u}}^{(J)}\|_{L^4(B_\varepsilon)} \leq c\varepsilon^{(n-1)/4}$, we obtain

$$\left| \int\limits_{B_\varepsilon} (\mathbf{u} \cdot \nabla)\mathbf{u} \cdot \widetilde{\mathbf{u}}^{(J)} dx \right| + \left| \int\limits_{B_\varepsilon} (\widetilde{\mathbf{u}}^{(J)} \cdot \nabla)\mathbf{u} \cdot \mathbf{u} dx \right| \leq 2\|\mathbf{u}\|_{L^4(B_\varepsilon)} \|\nabla \mathbf{u}\|_{L^2(B_\varepsilon)} \|\widetilde{\mathbf{u}}^{(J)}\|_{L^4(B_\varepsilon)}$$
$$\leq c\varepsilon^{\alpha + \frac{n-1}{4}} \|\nabla \mathbf{u}\|^2_{L^2(B_\varepsilon)} \leq c\varepsilon^{\frac{3}{4}} \|\nabla \mathbf{u}\|^2_{L^2(B_\varepsilon)},$$

where $\alpha = 1/2$ for $n = 2$ and $\alpha = 1/4$ for $n = 3$.

Furthermore, applying Poincaré's inequality, we bound the term related to $\mathbf{f}_1^{(J)}$:

$$\left| \int\limits_{B_\varepsilon} \mathbf{f}_1^{(J)} \cdot \mathbf{u} dx \right| \leq \varepsilon c \|\mathbf{f}_1^{(J)}\|_{L^2(B_\varepsilon)} \|\nabla \mathbf{u}\|_{L^2(B_\varepsilon)}.$$

Combining the above estimates, we arrive at

$$\nu \int\limits_{B_\varepsilon} |\nabla \mathbf{u}|^2 dx \le c\varepsilon^{\frac{2(J-1)+n}{2}} \sum_{l=N_1+1}^{N} |\widehat{a}_J^{[BLO_l,e]}| \|\nabla \mathbf{u}\|_{L^2(B_\varepsilon)}$$
$$+c\varepsilon^{\frac{3}{4}} \|\nabla \mathbf{u}\|^2_{L^2(B_\varepsilon)} + \varepsilon c \|\mathbf{f}_1^{(J)}\|_{L^2(B_\varepsilon)} \|\nabla \mathbf{u}\|_{L^2(B_\varepsilon)}.$$

This yields

$$\|\nabla \mathbf{u}\|_{L^2(B_\varepsilon)} \le \frac{C}{\nu - c\varepsilon^{3/4}} (\varepsilon^{\frac{2(J-1)+n}{2}} \sum_{l=N_1+1}^{N} |\widehat{a}_J^{[BLO_l,e]}| + \varepsilon \|\mathbf{f}_1^{(J)}\|_{L^2(B_\varepsilon)})$$
$$\le \frac{C}{\nu - c\varepsilon^{3/4}} (\varepsilon^{\frac{2(J-1)+n}{2}} \sum_{l=N_1+1}^{N} |\widehat{a}_J^{[BLO_l,e]}| + \varepsilon^{J-1}).$$

If $\nu - c\varepsilon^{3/4} > \nu/2$, then

$$\|\mathbf{u}\|_{W^{1,2}(B_\varepsilon)} = \|\mathbf{v} - \widetilde{\mathbf{u}}^{(J)}\|_{W^{1,2}(B_\varepsilon)} = O(\varepsilon^{J-1}). \tag{5.5.31}$$

By evaluating the norm of the difference between $\widetilde{\mathbf{u}}^{(J)}$ and $\widetilde{\mathbf{u}}^{(J+2)}$, we find

$$\|\widetilde{\mathbf{u}}^{(J)} - \widetilde{\mathbf{u}}^{(J+2)}\|_{W^{1,2}(B_\varepsilon)} = O(\varepsilon^{J+(n-1)/2}).$$

Replacing J by $J+2$ in (5.5.31), we deduce

$$\|\mathbf{v} - \widetilde{\mathbf{u}}^{(J+2)}\|_{W^{1,2}(B_\varepsilon)} = O(\varepsilon^{J+1}).$$

Applying the triangle inequality, we obtain the first estimate (5.5.30).

Define on $\widehat{W}^{1,2}(B_\varepsilon)$ the linear functional

$$\mathcal{N}(\eta) = \nu \int\limits_{B_\varepsilon} \nabla \mathbf{u} \cdot \nabla \eta dx + \int\limits_{B_\varepsilon} (\mathbf{u} \cdot \nabla)\mathbf{u} \cdot \eta dx + \int\limits_{B_\varepsilon} (\mathbf{u} \cdot \nabla)\widetilde{\mathbf{u}}^{(J)} \cdot \eta dx$$
$$+ \int\limits_{B_\varepsilon} (\widetilde{\mathbf{u}}^{(J)} \cdot \nabla)\mathbf{u} \cdot \eta dx - \int\limits_{B_\varepsilon} (\eta \cdot \nabla)\mathbf{u} \cdot \mathbf{u} dx$$
$$- \int\limits_{B_\varepsilon} (\eta \cdot \nabla)\mathbf{u} \cdot \widetilde{\mathbf{u}}^{(J)} dx - \int\limits_{B_\varepsilon} (\eta \cdot \nabla)\widetilde{\mathbf{u}}^{(J)} \cdot \mathbf{u} dx$$
$$+ \varepsilon^{J-1} \sum_{l=N_1+1}^{N} \widehat{a}_J^{[BLO_l,e]} \int\limits_{\gamma_\varepsilon^l} \eta \cdot \mathbf{n} dS - \int\limits_{B_\varepsilon} \mathbf{f}_1^{(J)} \cdot \eta dx$$

It is easy to see that it is bounded. Using the same reasoning as in the linear case (Stokes equation), we establish the existence of a unique function $\Psi \in L^2(B_\varepsilon)$ such that $\mathcal{N}(\eta) = \int_{B_\varepsilon} \Psi(x) \mathrm{div}\eta dx$ for all $\eta \in \widehat{W}^{1,2}(B_\varepsilon)$. Applying the same arguments as before, we represent the function Ψ in the form $\Psi = \Phi - \Phi^{(J)}$ and evaluate its norm via the norms $\|\mathbf{u}\|_{W^{1,2}(B_\varepsilon)}$ and $\|\mathbf{f}_1^{(J)}\|_{L^2(B_\varepsilon)}$ and get the estimate (5.5.30). □

5.5.6 MAPDD for Bernoulli Boundary Conditions

The weak formulation of the MAPDD method for the problem (5.5.2), with a vanishing external force $\mathbf{f} = 0$, is presented as follows:

Definition 5.5.2 Find a vector field $\mathbf{v}_d \in \widehat{\mathcal{H}}_{\mathrm{dec}}(B_\varepsilon, \delta)$ such that for all $\eta \in \widehat{\mathcal{H}}_{\mathrm{dec}}(B_\varepsilon, \delta)$ the following integral identity is satisfied:

$$\nu \int\limits_{B_\varepsilon} \nabla \mathbf{v}_d \cdot \nabla \eta \, dx + \int\limits_{B_\varepsilon} (\mathbf{v}_d \cdot \nabla)\mathbf{v}_d \cdot \eta \, dx - \int\limits_{B_\varepsilon} (\eta \cdot \nabla)\mathbf{v}_d \cdot \mathbf{v}_d \, dx$$
$$= - \sum_{j=N_1+1}^{N-1} p_j^* \int\limits_{\gamma_\varepsilon^j} \eta \cdot \mathbf{n} \, dS. \tag{5.5.32}$$

This formulation can also be equivalently expressed with functions defined only on $\bigcup\limits_{i=1}^{N} B_i^{\varepsilon,\delta}$:

Definition 5.5.3 Find a divergence-free vector field $\mathbf{v}_d \in \widehat{\widehat{W}}^{1,2}\left(\bigcup\limits_{i=1}^{N} B_i^{\varepsilon,\delta}\right)$ that satisfies, for all divergence-free $\eta \in \widehat{\widehat{W}}^{1,2}\left(\bigcup\limits_{i=1}^{N} B_i^{\varepsilon,\delta}\right)$, the integral identity:

$$\nu \sum_{i=1}^{N} \int\limits_{B_i^{\varepsilon,\delta}} \nabla \mathbf{v}_d \cdot \nabla \eta \, dx + \sum_{i=1}^{N} \int\limits_{B_i^{\varepsilon,\delta}} (\mathbf{v}_d \cdot \nabla)\mathbf{v}_d \cdot \eta \, dx$$
$$- \sum_{i=1}^{N} \int\limits_{B_i^{\varepsilon,\delta}} (\eta \cdot \nabla)\mathbf{v}_d \cdot \mathbf{v}_d \, dx + \nu \sum_{l=1}^{M} d_l \int\limits_{\sigma_\varepsilon^{(e_l)}} \nabla_{x^{(e_l)\prime}} \mathbf{v}_d \cdot \nabla_{x^{(e_l)\prime}} \eta \, dx^{(e_l)\prime} \tag{5.5.33}$$
$$= - \sum_{j=N_1+1}^{N-1} p_j^* \int\limits_{\gamma_\varepsilon^j} \eta \cdot \mathbf{n} \, dS.$$

It is important to note that the integrals over $\sigma_\varepsilon^{(e_l)}$ of the nonlinear terms vanish.

Here, $d_l = |e| - 2\delta$ represents the distance between the cross-sections S_{ij} and S_{ji} that intersect the segment e.

We can also consider an equivalent formulation involving both the divergence-free vector field $\mathbf{v}_d$ and the "MAPDD pressure" p_d:

Definition 5.5.4 Find a divergence-free vector field $\mathbf{v}_d \in \widehat{\widetilde{W}}^{1,2}\left(\bigcup_{i=1}^{N} B_i^{\varepsilon,\delta}\right)$ and the Bernoulli pressure field $\Phi_d \in L^2\left(\bigcup_{i=1}^{N} B_i^{\varepsilon,\delta}\right)$ that satisfy, for all $\eta \in \widehat{\widetilde{W}}^{1,2}\left(\bigcup_{i=1}^{N} B_i^{\varepsilon,\delta}\right)$, the following integral identity:

$$
\begin{aligned}
&\nu \sum_{i=1}^{N} \int_{B_i^{\varepsilon,\delta}} \nabla \mathbf{v}_d \cdot \nabla \eta \, dx + \sum_{i=1}^{N} \int_{B_i^{\varepsilon,\delta}} (\mathbf{v}_d \cdot \nabla) \mathbf{v}_d \cdot \eta \, dx \\
-&\sum_{i=1}^{N} \int_{B_i^{\varepsilon,\delta}} (\eta \cdot \nabla) \mathbf{v}_d \cdot \mathbf{v}_d \, dx + \nu \sum_{l=1}^{M} d_l \int_{\sigma_\varepsilon^{(e_l)}} \nabla_{x^{(e_l)\prime}} \mathbf{v}_d \cdot \nabla_{x^{(e_l)\prime}} \eta \, dx^{(e_l)\prime} \\
&= - \sum_{j=N_1+1}^{N} p_j \int_{\gamma_\varepsilon^j} \eta \cdot \mathbf{n} \, dS + \sum_{i=1}^{N} \int_{B_i^{\varepsilon,\delta}} \Phi_d \operatorname{div} \eta \, dx.
\end{aligned}
\tag{5.5.34}
$$

It is worth noting that in this definition the test functions η are not divergence-free.

Theorem 5.5.5 *Let δ be $C_{J+2n+1}\varepsilon|\ln\varepsilon|$, where the constant C_{J+2n+1} satisfies* (5.1.60) *with J replaced by $J' = J + 2n + 1$. Then, for sufficiently small ε, there exists a unique MAPDD solution $(\mathbf{v}_d, \Phi_d)$ that satisfies the following error estimate:*

$$
\|\mathbf{v} - \mathbf{v}_d\|_{W^{1,2}(B_\varepsilon)} = O(\varepsilon^{J+(n-1)/2}). \tag{5.5.35}
$$

Moreover, for the MAPDD Bernoulli pressure the following error estimate holds:

$$
\|\Phi - \Phi_d\|_{L^2(\bigcup_{i=1}^{N} B_i^{\varepsilon,\delta})} = O(\varepsilon^{J+(n-3)/2}). \tag{5.5.36}
$$

Proof

First, we demonstrate the existence of the divergence-free vector field $\mathbf{v}_d \in \widehat{\mathcal{H}}_{\text{dec}}(B_\varepsilon, \delta)$ which satisfies the identity (5.5.32). The proof closely resembles that of Theorem 5.5.1. We seek the solution $\mathbf{v}_d$ in the form $\mathbf{v}_d = \widetilde{\mathbf{u}}_d + \widetilde{\mathbf{u}}^{(J')}$. Then, the integral identity (5.5.32) can be rewritten as follows:

$$
\begin{aligned}
&\nu \int_{B_\varepsilon} \nabla \widetilde{\mathbf{u}}_d \cdot \nabla \eta dx + \int_{B_\varepsilon} ((\widetilde{\mathbf{u}}_d + \widetilde{\mathbf{u}}^{(J')}) \cdot \nabla) \widetilde{\mathbf{u}}_d \cdot \eta dx \\
&+ \int_{B_\varepsilon} (\widetilde{\mathbf{u}}_d \cdot \nabla) \widetilde{\mathbf{u}}^{(J')} \cdot \eta dx - \int_{B_\varepsilon} (\eta \cdot \nabla)(\widetilde{\mathbf{u}}_d + \widetilde{\mathbf{u}}^{(J')}) \cdot \widetilde{\mathbf{u}}_d dx - \int_{B_\varepsilon} (\eta \cdot \nabla) \widetilde{\mathbf{u}}_d \cdot \widetilde{\mathbf{u}}^{(J')} dx \\
&= -\varepsilon^{J'-1} \sum_{l=N_1+1}^{N} \widehat{a}_J^{[BLO_l,e]} \int_{\gamma_\varepsilon^l} \eta \cdot \mathbf{n} dS + \int_{B_\varepsilon} \mathbf{f}_1^{(J')} \cdot \eta dx,
\end{aligned}
\tag{5.5.37}
$$

where $\mathbf{f}_1^{(J')} = -(-\nu \Delta \widetilde{\mathbf{u}}^{(J')} + (\widetilde{\mathbf{u}}^{(J')} \cdot \nabla)\widetilde{\mathbf{u}}^{(J')} - (\widetilde{\mathbf{u}}^{(J')} \cdot (\nabla \widetilde{\mathbf{u}}^{(J')})^t + \nabla \Phi^{(J')})$. The identity (5.5.37) is equivalent to an operator equation in the space $\widehat{\mathcal{H}}_{\text{dec}}(B_\varepsilon, \delta)$:

$$
\widetilde{\mathbf{u}}_d = \mathcal{A} \widetilde{\mathbf{u}}_d, \tag{5.5.38}
$$

where $\mathcal{A}$ is a compact operator. Hence, the existence of the fixed point of (5.5.38) follows from the Leray–Schauder theorem if we establish an a priori estimate of the solution.

Taking $\eta = \widetilde{\mathbf{u}}_d$ in the identity (5.5.37) and applying Lemmas 2.3.11, 2.3.16, the Poincaré inequality (2.3.9), and the estimates (5.5.29) and $\|\nabla \widetilde{\mathbf{u}}^{(J')}\|_{L^2(B_\varepsilon)} \le c\varepsilon^{\frac{n-3}{2}}$, we obtain the following relations:

$$
\begin{aligned}
\nu \|\nabla \widetilde{\mathbf{u}}_d\|^2_{L^2(B_\varepsilon)} &= - \int_{B_\varepsilon} (\widetilde{\mathbf{u}}^{(J')} \cdot \nabla) \widetilde{\mathbf{u}}_d \cdot \widetilde{\mathbf{u}}_d dx + \int_{B_\varepsilon} (\widetilde{\mathbf{u}}_d \cdot \nabla) \widetilde{\mathbf{u}}_d \cdot \widetilde{\mathbf{u}}^{(J')} dx \\
&\quad -\varepsilon^{J'-1} \sum_{l=N_1+1}^{N} \widehat{a}_J^{[BLO_l,e]} \int_{\gamma_\varepsilon^l} \widetilde{\mathbf{u}}_d \cdot \mathbf{n} dS - \int_{B_\varepsilon} \mathbf{f}_1^{(J')} \cdot \widetilde{\mathbf{u}}_d dx \\
&\le \|\widetilde{\mathbf{u}}_d\|_{L^4(B_\varepsilon)} \|\nabla \widetilde{\mathbf{u}}_d\|_{L^2(B_\varepsilon)} \|\widetilde{\mathbf{u}}^{(J')}\|_{L^4(B_\varepsilon)} + c\varepsilon^{J'-1} \varepsilon^{(n-1)/2} \sum_{l=N_1+1}^{N} \|\widetilde{\mathbf{u}}_d\|_{L^2(\gamma_\varepsilon^l)} \\
&\quad + \|\mathbf{f}_1^{(J')}\|_{L^2(B_\varepsilon)} \|\widetilde{\mathbf{u}}_d\|_{L^2(B_\varepsilon)} \le c\varepsilon^{\alpha} \|\nabla \widetilde{\mathbf{u}}^{(J')}\|_{L^2(B_\varepsilon)} \|\nabla \widetilde{\mathbf{u}}_d\|^2_{L^2(B_\varepsilon)} \\
&\quad + c\varepsilon^{J'-1+n/2} \|\nabla \widetilde{\mathbf{u}}_d\|_{L^2(B_\varepsilon)} + c\varepsilon \|\mathbf{f}_1^{(J')}\|_{L^2(B_\varepsilon)} \|\nabla \widetilde{\mathbf{u}}_d\|_{L^2(B_\varepsilon)} \\
&\quad \le c_* \varepsilon^{1/2} \|\nabla \widetilde{\mathbf{u}}_d\|^2_{L^2(B_\varepsilon)} + c\varepsilon^{J'-1} \|\nabla \widetilde{\mathbf{u}}_d\|_{L^2(B_\varepsilon)}.
\end{aligned}
\tag{5.5.39}
$$

For $c_* \varepsilon^{1/2} < \nu$, the last inequality implies

$$
\|\mathbf{v}_d - \widetilde{\mathbf{u}}^{(J')}\|_{\widehat{\mathcal{H}}_{\text{dec}}(B_\varepsilon,\delta)} = \|\widetilde{\mathbf{u}}_d\|_{\widehat{\mathcal{H}}_{\text{dec}}(B_\varepsilon,\delta)} = \|\nabla \widetilde{\mathbf{u}}_d\|_{L^2(B_\varepsilon)} \le c\varepsilon^{J'-1} \le c\varepsilon^{J+2n}.
\tag{5.5.40}
$$

Hence, the problem (5.5.33) has a solution $\widetilde{\mathbf{u}}_d \in \widehat{\mathcal{H}}_{\text{dec}}(B_\varepsilon, \delta)$ that satisfies the estimate (5.5.40).

Now, consider the functional

$$f(\boldsymbol{\eta}) = \nu \sum_{i=1}^{N} \int\limits_{B_i^{\varepsilon,\delta}} \nabla \mathbf{v}_d \cdot \nabla \boldsymbol{\eta} dx + \sum_{i=1}^{N} \int\limits_{B_i^{\varepsilon,\delta}} (\mathbf{v}_d \cdot \nabla)\mathbf{v}_d \cdot \boldsymbol{\eta} dx$$

$$- \sum_{i=1}^{N} \int\limits_{B_i^{\varepsilon,\delta}} (\boldsymbol{\eta} \cdot \nabla)\mathbf{v}_d \cdot \mathbf{v}_d dx + \nu \sum_{l=1}^{M} d_l \int\limits_{\sigma_\varepsilon^{(e_l)}} \nabla_{x^{(e_l)'}} \mathbf{v}_d \cdot \nabla_{x^{(e_l)'}} \boldsymbol{\eta} dx^{(e_l)'}$$

$$+ \sum_{j=N_1+1}^{N} p_j \int\limits_{\gamma_\varepsilon^j} \boldsymbol{\eta} \cdot \mathbf{n} dS$$

defined for $\boldsymbol{\eta} \in \widehat{\widetilde{W}}^{1,2}\big(\bigcup_{i=1}^{N} B_i^{\varepsilon,\delta}\big)$. It is evident that $f(\boldsymbol{\eta})$ is a linear bounded functional, and from (5.5.33), it follows that $f(\boldsymbol{\eta})$ becomes zero for divergence-free $\boldsymbol{\eta}$. Thus, according to Theorem 5.4.4, it can be represented as

$$f(\boldsymbol{\eta}) = \sum_{i=1}^{N} \int\limits_{B_i^{\varepsilon,\delta}} \Phi_d \text{div} \boldsymbol{\eta} dx,$$

where $\Phi_d \in L^2(\bigcup_{i=1}^{N_1} B_i^{\varepsilon,\delta})$. Therefore, the pair $(\mathbf{v}_d, \Phi_d)$ satisfies the integral identity (5.5.34).

Substituting $\mathbf{v}_d = \widetilde{\mathbf{u}}_d + \widetilde{\mathbf{u}}^{(J')}$ and $\Phi_d = \widetilde{q}_d + \Phi^{(J')}$ into (5.5.34), and taking the test function $\boldsymbol{\eta} = \mathbf{U} \in \widehat{\mathcal{W}}_{\text{dec}}(B_\varepsilon, \delta)$, where $\text{div}\mathbf{U} = \widetilde{q}_d$ in $B_\varepsilon \backslash \bigcup_{i=1}^{N} B_i^{\varepsilon,\delta}$ and $\text{div}\mathbf{U} = 0$ in $B_\varepsilon \backslash \bigcup_{i=1}^{N} B_i^{\varepsilon,\delta}$, and

$$\|\mathbf{U}\|^2_{W^{1,2}(B_\varepsilon)} \le C\varepsilon^{-2n-3/2} \sum_{i=1}^{N} \|\widetilde{q}_d\|^2_{L^2(B_i^{\varepsilon,\delta})}, \tag{5.5.41}$$

with a constant C independent of ε (as given by Lemma 5.4.5), we can derive the following estimate:

$$\sum_{i=1}^{N} \int_{B_i^{\varepsilon,\delta}} |\widetilde{q}_d|^2 dx \le c\Big(\big(\|\nabla \widetilde{\mathbf{u}}_d\|_{L^2(B_\varepsilon)} + \varepsilon \|\mathbf{f}_1^{(J')}\|\big)\|\nabla \mathbf{U}\|_{L^2(B_\varepsilon)}$$

$$+\big(\|\widetilde{\mathbf{u}}_d\|_{L^4(B_\varepsilon)}\|\nabla \widetilde{\mathbf{u}}_d\|_{L^2(B_\varepsilon)} + \|\widetilde{\mathbf{u}}^{(J')}\|_{L^4(B_\varepsilon)}\|\nabla \widetilde{\mathbf{u}}^{(J')}\|_{L^2(B_\varepsilon)}\big)\|\mathbf{U}\|_{L^4(B_\varepsilon)}\Big)$$

$$\le c\Big(\varepsilon^{J'-1} + \varepsilon^{\alpha}\big(\|\nabla \widetilde{\mathbf{u}}_d\|^2_{L^2(B_\varepsilon)} + \|\nabla \widetilde{\mathbf{u}}^{(J')}\|_{L^2(B_\varepsilon)}\|\nabla \widetilde{\mathbf{u}}_d\|_{L^2(B_\varepsilon)}\big)\Big)\|\nabla \mathbf{U}\|_{L^2(B_\varepsilon)}$$

$$\le c\big(\varepsilon^{J'-1} + \varepsilon^{\alpha}\big(\varepsilon^{2J'-2} + \varepsilon^{J'-1+(n-3)/2}\big)\varepsilon^{-n-3/4}\big)\Big(\sum_{i=1}^{N} \|\widetilde{q}_d\|^2_{L^2(B_i^{\varepsilon,\delta})}\Big)^{1/2}$$

$$\le c\varepsilon^{J'-n-7/4}\Big(\sum_{i=1}^{N} \|\widetilde{q}_d\|^2_{L^2(B_i^{\varepsilon,\delta})}\Big)^{1/2}. \tag{5.5.42}$$

Taking into consideration that $J' = J + 2n + 1$, we obtain the following estimate for $\Phi_d - \Phi^{(J')}$:

$$\Big(\sum_{i=1}^{N} \|\Phi_d - \Phi^{(J')}\|^2_{L^2(B_i^{\varepsilon,\delta})}\Big)^{1/2} \le O(\varepsilon^{J+n-3/4}). \tag{5.5.43}$$

By utilizing (5.5.43) and the inequality (5.5.30) for $\Phi - \Phi^{(J')}$, we can deduce the subsequent estimate:

$$\Big(\sum_{i=1}^{N} \|\Phi - \Phi_d\|^2_{L^2(B_i^{\varepsilon,\delta})}\Big)^{1/2} \le O(\varepsilon^{J+n-3/4}), \tag{5.5.44}$$

which yields estimate (5.5.36). □

Note that now we can compute in $\bigcup_{i=1}^{N} B_i^{\varepsilon,\delta}$ the MAPDD pressure $p_d = \Phi_d - \frac{1}{2}|\mathbf{u}_d|^2$ and then extend p_d to the cylinders $B_{ij}^{dec,\varepsilon}$ as previously described (see Sect. 5.2):

$$p_d(x_n^{(e)}) = \langle p_d \rangle_{S_{ij}} + \frac{\langle p_d \rangle_{S_{ji}} - \langle p_d \rangle_{S_{ij}}}{|e| - 2\delta + \varepsilon}\Big(x_n^{(e)} - \delta + \frac{\varepsilon}{2}\Big). \tag{5.5.45}$$

Theorem 5.5.6 *The MAPDD pressure p_d satisfies the the error estimate:*

$$\|p - p_d\|_{L^2(B_\varepsilon)} = O(\varepsilon^{J+(n-3)/2}).$$

Theorems 5.5.5 and 5.5.6 justify the MAPDD for the Navier-Stokes equation in thin tube structure with the Bernoulli boundary conditions at the inflows/outflows.

5.6 Numerical Tests of the MAPDD for Stationary Navier–Stokes Equations in Tube Structure

The theoretical framework developed in the preceding chapter provides a rigorous foundation for the employment of the Method of Asymptotic Partial Decomposition of the Domain (MAPDD). By selecting an appropriate value for the parameter δ in the form $\delta = C_J\varepsilon|\ln\varepsilon|$, the resulting error has the order $O(\varepsilon^J)$. This theoretical validation remains applicable within the regime of moderate Reynolds numbers, typically of order one. In fluid dynamics, the Reynolds number is defined as the ratio vL/ν, where v and L represent characteristic velocity and characteristic linear dimensions of the fluid flow, respectively, while ν denotes the kinematic viscosity.

The present section aims to ascertain the practical limits of the method, as the theoretical constraints might be overly cautious. To this end, conducting numerical tests on the MAPDD becomes valuable. In the subsequent analysis, we subject the MAPDD approach to examination using a simplified Y-shaped thin tube structure. This structure incorporates additional cavities modeled as segments of disks, serving to simulate regions of clot formation within a network of vessels. By varying the flux at the inlet, we quantify the discrepancy between the numerical solution obtained through the full-dimensional model and the solution derived through the MAPDD technique. The comparative outcomes are presented in tabular format, outlining the relative errors in the velocity field with respect to the L^2-norm. Additionally, visual depictions of the computed velocities magnitude for both the full-dimensional model and the reduced geometry are provided in Figs. 5.6, 5.7, 5.8, 5.9, 5.10, and 5.11.

Section 6.11 extends analogous tests to the context of non-stationary Navier–Stokes equations within tube structures.

5.6.1 Numerical Setup

In this section, we outline the numerical setup for evaluating the MAPDD method in solving stationary Navier–Stokes equations within a simple Y-shaped thin tube structure $B_\varepsilon \subset \mathbb{R}^2$. The geometry is defined by the graph $\mathcal{B} = \overline{O_2O_1} \cup \overline{O_1O_3} \cup \overline{O_1O_4}$, where $O_1 = (0, 0)$, $O_2 = (-1, 0)$, $O_3 = (\frac{\sqrt{2}}{2}, \frac{\sqrt{2}}{2})$, and $O_4 = (\frac{\sqrt{2}}{2}, -\frac{\sqrt{2}}{2})$. The chosen value of ε is 0.05, which characterizes the thickness of different components within the structure: the thickness of the horizontal rectangle is equal to 4ε, while two other rectangles $\Pi_\varepsilon^{(e)}$ have thickness 2ε.

Denote $e_1 = \overline{O_1O_2}, e_2 = \overline{O_3O_1}, e_3 = \overline{O_4O_1}$ and choose the origins of the Cartesian local coordinates systems respectively at O_1 for e_1, O_3 for e_2 and O_4 for e_3. We also introduce the "auxiliary domains" ω_ε^i, so that finally we get the shape of B_ε as in Fig. 5.4. It simulates three vessels with clots.

Let us choose 1 cm as a unit, so that $\overline{O_1O_2} = 1\,\text{cm}$.

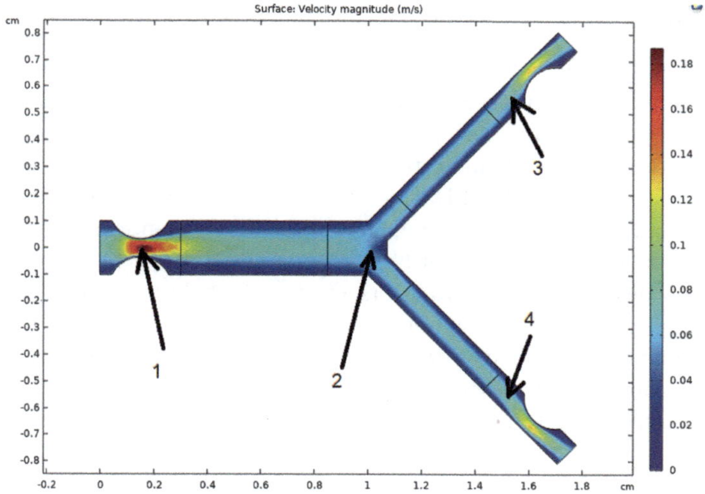

Fig. 5.4 Subdomain1: modulus of the velocity field computed in the whole domain for $F = 50$

The numerical setup, thus established, serves as the foundation for evaluating the MAPDD method efficacy in addressing stationary Navier–Stokes equations within this Y-shaped thin tube structure.

Consider the stationary Navier-Stokes equations:

$$\begin{cases} -\nu\Delta\mathbf{v} + (\mathbf{v}\cdot\nabla)\mathbf{v} + \nabla p = 0, \\ \qquad\qquad\qquad \operatorname{div}\mathbf{v} = 0. \end{cases} \tag{5.6.1}$$

These equations are defined within the domain B_ε. It is assumed that the pressures at the sections of the boundary γ_ε^j containing the points O_j, $j = 3, 4$ are set to be equal to zero. Additionally, on the lateral boundary Γ, a no-slip boundary condition is imposed

$$\mathbf{v} = 0.$$

At the "inflow" region ($x_1 = 0, x_2 \in (-2\varepsilon, 2\varepsilon)$), the normal velocity component is given by $-\frac{3F}{32\varepsilon^3}(x_2^2 - 4\varepsilon^2)$, while the tangential velocity is zero. Here, the parameter F, referred to as the flux, takes values 1, 5, 10, 50, 100, or 1000.

Despite the lack of C^2-smoothness in the neighborhood of the vertices O_1 and O_3, we consider the reduced geometry for the MAPDD. This involves dividing the domain B_ε into smaller segments labeled 1, 2, 3, and 4. These segments are truncated at distances of 6ε, 4ε, 8ε, and 8ε from O_1, O_2, O_3, and O_4 respectively.

It is noteworthy that in this simplified case, the graph equation provides exact flux values: F for the horizontal tube and $F/2$ for the two remaining tubes.

Consequently, the normal Poiseuille velocities in local coordinates can be explicitly calculated as $-\frac{3F}{32\varepsilon^3}(x_2^2 - 4\varepsilon^2)$ for the horizontal tube and $-\frac{3F}{8\varepsilon^3}((x_2^{(e)})^2 - \varepsilon^2)$ for the other two tubes (taking O_2 as the origin of the local coordinate system). These velocity values are then employed as boundary conditions at the truncation surfaces S_{ij} for the Navier–Stokes equations in the domains 1, 2, 3, and 4. Consequently, in this particular scenario, the MAPDD method reduces the full-dimensional problem to four independent Navier–Stokes problems in smaller domain sections. These four problems can be solved simultaneously.

Figure 5.5 displays the relative errors in these four domains for fluxes $F = 1, 5, 10$, comparing the exact solutions with those obtained using MAPDD, in the L^2-norm. Notably, the relative error increases as the inlet flux F (and consequently, the Reynolds number) increases. Despite this, the method maintains a high level of accuracy. This increase in error can be attributed to the presence of boundary

First domain		
F_inl	Absolute error	Relative error, %
1	1,25538*10^{-13}	0,04485
5	9,70348*10^{-12}	0,1424
10	7,6758*10^{-11}	0,2912

Second domain		
F_inl	Absolute error	Relative error, %
1	3,7353*10^{-13}	0,07375
5	1,37544*10^{-11}	1,048
10	1,04831*10^{-10}	2,213

Third domain		
F_inl	Absolute error	Relative error, %
1	3,67956*10^{-13}	0,04076
5	1,4371*10^{-11}	0,06192
10	8,20413*10^{-11}	0,08636

Fourth domain		
F_inl	Absolute error	Relative error, %
1	2,40171*10^{-13}	0,02684
5	7,28105*10^{-12}	0,03172
10	3,30086*10^{-11}	0,03509

Fig. 5.5 Relative error of the MAPDD for the fluxes $F = 1, 5, 10$

layers near nodes and vertices. For instance, near the node O_1, the Poiseuille velocities must be corrected by the exponentially decaying boundary layer term $\varepsilon^2 \mathbf{U}_0^{BLO_l}(x/\varepsilon)$. In the small boundary layer zone where the contribution of the boundary layer corrector $\varepsilon^2 \mathbf{U}_0^{BLO_l}(x/\varepsilon)$ is important, the solution differs from the Poiseuille flow, while out of the boundary layer zone it almost coincides with the Poiseuille flow. At the picture we see the Poiseuille flow zone as the parallel and symmetric level lines of the normal velocity. As the Reynolds number increases, the decay of the boundary layer term slows down, leading to a larger boundary layer region. For sufficiently high Reynolds numbers, the boundary layer region encompasses the entire rectangle $\Pi_\varepsilon^{(e)}$, and the asymptotic approximation is no more applicable (for $F = 1000$).

Figures 5.6, 5.7, 5.8, 5.9, 5.10, and 5.11 illustrate the disparity between velocity fields obtained from full-dimensional computations and those from MAPDD com-

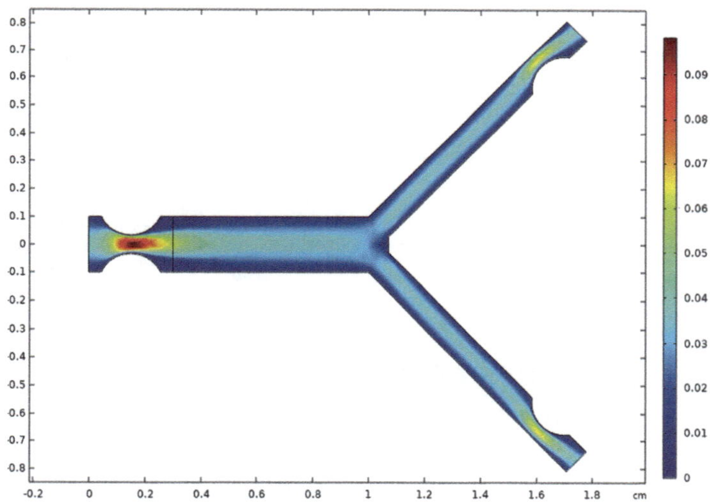

Fig. 5.6 Y-shaped thin tube structure with caverns: modulus of the velocity field computed in the whole domain for $F = 50$

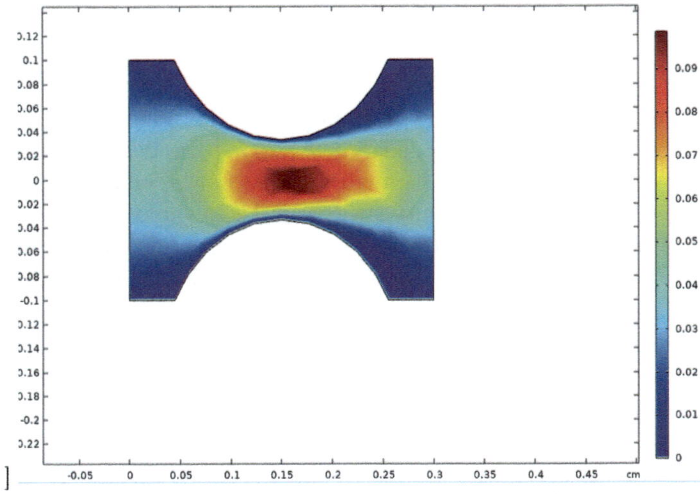

Fig. 5.7 Subdomain1: modulus of the velocity field computed by MAPDD for the inlet flux for $F = 50$

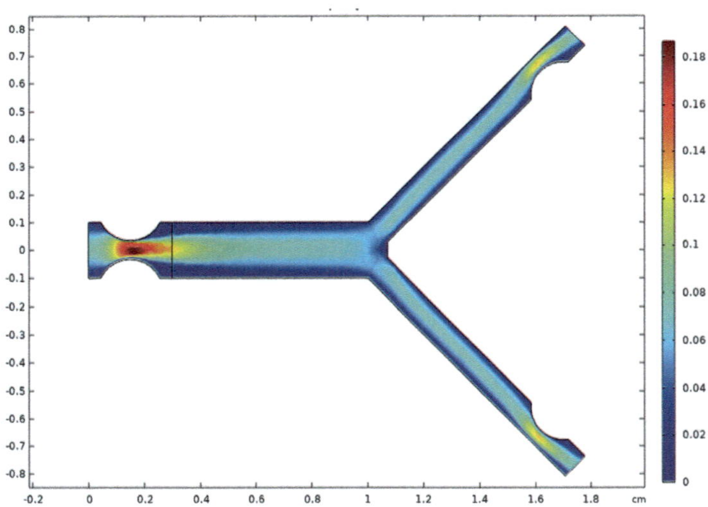

Fig. 5.8 Y-shaped thin tube structure with caverns: modulus of the velocity field computed in the whole domain for $F = 100$

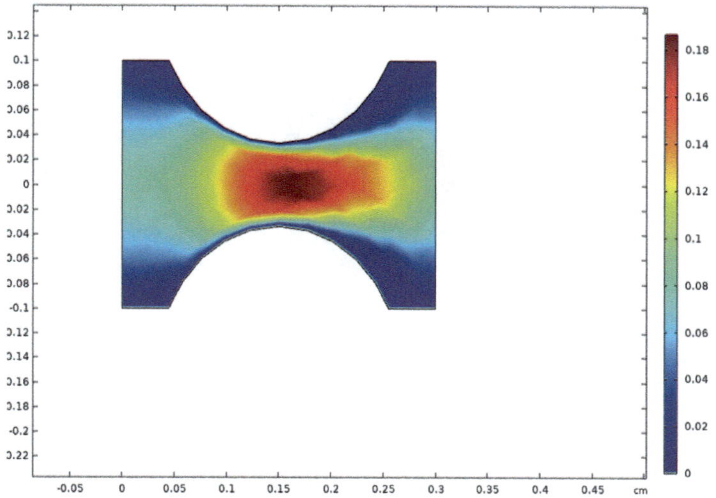

Fig. 5.9 Subdomain1: modulus of the velocity field computed by MAPDD for the inlet flux for $F = 100$

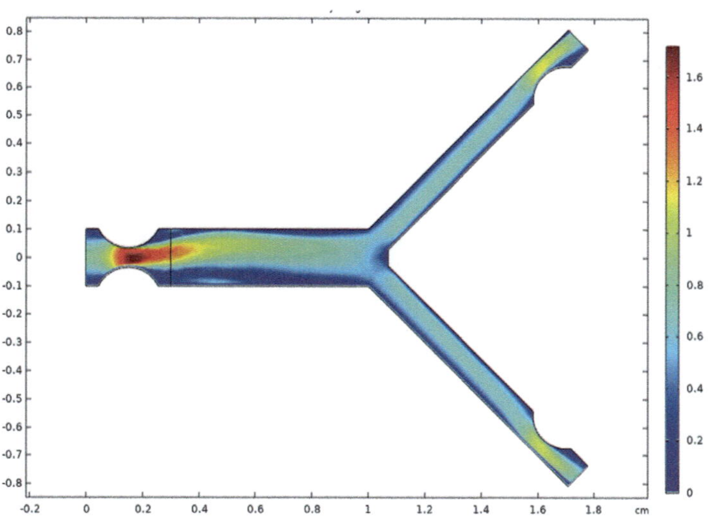

Fig. 5.10 Y-shaped thin tube structure with caverns: absolute velocity field computed in the whole domain for $F = 1000$

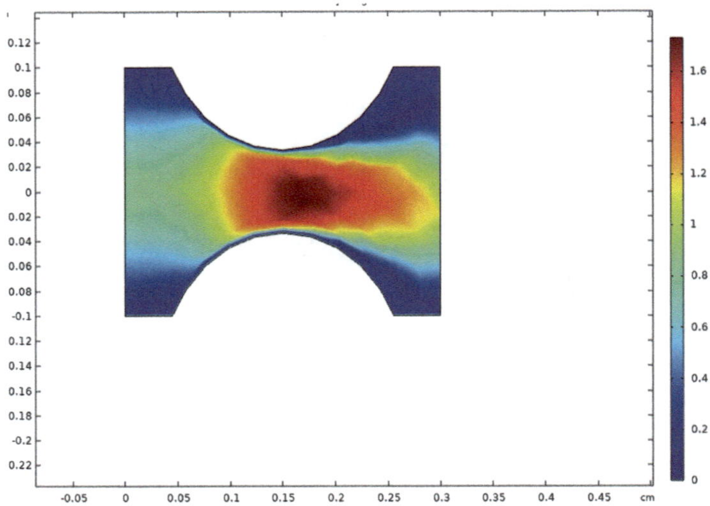

Fig. 5.11 Subdomain1: modulus of the velocity field computed by MAPDD for the inlet flux for $F = 1000$

putations, specifically for truncated subdomain 1. It is noticeable that for F values of 50 or even 100, the color patterns are quite similar, while for $F = 1000$, differences become discernible.

Chapter 6
Nonstationary Navier-Stokes Equations in Tube Structures

6.1 Formulation of the Problem

In this section, we present the formulation of the problem concerning the non-steady Navier–Stokes equations in a tube structure B_ε, defined based on Definition 1.4.2. The initial boundary value problem for the non-steady Navier–Stokes equations in B_ε is given by:

$$\begin{cases} \dfrac{\partial \mathbf{v}}{\partial t} - \nu\Delta\mathbf{v} + (\mathbf{v}\cdot\nabla)\mathbf{v} + \nabla p = 0,\ x \in B_\varepsilon^T, \\ \nabla\cdot\mathbf{v} = 0,\ x \in B_\varepsilon^T, \\ \mathbf{v} = \mathbf{g},\ x \in \partial B_\varepsilon \times (0,T), \\ \mathbf{v}(x,0) = 0,\ x \in B_\varepsilon, \end{cases} \tag{6.1.1}$$

where $\mathbf{g}$ stands for fluid velocity at the boundary ∂B_ε. The function $\mathbf{g}$ is assumed to satisfy the condition:

$$\mathbf{g}(x,0) = 0, \quad x \in \partial B_\varepsilon. \tag{6.1.2}$$

Additionally, $\mathbf{g}$ is equal to zero everywhere on ∂B_ε, except for sets $\gamma_\varepsilon^{N_1+1}, \dots, \gamma_\varepsilon^N$, which are part of the boundaries of domains $\omega_\varepsilon^{N_1+1}, \dots, \omega_\varepsilon^N$:

$$\begin{aligned} &\mathbf{g}(x,t)\big|_{\gamma_\varepsilon^j} = \mathbf{g}^j\left(\frac{x-O_j}{\varepsilon}, t\right)\Big|_{\gamma_\varepsilon^j}, \quad j = N_1+1, \dots, N, \\ &\mathbf{g}(x,t)\big|_{\partial B_\varepsilon \setminus \left(\bigcup\limits_{j=N_1+1}^{N} \gamma_\varepsilon^j\right)} = 0, \end{aligned} \tag{6.1.3}$$

G. Panasenko, K. Pileckas, *Multiscale Analysis of Viscous Flows in Thin Tube Structures*, Advances in Mathematical Fluid Mechanics,
https://doi.org/10.1007/978-3-031-54630-3_6

where $\mathbf{g}^j : \gamma^j \times [0, T] \to \mathbb{R}^n$, $\mathbf{g}$ belongs to the function space $C^{[\frac{J+1}{2}]+2}(0, T; W^{3/2,2}(\partial B_\varepsilon))$, with $[\alpha]$ representing the integer part of α. Here and below the space $C^l(0, T; W)$ with $W = W^{3/2,2}(\partial B_\varepsilon)$ or $W = W^{2,2}(B_\varepsilon)$ consists of functions such that $f(t, \cdot), \frac{\partial f}{\partial t}(t, \cdot), \ldots, \frac{\partial^l f}{\partial t^l}(t, \cdot)$ exist, are continuous on $[0, T]$ and belong to W. By $e = e_{O_j}$ and $x^{(e)}$, we denote the edge with the end O_j and the Cartesian coordinates corresponding to the origin O_j and the edge e, respectively, i.e., $x^{(e)} = \mathcal{P}^{(e)}(x - O_j)$, where $\mathcal{P}^{(e)}$ is the orthogonal matrix relating global coordinates x to local ones $x^{(e)}$.

Alongside the construction of the asymptotic expansion of the solution to (6.1.1), we establish the divergence-free extension of the boundary value function $\mathbf{g}$ denoted by the same symbol $\mathbf{g}$. This extension $\mathbf{g} \in C^{[\frac{J+1}{2}]+2}([0, T]; W^{2,2}(B_\varepsilon))$ and satisfies the following asymptotic estimates for all $t \in [0, T]$:

$$\begin{gathered}
\sup_{x \in B_\varepsilon} |\mathbf{g}(x, t)| \le c, \quad \|\mathbf{g}(\cdot, t)\|_{L^2(B_\varepsilon)} \le c\varepsilon^{\frac{n-1}{2}}, \\
\|\nabla \mathbf{g}(\cdot, t)\|_{L^2(B_\varepsilon)} \le c\varepsilon^{\frac{n-3}{2}}, \quad \|\mathbf{g}_t(\cdot, t)\|_{L^2(B_\varepsilon)} \le c\varepsilon^{\frac{n-1}{2}}, \\
\|\nabla^2 \mathbf{g}(\cdot, t)\|_{L^2(B_\varepsilon)} \le c\varepsilon^{\frac{n-5}{2}},
\end{gathered} \tag{6.1.4}$$

where c is a constant independent of ε and t.

The solution to problem (6.1.1) is sought in the form of $\mathbf{v} = \mathbf{u} + \mathbf{g}$. This leads to the following problem for $\mathbf{u}$ and p:

$$\begin{cases}
\mathbf{u}_t - \nu \Delta \mathbf{u} + (\mathbf{u} \cdot \nabla)\mathbf{u} + (\mathbf{g} \cdot \nabla)\mathbf{u} + (\mathbf{u} \cdot \nabla)\mathbf{g} + \nabla p = \mathbf{f}, \ x \in B_\varepsilon, \\
\qquad \operatorname{div} \mathbf{u} = 0, \ x \in B_\varepsilon, \\
\qquad \mathbf{u} = 0, \ x \in \partial B_\varepsilon, \\
\qquad \mathbf{u}(x, 0) = 0, \ x \in B_\varepsilon,
\end{cases} \tag{6.1.5}$$

where $\mathbf{f}$ represents the discrepancy of the extension $\mathbf{g}$, defined as:

$$\mathbf{f} = -\left(\mathbf{g}_t - \nu \Delta \mathbf{g} + (\mathbf{g} \cdot \nabla)\mathbf{g}\right).$$

Next, we introduce the concept of a weak solution for problem (6.1.1).

Definition 6.1.1 A weak solution of problem (6.1.1) is a divergence-free vector-valued function $\mathbf{v} = \mathbf{u} + \mathbf{g}$ satisfying the conditions $\mathbf{u}(x, 0) = 0$, $\mathbf{u} \in L^2(0, T; \mathring{H}^{1,2}(B_\varepsilon)) \cap L^\infty(0, T; \mathring{W}^{1,2}(B_\varepsilon))$, and $\mathbf{u}_t \in L^2(0, T; L^2(B_\varepsilon))$ and satisfying for arbitrary $\boldsymbol{\eta} \in \mathring{H}^{1,2}(B_\varepsilon)$ the integral identity:

$$\int_{B_\varepsilon} \left(\mathbf{u}_t \cdot \boldsymbol{\eta} + \nu \nabla \mathbf{u} \cdot \nabla \boldsymbol{\eta} - ((\mathbf{u} + \mathbf{g}) \cdot \nabla)\, \boldsymbol{\eta} \cdot \mathbf{u} - (\mathbf{u} \cdot \nabla)\, \boldsymbol{\eta} \cdot \mathbf{g}\right) dx = \int_{B_\varepsilon} \mathbf{f} \cdot \boldsymbol{\eta} dx. \tag{6.1.6}$$

The flow rates associated with the surfaces γ_ε^j, $j = N_1 + 1, \dots, N$, are defined as:

$$
\begin{aligned}
F_\varepsilon^j(t) &= \int\limits_{\gamma_\varepsilon^j} \mathbf{g}(x,t) \cdot \mathbf{n}(x)\, dS = \int\limits_{\gamma_\varepsilon^j} \mathbf{g}^j\left(\frac{x - O_j}{\varepsilon}, t\right) \cdot \mathbf{n}(x)\, dS, \\
&= \varepsilon^{n-1} \int\limits_{\gamma^j} g^j(y,t) \cdot \mathbf{n}(y)\, dS \equiv \varepsilon^{n-1} F^j(t),
\end{aligned}
\tag{6.1.7}
$$

where $\mathbf{n}$ is the unit outward (with respect to B_ε) normal vector to γ_ε^j.

The boundary value $\mathbf{g}(x,t)$ satisfies the compatibility condition:

$$
F^j(0) = 0, \qquad j = N_1 + 1, \dots, N. \tag{6.1.8}
$$

Due to the incompressibility of the fluid ($\nabla \cdot \mathbf{v} = 0$) and the no-slip condition on $\partial B_\varepsilon \setminus \bigcup\limits_{j=N_1+1}^{N} \gamma_\varepsilon^j$, the flow rates $F^j(t)$ are subject to the compatibility condition:

$$
\sum_{j=N_1+1}^{N} F^j(t) = 0, \quad \forall t \in [0, T]. \tag{6.1.9}
$$

Furthermore, the flow rates satisfy what are known as Kirchhoff-type junction conditions for all nodes O_l. Let us introduce flow rates within cylinders $\Pi_\varepsilon^{(e)}$, where e is an edge. Introduce a local coordinate system $O_l x^{(e)}$ and consider a cross section $\sigma_\varepsilon^{(e)}$ of the cylinder $\Pi_\varepsilon^{(e)}$. Denote the flow rate $F_\varepsilon^{(e)}$ through the cross section $\sigma_\varepsilon^{(e)}$: $F_\varepsilon^{(e)}(t) = \int_{\sigma_\varepsilon^{(e)}} \mathbf{v}(x^{(e)\prime}, t) \cdot \mathbf{n} dx^{(e)\prime}$, where $\mathbf{n}$ is colinear to the axis $O_l x^{(e)_n}$. Evidently:

$$
F_\varepsilon^{(e)}(t) = \varepsilon^{n-1} \int\limits_{\sigma^{(e)}} \mathbf{v}(\varepsilon y^{(e)\prime}, t) \cdot \mathbf{n} dy^{(e)\prime} = \varepsilon^{n-1} F^{(e)}(t).
$$

The Kirchhoff-type conditions can be expressed as:

$$
\sum_{e: O_l \in e} F^{(e)}(t) = 0, \quad l = 1, \dots, N_1. \tag{6.1.10}
$$

It is important to note that the last condition follows from the Stokes theorem, based on the incompressibility of the fluid ($\mathrm{div}\mathbf{v} = 0$), and it implies that the net flux is equal to zero.

6.2 Solvability of the Problem

6.2.1 The Case $n = 2$

This subsection investigates the existence of a unique weak solution to problem (6.1.1) in a tube structure $B_\varepsilon \subset \mathbb{R}^2$.

Theorem 6.2.1 *Consider a tube domain $B_\varepsilon \subset \mathbb{R}^2$, where the boundary ∂B_ε is C^2-regular. Let the extended function $\mathbf{g} \in C^{[\frac{J+1}{2}]+2}(0, T; W^{2,2}(B_\varepsilon))$*[1] *satisfy conditions* (6.1.3) *and* (6.1.4), *and let $\mathbf{f} \in L^2(0, T; L^2(B_\varepsilon))$. Then, for sufficiently small ε, the variational problem* (6.1.6) *has a solution $\mathbf{u}$ that satisfies the estimates:*

$$\sup_{t\in[0,T]} \|\mathbf{u}(\cdot, t)\|^2_{L^2(B_\varepsilon)} + \int_0^T \int_{B_\varepsilon} |\nabla \mathbf{u}(x,t)|^2 dx dt \le c\varepsilon^2 \int_0^T A_1(t) dt, \tag{6.2.1}$$

$$\sup_{t\in[0,T]} \|\nabla \mathbf{u}(\cdot, t)\|^2_{L^2(B_\varepsilon)} \le c e^{c\varepsilon^2 d_1 \int_0^T A_1(t)\,dt} \int_0^T A_1(t) dt, \tag{6.2.2}$$

where $d_1 = c_2\big(1 + \varepsilon^2 \int_0^T A_1(t)dt\big)$ and $A_1(t) = \int_{B_\varepsilon} |\mathbf{f}(x,t)|^2 dx$.

Furthermore, we have:

$$\int_0^T \int_{B_\varepsilon} |\nabla^2 \mathbf{u}|^2 dx dt + \int_0^T \int_{B_\varepsilon} |\mathbf{u}_t(x,t)|^2 dx dt \le C(A_1) \int_0^T A_1(t) dt, \tag{6.2.3}$$

where $C(A_1)$ is bounded as $\int_0^T A_1(t)dt$ tends to zero. The constants in the estimates are independent of ε.

[1] In this theorem, it suffices to impose the condition that the function $\mathbf{g}$ belongs to the space $C^1(0, T; W^{2,2}(B_\varepsilon))$. However, to construct the asymptotic decomposition of order J, we need a stronger condition $\mathbf{g} \in C^{[\frac{J+1}{2}]+2}(0, T; W^{2,2}(B_\varepsilon))$.

Remark 6.2.1 The constant $C(A_1)$ depends nonlinearly on $\int\limits_0^T A_1(t)dt$ and is bounded when this integral is bounded. However, we will use this result only for small values of $\int\limits_0^T A_1(t)dt$.

Proof
The existence of a solution to problem (6.1.6) is established using the Galerkin method (see [102, 203]). Our main focus is on obtaining suitable a priori estimates.

The Galerkin approximations of the solution to problem (6.1.6) are defined as follows:

$$\begin{gathered}\int\limits_{B_\varepsilon}\Big(\mathbf{u}_t^{(N)}\cdot\boldsymbol{\psi}_l+\nu\nabla\mathbf{u}^{(N)}\cdot\nabla\boldsymbol{\psi}_l-\big(\mathbf{u}^{(N)}+\mathbf{g}\big)\cdot\nabla\big)\boldsymbol{\psi}_l\cdot\mathbf{u}^{(N)}\\ -\big(\mathbf{u}^{(N)}\cdot\nabla\big)\boldsymbol{\psi}_l\cdot\mathbf{g}\Big)dx=\int\limits_{B_\varepsilon}\mathbf{f}\cdot\boldsymbol{\psi}_l dx,\\ \gamma_l^{(N)}(0)=0,\end{gathered}\tag{6.2.4}$$

where $l=1,..,N$, $\mathbf{u}^{(N)}(x,t)=\sum\limits_{k=1}^{N}\gamma_k^{(N)}(t)\boldsymbol{\psi}_k(x)$, and $\{\boldsymbol{\psi}_k\}_{k=1}^\infty$ is a basis in the space $\overset{\circ}{H}{}^{1,2}(B_\varepsilon)$. Without loss of generality, we may assume that $\{\boldsymbol{\psi}_k\}_{k=1}^\infty$ is orthonormal in $L^2(B_\varepsilon)$.

The equations (6.2.4) form a system of ordinary differential equations:

$$\begin{cases}\dfrac{d\gamma_l^{(N)}(t)}{dt}+\nu\sum\limits_{j=1}^{N}a_{lj}\gamma_j^{(N)}(t)+\sum\limits_{i,j=1}^{N}a_{lij}\gamma_i^{(N)}(t)\gamma_j^{(N)}(t)\\ \qquad+\sum\limits_{j=1}^{N}b_{lj}(t)\gamma_j^{(N)}(t)=f_l(t),\quad l=1,..,N,\\ \qquad\qquad\gamma_l^{(N)}(0)=0,\quad l=1,..,N,\end{cases}\tag{6.2.5}$$

where:

$$a_{lj}=\int\limits_{B_\varepsilon}\nabla\boldsymbol{\psi}_l\cdot\nabla\boldsymbol{\psi}_j dx,\quad a_{lij}=-\int\limits_{B_\varepsilon}(\boldsymbol{\psi}_i\cdot\nabla)\boldsymbol{\psi}_l\cdot\boldsymbol{\psi}_j dx,$$

$$b_{lj}(t)=-\int\limits_{B_\varepsilon}(\mathbf{g}\cdot\nabla)\boldsymbol{\psi}_l\cdot\boldsymbol{\psi}_j dx-\int\limits_{B_\varepsilon}(\boldsymbol{\psi}_j\cdot\nabla)\boldsymbol{\psi}_l\cdot\mathbf{g}dx,\quad f_l(t)=\int\limits_{B_\varepsilon}\mathbf{f}\cdot\boldsymbol{\psi}_l dx.$$

By multiplying (6.1.6) by $\gamma_k^{(N)}(t)$ and summing over $k=1,\ldots N$, we obtain:

$$\frac{1}{2}\frac{d}{dt}\int_{B_\varepsilon}|\mathbf{u}^{(N)}|^2dx+\nu\int_{B_\varepsilon}|\nabla\mathbf{u}^{(N)}|^2dx=\int_{B_\varepsilon}\left(\mathbf{u}^{(N)}\cdot\nabla\right)\mathbf{u}^{(N)}\cdot\mathbf{g}dx+\int_{B_\varepsilon}\mathbf{f}\cdot\mathbf{u}^{(N)}dx.$$

Utilizing inequality (2.3.7) and the Poincaré inequality (2.4.2) within the tube structure B_ε, we arrive at the estimate:

$$\begin{aligned}\left|\int_{B_\varepsilon}\left(\mathbf{u}^{(N)}\cdot\nabla\right)\mathbf{u}^{(N)}\cdot\mathbf{g}dx\right| &\le \|\mathbf{g}\|_{L^4(B_\varepsilon)}\|\mathbf{u}^{(N)}\|_{L^4(B_\varepsilon)}\|\nabla\mathbf{u}^{(N)}\|_{L^2(B_\varepsilon)}\\ &\le c_1\varepsilon^{\frac{1}{2}}\sup_{t\in[0,T]}\left(\|\mathbf{g}(\cdot,t)\|_{L^4(B_\varepsilon)}\right)\|\nabla\mathbf{u}^{(N)}\|^2_{L^2(B_\varepsilon)}.\end{aligned}\tag{6.2.6}$$

It is important to note that, as per (6.1.4):

$$\sup_{t\in[0,T]}\|\mathbf{g}(\cdot,t)\|_{L^4(B_\varepsilon)}\le c\varepsilon^{\frac{1}{4}}.$$

Thus, for sufficiently small ε (i.e., $c_1\varepsilon^{\frac{1}{2}}\sup\limits_{t\in[0,T]}\left(\|\mathbf{g}(\cdot,t)\|_{L^4(B_\varepsilon)}\right)\le\frac{\nu}{2}$), we have:

$$\frac{1}{2}\frac{d}{dt}\|\mathbf{u}^{(N)}(\cdot,t)\|^2_{L^2(B_\varepsilon)}+\frac{\nu}{2}\int_{B_\varepsilon}|\nabla\mathbf{u}^{(N)}|^2dx\le c\varepsilon^2\|\mathbf{f}\|^2_{L^2(B_\varepsilon)},$$

which further leads to:

$$\begin{aligned}&\frac{1}{2}\|\mathbf{u}^{(N)}(\cdot,t)\|^2_{L^2(B_\varepsilon)}+\frac{\nu}{2}\int_0^t\|\nabla\mathbf{u}^{(N)}(\cdot,\tau)\|^2_{L^2(B_\varepsilon)}d\tau\\ &\quad\le c\varepsilon^2\int_0^T\|\mathbf{f}(\cdot,t)\|^2_{L^2(B_\varepsilon)}dt\equiv c\varepsilon^2\int_0^T A_1(t)dt.\end{aligned}\tag{6.2.7}$$

Due to the orthonormality of the basis in $L^2(B_\varepsilon)$, it follows from (6.2.7) that:

$$\max_{t\in[0,T]}\|\mathbf{u}^{(N)}(\cdot,t)\|^2_{L^2(B_\varepsilon)}=\max_{t\in[0,T]}\sum_{l=1}^n|\gamma_l^{(N)}(t)|^2\le c\varepsilon^2\int_0^T A_1(t)dt.$$

The system (6.2.5) is analytically dependent on functions $\gamma_l^{(N)}(t)$. Therefore, from the a priori boundedness of the solution follows the unique solvability of the initial problem (6.2.5) on the interval $[0,\,T]$.

The estimate (6.2.7) holds for Galerkin approximations constructed using an arbitrary basis and for arbitrary bounded Lipschitz domains. To estimate the higher derivatives of $\mathbf{u}^{(N)}$, we need to assume that $\partial B_\varepsilon \in C^2$ and use the eigenfunctions of the Stokes operator $\widetilde{\Delta}$ within the tube structure B_ε. Therefore, let us assume that $\{\boldsymbol{\psi}_l\}$ is the system of eigenfunctions of the Stokes operator $\widetilde{\Delta}$ (see Sect. 4.3 of Chap. 4).

Multiplying Equation (6.2.4) by $\lambda_l \gamma_l^{(N)}(t)$ and summing over $l = 1, \dots, N$, we can reduce the equalities (6.2.4) to the following form:

$$\int_{B_\varepsilon} \Big(\sum_{l=1}^{N} \mathbf{u}_t^{(N)} \cdot \boldsymbol{\psi}_l \lambda_l \gamma_l^{(N)}(t) - \nu \sum_{l=1}^{N} \Delta \mathbf{u}^{(N)} \cdot \boldsymbol{\psi}_l \lambda_l \gamma_l^{(N)}(t)$$

$$- \sum_{l=1}^{N} ((\mathbf{u}^{(N)} + \mathbf{g}) \cdot \nabla) \lambda_l \gamma_l^{(N)}(t) \boldsymbol{\psi}_l \cdot \mathbf{u}^{(N)} - \sum_{l=1}^{N} (\mathbf{u}^{(N)} \cdot \nabla) \lambda_l \gamma_l^{(N)}(t) \boldsymbol{\psi}_l \cdot \mathbf{g} \Big) dx$$

$$= \int_{B_\varepsilon} \sum_{l=1}^{N} \mathbf{f} \cdot \boldsymbol{\psi}_l \lambda_k \gamma_l^{(N)}(t) dx, \quad l = 1, \dots, N.$$

Using the properties of the Stokes operator (Sect. 4.3 of Chap. 4), we can further rewrite the last equality as:

$$\int_{B_\varepsilon} \Big(\mathbf{u}_t^{(N)} \cdot \widetilde{\Delta} \mathbf{u}^{(N)} - \nu \Delta \mathbf{u}^{(N)} \cdot \widetilde{\Delta} \mathbf{u}^{(N)} \Big) dx$$

$$+ \int_{B_\varepsilon} \Big(((\mathbf{u}^{(N)} + \mathbf{g}) \cdot \nabla) \mathbf{u}^{(N)} \cdot \widetilde{\Delta} \mathbf{u}^{(N)} + (\mathbf{u}^{(N)} \cdot \nabla) \cdot \mathbf{g} \cdot \widetilde{\Delta} \mathbf{u}^{(N)} \Big) dx$$

$$= \int_{B_\varepsilon} \mathbf{f} \cdot \widetilde{\Delta} \mathbf{u}^{(N)} dx.$$

This expression can be further simplified to:

$$\nu \frac{d}{dt} \int_{B_\varepsilon} |\nabla \mathbf{u}^{(N)}|^2 dx + \int_{B_\varepsilon} |\widetilde{\Delta} \mathbf{u}^{(N)}|^2 dx$$

$$= - \int_{B_\varepsilon} ((\mathbf{u}^{(N)} + \mathbf{g}) \cdot \nabla) \mathbf{u}^{(N)} \cdot \widetilde{\Delta} \mathbf{u}^{(N)} dx \tag{6.2.8}$$

$$- \int_{B_\varepsilon} (\mathbf{u}^{(N)} \cdot \nabla) \mathbf{g} \cdot \widetilde{\Delta} \mathbf{u}^{(N)} dx + \int_{B_\varepsilon} \mathbf{f} \cdot \widetilde{\Delta} \mathbf{u}^{(N)} dx = \sum_{i=1}^{3} J_i,$$

where $J_i, i = 1, 2, 3$, denotes the three integrals in the right-hand side.

Now, we proceed with estimating the right-hand side of (6.2.8). Using the inequality (2.3.37), we obtain the following estimates:

$$\begin{aligned} |J_3| &\le \delta \int\limits_{B_\varepsilon} |\widetilde{\Delta}\mathbf{u}^{(N)}|^2 dx + c_\delta A_1(t), \\ |J_2| &\le \|\mathbf{u}^{(N)}\|_{L^4(B_\varepsilon)} \|\nabla \mathbf{g}\|_{L^4(B_\varepsilon)} \|\widetilde{\Delta}\mathbf{u}^{(N)}\|_{L^2(B_\varepsilon)} \\ &\le c\varepsilon^{\frac{1}{2}} \|\nabla \mathbf{u}^{(N)}\|_{L^2(B_\varepsilon)} \|\nabla \mathbf{g}\|_{L^4(B_\varepsilon)} \|\widetilde{\Delta}\mathbf{u}^{(N)}\|_{L^2(B_\varepsilon)} \\ &\le c_\delta \varepsilon \|\nabla \mathbf{u}^{(N)}\|^2_{L^2(B_\varepsilon)} \|\nabla \mathbf{g}\|^2_{L^4(B_\varepsilon)} + \delta \|\widetilde{\Delta}\mathbf{u}^{(N)}\|^2_{L^2(B_\varepsilon)}. \end{aligned} \tag{6.2.9}$$

By using inequality (2.3.39) and (6.1.4), we obtain:

$$\|\nabla \mathbf{g}\|^2_{L^4(B_\varepsilon)} \le c\varepsilon^{-1} \|\nabla \mathbf{g}\|_{L^2(B_\varepsilon)} \Big(\|\nabla \mathbf{g}\|^2_{L^2(B_\varepsilon)} + \varepsilon^2 \|\nabla^2 \mathbf{g}\|^2_{L^2(B_\varepsilon)} \Big)^{1/2} \le c\varepsilon^{-2}.$$

Thus, we can simplify the estimates for J_2 as follows:

$$|J_2| \le c_\delta \varepsilon^{-1} \|\nabla \mathbf{u}^{(N)}\|^2_{L^2(B_\varepsilon)} + \delta \|\widetilde{\Delta}\mathbf{u}^{(N)}\|^2_{L^2(B_\varepsilon)}. \tag{6.2.10}$$

Further, by using (6.1.4), we get:

$$\begin{aligned} \Big| \int\limits_{B_\varepsilon} (\mathbf{g} \cdot \nabla)\mathbf{u}^{(N)} \cdot \widetilde{\Delta}\mathbf{u}^{(N)} dx \Big| &\le \|\mathbf{g}\|_{L^\infty(B_\varepsilon)} \|\nabla \mathbf{u}^{(N)}\|_{L^2(B_\varepsilon)} \|\widetilde{\Delta}\mathbf{u}^{(N)}\|_{L^2(B_\varepsilon)} \\ &\le c_\delta \|\nabla \mathbf{u}^{(N)}\|^2_{L^2(B_\varepsilon)} + \delta \|\widetilde{\Delta}\mathbf{u}^{(N)}\|^2_{L^2(B_\varepsilon)}. \end{aligned} \tag{6.2.11}$$

Finally, using (2.3.37) and (2.3.39), we get the estimate:

$$\begin{aligned} &\Big| \int\limits_{B_\varepsilon} (\mathbf{u}^{(N)} \cdot \nabla)\mathbf{u}^{(N)} \cdot \widetilde{\Delta}\mathbf{u}^{(N)} dx \Big| \\ &\le \|\mathbf{u}^{(N)}\|_{L^4(B_\varepsilon)} \|\nabla \mathbf{u}^{(N)}\|_{L^4(B_\varepsilon)} \|\widetilde{\Delta}\mathbf{u}^{(N)}\|_{L^2(B_\varepsilon)} \\ &\le c_\delta \varepsilon^{-1} \|\mathbf{u}^{(N)}\|_{L^2(B_\varepsilon)} \|\nabla \mathbf{u}^{(N)}\|^2_{L^2(B_\varepsilon)} \times \\ &\times \Big(\|\nabla \mathbf{u}^{(N)}\|^2_{L^2(B_\varepsilon)} + \varepsilon^2 \|\nabla^2 \mathbf{u}^{(N)}\|^2_{L^2(B_\varepsilon)} \Big)^{\frac{1}{2}} + \frac{\delta}{2} \|\widetilde{\Delta}\mathbf{u}^{(N)}\|^2_{L^2(B_\varepsilon)} \\ &\le c_\delta \varepsilon^{-1} \|\mathbf{u}^{(N)}\|_{L^2(B_\varepsilon)} \|\nabla \mathbf{u}^{(N)}\|^3_{L^2(B_\varepsilon)} \\ &+ c_\delta \|\mathbf{u}^{(N)}\|_{L^2(B_\varepsilon)} \|\nabla \mathbf{u}^{(N)}\|^2_{L^2(B_\varepsilon)} \|\nabla^2 \mathbf{u}^{(N)}\|_{L^2(B_\varepsilon)} + \frac{\delta}{2} \|\widetilde{\Delta}\mathbf{u}^{(N)}\|^2_{L^2(B_\varepsilon)} \end{aligned}$$

$$\overset{(4.3.13)}{\leq} c_\delta\Big(1+\|\mathbf{u}^{(N)}\|^2_{L^2(B_\varepsilon)}\Big)\|\nabla\mathbf{u}^{(N)}\|^4_{L^2(B_\varepsilon)}+\delta\|\widetilde{\Delta}\mathbf{u}^{(N)}\|^2_{L^2(B_\varepsilon)}$$

$$\overset{(6.2.7)}{\leq} c_\delta\Big(1+\varepsilon^2\int\limits_0^T A_1(t)dt\Big)\,\|\nabla\mathbf{u}^{(N)}\|^4_{L^2(B_\varepsilon)}+\delta\|\widetilde{\Delta}\mathbf{u}^{(N)}\|^2_{L^2(B_\varepsilon)}. \tag{6.2.12}$$

By substituting the estimates (6.2.9)–(6.2.12) into (6.2.8) and taking $\delta=\frac{1}{8}$, we obtain the following inequality:

$$\nu\frac{d}{dt}\int\limits_{B_\varepsilon}|\nabla\mathbf{u}^{(N)}|^2dx+\int\limits_{B_\varepsilon}|\widetilde{\Delta}\mathbf{u}^{(N)}|^2dx\leq c_2\Big(1+\varepsilon^2\int\limits_0^T A_1(t)dt\Big)\|\nabla\mathbf{u}^{(N)}\|^4_{L^2(B_\varepsilon)}$$

$$+c_3\Big(1+\varepsilon^{-1}\Big)\|\nabla\mathbf{u}^{(N)}\|^2_{L^2(B_\varepsilon)}+c_4A_1(t). \tag{6.2.13}$$

Let us define $Y(t)=\int_{B_\varepsilon}|\nabla\mathbf{u}^{(N)}(x,t)|^2dx$ and $d_1=c_2\Big(1+\varepsilon^2\int\limits_0^T A_1(t)dt\Big)$. Then, we can express inequality (6.2.13) as follows:

$$Y'(t)\leq\frac{1}{\nu}d_1\|\nabla\mathbf{u}^{(N)}\|^2_{L^2(B_\varepsilon)}Y(t)$$

$$+c_3\frac{1}{\nu}(1+\varepsilon^{-1})\|\nabla\mathbf{u}^{(N)}\|^2_{L^2(B_\varepsilon)}+\frac{1}{\nu}c_4A_1(t).$$

Therefore:

$$\Big(Y(t)e^{-\frac{1}{\nu}\int\limits_0^t d_1\|\nabla\mathbf{u}^{(N)}(\cdot,\tau)\|^2_{L^2(B_\varepsilon)}d\tau}\Big)'$$

$$\leq c_5\Big((1+\varepsilon^{-1})\|\nabla\mathbf{u}^{(N)}\|^2_{L^2(B_\varepsilon)}+A_1(t)\Big)e^{-\frac{1}{\nu}\int\limits_0^t d_1\|\nabla\mathbf{u}^{(N)}(\cdot,\tau)\|^2_{L^2(B_\varepsilon)}d\tau}$$

$$\leq c_5\Big((1+\varepsilon^{-1})\|\nabla\mathbf{u}^{(N)}\|^2_{L^2(B_\varepsilon)}+A_1(t)\Big).$$

By integrating the last inequality and utilizing the fact that $Y(0)=0$, we arrive at:

$$Y(t)\leq$$

$$\leq e^{\frac{1}{\nu}\int\limits_0^t d_1\|\nabla\mathbf{u}^{(N)}(\cdot,\tau)\|^2_{L^2(B_\varepsilon)}d\tau}\Big[c(1+\varepsilon^{-1})\int\limits_0^T\|\nabla\mathbf{u}^{(N)}(\cdot,t)\|^2_{L^2(B_\varepsilon)}dt+c\int\limits_0^T A_1(t)dt\Big]$$

$$\leq ce^{c\varepsilon^2d_1\int\limits_0^T A_1(t)\,dt}\Big[\varepsilon^{-1}\int\limits_0^T\|\nabla\mathbf{u}^{(N)}(\cdot,t)\|^2_{L^2(B_\varepsilon)}dt+\int\limits_0^T A_1(t)dt\Big]. \tag{6.2.14}$$

From (6.2.7), we deduce the inequality:

$$\varepsilon^{-1}\int_0^T \|\nabla \mathbf{u}^{(N)}(\cdot,t)\|^2_{L^2(B_\varepsilon)}dt \le c\varepsilon \int_0^T A_1(t)dt. \tag{6.2.15}$$

Therefore, (6.2.14) implies:

$$Y(t) \le ce^{c\varepsilon^2 d_1 \int_0^T A_1(t)\,dt}\int_0^T A_1(t)dt, \tag{6.2.16}$$

and so, we obtain:

$$\sup_{t\in[0,T]} \|\nabla \mathbf{u}^{(N)}(\,\cdot\,,t)\|^2_{L^2(B_\varepsilon)} \le ce^{c\varepsilon^2 d_1 \int_0^T A_1(t)\,dt}\int_0^T A_1(t)dt. \tag{6.2.17}$$

By substituting (6.2.17) into (6.2.13), integrating by t from 0 to T, and applying estimates (6.2.7), (6.2.15), and (6.2.17), we obtain the following inequality:

$$\begin{aligned}\int_0^T\int_{B_\varepsilon} |\widetilde{\Delta}\mathbf{u}^{(N)}|^2 dxdt &\le c\Big(1+\varepsilon^2\int_0^T A_1(t)dt\Big)\int_0^T \|\nabla \mathbf{u}^{(N)}(\cdot,t)\|^4_{L^2(B_\varepsilon)}dt\\ &+c\Big(1+\varepsilon^{-1}\Big)\int_0^T \|\nabla \mathbf{u}^{(N)}(\cdot,t)\|^2_{L^2(B_\varepsilon)}dt + c\int_0^T A_1(t)dt\\ &\le c\Big(1+\varepsilon^2\int_0^T A_1(t)dt\Big)\sup_{t\in[0,T]}\|\nabla \mathbf{u}^{(N)}(\cdot,t)\|^2\int_0^T \|\nabla \mathbf{u}^{(N)}(\cdot,t)\|^2_{L^2(B_\varepsilon)}dt\\ &+c(1+\varepsilon)\int_0^T A_1(t)dt \le ce^{c\varepsilon^2 d_1\int_0^T A_1(t)\,dt}\Big[1+\int_0^T A_1(t)dt\Big]\Big(\int_0^T A_1(t)dt\Big)^2.\end{aligned} \tag{6.2.18}$$

From (6.2.18) and (4.3.13), we can deduce that:

$$\int_0^T\int_{B_\varepsilon} |\nabla^2\mathbf{u}^{(N)}|^2 dxdt \le C(A_1)\int_0^T A_1(t)dt, \tag{6.2.19}$$

where $C(A_1)$ is bounded as $\int_0^T A_1(t)dt$ tends to zero.

Now, let us estimate the norm of $\mathbf{u}_t^{(N)}$. We multiply relations (6.2.4) by $\frac{d}{dt}\gamma_k^{(N)}$ and sum up the obtained equalities over $k = 1, \ldots, N$:

$$\int\limits_{B_\varepsilon} |\mathbf{u}_t^{(N)}|^2 dx + \frac{\nu}{2}\frac{d}{dt}\int\limits_{B_\varepsilon} |\nabla \mathbf{u}^{(N)}|^2 dx = -\int\limits_{B_\varepsilon} \Big((\mathbf{u}^{(N)} + \mathbf{g}) \cdot \nabla \Big) \mathbf{u}^{(N)} \cdot \mathbf{u}_t^{(N)} dx$$
$$- \int\limits_{B_\varepsilon} (\mathbf{u}^{(N)} \cdot \nabla)\mathbf{g} \cdot \mathbf{u}_t^{(N)} dx + \int\limits_{B_\varepsilon} \mathbf{f} \cdot \mathbf{u}_t^{(N)} dx$$
$$\leq \Big(\|\mathbf{u}^{(N)}\|_{L^\infty(B_\varepsilon)} + \|\mathbf{g}\|_{L^\infty(B_\varepsilon)} \Big) \|\nabla \mathbf{u}^{(N)}\|_{L^2(B_\varepsilon)} \|\mathbf{u}_t^{(N)}\|_{L^2(B_\varepsilon)}$$
$$+ \|\mathbf{u}^{(N)}\|_{L^\infty(B_\varepsilon)} \|\nabla \mathbf{g}\|_{L^2(B_\varepsilon)} \|\mathbf{u}_t^{(N)}\|_{L^2(B_\varepsilon)} + \|\mathbf{f}\|_{L^2(B_\varepsilon)} \|\mathbf{u}_t^{(N)}\|_{L^2(B_\varepsilon)}$$
$$\leq \delta \|\mathbf{u}_t^{(N)}\|^2 + c_\delta \|\mathbf{u}^{(N)}\|^2_{L^\infty(B_\varepsilon)} \|\nabla \mathbf{u}^{(N)}\|^2_{L^2(B_\varepsilon)} + c_\delta \|\mathbf{g}\|^2_{L^\infty(B_\varepsilon)} \|\nabla \mathbf{u}^{(N)}\|^2_{L^2(B_\varepsilon)}$$
$$+ c_\delta \|\mathbf{u}^{(N)}\|^2_{L^\infty(B_\varepsilon)} \|\nabla \mathbf{g}\|^2_{L^2(B_\varepsilon)} + c_\delta A_1(t).$$

Taking sufficiently small δ and integrating over $[0, T]$, we deduce the inequality:

$$\int\limits_0^T \int\limits_{B_\varepsilon} |\mathbf{u}_t^{(N)}|^2 dx dt \leq c\Big(\int\limits_0^T A_1(t) dt$$
$$+ \int\limits_0^T \|\mathbf{u}^{(N)}\|^2_{L^\infty(B_\varepsilon)} \|\nabla \mathbf{u}^{(N)}\|^2_{L^2(B_\varepsilon)} dt + \int\limits_0^T \|\mathbf{g}\|^2_{L^\infty(B_\varepsilon)} \|\nabla \mathbf{u}^{(N)}\|^2_{L^2(B_\varepsilon)} dt \Big)$$
$$+ c \int\limits_0^T \|\nabla \mathbf{g}\|^2_{L^2(B_\varepsilon)} \|\mathbf{u}^{(N)}\|^2_{L^\infty(B_\varepsilon)} dt \overset{(4.3.15)}{\leq} c\Big(\int\limits_0^T A_1(t) dt$$
$$+ \varepsilon \sup_{t \in [0,T]} \Big(\|\nabla \mathbf{u}^{(N)}\|^2_{L^2(B_\varepsilon)} + \|\nabla \mathbf{g}\|^2_{L^2(B_\varepsilon)} \Big) \int\limits_0^T \|\nabla^2 \mathbf{u}^{(N)}\|^2_{L^2(B_\varepsilon)} dt$$
$$+ \sup_{t \in [0,T]} \|\mathbf{g}\|^2_{L^\infty(B_\varepsilon)} \int\limits_0^T \|\nabla \mathbf{u}^{(N)}\|^2_{L^2(B_\varepsilon)} dt \leq C(A_1) \int\limits_0^T A_1(t) dt, \tag{6.2.20}$$

where $C(A_1)$ is bounded as $\int\limits_0^T A_1(t)dt$ tends to zero. In the last estimate, we have applied inequalities (6.1.4), (6.2.7), (6.2.17), and (6.2.19).

Estimates (6.2.7), (6.2.17), (6.2.19), and (6.2.20) guarantee the existence of a subsequence $\{\mathbf{u}^{(N_l)}\}$ such that the sequences $\{\mathbf{u}^{(N_l)}\}$, $\{\nabla \mathbf{u}^{(N_l)}\}$, $\{\nabla^2 \mathbf{u}^{(N_l)}\}$, and $\{\mathbf{u}_t^{(N_l)}\}$ converge weakly in $L^2(0, T; L^2(B_\varepsilon))$ to $\mathbf{u}$, $\nabla \mathbf{u}$, $\nabla^2 \mathbf{u}$, and $\mathbf{u}_t$, respectively. It is evident that for the limit function $\mathbf{u}$, the estimates (6.2.7), (6.2.17), (6.2.19), and (6.2.20) remain valid.

Let us demonstrate that $\mathbf{u}$ is a weak solution of problem (6.1.6). By multiplying the equalities (6.2.4), written for the functions $\mathbf{u}^{(N_l)}$, with arbitrary continuous functions $d_l(t)$, we sum up the resulting relations for $l = 1, \ldots, M$ and integrate with respect to t from 0 to T:

$$\int_0^T \int_{B_\varepsilon} \Big(\mathbf{u}_t^{(N_l)} \cdot \mathbf{\Psi}^M + \nu \nabla \mathbf{u}^{(N_l)} \cdot \nabla \mathbf{\Psi}^M - \big((\mathbf{u}^{(N_l)} + \mathbf{g}) \cdot \nabla\big) \mathbf{\Psi}^M \cdot \mathbf{u}^{(N_l)} \\ - \big(\mathbf{u}^{(N_l)} \cdot \nabla\big) \mathbf{\Psi}^M \cdot \mathbf{g}\Big)\, dx\, dt = \int_0^T \int_{B_\varepsilon} \mathbf{f} \cdot \mathbf{\Psi}^M \, dx\, dt, \tag{6.2.21}$$

where:

$$\mathbf{\Psi}^M(x,t) = \sum_{j=1}^{M} d_j(t) \boldsymbol{\psi}_j(x). \tag{6.2.22}$$

It is evident that $\mathbf{\Psi}^M, \nabla \mathbf{\Psi}^M \in L^2(0, T; L^2(B_\varepsilon))$. Utilizing the compact embedding $W^{(1,1),2}(B_\varepsilon^T) \hookrightarrow L^4(B_\varepsilon^T)$, it is standard to demonstrate that we can take the limit as $N_l \to \infty$ in (6.2.21). This leads to the following expression:

$$\int_0^T \int_{B_\varepsilon} \Big(\mathbf{u}_t \cdot \mathbf{\Psi}^M + \nu \nabla \mathbf{u} \cdot \nabla \mathbf{\Psi}^M - \big((\mathbf{u} + \mathbf{g}) \cdot \nabla\big) \mathbf{\Psi}^M \cdot \mathbf{u} \\ - \big(\mathbf{u} \cdot \nabla\big) \mathbf{\Psi}^M \cdot \mathbf{g}\Big)\, dx\, dt = \int_0^T \int_{B_\varepsilon} \mathbf{f} \cdot \mathbf{\Psi}^M \, dx\, dt. \tag{6.2.23}$$

The linear combinations $\mathbf{\Psi}^M$ given by (6.2.22) with $d_j(t)$ being continuous are dense in the function space $\mathfrak{M}(B_\varepsilon^T)$ consisting of functions from $\mathring{W}_{\partial B_\varepsilon}^{(1,0),2}(B_\varepsilon^T)$ that depend continuously on t in $[0, T]$ in the norm of $\mathring{W}^{1,2}(B_\varepsilon)$ (see Theorem 2.3.8). Therefore, identity (6.2.23) remains valid for any $\mathbf{\Psi} \in \mathfrak{M}(B_\varepsilon^T)$:

$$\int_0^T \int_{B_\varepsilon} \Big(\mathbf{u}_t \cdot \mathbf{\Psi} + \nu \nabla \mathbf{u} \cdot \nabla \mathbf{\Psi} - \big((\mathbf{u} + \mathbf{g}) \cdot \nabla\big) \mathbf{\Psi} \cdot \mathbf{u} \\ - \big(\mathbf{u} \cdot \nabla\big) \mathbf{\Psi} \cdot \mathbf{g}\Big)\, dx\, dt = \int_0^T \int_{B_\varepsilon} \mathbf{f} \cdot \mathbf{\Psi} \, dx\, dt \quad \forall \mathbf{\Psi} \in \mathfrak{M}(B_\varepsilon^T). \tag{6.2.24}$$

By selecting $\mathbf{\Psi}(x,t) = d(t)\boldsymbol{\eta}(x)$ with $\boldsymbol{\eta} \in \overset{\circ}{H}{}^{1,2}(B_\varepsilon)$ and $d(t)$ being an arbitrary continuous function on $[0,T]$, we obtain the identity (6.1.6). Thus, $\mathbf{u}$ is a weak solution of (6.1.6). □

Now, we will prove the uniqueness of the solution.

Theorem 6.2.2 *The solution to problem* (6.1.6) *for $n = 2$ is unique.*

Proof
Suppose that there exist two solutions $\mathbf{u}_1$ and $\mathbf{u}_2$ of problem (6.1.6) that satisfy the conditions of Theorem 6.2.1. By subtracting the identity (6.1.6) for $\mathbf{u}_2$ from the one for $\mathbf{u}_1$, we obtain the following identity for the difference $\mathbf{w} = \mathbf{u}_1 - \mathbf{u}_2$:

$$\int_{B_\varepsilon} \Big(\mathbf{w}_t \cdot \boldsymbol{\eta} + \nu \nabla \mathbf{w} \cdot \nabla \boldsymbol{\eta} - (\mathbf{u}_1 \cdot \nabla)\boldsymbol{\eta} \cdot \mathbf{w} - (\mathbf{w} \cdot \nabla)\boldsymbol{\eta} \cdot \mathbf{u}_2 - (\mathbf{g} \cdot \nabla)\boldsymbol{\eta} \cdot \mathbf{w}$$
$$-(\mathbf{w} \cdot \nabla)\boldsymbol{\eta} \cdot \mathbf{g}\Big)dx = 0,$$

which holds for every vector field $\boldsymbol{\eta} \in \overset{\circ}{H}{}^{1,2}(B_\varepsilon)$. Taking $\boldsymbol{\eta} = \mathbf{w}$, we obtain:

$$\frac{d}{dt}\int_{B_\varepsilon} |\mathbf{w}|^2\, dx + \nu \int_{B_\varepsilon} |\nabla \mathbf{w}|^2\, dx = \int_{B_\varepsilon} (\mathbf{w} \cdot \nabla)\mathbf{w} \cdot \mathbf{u}_2\, dx + \int_{B_\varepsilon} (\mathbf{w} \cdot \nabla)\mathbf{w} \cdot \mathbf{g}\, dx$$
$$\leq \|\mathbf{w}\|_{L^4(B_\varepsilon)} \|\nabla \mathbf{w}\|_{L^2(B_\varepsilon)} \|\mathbf{u}_2\|_{L^4(B_\varepsilon)} + \sup_{x \in B_\varepsilon} |\mathbf{g}(x,t)| \|\mathbf{w}\|_{L^2(B_\varepsilon)} \|\nabla \mathbf{w}\|_{L^2(B_\varepsilon)}$$
$$\overset{(2.3.7),\,(6.1.4)}{\leq} c\Big(\|\mathbf{w}\|^{\frac{1}{2}}_{L^2(B_\varepsilon)} \|\nabla \mathbf{w}\|^{\frac{3}{2}}_{L^2(B_\varepsilon)} \|\nabla \mathbf{u}_2\|_{L^2(B_\varepsilon)} + \|\mathbf{w}\|_{L^2(B_\varepsilon)} \|\nabla \mathbf{w}\|_{L^2(B_\varepsilon)}\Big)$$
$$\overset{(2.1.2)}{\leq} c_\delta \Big(\|\mathbf{w}\|^2_{L^2(B_\varepsilon)} \big[\|\nabla \mathbf{u}_2\|^{\frac{4}{3}}_{L^2(B_\varepsilon)} + 1\big]\Big) + \delta \|\nabla \mathbf{w}\|^2_{L^2(B_\varepsilon)}.$$

Integrating the last inequality with respect to t, keeping in mind that $\mathbf{w}(x,0) = 0$, and taking $\delta = \nu/2$, we obtain the estimate:

$$\int_{B_\varepsilon} |\mathbf{w}(x,t)|^2\, dx \leq c\Big(\sup_{t \in [0,T]} \|\nabla \mathbf{u}_2\|^{\frac{4}{3}}_{L^2(B_\varepsilon)} + 1\Big) \int_0^t \int_{B_\varepsilon} |\mathbf{w}(x,\tau)|^2\, dx d\tau$$
$$\overset{(6.2.2)}{\leq} c\Big(e^{\frac{2}{3}c\varepsilon^2 d_1 \int_0^T A_1(t)\, dt} \Big[\int_0^T A_1(t) dt\Big]^{\frac{2}{3}} + 1\Big) \int_0^t \int_{B_\varepsilon} |\mathbf{w}(x,\tau)|^2\, dx d\tau.$$

Therefore, by Gronwall's lemma, we have $\mathbf{w} = 0$, which implies $\mathbf{u}_1 = \mathbf{u}_2$. □

We shall now proceed to recover and estimate the pressure function p, corresponding to the solution $\mathbf{u}$. The following theorem establishes the existence of the pressure function and provides estimates for it.

Theorem 6.2.3 *Assume that the conditions of Theorem* 6.2.1 *are satisfied. Then, there exists a function* $p \in L^2(0, T; L^2(B_\varepsilon))$ *such that the pair* $(\mathbf{u}, p)$ *satisfies the integral identity:*

$$\int\limits_{B_\varepsilon} \Big(\mathbf{u}_t \cdot \eta + \nu \nabla \mathbf{u} \cdot \nabla \eta - \big((\mathbf{u}+\mathbf{g}) \cdot \nabla\big)\eta \cdot \mathbf{u} - \big(\mathbf{u} \cdot \nabla\big)\eta \cdot \mathbf{g}\Big) dx = \int\limits_{B_\varepsilon} p \, div \, \boldsymbol{\eta} \, dx + \int\limits_{B_\varepsilon} \mathbf{f} \cdot \eta \, dx \quad \forall \eta \in \mathring{H}^{1,2}(B_\varepsilon). \tag{6.2.25}$$

Furthermore, the following estimate holds:

$$\int\limits_0^T \|p(\cdot, t)\|^2_{L^2(B_\varepsilon)} \, dt \le C(A_1)\varepsilon^{-1} \int\limits_0^T A_1(t) \, dt. \tag{6.2.26}$$

Moreover, there exists $\nabla p \in L^2(B_\varepsilon)$ *and:*

$$\int\limits_0^T \|\nabla p(\cdot, t)\|^2_{L^2(B_\varepsilon)} \, dt \le C(A_1) \int\limits_0^T A_1(t) \, dt. \tag{6.2.27}$$

Here $C(A_1)$ *is independent of* ε *and remains bounded as* $\int_0^T A_1(t) \, dt$ *tends to zero.*
The system (6.1.5) *is satisfied almost everywhere in* $B_\varepsilon \times (0, T)$.

Proof
Consider the functional:

$$L(\eta) = \int\limits_{B_\varepsilon} \Big(\mathbf{u}_t \cdot \eta + \nu \nabla \mathbf{u} \cdot \nabla \eta - \big((\mathbf{u}+\mathbf{g}) \cdot \nabla\big)\eta \cdot \mathbf{u} - \big(\mathbf{u} \cdot \nabla\big)\eta \cdot \mathbf{g}\Big) dx - \int\limits_{B_\varepsilon} \mathbf{f} \cdot \eta \, dx,$$

defined on the space $\mathring{W}^{1,2}(B_\varepsilon)$. It can be observed that:

$$\begin{aligned} \Big|L(\boldsymbol{\eta})\Big| &\le c\Big(\varepsilon\|\mathbf{u}_t\|_{L^2(B_\varepsilon)} + \|\nabla \mathbf{u}\|_{L^2(B_\varepsilon)} + \|\mathbf{u}\|^2_{L^4(B_\varepsilon)} \\ &+ \|\mathbf{u}\|_{L^4(B_\varepsilon)}\|\mathbf{g}\|_{L^4(B_\varepsilon)} + \varepsilon\|\mathbf{f}\|_{L^2(B_\varepsilon)}\Big)\|\nabla \boldsymbol{\eta}\|_{L^2(B_\varepsilon)} \\ &\le c\Big(\varepsilon\|\mathbf{u}_t\|_{L^2(B_\varepsilon)} + \|\nabla \mathbf{u}\|_{L^2(B_\varepsilon)} \\ &+ \varepsilon\|\nabla \mathbf{u}\|^2_{L^2(B_\varepsilon)} + \varepsilon\|\mathbf{f}\|_{L^2(B_\varepsilon)}\Big)\|\nabla \boldsymbol{\eta}\|_{L^2(B_\varepsilon)}. \end{aligned} \tag{6.2.28}$$

Hence, $L(\eta)$ is a bounded functional, and since $\mathbf{u}$ satisfies the identity (6.1.6), we obtain:

$$L(\eta) = 0 \ \text{ for } \ \text{ all } \ \eta \in \overset{\circ}{H}{}^{1,2}(B_\varepsilon).$$

Consequently, by Lemma 2.4.2, there exists a function $p(\cdot, t) \in L^2(B_\varepsilon)$ such that $\int_{B_\varepsilon} p(x, t)\, dx = 0$ for all $t \in (0, T)$ and:

$$L(\eta) = \int\limits_{B_\varepsilon} p \operatorname{div} \eta \, dx \quad \forall \eta \in \overset{\circ}{W}{}^{1,2}(B_\varepsilon). \tag{6.2.29}$$

This, along with the definition of $L(\eta)$, yields identity (6.2.25).

To establish estimate (6.2.26), we consider (6.2.29) with $\eta = \mathbf{w}$, where $\mathbf{w}$ is the solution of the problem:

$$\begin{cases} \operatorname{div}\mathbf{w} = p, \ x \in B_\varepsilon, \\ \quad \mathbf{w} = 0, \ \ x \in \partial B_\varepsilon. \end{cases}$$

By Lemma 2.4.14, a solution $\mathbf{w} \in \overset{\circ}{W}{}^{1,2}(B_\varepsilon)$ exists, satisfying the estimate:

$$\|\nabla \mathbf{w}\|_{L^2(B_\varepsilon)} \le c\varepsilon^{-3/2} \|p\|_{L^2(B_\varepsilon)}, \tag{6.2.30}$$

with the constant c independent of ε. This leads to:

$$\begin{aligned} &\|p\|^2_{L^2(B_\varepsilon)} = L(\mathbf{w}) \\ &\overset{(6.2.28)}{\le} c\Big(\varepsilon\|\mathbf{u}_t\|_{L^2(B_\varepsilon)} + \|\nabla\mathbf{u}\|_{L^2(B_\varepsilon)} + \varepsilon\|\nabla\mathbf{u}\|^2_{L^2(B_\varepsilon)} + \varepsilon\|\mathbf{f}\|_{L^2(B_\varepsilon)}\Big)\|\nabla\mathbf{w}\|_{L^2(B_\varepsilon)} \\ &\overset{(6.2.30)}{\le} c\varepsilon^{-3/2}\Big(\varepsilon\|\mathbf{u}_t\|_{L^2(B_\varepsilon)} + \|\nabla\mathbf{u}\|_{L^2(B_\varepsilon)} + \varepsilon\|\nabla\mathbf{u}\|^2_{L^2(B_\varepsilon)} + \varepsilon\|\mathbf{f}\|_{L^2(B_\varepsilon)}\Big)\|p\|_{L^2(B_\varepsilon)}, \end{aligned}$$

and using (6.2.1) and (6.2.3), we obtain estimate (6.2.26).

Rewriting identity (6.1.6) as:

$$\nu \int\limits_{B_\varepsilon} \nabla\mathbf{u} \cdot \nabla\eta dx = \int\limits_{B_\varepsilon} \Big(\mathbf{f} - \mathbf{u}_t - \big((\mathbf{u}+\mathbf{g})\cdot\nabla\big)\mathbf{u} - \big(\mathbf{u}\cdot\nabla\big)\mathbf{g}\Big) \cdot \eta dx \quad \forall \eta \in \overset{\circ}{H}{}^{1,2}(B_\varepsilon),$$

we can consider the solution $\mathbf{u}$ for almost all $t \in [0, T]$ as a weak solution of the Stokes problem with the right-hand side $\mathbf{f}_* = \mathbf{f} - \mathbf{u}_t - \big((\mathbf{u}+\mathbf{g})\cdot\nabla\big)\mathbf{u} - \big(\mathbf{u}\cdot\nabla\big)\mathbf{g}$. Since $\mathbf{u}_t \in L^2(B_\varepsilon)$ and $\mathbf{u} \in W^{2,2}(B_\varepsilon)$ for almost all $t \in [0, T]$, it follows from (6.1.4) that $\mathbf{f}_* \in L^2(B_\varepsilon)$. Consequently, by Theorem 4.1.3, the pressure function $p \in W^{1,2}(B_\varepsilon)$ and:

$$\|\nabla p\|^2_{L^2(B_\varepsilon)} \leq c\|\mathbf{f}_*\|^2_{L^2(B_\varepsilon)} \leq c\Big(\|\mathbf{f}\|^2_{L^2(B_\varepsilon)} + \|\mathbf{u}_t\|^2_{L^2(B_\varepsilon)}$$
$$+ \|\mathbf{u}\|^2_{\mathbf{W}^{2,2}(B_\varepsilon)}\|\nabla\mathbf{u}\|^2_{L^2(B_\varepsilon)} + \|\nabla\mathbf{u}\|^2_{L^2(B_\varepsilon)}\Big).$$

Integrating this inequality, we derive:

$$\int_0^T \|\nabla p(\cdot,t)\|^2_{L^2(B_\varepsilon)}\,dt \leq C(A_1)\int_0^T A_1(t)\,dt.$$

This concludes the proof. □

6.2.2 *The Case n = 3*

Theorem 6.2.4 *Let $B_\varepsilon \subset \mathbb{R}^3$, and the boundary ∂B_ε has C^2 smoothness. Assume that the extended function $\mathbf{g}$ belongs to $C^{[\frac{J+1}{2}]+2}(0,T;W^{2,2}(B_\varepsilon))$ and satisfies the conditions* (6.1.3) *and* (6.1.4). *There exists a constant γ_0 (independent of ε) such that if:*

$$\int_0^T A_1(t)dt \leq \gamma_0, \tag{6.2.31}$$

then for sufficiently small ε, the problem (6.1.6) *has a solution. The following estimates hold:*

$$\sup_{t\in[0,T]}\|\mathbf{u}(\cdot,t)\|^2_{L^2(B_\varepsilon)} + \int_0^T\int_{B_\varepsilon}|\nabla\mathbf{u}(x,t)|^2dxdt \leq c\varepsilon^2\int_0^T A_1(t)dt, \tag{6.2.32}$$

$$\sup_{t\in[0,T]}\|\nabla\mathbf{u}(\cdot,t)\|^2_{L^2(B_\varepsilon)} \leq c\int_0^T A_1(t)dt. \tag{6.2.33}$$

$$\int_0^T\|\nabla^2\mathbf{u}(\cdot,t)\|^2_{L^2(B_\varepsilon)}dt + \int_0^T\int_{B_\varepsilon}|\mathbf{u}_t|^2dxdt \leq c(A_1)\int_0^T A_1(t)dt, \tag{6.2.34}$$

where $c(A_1)$ is bounded as $\int_0^T A_1(t)dt$ tend to zero.

Proof

As in Theorem 6.2.1, we employ Galerkin approximations. First, by employing inequality (2.3.40) instead of (2.3.39), we establish, for sufficiently small ε, the existence of Galerkin approximations and the subsequent estimate:

$$\sup_{t\in[0,T]} \|\mathbf{u}^{(N)}(\cdot,t)\|^2_{L^2(B_\varepsilon)} + \int_0^T\int_{B_\varepsilon} |\nabla\mathbf{u}^{(N)}(x,t)|^2 dxdt \le c\varepsilon^2 \int_0^T A_1(t)dt. \tag{6.2.35}$$

To estimate the higher derivatives of $\mathbf{u}^{(N)}$, we employ a basis composed of eigenfunctions of the Stokes operator. By applying Galerkin approximations with the basis $\{\boldsymbol{\psi}_k\}_{k=1}^{\infty}$, multiplying them by $\lambda_k\gamma_k^{(N)}(t)$, summing up the resulting equations for $k = 1, \dots, N$, and utilizing properties of the Stokes operator, we derive:

$$\begin{aligned}
&\frac{\nu}{2}\frac{d}{dt}\int_{B_\varepsilon} |\nabla\mathbf{u}^{(N)}|^2 dx + \int_{B_\varepsilon} |\widetilde{\Delta}\mathbf{u}^{(N)}|^2 dx \\
&= -\int_{B_\varepsilon} ((\mathbf{u}^{(N)}+\mathbf{g})\cdot\nabla)\mathbf{u}^{(N)}\cdot\widetilde{\Delta}\mathbf{u}^{(N)} dx \\
&-\int_{B_\varepsilon} (\mathbf{u}^{(N)}\cdot\nabla)\mathbf{g}\cdot\widetilde{\Delta}\mathbf{u}^{(N)} dx + \int_{B_\varepsilon} \mathbf{f}\cdot\widetilde{\Delta}\mathbf{u}^{(N)} dx = \sum_{i=1}^{3} J_i.
\end{aligned} \tag{6.2.36}$$

We proceed to estimate the right-hand side of (6.2.36). Utilizing (4.3.14), (2.3.9), (2.3.42)–(2.3.43), (2.3.45), and (6.1.4) for $n = 3$, we obtain:

$$|J_3| \le \delta \int_{B_\varepsilon} |\widetilde{\Delta}\mathbf{u}^{(N)}|^2 dx + c_\delta A_1(t), \tag{6.2.37}$$

$$\begin{aligned}
|J_2| &= \left|\int_{B_\varepsilon} (\mathbf{u}^{(N)}\cdot\nabla)\mathbf{g}\cdot\widetilde{\Delta}\mathbf{u}^{(N)} dx\right| \le \|\mathbf{u}^{(N)}\|_{L^\infty(B_\varepsilon)}\|\nabla\mathbf{g}\|_{L^2(B_\varepsilon)}\|\widetilde{\Delta}\mathbf{u}^{(N)}\|_{L^2(B_\varepsilon)} \\
&\overset{(2.3.45),\,(6.1.4)}{\le} c\varepsilon^{\frac{1}{4}}\|\widetilde{\Delta}\mathbf{u}^{(N)}\|^2_{L^2(B_\varepsilon)}.
\end{aligned} \tag{6.2.38}$$

Let us subdivide J_1 into two integrals: $J_1 = J_{11} + J_{12}$. For $J_{11} = \int_{B_\varepsilon} (\mathbf{g}\cdot\nabla)\mathbf{u}^{(N)}\cdot\widetilde{\Delta}\mathbf{u}^{(N)} dx$, we have:

$$\begin{aligned}
|J_{11}| &= \left|\int_{B_\varepsilon} (\mathbf{g}\cdot\nabla)\mathbf{u}^{(N)}\cdot\widetilde{\Delta}\mathbf{u}^{(N)} dx\right| \\
&\overset{(2.3.42),\,(6.1.4)}{\le} c\varepsilon\|\nabla^2\mathbf{u}^{(N)}\|_{L^2(B_\varepsilon)}\|\widetilde{\Delta}\mathbf{u}^{(N)}\|_{L^2(B_\varepsilon)} \overset{(4.3.13)}{\le} c\varepsilon\|\widetilde{\Delta}\mathbf{u}^{(N)}\|^2_{L^2(B_\varepsilon)}.
\end{aligned} \tag{6.2.39}$$

For $J_{12} = \int_{B_\varepsilon} (\mathbf{u}^{(N)} \cdot \nabla)\mathbf{u}^{(N)} \cdot \widetilde{\Delta}\mathbf{u}^{(N)} dx$, we have:

$$
\begin{aligned}
|J_{12}| &= \Big| \int\limits_{B_\varepsilon} (\mathbf{u}^{(N)} \cdot \nabla)\mathbf{u}^{(N)} \cdot \widetilde{\Delta}\mathbf{u}^{(N)} dx \Big| \\
&\le c_\delta \|\mathbf{u}^{(N)}\|^2_{L^6(B_\varepsilon)} \|\nabla \mathbf{u}^{(N)}\|^2_{L^3(B_\varepsilon)} + \delta \int\limits_{B_\varepsilon} |\widetilde{\Delta}\mathbf{u}^{(N)}|^2 dx \overset{(2.3.9),\,(2.3.43)}{\le} \delta \int\limits_{B_\varepsilon} |\widetilde{\Delta}\mathbf{u}^{(N)}|^2 dx \\
&+ c_\delta \|\nabla \mathbf{u}^{(N)}\|^2_{L^2(B_\varepsilon)} \varepsilon^{-1} \|\nabla \mathbf{u}^{(N)}\|_{L^2(B_\varepsilon)} \Big(\|\nabla \mathbf{u}^{(N)}\|^2_{L^2(B_\varepsilon)} + \varepsilon^2 \|\nabla^2 \mathbf{u}^{(N)}\|^2_{L^2(B_\varepsilon)} \Big)^{\frac{1}{2}} \\
&\overset{(2.3.42)}{\le} \delta \int\limits_{B_\varepsilon} |\widetilde{\Delta}\mathbf{u}^{(N)}|^2 dx \\
&+ c_\delta \varepsilon^{-1} \|\nabla \mathbf{u}^{(N)}\|^3_{L^2(B_\varepsilon)} \Big(c\varepsilon^2 \|\nabla^2 \mathbf{u}^{(N)}\|^2_{L^2(B_\varepsilon)} + \varepsilon^2 \|\nabla^2 \mathbf{u}^{(N)}\|^2_{L^2(B_\varepsilon)} \Big)^{\frac{1}{2}} \\
&\overset{(4.3.13)}{\le} \delta \int\limits_{B_\varepsilon} |\widetilde{\Delta}\mathbf{u}^{(N)}|^2 dx + c_\delta \|\nabla \mathbf{u}^{(N)}\|^3_{L^2(B_\varepsilon)} \|\widetilde{\Delta}\mathbf{u}^{(N)}\|_{L^2(B_\varepsilon)} \\
&\le 2\delta \int\limits_{B_\varepsilon} |\widetilde{\Delta}\mathbf{u}^{(N)}|^2 dx + c_\delta \|\nabla \mathbf{u}^{(N)}\|^6_{L^2(B_\varepsilon)}.
\end{aligned} \tag{6.2.40}
$$

By substituting inequalities (6.2.37)–(6.2.40) into Eq. (6.2.36) and choosing $\delta = \frac{1}{12}$, we deduce the following estimate:

$$
\begin{aligned}
&\frac{\nu}{2} \frac{d}{dt} \int\limits_{B_\varepsilon} |\nabla \mathbf{u}^{(N)}|^2 \, dx + \left(\frac{3}{4} - c_1 \varepsilon^{1/4} \right) \int\limits_{B_\varepsilon} |\widetilde{\Delta}\mathbf{u}^{(N)}|^2 \, dx \\
&\le c_2 \|\nabla \mathbf{u}^{(N)}\|^6_{L^2(B_\varepsilon)} + c_3 A_1(t).
\end{aligned} \tag{6.2.41}
$$

For sufficiently small ε (specifically, when $\varepsilon^{1/4} \le \frac{1}{4c_1}$), we have:

$$
\frac{3}{4} - c_1 \varepsilon^{\frac{1}{4}} \ge \frac{1}{2},
$$

and from Eq. (6.2.41), we obtain the following result:

$$
\begin{aligned}
&\frac{\nu}{2} \frac{d}{dt} \int\limits_{B_\varepsilon} |\nabla \mathbf{u}^{(N)}|^2 \, dx + \frac{1}{2} \int\limits_{B_\varepsilon} |\widetilde{\Delta}\mathbf{u}^{(N)}|^2 \, dx \\
&\le c_2 \|\nabla \mathbf{u}^{(N)}\|^6_{L^2(B_\varepsilon)} + c_3 A_1(t).
\end{aligned} \tag{6.2.42}
$$

We denote $z(t) = \int_{B_\varepsilon} |\nabla \mathbf{u}(x,t)|^2\,dx$ and rewrite Eq. (6.2.42) as follows:

$$z'(t) \le \frac{2}{\nu} c_2 z(t)^3 + \frac{2}{\nu} c_3 A_1(t) \le \left(1 + z(t)^2\right)\left(\frac{2}{\nu} c_2 z(t) + \frac{2}{\nu} c_3 A_1(t)\right),$$

or equivalently:

$$\frac{z'(t)}{1+z(t)^2} \le \frac{2}{\nu} c_2 z(t) + \frac{2}{\nu} c_3 A_1(t). \tag{6.2.43}$$

Integrating Equation (6.2.43) with respect to t and utilizing Eq. (6.2.35), we obtain:

$$\begin{aligned}\arctan z(t) &\le \arctan z(0) + \frac{2c_2}{\nu}\int_0^T z(t)\,dt + \frac{2c_3}{\nu}\int_0^T A_1(t)\,dt \\ &\le \frac{2c_2}{\nu}\int_0^T\int_{B_e} |\nabla u^{(N)}|^2\,dx\,dt + c_4\int_0^T A_1(t)\,dt \qquad (6.2.44)\\ &\le c_5\int_0^T A_1(t)\,dt \overset{(6.2.31)}{\le} c_5\gamma_0.\end{aligned}$$

If the constant γ_0 in inequality (6.2.31) is sufficiently small (i.e., $c_5\gamma_0 < \frac{\pi}{2}$), then from inequality (6.2.44), we deduce the estimate:

$$z(t) = \int_{B_\varepsilon} |\nabla \mathbf{u}^{(N)}|^2\,dx \le \tan\left[c_5\gamma_0\right] \le c_6.$$

Hence, the relation (6.2.43) implies:

$$\frac{1}{1+c_6} z'(t) \le \frac{z'(t)}{1+z(t)^2} \le c_7 z(t) + c_8 A_1(t),$$

which further simplifies to:

$$z'(t) \le c_9 z(t) + c_{10} A_1(t),$$

and by applying Gronwall's lemma:

$$z(t) \le c_{11}\left(z(0) + \int_0^T A_1(t)\,dt\right) \le c_{11}\int_0^T A_1(t)\,dt.$$

Thus, we arrive at:

$$\sup_{t\in[0,T]} \|\nabla\mathbf{u}^{(N)}(\cdot,t)\|^2_{L^2(B_\varepsilon)} \le c\int_0^T A_1(t)\,dt. \tag{6.2.45}$$

Substituting Eq. (6.2.45) into Eq. (6.2.42) and integrating with respect to t yields:

$$\begin{gathered}\int_0^T \|\nabla^2\mathbf{u}^{(N)}(\cdot,t)\|^2_{L^2(B_\varepsilon)}\,dt \overset{(4.3.13)}{\le} c\int_0^T \|\widetilde{\Delta}\mathbf{u}^{(N)}(\cdot,t)\|^2_{L^2(B_\varepsilon)}\,dt \\ \le c\left(\left[\int_0^T A_1(t)\,dt\right]^3 + \int_0^T A_1(t)\,dt\right) \le c(A_1)\int_0^T A_1(t)\,dt,\end{gathered} \tag{6.2.46}$$

where $c(A_1)$ is bounded as $\int_0^T A_1(t)\,dt$ tends to zero.

Let us estimate the norm of $\mathbf{u}_t^{(N)}$. By considering $\boldsymbol{\eta} = \mathbf{u}_t^{(N)}$ in integral identity (6.1.6) (specifically, multiplying (6.2.4) by $\frac{d}{dt}\gamma_k^{(N)}$ and summing over $k = 1,\dots,N$), we obtain:

$$\begin{aligned}&\int_{B_\varepsilon} |\mathbf{u}_t^{(N)}|^2\,dx + \frac{\nu}{2}\frac{d}{dt}\int_{B_\varepsilon} |\nabla\mathbf{u}^{(N)}|^2\,dx \\ &= -\int_{B_\varepsilon} ((\mathbf{u}^{(N)}+\mathbf{g})\cdot\nabla)\mathbf{u}^{(N)}\cdot\mathbf{u}_t^{(N)}\,dx - \int_{B_\varepsilon} (\mathbf{u}^{(N)}\cdot\nabla)\mathbf{g}\cdot\mathbf{u}_t^{(N)}\,dx + \int_{B_\varepsilon} \mathbf{f}\cdot\mathbf{u}_t^{(N)}\,dx \\ &\le \|\mathbf{u}^{(N)}\|_{L^\infty(B_\varepsilon)}\|\nabla\mathbf{u}^{(N)}\|_{L^2(B_\varepsilon)}\|\mathbf{u}_t^{(N)}\|_{L^2(B_\varepsilon)} \\ &+\|\mathbf{g}\|_{L^\infty(B_\varepsilon)}\|\nabla\mathbf{u}^{(N)}\|_{L^2(B_\varepsilon)}\|\mathbf{u}_t^{(N)}\|_{L^2(B_\varepsilon)} \\ &+\|\mathbf{u}^{(N)}\|_{L^\infty(B_\varepsilon)}\|\nabla\mathbf{g}\|_{L^2(B_\varepsilon)}\|\mathbf{u}_t^{(N)}\|_{L^2(B_\varepsilon)} + \|\mathbf{f}\|_{L^2(B_\varepsilon)}\|\mathbf{u}_t^{(N)}\|_{L^2(B_\varepsilon)} \\ &\le 4\delta\|\mathbf{u}_t^{(N)}\|^2_{L^2(B_\varepsilon)} + c_\delta\|\mathbf{u}^{(N)}\|^2_{L^\infty(B_\varepsilon)}\|\nabla\mathbf{u}^{(N)}\|^2_{L^2(B_\varepsilon)} \\ &+c_\delta\|\mathbf{g}\|^2_{L^\infty(B_\varepsilon)}\|\nabla\mathbf{u}^{(N)}\|^2_{L^2(B_\varepsilon)} + c_\delta\|\mathbf{u}^{(N)}\|^2_{L^\infty(B_\varepsilon)}\|\nabla\mathbf{g}\|^2_{L^2(B_\varepsilon)} + c_\delta A(t).\end{aligned} \tag{6.2.47}$$

Integrating Equation (6.2.47) over $[0,T]$ and using conditions (6.1.4) and inequalities (4.3.14) and (6.2.46), we derive:

$$
\begin{aligned}
\int_0^T \int_{B_\varepsilon} |\mathbf{u}_t^{(N)}|^2\, dx\, dt &\le c \int_0^T \|\mathbf{u}^{(N)}\|^2_{L^\infty(B_\varepsilon)} \|\nabla \mathbf{u}^{(N)}\|^2_{L^2(B_\varepsilon)}\, dt \\
&\quad + c \sup_{t\in[0,T]} \|\mathbf{g}(\cdot,t)\|^2_{L^\infty(B_\varepsilon)} \int_0^T \|\nabla \mathbf{u}^{(N)}\|^2_{L^2(B_\varepsilon)}\, dt \\
&\quad + c \sup_{t\in[0,T]} \|\nabla \mathbf{g}(\cdot,t)\|^2_{L^2(B_\varepsilon)} \int_0^T \|\mathbf{u}^{(N)}\|^2_{L^\infty(B_\varepsilon)}\, dt + c\int_0^T A_1(t)\, dt \\
&\overset{(4.3.14),\,(6.1.4)}{\le} c\varepsilon^{1/2} \int_0^T \|\nabla^2 \mathbf{u}^{(N)}\|^2_{L^2(B_\varepsilon)} \|\nabla \mathbf{u}\|^2_{L^2(B_\varepsilon)}\, dt + c\int_0^T \|\nabla \mathbf{u}^{(N)}\|^2_{L^2(B_\varepsilon)}\, dt \\
&\quad + c\varepsilon^{1/2} \int_0^T \|\nabla^2 \mathbf{u}^{(N)}\|^2_{L^2(B_\varepsilon)}\, dt + c\int_0^T A_1(t)\, dt \le c(A_1) \int_0^T A_1(t)\, dt,
\end{aligned}
\tag{6.2.48}
$$

where $c(A_1)$ is bounded as $\int_0^T A_1(t)\, dt$ tends to zero.

Estimates (6.2.35), (6.2.46), and (6.2.48) ensure the existence of a subsequence $\{\mathbf{u}^{(N_l)}\}$ such that $\{\mathbf{u}^{(N_l)}\}$, $\{\nabla \mathbf{u}^{(N_l)}\}$, $\{\nabla^2 \mathbf{u}^{(N_l)}\}$, and $\{\mathbf{u}_t^{(N_l)}\}$ converge weakly in $L^2(0,T; L^2(B_\varepsilon))$ to $\mathbf{u}$, $\nabla \mathbf{u}$, $\nabla^2 \mathbf{u}$, and $\mathbf{u}_t$, respectively. Since in the three-dimensional case embedding $W^{(1,1),2}(B_\varepsilon^T) \hookrightarrow L^4(B_\varepsilon^T)$ is compact, it can be demonstrated that it is possible to pass to the limit as $N_l \to \infty$ in the equality (6.2.21) (which is the same for two- and three-dimensional cases). The rest of the proof that $\mathbf{u}$ is a weak solution of problem (6.1.1) for $n = 3$ repeats the arguments of Theorem 6.2.1 for the two-dimensional case. □

Theorem 6.2.5 *The solution of problem* (6.1.6) *for the dimension* $n = 3$ *is unique.*

Proof
Drawing parallels to the reasoning in Theorem 6.2.2, consider the difference $\mathbf{w} = \mathbf{u}_1 - \mathbf{u}_2$ between two solutions of problem (6.1.6). This yields the subsequent equality:

$$
\begin{aligned}
&\frac{d}{dt} \int_{B_\varepsilon} |\mathbf{w}|^2\, dx + \nu \int_{B_\varepsilon} |\nabla \mathbf{w}|^2\, dx \\
&= \int_{B_\varepsilon} (\mathbf{w}\cdot\nabla)\mathbf{w}\cdot \mathbf{u}_2\, dx + \int_{B_\varepsilon} (\mathbf{w}\cdot\nabla)\mathbf{w}\cdot \mathbf{g}\, dx.
\end{aligned}
$$

By estimating the right-hand side utilizing (2.3.8) and (6.1.4), the ensuing inequality is obtained:

$$\frac{d}{dt}\int\limits_{B_\varepsilon}|\mathbf{w}|^2\,dx+\nu\int\limits_{B_\varepsilon}|\nabla\mathbf{w}|^2\,dx$$
$$\leq\|\mathbf{w}\|_{L^4(B_\varepsilon)}\|\nabla\mathbf{w}\|_{L^2(B_\varepsilon)}\|\mathbf{u}_2\|_{L^4(B_\varepsilon)}+\sup_{x\in B_\varepsilon}|\mathbf{g}(x,t)|\|\mathbf{w}\|_{L^2(B_\varepsilon)}\|\nabla\mathbf{w}\|_{L^2(B_\varepsilon)}$$
$$\leq c\Big(\|\mathbf{w}\|^{\frac14}_{L^2(B_\varepsilon)}\|\nabla\mathbf{w}\|^{\frac74}_{L^2(B_\varepsilon)}\|\nabla\mathbf{u}_2\|_{L^2(B_\varepsilon)}+\|\mathbf{w}\|_{L^2(B_\varepsilon)}\|\nabla\mathbf{w}\|_{L^2(B_\varepsilon)}\Big)$$
$$\leq c_\delta\Big(\|\mathbf{w}\|^2_{L^2(B_\varepsilon)}\big[\|\nabla\mathbf{u}_2\|^8_{L^2(B_\varepsilon)}+1\big]\Big)+\delta\|\nabla\mathbf{w}\|^2_{L^2(B_\varepsilon)}.$$

Integrating the last inequality over time and setting $\delta=\nu/2$ yields the estimate:

$$\int\limits_{B_\varepsilon}|\mathbf{w}(x,t)|^2\,dx$$
$$\leq c\Big(\sup_{t\in[0,T]}\|\nabla\mathbf{u}_2\|^8_{L^2(B_\varepsilon)}+1\Big)\int\limits_0^t\int\limits_{B_\varepsilon}|\mathbf{w}(x,\tau)|^2\,dxd\tau$$
$$\overset{(6.2.32)}{\leq}c\Big(\Big(\int\limits_0^T A_1(t)\,dt\Big)^4+1\Big)\int\limits_0^t\int\limits_{B_\varepsilon}|\mathbf{w}(x,\tau)|^2\,dxd\tau.$$

Hence, applying Gronwall's lemma, it follows that $\mathbf{w}=0$, thus implying $\mathbf{u}_1=\mathbf{u}_2$. □

Theorem 6.2.6 *Under the assumptions of Theorem* 6.2.4, *there exists a unique function* $p\in L^2(B_\varepsilon)$ *such that* $\int_{B_\varepsilon}p(x,t)dx=0$, *and the pair* $(\mathbf{u},p)$ *satisfies the integral identity:*

$$\int\limits_{B_\varepsilon}\Big(\mathbf{u}_t\cdot\eta+\nu\nabla\mathbf{u}\cdot\nabla\eta-\big((\mathbf{u}+\mathbf{g})\cdot\nabla\big)\eta\cdot\mathbf{u}-\big(\mathbf{u}\cdot\nabla\big)\eta\cdot\mathbf{g}\Big)dx$$
$$=\int\limits_{B_\varepsilon}p\,\mathrm{div}\eta dx+\int\limits_{B_\varepsilon}\mathbf{f}\cdot\eta dx\quad\forall\eta\in\mathring{W}^{1,2}(B_\varepsilon).\tag{6.2.49}$$

Additionally, the following estimate holds:

$$\int\limits_0^T\|p(\cdot,t)\|^2_{L^2(B_\varepsilon)}dt\leq c(A_1)\varepsilon^{-1}\int\limits_0^T A_1(t)dt.\tag{6.2.50}$$

Furthermore, there exists $\nabla p\in L^2(B_e)$ *satisfying the estimate:*

$$\int\limits_0^T\|\nabla p(\cdot,t)\|^2_{L^2(B_e)}dt\leq c(A_1)\int\limits_0^T A_1(t)dt.\tag{6.2.51}$$

The quantity $c(A_1)$ *remains bounded as* $\int\limits_0^T A_1(t)dt$ *approaches zero.*

The system (6.1.5) *is satisfied almost everywhere in* $B_\varepsilon \times (0, T)$.

Proof

The proof parallels that of the two-dimensional case, with the exception that inequality (2.3.37) is replaced with inequality (2.3.38), and estimates from Theorem 6.2.4 are utilized in place of the two-dimensional estimates from Theorem 6.2.3. □

6.3 Construction of the Asymptotic Expansion: Case of Small-Time Vanishing Data

In this section, we outline the procedure for constructing an asymptotic expansion of the solution to problem (6.1.1). We first focus on the scenario where the data vanishes for small time intervals. Specifically, we assume that $\mathbf{g}(x, t)$ is identically zero in a neighborhood of the origin $t = 0$, meaning there exists a positive t^* such that for all $t \in [0, t^*]$, all components of $\mathbf{g}^j$ are zero:

$$\mathbf{g}^j(\cdot, t) = 0 \;\; \forall\, t \in [0, t^*]. \tag{6.3.1}$$

In this case, the time-dependent boundary layer does not arise, simplifying the construction of the asymptotic expansion compared to the general case.

Given that the boundary velocity $\mathbf{g}$ is zero in a neighborhood of $t = 0$ implies:

$$F_\varepsilon^j(t) = 0, \quad j = N_1 + 1, \dots, N, \tag{6.3.2}$$

for $t \in [0, t^*]$.

6.3.1 *Tree-Type Tube Structures*

In the case where the graph $\mathcal{G}$ of the tube structure B_ε forms a tree, the flow rates in cylinders $\Pi_\varepsilon^{(e)}$ can be defined, taking into account the orientation of the edges from the lower level to the higher one. We denote:

$$F_\varepsilon^{(e)}(t) = \int\limits_{\sigma_\varepsilon^{(e)}} \mathbf{v}(x, t) \cdot \mathbf{e}\, dx, \tag{6.3.3}$$

the flow rate in the cylinder $\Pi_\varepsilon^{(e)}$, where $\mathbf{e}$ is the direction vector of the local coordinate axis $x_n^{(e)}$ and $\sigma_\varepsilon^{(e)}$ is an orthogonal cross section of the cylinder $\Pi_\varepsilon^{(e)}$.

This flow rate is constant with respect to $x_n^{(e)}$. In accordance with the definition of the flow rates $F_\varepsilon^j(t)$, the relationship between $F_\varepsilon^{(e)}(t)$ and $F_\varepsilon^j(t)$ for $N_1+1 \le j < N$ and any edge e having its end in the vertex O_j is given by:

$$F_\varepsilon^{(e)}(t) = F_\varepsilon^j(t). \tag{6.3.4}$$

For the root vertex O_N, the relation is reversed:

$$F_\varepsilon^{(e_1)}(t) = -F_\varepsilon^N(t). \tag{6.3.5}$$

As in the stationary case, for a tree structure, all flow rates $F_\varepsilon^{(e)}$ can be uniquely determined using relations (6.3.4) and (6.1.10) and (6.1.9). The determination process starts from the edges at the highest level, where relations (6.1.7) are applied, and then proceeds to descend from higher-level nodes to lower-level ones. It is worth noting that in this algorithm, each relation (6.1.10) involves precisely one unknown flow rate corresponding to the edge e_s of the previous level.

The steps in the algorithm for constructing the asymptotic expansion of the solution to (6.1.1) are as follows:

Step 1 As previously indicated, for each edge, the flow rate $F_\varepsilon^{(e)}(t)$ is known. This enables the search for the regular component of the asymptotic expansion (in local coordinates) in the form of a non-steady Poiseuille-type solution:

$$\mathbf{u}^{(e)}(x^{(e)}, t) = (0, \ldots, u^{(e)}(x^{(e)\prime}, t))^*, \ \ p^{(e)}(x^{(e)}, t) = -s^{(e)}(t)x_n^{(e)} + a^{(e)}(t), \tag{6.3.6}$$

which possesses the given flow rate $F_\varepsilon^{(e)}(t)$ (see Chap. 3, Sect. 3.5). By employing such Poiseuille-type solutions for all edges, it is observable that the pressure function p from (6.3.6) is defined across the graph, hence referred to as the *macroscopic pressure*. The asymptotic expansion for $u^{(e)}(x^{(e)\prime}, t)$ takes the form (see Sect. 3.5 of Chap. 3):

$$\begin{aligned}
u^{(e,J)}(x^{(e)\prime}, t) &= \sum_{k=0}^{J/2} \varepsilon^{2k} \widetilde{U}_{2k}^{(e)}\left(\frac{x^{(e)\prime}}{\varepsilon}, t\right),\\
p^{(e,J)}(x, t) &= -s^{(e,J)}(t)x_n^{(e)} + a^{(e,J)}(t),\\
s^{(e,J)}(t) &= \sum_{k=0}^{J/2} \varepsilon^{2k-2} s_{2k}^{(e)}(t), \quad J = 2K,
\end{aligned} \tag{6.3.7}$$

where K is an integer:

$$\widetilde{U}_{2k}^{(e)}(y^{(e)'},t) = s_{2k}^{(e)}(t)U_0^{(e)}(y^{(e)\prime}) + \frac{ds_{2k-2}^{(e)}(t)}{dt}U_2^{(e)}(y^{(e)'}) \\ \dots + \frac{d^k s_0^{(e)}(t)}{dt^k}U_{2k}^{(e)}(y^{(e)'}), \tag{6.3.8}$$

$$s_0^{(e)}(t) = \frac{1}{\kappa_0}F^{(e)}(t), \\ s_{2k}^{(e)}(t) = -\kappa_0^{-1}\kappa_2\frac{ds_{2k-2}^{(e)}(t)}{dt} - \dots - \kappa_0^{-1}\kappa_{2k}\frac{d^k s_0^{(e)}(t)}{dt^k}, \ k \geq 1, \tag{6.3.9}$$

$y^{(e)} = \frac{x^{(e)}}{\varepsilon}$, and $U_{2r}^{(e)}(y^{(e)\prime})$ are recursively defined, starting from $U_0^{(e)}$, as the solutions of the following elliptic problems on $\sigma^{(e)}$:

$$\begin{cases} -\nu\Delta' U_0^{(e)}(y^{(e)\prime}) = 1, \ y^{(e)\prime} \in \sigma^{(e)}, \\ U_0^{(e)}(y^{(e)\prime}) = 0, \ y^{(e)\prime} \in \partial\sigma^{(e)}; \end{cases} \tag{6.3.10}$$

$$\begin{cases} -\nu\Delta' U_{2r}^{(e)}(y^{(e)\prime}) = -U_{2r-2}^{(e)}(y^{(e)\prime}), \ y^{(e)\prime} \in \sigma^{(e)}, \\ U_{2r}^{(e)}(y^{(e)\prime}) = 0, \qquad y^{(e)\prime} \in \partial\sigma^{(e)}; \ r \geq 1; \end{cases} \tag{6.3.11}$$

and:

$$\kappa_{2s} = \int\limits_{\sigma^{(e)}} U_{2s}^{(e)}(y^{(e)\prime})dy^{(e)\prime} \neq 0, \ \ s \geq 0.$$

It should be noted that changing the orientation of the local coordinate $x_n^{(e)}$ to the opposite (it corresponds to the alternative choice of the origin of the local coordinate system) also leads to a change in sign for all functions $s_j^{(e)}$.

The terms $a^{(e,J)}(t)$ in (6.3.7) will be defined in full at a later point. Currently, their "temporary values" $\widehat{a}^{(e,J)}(t)$ are defined in a manner that ensures the corresponding macroscopic pressure $\widehat{p}^{(e,J)}$ to be continuous on the graph $\mathcal{G}$. Starting from the root vertex O_N, $\widehat{a}^{(e_1,J)}(t)$ is set to 0, and subsequently, $\widehat{a}^{(e,J)}(t)$ are defined for edges of increasing levels through the continuous extension of $\widehat{p}^{(e,J)}$. Specifically, $\widehat{a}^{(e,J)}(t)$ for edges of level k is set equal to the value of the macroscopic pressure $\widehat{p}^{(e,J)}$ at the end of the "parent edge" of the preceding level (where e_α of level $k-1$ is considered the parent of the edge e_β of level k if they share a common end). The expansion of $\widehat{a}^{(e,J)}(t)$ in terms of ε is given by:

$$\widehat{a}^{(e,J)}(t) = \sum_{j=0}^{J} \varepsilon^{j-2} \widehat{a}_j^{(e)}(t). \tag{6.3.12}$$

Consequently, the macroscopic pressure takes the form:

$$\widehat{p}^{(e,J)}(x,t) = \sum_{j=0}^{J} \varepsilon^{j-2} \widehat{p}_j^{(e)}(x,t),$$

where:

$$\widehat{p}_j^{(e)}(x,t) = -s_j^{(e)}(t) x_n^{(e)} + \widehat{a}_j^{(e)}(t).$$

Remark 6.3.1 It is worth noting that the following formulas are valid (see Sect. 3.5 of Chap. 3):

$$\widetilde{U}_0^{(e)}(y',t) = \frac{1}{\kappa_0} F^{(e)}(t) U_0^{(e)}(y^{(e)\prime}), \quad s_0^{(e)}(t) = \frac{1}{\kappa_0} F^{(e)}(t), \tag{6.3.13}$$

and:

$$\begin{aligned} \widetilde{U}_{2k}^{(e)}(y^{(e)\prime},t) &= \frac{d^k F^{(e)}(t)}{dt^k} \sum_{i=0}^{k} \alpha_{2k-2i} U_{2i}^{(e)}(y^{(e)\prime}), \\ s_{2k}^{(e)}(t) &= \alpha_{2k} \frac{d^k F^{(e)}(t)}{dt^k}, \end{aligned} \tag{6.3.14}$$

for $k = 1, \ldots, K$, where the coefficients a_{2r} are defined recursively:

$$\alpha_0 = \frac{1}{\kappa_0}, \quad \alpha_{2r} = \frac{1}{\kappa_0} \Big(\sum_{s=1}^{r-1} \kappa_{2s} \alpha_{2r-2s} - \kappa_{2r} \alpha_0 \Big). \tag{6.3.15}$$

We see here that the regularity of $\mathbf{g}^j$ in time implies that all functions s_j, a_j, $\widetilde{U}_j^{(e)}$, and $\mathbf{u}^{(e,J)}$ are twice differentiable in time.

Step 2 The second phase of the procedure involves multiplying the regular component of the asymptotic expansion by cutoff functions. These functions are equal to one everywhere except for spherical areas centered at nodes, with a radius $r\varepsilon$. The parameter r denotes the maximal diameter of the domains ω^j. This multiplication results in a residual term on the right-hand side of the Navier–Stokes equations (6.1.1). The support of this residual is contained around the nodes and will be subsequently compensated for in the third step by solving boundary layer problems within unbounded domains. These problems have solutions that tend to zero exponentially as the scaled variables $x_n^{(e)}/\varepsilon$ tend to infinity. The boundary layer solutions are then multiplied by a second cutoff function to eliminate contributions from boundary layers far from the nodes. This ensures that there is no overlap between boundary layers associated with different nodes.

Let $\mathbf{U}^{[e,J]}$ represent the regular portion of the asymptotic approximation of the velocity field in the global coordinate system. Specifically, $\mathbf{U}^{[e,J]} = (\mathcal{P}^{(e)})^*(0, \ldots, u^{(e,J)})^*$. Similarly, set $\mathbf{U}_j^{[e]} = (\mathcal{P}^{(e)})^*(0, \ldots, \widetilde{U}_j^{(e)})^*$. Here, $\mathcal{P}^{(e)}$ denotes the orthogonal matrix that relates global coordinates x to local coordinates $x^{(e)} = \mathcal{P}^{(e)}(x - O_l)$ (where O_l is both the end of e and the origin of the local coordinate system). The sought asymptotic expansion takes the form:

$$\begin{aligned}\mathbf{u}^{(J)}(x,t) = \sum_{i=1}^{M} \zeta\left(\frac{x_n^{(e_i)}}{3r\varepsilon}\right)\zeta\left(\frac{|e_i| - x_n^{(e_i)}}{3r\varepsilon}\right)\mathbf{U}^{[e_i,J]}\left(\frac{x^{(e_i)\prime}}{\varepsilon}, t\right) \\ + \sum_{l=1}^{N}(1 - \zeta\left(\frac{|x - O_l|}{|e|_{min}}\right)\mathbf{U}^{[BLO_l,J]}\left(\frac{x - O_l}{\varepsilon}, t\right),\end{aligned} \tag{6.3.16}$$

where:

$$\mathbf{U}^{[e_i,J]}(y^{(e)\prime}, t) = \sum_{j=0}^{J} \varepsilon^j \mathbf{U}^{(e)}(y^{(e)\prime}, t),$$

$$\mathbf{U}^{(e)}(y^{(e)\prime}, t) = (\mathcal{P}^{(e)})^*(0, \ldots, \widetilde{U}_j^{(e)}(y^{(e)\prime}, t)).$$

In this context, as before, $\zeta(\tau)$ denotes a smooth cutoff function that is independent of ε, with $\zeta(\tau) = 0$ for $\tau \leq 1/3$ and $\zeta(\tau) = 1$ for $\tau \geq 2/3$. If an edge e_i terminates at a vertex, the cutoff function ζ in the first sum (which vanishes at the vertex) should be omitted. Additionally, $|e|_{min}$ represents the minimum length among the edges. The boundary layer term $\mathbf{U}^{[BLO_l,J]}(y, t)$ decays exponentially as $|y|$ approaches infinity, and it is represented as:

$$\mathbf{U}^{[BLO_l,J]}(y, t) = \sum_{j=-1}^{J} \varepsilon^j \mathbf{U}_j^{[BLO_l]}(y, t). \tag{6.3.17}$$

The expansion of the pressure follows a similar form:

$$\begin{aligned}p^{(J)}(x,t) = \sum_{i=1}^{M} \zeta\left(\frac{x_n^{(e_i)}}{3r\varepsilon}\right)\zeta\left(\frac{|e_i| - x_n^{(e_i)}}{3r\varepsilon}\right)\left(-s^{(e_i,J)}(t)x_n^{(e_i)} + \widehat{a}^{(e_i,J)}(t)\right) \\ + \frac{1}{\varepsilon}\sum_{l=0}^{N}\left(1 - \zeta\left(\frac{|x - O_l|}{|e|_{min}}\right)\right)p^{[BLO_l,J]}\left(\frac{x - O_l}{\varepsilon}, t\right),\end{aligned} \tag{6.3.18}$$

with:

$$P^{[BLO_l,J]}(y,t) = \sum_{j=-1}^{J} \varepsilon^j P_j^{[BLO_l]}(y,t). \tag{6.3.19}$$

Here, $\widehat{a}^{(e,J)}(t)$ is determined by Eq. (6.3.12). The boundary layer functions are defined within dilated domains, which are the union of semi-infinite cylinders sharing a common node O_l and the corresponding ω^l. Specifically, if O_l is a node and also serves as a common end for the edges $e_{i_1}, \dots, e_{i_m}$ in a bundle $\mathcal{B}_l$, then the semi-infinite cylinders $\Pi_\infty^{(e_j)}$ are defined by:

$$\Pi_\infty^{(e_j)} = \{y \in \mathbb{R}^n : \mathcal{P}^{(e_j)} y \in \sigma^j \times (0, +\infty)\}.$$

As before, the domain Ω_l, featuring m outlets to infinity corresponding to the node O_l, is given by:

$$\Omega_l = \Big(\bigcup_{j=1}^{m} \Pi_\infty^{(e_{i_j})}\Big) \bigcup \omega^l$$

(see Fig. 1.7).

Similarly, if O_l is a vertex and the end of the edge e_i, then the domain Ω_l is a union of a semi-infinite cylinder and the domain ω^l:

$$\Omega_l = \{y \in \mathbb{R}^n : \mathcal{P}^{(e_i)} y \in \sigma^i \times (0, +\infty)\} \bigcup \omega^l.$$

Step 3 The solution for the boundary layer consists of a pair of functions: a vector-valued function $\mathbf{U}_j^{[BLO_l]}$ and a scalar function $P_j^{[BLO_l]}$. By substituting the expansion given in Eqs. (6.3.16)–(6.3.19) into the Navier–Stokes equations and the associated boundary conditions, while neglecting the influence of the cutoff functions $\zeta(\frac{|e_i| - x_n^{(e_i)}}{3r\varepsilon})$ and $\zeta(\frac{|x - O_l|}{|e|_{min}})$, we obtain for every node O_l the following Stokes system in the domain Ω_l:

$$\begin{cases} -\nu \Delta_y \mathbf{U}_j^{[BLO_l]} + \nabla_y P_j^{[BLO_l]} = \mathbf{f}_j^{[REGO_l]} + \mathbf{f}_j^{[BLO_l]}, & y^{(e)} \in \Omega_l, \\ \operatorname{div}_y \mathbf{U}_j^{[BLO_l]} = h_j^{[REGO_l]}, & y^{(e)} \in \Omega_l, \\ \mathbf{U}_j^{[BLO_l]} = 0, & y^{(e)} \in \partial\Omega_l, \end{cases} \tag{6.3.20}$$

where $j = -1, 0, \dots, J$ and:

$$\mathbf{f}_j^{[REGO_l]}(y^{(e)},t) = -\sum_{e:O_l\in e}\Big\{-\nu\Delta_y\big[\zeta(\frac{y_n^{(e)}}{3r})\mathbf{U}_j^{(e)}(y^{(e)\prime},t)\big]$$
$$+\nabla_y\big[\zeta(\frac{y_n^{(e)}}{3r})(-s_j^{(e)}(t)y_n^{(e)})\big]+\zeta(\frac{y_n^{(e)}}{3r})\frac{\partial\mathbf{U}_{j-2}^{(e)}(y^{(e)\prime},t)}{\partial t}$$
$$+\sum_{p+r=j-1}\zeta(\frac{y_n^{(e)}}{3r})\big(\mathbf{U}_p^{(e)}(y^{(e)\prime},t)\cdot\nabla_y\big)\big(\zeta(\frac{y_n^{(e)}}{3r})\mathbf{U}_r^{(e)}(y^{(e)\prime},t)\big)$$
$$+\widehat{a}_{j+1}^{(e)}(t)\,\nabla_y\zeta(\frac{y_n^{(e)}}{3r})\Big\}, \tag{6.3.21}$$

with $\widehat{a}_{j+1}^{(e)}(t)$ omitted when $j=J$. Additionally:

$$\mathbf{f}_j^{[BLO_l]}(y^{(e)},t) = -\sum_{e:O_l\in e}\Big\{\sum_{p+r=j-1}\zeta(\frac{y_n^{(e)}}{3r})\big(\mathbf{U}_p^{(e)}(y^{(e)\prime},t)\cdot\nabla_y\big)\mathbf{U}_r^{[BLO_l]}(y,t)$$
$$+\sum_{p+r=j-1}\big(\mathbf{U}_p^{[BLO_l]}(y,t)\cdot\nabla_y\big)\big(\zeta(\frac{y_n^{(e)}}{3r})\mathbf{U}_r^{(e)}(y^{(e)\prime},t)\big)\Big\}$$
$$-\sum_{p+r=j-1}\big(\mathbf{U}_p^{[BLO_l]}(y,t)\cdot\nabla_y\big)\mathbf{U}_r^{[BLO_l]}(y,t)\frac{\partial\mathbf{U}_{j-2}^{[BLO_l]}}{\partial t}(y,t), \tag{6.3.22}$$

and:

$$h_j^{[REGO_l]}(y,t) = -\sum_{e:O_l\in e}\operatorname{div}_y\Big(\zeta(\frac{y_n^{(e)}}{3r})\mathbf{U}_j^{(e)}(y^{(e)\prime},t)\Big)$$
$$= -\operatorname{div}_y\Big(\sum_{e:O_l\in e}\Big\{s_j^{(e)}(t)\zeta(\frac{y_n^{(e)}}{3r})(\mathcal{P}^{(e)})^*(0,\ldots,U_0^{(e)}(y^{(e)\prime}))^*$$
$$+\sum_{k:0<2k\le j}\frac{d^k s_{j-2k}^{(e)}(t)}{dt^k}(t)\zeta(\frac{y_n^{(e)}}{3r})(\mathcal{P}^{(e)})^*(0,\ldots,U_{2k}^{(e)}(y^{(e)\prime}))^*\Big\}\Big). \tag{6.3.23}$$

Here, we consider the sum $\sum_{e:O_l\in e}$ over all edges e that have endpoints at the node O_l. As a convention, the terms with negative subscripts j, except for $P_{-1}^{[BLO_l]}$, are assigned a value of zero. In this subsection, the terms $\mathbf{U}_j^{(e)}$ are only present for even values of j. For the sake of consistency, all local coordinates $x^{(e)}$ and $y^{(e)}$ are taken with respect to the origin O_l. This corresponds to the selection of local coordinates $x^{(e)}$ when transitioning "from a lower level to a higher one" for all edges e in the bundle $\mathcal{B}^{(l)}$, excluding the chosen edge e_s. It is also worth noting that in this context, the functions $\widehat{a}_{j+1}^{(e)}(t)$ are the same for the node O_l due to the continuity of the macroscopic pressure $\widehat{p}^{(e,J)}$ in this node. These functions are denoted collectively as $\widehat{p}_j(O_l,t)$.

The right-hand sides $\mathbf{f}_j^{[REGO_l]}$ and $\mathbf{h}_j^{[REGO_l]}$ have compact supports, disappearing beyond the sphere of radius $2r$ centered at the origin. This property arises from the cutoff functions being equal to one outside this sphere. Additionally, the Poiseuille-type solutions $\big(\mathbf{U}_j^{(e)}(x^{(e)\prime},t), -s_j^{(e)}(t)x_n^{(e)} + const\big)$ satisfy the relationships:

$$-\nu\Delta\mathbf{U}_j^{(e)} + \nabla\big(-s_j^{(e)}(t)x_n^{(e)} + const\big) + \frac{\partial\mathbf{U}_{j-2}^{(e)}}{\partial t} = 0, \quad \mathrm{div}\mathbf{U}_j^{(e)} = 0.$$

Notably, the function $\mathbf{f}_j^{[REGO_l]}$ also vanishes within the sphere of radius r, due to the vanishing of the cutoff function within that region. Conversely, $\mathbf{f}_j^{[BLO_l]}$ may have support across the entire domain $\overline{\Omega}_l$, yet it exponentially tends to zero as $|y| \to +\infty$. This property is a consequence of inductive reasoning based on the outcomes of Sect. 4.4.

Similarly, at each vertex O_l for $l = N_1+1,\dots,N$, the Stokes problem is derived for $\big(\mathbf{U}_j^{[BLO_l]}, P_j^{[BLO_l]}\big)$ within the semi-infinite dilated cylinder Ω_l:

$$\begin{cases} -\nu\Delta_y\mathbf{U}_j^{[BLO_l]} + \nabla_y P_j^{[BLO_l]} = \mathbf{f}_j^{[REGO_l]} + \mathbf{f}_j^{[BLO_l]}, & y \in \Omega_l, \\ \mathrm{div}_y\mathbf{U}_j^{[BLO_l]} = h_j^{[REGO_l]}, & y \in \Omega_l, \\ \mathbf{U}_j^{[BLO_l]}\big|_{\partial\Omega_l\setminus\gamma^l} = 0, & \\ \mathbf{U}_j^{[BLO_l]}\big|_{\gamma^l} = \mathbf{g}^l(y,t)\delta_{j,0}, & \end{cases} \tag{6.3.24}$$

where δ_{jm} is Kronecker's delta, and the right-hand sides are defined by (6.3.21), (6.3.22), and (6.3.23), where the sum $\sum\limits_{e:O_l\in e}$ contains only one term.

The problem (6.3.20) possesses a unique solution, up to an additive constant in the pressure field, such that $\mathbf{U}_j^{[BLO_l]}(\cdot,t) \in \mathring{W}^{1,2}(\Omega_l)$ and $P_j^{[BLO_l]}(\cdot,t) \in L^2_{loc}(\Omega_l)$ where t is a parameter. The existence of such a solution is contingent upon the condition that the integral over Ω_l of the right-hand side of the divergence equation $(6.3.20)_2$ vanishes, which can be expressed as follows:

$$\int\limits_{\Omega_l} h_j^{[REGO_l]}(y,t)dy = 0. \tag{6.3.25}$$

This stipulation is indeed met. Specifically, due to the Kirchhoff conditions for the fluxes at the nodes:

$$\int_{\Omega_l} h_j^{[REGO_l]}(y,t)dy = -\sum_{e:O_l\in e}\int_{\Omega_l} \mathrm{div}_y\Big(\zeta(\frac{y_n^{(e)}}{3r})\mathbf{U}_j^{(e)}(y^{(e)\prime},t)\Big)dy^{(e)\prime}$$

$$= -\sum_{e:O_l\in e}\int_{\sigma^{(e)}} \mathbf{U}_j^{(e)}(y^{(e)\prime},t)\cdot \mathbf{n}\,dy^{(e)\prime} = 0.$$

In the above equations, condition (6.1.10) is utilized, alongside the fact that the functions $\mathbf{U}_j^{(e)}$ satisfy the flow rate conditions specified in (6.3.3). By virtue of (6.3.8), these relations can be rewritten for all j in the subsequent manner:

$$\sum_{e:O_l\in e}\int_{\sigma^{(e)}} U_0^{(e)}(y^{(e)\prime})dy^{(e)\prime}s_j^{(e)}(t)$$

$$= -\sum_{e:O_l\in e}\ \sum_{k:0<2k\le j}\frac{d^k s_{j-2k}^{(e)}(t)}{dt^k}(t)\int_{\sigma^{(e)}} U_{2k}^{(e)}(y^{(e)\prime})dy^{(e)\prime}.$$

The boundary layer functions are recursively defined with respect to the parameter j. Let us consider the base case of the induction. For $j=-1$, all terms on the right-hand side of Eq. (6.3.20) disappear, excluding the final term of $\mathbf{f}_{-1}^{[REGO_l]}$. This implies that $\mathbf{f}_{-1}^{[REGO_l]}(y^{(e)},t) = -\widehat{a}_0^{(e)}(t)\,\nabla_y(\zeta(\frac{y_n^{(e)}}{3r}))$. It is pertinent to recall that all functions $\widehat{a}_0^{(e)}(t)$ are equal to the value of $\widehat{p}_0(O_l,t)$. Consequently, for $j=-1$, the problem (6.3.20) is characterized by an evident analytical solution with a compact support: the velocity $\mathbf{U}_{-1}^{[BLO_l]}(y,t)$ is identically zero, and the pressure is given by:

$$P_{-1}^{[BLO_l]}(y,t) = -\sum_{e:O_l\in e}\big(\zeta(\frac{y_n^{(e)}}{3r})-1\big)\widehat{p}_0(O_l,t).$$

Utilizing Theorem 4.4.2, we establish, using an induction-based approach, that the expressions on the right-hand side, namely, $\mathbf{f}_j^{[REGO_l]}(y^{(e)},t)$, $\mathbf{f}_j^{[BLO_l]}(y^{(e)},t)$, and $h_j^{[REGO_l]}(y,t)$, within the problems for boundary layer correctors, depend smoothly on time t (as a parameter). For each j, these expressions are $[\frac{J-j}{2}]+2$ times differentiable in t. Consequently, the solution $(\mathbf{U}_j^{[BLO_l]}, P_j^{[BLO_l]})$ inherits analogous regularity characteristics. Moreover, applying Theorem 4.4.2 in conjunction with Lemma 4.4.2, we establish that $\widehat{a}_j^{[BLO_l,e]}(t)$ have the same regularity in t.

Step 4 The solutions $\mathbf{U}_j^{[BLO_l]} \in \mathring{W}^{1,2}(\Omega_l)$ and $P_j^{[BLO_l]} \in L^2_{loc}(\Omega_l)$ of the boundary layer problems stated in Eq. (6.3.20) have an important property: the velocity $\mathbf{U}_j^{[BLO_l]}$ and its derivatives exponentially vanish as $|y| \to +\infty$, where y is the spatial coordinate within the domain Ω_l. This phenomenon is observed for all outlets $\Pi^{(e)}$ of the domain Ω_l. Simultaneously, the pressure component $P_j^{[BLO_l]}$ stabilizes toward a constant value denoted as $\widehat{a}_j^{[BLO_l,e]}(t)$. These constants depend on the parameter t and may vary between different outlets. Without loss of generality, one of these constants can be arbitrarily selected. To standardize the analysis, we fix the constant within the outlet corresponding to the chosen (parent) edge of the bundle, setting it equal to zero (we refer to Chap. 4 for further details).

All the assertions made in Steps 2–4 pertaining to the solution of the problem (6.3.20) remain valid for the boundary layer (6.3.24), corresponding to the vertices O_l with $l = N_1+1, \dots, N$. The primary distinction lies in the solvability condition. Specifically, problem (6.3.24) possesses a unique solution if the following condition is met:

$$\delta_{j0} \int\limits_{\gamma^l} \mathbf{g}^l(y,t) \cdot \mathbf{n}(y) dS + \int\limits_{\Omega_l} \mathrm{div}_y \Big(\zeta(\frac{y_n^{(e)}}{3r}) \mathbf{U}_j^{(e)}(y^{(e)\prime}, t) \Big) dy^{(e)\prime} = 0.$$

The latter condition is satisfied, due to relations (6.3.4) and (6.3.5).

Step 5 Subsequently, we redefine the macroscopic pressure $p^{(e)}$ for each edge. This redefinition follows the procedure of iteratively determining the constants $a^{(e,J)}(t)$ from lower levels to higher ones. We initiate the process at the root vertex O_N, where we set $a_j^{(e_1)}(t) = \widehat{a}_j^{(e_1)}(t) = 0$. Assume that $a_j^{(e)}(t)$ has already been assigned for edges at level k. Progressing to level $k+1$, we consider the bundle $\mathcal{B}_l$ encompassing the chosen (parent) edge e_s and the edges $e_{\alpha_1}, \dots, e_{\alpha_r}$ at level $k+1$. We proceed to set $a_j^{(e_{\alpha_m})}(t) = a_j^{(e_s)}(t) + \widehat{a}_{j-1}^{[BLO_l,e_{\alpha_m}]}(t)$ for $m = 1, \dots, r$ and denote $a^{(e,J)} = \frac{1}{\varepsilon^2} \sum\limits_{j=0}^{J} \varepsilon^j a_j^{(e)}$. Next, we redefine the boundary layer pressures through the introduction of a new formulation:

$$P_j^{[BLO_l,new]}(y,t) = P_j^{[BLO_l]}(y,t) - \sum_{m=1}^{r} \zeta(\frac{y_n^{(e_{\alpha_m})}}{3r}) \widehat{a}_j^{[BLO_l,e_{\alpha_m}]}(t)$$
$$- \sum_{e:O_l \in e} \Big(\zeta(\frac{y_n^{(e)}}{3r}) - 1 \Big) \Big(a_{j+1}^{(e)}(t) - \widehat{a}_j^{(e_s)}(t) \Big) \Big).$$

In the case where $j = J$, the last sum above is absent. Take in formulas (6.3.18) and (6.3.19) of asymptotic expansion these newly defined pressures $P_j^{[BLO_l,new]}(y,t)$ instead of the former ones $P_j^{[BLO_l]}(y,t)$ and keep for them the old notation $P_j^{[BLO_l]}(y,t)$. After such redefinition, the boundary layer pressures

$p_j^{[BLO_l]}$ become tending to zero at infinity, and they evidently still satisfy the boundary layer problems (6.3.20) with the unchanged velocity fields $\mathbf{U}_j^{[BLO_l]}$ and the new values of constants $a_{j+1}^{(e)}(t)$ in the right-hand sides $\mathbf{f}_j^{[REGO_l]}$ (instead of the former temporary values $\widehat{a}_{j+1}^{(e)}(t)$). Notably, the redefined macroscopic pressure p may potentially be discontinuities at nodes, where the gaps are known.

6.3.2 General Tube Structure

In the context of a general tube structure (see Fig. 1.6b), a distinguishing feature when compared to tree-type configurations lies in the potential ambiguity regarding the flow rates within the cylinders $\Pi_\varepsilon^{(e)}$. In such cases, a priori knowledge of the uniqueness of the solution to the algebraic equation system (6.1.10) is lacking. Consequently, the determination of functions $s_j^{(e)}(t)$ becomes nontrivial and cannot be straightforwardly derived from the relations (6.1.10) and (6.1.9). Instead, for each $j = 0, \ldots, J$, these functions need to be successively ascertained along with the functions $\widehat{a}_j^{(e)}(t)$ via the conductivity problem on the graph $\mathcal{B}$ for the macroscopic pressure $\widehat{p}_j^{(e)}$. As deduced from the constructions of the preceding subsection, in the case of a tree, the macroscopic pressure $\widehat{p}^{(e)}(x_n^{(e)}, t) = \sum_{j=0}^{J} \varepsilon^{j-2} \widehat{p}_j^{(e)}(x_n^{(e)}, t)$, defined for every edge e, takes the form of the sum $\sum_{j=0}^{J} \varepsilon^{j-2}(s_j^{(e)}(t) x_n^{(e)} + \widehat{a}_j^{(e)}(t))$ and satisfies the following relations on the graph $\mathcal{B}$:

(i) Second-order differential equations are valid across all edges:

$$- \kappa_e \frac{\partial^2 \widehat{p}_j^{(e)}}{\partial x_n^{(e)2}}(x_n^{(e)}, t) = 0, \quad x_n^{(e)} \in (0, |e|), \quad j = 0, 1, \ldots, J,$$

where $\kappa_e = \int_{\sigma^{(e)}} U_0(y')dy'$. This equation stems from the linearity of pressure along each edge e, with κ_e representing the steady-state permeability of the tube, serving as the proportionality coefficient in Darcy's law that relates pressure drop to flow rate.

(ii) Junction conditions are established at nodes and vertices O_l, with $l = 1, \ldots, N$.

(ii)$_1$ Kirchhoff-type conditions at nodes:

$$- \sum_{e: O_l \in e} \kappa_e \frac{\partial \widehat{p}_j^{(e)}(0, t)}{\partial x_n^{(e)}} = \Psi_{l,j}(t), \quad l = 1, \ldots, N_1,$$

and Neumann conditions at vertices:

$$-\kappa_{e_l}\frac{\partial \widehat{p}_j^{(e_l)}}{\partial x_n^{(e_l)}}(0,t)=\Psi_{l,j}(t),\quad l=N_1+1,\dots,N,$$

where:

$$\Psi_{l,j}(t)=\sum_{e:O_l\in e}\ \sum_{k:0<2k\le j}\frac{\partial^{k+1}\widehat{p}_{j-2k}^{(e)}(0,t)}{\partial t^k\partial x_n^{(e)}}\int_{\sigma^{(e)}}U_{2k}^{(e)}(y^{(e)\prime})dy^{(e)\prime}$$

for $l=1,\dots,N_1$, and:

$$\Psi_{l,j}(t)=\sum_{k:0<2k\le j}\frac{\partial^{k+1}\widehat{p}_{j-2k}^{(e_l)}(0,t)}{\partial t^k\partial x_n^{(e_l)}}\int_{\sigma^{(e_l)}}U_{2k}^{(e_l)}(y^{(e_l)\prime})dy^{(e_l)\prime}-F^l(t)\delta_{j0},$$

where δ_{jm} denotes Kronecker's delta, for $l=N_1+1,\dots,N$. These conditions arise from asymptotic formulas (6.3.8) and (6.3.9) and the solvability condition (6.3.25) for the boundary layer problem (6.3.20). Notably, the local coordinates $x^{(e)}$ have been redefined such that they all share the same origin O_l. In accordance with the coordinate systems outlined in the previous subsection for trees (from lower levels to higher ones), the sign of each term in the summation depends on the direction of the local coordinate $x_n^{(e)}$: it is negative for the selected edge e within the bundle and positive for all others.

For the conditions corresponding to $l=N_1+1,\dots,N$ and $j=0$, the formulation takes the shape:

$$-\kappa_{e_l}\frac{\partial \widehat{p}_0^{(e_l)}}{\partial x_n^{(e_l)}}(0,t)=-F^l(t),$$

where $F_l(t)$ represents the known flow rate linked to vertex O_l. For $j=2k,\ k\ge 1$, these conditions can be expressed as:

$$-\kappa_{e_l}\frac{\partial \widehat{p}_{2k}^{(e_l)}}{\partial x_n^{(e_l)}}(0,t)=-\kappa_0\alpha_{2k}\frac{d^kF^l(t)}{dt^k}$$

(see (6.3.14) and (6.3.15)).

In instances involving nodes O_l, i.e., when $l=1,\dots,N_1$, the flow rates associated with this node are generally unknown. The condition for $j=0$ and $l=1,\dots,N_1$ signifies that the total flux across all edges equals zero (see (6.1.9) and (6.1.10)).

(ii)$_2$ Continuity conditions on the graph:

$$\widehat{p}_j^{(e_\alpha)}(0,t) = \widehat{p}_j^{(e_\beta)}(0,t) \ \forall e_\alpha, e_\beta \in \mathcal{B}_l;$$

The redefined terms for macroscopic pressure, denoted as p_j, satisfy the same problem on the graph but with the condition of known gaps in the nodes:

$$p_j^{(e)}(0,t) - p_j^{(e_s)}(0,t) = \varphi_{l,j}^{(e)}(t), \quad \forall e \in \mathcal{B}_l, e \neq e_s,$$

instead of the continuity condition $(ii)_2$. Here, e_s designates the selected edge within the bundle and $\varphi_{l,j}^{(e)}(t) = \widehat{a}_{j-1}^{[BLO_l,e]}(t)$. In the general scenario, any edge O_l within the bundle can be chosen as e_s (while bearing in mind that, for tree-like graphs, the parent edge was designated as the selected one).

We solve such problem for the macroscopic pressure at every passage from j to $j+1$. Subsequently, we establish a recurrent series of issues on the graph. Each such problem has the following form. Let p be a function defined on this graph. For every edge e, it can be considered as a function of the local coordinate $x_n^{(e)}$, namely, $p(x,t) = p^{(e)}(x_n^{(e)}, t)$, where $x_n^{(e)}$ is the last coordinate of $x^{(e)} = \mathcal{P}^{(e)}(x - O_l)$ and O_l is an end of e, the origin of the local coordinate system.

Consider the graph $\mathcal{B}$ and the following conductivity problem on it.

(i) The second-order differential equation across the edges:

$$-\kappa_e \frac{\partial^2 p^{(e)}}{\partial x_n^{(e)2}}(x_n^{(e)}, t) = f^{(e)}(x_n^{(e)}, t), \quad x_n^{(e)} \in (0, |e|), \tag{6.3.26}$$

where κ_e represents positive constants.

(ii) The junction conditions at nodes and vertices O_l, where $l = 1, \ldots, N$:

(ii)$_1$ The Kirchhoff-type conditions at nodes:

$$-\sum_{e:O_l \in e} \kappa_e \frac{\partial p^{(e)}}{\partial x_n^{(e)}}(0,t) = \Psi_l(t), \quad l = 1, \ldots, N_1, \tag{6.3.27}$$

and Neumann conditions at vertices:

$$-\kappa_{e_l} \frac{\partial p^{(e_l)}}{\partial x_n^{(e_l)}}(0,t) = \Psi_l(t), \quad l = N_1 + 1, \ldots, N. \tag{6.3.28}$$

(ii)$_2$ The prescribed gaps of the pressure at the nodes O_l:

$$p^{(e)}(0,t) - p^{(e_{s,l})}(0,t) = \varphi_l^{(e)}(t) \ \forall e \in \mathcal{B} : e \subset \mathcal{B}_l, e \neq e_{s,l}; \tag{6.3.29}$$

Here, $e_{s,l}$ designates the selected edge from bundle $\mathcal{B}_l$, and the local coordinate system $x^{(e)}$ is redefined so that its origin is O_l. Ψ_l and $\varphi_l^{(e)}$ are real numbers (dependent on the parameter t), $f^{(e)}(\cdot, t) \in L^2\big((0, |e|)\big)$. As it is usual for the Neumann boundary value problem, the compatibility condition appears as follows:

$$\sum_{i=1,\dots,M} \int_0^{|e_i|} f^{(e_i)}(x_n^{(e_i)}, t) dx_n^{(e_i)} + \sum_{l=1}^{N} \Psi_l(t) = 0, \qquad (6.3.30)$$

and the solution remains unique apart from an additive constant.

Condition (6.3.30) emerges through integrating Eq. (6.3.26) over graph $\mathcal{B}$, incorporating the orientation of local coordinates within each bundle $\mathcal{B}_l$.

Our aim is to reduce problems (6.3.26)–(6.3.29) into an analogous problem with $\varphi_l^{(e)}(t) = 0$. This would mean replacing the prescribed gaps in nodes with a continuity requirement on the unknown function across the nodes. For this purpose, we introduce the following transformation on the unknown function. For each node O_l and edge e such that $e \subset \mathcal{B}_l$, excluding e_s, we define:

$$\widehat{p}^{(e)}(x_n^{(e)}, t) = p^{(e)}(x_n^{(e)}, t) - (1 - \zeta(\frac{2x_n^{(e)}}{|e|}))\varphi_l^{(e)}(t), \quad x_n^{(e)} \in [0, \frac{|e|}{2}]. \qquad (6.3.31)$$

Here, ζ signifies the cutoff function, as defined after formula (6.3.16). It is noteworthy that for $x_n^{(e)} \in [\frac{|e|}{2}, |e|]$, a similar transformation is applied, corresponding to the second end of e. However, if the second end of the edge is a vertex or $e = e_{s,l}$, the change of the unknown function is not necessary. For $\widehat{p}^{(e)}$, we obtain the following equations:

$$-\kappa_e \frac{\partial^2 \widehat{p}^{(e)}}{\partial x_n^{(e)2}}(x_n^{(e)}, t) = \widehat{f}^{(e)}(x_n^{(e)}, t), \quad x_n^{(e)} \in (0, |e|), \qquad (6.3.32)$$

along with the junction conditions at nodes and vertices:

$$- \sum_{e: O_l \in e} \kappa_e \frac{\partial \widehat{p}^{(e)}}{\partial x_n^{(e)}}(0, t) = \Psi_l(t), \quad l = 1, \dots, N_1, \qquad (6.3.33)$$

$$-\kappa_{e_l} \frac{\partial \widehat{p}^{(e_l)}}{\partial x_n^{(e_l)}}(0, t) = \Psi_l(t), \quad l = N_1 + 1, \dots, N, \qquad (6.3.34)$$

and the continuity conditions at nodes:

$$\widehat{p}^{(e_\alpha)}(0,t) = \widehat{p}^{(e_\beta)}(0,t), \ \ \forall e_\alpha, e_\beta \subset \mathcal{B}_l, \tag{6.3.35}$$

where $\widehat{f}^{(e)}$ is defined for $x_n^{(e)} \in [0, \frac{|e|}{2}]$ by the formula:

$$\widehat{f}^{(e)}(x_n^{(e)}, t) = f^{(e)}(x_n^{(e)}, t) + \kappa_e \varphi_l^{(e)} \zeta''(\frac{2x_n^{(e)}}{|e|})(\frac{2}{|e|})^2. \tag{6.3.36}$$

Problems (6.3.32)–(6.3.35) are similar to problem (5.1.11) associated with the stationary case. The sole distinction lies in the fact that, at present, the data and unknown functions depend upon the variable t serving as the parameter. Similarly to Sect. 5 (cf. Theorem 5.1.1), we prove the following assertion.

Theorem 6.3.1 *Problem* (6.3.32)–(6.3.35) *possess a unique weak solution* $\widehat{p} \in W^{1,2}_{0_N}(\mathcal{B})$ *if and only if condition* (6.3.30) *holds. Moreover, the solution* $\widehat{p}$ *satisfies the inequality:*

$$\|\widehat{p}\|_{W^{1,2}(\mathcal{B})} \leq c\Big(\|\widehat{f}\|_{L^2(\mathcal{B})} + \big(\sum_{l=1}^{N-1} |\Psi_l|^2\big)^{1/2}\Big), \tag{6.3.37}$$

where the constant c *does not depend on* t *and* $\|\widehat{f}\|^2_{L^2(\mathcal{B})} = \sum_{i=1}^{M} \int_0^{|e_i|} |\widehat{f}^{(e_i)}|^2 dx_n^{(e_i)}$.

Corollary 6.3.1 *If condition* (6.3.30) *holds, then problem* (6.3.26)–(6.3.29) *possess a solution* p. *This solution is unique, up to an additive constant. The solution, normalized by the condition* $p(O_N) = 0$, *satisfies the following estimate:*

$$\|p\|^2_{W^{1,2}(\mathcal{B})} \leq c\left(\|\widehat{f}\|^2_{L^2(\mathcal{B})} + \sum_{l=1}^{N-1} |\Psi_l|^2 + \sum_{l=1}^{N_1} \sum_{e:O_l \in e} |\varphi_l^{(e)}|^2\right). \tag{6.3.38}$$

Corollary 6.3.2 *Assuming that* $f^{(e_i)} = 0$ *for all* e_i, *where* $i = 1, \ldots, M$, *and the compatibility condition* (6.3.30) *is satisfied, the solution* p *of problem* (6.3.26)–(6.3.29), *with the normalization constraint* $p(O_N) = 0$, *is linear along each edge* e:

$$p^{(e)}(x_n^{(e)}, t) = -s^{(e)}(t)x_n^{(e)} + a^{(e)}(t). \tag{6.3.39}$$

Furthermore, the subsequent inequalities hold:

$$\sum_{i=1}^{M} |s^{(e_i)}(t)|^2 \leq c\left(\sum_{l=1}^{N-1} |\Psi_l(t)|^2 + \sum_{l=1, e:O_l \in e}^{N_1} |\varphi_l^{(e)}(t)|^2\right), \tag{6.3.40}$$

$$\sum_{i=1}^{M} |a^{(e_i)}(t)|^2 \leq c \left(\sum_{l=1}^{N-1} |\Psi_l(t)|^2 + \sum_{l=1, e: O_l \in e}^{N_1} |\varphi_l^{(e)}(t)|^2 \right), \tag{6.3.41}$$

Remark 6.3.2 When $f^{(e_i)} = 0$ for all e_i, $i = 1, \ldots, M$, problems (6.3.26)–(6.3.29) can be reformulated as a system of linear algebraic equations involving $s^{(e)}(t)$ and $a^{(e)}(t)$:

$$\begin{cases} \sum_{e: O_l \in e} \kappa_e s^{(e)}(t) = \Psi_l(t), \quad l = 1, \ldots, N_1, \\ \kappa_e s^{(e)}(t) = \Psi_l(t), \quad l = N_1 + 1, \ldots, N, \\ a^{(e)}(t) - a^{(e_{s,l})}(t) = \varphi_l^{(e)}(t), \forall e \in \mathcal{B}_l, \ e \neq e_{s,l}, \ l = 1, \ldots, N_1, \end{cases} \tag{6.3.42}$$

completed for each edge $e = \overline{O_i O_j}$ by introducing relationships between the coefficient values $s^{(e)}$ and $a^{(e)}$ in two coordinate systems: one with the origin at endpoint O_i and the other at O_j. These coefficients are represented as $s^{(e,O_i)}$, $a^{(e,O_i)}$ in the first system and $s^{(e,O_j)}$, $a^{(e,O_j)}$ in the second. The two relations connecting these coefficients are:

$$s^{(e,O_i)}(t) = -s^{(e,O_j)}(t), \quad a^{(e,O_j)}(t) - a^{(e,O_i)}(t) = s^{(e,O_i)}(t)|e|. \tag{6.3.43}$$

Adding an additional equation:

$$a^{(e_N)}(t) = 0, \tag{6.3.44}$$

we get the system with an invertible matrix, and thus, this system admits a unique solution (this follows from Theorem 6.3.1).

6.3.3 Construction of the Asymptotic Expansion for General Graphs

In this subsection, we outline the methodology for constructing the asymptotic expansion applicable to general graphs. The objective is to derive an asymptotic expansion for the velocity field and pressure within this context. The expansion for the velocity field is structured similarly to that for tree-type graphs and is expressed as follows:

$$\begin{aligned} \mathbf{u}^{(J)}(x,t) = &\sum_{i=1}^{M} \zeta\left(\frac{x_n^{(e_i)}}{3r\varepsilon}\right) \zeta\left(\frac{|e| - x_n^{(e_j)}}{3r\varepsilon}\right) \mathbf{U}^{[e_i,J]}\left(\frac{x^{(e_i)\prime}}{\varepsilon}, t\right) \\ &+ \sum_{l=1}^{N} \left(1 - \zeta\left(\frac{|x - O_l|}{|e|_{min}}\right)\right) \mathbf{U}^{[BLO_l,J]}\left(\frac{x - O_l}{\varepsilon}, t\right), \end{aligned} \tag{6.3.45}$$

where:

$$\mathbf{U}^{[e,J]} = \left(P^{(e)}\right)^* (0, \ldots, 0, \widetilde{U}^{[e,J]})^*,$$

$$\mathbf{U}_j^{(e)} = \left(P^{(e)}\right)^* (0, \ldots, 0, \widetilde{U}_j^{(e)})^*, \quad j = 0, 1, \ldots, J,$$

$$\widetilde{U}^{[e,J]}(y^{(e)\prime}, t) = \sum_{j=0}^{J} \varepsilon^j \widetilde{U}_j^{(e)}(y^{(e)\prime}, t),$$
$$\mathbf{U}^{[BLO_l,J]}(y, t) = \sum_{j=0}^{J} \varepsilon^j \mathbf{U}_j^{[BLO_l]}(y, t). \tag{6.3.46}$$

It is worth noting that the primary difference between (6.3.46) and the formula ($6.3.7_1$) used for tree-type graphs is the broader range of natural powers of ε considered in (6.3.46). This stems from the possibility of non-vanishing $s_1^{(e)}(t)$ and, consequently, non-vanishing odd-order terms $s_j^{(e)}(t)$ in the general graph scenario. It is essential to remember that functions $s_j^{(e)}(t)$ remain unknown for general graphs.

Similarly to the velocity field expansion, the asymptotic expansion for the pressure can be expressed as:

$$p^{(J)}(x, t) = \sum_{i=1}^{M} \zeta\left(\frac{x_n^{(e_i)}}{3r\varepsilon}\right) \zeta\left(\frac{|e| - x_n^{(e_j)}}{3r\varepsilon}\right) \left(-s^{e_i}(t) x_n^{(e_i)} + a^{e_i}(t)\right)$$
$$+ \frac{1}{\varepsilon} \sum_{l=0}^{N} (1 - \zeta\left(\frac{|x - O_l|}{|e|_{min}}\right)) P^{[BLO_l,J]}\left(\frac{x - O_l}{\varepsilon}, t\right), \tag{6.3.47}$$

where:

$$s^{(e)}(t) = \frac{1}{\varepsilon^2} \sum_{j=0}^{J} \varepsilon^j s_j^{(e)}(t),$$
$$a^{(e)}(t) = \frac{1}{\varepsilon^2} \sum_{j=0}^{J} \varepsilon^j a_j^{(e)}(t), \tag{6.3.48}$$

and:

$$P^{[BLO_l,J]}(y, t) = \sum_{j=-1}^{J} \varepsilon^j P_j^{[BLO_l,J]}(y, t). \tag{6.3.49}$$

The step-by-step algorithm for determining the terms in the asymptotic expansions (6.3.16) and (6.3.47) is outlined as follows:

Base Case Solve the conductivity problem on the graph to find p_0:

$$\begin{cases} -\kappa_e \dfrac{\partial^2 p_0^{(e)}}{\partial x_n^{(e)2}}(x_n^{(e)}, t) = 0, & x_n^{(e)} \in (0, |e|), \\ -\sum_{e:O_l\in e} \kappa_e \dfrac{\partial p_0^{(e)}}{\partial x_n^{(e)}}(0, t) = 0, & l = 1, \dots, N_1, \\ -\kappa_e \dfrac{\partial p_0^{(e)}}{\partial x_n^{(e)}}(0, t) = -F^{(j)}(t), & l = N_1 + 1, \dots, N, \\ p_0^{(e)}(0, t) = p_0^{(e_s)}(0, t), \ \forall e \subset \mathcal{B}_l. \end{cases}$$

Solving the above problem, we find functions $s_0^{(e)}(t)$ and $a_0^{(e)}(t)$, yielding:

$$p_0^{(e)}(x^{(e)}, t) = -s_0^{(e)}(t)x_n^{(e)} + a_0^{(e)}(t).$$

It is important to observe that this problem on the graph and subsequent issues related to problems on the graph depend on the variable t serving as the parameter. By employing Theorem 6.3.1 in conjunction with Corollaries 6.3.1 and 6.3.2, we establish the regularity of their solutions, $p_j^{(e)}$, and, subsequently, the regularity of the coefficients $s_j^{(e)}(t)$ and $a_j^{(e)}(t)$.

Also, we derive the velocity:

$$\widetilde{U}_0^{(e)}(y^{(e)\prime}, t) = s_0^{(e)}(t)U_0^{(e)}(y^{(e)\prime}), \ \ \mathbf{U}_0^{(e)} = \left(\mathcal{P}^{(e)}\right)^* (0, \dots, 0, \widetilde{U}_0^{(e)})^*. \tag{6.3.50}$$

Next, introduce the boundary layer pressure of rank -1 (see (6.3.49)):

$$P_{-1}^{[BLO_l]}(y, t) = -\left(\sum_{e:O_l\in e} \zeta\left(\frac{y_n^{(e)}}{3r}\right) - 1\right) p_0(O_l, t).$$

Here p_0 is a continuous function on $\mathcal{B}$ without jumps at the nodes, so that $p_0(O_l, t)$ is well-defined.

The velocity of rank -1 is zero: $\mathbf{U}_{-1}^{[BLO_l]}(y, t) = 0$.

The boundary layer problem for $(\mathbf{U}_0^{[BLO_l]}(y, t), P_0^{[BLO_l]}(y, t))$ is the following:

$$\begin{cases} -\nu\Delta_y \mathbf{U}_0^{[BLO_l]} + \nabla_y P_0^{[BLO_l]} = \mathbf{f}_0^{[REGO_l]} + \mathbf{f}_0^{[BLO_l]}, & y \in \Omega_l, \\ \mathrm{div}_y \mathbf{U}_0^{BLO_l} = h_0^{[REGO_l]}, & y \in \Omega_l, \\ \mathbf{U}_0^{[BLO_l]} = 0, & y \in \partial\Omega_l, \end{cases} \tag{6.3.51}$$

where:

$$\mathbf{f}_0^{[REGO_l]}(y,t) = -\sum_{e:O_l\in e} \Big\{ s_0^{(e)}(t)\Big(\nu\Delta_y\big(\zeta(\frac{y_n^{(e)}}{3r})(\mathcal{P}^{(e)})^*(0,\ldots,0,U_0^{(e)}(y^{(e)\prime}))^*\big) - \nabla_y\big(\zeta(\frac{y_n^{(e)}}{3r})y_n^{(e)}\big)\Big) + a_1^{(e)}(t)\,\nabla_y\big(\zeta(\frac{y_n^{(e)}}{3r})\big)\Big\}, \tag{6.3.52}$$

$$\mathbf{f}_0^{[BLO_l]}(y,t) = 0, \tag{6.3.53}$$

$$h_0^{[REGO_l]}(y,t) = -\mathrm{div}_y \sum_{e:O_l\in e} \Big\{ s_0^{(e)}(t)\zeta(\frac{y_n^{(e)}}{3r})(\mathcal{P}^{(e)})^*(0,\ldots,0,U_0^{(e)}(y^{(e)\prime}))^*\Big\}. \tag{6.3.54}$$

Here the quantity $a_1^{(e)}(t)$ in the right-hand side is unknown. Let us denote by $(\mathbf{U}_0^{[BLO_l]}(y,t), \widehat{P}_0^{[BLO_l]}(y,t))$ the solution of problem (6.3.51) without the last term $a_1^{(e)}(t)\nabla_y\big(\zeta(\frac{y_n^{(e)}}{3r})\big)$ in $\mathbf{f}_0^{[REGO_l]}(y,t)$ (since this term is of gradient form, the solutions differ only by the pressure component). The right-hand sides of system (6.3.51) have compact supports. Therefore, the pressure $\widehat{P}_0^{[BLO_l]}(y,t)$ exponentially stabilizes in any outlet (corresponding to the edge e) to a constant, dependent on the parameter t, say: $\widehat{a}_0^{[BLO_l,e]}(t)$,

$$\lim_{y_n^{(e)}\to+\infty} \widehat{P}_0^{[BLO_l]}(y,t) = \widehat{a}_0^{[BLO_l,e]}(t).$$

Moving forward, we solve the conductivity problem of rank 1 for p_1:

$$\begin{cases} -\kappa_e \dfrac{\partial^2 p_1^{(e)}}{\partial x_n^{(e)2}}(x_n^{(e)},t) = 0, & x_n^{(e)} \in (0,|e|),\\ -\displaystyle\sum_{e:O_l\in e} \kappa_e \dfrac{\partial p_1^{(e)}}{\partial x_n^{(e)}}(0,t) = 0, & l = 1,\ldots,N_1,\\ -\kappa_e \dfrac{\partial p_1^{(e)}}{\partial x_n^{(e)}}(0,t) = 0, & l = N_1+1,\ldots,N,\\ p_1^{(e)}(0,t) - p_1^{(e_s)}(0,t) = \widehat{a}_0^{[BLO_l,e]}(t),\ \forall e \subset \mathcal{B}_l, e \neq e_s. \end{cases}$$

where e_s is the selected edge of the bundle.

This problem possesses a solution p_1, which is unique up to a constant (t-dependent function).

The functions $s_1^{(e)}(t)$ and $a_1^{(e)}(t)$ are established. The expression for $p_1^{(e)}(x_n^{(e)},t)$ is given by:

$$p_1^{(e)}(x_n^{(e)},t) = -s_1^{(e)}(t)x_n^{(e)} + a_1^{(e)}(t).$$

Consequently, we can completely determine the pressure in the boundary layer problem (6.3.51):

$$P_0^{[BLO_l]}(y,t) = \widehat{P}_0^{[BLO_l]}(y,t) - \sum_{e:O_l\in e,\, e\neq e_s} \zeta(\frac{y_n^{(e)}}{3r})\widehat{a}_0^{[BLO_l,e]}(t)$$

$$-\Big(\sum_{e:O_l\in e} \zeta(\frac{y_n^{(e)}}{3r}) - 1\Big) p_1^{(e_s)}(0,t).$$

The boundary layer functions in the vertex-associated unbounded domains Ω_l can be readily determined by solving the boundary value problem outlined below:

$$\begin{cases} -\nu\Delta_y \mathbf{U}_0^{[BLO_l]} + \nabla_y P_0^{[BLO_l]} = 0, & y\in\Omega_l, \\ \mathrm{div}_y \mathbf{U}_0^{BLO_l} = 0, & y\in\Omega_l, \\ \mathbf{U}_0^{[BLO_l]} = 0, & y\in\partial\Omega_l\setminus\gamma^l, \\ \mathbf{U}_0^{[BLO_l]} = \mathbf{g}^l(y,t), & y\in\gamma^l. \end{cases} \quad (6.3.55)$$

Assuming that all terms within the expansions (6.3.45)–(6.3.49) corresponding to a rank less than or equal to $j-1$ are known, and the macroscopic pressure on the graph p_j is also known, the transition from rank $j-1$ to rank j can be described.

Step 1 As the macroscopic pressure p_j on the graph is known, functions $s_j^{(e)}(t)$ and $a_j^{(e)}(t)$ are introduced for every edge e as follows:

$$p_j^{(e)}(x^{(e)},t) = -s_j^{(e)}(t)x_n^{(e)} + a_j^{(e)}(t)$$

and:

$$\widetilde{U}_j^{(e)}(y^{(e)\prime},t) = s_j^{(e)}(t)U_0^{(e)}(y^{(e)\prime}) + \frac{ds_{j-2}^{(e)}(t)}{dt}U_2^{(e)}(y^{(e)\prime}) +$$

$$\ldots + \frac{d^k s_0^{(e)}(t)}{dt^k}U_{2k}^{e}(y^{(e)\prime}), \quad (6.3.56)$$

$$\mathbf{U}_j^{(e)} = (\mathcal{P}^{(e)})^*(0,\ldots,0,\widetilde{U}_j^{(e)})^*,$$

where $k=[j/2]$, i.e., k denotes the integer part of $j/2$.

Step 2 The boundary layer solution in Ω_l, $l = 1,\ldots,N_1$, is a pair $\big(\mathbf{U}_j^{[BLO_l]}, P_j^{[BLO_l]}\big)$, which solves the problem (6.3.20). Initially, we determine the pair $\big(\mathbf{U}_j^{[BLO_l]}, \widehat{P}_j^{[BLO_l]}\big)$ as the solution to the same problem (6.3.20), excluding the final term in the formulation of $\mathbf{f}_j^{[REGO_l]}$ (see Eq. (6.3.21)). It can be established via induction that the vector field $\mathbf{U}_j^{[BLO_l]}$ exponentially approaches zero as $|y| \to \infty$. Concurrently, the corresponding pressure function $\widehat{P}_j^{[BLO_l]}$ converges to constants $\widehat{a}_j^{BLO_l,e}(t)$ at the outlets to infinity. Notably, these constants may differ across various outlets. Since the pressure function's definition admits an additive constant, we can set the limit constant to zero for the outlet corresponding to the selected edge e_s. We introduce the notation $\varphi_{l,j+1}^{(e)}(t) = \widehat{a}_j^{BLO_l,e}(t)$.

For the vertices $O_l,\ l = 1, \dots, N$, we solve problem (6.3.55).

Step 3 The conductivity problem on the graph is solved for the function $p_{j+1}^{(e)}$ ($j < J$), governed by the following equations:

$$\begin{cases} -\kappa_e \dfrac{\partial^2 p_{j+1}^{(e)}}{\partial x_n^{(e)2}}(x_n^{(e)}, t) = 0, & x_n^{(e)} \in (0, |e|), \\ -\sum\limits_{e:O_l\in e} \kappa_e \dfrac{\partial p_{j+1}^{(e)}}{\partial x_n^{(e)}}(0, t) = \Psi_{l,j+1}(t),\ l = 1, \dots, N_1, \\ -\kappa_e \dfrac{\partial p_{j+1}^{(e)}}{\partial x_n^{(e)}}(0, t) = \Psi_{l,j+1}(t),\ l = N_1 + 1, \dots, N, \\ p_{j+1}^{(e)}(0, t) - p_{j+1}^{(e_s)}(0, t) = \varphi_{l,j+1}^{(e)}(t),\ \forall e \subset \mathcal{B}_l, e \neq e_s, \end{cases}$$

where:

$$\Psi_{l,j}(t) = \sum_{e:O_l\in e} \sum_{k:0<2k\le j} \frac{\partial^{k+1} p_{j-2k}^{(e)}}{\partial t^k \partial x_n^{(e)}}(0, t) \int_{\sigma^{(e)}} U_{2k}^{(e)}(y^{(e)\prime}) dy^{(e)\prime},$$

and the local coordinates $x^{(e)}$ are redefined so that all of them have the same origin O_l.

Step 4 Finally, we redefine the pressure $P_j^{[BLO_l]}(y, t)$ in the boundary layer problems (6.3.20) and (6.3.21):

$$P_j^{[BLO_l]}(y, t) = \widehat{P}_j^{[BLO_l]}(y, t) - \sum_{e:O_l\in e,\ e\neq e_s} \zeta\Big(\frac{y_n^{(e)}}{3r}\Big)\widehat{a}_j^{[BLO_l,e]}(t)$$
$$-\Big(\sum_{e:O_l\in e} \zeta\Big(\frac{y_n^{(e)}}{3r}\Big) - 1\Big) p_{j+1}^{(e_s)}(0, t).$$

For the specific case of $j = J$, the last summation term is omitted.

This concludes the iterative progression from the jth to the $(j + 1)$th step.

6.4 Residual in the Case of Data Vanishing for Small Times

By construction, we have:

$$\begin{aligned}\mathbf{u}^{(J)} &\in L^2(0,T;W^{2,2}(B_\varepsilon))\cap L^\infty(0,T;W^{1,2}(B_\varepsilon)),\\ \frac{\partial \mathbf{u}^{(J)}}{\partial t} &\in L^2(0,T;W^{1,2}(B_\varepsilon))\cap L^\infty(0,T;L^2(B_\varepsilon)),\\ \frac{\partial^2 \mathbf{u}^{(J)}}{\partial t^2} &\in L^2(0,T;L^2(B_\varepsilon)),\\ \nabla p^{(J)} &\in L^2(0,T;L^2(B_\varepsilon)).\end{aligned} \tag{6.4.1}$$

Remark 6.4.1 It can be demonstrated that $\|\mathbf{u}^{(J)}\|^4_{L^4(B_\varepsilon)} \le c\varepsilon^{n-1}$. This result is established by observing that the Poiseuille component of $\mathbf{u}^{(J)}$ satisfies this inequality (see Chap. 3). The $\|\cdot\|_{L^4(B_\varepsilon^{(i)})}$ norm of the boundary layer functions in each bundle $B_\varepsilon^{(i)}$ can be estimated by the L^4 norm in the unbounded dilated domain Ω_i multiplied by $\varepsilon^{n/4}$. Accounting for an exponential decay of the boundary layers, the desired estimate follows.

Throughout the construction of the asymptotic expansion, we have indicated the regularity of all terms within the expansion. This yields the regularity of the entire asymptotic expansion $\mathbf{u}^{(J)}$, implying that it is twice differentiable with respect to time.

Unfortunately, the constructed function $\mathbf{u}^{(J)}$ is not divergence-free due to the multiplication of the boundary layer terms $\mathbf{U}^{[BLO_l,J]}$ in each bundle of cylinders by the cutoff factor $\zeta(\frac{|x-O_l|}{|e|_{min}})$. To address this, we introduce a corrective term to the approximation $\mathbf{u}^{(J)}$. In particular, we have:

$$\mathrm{div}\mathbf{u}^{(J)}(x,t) = h^{(J)}(x,t),$$

where:

$$h^{(J)}(x,t) = -\sum_{l=1}^{N} \nabla\zeta(\frac{|x-O_l|}{|e|_{min}})\cdot \mathbf{U}^{[BLO_l,J]}(\frac{x-O_l}{\varepsilon},t).$$

Clearly, $h^{(J)} \in L^2(0,T;W^{2,2}\cap \mathring{W}^{1,2}(B_\varepsilon))$, $h_t^{(J)} \in L^2(0,T;\mathring{W}^{1,2}(B_\varepsilon))$, and $h_{tt}^{(J)} \in L^2(0,T;L^2(B_\varepsilon))$. Since the support of the function $\nabla\zeta(\dfrac{|x-O_l|}{|e|_{min}})$ belongs to the middle third of every cylinder, we deduce the following relations:

$$\|h^{(J)}\|_{L^2(0,T;W^{2,2}(B_\varepsilon))} = O(e^{-c_1/\varepsilon}),$$
$$\left\|\frac{\partial h^{(J)}}{\partial t}\right\|_{L^2(0,T;W^{1,2}(B_\varepsilon))} = O(e^{-c_1/\varepsilon}), \tag{6.4.2}$$

and:

$$\left\|\frac{\partial^2 h^{(J)}}{\partial t^2}\right\|_{L^2(0,T;L^2(B_\varepsilon))} = O(e^{-c_1/\varepsilon}),$$

where $c_1 > 0$.

Now, let us construct the corrective term $\mathbf{w}$ such that:

$$\mathbf{w}^{(J)} \in L^2(0,T;\overset{\circ}{W}{}^{2,2}(B_\varepsilon)), \quad \mathbf{w}_t^{(J)} \in L^2(0,T;\overset{\circ}{W}{}^{2,2}(B_\varepsilon)),$$
$$\mathbf{w}_{tt}^{(J)} \in L^2(0,T;\overset{\circ}{W}{}^{1,2}(B_\varepsilon)),$$

and:

$$\operatorname{div}\mathbf{w}^{(J)}(x,t) = -h^{(J)}(x,t).$$

It is easy to verify that:

$$\int\limits_{B_\varepsilon} h^{(J)}(y,t)dy = 0.$$

Therefore, due to Lemma 2.4.14, such a function $\mathbf{w}^{(J)}$ exists. Moreover, the following estimates hold:

$$\|\mathbf{w}^{(J)}(\cdot,t)\|_{W^{2,2}(B_\varepsilon)} \le \varepsilon^{-5/2}c\|h^{(J)}(\cdot,t)\|_{W^{1,2}(B_\varepsilon)},$$
$$\|\mathbf{w}_t^{(J)}\|_{L^2(0,T;W^{2,2}(B_\varepsilon))} \le \varepsilon^{-5/2}c\|h_t^{(J)}\|_{L^2(0,T;W^{1,2}(B_\varepsilon))}, \tag{6.4.3}$$
$$\|\mathbf{w}_{tt}^{(J)}\|_{L^2(0,T;W^{1,2}(B_\varepsilon))} \le \varepsilon^{-5/2}c\|h_{tt}^{(J)}\|_{L^2(0,T;L^2(B_\varepsilon))},$$

where c is a constant independent of t and ε.

Notably, $h^{(J)}(x,t) = 0$ in the neighborhood of the point $t = 0$. Then, the construction of Lemma 2.4.14 implies that also $\mathbf{w}^{(J)}(x,t) = 0$ for small t.

Denote $\widetilde{\mathbf{u}}^{(J)} = \mathbf{u}^{(J)} + \mathbf{w}^{(J)}$ and calculate $\mathcal{L}(\widetilde{\mathbf{u}}^{(J)}, p^{(J)})$ in a half-bundle HB_{O_l}, $l = 1,\dots,N$:

$$-\mathbf{f}^{(J)}(x,t) = \left(\widetilde{\mathbf{u}}_t^{(J)} - \nu\Delta\widetilde{\mathbf{u}}^{(J)} + (\widetilde{\mathbf{u}}^{(J)}\cdot\nabla)\widetilde{\mathbf{u}}^{(J)} - \nabla p^{(J)}\right) = \mathcal{L}(\mathbf{u}^{(J)} + \mathbf{w}^{(J)}, p^{(J)})$$

$$= \varepsilon^{J-1} \frac{\partial (\mathbf{U}^{(e)}_{J-1}(y^{(e)\prime}, t)\zeta(\frac{x_n^{(e)}}{3r\varepsilon})\zeta(\frac{|e|-x_n^{(e)}}{3r\varepsilon}))}{\partial t} + \varepsilon^{J} \frac{\partial (\mathbf{U}^{(e)}_{J}(y^{(e)\prime}, t)\zeta(\frac{x_n^{(e)}}{3r\varepsilon})\zeta(\frac{|e|-x_n^{(e)}}{3r\varepsilon}))}{\partial t}$$

$$+ \varepsilon^{J-1} \frac{\partial \mathbf{U}^{[BLO_l]}_{J-1}}{\partial t}(y,t) + \varepsilon^{J} \frac{\partial \mathbf{U}^{[BLO_l]}_{J}}{\partial t}(y,t)$$

$$+ \sum_{J+1\le j\le 2J} \varepsilon^{j-2} \Big(\sum_{p+r=j-1} \Big(\mathbf{U}^{(e)}_p(y^{(e)\prime}, t)\zeta\Big(\frac{y_n^{(e)}}{3r}\Big) \cdot \nabla_y\Big)\mathbf{U}^{(e)}_r(y^{(e)\prime}, t)\zeta\Big(\frac{y_n^{(e)}}{3r}\Big)$$

$$+ \sum_{p+r=j-1} \zeta\Big(\frac{y_n^{(e)}}{3r}\Big)\big(\mathbf{U}^{(e)}_p(y^{(e)\prime}, t)\cdot\nabla_y\big)\mathbf{U}^{[BLO_l]}_r(y,t)$$

$$+ \sum_{p+r=j-1} \big(\mathbf{U}^{[BLO_l]}_p(y,t)\cdot\nabla_y\big)\Big(\zeta\Big(\frac{y_n^{(e)}}{3r}\Big)\mathbf{U}^{(e)}_r(y^{(e)\prime}, t)\Big)$$

$$+ \sum_{p+r=j-1} \big(\mathbf{U}^{[BLO_l]}_p(y,t)\cdot\nabla_y\big)\mathbf{U}^{[BLO_l]}_r(y,t)\Big)$$

$$- \Big(\frac{\partial \mathbf{U}^{[BLO_l,J]}(y,t)}{\partial t}\zeta(\frac{|x-O_l|}{|e|_{min}}) - \nu\Delta\Big(\mathbf{U}^{[BLO_l,J]}(y,t)\zeta(\frac{|x-O_l|}{|e|_{min}})\Big)\Big)\chi(x)$$

$$+ \Big((\mathbf{U}^{[BLO_l,J]}(y,t)\big(1-\zeta(\frac{|x-O_l|}{|e|_{min}})\big)\cdot\nabla)(\mathbf{U}^{[BLO_l,J]}(y,t)\Big(1-\zeta(\frac{|x-O_l|}{|e|_{min}})\Big)$$

$$- (\mathbf{U}^{[BLO_l,J]}(y,t)\cdot\nabla)\mathbf{U}^{[BLO_l,J]}(y,t)\Big)\chi(x)$$

$$- \Big(\mathbf{U}^{[BLO_l,J]}(y,t)\zeta(\frac{|x-O_l|}{|e|_{\min}})\cdot\nabla\Big)\mathbf{U}^{[e,J]}(y^{(e)\prime}, t)$$

$$- \Big(\mathbf{U}^{[e,J]}(y^{(e)\prime}, t)\cdot\nabla\Big)\Big(\mathbf{U}^{[BLO_l,J]}(y,t)\zeta(\frac{|x-O_l|}{|e|_{\min}})\Big)$$

$$+ \frac{\partial \mathbf{w}^{(J)}}{\partial t} - \nu\Delta\mathbf{w}^{(J)} + (\mathbf{w}^{(J)}\cdot\nabla)\mathbf{w}^{(J)} + (\mathbf{u}^{(J)}\cdot\nabla)\mathbf{w}^{(J)} + (\mathbf{w}^{(J)}\cdot\nabla)\mathbf{u}^{(J)}$$

$$+ \varepsilon^{J-2}\widehat{a}_J^{[BLO_l,e]}(t)\nabla_y\zeta\Big(\frac{y_n^{(e)}}{3r}\Big) - \nabla\big(\zeta(\frac{|x-O_l|}{|e|_{min}})P^{[BLO_l,J]}(y,t)\big)\chi(x).$$

Here $y = \dfrac{x - O_l}{\varepsilon}$; $y^{(e)} = \dfrac{x^{(e)}}{\varepsilon}$; $\chi = \chi_{\mathrm{supp}\zeta\left(\frac{|x-O_l|}{|e|_{min}}\right)}$ is the characteristic function of the set $\mathrm{supp}\zeta\left(\frac{|x-O_l|}{|e|_{min}}\right)$. As before, the terms of the sums $\sum\limits_{e:O_l\in e}$ are extended by zero out of cylinders $\Pi_\varepsilon^{(e)}$.

In the presented formula, the initial four terms are the part of the residual coming from the time derivative of the velocity, and the next four terms originate from the inertial term, encompassing all possible combinations of $\mathbf{U}_\alpha^{(e)}$ and $\mathbf{U}_\beta^{[BLO_l]}$ with powers of ε higher than $J - 2$. The subsequent five lines represent the residual resulting from the multiplication of the boundary layer correctors by the cutoff function $1 - \zeta\left(\frac{|x-O_l|}{|e|_{min}}\right)$. Notably, these terms vanish exponentially due to their exponential decay within the set $\mathrm{supp}\zeta\left(\frac{|x-O_l|}{|e|_{min}}\right)$ (where $\chi \neq 0$). Specifically, the order of this term in the L^2 norm diminishes as $O(e^{-c_1/\varepsilon})$, where c_1 is a positive constant. The next five terms are generated by the addition of the "divergence corrector" $\mathbf{w}^{(J)}$ to the expansion $\mathbf{u}^{(J)}$, and their support also belongs to $\mathrm{supp}\zeta\left(\frac{|x-O_l|}{|e|_{min}}\right)$. The last two terms come from the pressure gradient term: the first one remains uncompensated by the boundary layer-in-space problems, while the second one results from the multiplication of the boundary layer correctors by the cutoff function $1 - \zeta\left(\frac{|x-O_l|}{|e|_{min}}\right)$ and so diminishes exponentially.

From the derived expressions, it can be deduced that $\mathbf{f}^{(J)} \in L^2(0, T; L^2(B_\varepsilon))$ and:

$$\|\mathbf{f}^{(J)}\|_{L^2(0,T;L^2(B_\varepsilon))} = O(\varepsilon^{J-2}). \tag{6.4.4}$$

Furthermore, all terms of this sum are differentiable in time, so that:

$$\frac{\partial \mathbf{f}^{(J)}}{\partial t} \in L^2(0, T; L^2(B_\varepsilon))$$

and:

$$\left\|\frac{\partial \mathbf{f}^{(J)}}{\partial t}\right\|_{L^2(0,T;L^2(B_\varepsilon))} = O(\varepsilon^{J-2}). \tag{6.4.5}$$

6.5 Justification of the Asymptotic Expansion

In this section, we justify the constructed asymptotic expansion of the solution to the Navier–Stokes problem (6.1.1). Let $\mathbf{v}$ be the solution to (6.1.1). As an extension of the boundary value $\mathbf{g}$, we consider the asymptotic approximation $\widetilde{\mathbf{u}}^{(J)}$ derived in the preceding sections. We represent $\mathbf{v}$ and the pressure p as the sums $\mathbf{v} = \mathbf{u} + \widetilde{\mathbf{u}}^{(J)}$ and $p = q + p^{(J)}$. Here, $\mathbf{u} \in L^2(0, T; \mathring{W}^{1,2}(B_\varepsilon))$, $\mathbf{u}_t \in L^2(0, T; L^2(B_\varepsilon))$, $\widetilde{\mathbf{u}}^{(J)} \in$

$L^2(0, T; \mathring{W}^{1,2}(B_\varepsilon))$, and $\widetilde{\mathbf{u}}_t^{(J)} \in L^2(0, T; \mathring{W}^{1,2}(B_\varepsilon))$. The difference $\mathbf{u} = \mathbf{v} - \widetilde{\mathbf{u}}^{(J)}$ satisfies the initial condition $\mathbf{u}(x, 0) = 0$, the boundary condition $\mathbf{u}(x, t)|_{\partial B_\varepsilon} = 0$, and the integral identity:

$$\int\limits_{B_\varepsilon} \Big(\mathbf{u}_t \cdot \boldsymbol{\eta} + \nu \nabla \mathbf{u} \cdot \nabla \boldsymbol{\eta} - \big((\mathbf{u} + \widetilde{\mathbf{u}}^{(J)}) \cdot \nabla\big)\boldsymbol{\eta} \cdot \mathbf{u} \\ - \big(\mathbf{u} \cdot \nabla\big)\boldsymbol{\eta} \cdot \widetilde{\mathbf{u}}^{(J)}\Big) dx = \int\limits_{B_\varepsilon} \mathbf{f}^{(J)} \cdot \boldsymbol{\eta} dx \quad \forall \boldsymbol{\eta} \in \mathring{H}^{1,2}(B_\varepsilon), \tag{6.5.1}$$

where $\mathbf{f}^{(J)} = -\big(\widetilde{\mathbf{u}}_t^{(J)} - \nu \Delta \widetilde{\mathbf{u}}^{(J)} + (\widetilde{\mathbf{u}}^{(J)} \cdot \nabla)\widetilde{\mathbf{u}}^{(J)} - \nabla p^{(J)}\big)$.

We state and prove the following theorem:

Theorem 6.5.1 *Let $n = 2$ or $n = 3$. Suppose $\partial B_\varepsilon \in C^2$ and $\mathbf{g} \in C^{[\frac{(J+1)}{2}]+2}(0, T; W^{3/2,2}(\partial B_\varepsilon))$. Then, for sufficiently small ε, the following estimates hold:*

$$\sup_{t \in [0,T]} \|\mathbf{u}(\cdot, t)\|_{L^2(B_\varepsilon)} + \|\nabla \mathbf{u}\|_{L^2(0,T;L^2(B_\varepsilon))} \leq c\varepsilon^{J-1}, \tag{6.5.2}$$

$$\sup_{t \in [0,T]} \|\nabla \mathbf{u}(\cdot, t)\|_{L^2(B_\varepsilon)} + \|\mathbf{u}_t\|_{L^2(0,T;L^2(B_\varepsilon))} \\ + \|\nabla^2 \mathbf{u}\|_{L^2(0,T;L^2(B_\varepsilon))} \leq c\varepsilon^{J-2}. \tag{6.5.3}$$

Furthermore, there exists a pressure function $q \in L^2(0, T; L^2(B_\varepsilon))$ such that $\int_{B_\varepsilon} q(x, t) dx = 0$ and the following integral identity is valid:

$$\int\limits_{B_\varepsilon} \Big(\mathbf{u}_t \cdot \boldsymbol{\eta} + \nu \nabla \mathbf{u} \cdot \nabla \boldsymbol{\eta} - \big((\mathbf{u} + \widetilde{\mathbf{u}}^{(J)}) \cdot \nabla\big)\boldsymbol{\eta} \cdot \mathbf{u} - \big(\mathbf{u} \cdot \nabla\big)\boldsymbol{\eta} \cdot \widetilde{\mathbf{u}}^{(J)}\Big) dx \\ = \int\limits_{B_\varepsilon} q \operatorname{div} \boldsymbol{\eta} dx + \int\limits_{B_\varepsilon} \mathbf{f}^{(J)} \cdot \boldsymbol{\eta} dx \quad \forall \boldsymbol{\eta} \in \mathring{W}^{1,2}(B_\varepsilon), \tag{6.5.4}$$

along with the pressure estimate:

$$\|q\|_{L^2(0,T;L^2(B_\varepsilon))} \leq c\varepsilon^{J-5/2}. \tag{6.5.5}$$

In the case $n = 2$, the estimates of Theorem 6.5.1 follow from Theorems 6.2.1 and 6.2.3 if we set $\mathbf{g} = \widetilde{\mathbf{u}}^{(J)}$ and take into account inequalities (6.4.4). Similarly, for $n = 3$, this theorem follows from Theorems 6.2.4 and 6.2.6 and inequalities (6.4.4) and (6.4.5). Notice that in view of (6.4.4), for sufficiently small ε and $J > 2$, the condition (6.2.31) is fulfilled.

If the boundary value $\mathbf{g}$ is more regular, the obtained estimates can be improved. Assume that $\mathbf{g} \in C^{[\frac{J+1}{2}]+3}([0, T]; W^{3/2,2}(\sigma))$. Then we can construct the asymptotic approximation $\mathbf{u}^{(J+2)}$ and estimate (6.5.2) takes the following form:

$$\sup_{t\in[0,T]} \|\mathbf{v}(\cdot,t)-\widetilde{\mathbf{u}}^{(J+2)}(\cdot,t)\|_{L^2(B_\varepsilon)}+\|\nabla\mathbf{v}-\nabla\widetilde{\mathbf{u}}^{(J+2)}\|_{L^2(0,T;L^2(B_\varepsilon))}$$

$$\le c\varepsilon^{J+1}\le c\varepsilon^J\sqrt{\mathrm{mes}(B_\varepsilon)},$$

where $\mathrm{mes}(B_\varepsilon)\sim\varepsilon^{n-1}$. Comparing $\widetilde{\mathbf{u}}^{(J)}$ and $\widetilde{\mathbf{u}}^{(J+2)}$, we notice that:

$$\sup_{t\in[0,T]} \|\widetilde{\mathbf{u}}^{(J)}(\cdot,t)-\widetilde{\mathbf{u}}^{(J+2)}(\cdot,t)\|_{L^2(B_\varepsilon)}+\|\nabla\widetilde{\mathbf{u}}^{(J)}-\nabla\widetilde{\mathbf{u}}^{(J+2)}\|_{L^2(0,T;L^2(B_\varepsilon))}$$

$$\le c\varepsilon^J\sqrt{\mathrm{mes}(B_\varepsilon)}.$$

Thus, by the triangle inequality, it follows that:

$$\sup_{t\in[0,T]} \|\mathbf{v}(\cdot,t)-\widetilde{\mathbf{u}}^{(J)}(\cdot,t)\|_{L^2(B_\varepsilon)}+\|\nabla\mathbf{v}-\nabla\widetilde{\mathbf{u}}^{(J)}\|_{L^2(0,T;L^2(B_\varepsilon))}\le c\varepsilon^J\sqrt{\mathrm{mes}(B_\varepsilon)}.$$

Analogous improvements can be obtained for estimates (6.5.3) and (6.5.5). This refinement yields the following result:

Theorem 6.5.2 *If* $\mathbf{g}\in C^{[\frac{J+1}{2}]+3}(0,T;W^{3/2,2}(\partial B_\varepsilon))$, *then the following improved estimates hold:*

$$\sup_{t\in[0,T]} \|\mathbf{v}(\cdot,t)-\widetilde{\mathbf{u}}^{(J)}(\cdot,t)\|_{L^2(B_\varepsilon)}+\|\nabla\mathbf{v}-\nabla\widetilde{\mathbf{u}}^{(J)}\|_{L^2(0,T;L^2(B_\varepsilon))}$$
$$\le c\varepsilon^J\sqrt{\mathrm{mes}(B_\varepsilon)}=O(\varepsilon^{J+\frac{n-1}{2}}), \tag{6.5.6}$$

$$\sup_{t\in[0,T]} \|\nabla\mathbf{v}(\cdot,t)-\nabla\widetilde{\mathbf{u}}^{(J)}(\cdot,t)\|_{L^2(B_\varepsilon)}+\|\mathbf{v}_t-\widetilde{\mathbf{u}}_t^{(J)}\|_{L^2(0,T;L^2(B_\varepsilon))}$$
$$+\|\nabla^2\mathbf{v}-\nabla^2\widetilde{\mathbf{u}}^{(J)}\|_{L^2(0,T;L^2(B_\varepsilon))}\le c\varepsilon^{J-1}\sqrt{\mathrm{mes}(B_\varepsilon)}=O(\varepsilon^{J+\frac{n-3}{2}}), \tag{6.5.7}$$

along with an estimate for the pressure:

$$\|p-p^{(J)}\|_{L^2(0,T;L^2(B_\varepsilon))}\le c\varepsilon^{J-1}\sqrt{\mathrm{mes}(B_\varepsilon)}=O(\varepsilon^{J+\frac{n-3}{2}}). \tag{6.5.8}$$

Remark 6.5.1 As in the stationary case, we will modify the asymptotic expansion (5.1.5)–(5.1.9), without loss of accuracy, in a way that it belongs to the space $\mathcal{H}_{\mathrm{dec}}(B_\varepsilon,\delta)$ for almost all $t\in(0,T)$. Specifically, the argument $\frac{|x-O_l|}{|e|_{\min}}$ of the cutoff function ζ can be substituted with $\frac{|x-O_l|}{\delta}$, where $\delta=C_J\varepsilon|\ln\varepsilon||e|_{\min}$, and the constant C_J is chosen such that:

$$E_{\delta/\varepsilon}(t) = \|\mathbf{U}^{[BLO_l,J]}(\cdot,t)\|_{W^{2,2}(\Omega_{l,\delta/\varepsilon})} + \|\nabla P^{[BLO_l,J]}(\cdot,t)\|_{L^2(\Omega_{l,\delta/\varepsilon})} \le c\varepsilon^{J+(n-1)/2+2} \tag{6.5.9}$$

with the constant c independent of t. Here, as before, $\Omega_{l,R} = \Omega_l \cap \{|y| > R\}$.

Indeed, by Theorem 4.4.3, $\mathbf{U}^{[BLO_l,J]}(y,t)$ and $P^{[BLO_l,J]}(y,t)$ tend to zero exponentially as $|y| \to \infty$, leading to the inequality:

$$E_R(t) \le c_1 e^{-c_2 R}, \quad c_1, c_2 > 0.$$

Consequently, the estimate $E_{\delta/\varepsilon}(t) = O(\varepsilon^{J+(n-1)/2+2})$ is valid for δ such that:

$$c_2\delta/\varepsilon \ge (J + (n-1)/2 + 2)|\ln\varepsilon|,$$

implying $\delta = C_J\varepsilon|\ln\varepsilon|$. Notably, the constant $C_J = c_3 J + c_4$, where c_3 and c_4 are independent of J. This adjustment along with (5.1.52) and (5.1.60) and the change of variables $\nabla_x = \varepsilon^{-1}\nabla_y$ yields that $W^{2,2}(B_\varepsilon)$ norm of the difference between the constructed asymptotic expansion $\widetilde{\mathbf{u}}^{(J)}$ and the modified one is of order $O(\varepsilon^{J+(n-1)/2})$.

6.6 MAPDD in the Case of Data Vanishing for Small Times

The method of asymptotic partial decomposition of the domain (MAPDD) is applied to the nonstationary Navier–Stokes equations in a manner analogous to its application in the stationary case. Specifically, the function space $\overset{\circ}{H}{}^{1,2}(B_\varepsilon)$, consisting of divergence-free vector functions used in the variational formulation, is substituted with the subspace $\mathcal{H}_{\text{dec}}(B_\varepsilon,\delta)$. Consequently, the weak formulation of MAPDD for the nonstationary Navier–Stokes equations is as follows:

Definition 6.6.1 Find $\mathbf{v}_d \in L^2(0,T;\mathcal{H}_{\text{dec}}(B_\varepsilon,\delta))$ such that $\dfrac{\partial \mathbf{v}_d}{\partial t} \in L^2(0,T;$ $L^2(B_\varepsilon))$, $\mathbf{v}_d|_{t=0} = 0$, $\mathbf{v}_d|_{\partial B_\varepsilon} = \mathbf{g}(x,t)$, and such that for almost all $t \in [0,T)$ and for all $\boldsymbol{\eta} \in \mathcal{H}_{0,\text{dec}}(B_\varepsilon,\delta)$, the following integral identity holds:

$$\int_{B_\varepsilon} \frac{\partial \mathbf{v}_d}{\partial t}\cdot\boldsymbol{\eta} + \nu\nabla\mathbf{v}_d\cdot\nabla\boldsymbol{\eta} + (\mathbf{v}_d\cdot\nabla)\mathbf{v}_d\cdot\boldsymbol{\eta}\,dx = 0. \tag{6.6.1}$$

For this formulation, we prove the existence and uniqueness of the solution and derive a corresponding estimate analogous to (5.1.65).

It is important to note that this formulation is equivalent to the subsequent one:

Definition 6.6.2 Find $\mathbf{u}_d \in L^2(0, T; \mathcal{H}_{0,\text{dec}}(B_\varepsilon, \delta))$ such that $\dfrac{\partial \mathbf{u}_d}{\partial t} \in L^2(0, T; L^2(B_\varepsilon))$, $\mathbf{u}_d|_{t=0} = 0$, and such that for almost all $t \in [0, T)$ and for all $\boldsymbol{\eta} \in \mathcal{H}_{0,\text{dec}}(B_\varepsilon, \delta)$, the integral identity is satisfied:

$$\int_{B_\varepsilon} \left(\frac{\partial \mathbf{u}_d}{\partial t} \cdot \boldsymbol{\eta} + \nu \nabla \mathbf{u}_d \cdot \nabla \boldsymbol{\eta} - ((\mathbf{u}_d + \widetilde{\mathbf{u}}^{(J)}) \cdot \nabla) \boldsymbol{\eta} \cdot \mathbf{u}_d - (\mathbf{u}_d \cdot \nabla) \boldsymbol{\eta} \cdot \widetilde{\mathbf{u}}^{(J)} \right) dx$$

$$= - \int_{B_e} \left(\frac{\partial \widetilde{\mathbf{u}}^{(J)}}{\partial t} - \nu \Delta \widetilde{\mathbf{u}}^{(J)} + (\widetilde{\mathbf{u}}^{(J)} \cdot \nabla) \widetilde{\mathbf{u}}^{(J)} \right) \cdot \boldsymbol{\eta} \, dx. \tag{6.6.2}$$

Here, $\mathbf{v}_d = \mathbf{u}_d + \widetilde{\mathbf{u}}^{(J)}$. It is worth noting that the pressure term $\int_{B_\varepsilon} \nabla p^{(J)} \cdot \boldsymbol{\eta} \, dx$ can be added to the left-hand side since it equals zero.

As previously indicated, formulations "without pressure" are not conducive to numerical implementation due to the space of test functions being "divergence-free," making the construction of a suitable basis for finite element methods challenging. Consequently, we introduce an alternative "with pressure" formulation:

Definition 6.6.3 Find $\mathbf{u}_d \in L^2(0, T; \mathcal{H}_{0,\text{dec}}(B_\varepsilon, \delta))$ and the "MAPDD pressure" $q_d \in L^2(0, T; L^2(\bigcup_{i=1}^{N} B_i^{\varepsilon,\delta}))$, such that $\dfrac{\partial \mathbf{u}_d}{\partial t} \in L^2(0, T; L^2(B_\varepsilon))$, $\mathbf{u}_d|_{t=0} = 0$, and for almost all $t \in [0, T)$ and for all $\boldsymbol{\eta} \in \mathcal{W}_{0,\text{dec}}(B_\varepsilon, \delta)$, the following integral identity is valid:

$$\int_{B_\varepsilon} \left(\frac{\partial \mathbf{u}_d}{\partial t} \cdot \boldsymbol{\eta} + \nu \nabla \mathbf{u}_d \cdot \nabla \boldsymbol{\eta} - ((\mathbf{u}_d + \widetilde{\mathbf{u}}^{(J)}) \cdot \nabla) \boldsymbol{\eta} \cdot \mathbf{u}_d - (\mathbf{u}_d \cdot \nabla) \boldsymbol{\eta} \cdot \widetilde{\mathbf{u}}^{(J)} \right) dx$$

$$= \sum_{i=1}^{N} \int_{B_i^{\varepsilon,\delta}} q_d \operatorname{div} \boldsymbol{\eta} \, dx + \int_{B_e} \mathbf{f}^{(J)} \cdot \boldsymbol{\eta} \, dx. \tag{6.6.3}$$

Recall that $\mathbf{f}^{(J)}(x, t)$ is defined as follows:

$$\begin{aligned} \mathbf{f}^{(J)}(x, t) &= -\mathcal{L}(\widetilde{\mathbf{u}}^{(J)}, p^{(J)}) \\ &= -\left(\tfrac{\partial \widetilde{\mathbf{u}}^{(J)}(x,t)}{\partial t} - \nu \Delta \widetilde{\mathbf{u}}^{(J)}(x, t) + (\widetilde{\mathbf{u}}^{(J)} \cdot \nabla) \widetilde{\mathbf{u}}^{(J)}(x, t) + \nabla p^{(J)}(x, t) \right). \end{aligned}$$

It is notable to mention that the preceding formulation is equivalent to the subsequent one:

Definition 6.6.4 Find the divergence-free vector field $\mathbf{v}_d \in L^2(0,T;\mathcal{H}_{\mathrm{dec}}(B_\varepsilon,\delta))$ and the "MAPDD pressure" $p_d \in L^2(0,T;L^2(\bigcup_{i=1}^N B_i^{\varepsilon,\delta}))$ such that $\mathbf{v}_d(x,t)|_{\partial B_\varepsilon} = \mathbf{g}(x,t)$, $\frac{\partial \mathbf{v}_d}{\partial t} \in L^2(0,T;L^2(B_\varepsilon))$, $\mathbf{v}_d|_{t=0}=0$, and for almost all $t \in [0,T)$ and for all $\boldsymbol{\eta} \in \mathcal{W}_{0,\mathrm{dec}}(B_\varepsilon,\delta)$, the following integral identity holds:

$$\sum_{i=1}^N \int_{B_i^{\varepsilon,\delta}} \Big(\frac{\partial \mathbf{v}_d}{\partial t}\cdot\boldsymbol{\eta} + \nu\nabla\mathbf{v}_d\cdot\nabla\boldsymbol{\eta} + (\mathbf{v}_d\cdot\nabla)\mathbf{v}_d\cdot\boldsymbol{\eta}_d\Big)\,dx$$

$$+\sum_{l=1}^M d_l \int_{\sigma_\varepsilon^{(e_l)}} \Big(\frac{\partial \mathbf{v}_d}{\partial t}\cdot\boldsymbol{\eta} + \nu\nabla_{x^{(e_l)\prime}}\mathbf{v}_d\cdot\nabla_{x^{(e_l)\prime}}\boldsymbol{\eta}dx^{(e_l)\prime}\Big)\,dx = \sum_{i=1}^N \int_{B_i^{\varepsilon,\delta}} p_d \mathrm{div}\boldsymbol{\eta}\,dx. \tag{6.6.4}$$

In this equation, $d_l = |e| - 2\delta$ represents the distance between the cross sections S_{ij} and S_{ji} intersecting the segment e.

Theorem 6.6.1 *Suppose $B_\varepsilon \subset \mathbb{R}^3$. For sufficiently small ε, there exists a solution $\mathbf{u}_d \in L^2(0,T;\mathcal{H}_{0,\mathrm{dec}}(B_\varepsilon,\delta))$ that satisfies the integral identity* (6.6.2). *It possesses the time derivative $\mathbf{u}_{d,t} \in L^2(0,T;\mathcal{H}_{0,\mathrm{dec}}(B_\varepsilon,\delta))$. The subsequent estimates hold:*

$$\sup_{t\in[0,T]} \|\mathbf{u}_d(\cdot,t)\|^2_{L^2(B_\varepsilon)} + \int_0^T \|\nabla\mathbf{u}_d(\cdot,\tau)\|^2_{L^2(B_\varepsilon)}d\tau \le c\varepsilon^2 \int_0^T \|\mathbf{f}^{(J)}(\cdot,t)\|^2_{L^2(B_\varepsilon)}\,dt, \tag{6.6.5}$$

$$\sup_{t\in[0,T]} \|\mathbf{u}_{d,t}(\cdot,t)\|^2_{L^2(B_\varepsilon)} + \int_0^T \|\nabla\mathbf{u}_{d,t}(\cdot,t)\|^2_{L^2(B_\varepsilon)}\,dt + \varepsilon^{-2}\int_0^T \|\mathbf{u}_{d,t}(\cdot,t)\|^2_{L^2(B_\varepsilon)}\,dt$$

$$\le c\int_0^T \Big(\|\mathbf{f}^{(J)}(\cdot,t)\|^2_{L^2(B_\varepsilon)} + \|\mathbf{f}_t^{(J)}(\cdot,t)\|^2_{L^2(B_\varepsilon)}\Big)\,dt, \tag{6.6.6}$$

where the constant c depends on $\int_0^T \|\mathbf{f}^{(J)}(\cdot,t)\|^2_{L^2(B_\varepsilon)}dt$ and remains bounded as this integral tends to zero; furthermore, c is independent of ε for ε small enough.

The solution to problem (6.6.2) *is unique.*

Proof

We establish the solvability of the problem (6.6.2) utilizing the Galerkin method (e.g., [102, 203]). We follow the scheme proposed by O. Ladyzhenskaya [102] (see also [151] where this method was applied to prove the existence for the Navier–Stokes equations in the tube structure B_ε). Thus, the main objective is to obtain suitable a priori estimates.

1. First, we prove estimate (6.6.5) for the Galerkin approximations. The Galerkin approximations for solving problem (6.6.2) are defined by the following equations:

$$\int\limits_{B_\varepsilon}\Big(\frac{\partial \mathbf{u}_d^{(N)}}{\partial t}\cdot\boldsymbol{\xi}_l+\nu\nabla\mathbf{u}_d^{(N)}\cdot\nabla\boldsymbol{\xi}_l-\big(\mathbf{u}_d^{(N)}+\widetilde{\mathbf{u}}^{(J)}\big)\cdot\nabla)\boldsymbol{\xi}_l\cdot\mathbf{u}_d^{(N)}$$
$$-\big(\mathbf{u}_d^{(N)}\cdot\nabla\big)\boldsymbol{\xi}_l\cdot\widetilde{\mathbf{u}}^{(J)}\Big)dx=\int\limits_{B_\varepsilon}\mathbf{f}^{(J)}\cdot\boldsymbol{\xi}_l dx, \tag{6.6.7}$$
$$\gamma_l^{(N)}(0)=0,$$

where $l=1,..,N,\ \mathbf{u}_d^{(N)}(x,t)=\sum\limits_{k=1}^{N}\gamma_k^{(N)}(t)\boldsymbol{\xi}_k(x)$, and $\{\boldsymbol{\xi}_k\}_{k=1}^{\infty}$ constitutes a basis within the space $\mathcal{H}_{0,\mathrm{dec}}(B_\varepsilon,\delta)$. Without loss of generality, it is assumed that $\{\boldsymbol{\xi}_k\}_{k=1}^{\infty}$ is orthonormal in $L^2(B_\varepsilon)$.

Evidently, the relations (6.6.7) are equivalent to the following system of ordinary differential equations:

$$\begin{cases}\dfrac{d\gamma_l^{(N)}(t)}{dt}+\nu\sum\limits_{j=1}^{N}a_{lj}\gamma_j^{(N)}(t)+\sum\limits_{i,j=1}^{N}a_{lij}\gamma_i^{(N)}(t)\gamma_j^{(N)}(t)+\\ \qquad+\sum\limits_{j=1}^{N}b_{lj}^{(1)}(t)\gamma_j^{(N)}(t)+\sum\limits_{j=1}^{N}b_{lj}^{(2)}(t)\gamma_j^{(N)}(t)=f_l(t),\\ \qquad\qquad\gamma_l^{(N)}(0)=0,\end{cases} \tag{6.6.8}$$

where the coefficients are defined as follows:

$$a_{lj}=\int\limits_{B_\varepsilon}\nabla\boldsymbol{\xi}_l\cdot\nabla\boldsymbol{\xi}_j dx,\quad a_{lij}=-\int\limits_{B_\varepsilon}(\boldsymbol{\xi}_i\cdot\nabla)\boldsymbol{\xi}_l\cdot\boldsymbol{\xi}_j dx,$$
$$b_{lj}^{(1)}(t)=-\int\limits_{B_\varepsilon}(\widetilde{\mathbf{u}}^{(J)}\cdot\nabla)\boldsymbol{\xi}_l\cdot\boldsymbol{\xi}_j dx,\ b_{lj}^{(2)}(t)=-\int\limits_{B_\varepsilon}(\boldsymbol{\xi}_j\cdot\nabla)\boldsymbol{\xi}_l\cdot\widetilde{\mathbf{u}}^{(J)}dx,$$
$$f_l(t)=\int\limits_{B_\varepsilon}\mathbf{f}^{(J)}\cdot\boldsymbol{\xi}_l dx,\quad l=1,..,N.$$

By multiplying relations (6.6.7) with $\gamma_l^{(N)}(t)$ and summing over $l=1,\dots N$, we obtain:

$$\frac{1}{2}\frac{d}{dt}\int\limits_{B_\varepsilon}|\mathbf{u}_d^{(N)}|^2dx+\nu\int\limits_{B_\varepsilon}|\nabla\mathbf{u}_d^{(N)}|^2dx=\int\limits_{B_\varepsilon}(\mathbf{u}_d^{(N)}\cdot\nabla)\mathbf{u}_d^{(N)}\cdot\widetilde{\mathbf{u}}^{(J)}dx$$

Based on this equation and utilizing the estimate:

$$\sup_{t\in[0,T]} \|\widetilde{\mathbf{u}}^{(J)}(\cdot,t)\|_{L^4(B_\varepsilon)} + \sup_{t\in[0,T]} \|\widetilde{\mathbf{u}}_t^{(J)}(\cdot,t)\|_{L^4(B_\varepsilon)} \leq c\varepsilon^{\frac{(n-1)}{4}}, \tag{6.6.9}$$

we deduce that for sufficiently small ε, the inequality holds:

$$\frac{1}{2}\|\mathbf{u}_d^{(N)}(\cdot,t)\|^2_{L^2(B_\varepsilon)} + \frac{\nu}{2}\int_0^t \|\nabla\mathbf{u}_d^{(N)}(\cdot,\tau)\|^2_{L^2(B_\varepsilon)}d\tau \leq c\varepsilon^2 \int_0^T \|\mathbf{f}^{(J)}(\cdot,t)\|^2_{L^2(B_\varepsilon)}dt. \tag{6.6.10}$$

This inequality is the Galerkin analogue for (6.6.5).

Due to the orthonormality of the basis $\{\boldsymbol{\xi}_l\}$ in $L^2(B_\varepsilon)$, we have from (6.6.10):

$$\max_{t\in[0,T]} \|\mathbf{u}_d^{(N)}(\cdot,t)\|^2_{L^2(B_\varepsilon)} = \max_{t\in[0,T]} \sum_{l=1}^n |\gamma_l^{(N)}(t)|^2 \leq c\varepsilon^2 \int_0^T \|\mathbf{f}^{(J)}(\cdot,t)\|^2_{L^2(B_\varepsilon)}dt.$$

The system (6.6.8) depends analytically on the functions $\gamma_l^{(N)}(t)$. Thus, the a priori boundedness of the solution implies the unique solvability of the initial problem (6.6.8) on the interval $[0, T]$. Additionally, the right-hand side $\mathbf{f}^{(J)}$ possesses a derivative $\mathbf{f}_t^{(J)} \in L^2(B_\varepsilon)$, which implies that the functions $\gamma_l^{(N)}(t)$ have second-order derivatives $\dfrac{d^2\gamma_l^{(N)}(t)}{dt^2}$.

2. Let us prove the Galerkin analogue of estimate (6.6.6). Upon differentiating the integral identity (6.6.7) with respect to t, we derive:

$$\begin{aligned}
&\int_{B_\varepsilon} \Big(\mathbf{u}_{d,tt}^{(N)} \cdot \boldsymbol{\xi}_l + \nu\nabla\mathbf{u}_{d,t}^{(N)} \cdot \nabla\boldsymbol{\xi}_l - \big((\mathbf{u}_{d,t}^{(N)} + \widetilde{\mathbf{u}}_t^{(J)}) \cdot \nabla\big)\boldsymbol{\xi}_l \cdot \mathbf{u}_d^{(N)} \\
&-\big((\mathbf{u}_d^{(N)} + \widetilde{\mathbf{u}}^{(J)}) \cdot \nabla\big)\boldsymbol{\xi}_l \cdot \mathbf{u}_{d,t}^{(N)} - \big(\mathbf{u}_{d,t}^{(N)} \cdot \nabla\big)\boldsymbol{\xi}_l \cdot \widetilde{\mathbf{u}}^{(J)} - \big(\mathbf{u}_d^{(N)} \cdot \nabla\big)\boldsymbol{\xi}_l \cdot \widetilde{\mathbf{u}}_t^{(J)}\Big)dx \\
&= \int_{B_\varepsilon} \mathbf{f}_t^{(J)} \cdot \boldsymbol{\xi}_l dx,
\end{aligned} \tag{6.6.11}$$

where $\mathbf{u}_{d,t}^{(N)} = \dfrac{\partial\mathbf{u}_d^{(N)}}{\partial t}$. By multiplying relations (6.6.11) with $\gamma_l^{(N)\prime}(t)$ and summing over $l = 1, \ldots N$, we obtain:

$$
\begin{aligned}
&\frac{1}{2}\frac{d}{dt}\int\limits_{B_\varepsilon}|\mathbf{u}_{d,t}^{(N)}|^2dx+\nu\int\limits_{B_\varepsilon}|\nabla\mathbf{u}_{d,t}^{(N)}|^2dx\\
&=\int\limits_{B_\varepsilon}((\mathbf{u}_{d,t}^{(N)}+\widetilde{\mathbf{u}}_t^{(J)})\cdot\nabla)\mathbf{u}_{d,t}^{(N)}\cdot\mathbf{u}_d^{(N)}dx+\int\limits_{B_\varepsilon}(\mathbf{u}_{d,t}^{(N)}\cdot\nabla)\mathbf{u}_{d,t}^{(N)}\cdot\widetilde{\mathbf{u}}^{(J)}dx\\
&\quad+\int\limits_{B_\varepsilon}(\mathbf{u}_d^{(N)}\cdot\nabla)\mathbf{u}_{d,t}^{(N)}\cdot\widetilde{\mathbf{u}}_t^{(J)}dx+\int\limits_{B_\varepsilon}\mathbf{f}_t^{(J)}\cdot\mathbf{u}_{d,t}^{(N)}dx=\sum_{k=1}^{4}J_k.
\end{aligned}
\tag{6.6.12}
$$

We proceed to estimate the terms on the right-hand side of (6.6.12):

$$
\begin{aligned}
|J_1|&\le\big(\|\mathbf{u}_{d,t}^{(N)}\|_{L^4(B_\varepsilon)}+\|\widetilde{\mathbf{u}}_t^{(J)}\|_{L^4(B_\varepsilon)}\big)\|\nabla\mathbf{u}_{d,t}^{(N)}\|_{L^2(B_\varepsilon)}\|\mathbf{u}_d^{(N)}\|_{L^4(B_\varepsilon)}\\
&\le c_1\varepsilon^{\frac{1}{2}}\|\nabla\mathbf{u}_{d,t}^{(N)}\|^2_{L^2(B_\varepsilon)}\|\nabla\mathbf{u}_d^{(N)}\|_{L^2(B_\varepsilon)}\\
&\quad+c\varepsilon^{\frac{1}{4}}\|\widetilde{\mathbf{u}}_t^{(J)}\|_{L^4(B_\varepsilon)}\|\nabla\mathbf{u}_{d,t}^{(N)}\|_{L^2(B_\varepsilon)}\|\nabla\mathbf{u}_d^{(N)}\|_{L^2(B_\varepsilon)}\\
&\le c_2\varepsilon^{\frac{1}{2}}\|\nabla\mathbf{u}_{d,t}^{(N)}\|^2_{L^2(B_\varepsilon)}\|\nabla\mathbf{u}_d^{(N)}\|_{L^2(B_\varepsilon)}+c\|\widetilde{\mathbf{u}}_t^{(J)}\|^2_{L^4(B_\varepsilon)}\|\nabla\mathbf{u}_d^{(N)}\|_{L^2(B_\varepsilon)};
\end{aligned}
$$

$$|J_2|\le c\varepsilon^{\frac{1}{4}}\|\widetilde{\mathbf{u}}^{(J)}\|_{L^4(B_\varepsilon)}\|\nabla\mathbf{u}_{d,t}^{(N)}\|^2_{L^2(B_\varepsilon)};$$

$$|J_3|\le\varepsilon^{\frac{1}{2}}\|\nabla\mathbf{u}_{d,t}^{(N)}\|^2_{L^2(B_\varepsilon)}+c\|\widetilde{\mathbf{u}}_t^{(J)}\|^2_{L^4(B_\varepsilon)}\|\nabla\mathbf{u}_d^{(N)}\|^2_{L^2(B_\varepsilon)};$$

$$|J_4|\le\varepsilon^{\frac{1}{2}}\|\nabla\mathbf{u}_{d,t}^{(N)}\|^2_{L^2(B_\varepsilon)}+c\varepsilon^{\frac{3}{2}}\|\mathbf{f}_t^{(J)}\|^2_{L^2(B_\varepsilon)}.$$

In these estimates, we have employed the inequality (2.3.38).

Substituting the aforementioned estimates into Eq. (6.6.12) and utilizing inequality (6.6.9), we deduce the following estimate:

$$
\begin{aligned}
&\frac{1}{2}\frac{d}{dt}\int\limits_{B_\varepsilon}|\mathbf{u}_{d,t}^{(N)}|^2dx+\nu\int\limits_{B_\varepsilon}|\nabla\mathbf{u}_{d,t}^{(N)}|^2dx\\
&\le\Big(c_1\varepsilon^{\frac{1}{2}}\|\nabla\mathbf{u}_d^{(N)}\|_{L^2(B_\varepsilon)}+c_2\varepsilon^{\frac{1}{2}}\Big)\int\limits_{B_\varepsilon}|\nabla\mathbf{u}_{d,t}^{(N)}|^2dx+c\varepsilon^{\frac{3}{2}}\|\mathbf{f}_t^{(J)}\|^2_{L^2(B_\varepsilon)}.
\end{aligned}
\tag{6.6.13}
$$

Suppose that ε is sufficiently small, specifically such that $c_2\varepsilon^{\frac{1}{2}}\le\frac{\nu}{2}$. Then, from the inequality (6.6.13), we infer:

$$\frac{1}{2}\frac{d}{dt}\int\limits_{B_\varepsilon} |\mathbf{u}_{d,t}^{(N)}|^2 dx + \left(\frac{\nu}{2} - c_1\varepsilon^{\frac{1}{2}}\|\nabla \mathbf{u}_d^{(N)}\|_{L^2(B_\varepsilon)}\right)\int\limits_{B_\varepsilon} |\nabla \mathbf{u}_{d,t}^{(N)}|^2 dx \le c\varepsilon^{\frac{3}{2}}\|\mathbf{f}_t^{(J)}\|^2_{L^2(B_\varepsilon)}. \tag{6.6.14}$$

This inequality is the main step to prove estimate (6.6.6); however, we need to prove that the coefficient $\left(\frac{\nu}{2} - c_1\varepsilon^{\frac{1}{2}}\|\nabla \mathbf{u}_d^{(N)}\|_{L^2(B_\varepsilon)}\right)$ is positive.

2.1. Based on Eq. (6.6.7), it follows that:

$$\nu\|\nabla \mathbf{u}_d^{(N)}\|^2_{L_2(B_\varepsilon)} = -\int\limits_{B_\varepsilon} \mathbf{u}_{d,t}^{(N)}\cdot \mathbf{u}_d^{(N)} dt + \int\limits_{B_\varepsilon} \mathbf{f}^{(J)}\cdot \mathbf{u}_d^{(N)} dx - \int\limits_{B_\varepsilon} (\mathbf{u}_d^{(N)}\cdot\nabla)\mathbf{u}_d^{(N)}\cdot\widetilde{\mathbf{u}}^{(J)} dx$$

$$\le \|\mathbf{u}_d^{(N)}\|_{L^2(B_\varepsilon)}\left(\|\mathbf{u}_{d,t}^{(N)}\|_{L^2(B_\varepsilon)} + \|\mathbf{f}^{(J)}\|_{L^2(B_\varepsilon)}\right) + \|\mathbf{u}_d^{(N)}\|_{L^4(B_\varepsilon)}\|\widetilde{\mathbf{u}}^{(J)}\|_{L^4(B_\varepsilon)}\|\nabla \mathbf{u}_d^{(N)}\|_{L^2(B_\varepsilon)}$$

$$\le c_3\varepsilon\|\nabla \mathbf{u}_d^{(N)}\|_{L^2(B_\varepsilon)}\left(\|\mathbf{u}_{d,t}^{(N)}\|_{L^2(B_\varepsilon)} + \|\mathbf{f}^{(J)}\|_{L^2(B_\varepsilon)}\right) + c_4\varepsilon^{\frac{3}{4}}\|\nabla \mathbf{u}_d^{(N)}\|^2_{L^2(B_\varepsilon)}.$$

If $c_4\varepsilon^{\frac{3}{4}} \le \frac{\nu}{2}$, the last inequality implies:

$$\|\nabla \mathbf{u}_d^{(N)}\|_{L^2(B_\varepsilon)} \le \frac{2c_3\varepsilon}{\nu}\left(\|\mathbf{u}_{d,t}^{(N)}\|_{L^2(B_\varepsilon)} + \|\mathbf{f}^{(J)}\|_{L^2(B_\varepsilon)}\right). \tag{6.6.15}$$

Consequently:

$$\frac{\nu}{2} - c_1\varepsilon^{\frac{1}{2}}\|\nabla \mathbf{u}_d^{(N)}(\cdot,t)\|_{L^2(B_\varepsilon)} \ge \frac{\nu}{2} - \frac{2c_1c_3}{\nu}\varepsilon^{\frac{3}{2}}\left(\|\mathbf{u}_{d,t}^{(N)}(\cdot,t)\|_{L^2(B_\varepsilon)} + \|\mathbf{f}^{(J)}(\cdot,t)\|_{L^2(B_\varepsilon)}\right). \tag{6.6.16}$$

Particularly, when $t = 0$, it holds that $\|\nabla \mathbf{u}_d^{(N)}(\cdot,0)\|_{L^2(B_\varepsilon)} = 0$, thus resulting in:

$$\frac{\nu}{2} - c_1\varepsilon^{\frac{1}{2}}\|\nabla \mathbf{u}_d^{(N)}(\cdot,0)\|_{L^2(B_\varepsilon)} \ge \frac{\nu}{2} - \frac{2c_1c_3}{\nu}\varepsilon^{\frac{3}{2}}\left(\|\mathbf{u}_{d,t}^{(N)}(\cdot,0)\|_{L^2(B_\varepsilon)} + \|\mathbf{f}^{(J)}(\cdot,0)\|_{L^2(B_\varepsilon)}\right).$$

Consider the function $y(t) = \frac{\nu}{4} - c_1\varepsilon^{\frac{1}{2}}\|\nabla \mathbf{u}_d^{(N)}(\cdot,t)\|_{L^2(B_\varepsilon)}$. This function is continuous for $t \ge 0$, and as $\nabla \mathbf{u}_d^{(N)}(x,0) = 0$, we have $y(0) = \frac{\nu}{4} > 0$. Therefore, either there exists a point $t_1 \in [0, T]$ such that $y(t) > 0$ for $t < t_1$ and $y(t_1) = 0$ or $y(t) > 0$ for all $t \in [0, T]$. It can be demonstrated that the first case is impossible. Assume this to be the case. Then, for $t \in [0, t_1)$, the following estimate is valid:

$$\frac{1}{2}\frac{d}{dt}\int\limits_{B_\varepsilon} |\mathbf{u}_{d,t}|^2 dx \leq c\varepsilon^{\frac{3}{2}} \|\mathbf{f}_t^{(J)}\|^2_{L^2(B_\varepsilon)}. \tag{6.6.17}$$

(see (6.6.14)). Moreover:

$$\frac{1}{2}\int\limits_{B_\varepsilon} |\mathbf{u}_{d,t}^{(N)}(x,t)|^2 dx$$

$$\leq \frac{1}{2}\int\limits_{B_\varepsilon} |\mathbf{u}_{d,t}^{(N)}(x,0)|^2 dx + c\varepsilon^{\frac{3}{2}} \int\limits_0^T \|\mathbf{f}_t^{(J)}(\cdot,t)\|^2_{L^2(B_\varepsilon)} dt \quad \forall t \in [0, t_1). \tag{6.6.18}$$

Taking $t = 0$ in Eq. (6.6.7) and subsequently multiplying the resultant equations by $\gamma_l^{(N)\prime}(0)$ for $l = 1, \ldots, N$, the equations are summed over l to derive the ensuing expression:

$$\int\limits_{B_\varepsilon} |\mathbf{u}_{d,t}^{(N)}(x,0)|^2 dx = \int\limits_{B_\varepsilon} \mathbf{f}^{(J)}(x,0) \cdot \mathbf{u}_{d,t}^{(N)}(x,0) dx.$$

Here the conditions $\mathbf{u}_d^{(N)}(x,0) = 0$ and $\mathbf{u}^{(J)}(x,0) = 0$ have been utilized.

Thus:

$$\|\mathbf{u}_{d,t}^{(N)}(\cdot,0)\|_{L^2(B_\varepsilon)} \leq \|\mathbf{f}^{(J)}(\cdot,0)\|_{L^2(B_\varepsilon)}, \tag{6.6.19}$$

and using Eq. (6.6.18), we further obtain:

$$\frac{1}{2}\int\limits_{B_\varepsilon} |\mathbf{u}_{d,t}^{(N)}(x,t)|^2 dx \leq c\Big(\|\mathbf{f}^{(J)}(\cdot,0)\|^2_{L^2(B_\varepsilon)} + \varepsilon^{\frac{3}{2}} \int\limits_0^T \|\mathbf{f}_t^{(J)}(\cdot,t)\|^2_{L^2(B_\varepsilon)} dt\Big) \ \forall t \in [0, t_1). \tag{6.6.20}$$

By continuity, the above inequality remains valid for $t = t_1$. Substituting (6.6.20) for $t = t_1$ into (6.6.16) yields:

$$\begin{aligned}
&\frac{\nu}{4} - c_1\varepsilon^{\frac{1}{2}}\|\nabla \mathbf{u}_d^{(N)}(\cdot,t_1)\|_{L^2(B_\varepsilon)} \\
&\quad\geq \frac{\nu}{4} - \frac{4c_1c_3}{\nu}\varepsilon^{\frac{3}{2}}\Big(\|\mathbf{u}_{d,t}^{(N)}(\cdot,t_1)\|_{L^2(B_\varepsilon)} + \|\mathbf{f}^{(J)}(\cdot,t_1)\|_{L^2(B_\varepsilon)}\Big) \\
&\quad\geq \frac{\nu}{4} - c\varepsilon^{\frac{3}{2}}\Big(\sup_{t\in[0,t]}\|\mathbf{f}^{(J)}(\cdot,t)\|_{L^2(B_\varepsilon)} + \varepsilon^{\frac{3}{4}}\|\mathbf{f}_t^{(J)}\|_{L^2(0,T;L^2(B_\varepsilon))}\Big) \\
&\quad\geq \frac{\nu}{4} - c\varepsilon^{\frac{3}{2}}\Big(\|\mathbf{f}^{(J)}\|_{L^2(0,T;L^2(B_\varepsilon))} + \|\mathbf{f}_t^{(J)}\|_{L^2(0,T;L^2(B_\varepsilon))}\Big).
\end{aligned} \tag{6.6.21}$$

Here has been employed the inequality[2]:

$$\sup_{t\in[0,t]}\|\mathbf{f}^{(J)}(\cdot,t)\|_{L^2(B_\varepsilon)} \leq c\Big(\|\mathbf{f}^{(J)}\|_{L^2(0,T;L^2(B_\varepsilon))} + \|\mathbf{f}_t^{(J)}\|_{L^2(0,T;L^2(B_\varepsilon))}\Big). \tag{6.6.22}$$

For sufficiently small ε, the inequality above implies that $y(t_1) > 0$, leading to a contradiction with the initial assumption. Therefore, for all $t \in [0, T]$, the inequality $y(t) = \frac{\nu}{4} - c_1\varepsilon^{\frac{1}{2}}\|\nabla \mathbf{u}^{(N)}(\cdot,t)\|_{L^2(B_\varepsilon)} > 0$ holds. This consequently gives rise to the following conclusion:

$$\frac{\nu}{2} - c_1\varepsilon^{\frac{1}{2}}\|\nabla \mathbf{u}^{(N)}(\cdot,t)\|_{L^2(B_\varepsilon)} > \frac{\nu}{4}, \tag{6.6.23}$$

and by utilizing the results of (6.6.14) and (6.6.19), the following inequality can be deduced:

---

[2] To prove (6.6.22), we use the continuous embedding $W^{1,2}(0,T)$ into $C(0,T)$ for the function $\|\mathbf{f}^{(J)}(\cdot,t)\|_{L^2(B_\varepsilon)}$:

$$\sup_{t\in[0,T]}\|\mathbf{f}^{(J)}(\cdot,t)\|^2_{L^2(B_\varepsilon)} \leq c\Big(\int_0^T \|\mathbf{f}^{(J)}(\cdot,t)\|^2_{L^2(B_\varepsilon)}dt + \int_0^T \Big|\frac{d}{dt}\|\mathbf{f}^{(J)}(\cdot,t)\|_{L^2(B_\varepsilon)}\Big|^2 dt\Big).$$

Since:

$$\frac{d}{dt}\Big(\int_{B_\varepsilon} |\mathbf{f}^{(J)}(x,t)|^2dx\Big)^{1/2} = \frac{\int_{B_\varepsilon}\mathbf{f}^{(J)}(x,t)\mathbf{f}_t^{(J)}(x,t)dx}{\big(\int_{B_\varepsilon}|\mathbf{f}^{(J)}(x,t)|^2dx\big)^{1/2}} \leq \Big(\int_{B_\varepsilon}|\mathbf{f}_t^{(J)}(x,t)|^2dx\Big)^{1/2},$$

from the latter inequality follows estimate (6.6.22).

$$\frac{1}{2}\int\limits_{B_\varepsilon} |\mathbf{u}_{d,t}^{(N)}(x,t)|^2 dx + \frac{\nu}{4}\int\limits_0^t\int\limits_{B_\varepsilon} |\nabla \mathbf{u}_{d,t}^{(N)}(x,t)|^2 dxdt$$
$$\leq c\int\limits_0^T \Big(\|\mathbf{f}^{(J)}(\cdot,t)\|^2_{L^2(B_\varepsilon)} + \|\mathbf{f}_t^{(J)}(\cdot,t)\|^2_{L^2(B_\varepsilon)}\Big)dt.$$

Finally, applying the Poincaré inequality leads to:

$$\int\limits_0^T\int\limits_{B_\varepsilon} |\mathbf{u}_{d,t}^{(N)}(x,t)|^2 dxdt \leq c\varepsilon^2 \int\limits_0^T\int\limits_{B_\varepsilon} |\nabla \mathbf{u}_{d,t}^{(N)}(x,t)|^2 dxdt.$$

Thus, the estimate (6.6.6) for Galerkin approximation can be concluded.

3. The estimates (6.6.5) and (6.6.6) for Galerkin approximation guarantee the existence of a subsequence $\{\mathbf{u}_d^{(N_l)}\}$ such that the corresponding sequences $\{\mathbf{u}_d^{(N_l)}\}$, $\{\nabla \mathbf{u}_d^{(N_l)}\}$, and $\{\mathbf{u}_{d,t}^{(N_l)}\}$ converge weakly in $L^2(0,T;L^2(B_\varepsilon))$ to $\mathbf{u}_d$, $\nabla \mathbf{u}_d$, and $\mathbf{u}_{d,t}$, respectively. This implies that $\mathbf{u}_d \in L^2(0,T;\mathcal{H}_{0,\mathrm{dec}}(B_\varepsilon,\delta))$, $\mathbf{u}_{d,t} \in L^2(0,T;L^2(B_\varepsilon))$, and the estimates (6.6.5) and (6.6.6) still hold for $\mathbf{u}_d$. By following a similar reasoning to that presented in the proof of Theorem 6.2.1, it can be shown that $\mathbf{u}_d$ satisfies the integral identity:

$$\int\limits_0^T\int\limits_{B_\varepsilon} \Big(\mathbf{u}_{d,t}\cdot \Xi^M + \nu\nabla \mathbf{u}_d \cdot \nabla \Xi^M - \big(\mathbf{u}_d + \widetilde{\mathbf{u}}^{[J]}\big)\cdot\nabla\big)\Xi^M \cdot \mathbf{u}_d$$
$$-\big(\mathbf{u}_d\cdot\nabla\big)\Xi^M\cdot\widetilde{\mathbf{u}}^{[J]}\Big)dxdt = \int\limits_0^T\int\limits_{B_\varepsilon} \mathbf{f}^{[J]}\cdot\Xi^M dxdt \tag{6.6.24}$$

for any Ξ^M having the form $\Xi^M(x,t) = \sum\limits_{j=1}^{M} d_j(t)\boldsymbol{\xi}_j(x)$, where $d_j(t)$ are continuous functions. Due to Theorem 2.3.8, the set of such linear combinations is dense in the space $L^2(0,T;\mathcal{H}_{0,\mathrm{dec}}(B_\varepsilon,\delta))$. Therefore, the identity (6.6.24) is valid for any $\Xi \in L^2(0,T;\mathcal{H}_{0,\mathrm{dec}}(B_\varepsilon,\delta))$. Taking in (6.6.24) $\Xi(x,t) = d(t)\boldsymbol{\eta}(x)$ with $\boldsymbol{\eta} \in \mathcal{H}_{0,\mathrm{dec}}(B_\varepsilon,\delta))$ and an arbitrary continuous on $[0,T]$ function $d(t)$, we can deduce that $\mathbf{u}_d$ satisfies the identity (6.6.2), and thus, we have proved the existence of a weak solution.

4. Now, let us prove the uniqueness. For the difference $\mathbf{w}_d = \mathbf{u}_d^{(1)} - \mathbf{u}_d^{(2)}$ between two solutions to problem (6.6.2), we deduce the equality:

$$\frac{1}{2}\frac{d}{dt}\int\limits_{B_\varepsilon} |\mathbf{w}_d|^2\,dx + \nu\int\limits_{B_\varepsilon} |\nabla\mathbf{w}_d|^2\,dx$$
$$= \int\limits_{B_\varepsilon} (\mathbf{w}_d\cdot\nabla)\mathbf{w}_d\cdot\mathbf{u}_d^{(2)}\,dx + \int\limits_{B_\varepsilon} (\mathbf{w}_d\cdot\nabla)\mathbf{w}_d\cdot\widetilde{\mathbf{u}}^{(J)}\,dx.$$

By leveraging the inequalities (2.3.8) and (6.6.9), we can derive the following estimate:

$$\begin{aligned}
&\frac{1}{2}\frac{d}{dt}\int\limits_{B_\varepsilon} |\mathbf{w}_d|^2\,dx + \nu\int\limits_{B_\varepsilon} |\nabla\mathbf{w}_d|^2\,dx\\
&\le \|\mathbf{w}_d\|_{L^4(B_\varepsilon)}\|\nabla\mathbf{w}_d\|_{L^2(B_\varepsilon)}\Big(\|\mathbf{u}_d^{(2)}\|_{L^4(B_\varepsilon)} + \sup_{t\in(0,T)}\|\widetilde{\mathbf{u}}^{[J]}(x,t)\|_{L^4(B_\varepsilon)}\Big)\\
&\le c\|\mathbf{w}_d\|^{\frac{1}{4}}_{L^2(B_e)}\|\nabla\mathbf{w}_d\|^{\frac{7}{4}}_{L^2(B_\varepsilon)}\Big(\|\nabla\mathbf{u}_d^{(2)}\|_{L^2(B_\varepsilon)} + \varepsilon^{\frac{1}{2}}\Big)\\
&\le c_\delta\Big(\|\mathbf{w}\|^2_{L^2(B_\varepsilon)}\big[\|\nabla\mathbf{u}_d^{(2)}\|^8_{L^2(B_\varepsilon)} + \varepsilon^4\big]\Big) + \delta\|\nabla\mathbf{w}\|^2_{L^2(B_\varepsilon)}.
\end{aligned} \tag{6.6.25}$$

Taking in the latter inequality $\delta = \nu/2$ and integrating by t, we obtain the estimate:

$$\int\limits_{B_\varepsilon} |\mathbf{w}_d(x,t)|^2\,dx \le c\Big(\sup_{t\in[0,T]}\|\nabla\mathbf{u}_d^{(2)}\|^8_{L^2(B_\varepsilon)} + 1\Big)\int\limits_0^t\int\limits_{B_\varepsilon} |\mathbf{w}_d(x,\tau)|^2\,dxd\tau.$$

Consequently, through the application of Gronwall's lemma, it follows that $\mathbf{w}_d = 0$, leading to the conclusion $\mathbf{u}_d^{(1)} = \mathbf{u}_d^{(2)}$. This completes the proof. □

Remark 6.6.1 The propositions stated in Theorem 6.6.1 also hold true for two-dimensional tube structures $B_\varepsilon \subset \mathbb{R}^2$. In order to establish the validity of estimates (6.6.5) and (6.6.6), it suffices to use two-dimensional multiplicative inequalities (2.3.7) and (2.3.37) instead of their three-dimensional counterparts (2.3.8) and (2.3.38).

Remark 6.6.2 Estimates (6.6.6) and (6.6.15) yield:

$$\|\nabla\mathbf{u}_d\|_{L^2(B_\varepsilon)} \le c\varepsilon\Big\{\int\limits_0^T \Big(\|\mathbf{f}^{(J)}(\cdot,t)\|^2_{L^2(B_\varepsilon)} + \|\mathbf{f}_t^{(J)}(\cdot,t)\|^2_{L^2(B_\varepsilon)}\Big)dt\Big\}^{1/2}. \tag{6.6.26}$$

Let us proceed to estimate the MAPDD pressure p_d.

Theorem 6.6.2 *Suppose $B_\varepsilon \subset \mathbb{R}^n, n = 2, 3$, $\partial B_\varepsilon \in C^2$. There exists a unique $q_d \in L^2(0,T;\widehat{L}^2_{dec}(B_\varepsilon))$, which, together with $\mathbf{u}_d$, satisfies the integral identity* (6.6.3). *The following estimate holds:*

$$\int_0^T \sum_{i=1}^N \|q_d(\cdot,t)\|^2_{L^2(B_i^{\varepsilon,\delta})} dt \le c\varepsilon^{-2n+1/2} \int_0^T \Big(\|\mathbf{f}^{(J)}(\cdot,t)\|^2_{L^2(B_\varepsilon)} + \|\mathbf{f}_t^{(J)}(\cdot,t)\|^2_{L^2(B_\varepsilon)} \Big) dt. \tag{6.6.27}$$

Here the constant c depends on $\int_0^T \|\mathbf{f}^{(J)}(\cdot,t)\|^2_{L^2(B_\varepsilon)} dt$ *and remains bounded as this integral tends to zero; furthermore, c is independent of* ε *for* ε *small enough.*

Proof

We begin by introducing the linear bounded functional defined on $\mathcal{W}_{0,\mathrm{dec}}(B_\varepsilon,\delta)$:

$$f(\eta) = \int_{B_\varepsilon} \frac{\partial \mathbf{v}_d}{\partial t} \cdot \eta + \nu \nabla \mathbf{v}_d \cdot \nabla \eta + (\mathbf{v}_d \cdot \nabla)\mathbf{v}_d \cdot \eta dx, \tag{6.6.28}$$

where $\mathbf{v}_d = \mathbf{u}_d + \widetilde{\mathbf{u}}^{(J)}$. This functional vanishes on the space $\mathcal{H}_{0,\mathrm{dec}}(B_\varepsilon,\delta)$. Thus, by Theorem 5.2.1, there exists a function $q_d(\cdot,t) \in \widehat{L}^2_{dec}(B_\varepsilon)$ (recall that q_d is zero on all cylinders B^ε_{ij} and $\int_{B_\varepsilon} q_d(x,t)dx = 0$) such that:

$$f(\eta) = \sum_{i=1}^N \int_{B_i^{\varepsilon,\delta}} q_d \mathrm{div}\eta dx \quad \forall \eta \in \mathcal{W}_{0,\mathrm{dec}}(B_\varepsilon,\delta). \tag{6.6.29}$$

By comparing (6.6.28) and (6.6.29), it is clear that $(\mathbf{u}_d, q_d)$ satisfies the integral identity (6.6.3).

According to Lemma 5.2.2, there exists a function $\mathbf{U} \in \mathcal{W}_{0,\mathrm{dec}}(B_\varepsilon,\delta)$ such that:

$$\mathrm{div}\mathbf{U}(x,t) = \begin{cases} q_d(x,t), \ x \in \bigcup_{i=1}^N B_i^{\varepsilon,\delta}, \\ 0, \ x \in B_\varepsilon \setminus \bigcup_{i=1}^N B_i^{\varepsilon,\delta}, \end{cases}$$

and:

$$\|\mathbf{U}(\cdot,t)\|^2_{W^{1,2}(B_\varepsilon)} \le C\varepsilon^{-2n-3/2} \sum_{i=1}^N \|p_d(\cdot,t)\|^2_{L^2(B_i^{\varepsilon,\delta})} \tag{6.6.30}$$

where C is a constant independent of ε and δ. To prove the estimate (6.6.27), we take $\eta = \mathbf{U}$ in the integral identity (6.6.3):

$$\sum_{i=1}^{N}\int\limits_{B_i^{\varepsilon,\delta}}|q_d|^2dx=\sum_{i=1}^{N}\int\limits_{B_i^{\varepsilon,\delta}}p_d\mathrm{div}\mathbf{U}dx$$
$$=\int\limits_{B_\varepsilon}\Big(\frac{\partial\mathbf{u}_d}{\partial t}\cdot\mathbf{U}+\nu\nabla\mathbf{u}_d\cdot\nabla\mathbf{U}-((\mathbf{u}_d+\widetilde{\mathbf{u}}^{(J)})\cdot\nabla)\mathbf{U}\cdot\mathbf{u}_d-(\mathbf{u}_d\cdot\nabla)\mathbf{U}\cdot\widetilde{\mathbf{u}}^{(J)}\Big)dx$$
$$-\int\limits_{B_e}\mathbf{f}^{(J)}\cdot\mathbf{U}dx\le c\Big(\varepsilon\|\mathbf{u}_{d,t}\|_{L^2(B_\varepsilon)}+\|\nabla\mathbf{u}_d\|_{L^2(B_\varepsilon)}+\|\mathbf{u}_d\|^2_{L^4(B_\varepsilon)}$$
$$+\|\mathbf{u}_d\|_{L^4(B_\varepsilon)}\|\widetilde{\mathbf{u}}^{(J)}\|_{L^4(B_\varepsilon)}+\varepsilon\|\mathbf{f}^{(J)}\|_{L^2(B_\varepsilon)}\Big)\|\nabla\mathbf{U}\|_{L^2(B_\varepsilon)}$$
$$\le c\Big(\varepsilon\|\mathbf{u}_{d,t}\|_{L^2(B_\varepsilon)}+\|\nabla\mathbf{u}_d\|_{L^2(B_\varepsilon)}+\varepsilon^{1/2}\|\nabla\mathbf{u}_d\|^2_{L^2(B_\varepsilon)}$$
$$+\varepsilon\|\mathbf{f}^{(J)}\|_{L^2(B_\varepsilon)}\Big)\varepsilon^{-n-3/4}\sum_{i=1}^{N}\|q_d(\cdot,t)\|_{L^2(B_i^{\varepsilon,\delta})}. \tag{6.6.31}$$

So:

$$\Big(\sum_{i=1}^{N}\int\limits_{B_i^{\varepsilon,\delta}}|q_d|^2dx\Big)^{1/2}\le c\varepsilon^{-n-3/4}\Big(\varepsilon\|\mathbf{u}_{d,t}\|_{L^2(B_\varepsilon)}$$
$$+\|\nabla\mathbf{u}_d\|_{L^2(B_\varepsilon)}+\varepsilon^{1/2}\|\nabla\mathbf{u}_d\|^2_{L^2(B_\varepsilon)}+\varepsilon\|\mathbf{f}^{(J)}\|_{L^2(B_\varepsilon)}\Big).$$

Thus, by utilizing inequalities (6.6.6) and (6.6.26) and integrating over t, we obtain:

$$\int\limits_0^T\sum_{i=1}^{N}\|q_d(\cdot,t)\|^2_{L^2(B_i^{\varepsilon,\delta})}dt$$
$$\le c\varepsilon^{-2n-3/2}\Big(\varepsilon^2\int\limits_0^T\Big(\|\mathbf{f}^{(J)}(\cdot,t)\|^2_{L^2(B_\varepsilon)}+\|\mathbf{f}_t^{(J)}(\cdot,t)\|^2_{L^2(B_\varepsilon)}\Big)dt$$
$$+\varepsilon\sup_{t\in[0,T]}\|\nabla\mathbf{u}_d(\cdot,t)\|^2_{L^2(B_\varepsilon)}\int\limits_0^T\|\nabla\mathbf{u}_d(\cdot,t)\|^2_{L^2(B_\varepsilon)}dt\Big) \tag{6.6.32}$$
$$\le c\varepsilon^{-2n+1/2}\int\limits_0^T\Big(\|\mathbf{f}^{(J)}(\cdot,t)\|^2_{L^2(B_\varepsilon)}+\|\mathbf{f}_t^{(J)}(\cdot,t)\|^2_{L^2(B_\varepsilon)}\Big)dt.$$

This concludes the proof. □

Theorem 6.6.3 *Assume that* $\mathbf{g} \in C^{[\frac{J+1}{2}]+3}(0, T; W^{3/2,2}(\partial B_\varepsilon))$, *and let* δ *be* $C_{J+2}\varepsilon|\ln\varepsilon|$, *where the constant* C_{J+2} *satisfies* (6.5.9) *with* J *replaced by* $J+2$. *Then the following error estimates hold:*

$$\sup_{t\in[0,T]} \|\mathbf{v}-\mathbf{v}_d\|_{L^2(B_\varepsilon)} + \|\nabla(\mathbf{v}-\mathbf{v}_d)\|_{L^2(B_\varepsilon)\times(0,T)} = O(\varepsilon^{J+(n-1)/2}), \qquad (6.6.33)$$

$$\sup_{t\in[0,T]} \|\mathbf{v}_t-\mathbf{v}_{d,t}\|_{L^2(B_\varepsilon)}+\|\nabla(\mathbf{v}_t-\mathbf{v}_{d,t})\|_{L^2(B_\varepsilon)\times(0,T)} = O(\varepsilon^{J+(n-3)/2}), \qquad (6.6.34)$$

$$\|\mathbf{v}_t - \mathbf{v}_{d,t}\|_{L^2(B_\varepsilon)\times(0,T)} = O(\varepsilon^{J+(n-1)/2}). \qquad (6.6.35)$$

Proof
Denote $\mathbf{u}_d = \mathbf{v}_d - \widetilde{\mathbf{u}}^{(J')}$, where $J' = J+2$. Taking into account that the right-hand side of (6.6.5) has the order $O(\varepsilon^{2J'-2}) = O(\varepsilon^{2J+2})$ and the right-hand side of (6.6.6) has the order $O(\varepsilon^{2J'-4}) = O(\varepsilon^{2J})$, we obtain:

$$\sup_{t\in[0,T]} \|\widetilde{\mathbf{u}}^{(J')} - \mathbf{v}_d\|_{L^2(B_\varepsilon)} + \|\nabla(\tilde{\mathbf{u}}^{(J')} - \mathbf{v}_d)\|_{L^2(B_\varepsilon)\times(0,T)} = O(\varepsilon^{J+1}),$$

$$\sup_{t\in[0,T]} \|\widetilde{\mathbf{u}}_t^{(J')} - \mathbf{v}_{d,t}\|_{L^2(B_\varepsilon)} + \|\nabla(\widetilde{\mathbf{u}}_t^{(J')} - \mathbf{v}_{d,t})\|_{L^2(B_\varepsilon)\times(0,T)} = O(\varepsilon^{J}),$$

$$\|\widetilde{\mathbf{u}}_t^{(J')} - \mathbf{v}_{d,t}\|_{L^2(B_\varepsilon)\times(0,T)} = O(\varepsilon^{J+1}).$$

Using estimates (6.5.2) and (6.5.3) and estimates (6.5.6), (6.5.7), and (6.5.8) and applying the triangle inequality, we prove estimates (6.6.33), (6.6.34), and (6.6.35). The higher regularity of the boundary value function $\mathbf{g}$ is related to the construction of the asymptotic approximation $\widetilde{\mathbf{u}}^{(J')}$ with $J' = J+2$. □

Theorem 6.6.4 *Assume that* $\mathbf{g} \in C^{[\frac{J+1}{2}]+n+3}(0, T; W^{3/2,2}(\partial B_\varepsilon))$, *and let* δ *be* $C_{J+2n+2}\varepsilon|\ln\varepsilon|$, *where the constant* C_{J+2n+2} *satisfies* (6.5.9) *with* J *replaced by* $J+2$. *Then there exists the pressure* p, *solution to problem* (6.1.1), *such that the following error estimate holds:*

$$\Big(\int_0^T \sum_{i=1}^N \|p(\cdot,t) - p_d(\cdot,t)\|^2_{L^2(B_i^{\varepsilon,\delta})} dt\Big)^{1/2} = O(\varepsilon^{J+n/2+(n-3)/2}), \qquad (6.6.36)$$

Proof
Denote $q_d = p_d - p^{(J')}$, where $J' = J+2n+2$. Taking into account that the right-hand side of (6.6.27) has the order $O(\varepsilon^{-2n+1/2}\varepsilon^{2J'-4}) = O(\varepsilon^{2J+2n+1/2})$, we obtain:

$$\Big(\int_0^T \sum_{i=1}^N \|p^{(J')}(\cdot,t) - p_d(\cdot,t)\|^2_{L^2(B_i^{\varepsilon,\delta})} dt\Big)^{1/2} = O(\varepsilon^{J+n}) = O(\varepsilon^{J+n/2+(n-3)/2}).$$

Using estimate (6.5.8) and applying the triangle inequality, we prove estimate (6.6.36). The higher regularity of the boundary value function $\mathbf{g}$ is related to the construction of the asymptotic approximation $\widetilde{\mathbf{u}}^{(J')}$ with $J' = J + 2n + 2$. □

Now, we extend p_d to the cylinders $B_{ij}^{dec,\varepsilon}$ as previously described (see Sect. 5.2):

$$p_d(x_n^{(e)}, t) = \langle p_d \rangle_{S_{ij}} + \frac{\langle p_d \rangle_{S_{ji}} - \langle p_d \rangle_{S_{ij}}}{|e| - 2\delta + \varepsilon} \Big(x_n^{(e)} - \delta + \frac{\varepsilon}{2}\Big). \tag{6.6.37}$$

As in Sect. 5.2, for so extended p_d, we prove that:

$$\|p - p_d\|_{L^2(B_\varepsilon \times (0,T))} = O(\varepsilon^{J+(n-3)/2}).$$

These two theorems justify the MAPDD for the nonstationary Navier–Stokes equation in thin tube structure in the case of vanishing data for small times.

6.7 The Case of Non-vanishing for Small Time Data

In the general scenario where data is non-vanishing in a neighborhood of the point $t = 0$, the asymptotic expansion established in the preceding sections produces a residual of order ε^2 for the initial condition. In order to compensate this residual, we construct two additional sums of the expansion: firstly, the boundary layer-in-time correctors for each edge (these terms are the analogues for the regular expansion in Sect. 6.3) and, secondly, the boundary layer-in-time-and-in-space correctors localized around all nodes and vertices (analogous to the boundary layer-in-space expansion in Sect. 6.3). Notably, all these newly introduced terms exhibit exponential decay with respect to the fast time $\tau = t/\varepsilon^2$. The flux distribution is described by the equation for the macroscopic fast-time pressure on the graph, but now this problem becomes non-steady. Consequently, the theorems concerning solution existence, uniqueness, and exponential decay with respect to the fast time become considerably intricate.

The construction of the boundary layer-in-time closely parallels that of the previous section's expansion. To elucidate, the initial step involves creating the boundary layer-in-time correctors for nonstationary flows of a Poiseuille type along each edge. The velocity of these flows solely relies upon the transverse spatial variables $x^{(e)\prime}$ of the tube, while pressure is a linear function of the longitudinal variable $x_n^{(e)}$. Subsequently, these edge-associated flows undergo multiplication by the cutoff function ζ, which vanishes in the neighborhoods of nodes. This multiplication generates the discrepancy in the right-hand side of the equation, which has compact support within the neighborhoods of the nodes. To counterbalance this discrepancy, the boundary layer-in-space are constructed (analogously to the preceding sections). Notably, these boundary layer-in-time-and-in-space have the solvability condition of the same type as (6.3.30). These conditions yield junction criteria for the macroscopic pressure, stipulating that the total flux of each node must be zero.

On the other hand, the pressure within the solution of the boundary layer-in-space converges in outlets to infinity toward certain time-dependent constants. These constants are then compensated as previously elucidated, leading to a redefined macroscopic pressure characterized by specified discontinuities at the nodes. Consequently, we obtain for the macroscopic pressure a conductivity problem on the graph similar to the previous one. The fundamental distinction between these conductivity problems resides in the fact that in the case of the fast-time boundary layer, the relation between flow rate and pressure drop is more complicated than the classical Darcy law: it becomes non-local in time. Nevertheless, all recursive conductivity problems on the graph admit a unique solution, thereby rendering the algorithm for constructing boundary layer-in-time correctors akin to the prior considerations.

A salient contrast in the constructions detailed in Sect. 6.3 is that the inertial term $(\mathbf{v} \cdot \nabla)\mathbf{v}$ in the Navier–Stokes equations mixes the terms depending on the slow time t with the terms depending on the fast time $\tau = t/\varepsilon^2$. Specifically, $(\mathbf{v} \cdot \nabla)\mathbf{v}$ is a finite sum of products of terms of these two types. Functions reliant on τ vanish exponentially as τ tends to infinity, while the products where both factors depend on t have been already compensated in the previous section by constructing the corresponding boundary layer correctors in the space variables (the BLO_l terms).

Consequently, the current focus centers on compensating the sums of products featuring at least one factor depending on τ (and, thus, exponentially vanishing as τ tends to infinity). This property allows to expand all t-dependent factors through Taylor series expansions with respect to the variable $\varepsilon^2\tau$. The growth of powers of τ is compensated here by the exponentially vanishing factors. As a result, we get the boundary layer-in-time-and-in-space problems with the right-hand sides depending only on y and τ.

6.7.1 *Boundary Layer with Respect to the Fast Time*

We commence by observing that, due to the condition $\mathbf{g}(x, 0) = 0$, the terms of orders ε^0 and ε^1 within the expansion (6.3.16) become zero. Consequently, for $t = 0$, all flow rates are equal to zero. The boundary layer-in-time has the following structure:

$$\mathbf{u}^{(BLt,J)}(x,t) = \sum_{i=1}^{M} \zeta\left(\frac{x_n^{(e_i)}}{3r\varepsilon}\right)\zeta\left(\frac{|e_i| - x_n^{(e_j)}}{3r\varepsilon}\right)\mathbf{U}^{[BLt,(e_i),J]}\left(\frac{x^{(e_i)\prime}}{\varepsilon}, \frac{t}{\varepsilon^2}\right)$$
$$+\sum_{l=1}^{N}(1-\zeta\left(\frac{|x-O_l|}{|e|_{min}}\right))\mathbf{U}^{[BLt,BLO_l,J]}\left(\frac{x-O_l}{\varepsilon}, \frac{t}{\varepsilon^2}\right). \tag{6.7.1}$$

The terms $\mathbf{U}^{[BLt,(e_i),J]}(y^{(e_i)\prime},\tau)$ vanish exponentially as $\tau = \frac{t}{\varepsilon^2}$ tends to infinity. These terms remain consistent with the previous section. The new boundary layer terms $\mathbf{U}^{[BLt,BLO_l,J]}(y,\tau)$ also vanish exponentially as $|y|$ and τ tend to infinity (see Theorems 4.5.3 and 4.5.5). The following representations are valid:

$$\mathbf{U}^{[BLt,(e_i),J]}(y^{(e)\prime},\tau) = \sum_{j=-1}^{J} \varepsilon^j \mathbf{U}_j^{[BLt,(e_i)]}(y^{(e)\prime},\tau),$$
$$\mathbf{U}_j^{[BLt,(e_i)]}(y^{(e)\prime},\tau) = \mathcal{P}^{(e)*}\big(0,\ldots,0,U_j^{[BLt,(e_i)]}(y^{(e)\prime},\tau)\big); \tag{6.7.2}$$

$$\mathbf{U}^{[BLt,BLO_l,J]}(y,\tau) = \sum_{j=-1}^{J} \varepsilon^j \mathbf{U}_j^{[BLt,BLO_l]}(y,\tau); \tag{6.7.3}$$

$$p^{(BLt,J)}(x,t) = \sum_{i=1}^{M} \zeta\left(\frac{x_n^{(e_i)}}{3r\varepsilon}\right)\zeta\left(\frac{|e_i| - x_n^{(e_j)}}{3r\varepsilon}\right) p^{[BLt,(e_i),J]}(x_n^{(e_i)}, \frac{t}{\varepsilon^2})$$
$$+\sum_{l=0}^{N}(1-\zeta\left(\frac{|x-O_l|}{|e|_{min}}\right))P^{[BLt,BLO_l,J]}(\frac{x-O_l}{\varepsilon}, \frac{t}{\varepsilon^2}), \tag{6.7.4}$$

$$p^{[BLt,(e_i),J]} = \sum_{j=0}^{J} \varepsilon^{j-2} p_j^{[BLt,(e_i)]},$$
$$p_j^{[BLt,(e_i)]}(x_n^{(e_i)},\tau) = -s_j^{[BLt,(e_i)]}(\tau)x_n^{(e_i)} + a_j^{[BLt,(e_i)]}(\tau), \tag{6.7.5}$$

$$P^{[BLt,BLO_l,J]}(y,\tau) = \frac{1}{\varepsilon}\sum_{j=-1}^{J} \varepsilon^j P_j^{[BLt,BLO_l]}(y,\tau). \tag{6.7.6}$$

Expanding approximation (6.3.45) in Taylor series with respect to the time variable t and replacing t in this expansion with the product $\varepsilon^2\tau$, retaining terms of order ε^J for the velocity and applying the Taylor–Lagrange formula:

$$G(t) = G(0) + \varepsilon^2\tau G'(0) + \ldots + \varepsilon^{2\alpha}\tau^\alpha G^{(\alpha)}(0) + \varepsilon^{2\alpha+2}\tau^{\alpha+1}G^{(\alpha+1)}(\varepsilon^2\theta),\ \ \theta \in (0,\tau),$$

we deduce a new representation for $\mathbf{u}^{(J)}$:

$$\begin{aligned}\mathbf{u}^{(J)}(x,\varepsilon^2\tau) = \sum_{i=1}^{M} \zeta(\frac{x_n^{(e_i)}}{3r\varepsilon})\zeta(\frac{|e_i| - x_n^{(e_j)}}{3r\varepsilon})\Big\{\sum_{j=0}^{J}\varepsilon^j\Big(\mathbf{U}_j^{(e_i)}(y^{(e_i)\prime},0) \\ + \sum_{2\alpha+q=j,\alpha\neq 0} \tau^\alpha \frac{\partial^\alpha \mathbf{U}_q^{(e_i)}}{\partial t^\alpha}(y^{(e_i)\prime},0)\Big) \\ +\varepsilon^{J+1} \sum_{2\alpha+q=J+1,\alpha\neq 0} \tau^\alpha \frac{\partial^\alpha \mathbf{U}_q^{(e_i)}}{\partial t^\alpha}(y^{(e_i)\prime},\varepsilon^2\theta_{1,q})\Big\} \\ +\sum_{l=1}^{N}(1-\zeta(\frac{|x-O_l|}{|e|_{min}}))\Big\{\sum_{j=0}^{J}\varepsilon^j\Big(\mathbf{U}_j^{[BLO_l]}(y,0) \\ + \sum_{2\alpha+q=j,\alpha\neq 0} \tau^\alpha \frac{\partial^\alpha \mathbf{U}_q^{[BLO_l]}}{\partial t^\alpha}(y,0)\Big) \\ +\varepsilon^{J+1} \sum_{2\alpha+q=J+1,\alpha\neq 0} \tau^\alpha \frac{\partial^\alpha \mathbf{U}_q^{[BLO_l]}}{\partial t^\alpha}(y,\varepsilon^2\theta_{2,q})\Big\}.\end{aligned} \tag{6.7.7}$$

Here, $\theta_{1,q}$ and $\theta_{2,q} \in (0,\tau)$.

Substituting expansions (6.3.45) and (6.7.1)–(6.7.7) into the Navier–Stokes problem (6.1.1) and grouping terms with the same power of ε, we derive the algorithm for the successive determination for the terms of expansions (6.7.1)–(6.7.7). To determine the boundary layer-in-time terms $\mathbf{U}_j^{[BLt,(e_i)]}(y^{(e_i)\prime},\tau)$, we must solve a special problem for the macroscopic pressure at each step j. The notation below is introduced for this purpose. For each edge e, the following problem is considered: given $\mathcal{H} \in W^{1,2}(0,+\infty)$, find $\mathcal{S} \in L^2(0,+\infty)$ and $\mathcal{V} \in L^2(0,+\infty;\mathring{W}^{1,2}(\sigma^{(e)}))$ such that:

$$\begin{cases} \dfrac{\partial \mathcal{V}}{\partial \tau}(y^{(e)\prime},\tau) - \nu\Delta'_{y^{(e)\prime}}\mathcal{V}(y^{(e)\prime},\tau) = \mathcal{S}(\tau), & y^{(e)\prime} \in \sigma^{(e)}, \tau > 0, \\ \mathcal{V}(y^{(e)\prime},\tau)|_{\partial\sigma^{(e)}} = 0, & y^{(e)\prime} \in \sigma^{(e)}, \tau > 0, \\ \mathcal{V}(y^{(e)\prime},0) = 0, & y^{(e)\prime} \in \sigma^{(e)} \\ \displaystyle\int_{\sigma^{(e)}} \mathcal{V}(y^{(e)\prime},\tau)dy^{(e)\prime} = \mathcal{H}(\tau), \ \tau \geq 0. \end{cases} \tag{6.7.8}$$

The relationship $\mathcal{S}(\tau) = (L^{(e)})^{-1}\mathcal{H}(\tau)$ is demonstrated in Sect. 3.2 of Chap. 3, where it is shown that $(L^{(e)})^{-1}$ is a bounded linear operator from $W^{1,2}(0,+\infty)$ to $L^2(0,+\infty)$. In a similar way, for any $\mathcal{S} \in L^2(0,+\infty)$, the initial boundary value problem for the heat equation can be solved:

$$\begin{cases} \dfrac{\partial \mathcal{V}}{\partial \tau}(y^{(e)\prime}, \tau) - \nu \Delta'_{y^{(e)\prime}} \mathcal{V}(y^{(e)\prime}, \tau) = \mathcal{S}(\tau), \ y^{(e)\prime} \in \sigma^{(e)}, \tau > 0, \\ \mathcal{V}(y^{(e)\prime}, \tau)|_{\partial\sigma^{(e)}} = 0, \qquad y^{(e)\prime} \in \sigma^{(e)}, \tau > 0, \\ \mathcal{V}(y^{(e)\prime}, 0) = 0, \qquad y^{(e)\prime} \in \sigma^{(e)}, \end{cases} \tag{6.7.9}$$

and set:

$$L^{(e)}\mathcal{S}(\tau) = \int_{\sigma^{(e)}} \mathcal{V}(y^{(e)\prime}, \tau) dy^{(e)\prime}. \tag{6.7.10}$$

Evidently, $L^{(e)}$ is a bounded linear operator acting from $L^2(0, +\infty)$ to $W^{1,2}(0, +\infty)$. Moreover, the operators $L^{(e)}$ and $(L^{(e)})^{-1}$ are inverses of each other. An important inequality to note is:

$$c_1 \|\mathcal{Q}\|_{L^2(0,\hat{T})} \le \|L^{(e)}\mathcal{Q}\|_{W^{1,2}(0,\hat{T})} \le c_2 \|\mathcal{Q}\|_{L^2(0,\hat{T})} \quad \forall \mathcal{Q} \in L^2(0, \hat{T}), \tag{6.7.11}$$

where c_1 and c_2 are constants independent of $\hat{T}$.

The procedure for successive determination of the terms in the asymptotic expansion (6.7.1)–(6.7.7) closely resembles that of previous sections.

The Base Case We commence by addressing the conductivity problem on the graph for the function $p_0^{[BLt]}$:

$$\begin{cases} -\dfrac{\partial}{\partial x_n^{(e)}}\Big(L^{(e)} \dfrac{\partial p_0^{[BLt,(e)]}}{\partial x_n^{(e)}}(x_n^{(e)}, \tau)\Big) = 0, & x_n^{(e)} \in (0, |e|), \\ -\displaystyle\sum_{e: O_l \in e} \Big(L^{(e)} \dfrac{\partial p_0^{[BLt,(e)]}}{\partial x_n^{(e)}}\Big)(0, \tau) = 0, & l = 1, \dots, N_1, \\ -\Big(L^{(e)} \dfrac{\partial p_0^{[BLt,(e)]}}{\partial x_n^{(e)}}\Big)(0, \tau) = 0, & l = N_1 + 1, \dots, N, \\ p_0^{[BLt,(e)]}(0, \tau) = p_0^{[BLt,(e_s)]}(0, \tau), \ \forall e \in \mathcal{B}_l, & \end{cases}$$

where the local coordinates $x^{(e)}$ have been redefined to ensure a common origin O_l.

It is evident that this problem possesses a solution depending only on time. By normalizing it by the condition $p_j^{[BLt]}(O_N, \tau) = 0$, we deduce that $p_0^{[BLt]} = 0$. For each edge e, let us define functions $s_0^{[BLt,(e)]}(\tau) = 0$ and $a_0^{[BLt,(e)]}(\tau) = 0$, resulting in:

$$p_0^{[BLt,(e)]}(x^{(e)}, \tau) = -s_0^{[BLt,(e)]}(\tau) x_n^{(e)} + a_0^{[BLt,(e)]}(\tau) = 0$$

Furthermore, given that the terms of order ε^0 and ε^1 in the expansion (6.3.16) vanish, we can conclude:

$$P_{-1}^{[BLt,BLO_l]}(y,\tau)=0,\quad \mathbf{U}_{-1}^{[BLt,BLO_l]}(y,\tau)=0,\quad \mathbf{U}_0^{[BLt,(e)]}(y^{(e)\prime},\tau)=0,$$

$$s_1^{[BLt,(e)]}(\tau)=0,\quad a_1^{[BLt,(e)]}(\tau)=0.$$

The rank 0 boundary layer functions within the vertex-associated unbounded domains Ω_l are identically zero, being the solution of the following problem:

$$\begin{cases} -\nu\Delta_y\mathbf{U}_0^{[BLt,BLO_l]}+\nabla_y P_0^{[BLt,BLO_l]}=0,\ y\in\Omega_l,\\ \operatorname{div}_y\mathbf{U}_0^{[BLt,BLO_l]}=0,\ y\in\Omega_l,\\ \mathbf{U}_0^{[BLt,BLO_l]}(y)=0,\ y\in\partial\Omega_l\backslash\gamma^l,\\ \mathbf{U}_0^{[BLt,BLO_l]}(y)=0\ \ y\in\gamma^l. \end{cases} \tag{6.7.12}$$

Now, let us assume that all terms in the expansion (6.7.1)–(6.7.7) up to rank $j-1$ are known, along with the macroscopic pressure on the graph $p_j^{[BLt]}$. We describe the transition from rank $j-1$ to rank j.

Step 1 By inductive assumption, we know the macroscopic pressure on the graph $p_j^{[BLt]}$. For each edge e, let us introduce functions $s_j^{[BLt,(e)]}(\tau)$ and $a_j^{[BLt,(e)]}(\tau)$, which satisfy:

$$p_j^{[BLt,(e)]}(x^{(e)},\tau)=-s_j^{[BLt,(e)]}(\tau)x_n^{(e)}+a_j^{[BLt,(e)]}(\tau),$$

and define:

$$\mathbf{U}_j^{[BLt,(e)]}=\left(P^{(e)}\right)^*(0,\dots,0,\widetilde{U}_j^{[BLt,(e)]})^*,$$

where $\widetilde{U}_j^{BLt,(e)}$ is the solution to the heat equation:

$$\begin{cases} \dfrac{\partial\widetilde{U}_j^{[BLt,(e)]}}{\partial\tau}-\nu\Delta'_{y^{(e)\prime}}\widetilde{U}_j^{[BLt,(e)]}=s_j^{[BLt,(e)]}(\tau), & y^{(e)\prime}\in\sigma^{(e)},\tau>0,\\ \widetilde{U}_j^{[BLt,(e)]}=0, & y^{(e)\prime}\in\sigma^{(e)},\tau>0,\\ \widetilde{U}_j^{[BLt,(e)]}(y^{(e)\prime},0)=-U_j^{(e)}(y^{(e)\prime},0), & y^{(e)\prime}\in\sigma^{(e)}. \end{cases} \tag{6.7.13}$$

Step 2 Now, we find the boundary layer-in-time solution, which consists of a vector-valued function $\mathbf{U}_j^{[BLt,BLO_l]}$ and a scalar function $P_j^{[BLt,BLO_l]}$ satisfying the following problem:

$$\begin{cases} \dfrac{\partial \mathbf{U}_j^{[BLt,BLO_l]}}{\partial \tau} - \nu \Delta_y \mathbf{U}_j^{[BLt,BLO_l]} + \nabla_y P_j^{[BLt,BLO_l]} = \mathbf{f}_j^{[BLt,REGO_l]} + \mathbf{f}_j^{[BLt,BLO_l]}, \\ \mathrm{div}_y \mathbf{U}_j^{[BLt,BLO_l]} = -\mathrm{div}_y \Big\{ \sum_{e:O_l \in e} \mathbf{U}_j^{[BLt,(e)]} \zeta(\frac{y_n^{(e)}}{3r}) \Big\}, \\ \mathbf{U}_j^{[BLt,BLO_l]}\big|_{\partial\Omega_l} = 0, \\ \mathbf{U}_j^{[BLt,BLO_l]}(y,0) = -\mathbf{U}_j^{[BLO_l]}(y,0), \end{cases} \tag{6.7.14}$$

where:

$$\begin{aligned} \mathbf{f}_j^{[BLt,REGO_l]}(y,\tau) = \sum_{e:O_l\in e} \Big\{ \nu \Delta_y \big(\zeta(\frac{y_n^{(e)}}{3r}) \mathbf{U}_j^{[BLt,(e)]}(y^{(e)\prime},\tau) \big) \\ + \nabla_y \big(\zeta(\frac{y_n^{(e)}}{3r}) s_j^{[BLt,(e)]}(\tau) \big) y_n^{(e)} - a_{j+1}^{[BLt,(e)]}(\tau) \nabla_y \big(\zeta(\frac{y_n^{(e)}}{3r}) \big) \Big\} \\ - \sum_{e:O_l\in e} \sum_{\substack{\gamma,\delta\in\{1,2\},\\ (\gamma,\delta)\neq(1,1)}} \sum_{\alpha+\beta=j-1} (\mathbf{U}_\alpha^{(e),\gamma}(y,\tau)\cdot\nabla) \mathbf{U}_\beta^{(e),\delta}(y,\tau), \end{aligned} \tag{6.7.15}$$

$$\begin{aligned} \mathbf{U}_m^{(e),1}(y,\tau) = \Big\{ \mathbf{U}_m^{(e)}(y^{(e)\prime},0) + \sum_{2\alpha+q=m,\alpha\neq 0} \tau^\alpha \frac{\partial^\alpha \mathbf{U}_q^{(e)}}{\partial t^\alpha}(y^{(e)\prime},0) \Big\} \zeta(\frac{y_n^{(e)}}{3r}), \\ \mathbf{U}_m^{(e),2}(y,\tau) = \mathbf{U}_m^{[BLt,(e)]}(y^{(e)\prime},\tau) \zeta(\frac{y_n^{(e)}}{3r}), \end{aligned} \tag{6.7.16}$$

and:

$$\begin{aligned} \mathbf{f}_j^{[BLt,BLO_l]}(y,\tau) = - \sum_{e:O_l\in e} \sum_{\substack{\gamma\in\{1,2\},\delta\in\{3,4\},\\ (\gamma,\delta)\neq(1,3)}} \sum_{\alpha+\beta=j-1} (\mathbf{U}_\alpha^{(e),\gamma}(y,\tau)\cdot\nabla) \mathbf{U}_\beta^{(O_l),\delta}(y,\tau) \\ - \sum_{e:O_l\in e} \sum_{\substack{\gamma\in\{3,4\},\delta\in\{1,2\},\\ (\gamma,\delta)\neq(3,1)}} \sum_{\alpha+\beta=j-1} (\mathbf{U}_\alpha^{(O_l),\gamma}(y,\tau)\cdot\nabla) \mathbf{U}_\beta^{(e),\delta}(y,\tau) \\ - \sum_{\substack{\gamma,\delta\in\{3,4\},\\ (\gamma,\delta)\neq(3,3)}} \sum_{\alpha+\beta=j-1} (\mathbf{U}_\alpha^{(O_l),\gamma}(y,\tau)\cdot\nabla) \mathbf{U}_\beta^{(O_l),\delta}(y,\tau), \end{aligned} \tag{6.7.17}$$

$$\begin{aligned} \mathbf{U}_m^{(O_l),3}(y,\tau) = \Big\{ \mathbf{V}_m^{[BLO_l]}(y,0) + \sum_{2\alpha+q=m,\alpha\neq 0} \tau^\alpha \frac{\partial^\alpha \mathbf{U}_q^{[BLO_l]}}{\partial t^\alpha}(y,0) \Big\}, \\ \mathbf{U}_m^{(O_l),4}(y,\tau) = \mathbf{U}_m^{[BLt,BLO_l]}(y,\tau). \end{aligned} \tag{6.7.18}$$

The terms of the form $(\mathbf{U}_\alpha^{(.),\gamma} \cdot \nabla)\mathbf{U}_\beta^{(.),\delta}$ stem from substituting the velocity expansion into the inertial term in the Navier–Stokes equations. In this context, the superscripts 1, 2, 3, or 4 delineate distinct categories of terms in the expansion: superscript 1 pertains to the Taylor expansion with respect to the time for the regular Poiseuille-type function $\mathbf{U}_j^{(e)}$ on e, multiplied by the cutoff function ζ; superscript 2 corresponds to the boundary layer-in-time $\mathbf{U}_j^{[BLt,(e)]}$ (fast-time Poiseuille-type flow; see Sect. 3.5 of Chap. 3), also multiplied by the cutoff function ζ; superscript 3 signifies the Taylor expansion with respect to time for the boundary layer-in-space corrector $\mathbf{U}_j^{[BLO_l]}$, constructed in the prior section; and superscript 4 designates the boundary layer-in-time-and-in-space correctors, solutions to problem (6.7.14) with indices lower than j.

The terms $(\mathbf{U}_\alpha^{(e),\gamma} \cdot \nabla)\mathbf{U}_\beta^{(e),\delta}$ with $\gamma = \delta = 1$ have been taken into consideration during the construction of the spatial boundary layer correctors in the preceding section; hence, they do not appear in the last sum of (6.7.15). Similar reasoning explains the absence of products with indices $(\gamma, \delta) = (1, 3), (3, 1), (3, 3)$ in (6.7.17).

The part (6.7.17) on the right-hand side encompasses only products that possess at least one factor causing exponential decay as $|y| \to \infty$, while the part (6.7.15) encompasses only Poiseuille-type factors multiplied by the cutoff function.

It is noteworthy that the inertial terms vanish for the Poiseuille-type flows, i.e.:

$$(\mathbf{U}_\alpha^{(e),\gamma} \cdot \nabla)\mathbf{U}_\beta^{(e),\delta} = 0$$

for $\gamma, \delta \in \{1, 2\}$, in the subdomain where the cutoff function is equal to 1 or zero. Thus, the terms $(\mathbf{U}_\alpha^{(e),\gamma} \cdot \nabla)\mathbf{U}_\beta^{(e),\delta}$ have the compact supports.

In the subsequent steps, we initially solve this problem while excluding the term $a_{j+1}^{BLt,(e)}(\tau)\nabla_y\zeta(\frac{y_n^{(e)}}{3r})$ from the right-hand side. Through this process, we find the pressure $\widehat{P}_j^{[BLt,BLO_l]}(y, \tau)$, which approaches a constant value $\widehat{a}_j^{[BLt,(e)]}(\tau)$ as $|y| \to +\infty$ in the outlet associated with edge e.

Step 3 Consider the solution $\widetilde{U}_{j+1,ht}^{(e)}$ of the heat equation:

$$\begin{cases} \dfrac{\partial \widetilde{U}_{j+1,ht}^{(e)}}{\partial \tau} - \nu\Delta'_{y'}\widetilde{U}_{j+1,ht}^{(e)} = 0, & y^{(e)\prime} \in \sigma^{(e)}, \tau > 0, \\ \widetilde{U}_{j+1,ht}^{(e)}(y^{(e)\prime}, t) = 0, & y^{(e)\prime} \in \partial\sigma^{(e)}, \tau > 0, \\ \widetilde{U}_{j+1,ht}^{(e)}(y^{(e)\prime}, 0) = -U_{j+1}^{(e)}(y^{(e)\prime}, 0), & y^{(e)\prime} \in \sigma^{(e)}. \end{cases} \tag{6.7.19}$$

Calculate the flow rate $\mathcal{H}_{j+1}^{(e)}(\tau)$:

$$\mathcal{H}_{j+1}^{(e)}(\tau) = \int_{\sigma^{(e)}} \widetilde{U}_{j+1,ht}^{(e)}(y^{(e)\prime}, \tau)\, dy^{(e)\prime}.$$

6.7.1.1 Nonstationary Conductivity Problem on the Graph

Now, we consider the conductivity problem on the graph for the function $p_{j+1}^{[BLt,(e)]}$ according to:

$$\begin{cases} -\dfrac{\partial}{\partial x_n^{(e)}}\Big(L^{(e)} \dfrac{\partial p_{j+1}^{[BLt,(e)]}}{\partial x_n^{(e)}}(x_n^{(e)}, \tau)\Big) = 0, \quad x_n^{(e)} \in (0, |e|), \\ -\sum\limits_{e:O_l \in e} \Big(L^{(e)} \dfrac{\partial p_{j+1}^{[BLt,(e)]}}{\partial x_n^{(e)}}\Big)(0, \tau) = -\sum\limits_{e:O_l \in e} \mathcal{H}_{j+1}^{(e)}(\tau), \quad l = 1, \ldots, N_1, \\ -\Big(L^{(e)} \dfrac{\partial p_{j+1}^{[BLt,(e)]}}{\partial x_n^{(e)}}\Big)(0, \tau) = -\mathcal{H}_{j+1}^{(e)}(\tau), \quad l = N_1 + 1, \ldots, N, \\ p_{j+1}^{[BLt,(e)]}(0, \tau) = p_{j+1}^{[BLt,(e_s)]}(0, \tau) + \widehat{a}_j^{[BLt,(e)]}(\tau), \forall e \subset \mathcal{B}_l, \end{cases} \tag{6.7.20}$$

where e_s is the selected edge of the bundle, and the local coordinates $x^{(e)}$ have been redefined so that all share the common origin O_l.

It is important to note that:

$$\sum_{l=1}^{N} \sum_{e:O_l \in e} \mathcal{H}_{j+1}^{(e)}(\tau) = 0, \tag{6.7.21}$$

holds true, with the local coordinates $x^{(e)}$ redefined to share the origin O_l. For each edge e, the flow rate remains constant with respect to the longitudinal variable $x_n^{(e)}$. Changing the orientation of this variable to its opposite results in a change of sign in the flow rate. Hence, the value of $\mathcal{H}_{j+1}^{(e)}(\tau)$ contributes twice to the sum (6.7.21), once positively and once negatively.

We will establish the existence and uniqueness of the solution $p_{j+1}^{[BLt,(e)]}$, which is constrained to vanish at some point O_N. To achieve this, we first transform (6.7.20) to an equivalent problem with a modified right-hand side in the first equation and the elimination of $\widehat{a}_j^{[BLt,(e)]}(\tau)$ in the last condition. Specifically, we introduce a change of the unknown function $\widehat{p}^{(e)}$ defined as:

$$\widehat{p}^{(e)} = p_{j+1}^{[BLt,(e)]}(x_n^{(e)}, \tau) - (1 - \zeta(\frac{2x_n^{(e)}}{|e|}))\hat{a}_j^{[BLt,(e)]}(\tau),$$

and set:

$$F^{(e)}(x_n^{(e)},\tau)=L^{(e)}\hat{a}_j^{[BLt,(e)]}(\tau)\zeta''(\frac{2x_n^{(e)}}{|e|})(\frac{2}{|e|})^2.$$

As a result of these transformations, $\widehat{p}^{(e)}$ now satisfies the following equations:

$$\begin{cases} -\dfrac{\partial}{\partial x_n^{(e)}}\Big(L^{(e)}\dfrac{\partial \widehat{p}^{(e)}}{\partial x_n^{(e)}}(x_n^{(e)},\tau)\Big)=F^{(e)}(x_n^{(e)},\tau), & x_n^{(e)}\in(0,|e|),\\ -\sum\limits_{e:O_l\in e}\Big(L^{(e)}\dfrac{\partial \widehat{p}^{(e)}}{\partial x_n^{(e)}}\Big)(0,\tau)=-\sum\limits_{e:O_l\in e}\mathcal{H}_{j+1}^{(e)}(\tau),\ l=1,\dots,N_1, & \\ -\Big(L^{(e)}\dfrac{\partial \widehat{p}^{(e)}}{\partial x_n^{(e)}}\Big)(0,\tau)=-\mathcal{H}_{j+1}^{(e)}(\tau), & l=N_1+1,\dots,N,\\ \widehat{p}^{(e)}(0,\tau)=\widehat{p}^{(e_s)}(0,\tau), & \forall e\subset\mathcal{B}_l. \end{cases} \tag{6.7.22}$$

Next, we differentiate three first equations of problem (6.7.22) with respect to τ:

$$\begin{cases} -\dfrac{\partial}{\partial x_n^{(e)}}\Big(L^{(e)}\dfrac{\partial \widehat{p}^{(e)}}{\partial x_n^{(e)}}(x_n^{(e)},\tau)\Big)_\tau=(F^{(e)})_\tau(x_n^{(e)},\tau), & x_n^{(e)}\in(0,|e|),\\ -\sum\limits_{e:O_l\in e}\Big(L^{(e)}\dfrac{\partial \widehat{p}^{(e)}}{\partial x_n^{(e)}}\Big)_\tau(0,\tau)=-\sum\limits_{e:O_l\in e}(\mathcal{H}_{j+1}^{(e)})_\tau(\tau),\ l=1,\dots,N_1, & \\ -\Big(L^{(e)}\dfrac{\partial \widehat{p}^{(e)}}{\partial x_n^{(e)}}\Big)_\tau(0,\tau)=-(\mathcal{H}_{j+1}^{(e)})_\tau(\tau), & l=N_1+1,\dots,N,\\ \widehat{p}^{(e)}(0,\tau)=\widehat{p}^{(e_s)}(0,\tau), & \forall e\subset\mathcal{B}_l, \end{cases} \tag{6.7.23}$$

where $g_\tau=\dfrac{\partial g}{\partial\tau}$. Note that problems (6.7.22) and (6.7.23) are equivalent, because fluxes at vertices are equal to zero for $t=0$, also algebraic sum of fluxes at the nodes is equal to zero for $t=0$; $\sum\limits_{e:O_l\in e}\mathcal{H}_{j+1}^{(e)}(0)=0$ and $\sum\limits_{e:O_l\in e}L^{(e)}\dfrac{\partial\widehat{p}^{(e)}}{\partial x_n^{(e)}}(x_n^{(e)},0)=0$ (see (6.7.8) and (6.7.9)).

Let $\widehat{T}>0$. Then, the weak solution $\widehat{p}^{(e)}$ can be determined by solving the following variational problem:

$$a_{\widehat{T}}(\widehat{p},\psi)=b_{\widehat{T}}(\psi),\ \ \forall\psi\in L^2(0,\widehat{T};W^{1,2}(\mathcal{B})), \tag{6.7.24}$$

where:

$$a_{\widehat{T}}(p,\psi)=\sum_{i=1}^{M}\int_0^{\widehat{T}}\int_0^{|e_i|}\Big(L^{(e_i)}\frac{\partial p}{\partial x_n^{(e_i)}}\Big)_\tau\frac{\partial\psi}{\partial x_n^{(e_i)}}dx_n^{(e_i)}d\tau$$

and:

$$b_{\widehat{T}}(\psi)=\sum_{i=1}^{M}\int_0^{\widehat{T}}\int_0^{|e_i|}(F^{(e_i)})_\tau\psi dx_n^{(e_i)}d\tau+\sum_{l=1}^{N}\int_0^{\widehat{T}}\sum_{e:O_l\in e}(\mathcal{H}_{j+1}^{(e)})_\tau(\tau)\psi(O_l,\tau)d\tau.$$

We establish the existence and uniqueness of a weak solution to problem (6.7.23) under the compatibility condition:

$$\sum_{i=1}^{M}\int_0^{|e_i|}(F^{(e_i)})_\tau dx_n^{(e_i)}+\sum_{l=1}^{N}\sum_{e:O_l\in e}(\mathcal{H}_{j+1}^{(e)})_\tau(\tau)=0. \tag{6.7.25}$$

Since the derivative of the cutoff function vanishes in the neighborhood of the ends of each edge, we have:

$$\int_0^{|e_i|}(F^{(e_i)})_\tau dx_n^{(e_i)}=\left(\frac{2}{|e|}\right)^2(L^{(e)}\varphi_l^{(e)})_\tau\int_0^{|e_i|}\zeta''\left(\frac{2x_n^{(e)}}{|e|}\right)dx_n^{(e_i)}=0.$$

Thus, the validity of the compatibility condition is a direct consequence of relations (6.7.21).

Let $L^2(0,\widehat{T};W^{1,2}_{O_N}(\mathcal{B}))$ be the subset of functions $\psi\in L^2(0,\widehat{T};W^{1,2}(\mathcal{B}))$ that satisfy the condition $\psi(O_N,\tau)=0$.

Theorem 6.7.1 *Let $\mathcal{H}_l\in\mathring{W}^{1,2}(0,+\infty)$, $l=1,\dots,N$, and $F^{(e_i)},F_\tau^{(e_i)}\in L^2(0,+\infty;W^{1,2}(\mathcal{B}))$, $F^{(e_i)}|_{\tau=0}=0, i=1,\dots,M$. The problem (6.7.24) possesses a unique solution $\widehat{p}\in L^2(0,\hat{T};W^{1,2}_{O_N}(\mathcal{B}))$, which vanishes at the vertex O_N, if and only if condition (6.7.25) is satisfied for almost all $\tau\in(0,\hat{T})$.*

Proof
The proof revolves around the application of the Lax–Milgram representation lemma (see Lemma 2.2.1) along with the utilization of inequality (6.7.11) on each edge e. Consider the subspace $W^{1,2}_{O_N}(\mathcal{B})$ consisting of functions in $W^{1,2}(\mathcal{B})$ that vanish at the vertex O_N. By direct inspection using (6.7.11), it is established that the bilinear form $a_{\hat{T}}$ is continuous with respect to the norm of $L^2(0,\hat{T};W^{1,2}_{O_N}(\mathcal{B}))$ and that $b_{\hat{T}}$ is a bounded linear functional. The coerciveness of $a_{\hat{T}}$ is a consequence of the following estimates, which hold for any edge e and the solution $(\mathcal{V},\mathcal{S})$ to problem (6.7.8):

$$\int_0^{\hat{T}}(L^{(e)}\mathcal{S})_\tau\mathcal{S}d\tau=\int_0^{\hat{T}}\int_{\sigma^{(e)}}(\mathcal{V}(y^{(e)\prime},\tau))_\tau\mathcal{S}(\tau)\,dy^{(e)\prime}d\tau$$

$$= \int_0^{\hat{T}} \int_{\sigma^{(e)}} \left\{ |(\mathcal{V}(y^{(e)\prime}, \tau))_\tau|^2 + \frac{\nu}{2} \frac{\partial}{\partial \tau} \left| \nabla_{y^{(e)\prime}} \mathcal{V}(y^{(e)\prime}, \tau) \right|^2 \right\} dy^{(e)\prime} d\tau$$

$$\geq \int_0^{\hat{T}} \int_{\sigma^{(e)}} |(\mathcal{V}(y^{(e)\prime}, \tau))_\tau|^2 dy^{(e)\prime} d\tau$$

$$\geq \frac{1}{|\sigma^{(e)}|} \int_0^{\hat{T}} \Big(\int_{\sigma^{(e)}} (\mathcal{V}(y^{(e)\prime}, \tau))_\tau dy^{(e)\prime} \Big)^2 d\tau = \frac{1}{|\sigma^{(e)}|} \int_0^{\hat{T}} |(L^{(e)} \mathcal{S})_\tau|^2 d\tau$$

$$\geq \frac{1}{2|\sigma^{(e)}|} \int_0^{\hat{T}} |(L^{(e)} \mathcal{S})_\tau|^2 d\tau + \frac{1}{\hat{T}^2 |\sigma^{(e)}|} \int_0^{\hat{T}} |L^{(e)} \mathcal{S}|^2 d\tau$$

$$\geq c \min\left(\frac{1}{2}, \frac{1}{\hat{T}^2}\right) \frac{1}{|\sigma^{(e)}|} \int_0^{\hat{T}} |\mathcal{S}(\tau)|^2 d\tau.$$

Thus:

$$a_{\hat{T}}(p, p) \geq \min\left(\frac{1}{2}, \frac{1}{\hat{T}^2}\right) \frac{c}{\min\limits_{i=1,\dots,M} |\sigma^{(e_i)}|} (p, p)_{\hat{T}}, \tag{6.7.26}$$

where:

$$(p, \psi)_{\hat{T}} = \sum_{i=1}^{M} \int_0^{\hat{T}} \int_0^{|e_i|} \frac{\partial p}{\partial x_n^{(e_i)}} \frac{\partial \psi}{\partial x_n^{(e_i)}} dx_n^{(e_i)} d\tau.$$

Hence, applying the Lax–Milgram lemma (see Lemma 2.2.1) in the Hilbert space $L^2(0, \hat{T}; W^{1,2}_{O_N}(\mathcal{B}))$, the existence and uniqueness of the solution to problem (6.7.24) projected onto the subspace $L^2(0, \hat{T}; W^{1,2}_{O_N}(\mathcal{B}))$ is demonstrated. Moreover, it can be checked directly that if the condition (6.7.25) holds, then (6.7.24) still holds for any test function from the larger space $L^2(0, \hat{T}; W^{1,2}(\mathcal{B}))$. This is achieved by decomposing $\psi \in L^2(0, \hat{T}; W^{1,2}(\mathcal{B}))$ into a sum $\psi(x_n^{(e_i)}, \tau) = \tilde{\psi}(x_n^{(e_i)}, \tau) + \psi(O_N, \tau)$, where $\tilde{\psi} \in L^2(0, \hat{T}; W^{1,2}_{O_N}(\mathcal{B}))$. Taking into account that $\tilde{\psi}$ satisfies (6.7.24); that the $x_n^{(e_i)}$ derivative of $\psi(O_N, \tau)$ is equal to zero, so that $a_{\hat{T}}(p, \psi(O_N, \tau)) = 0$; and that, due to (6.7.25), $b_{\hat{T}}(\psi(O_N, \tau)) = 0$, we get

(6.7.24) for all $\psi \in L^2(0, \hat{T}; W^{1,2}_{O_N}(\mathcal{B}))$. The necessity of condition (6.7.25) follows from the identity (6.7.24) written for $\psi = \psi(\tau)$, an arbitrary function of $L^2(0, \hat{T})$. □

Remark 6.7.1 If the condition $\widehat{p}(O_N) = 0$ is omitted, a solution to (6.7.22) still exists and remains unique up to an additive function of τ.

Remark 6.7.2 The estimate (6.7.26) leads to the inequality:

$$\sqrt{(p, p)_{\hat{T}}} \leq C_* \hat{T}^2 \quad \forall \hat{T} > 1, \tag{6.7.27}$$

where the constant $C_* \geq 0$ is independent of $\hat{T}$.

Now, let us assume that the data of problem (6.7.22) decays exponentially as $\tau \to \infty$ and show that then the solution of (6.7.24) also exponentially vanishes as $\tau \to \infty$. To clarify the notation, we recall the following definitions:

$\mathbb{L}^2_\beta(0, +\infty)$ is the space of functions from $L^2(0, +\infty)$ such that:

$$\int_0^{+\infty} |f(\tau)|^2 \exp\{2\beta\tau\} d\tau < \infty,$$

$\mathbb{W}^{1,2}_\beta(0, +\infty)$ is the space of functions $f \in W^{1,2}(0, +\infty)$ such that $f, f' \in \mathbb{L}^2_\beta(0, +\infty)$.

Additionally, let $\mathbb{W}^{1,2}_{0,\beta}(0, +\infty)$ denote the subset of $\mathbb{W}^{1,2}_\beta(0, +\infty)$ consisting of functions that vanish at $\tau = 0$.

Theorem 6.7.2 *Let p denote the solution to problem* (6.7.24) *for arbitrary $\hat{T} > 0$ and let $\mathcal{H}_l \in \mathbb{W}^{1,2}_{0,\gamma}(0, +\infty)$ for $l = 1, \ldots, N$ and $F^{(e_i)} \in \mathbb{W}^{1,2}_{0,\gamma}(0, +\infty; L^2(e_i))$ for $i = 1, \ldots, M$. Then $p \in \mathbb{L}^2_{\beta_*}(0, +\infty; W^{1,2}(\mathcal{B}))$ with a certain positive $\beta_* < \gamma$.*

Proof

Let $\beta < \gamma/2$. The proof of this theorem relies on establishing uniform bounds with respect to $\hat{T}$ for the bilinear and linear forms:

$$a_{\hat{T},\beta}(p, \psi) = \sum_{i=1}^{M} \int_0^{\hat{T}} \exp\{2\beta\tau\} \int_0^{|e_i|} \frac{\partial\{(L^{(e_i)}p)_\tau + \beta L^{(e_i)}p\}}{\partial x_n^{(e_i)}} \frac{\partial \psi}{\partial x_n^{(e_i)}} dx_n^{(e_i)} d\tau,$$

and:

$$b_{\hat{T},\beta}(\psi) = \sum_{i=1}^{M} \int_0^{\hat{T}} \exp\{2\beta\tau\} \int_0^{|e_i|} \{(F^{(e_i)})_\tau + \beta F^{(e_i)})\} \psi dx_n^{(e_i)} d\tau$$

$$+\sum_{l=1}^{N}\int_{0}^{\hat{T}}\exp\{2\beta\tau\}\{(\mathcal{H}_l)_\tau(\tau)+\beta\mathcal{H}_l(\tau)\}\psi(O_l,\tau)d\tau.$$

Note that p satisfies the relation:

$$a_{\hat{T},\beta}(p,p)=b_{\hat{T},\beta}(p). \tag{6.7.28}$$

This result is derived by multiplying equations (6.7.22) and (6.7.23) by p, followed by integration over the domain $\mathcal{B}$. Afterward, integration by parts is applied, and the resulting equation is multiplied by the exponential term $\exp\{2\beta\tau\}$. Finally, integration over the interval $(0,\hat{T})$ with respect to τ is performed.

Consider the bilinear form $a_{\hat{T},\beta}(p,p)$, where we change the order of integration. Let us evaluate the terms of the sum over i. The edge e_i we denote further as e. Note that, for any point $x_n^{(e)}$, belonging to the edge e, the equality holds:

$$\mathcal{S}(\tau)=\frac{\partial p}{\partial x_n^{(e)}}(x_n^{(e)},\tau).$$

So, for any $\mathcal{S}\in L^2(0,\hat{T})$, let us consider the integral (a term of $a_{\hat{T},\beta}(p,p)$):

$$I^{(e)}(\mathcal{S})=\int_0^{\hat{T}}\Big((L^{(e)}\mathcal{S})_\tau(\tau)\mathcal{S}(\tau)+\beta(L^{(e)}\mathcal{S})(\tau)\mathcal{S}(\tau)\Big)\exp\{2\beta\tau\}d\tau.$$

Using the definition of $L^{(e)}\mathcal{S}$ as the flux of the solution of the inverse problem to the heat equation with the right-hand side $\mathcal{S}$ (see Eq. (6.7.9) for $(\mathcal{V},\mathcal{S})$ and formula (6.7.10) for the definition of $L^{(e)}\mathcal{S}$), we replace $L^{(e)}\mathcal{S}(\tau)$ by the integral $\int_{\sigma^{(e)}}\mathcal{V}(y^{(e)\prime},\tau)dy^{(e)\prime}$. This furnishes:

$$I^{(e)}(\mathcal{S})=\int_0^{\hat{T}}\int_{\sigma^{(e)}}\Big\{(\mathcal{V}(y^{(e)\prime},\tau))_\tau+\beta\mathcal{V}(y^{(e)\prime},\tau)\Big\}dy^{(e)\prime}\mathcal{S}(\tau)\exp\{2\beta\tau\}d\tau.$$

Next, utilizing the property that $\mathcal{S}(\tau)$ is the right-hand side of the heat equation (see (6.7.9)), we replace $\mathcal{S}(\tau)$ by $\frac{\partial}{\partial\tau}\mathcal{V}(y^{(e)\prime},\tau)-\nu\Delta\mathcal{V}(y^{(e)\prime},\tau)$ and integrate by parts the Laplacian:

$$I^{(e)}(\mathcal{S})=\int_0^{\hat{T}}\int_{\sigma^{(e)}}\Big\{|(\mathcal{V}(y^{(e)\prime},\tau))_\tau|^2+\frac{\nu}{2}\frac{\partial}{\partial\tau}(\nabla_{y^{(e)\prime}}\mathcal{V}(y^{(e)\prime},\tau))^2$$

$$+\beta\nu|\nabla_{y^{(e)\prime}}\mathcal{V}(y^{(e)\prime},\tau)|^2+\frac{\beta}{2}\frac{\partial}{\partial\tau}\left|\mathcal{V}(y^{(e)\prime},\tau)\right|^2\Big\}dy^{(e)\prime}\exp\{2\beta\tau\}d\tau.$$

Integrating by parts with respect to time variable in the second and the fourth terms, we deduce:

$$I^{(e)}(\mathcal{S})=\int_0^{\hat{T}}\int_{\sigma^{(e)}}\Big\{|(\mathcal{V}(y^{(e)\prime},\tau))_\tau|^2-\nu\beta|\nabla_{y^{(e)\prime}}\mathcal{V}(y^{(e)\prime},\tau)|^2$$

$$+\beta\nu|\nabla_{y^{(e)\prime}}\mathcal{V}(y^{(e)\prime},\tau)|^2-\beta^2|\mathcal{V}(y^{(e)\prime},\tau)|^2\Big\}dy^{(e)\prime}\exp\{2\beta\tau\}d\tau$$

$$+\int_{\sigma^{(e)}}\frac{\nu}{2}|\nabla_{y^{(e)\prime}}\mathcal{V}(y^{(e)\prime},\hat{T})|^2dy^{(e)\prime}\exp\{2\beta\hat{T}\}+\int_{\sigma^{(e)}}\frac{\beta}{2}|\mathcal{V}(y^{(e)\prime},\hat{T})|^2dy^{(e)\prime}\exp\{2\beta\hat{T}\}.$$

Simplifying the second and the third terms, we get:

$$I^{(e)}(\mathcal{S})=\int_0^{\hat{T}}\int_{\sigma^{(e)}}\Big\{|(\mathcal{V}(y^{(e)\prime},\tau))_\tau|^2-\beta^2|\mathcal{V}(y^{(e)\prime},\tau)|^2\Big\}dy^{(e)\prime}\exp\{2\beta\tau\}d\tau$$
$$+\int_{\sigma^{(e)}}\frac{\nu}{2}|\nabla_{y^{(e)\prime}}\mathcal{V}(y^{(e)\prime},\hat{T})|^2dy^{(e)\prime}\exp\{2\beta\hat{T}\}+\int_{\sigma^{(e)}}\frac{\beta}{2}|\mathcal{V}(y^{(e)\prime},\hat{T})|^2dy^{(e)\prime}\exp\{2\beta\hat{T}\}.\tag{6.7.29}$$

Applying estimate (2.3.16) to the first term yields:

$$I^{(e)}(\mathcal{S})\geq\beta^2\int_0^{\hat{T}}\int_{\sigma^{(e)}}|\mathcal{V}(y^{(e)\prime},\tau)|^2dy^{(e)\prime}\exp\{2\beta\tau\}d\tau$$

$$-\beta\int_{\sigma^{(e)}}|\mathcal{V}(y^{(e)\prime},\hat{T})|^2dy^{(e)\prime}\exp\{2\beta\hat{T}\}-\beta^2\int_0^{\hat{T}}\int_{\sigma^{(e)}}|\mathcal{V}(y^{(e)\prime},\tau)|^2dy^{(e)\prime}\exp\{2\beta\tau\}d\tau$$

$$+\frac{\nu}{2}\int_{\sigma^{(e)}}|\nabla_{y^{(e)\prime}}\mathcal{V}(y^{(e)\prime},\hat{T})|^2dy^{(e)\prime}\exp\{2\beta\hat{T}\}+\frac{\beta}{2}\int_{\sigma^{(e)}}|\mathcal{V}(y^{(e)\prime},\hat{T})|^2dy^{(e)\prime}\exp\{2\beta\hat{T}\}$$

$$=\frac{\nu}{2}\int_{\sigma^{(e)}}|\nabla_{y^{(e)\prime}}\mathcal{V}(y^{(e)\prime},\hat{T})|^2dy^{(e)\prime}\exp\{2\beta\hat{T}\}-\frac{\beta}{2}\int_{\sigma^{(e)}}|\mathcal{V}(y^{(e)\prime},\hat{T})|^2dy^{(e)\prime}exp\{2\beta\hat{T}\}.$$

Applying now the Poincaré inequality in the domain $\sigma^{(e)}$ with constant C_P, we get for $\beta < \frac{\nu}{2C_P}$:

$$I^{(e)}(\mathcal{S}^{(e)}) \geq \frac{\nu}{4} \int\limits_{\sigma^{(e)}} |\nabla_{y^{(e)\prime}} \mathcal{V}(y^{(e)\prime}, \hat{T})|^2 dy^{(e)\prime} \exp\{2\beta\hat{T}\} \tag{6.7.30}$$

and:

$$I^{(e)}(\mathcal{S}^{(e)}) \geq \frac{\beta}{2} \int\limits_{\sigma^{(e)}} |\mathcal{V}^{(e)}(y^{(e)\prime}, \hat{T})|^2 dy^{(e)\prime} \exp\{2\beta\hat{T}\}. \tag{6.7.31}$$

Using the Cauchy-Schwarz-Bunyakovsky inequality, from here, we deduce:

$$\begin{aligned} I^{(e)}(\mathcal{S}^{(e)}) &\geq \frac{\beta}{2} \frac{1}{|\sigma^{(e)}|} \Big(\int\limits_{\sigma^{(e)}} \mathcal{V}^{(e)}(y^{(e)\prime}, \hat{T}) dy^{(e)\prime} \Big)^2 \exp\{2\beta\hat{T}\} \\ &= \frac{\beta}{2} \frac{1}{|\sigma^{(e)}|} |L^{(e)} \mathcal{S}^{(e)}(\hat{T})|^2 \exp\{2\beta\hat{T}\}. \end{aligned}$$

Here we used the definition of the operator $L^{(e)}$. Taking in the last inequality $S^{(e)} = \frac{\partial p}{\partial x_n^{(e_i)}}$, $i = 1, \ldots M$ (here we consider p as the function of t depending on $x_n^{(e_i)}$ as a parameter), we obtain:

$$I^{(e)}\Big(\frac{\partial p}{\partial x_n^{(e_i)}}\Big) \geq \frac{\beta}{2} \frac{1}{|\sigma^{(e)}|} |L^{(e)}\Big(\frac{\partial p}{\partial x_n^{(e_i)}}\Big)(\hat{T})|^2 \exp\{2\beta\hat{T}\}. \tag{6.7.32}$$

So, for all $\hat{T} > 1$:

$$\begin{aligned} a_{\hat{T},\beta}(p, p) &= \sum_{i=1}^{M} \int\limits_0^{|e_i|} I^{(e_i)}\Big(\frac{\partial p}{\partial x_n^{(e_i)}}\Big) dx_n^{(e_i)} \\ &\geq \frac{\beta}{2} \frac{1}{\max_e |\sigma^{(e)}|} \exp\{2\beta\hat{T}\} \sum_{i=1}^{M} \int\limits_0^{|e_i|} \Big|\frac{\partial (L^{(e_i)} p)}{\partial x_n^{(e_i)}}\Big|^2 dx_n^{(e_i)} \end{aligned} \tag{6.7.33}$$

(the operator $L^{(e_i)}$ commutes with the derivative $\frac{\partial}{\partial x_n^{(e_i)}}$).

By the Cauchy-Schwarz-Bunyakovsky inequality, for any positive $\delta < (\gamma - 2\beta)/2$:

$$b_{\hat{T},\beta}(p) \le \sqrt{\int_0^{\hat{T}} \exp\{4(\delta+\beta)\tau\} \sum_{i=1}^{M} \int_0^{|e_i|} \left((F^{(e_i)})_\tau(\tau) + \beta F^{(e_i)}(\tau)\right)^2 dx_n^{(e_i)} d\tau} \times$$

$$\times \sqrt{\int_0^{\hat{T}} \exp\{-4\delta\tau\} \sum_{i=1}^{M} \int_0^{|e_i|} \left|\frac{\partial p}{\partial x_n^{(e_i)}}\right|^2 dx_n^{(e_i)} d\tau} \; +$$

$$+ \sqrt{\int_0^{\hat{T}} \exp\{4(\delta+\beta)\tau\} \sum_{l=1}^{N} \left((\mathcal{H}_l(O_l,\tau))_\tau + \beta\mathcal{H}_l(O_l,\tau)\right)^2 d\tau} \; \times$$

$$\times \sqrt{\int_0^{\hat{T}} \exp\{-4\delta\tau\} \sum_{l=1}^{N} |p(O_l,\tau)|^2 d\tau}.$$

Hence, due to (6.7.27) and Lemma 2.3.6, $b_{\hat{T},\beta}(p)$ is bounded uniformly with respect to $\hat{T}$, and so, (6.7.28) and (6.7.33) yield, that for any edge $e_i, i = 1,\ldots,M,$:

$$\exp\{2\beta\hat{T}\} \sum_{i=1}^{M} \int_0^{|e_i|} \left|\frac{\partial(L^{(e_i)}p^{(e_i)})}{\partial x_n^{(e_i)}}\right|^2 dx_n^{(e_i)} \le a_{\hat{T},\beta}(p,p) = b_{\hat{T},\beta}(p)$$

is bounded uniformly. This implies that for any edge $e_i, i = 1,\ldots,M$, the inclusion:

$$\frac{\partial(L^{(e_i)}p^{(e_i)})}{\partial x_n^{(e_i)}} \in \mathbb{L}^2_{\beta_1}(0,+\infty; L^2(e_i)), \tag{6.7.34}$$

holds for every $\beta_1 < \beta$. This is the first exponential estimate for $L^{(e_i)}p^{(e_i)}$.

Now, let us prove the exponential estimate for the time derivative $(L^{(e_i)}p^{(e_i)})_\tau$. Note that $a_{\hat{T},\beta}(p,p) = \sum_{i=1}^{M} \int_0^{|e_i|} I^{(e_i)}(S^{(e_i)})dx_n^{(e)}$, where $S^{(e_i)} = \dfrac{\partial p}{\partial x_n^{(e_i)}}$ and all terms $I^{(e_i)}(S^{(e_i)})$ are non-negative. So, for any edge e, $a^{(\beta)}_{\hat{T},\beta}(p,p) \ge 0$.

Earlier, we have proved inequalities (6.7.30) and (6.7.31), which yield:

$$\int_{\sigma^{(e)}} |\mathcal{V}^{(e)}(y^{(e)\prime}, \hat{T})|^2 dy^{(e)\prime} + \int_{\sigma^{(e)}} |\nabla_{y^{(e)\prime}} \mathcal{V}(y^{(e)\prime}, \hat{T})|^2 dy^{(e)\prime} \in \mathbb{L}^2_{\beta_1}(0, +\infty; L^2(e_i)). \tag{6.7.35}$$

On the other hand, replacing β by β_1 ($\beta_1 < \beta$) in the integral $I^{(e)}(\mathcal{S})$ (see (6.7.29)), we see that:

$$I^{(e)}(\mathcal{S}) = \int_0^{\hat{T}} \int_{\sigma^{(e)}} \left\{ |(\mathcal{V}(y^{(e)\prime}, \tau))_\tau|^2 - \beta_1{}^2 |\mathcal{V}(y^{(e)\prime}, \tau)|^2 \right\} dy^{(e)\prime} \exp\{2\beta_1\tau\} d\tau$$

$$+ \int_{\sigma^{(e)}} \frac{\nu}{2} |\nabla_{y^{(e)\prime}} \mathcal{V}(y^{(e)\prime}, \hat{T})|^2 dy^{(e)\prime} \exp\{2\beta_1\hat{T}\} + \int_{\sigma^{(e)}} \frac{\beta_1}{2} |\mathcal{V}(y^{(e)\prime}, \hat{T})|^2 dy^{(e)\prime} \exp\{2\beta_1\hat{T}\}$$

is bounded by a constant independent of $\hat{T}$ (because $a_{\hat{T},\beta_1}(p, p)$ is its upper bound and it is bounded). Taking into consideration that the second, the third, and the fourth terms are bounded (see (6.7.35)), we get the boundedness of the integral:

$$\int_0^{\hat{T}} \int_{\sigma^{(e)}} |(\mathcal{V}(y^{(e)\prime}, \tau))_\tau|^2 dy^{(e)\prime} \exp\{2\beta_1\tau\} d\tau.$$

By Cauchy-Schwarz-Bunyakovsky inequality, this integral is larger than:

$$|\sigma^{(e)}|^{-1} \int_0^{\hat{T}} (L^{(e)} s^{(e)})_\tau^2 \exp\{2\beta_1\tau\} d\tau.$$

This yields:

$$\frac{\partial (L^{(e_i)} p^{(e_i)})_\tau}{\partial x_n^{(e_i)}} \in \mathbb{L}^2_{\beta_1}(0, +\infty; L^2(e_i)). \tag{6.7.36}$$

Consequently, by virtue of Theorem 3.2.7, relations (6.7.36) and (6.7.34) imply that there exists a positive $\beta_* \le \beta_1$ such that:

$$\frac{\partial p^{(e_i)}}{\partial x_n^{(e_i)}} \in \mathbb{L}^2_{\beta_*}(0, +\infty; L^2(e_i)).$$

Finally, by (2.3.17), $p \in \mathbb{L}^2_{\beta_*}(0, +\infty; L^2(e_i))$, and so, $p \in \mathbb{L}^2_{\beta_*}(0, +\infty; W^{1,2}(\mathcal{B}))$.

□

Corollary 6.7.1 *Let $\mathcal{H}_l$ and $(\mathcal{H}_l)_\tau$ belong to $\mathbb{L}^2_\gamma(0,+\infty)$, $\mathcal{H}_l(0)=0$, and let $F^{(e_i)}$ and their time derivatives belong to $\mathbb{L}^2_\gamma(0,+\infty;W^{1,2}(\mathcal{B}))$. Then there exists $\beta_*<\gamma$, $\beta_*>0$, such that the following estimate holds:*

$$\int\limits_0^{+\infty}\exp\{2\beta_*\tau\}\sum_{i=1}^M\int\limits_0^{|e_i|}\Big|\frac{\partial p}{\partial x_n^{(e_i)}}\Big|^2dx_n^{(e_i)}d\tau$$

$$\le c\Big\{\int\limits_0^{+\infty}\exp\{2\gamma\tau\}\sum_{i=1}^M\int\limits_0^{|e_i|}|(F^{(e_i)})_\tau(\tau)|^2+|F^{(e_i)}(\tau)|^2dx_n^{(e_i)}d\tau$$

$$+\int\limits_0^{+\infty}\exp\{2\gamma\tau\}\sum_{l=1}^N|(\mathcal{H}_l)_\tau(O_l,\tau))|^2+|\mathcal{H}_l(O_l,\tau)|^2d\tau\Big\}.$$

Proof
Due to estimate (6.7.33), we get the inequality:

$$\exp\{2\beta\hat{T}\}\sum_{i=1}^M\int\limits_0^{|e_i|}\Big|\frac{\partial(L^{(e_i)}p^{(e_i)})}{\partial x_n^{(e_i)}}\Big|^2dx_n^{(e_i)}\le a_{\hat{T},\beta}(p,p)=b_{\hat{T},\beta}(p)$$

$$\le\sqrt{\int\limits_0^{\hat{T}}\exp\{4(\delta+\beta)\tau\}\sum_{i=1}^M\int\limits_0^{|e_i|}\big((F^{(e_i)})_\tau(\tau)+\beta F^{(e_i)}(\tau)\big)^2dx_n^{(e_i)}d\tau}$$

$$\times\sqrt{\int\limits_0^{\hat{T}}\exp\{-4\delta\tau\}\sum_{i=1}^M\int\limits_0^{|e_i|}\Big|\frac{\partial p}{\partial x_n^{(e_i)}}\Big|^2dx_n^{(e_i)}d\tau}\;+$$

$$+\sqrt{\int\limits_0^{\hat{T}}\exp\{4(\delta+\beta)\tau\}\sum_{l=1}^N\big((\mathcal{H}_l(O_l,\tau))_\tau+\beta\mathcal{H}_l(O_l,\tau)\big)^2d\tau}$$

$$\times\sqrt{\int\limits_0^{\hat{T}}\exp\{-4\delta\tau\}\sum_{l=1}^N|p(O_l,\tau)|^2d\tau}$$

$$\leq c\Bigg(\sqrt{\int_0^\infty \exp\{4(\delta+\beta)\tau\}\sum_{i=1}^{M}\int_0^{|e_i|}\left((F^{(e_i)})_\tau(\tau)+\beta F^{(e_i)}(\tau)\right)^2 dx_n^{(e_i)}\,d\tau}$$

$$+\sqrt{\int_0^\infty \exp\{4(\delta+\beta)\tau\}\sum_{l=1}^{N}\left((\mathcal{H}_l(O_l,\tau))_\tau+\beta\mathcal{H}_l(O_l,\tau)\right)^2 d\tau}\Bigg)$$

$$\times\sqrt{\int_0^\infty \exp\{-4\delta\tau\}\sum_{i=1}^{M}\int_0^{|e_i|}\left|\frac{\partial p}{\partial x_n^{(e_i)}}\right|^2 dx_n^{(e_i)}\,d\tau}.$$

So:

$$\int_0^\infty \exp\{\beta t\}\sum_{i=1}^{M}\int_0^{|e_i|}\left|\frac{\partial(L^{(e_i)}p^{(e_i)})}{\partial x_n^{(e_i)}}\right|^2 dx_n^{(e_i)}\,dt$$

$$\leq (c/\beta)\Bigg(\sqrt{\int_0^\infty \exp\{4(\delta+\beta)\tau\}\sum_{i=1}^{M}\int_0^{|e_i|}\left((F^{(e_i)})_\tau(\tau)+\beta F^{(e_i)}(\tau)\right)^2 dx_n^{(e_i)}\,d\tau}$$

$$+\sqrt{\int_0^\infty \exp\{4(\delta+\beta)\tau\}\sum_{l=1}^{N}\left((\mathcal{H}_l(O_l,\tau))_\tau+\beta\mathcal{H}_l(O_l,\tau)\right)^2 d\tau}\Bigg)$$

$$\times\sqrt{\int_0^\infty \exp\{-4\delta\tau\}\sum_{i=1}^{M}\int_0^{|e_i|}\left|\frac{\partial p}{\partial x_n^{(e_i)}}\right|^2 dx_n^{(e_i)}\,d\tau}.$$

The last estimate used estimate (5.1.15). On the other hand, according to Theorem 3.2.7, there exists $\beta_* > 0$, such that:

$$\int_0^\infty \exp\{\beta_* t\}\sum_{i=1}^{M}\int_0^{|e_i|}\left|\frac{\partial p^{(e_i)}}{\partial x_n^{(e_i)}}\right|^2 dx_n^{(e_i)}\,dt$$

$$\leq \int_0^\infty \exp\{\beta t\}\sum_{i=1}^{M}\int_0^{|e_i|}\left|\frac{\partial(L^{(e_i)}p^{(e_i)})}{\partial x_n^{(e_i)}}\right|^2 dx_n^{(e_i)}\,dt.$$

Also:

$$\sqrt{\int_0^\infty \exp\{-4\delta\tau\}\sum_{i=1}^M \int_0^{|e_i|} \Big|\frac{\partial p}{\partial x_n^{(e_i)}}\Big|^2 dx_n^{(e_i)} d\tau}$$

$$\leq \sqrt{\int_0^\infty \exp\{\beta_*\tau\}\sum_{i=1}^M \int_0^{|e_i|} \Big|\frac{\partial p}{\partial x_n^{(e_i)}}\Big|^2 dx_n^{(e_i)} d\tau}.$$

Thus, taking into account that $4(\delta+\beta) < 2\gamma$, we conclude:

$$\sqrt{\int_0^\infty \exp\{\beta_* t\}\sum_{i=1}^M \int_0^{|e_i|} \Big|\frac{\partial p^{(e_i)}}{\partial x_n^{(e_i)}}\Big|^2 dx_n^{(e_i)} dt}$$

$$\leq (c/\beta)\Bigg(\sqrt{\int_0^\infty \exp\{2\gamma\tau\}\sum_{i=1}^M \int_0^{|e_i|} \big((F^{(e_i)})_\tau(\tau) + \beta F^{(e_i)}(\tau)\big)^2 dx_n^{(e_i)} d\tau}$$

$$+\sqrt{\int_0^\infty \exp\{2\gamma\tau\}\sum_{l=1}^N \big((\mathcal{H}_l(O_l,\tau))_\tau + \beta\mathcal{H}_l(O_l,\tau)\big)^2 d\tau}\Bigg).$$

The proposition is proved. □

6.7.1.2 Final Step of Construction of the Asymptotic Expansion

Step 4 Next, define $s_{j+1}^{[BLt,(e)]}(\tau)$ and $a_{j+1}^{[BLt,(e)]}(\tau)$ as follows:

$$p_{j+1}^{[BLt,(e)]}(x_n^{(e)},\tau) = -s_{j+1}^{[BLt,(e)]}(\tau)x_n^{(e)} + a_{j+1}^{[BLt,(e)]}(\tau),$$

and define

$$U_{j+1}^{[BLt,(e)]}\left(y^{(e)'},\tau\right) = \tilde{U}_{j+1,ht}^{(e)}\left(y^{(e)'},\tau\right) + \mathcal{V}_{j+1}^{(e)}\left(y^{(e)'},\tau\right),$$

where $\mathcal{V}^{(e)}_{j+1}$ is the solution of problem (6.7.9) corresponding to $S(\tau) = s^{[BLt,(e)]}_{j+1}(\tau)$, and $\tilde{U}^{(e)}_{j+1,ht}\left(y^{(e)'}, \tau\right)$ is the solution of problem (6.7.19).

We then determine $P^{[BLt,BLO_l]}_j(y,\tau)$, the pressure in the boundary layer problems (6.7.14) and (6.7.15), by employing the following formula:

$$P^{[BLt,BLO_l]}_j(y,\tau)=\widetilde{P}^{[BLt,BLO_l]}_j(y,\tau)-\sum_{e:O_l\in e,\ e\neq e_s}\zeta\left(\frac{y^{(e)}_n}{3r}\right)\widehat{a}^{[BLt,BLO_l,(e)]}_j(\tau)$$
$$-\left(\sum_{e:O_l\in e}\zeta\left(\frac{y^{(e)}_n}{3r}\right)-1\right)a^{[BLt,(e_s)]}_{j+1}(\tau).$$

Function $\sum_{e:O_l\in e}\zeta\left(\frac{y^{(e)}_n}{3r}\right)$ is equal to $\zeta\left(\frac{y^{(e)}_n}{3r}\right)$ on each edge of the bundle $\mathcal{B}^l$. This step finalizes the passage from j to $j+1$. It should be noted that $p^{[BLt,(e_s)]}_{j+1}(0,\tau) = a^{[BLt,(e_s)]}_{j+1}(\tau)$.

6.8 Residual

The constructed function $\mathbf{u}^{[J]}$ is not divergence-free because of the multiplication of the boundary layer functions in each bundle of cylinders by the cutoff factor $1-\zeta(\frac{|x-O_l|}{|e|_{min}})$. Namely,

$$\operatorname{div}\mathbf{u}^{[J]}(x,t) = h^{[J]}(x,t),$$

where:

$$h^{[J]}(x,t) = -\sum_{l=1}^{N}\nabla\zeta(\frac{|x-O_l|}{|e|_{min}})\cdot(\mathbf{U}^{[BLO_l,J]}(\frac{x-O_l}{|e|_{min}},t)$$
$$+\mathbf{U}^{[BLt,BLO_l,J]}(\frac{x-O_l}{|e|_{min}},t/\varepsilon^2)).$$

Obviously, $h^{[J]}\in L^2(0,T;\mathring{W}^{1,2}(B_\varepsilon))$ and $h^{[J]}_t\in L^2(0,T;L^2(B_\varepsilon))$. Since the support of the function $\nabla\zeta(\dfrac{|x-O_l|}{|e|_{min}})$ belongs to the middle third of every cylinder, there hold the relations:

$$\|h^{[J]}\|_{L^2(0,T;W^{1,2}(B_\varepsilon))} = O(e^{-c_1/\varepsilon}),\qquad \left\|\frac{\partial h^{[J]}}{\partial t}\right\|_{L^2(0,T;L^2(B_\varepsilon))} = O(e^{-c_1/\varepsilon}), \tag{6.8.1}$$

with some $c_1 > 0$. So, let us construct the corrector $\mathbf{w}^{[J]} \in W^{1,2}(0,T; W^{2,2}(B_\varepsilon) \cap \mathring{W}^{1,2}(B_\varepsilon))$, such that:

$$\mathrm{div}\mathbf{w}^{[J]}(x,t) = -h^{[J]}(x,t).$$

It is easy to see that:

$$\int\limits_{B_\varepsilon} h^{[J]}(x,t)dx = 0,$$

(moreover, $\int_{\Pi_\varepsilon^{(e)}} h^{[J]}(x,t)dx = 0$ for any cylinder $\Pi_\varepsilon^{(e)}$ composing B_ε). By Lemma 2.4.13, such function $\mathbf{w}^{[J]} \in L^2(0,T; \mathring{W}^{2,2}(B_\varepsilon))$ with $\mathbf{w}_t^{[J]} \in L^2(0,T; \mathring{W}^{1,2}(B_\varepsilon))$ exists. Moreover, the following estimates:

$$\|\mathbf{w}^{[J]}(\cdot,t)\|_{L^2(0,T;W^{2,2}(B_\varepsilon))} \le \varepsilon^{-2} c \|h^{[J]}(\cdot,t)\|_{L^2(0,T;W^{1,2}(B_\varepsilon))}, \tag{6.8.2}$$

$$\|\mathbf{w}_t^{[J]}\|_{L^2(0,T;W^{1,2}(B_\varepsilon))} \le \varepsilon^{-1} c \|h_t^{[J]}\|_{L^2(0,T;L^2(B_\varepsilon))} \tag{6.8.3}$$

hold (see Lemma 2.4.13). Here the constant c is independent of t and ε.

Set $\widetilde{\mathbf{u}}^{[J]} = \mathbf{u}^{[J]} + \mathbf{w}^{[J]}$. Then $\mathrm{div}\widetilde{\mathbf{u}}^{[J]} = 0$, and $\widetilde{\mathbf{u}}^{[J]}$ satisfies the boundary conditions and the initial condition.

Now we are in position to compute the residual $\mathcal{L}(\widetilde{\mathbf{u}}^{[J]}, p^{[J]})$ in a half-bundle HB_{O_l}. We obtain:

$$
\begin{aligned}
-\mathbf{f}^{[J]}(x,t) &= \mathcal{L}(\mathbf{u}^{[J]} + \mathbf{w}^{[J]}, p^{[J]}) \\
&= \varepsilon^{J-1}\frac{\partial \mathbf{U}_{J-1}^{(e)}}{\partial t}(y^{(e)\prime},t)\zeta\Big(\frac{x_n^{(e)}}{3r\varepsilon}\Big) + \varepsilon^{J}\frac{\partial \mathbf{U}_{J}^{(e)}}{\partial t}(y^{(e)\prime},t)\zeta\Big(\frac{x_n^{(e)}}{3r\varepsilon}\Big) \\
&\quad + \varepsilon^{J-1}\frac{\partial \mathbf{U}_{J-1}^{[BLO_l]}}{\partial t}(y,t) + \varepsilon^{J}\frac{\partial \mathbf{U}_{J}^{[BLO_l]}}{\partial t}(y,t) \\
+ \sum_{J+1\le j\le 2J} \varepsilon^{j-2}\Bigg\{&\Bigg(\sum_{p+r=j-1}\Big(\mathbf{U}_p^{(e)}(y^{(e)\prime},t)\zeta\Big(\frac{y_n^{(e)}}{3r}\Big)\cdot\nabla_y\Big)\Big(\mathbf{U}_r^{(e)}(y^{(e)\prime},t)\zeta\Big(\frac{y_n^{(e)}}{3r}\Big)\Big)\Bigg) \\
&\quad + \sum_{p+r=j-1}\zeta\Big(\frac{y_n^{(e)}}{3r}\Big)(\mathbf{U}_p^{(e)}(y^{(e)\prime},t)\cdot\nabla_y)\mathbf{U}_r^{[BLO_l]}(y,t) \\
&\quad + \sum_{p+r=j-1}(\mathbf{U}_p^{[BLO_l]}(y,t)\cdot\nabla_y)\Big(\zeta\Big(\frac{y_n^{(e)}}{3r}\Big)\mathbf{U}_r^{(e)}(y^{(e)\prime},t)\Big)
\end{aligned}
$$

$$+ \sum_{p+r=j-1} (\mathbf{U}_p^{[BLO_l]}(y,t)\cdot\nabla_y)\mathbf{U}_r^{[BLO_l]}(y,t)$$

$$+ \sum_{\substack{\gamma,\delta\in\{1,2\},\\ (\gamma,\delta)\neq(1,1)}} \sum_{\alpha+\beta=j-1} (\mathbf{U}_\alpha^{(e),\gamma}(y,\tau)\cdot\nabla_y)\mathbf{U}_\beta^{(e),\delta}(y,\tau)$$

$$+ \sum_{\substack{\gamma\in\{1,2\},\delta\in\{3,4\},\\ (\gamma,\delta)\neq(1,3)}} \sum_{\alpha+\beta=j-1} (\mathbf{U}_\alpha^{(e),\gamma}(y,\tau)\cdot\nabla)\mathbf{U}_\beta^{(O_l),\delta}(y,\tau)$$

$$+ \sum_{\substack{\gamma\in\{3,4\},\delta\in\{1,2\},\\ (\gamma,\delta)\neq(3,1)}} \sum_{\alpha+\beta=j-1} (\mathbf{U}_\alpha^{(O_l),\gamma}(y,\tau)\cdot\nabla)\mathbf{U}_\beta^{(e),\delta}(y,\tau)$$

$$+ \sum_{\substack{\gamma,\delta\in\{3,4\},\\ (\gamma,\delta)\neq(3,3)}} \sum_{\alpha+\beta=j-1} (\mathbf{U}_\alpha^{(O_l),\gamma}(y,\tau)\cdot\nabla)\mathbf{U}_\beta^{(O_l),\delta}(y,\tau)\Big)\Big\}$$

$$+\mathcal{L}\big(1-\zeta(\frac{|x-O_l|}{|e|_{min}}))(\mathbf{U}^{[BLO_l,J]}(y,t)+\mathbf{U}^{[BLt,BLO_l,J]}(y,\tau)),$$

$$(1-\zeta(\frac{|x-O_l|}{|e|_{min}}))(P^{[BLO_l,J]}(y,t)+P^{[BLt,BLO_l,J]}(y,\tau))\big)\chi(x)$$

$$-\mathcal{L}(\mathbf{U}^{[BLO_l,J]}(y,t)+\mathbf{U}^{[BLt,BLO_l,J]}(y,\tau),\ P^{[BLO_l,J]}(y,t)+P^{[BLt,BLO_l,J]}(y,\tau))\chi(x)$$

$$+\frac{\partial\mathbf{w}^{[J]}}{\partial t} - \nu\Delta\mathbf{w}^{[J]} + (\mathbf{w}^{[J]}\cdot\nabla)\mathbf{w}^{[J]}$$

$$+(\mathbf{u}^{[J]}\cdot\nabla)\mathbf{w}^{[J]} + (\mathbf{w}^{[J]}\cdot\nabla)\mathbf{u}^{[J]}$$

$$+\varepsilon^{J-2}(\widehat{a}_J^{[BLO_l,(e)]}+\widehat{a}_J^{[BLt,BLO_l,(e)]})\nabla_y\zeta(\frac{y_n^{(e)}}{3r})).$$

Here $y=\dfrac{x-O_l}{\varepsilon}$; $y^{(e)}=\dfrac{x^{(e)}}{\varepsilon}$; $\chi=\chi_{supp\ \zeta(\frac{|x-O_l|}{|\mathbf{U}|_{min}}}$ is the characteristic function of the set supp $\zeta(\dfrac{|x-O_l|}{|e|_{min}})$, and $\mathbf{U}_\alpha^{(e),1}$ and $\mathbf{U}_\alpha^{(O_l),3}$ are the same as before (see (6.7.16) and (6.7.18)), except for:

$$\mathbf{U}_{J+1}^{(e),1}(y,\tau)=\sum_{2\alpha+q=J+1}\tau^\alpha\frac{\partial^\alpha\mathbf{U}_q^{(e)}}{\partial t^\alpha}(y^{(e)\prime},\theta_{1,q}\varepsilon)\zeta(\frac{y_n^{(e)}}{3r}),$$

and:

$$\mathbf{U}_{J+1}^{(O_l),3}(y,\tau) = \sum_{2\alpha+q=J+1} \tau^{\alpha} \frac{\partial^{\alpha}\mathbf{U}_q^{[BLO_l]}}{\partial t^{\alpha}}(y^{(e)\prime}, \theta_{3,q}\varepsilon).$$

In the presented formulas, the initial four terms are the part of the residual coming from the time derivative of the velocity, and the next four terms originate from the inertial term, encompassing all possible combinations of $\mathbf{U}_{\beta}^{(e)}$ and $\mathbf{U}_{\beta}^{[BLO_l]}$ with orders higher than $J-2$. These terms appeared earlier in the residual in the case of data vanishing for small times. The next four terms complete the list of terms originated from $(\widetilde{\mathbf{u}}^{[J]}\cdot\nabla)\widetilde{\mathbf{u}}^{[J]}$. The subsequent two terms represent the residual resulting from the multiplication of the boundary layer correctors by the cutoff function $1-\zeta(\frac{|x-O_l|}{|e|_{min}})$. Notably, these terms vanish exponentially due to their exponential decay within the set $\text{supp}\zeta(\frac{|x-O_l|}{|e|_{min}})$ (where $\chi \neq 0$). Specifically, the order of these terms in the L^2 norm diminishes as $O(e^{-c_1/\varepsilon})$, where c_1 is a positive constant. The next three terms are generated by the addition of the "divergence corrector" $\mathbf{w}^{[J]}$ to the expansion $\mathbf{u}^{[J]}$, and their support also belongs to $\text{supp}\zeta\left(\frac{|x-O_l|}{|e|_{min}}\right)$. So, these terms as well diminish exponentially. The last line comes from the pressure gradient term, which remains uncompensated by the boundary layer problems.

Clearly:

$$\|\mathbf{f}^{[J]}\|_{L^2(0,T;L^2(B_\varepsilon))} = O(\varepsilon^{J-2}). \tag{6.8.4}$$

6.9 Justification of the Asymptotic Expansion in the Case of Non-vanishing for Small Time Data

The considerations within this section parallel those of Sect. 6.5. It is sufficient to define the extension of the boundary value $\mathbf{g}$ as the derived asymptotic expansion $\widetilde{\mathbf{u}}^{[J]}$. By subsequently applying Theorems 6.2.1 and 6.2.3 (for the two-dimensional case) and Theorems 6.2.4 and 6.2.6 (for the case $n=3$) and applying estimate (6.8.4) all in a manner analogous to the approach outlined in Sect. 6.5, we arrive at the ensuing result.

Theorem 6.9.1 *If* $\mathbf{g} \in C^{[\frac{J+1}{2}]+3}(0,T;W^{3/2,2}(\partial B_\varepsilon))$, *then the following estimates hold:*

$$\sup_{t\in[0,T]} \|\mathbf{v}(\cdot,t)-\widetilde{\mathbf{u}}^{[J]}(\cdot,t)\|_{L^2(B_\varepsilon)} + \|\nabla\mathbf{v}-\nabla\widetilde{\mathbf{u}}^{[J]}\|_{L^2(0,T;L^2(B_\varepsilon))} = O(\varepsilon^{J+\frac{n-1}{2}}), \tag{6.9.1}$$

$$\sup_{t\in[0,T]} \|\nabla \mathbf{v}(\cdot,t) - \nabla\widetilde{\mathbf{u}}^{[J]}(\cdot,t)\|_{L^2(B_\varepsilon)} + \|\mathbf{v}_t - \widetilde{\mathbf{u}}_t^{[J]}\|_{L^2(0,T;L^2(B_\varepsilon))}$$
$$+ \|\nabla^2\mathbf{v} - \nabla^2\widetilde{\mathbf{u}}^{[J]}\|_{L^2(0,T;L^2(B_\varepsilon))} = O(\varepsilon^{J+\frac{n-3}{2}}). \tag{6.9.2}$$

If $\mathbf{g} \in C^{[\frac{J+1}{2}]+n+3}(0,T;\, W^{3/2,2}(\partial B_\varepsilon))$*, then the following estimates hold:*

$$\|p - p^{[J]}\|_{L^2(0,T;L^2(B_\varepsilon))} = O(\varepsilon^{J+\frac{n-3}{2}}). \tag{6.9.3}$$

Remark 6.9.1 The argument $\frac{|x-O_l|}{|e|_{\min}}$ of the cutoff function ζ can be substituted with $\frac{|x-O_l|}{\delta}$, where $\delta = C_J\varepsilon|\ln\varepsilon||e|_{\min}$, and the constant C_J is chosen such that:

$$E_{\delta/\varepsilon}(t) = \|\mathbf{U}^{[BLO_l,J]}(\cdot,t)\|_{W^{2,2}(\Omega_{l,\delta/\varepsilon})} + \|\nabla P^{[BLO_l,J]}(\cdot,t)\|_{L^2(\Omega_{l,\delta/\varepsilon})}$$
$$\le c\varepsilon^{J+(n-1)/2+2} \tag{6.9.4}$$

with c independent of t and:

$$\widetilde{E}_{\delta/\varepsilon} = \|\mathbf{U}^{[BLt,BLO_l,J]}\|_{W^{(2,1),2}(\Omega_{l,\delta/\varepsilon}\times(0,+\infty))} + \|\nabla P^{[BLt,BLO_l,J]}\|_{L^2(\Omega_{l,\delta/\varepsilon}\times(0,+\infty))}$$
$$= O(\varepsilon^{J+(n-1)/2+2}). \tag{6.9.5}$$

Here, as before, $\Omega_{l,R} = \Omega_l \cup \{|y| > R\}$.

By Theorem 4.4.3, $\mathbf{U}^{[BLO_l,J]}(y,t)$ and $P^{[BLO_l,J]}(y,t)$ tend to zero exponentially as $|y| \to \infty$, leading to the inequality:

$$E_R(t) \le c_1 e^{-c_2 R}, \quad c_1, c_2 > 0.$$

Similarly, by Theorem 4.5.5, $\mathbf{U}^{[BLt,BLO_l,J]}(y,t)$ and $P^{[BLt,BLO_l,J]}(y,t)$ tend to zero exponentially as $|y| \to \infty$, leading to the inequality:

$$\widetilde{E}_R(t) \le c_1 e^{-c_2 R}, \quad c_1, c_2 > 0.$$

Consequently, the estimates $E_{\delta/\varepsilon}(t)$, $\widetilde{E}_{\delta/\varepsilon} = O(\varepsilon^{J+(n-1)/2+2})$ are valid for δ such that:

$$c_2\delta/\varepsilon \ge (J + (n-1)/2 + 2)|\ln\varepsilon|,$$

implying $\delta = C_J\varepsilon|\ln\varepsilon|$. So modified asymptotic expansion satisfies the same estimates as the pair $\widetilde{\mathbf{u}}^{[J]}$, $p^{[J]}$ (see Theorem 6.9.1).

6.10 MAPDD in the Case of Non-vanishing for Small Time Data

The method of asymptotic partial decomposition of the domain (MAPDD) is investigated in case where data does not vanish for small time. In this context, we can establish the existence and estimates for the MAPDD velocity $\mathbf{u}_d$, which belongs to the space $L^2(0,T;\mathcal{H}_{0,\mathrm{dec}}(B_\varepsilon,\delta)) \cap L^\infty(0,T;L^2(B_\varepsilon))$. However, to prove the existence of its time derivative $\mathbf{u}_{d,t} \in L^2(0,T;L^2(B_\varepsilon,\delta))$, as well as to establish the existence and estimates of the MAPDD pressure p_d, remains an unresolved challenge.

To address this, a different definition of the MAPDD solution is introduced, which defines weaker solution compared to the previously used one. This new definition aligns with the concept of Leray–Hopf solutions for the Navier–Stokes equations (see [102] for more details).

Definition 6.10.1 Find $\mathbf{v}_d \in L^2(0,T;\mathcal{H}_{\mathrm{dec}}(B_\varepsilon,\delta)) \bigcap L^\infty(0,T;L^2(B_\varepsilon))$, such that $\mathbf{v}_d|_{t=0} = 0$, $\mathbf{v}_d(x,t) = \mathbf{g}(x,t)$ on ∂B_ε, and for all test functions $\boldsymbol{\eta} \in \mathcal{M}(B_\varepsilon,\delta) = \big\{\mathbf{w}_t \in L^2(0,T;\mathcal{H}_{0,dec}(B_\varepsilon,\delta)) | \mathbf{w} \in L^2(0,T;L^2(B_\varepsilon)), \mathbf{w}(x,T) = 0\big\}$, the following integral identity is valid:

$$\int_0^T \int_{B_\varepsilon} \Big(- \mathbf{v}_d \cdot \frac{\partial \boldsymbol{\eta}}{\partial t} + \nu \nabla \mathbf{v}_d \cdot \nabla \boldsymbol{\eta} + (\mathbf{v}_d \cdot \nabla)\mathbf{v}_d \cdot \boldsymbol{\eta} \Big)\, dxdt = 0. \tag{6.10.1}$$

Introduce $\mathbf{u}_d = \mathbf{v}_d - \widetilde{\mathbf{u}}^{[J]}$. Then this formulation is equivalent to the following one.

Definition 6.10.2 Find $\mathbf{u}_d \in L^2(0,T;\mathcal{H}_{0,\mathrm{dec}}(B_\varepsilon,\delta)) \bigcap L^\infty(0,T;L^2(B_\varepsilon))$, such that $\mathbf{u}_d|_{t=0} = 0$, and for all test functions $\boldsymbol{\eta} \in \mathcal{M}(B_\varepsilon,\delta)$ with $\boldsymbol{\eta}(x',T) = 0$, the following integral identity is valid:

$$\begin{aligned} &\int_0^T \int_{B_\varepsilon} \Big(- \mathbf{u}_d \cdot \frac{\partial \boldsymbol{\eta}}{\partial t} + \nu \nabla \mathbf{u}_d \cdot \nabla \boldsymbol{\eta} - ((\mathbf{u}_d + \widetilde{\mathbf{u}}^{[J]}) \cdot \nabla)\boldsymbol{\eta} \cdot \mathbf{u}_d \\ &\quad - (\mathbf{u}_d \cdot \nabla)\boldsymbol{\eta} \cdot \widetilde{\mathbf{u}}^{[J]} \Big)\, dxdt = \int_0^T \int_{B_e} \mathbf{f}^{[J]} \cdot \boldsymbol{\eta}\, dxdt. \end{aligned} \tag{6.10.2}$$

Theorem 6.10.1 *Suppose $B_\varepsilon \subset \mathbb{R}^n$, $n = 2, 3$. For sufficiently small ε, there exists a solution $\mathbf{u}_d \in L^2(0,T;\mathcal{H}_{0,\mathrm{dec}}(B_\varepsilon,\delta))$ that satisfies the integral identity* (6.6.2). *The subsequent estimates hold:*

$$\sup_{t\in[0,T]} \|\mathbf{u}_d(\cdot,t)\|^2_{L^2(B_\varepsilon)} + \int_0^T \|\nabla \mathbf{u}_d(\cdot,t)\|^2_{L^2(B_\varepsilon)} dt \le c\varepsilon^2 \int_0^T \|\mathbf{f}^{[J]}(\cdot,t)\|^2_{L^2(B_\varepsilon)} dt, \tag{6.10.3}$$

where the constant c is independent of ε.

Proof
The existence of the solution to the problem (6.10.2) is established using the Galerkin method, similar to the proof of Theorem 6.6.1. The Galerkin approximations are defined by the equations:

$$\begin{gathered}\int_{B_\varepsilon} \Big(\frac{\partial \mathbf{u}_d^{(N)}}{\partial t} \cdot \boldsymbol{\xi}_l + \nu \nabla \mathbf{u}_d^{(N)} \cdot \nabla \boldsymbol{\xi}_l - \big(\mathbf{u}_d^{(N)} + \widetilde{\mathbf{u}}^{[J]}\big) \cdot \nabla\big)\boldsymbol{\xi}_l \cdot \mathbf{u}_d^{(N)} \\ - \big(\mathbf{u}_d^{(N)} \cdot \nabla\big)\boldsymbol{\xi}_l \cdot \mathbf{u}^{(J)} \Big) dx = \int_{B_\varepsilon} \mathbf{f}^{[J]} \cdot \boldsymbol{\xi}_l dx, \\ \mathbf{u}_d^{(N)}(x,0) = 0,\end{gathered} \tag{6.10.4}$$

where $l = 1,.., N$, $\mathbf{u}_d^{(N)}(x,t) = \sum_{k=1}^{N} \gamma_k^{(N)}(t)\boldsymbol{\xi}_k(x)$, and $\{\boldsymbol{\xi}_k\}_{k=1}^{\infty}$ is a basis within the space $\mathcal{H}_{0,\mathrm{dec}}(B_\varepsilon,\delta)$ which we assume to be orthonormal in $L^2(B_\varepsilon)$. By multiplying relations (6.6.7) with $\gamma_l^{(N)}(t)$ and summing over $l = 1,\dots N$, we obtain:

$$\begin{aligned}\frac{1}{2}\frac{d}{dt}\int_{B_\varepsilon} |\mathbf{u}_d^{(N)}|^2 dx + \nu \int_{B_\varepsilon} |\nabla \mathbf{u}_d^{(N)}|^2 dx = \int_{B_\varepsilon} \big(\mathbf{u}_d^{(N)} \cdot \nabla\big)\mathbf{u}_d^{(N)} \cdot \widetilde{\mathbf{u}}^{[J]} dx \\ + \int_{B_\varepsilon} \mathbf{f}^{[J]} \cdot \mathbf{u}_d^{(N)} dx.\end{aligned}$$

Based on this equation, we deduce that for sufficiently small ε, the following inequality holds:

$$\sup_{t\in(0,T)} \|\mathbf{u}_d^{(N)}(\cdot,t)\|^2_{L^2(B_\varepsilon)} + \int_0^T \|\nabla \mathbf{u}_d^{(N)}(\cdot,\tau)\|^2_{L^2(B_\varepsilon)} d\tau \le c\varepsilon^2 \int_0^T \|\mathbf{f}^{[J]}(\cdot,t)\|^2_{L^2(B_\varepsilon)} dt. \tag{6.10.5}$$

Estimate (6.10.5) for the Galerkin approximation ensures the existence of a subsequence $\mathbf{u}_d^{(N_l)}$ such that the corresponding sequences of $\mathbf{u}_d^{(N_l)}$ and $\nabla \mathbf{u}_d^{(N_l)}$ converge weakly in $L^2(0,T; L^2(B_\varepsilon))$ to $\mathbf{u}_d$ and $\nabla \mathbf{u}_d$, respectively. Furthermore, using arguments from the book of O.A. Ladyzhenskaya [102] (see Sect. 6.7), it can be established that $\mathbf{u}_d$ is uniformly continuous on $[0, T]$ in a weak topology of $L^2(B_\varepsilon)$, and the sequence $\mathbf{u}_d^{(N_l)}$ converges strongly in $L^2(0, T; L^2(B_\varepsilon))$.

This information is sufficient to pass to the limit in the integral identity. We proceed by multiplying the equalities (6.10.4) by arbitrary functions $c_l \in C^1(0, T)$ that adhere to the condition $c_l(T) = 0$. Following this, we integrate the resulting relationships across the interval $(0, T)$ and subsequently apply integration by parts with respect to the time variable:

$$\int_0^T \int_{B_\varepsilon} \Big(-\mathbf{u}_d^{(N)} \cdot \Xi_t^M + \nu \nabla \mathbf{u}_d^{(N)} \cdot \nabla \Xi^M - \big(\mathbf{u}_d^{(N)} + \widetilde{\mathbf{u}}^{[J]}\big) \cdot \nabla\big) \Xi^M \cdot \mathbf{u}_d^{(N)}$$
$$-\big(\mathbf{u}_d^{(N)} \cdot \nabla\big) \Xi^M \cdot \widetilde{\mathbf{u}}^{[J]}\Big) dxdt = \int_0^T \int_{B_\varepsilon} \mathbf{f}^{[J]} \cdot \Xi^M dxdt \tag{6.10.6}$$

for any $\Xi^M(x,t) = \sum_{j=1}^{M} c_j(t) \boldsymbol{\xi}_j(x)$ with arbitrary M. By taking the limit as N_l approaches infinity in (6.10.6), we establish the identity (6.10.2) for the limit function $\mathbf{u}_d$. This identity holds for test functions of the form $\Xi^M(x,t) = \sum_{j=1}^{M} c_j(t) \boldsymbol{\xi}_j(x)$. This class of functions forms a dense set within the space $\mathcal{M}(B_\varepsilon, \delta)$, as indicated by Theorem 2.3.8. Consequently, the identity holds for any $\boldsymbol{\eta} \in \mathcal{M}(B_\varepsilon, \delta)$ with the condition $\boldsymbol{\eta}(x, T) = 0$. This establishes the existence of a weak solution. □

Remark 6.10.1 In the three-dimensional case, the uniqueness of Leray–Hopf-type solutions for the Navier–Stokes equations remains an open question. However, in the two-dimensional case, such solutions are known to be unique (see [102]).

Theorem 6.10.2 *Assume that* $\mathbf{g} \in C^{[\frac{J+1}{2}]+3}(0, T; W^{2,2}(B_\varepsilon))$, *and let* δ *be* $C_{J+2}\varepsilon|\ln \varepsilon|$, *where the constant* C_{J+2} *satisfies* (6.9.4) *and* (6.9.5) *with* J *replaced by* $J+2$. *Then there exists a solution* $\mathbf{v}_d$ *to problem* (6.10.1) *satisfying the following error estimate:*

$$\sup_{t\in[0,T]} \|\mathbf{v}_d - \mathbf{v}\|_{L^2(B_\varepsilon)} + \|\nabla \mathbf{v}_d - \nabla \mathbf{v}\|_{L^2(B_\varepsilon)\times(0,T)} = O(\varepsilon^{J+(n-1)/2}). \tag{6.10.7}$$

Proof

Denote now $\mathbf{u}_d = \mathbf{v}_d - \widetilde{\mathbf{u}}^{(J')}$, where $J' = J+2$. Taking into account estimate (6.8.4) with J replaced by J' and (6.10.3), along with the estimate (6.9.1) of Theorem 6.9.1:

$$\sup_{t\in[0,T]} \|\tilde{\mathbf{u}}^{[J]} - \mathbf{v}\|_{L^2(B_\varepsilon)} + \|\nabla(\tilde{\mathbf{u}}^{[J]} - \mathbf{v})\|_{L^2(B_\varepsilon)\times(0,T)} = O(\varepsilon^{J+(n-1)/2}),$$

and applying the triangle inequality, we get (6.10.7). □

6.11 Numerical Tests of the MAPDD for Nonstationary Navier–Stokes Equations in Tube Structures

This section presents the numerical tests of the MAPDD for the nonstationary Navier–Stokes equations within tube structures. The test cases feature an idealized geometry resembling the bifurcation of arteries. The objective is to investigate the maximal Reynolds numbers for which the solutions obtained through the MAPDD approach remain suitable in comparison to solutions derived from the full-dimensional model. It is noteworthy that in the work [55], real-life arterial bifurcation geometry was considered, and the performance of the MAPDD method was compared with results obtained through direct numerical computations.

The nonstationary Navier–Stokes equations are expressed as follows:

$$\begin{cases} \dfrac{\partial \mathbf{v}}{\partial t} - \nu \Delta \mathbf{v} + (\mathbf{v} \cdot \nabla)\mathbf{v} + \nabla p = 0, \\ \text{div } \mathbf{v} = 0, \end{cases} \tag{6.11.1}$$

with the initial condition $\mathbf{v}(x, 0) = 0$ and the boundary conditions prescribing the inlet velocity and zero pressure at the outflow.

The convergence of the MAPDD model for a sequence of decreasing ε values is examined in relation to error estimates for both velocity and pressure components.

6.11.1 Numerical Setup

Consider the planar layout of the domain B_ε, as depicted in Fig. 6.1.

Two interconnected junction regions, as ω_ε^1 and ω_ε^2 (white color), are linked by two straight tubes, namely, $B_{1,2}^{dec,\varepsilon}$ (upper one) and $\widetilde{B}_{1,2}^{dec,\varepsilon}$ (lower one, gray color), alongside corresponding buffer regions $B_i^{\varepsilon,\delta}$ and $\widetilde{B}_i^{\varepsilon,\delta}$, where $i = 1, 2$ (blue color). The inlet boundary γ_{in} (on the left) serves as the entry point for the flow into the

Fig. 6.1 Simplified thin tube structure

first junction domain, and the exit is through γ_{out} in the second junction domain (on the right).

The tubes $B_{1,2}^{dec,\varepsilon}$ and $\widetilde{B}_{1,2}^{dec,\varepsilon}$ are included in the complete reference model but truncated in the reduced MAPDD model. In the latter case, the MAPDD interface conditions are enforced on the boundaries S_{12}, S_{21}, $\widetilde{S}_{12}$, and $\widetilde{S}_{21}$ (black straight lines crossing the tubes).

The radius R of the tubes scales proportionally to ε; specifically, we set $R = \varepsilon$. For each value of ε, the junction regions contract homothetically by a factor of ε with respect to center points indicated by crosses in Fig. 6.1. The distance L between these points remains constant across various ε values, with $L = 7$ chosen arbitrarily for the experiments. We consider a sequence of ε values given by $\varepsilon = 2^{-k}$, where k ranges from 2 to 6.

In accordance with the developed theory, the parameter δ, representing the distance from the junction region centers to the interfaces S_{ij} and $\widetilde{S}_{ij}$, is defined as:

$$\delta = C_0 J \varepsilon |\ln(\varepsilon)|, \tag{6.11.2}$$

where C_0 is a constant independent of ε and J; here J is the order of the admissible error, which is given as a power of ε. Here, $\widetilde{C} = C_0 J$, introduced as a user-defined parameter, must satisfy the condition $4\varepsilon < \delta < \frac{L}{2}$ to ensure the feasibility of the selected geometry, with no intersection of junction regions and non-emptiness of buffer regions. With the choice $\widetilde{C} = C/\ln(2)$, admissible values for C fall in the range $2 < C < 7$ for the present geometry and chosen ε values.

The impact of the constant C on model accuracy is investigated over the range $C = \{2.5;\ 3;\ \ldots;\ 6.5\}$ and extremes $C = 2.1$ and $C = 6.9$, aimed at providing insights into optimal C selection. Furthermore, an examination of error convergence with respect to ε and the influence of Reynolds number is carried out within the subset $C = \{3;\ 4;\ 5\}$.

Both the full reference domain and the MAPDD formulation consider the incompressible, time-dependent Navier–Stokes equations (6.11.1) (and their MAPDD variant (6.6.1)) over a time interval $t \in (0,\ T]$, with $T = 0.4\,\mathrm{s}$. Parabolic velocity profiles are imposed through Dirichlet boundary conditions on γ_{in} and γ_{out}, represented as:

$$\mathbf{g} = \left(U_0 \left(1 - \frac{(x_2 - c_0)^2}{\varepsilon^2} \right) \sin\left(\pi \frac{t}{T} \right),\ 0 \right)^t,$$

where c_0 signifies the x_2 coordinate of the boundary's center. The initial velocity is set as $\mathbf{u}_\varepsilon(x, 0) = 0$. No-slip conditions are applicable to the remaining boundaries, except the truncated domain interfaces. To address the indeterminate pressure constant of the Dirichlet problem and ensure solvability of the discretized system, a zero-pressure condition is enforced on a single degree of freedom at the outlet. Periodic boundary conditions for velocity are applied between the interfaces S_{ij}

and $\tilde{S}_{ij}$ in accordance with MAPDD conditions (i.e., the profile of normal velocity is constant along the tube).

Two distinct setups are investigated. Firstly, numerical verification of error estimates is conducted for velocity:

$$\sup_{t\in(0,T)} \|\mathbf{u}_{\varepsilon,\delta} - \mathbf{u}_\varepsilon\|_{L^2(B_\varepsilon)} + \|\nabla(\mathbf{u}_{\varepsilon,\delta} - \mathbf{u}_\varepsilon)\|_{L^2((0,T);L^2(B_\varepsilon))} = O(\varepsilon^J), \tag{6.11.3}$$

and for pressure:

$$\|p_{\varepsilon,\delta} - p_\varepsilon\|_{L^2(0,T;L^2(B_\varepsilon))} = O(\varepsilon^J). \tag{6.11.4}$$

Subsequently, the convergence rate with respect to ε is estimated, and the influence of constant C is studied. As the theoretical framework assumes $\mathbf{g}_\varepsilon = \mathcal{O}(1)$, we set $U_0 = 1$.

Additionally, in accordance with the low Reynolds number assumption, defined as $Re(\varepsilon) = \frac{4}{3}\frac{\varepsilon}{\nu}U_0$, viscosity is chosen as $\nu = 0.01$. For the largest ε value, the resulting Re is approximately 33.3.

Next, we delve into the investigation of error behaviors with respect to the Reynolds number. We examine fixed Reynolds numbers within the range of $1 \leq Re(\varepsilon) \leq 300$, where ν is set to 0.035. The variable inflow U_0 is calculated from the Reynolds number Re.

Additionally, we conduct simulations in the convection-dominated regime by employing a Reynolds number of $Re = 1000$, with a chosen value of $C = 3$.

6.11.2 Numerical Discretization

The model equations are discretized through a monolithic mixed finite element method utilizing Taylor–Hood basis functions (quadratic basis functions for velocity and linear continuous ones for pressure). Unstructured triangular meshes are employed, with element sizes of $h = \varepsilon/10$ for $Re \leq 300$ and $h = \varepsilon/20$ for $Re = 1000$. It is worth noting that Taylor–Hood elements are recognized for their inf-sup stability for standard Navier–Stokes problems, though the formal proof for their stability in the MAPDD formulations remains a topic for future exploration. However, our numerical observations show no signs of instability, offering supporting evidence for the assumed inf-sup stability.

An implicit Euler scheme, unconditionally stable, discretizes the time derivative with a time step size of $\Delta t = 0.01$ s. To handle the convection term, a semi-implicit approach is employed, utilizing a skew-symmetric formulation.

The computational implementation and solution are executed using the FEniCS finite element library [8].

6.11.3 Comparison of Full-Dimension and MAPDD Solutions

We proceed to compare the solutions of the full-dimensional problem with those obtained using the MAPDD approach. For this purpose, we evaluate the error norms (6.11.3) and (6.11.4) for both reference and MAPDD results, considering the prescribed sequences of ε and of the constant $C \in (2, 7)$. This parameter C governs the extension of the straight section of the junction domain as described in (6.11.2).

Figure 6.2 illustrates the dependence of resulting errors on the values of C for the four largest ε values. Notably, as C increases, the errors generally diminish. Particularly, for the largest ε value ($\varepsilon = 2^{-2}$), small C values result in a notable rise in errors. Within the intermediate range ($3 \leq C \leq 6.5$ for velocity and $3.5 \leq C \leq 6$ for pressure), the impact on errors is relatively mild. However, as C exceeds 6 (indicating almost-touching interfaces), a significant drop in errors is observed. As ε decreases, the errors show reduced sensitivity to variations in C. Specifically,

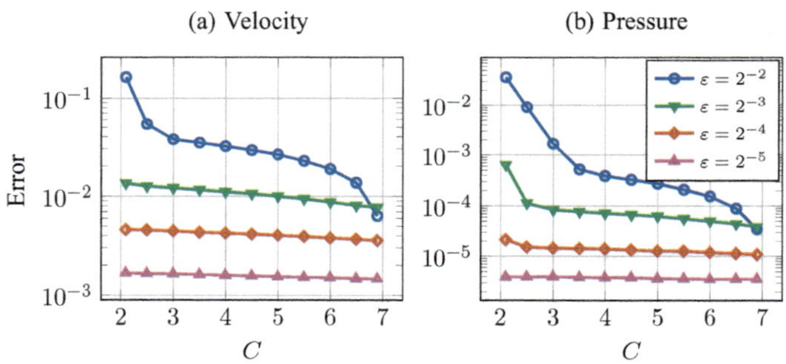

Fig. 6.2 Errors of the MAPDD model in (**a**) the velocity (Eq. 6.11.3) and (**b**) the pressure (Eq. 6.11.4) over C for fixed ε

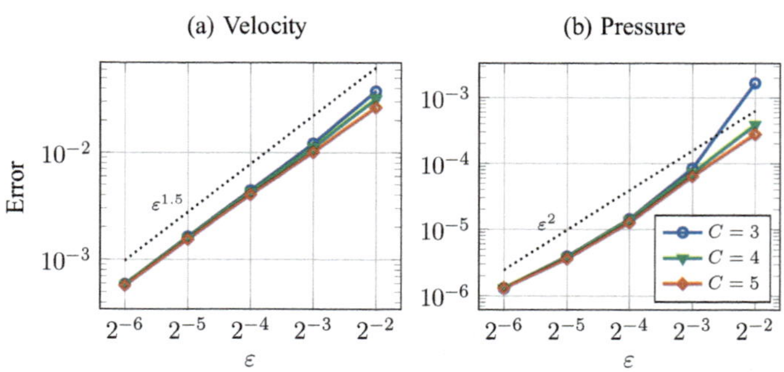

Fig. 6.3 Errors of the MAPDD model in (**a**) the velocity (Eq. 6.11.3) and (**b**) the pressure (Eq. 6.11.4) for different values of C over ε. The dotted lines indicate ε^J with (**a**) $J = 1.5$ and (**b**) $J = 2$ (cf. Table 6.1)

for $\varepsilon = 2^{-3}$ and 2^{-4}, selecting $C \geq 2.5$ seems to significantly enhance accuracy, as evident from the pressure error.

The dependence of errors on ε is illustrated in Fig. 6.3, focusing on the selected range $C = \{3;\ 4;\ 5\}$. The velocity error shows a clear exponential decay pattern as ε decreases, maintaining a near-constant rate. Higher values of C yield marginally smaller errors but with a slightly slower convergence. A similar behavior is observed for the pressure error. Notably, for the largest ε, a choice of $C = 3$ leads to a pressure error nearly one order of magnitude larger compared to using $C = 4$ or 5.

At large values of ε, the rate of convergence appears to be faster than in the lower range of ε, where the slope is similar to the velocity error. For reference, the figures include lines indicating the slopes of $J = 1.5$ (velocity) and $J = 2$ (pressure).

Finally, Table 6.1 presents empirically estimated average convergence rates denoted by J. It is apparent from the table that larger C values slightly mitigate the convergence rate.

Figure 6.4 illustrates the time evolution of velocity and pressure errors measured in the $L^2(B_\varepsilon)$ norm. The focus is on the case of $\varepsilon = 2^{-3}$ while considering different values of C. Notably, the pressure error is depicted with a logarithmic scale for the y axis due to the wide-ranging variations spanning two orders of magnitude. It is clearly dominated by the large errors at the first time step, which sets in motion initially quiescent flow.

The time profile confirms the observation (Fig. 6.4) that for $\varepsilon = 2^{-3}$, the improvement in accuracy between $C = 3, 4, 5$ is very small. The minimum velocity error at $t/T = 0.5$ indicates that the assumption of parallel flow through the straight

Table 6.1 Estimated rates of convergence J for different values of C

	Rate of convergence J	
	Velocity	Pressure
$C = 3$	1.50	2.58
$C = 4$	1.44	2.06
$C = 5$	1.38	1.93

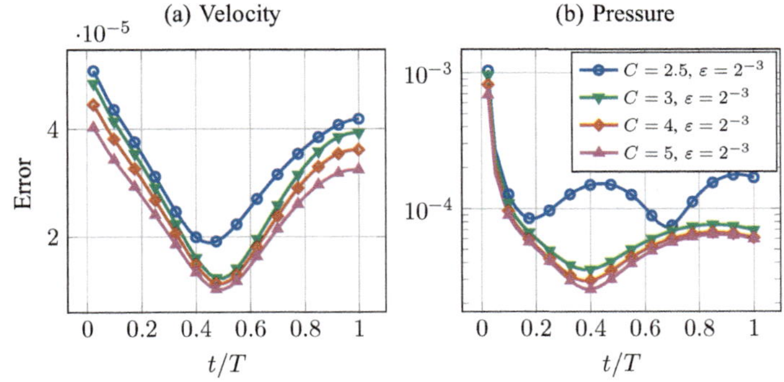

Fig. 6.4 Time evolution of the $L^2(B_\varepsilon)$ errors of the MAPDD model in (**a**) the velocity and (**b**) the pressure (logarithmic y axis), for $\varepsilon = 2^{-3}$ and different values of C

sections is a more accurate approximation during the peak flow. With $C = 2.5$, a slightly elevated velocity error is observed close to $t/T = 0.5$, the time of maximum flow rate. The pressure error using $C = 2.5$ clearly deviates from and exceeds the errors obtained with $C \geq 3$. In Fig. 6.4b, the corresponding total error weakly exceeded the trend with constant rate observed for $C \geq 3$.

Relative errors, computed for varying Reynolds numbers ($Re = 1$, 10, 100, and 300), are shown in Fig. 6.5 for both velocity and pressure. It is important to note that the errors (6.11.3) and (6.11.4) are scaled by the corresponding full reference solutions evaluated using the same norms. The magnitude of these relative errors is notably minuscule. Even under the most challenging scenario (at $Re = 300$), the errors barely exceed 10% for velocity and 1% for pressure. For situations with lower Reynolds numbers, such as $Re = 1$, 10, the relative errors reach values as low as 10^{-5} for velocity and 10^{-8} for pressure. These values are associated with the smallest considered ε. Remarkably, for velocity, the outcomes for $Re = 1$, 10, and 100 exhibit substantial similarity. The only exception is when $Re = 100$ and $C = 3$, which displays a noteworthy increase in error—a trend that mitigates as ε decreases. Once again, it is worth highlighting that adopting higher values of C consistently leads to diminished errors across all cases. In the case of $Re = 300$, the errors surge significantly and are remarkably sensitive to the chosen value of C. Particularly, the errors obtained with $C = 3$, 4, and 5 differ by approximately half an order of magnitude.

For $\varepsilon \leq 2^{-3}$, the error can be seen to decrease exponentially with a rate comparable to the lower Reynolds numbers.

The relative pressure error, at Reynolds numbers $Re = 1$ and $Re = 10$, displays almost identical behavior. However, at $Re = 100$, it exhibits increased sensitivity to the selection of C compared to the velocity error. When $C = 5$, the relative error coincides for $Re \in \{1; 10; 100\}$. Conversely, with $C = 4$, and even more so

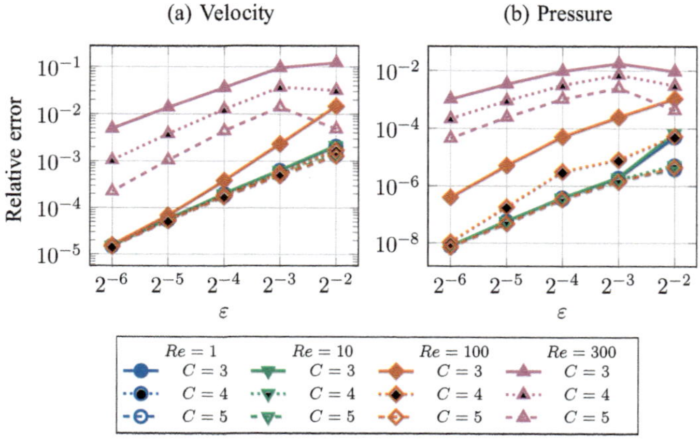

Fig. 6.5 Relative velocity errors of the Navier–Stokes MAPDD model w.r.t. the full solution for different Reynolds numbers and C over ε

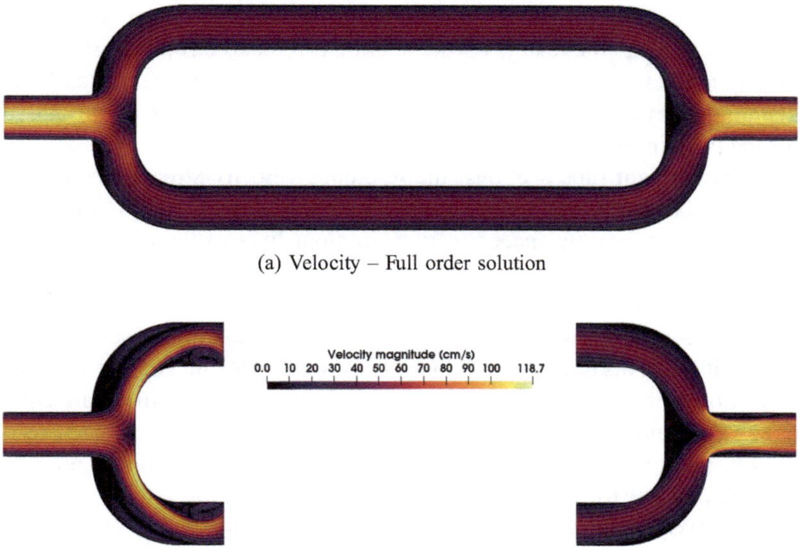

(a) Velocity – Full order solution

(b) Velocity – MAPDD solution

Fig. 6.6 Velocity magnitude and streamlines for $\varepsilon = 0.25$, $Re = 100$, at peak time $t = T/2$: full reference model (**a**) versus MAPDD model (**b**)

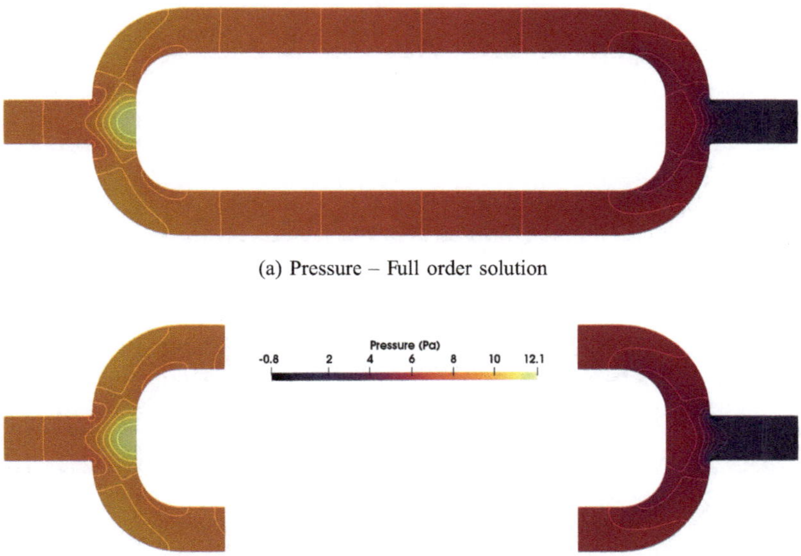

(a) Pressure – Full order solution

(b) Pressure – MAPDD solution

Fig. 6.7 Pressure field and contours for $\varepsilon = 0.25$, $Re = 100$, at peak time $t = T/2$: full reference model (**a**) versus MAPDD model (**b**)

with $C = 3$, the error escalates significantly. Under the scenario of $Re = 300$, the pressure error mirrors the features observed in the velocity error analysis.

Velocity magnitude and pressure fields of both the full and the reduced MAPDD models are displayed in Figs. 6.6, 6.7 for the case of $Re = 100$, $\varepsilon = 0.25$, at the time of the peak inflow $t = T/2$.

Inside the straight tube sections, the reference velocity streamlines are highly parallel, in accordance with the Womersley assumption of the MAPDD model. Likewise, the equidistantly spaced, vertical contour lines of the reference pressure (Fig. 6.7) indicate a nearly linear pressure along the tubes. The velocity field obtained with the MAPDD model in the reduced domain is visually indiscernible from the reference solution. The MAPDD pressure solution also matches the reference with high accuracy. Small differences in the pressure contour lines appear at the position of the upstream interfaces, where Womersley conditions are enforced by the MAPDD model.

In addition, Figs. 6.8, 6.9 show the velocity and pressure fields for an elevated Reynolds number of $Re = 1000$.

Strong convective effects, flow separation, and recirculation are present, mostly in the bifurcations of the upstream junction domains. Also this more complex flow is recovered with high fidelity by the MAPDD model. At the axial position of the upstream interfaces, the reference solution exhibits recirculation zones near the inner walls. In the MAPDD solution, with the presence of the interfaces, imposing

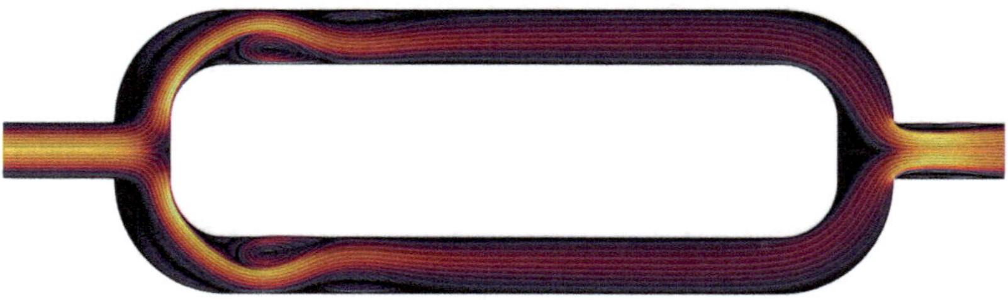

(a) Velocity – Full order solution

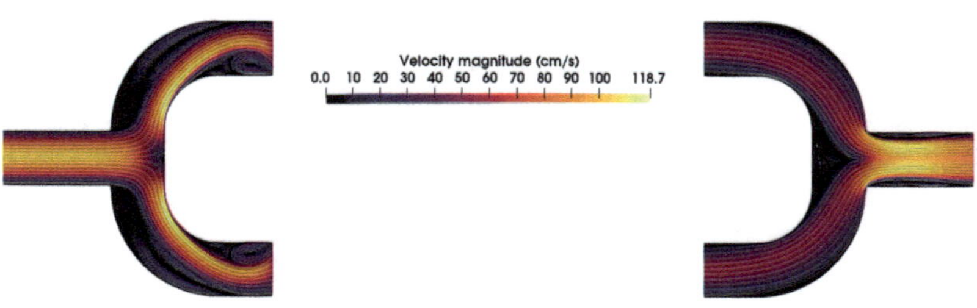

(b) Velocity – MAPDD solution

Fig. 6.8 Velocity magnitude and streamlines for $\varepsilon = 0.25$, $Re = 1000$, at peak time $t = T/2$: full reference model (**a**) versus MAPDD model (**b**)

(a) Pressure – Full order solution

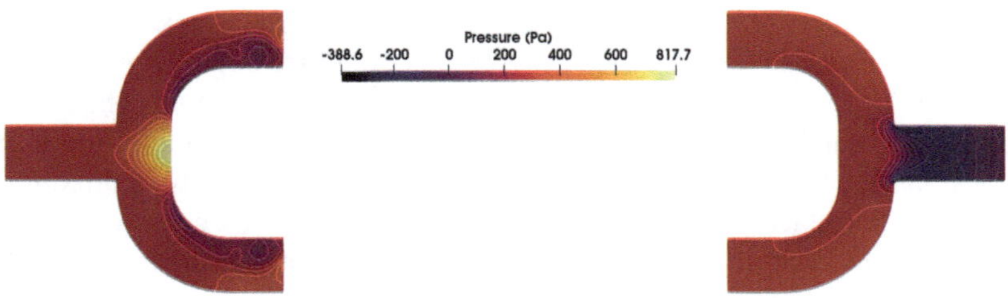

(b) Pressure – MAPDD solution

Fig. 6.9 Pressure field and contours for $\varepsilon = 0.25$, $Re = 1000$, at peak time $t = T/2$: full reference model (**a**) versus MAPDD model (**b**)

parallel-flow conditions, the recirculating flow is slightly modified with respect to the reference. In the junction domain on the downstream side, the full-order reference flow is recovered with excellent accuracy. Here, the interference of the interface conditions is minimal, since the straight tube sections have a parallelizing effect on the flow. Comparison of the pressure fields of the reference and the MAPDD solution (Fig. 6.9) reveals similar features. The most pronounced characteristics of the pressure are the stagnation point in the upstream junction domain and pressure minima close to the upstream interfaces, matching the positions of the recirculation zones, and a low pressure region in the outflow section of the downstream junction domain. All these features are accurately represented by the MAPDD solution. Again, only near the upstream interfaces, small differences can be observed, where the magnitude and extent of the MAPDD pressure minima is smaller than in the reference case. Elsewhere the pressure contours match almost exactly.

The numerical experiments confirm the theoretical findings for a simple domain of two junction domains connected by two straight tubes, for small Reynolds numbers, when the Womersley flow assumption for the truncated tubes is accurate. The reduced geometry includes straight buffer zone extension of the junction domains at the interfaces of the truncated tubes. The length of these extensions is controlled by a user parameter, C. Longer extensions help to counteract the departure from the Womersley regime in the truncated tubes for elevated Reynolds numbers.

The choice of the parameter C is a trade-off of savings in computational time (smaller C) and accuracy (larger C). It should be chosen as small as possible while the maximum admissible error is not exceeded or can be determined, for instance, by increasing C from a small initial value until the error is insensitive to changes in C with a given tolerance.

The numerical results show that the velocity MAPDD solution converges to the full solution with order ≈ 1.5 under ε, as indicated by the theoretical error bound. The pressure error appears to converge faster, but at non-constant rate, suggesting that for the selected parameters, the estimate is not sharp. Very small relative errors indicate that the MAPDD model is very accurate for low Reynolds numbers. In particular, for larger Re, higher C could significantly improve the accuracy.

The MAPDD model delivered excellent results even for a large Reynolds number of $Re = 1000$, for which the theoretical error estimates do not hold. The MAPDD model results matched the reference velocity and pressure distributions with high accuracy and succeeded in recovering complex flow features. Only in proximity to the upstream interfaces, for MAPDD, the interface conditions slightly altered the flow in order to accommodate the model assumptions of a Womersley-like flow inside the truncated tubes. These results are obtained in [26].

Chapter 7
Time-Periodic Case

The time-periodic setting is motivated by the challenge of computer modeling within the blood vessel network. Such modeling plays a crucial role in optimizing the selection of surge strategies in cases involving cardiovascular or cerebrovascular issues. As it was mentioned in previous chapters, the existing one-dimensional models and codes, fail to provide the necessary accuracy in proximity to regions where clot formation occurs, as well as around stents and vessel bifurcations, while full three-dimensional simulations are currently highly time-consuming and applicable only to limited segments of the circulatory system. Consequently, we propose a hybrid-dimensional approach that combines one-dimensional simplification within regular regions with three-dimensional refinements in localized areas exhibiting singular behavior.

This approach has previously been applied to nonstationary initial boundary value problems for the Navier–Stokes equations in slender tube structures, as discussed in the preceding chapter. However, for hemodynamic modeling, it is more natural to consider time-periodic scenarios. Thus, we examine the time-periodic Navier–Stokes equation. We establish theorems concerning the existence and uniqueness of solutions within this framework, and we construct an asymptotic expansion of the solution with respect to the small parameter ε, representing the ratio of vessel diameter to length.

Our focus pertains to the Navier–Stokes equations in tube structures. We address the problem under two distinct scalings: one analogous to the previous context and the second scaling introduces a big parameter ε^{-2} in front of the time derivative of velocity. The derived asymptotic expansion is subsequently employed to address a specialized numerical approach that combines one-dimensional descriptions with three-dimensional refinements (MAPDD).

G. Panasenko, K. Pileckas, *Multiscale Analysis of Viscous Flows in Thin Tube Structures*, Advances in Mathematical Fluid Mechanics,
https://doi.org/10.1007/978-3-031-54630-3_7

7.1 Problem Formulation

In this section, we describe two distinct scalings of the Navier–Stokes equations applied to small and very small arterioles or capillaries. The experimental data vary depending on factors such as the health status of the human or animal blood system under consideration. Averaged data are taken from sources [116] and [119]. The characteristic time period is 1 s, with a characteristic velocity of approximately 0.5×10^{-3} m/s. Two scaling scenarios are examined for the characteristic length and vessel diameter: (1) length of 10^{-3} m and characteristic diameter of 10^{-4} m and (2) length of 10^{-2} m and characteristic diameter of 10^{-3} m (in both cases, denoted as $\varepsilon = 0.1$).

For the first case, by introducing the spatial variable transformation $X = 10^{-3}x$, velocity transformation $\mathbf{v} = 10^{-3}\mathbf{V}$, and pressure transformation $p = 10^3 P$, accounting for the dynamic viscosity of blood (4×10^{-3} Pa·s) and its density (10^3 kg/m), the Navier–Stokes equations are obtained in new variables:

$$\begin{cases} \dfrac{\partial \mathbf{V}}{\partial t} - 4\Delta_X \mathbf{V} + 0.5(\mathbf{V} \cdot \nabla_X)\mathbf{V} + \nabla_X P = 0, \\ \operatorname{div}\mathbf{V} = 0. \end{cases}$$

In the second case, employing the velocity transformation $\mathbf{v} = 10^{-4}\mathbf{V}$ and the pressure transformation $p = 10^1 P$, the Navier–Stokes equations in new variables are:

$$\begin{cases} 10^2 \dfrac{\partial \mathbf{V}}{\partial t} - 4\Delta_X \mathbf{V} + 5(\mathbf{V} \cdot \nabla_X)\mathbf{V} + \nabla_X P = 0, \\ \operatorname{div}\mathbf{V} = 0. \end{cases}$$

The investigation proceeds with two distinct settings: one with the factor ε^{-2} before the time derivative term and another without it. The study focuses on a time-periodic boundary value problem within a tube structure B_ε. The problem involves the Navier–Stokes equations:

$$\begin{cases} \dfrac{1}{\varepsilon^\beta}\mathbf{v}_t - \nu\Delta\mathbf{v} + (\mathbf{v} \cdot \nabla)\mathbf{v} + \nabla p = 0, & x \in B_\varepsilon,\ t \in (0, 2\pi), \\ \operatorname{div}\mathbf{v} = 0, & x \in B_\varepsilon,\ t \in (0, 2\pi), \\ \mathbf{v}(x,t) = \mathbf{g}, & x \in \partial B_\varepsilon,\ t \in (0, 2\pi), \\ \mathbf{v}(x,t) = \mathbf{v}(x, t + 2\pi), & x \in B_\varepsilon,\ t \in (0, 2\pi), \end{cases} \tag{7.1.1}$$

with parameter options $\beta = 0$ or $\beta = 2$.

As in the previous section, we assume that the fluid velocity $\mathbf{g}$ at the boundary ∂B_ε has the following structure: $\mathbf{g} = 0$ everywhere on ∂B_ε except for the set $\gamma_\varepsilon^{N_1+1}, \ldots, \gamma_\varepsilon^N$, where $\gamma_\varepsilon^j = \partial B_\varepsilon \cap \partial\omega_\varepsilon^j$, $j = N_1+1, \ldots, N$, i.e.:

$$\mathbf{g}(x,t)\big|_{\gamma_\varepsilon^j} = \mathbf{g}^j\left(\frac{x - O_j}{\varepsilon}, t\right)\Big|_{\gamma_\varepsilon^j}, \quad j = N_1+1, \ldots, N,$$
$$\mathbf{g}(x,t)\Big|_{\partial B_\varepsilon \setminus (\bigcup\limits_{j=N_1+1}^{N} \gamma_\varepsilon^j)} = 0,$$

where $\mathbf{g}^j \in C^{[\frac{J+2}{2}]+1}(0, 2\pi; W^{3/2,2}(\gamma^j))$, $[\alpha]$ is the integer part of α, $\gamma^j = \varepsilon^{-1}(\gamma_\varepsilon^j - O_j)$ is the corresponding dilated part of the boundary, and $\mathbf{g} \in C^{[\frac{J+1}{2}]+1}(0, 2\pi; W^{3/2,2}(\partial B_\varepsilon))$.

As before:

$$\widetilde{F}^j(t) = \int\limits_{\gamma_\varepsilon^j} \mathbf{g}(x,t)\cdot \mathbf{n}(x)dS = \int\limits_{\gamma_\varepsilon^j} \mathbf{g}^j\left(\frac{x - O_j}{\varepsilon}, t\right)\cdot \mathbf{n}(x)dS$$
$$\equiv \varepsilon^{n-1}F^j(t), \quad j = N_1+1, \ldots, N, \tag{7.1.2}$$

where $\mathbf{n}$ is the unit outward (with respect to B_ε) normal vector to γ_ε^j. Since $\mathbf{g}(x,t)$ is time-periodic, $F^j(t)$ also must be time-periodic. Moreover, since we will need the divergence-free extension of $\mathbf{g}$, we assume the compatibility condition for the flow rates $F^j(t)$:

$$\sum_{j=N_1+1}^{N} F^j(t) = 0 \ \ \forall t \in [0, 2\pi]. \tag{7.1.3}$$

Let $\mathbf{g}$ be the divergence-free time-periodic extension of the boundary function $\mathbf{g}$ (which we denote by the same symbol $\mathbf{g}$, $\mathbf{g} \in C^{[\frac{J+1}{2}]+2}([0, 2\pi]; W^{2,2}(B_\varepsilon)))$. Suppose that $\mathbf{g}$ satisfies for all $t \in [0, 2\pi]$ the following asymptotic estimates:

$$\sup_{x\in B_\varepsilon} |\mathbf{g}(x,t)| \le c, \quad \|\nabla \mathbf{g}\|_{L^2(B_\varepsilon)} \le c\varepsilon^{\frac{n-3}{2}},$$
$$\|\mathbf{g}_t\|_{L^2(B_\varepsilon)} \le c\varepsilon^{\frac{n-1}{2}}, \quad \|\nabla^2 \mathbf{g}\|_{L^2(B_\varepsilon)} \le c\varepsilon^{\frac{n-5}{2}}, \quad n = 2, 3, \tag{7.1.4}$$

where the constant c is independent of ε.

As in Chap. 6, we will construct a special extension $\mathbf{g}$, which is the velocity part of the asymptotic representation of the solution. It is worth noting that such extension $\mathbf{g}$ and the corresponding asymptotic representation for the pressure generate a "small" discrepancy in Eqs. (7.1.1).

As we have a divergence-free time-periodic extension $\mathbf{g}$, satisfying estimates (7.1.4), a new unknown function $\mathbf{u}$ can be introduced, representing $\mathbf{v}$ as the sum $\mathbf{v} = \mathbf{u} + \mathbf{g}$. For $\mathbf{u}$, we get the following problem:

$$\begin{cases} \frac{1}{\varepsilon^\beta}\mathbf{u}_t - \nu\Delta\mathbf{u} + (\mathbf{u}\cdot\nabla)\mathbf{g} + (\mathbf{g}\cdot\nabla)\mathbf{u} + \\ \qquad +(\mathbf{u}\cdot\nabla)\mathbf{u} + \nabla p = \mathbf{f}, & x \in B_\varepsilon,\ t \in (0, 2\pi), \\ \qquad \mathrm{div}\mathbf{u} = 0, & x \in B_\varepsilon,\ t \in (0, 2\pi), \\ \qquad \mathbf{u}(x,t) = 0, & x \in \partial B_\varepsilon,\ t \in (0, 2\pi), \\ \qquad \mathbf{u}(x,t) = \mathbf{u}(x, t+2\pi), & x \in B_\varepsilon,\ t \in (0, 2\pi), \end{cases} \tag{7.1.5}$$

where $\mathbf{f} = -\Big(\frac{1}{\varepsilon^\beta}\mathbf{g}_t - \nu\Delta\mathbf{g} + (\mathbf{g}\cdot\nabla)\mathbf{g}\Big)$.

The variational formulation of the problem (7.1.5) is the following:

Definition 7.1.1 Find a time-periodic vector field $\mathbf{u} \in \overset{\circ}{W}{}^{(1,1),2}_{\partial B_\varepsilon}(B_\varepsilon \times (0, 2\pi))$, $\mathbf{u}(\cdot, t) \in \overset{\circ}{H}{}^{1,2}(B_\varepsilon)$ satisfying, for every time-periodic $\eta \in \overset{\circ}{W}{}^{(1,0),2}_{\partial B_\varepsilon}(B_\varepsilon \times (0, 2\pi))$, $\eta(\cdot, t) \in \overset{\circ}{H}{}^{1,2}(B_\varepsilon)$, the integral identity:

$$\int_0^{2\pi}\int_{B_\varepsilon}\Big(\frac{1}{\varepsilon^\beta}\mathbf{u}_t\cdot\eta + \nu\nabla\mathbf{u}\cdot\nabla\eta - ((\mathbf{u}+\mathbf{g})\cdot\nabla)\eta\cdot\mathbf{u} - (\mathbf{u}\cdot\nabla)\eta\cdot\mathbf{g}\Big)\,dxdt$$
$$= \int_0^{2\pi}\int_{B_\varepsilon}\mathbf{f}\cdot\eta\,dxdt, \qquad \forall t \in [0, 2\pi]. \tag{7.1.6}$$

7.2 Existence and Uniqueness of the Solution to the Problem

7.2.1 Solvability of the Problem: Case n = 2

Let $B_\varepsilon \subset \mathbb{R}^2$. In this section, we establish the theorems concerning the existence and uniqueness of solutions for the time-periodic Navier–Stokes equations in B_ε.

Theorem 7.2.1 *Let $B_\varepsilon \subset \mathbb{R}^2$ where $\partial B_\varepsilon \in C^2$ and $\mathbf{f} \in L^2(B_\varepsilon \times (0, 2\pi))$. Suppose that the extended function $\mathbf{g}$ satisfies conditions* (7.1.2)–(7.1.4) *and lies in $C^{[\frac{J+1}{2}]+2}(0, 2\pi; W^{2,2}(B_\varepsilon))$*[1]*. Then, for sufficiently small ε, the variational problem* (7.1.6) *has a solution $\mathbf{u}$ in $\overset{\circ}{W}{}^{(2,1),2}_{\partial B_\varepsilon}(B_\varepsilon^T)$ that satisfies the following estimates:*

[1] Here and below $C^l(0, 2\pi; W)$ is the space of 2π-periodic functions from $C^l(\mathbb{R}; W)$.

$$\sup_{t\in[0,2\pi]} \|\mathbf{u}(\cdot\,,t)\|^2_{L^2(B_\varepsilon)} + \varepsilon^\beta \int_0^{2\pi}\int_{B_\varepsilon} |\nabla \mathbf{u}(x,t)|^2 dxdt \leq c\varepsilon^{2+\beta}\int_0^{2\pi} A_1(t)dt, \tag{7.2.1}$$

$$\sup_{t\in[0,2\pi]} \|\nabla\mathbf{u}(\cdot\,,t)\|^2_{L^2(B_\varepsilon)} + \int_0^{2\pi}\int_{B_\varepsilon} |\mathbf{u}_t(x,t)|^2 dxdt + \varepsilon^\beta \int_0^{2\pi}\int_{B_\varepsilon} |\nabla^2 \mathbf{u}(x,t)|^2 dxdt \leq c\varepsilon^{\beta}\int_0^{2\pi} A_1(t)dt, \tag{7.2.2}$$

where $A_1(t) = \|\mathbf{f}(\cdot,t)\|^2_{L^2(B_\varepsilon)}$ *and the constant* c *is independent of* ε.

Proof

We establish the solvability of problem (7.1.5) applying the Galerkin method, using as a basis the eigenfunctions of the Stokes operator within B_ε.

1. We introduce Galerkin approximations $\mathbf{u}^{(N)}(x,t) = \sum\limits_{k=1}^{N} \gamma_k^{(N)}(t)\boldsymbol{\psi}_k(x)$ for the solution $\mathbf{u}(x,t)$ to problem (7.1.5), where the coefficients $\gamma_k^{(N)}(t)$ are defined by the system of ordinary differential equations:

$$\int_{B_\varepsilon} \Big(\frac{1}{\varepsilon^\beta}\mathbf{u}_t^{(N)}\cdot\boldsymbol{\psi}_l + \nu\nabla\mathbf{u}^{(N)}\cdot\nabla\boldsymbol{\psi}_l - \Big((\mathbf{u}^{(N)}+\mathbf{g})\cdot\nabla\Big)\boldsymbol{\psi}_l\cdot\mathbf{u}^{(N)} - (\mathbf{u}^{(N)}\cdot\nabla)\boldsymbol{\psi}_l\cdot\mathbf{g}\Big)dx = \int_{B_\varepsilon}\mathbf{f}\cdot\boldsymbol{\psi}_l dx, \quad l = 1,..,N, \tag{7.2.3}$$

with the initial condition:

$$\mathbf{u}^{(N)}(x,0) = \mathbf{u}_0^{(N)}(x). \tag{7.2.4}$$

Here, $\mathbf{u}_0^{(N)}(x) = \sum\limits_{k=1}^{N}\gamma_{0k}^{(N)}\boldsymbol{\psi}_k(x)$, and $\{\boldsymbol{\psi}_k\}_{k=1}^\infty$ is a basis in the space $\mathring{H}^{1,2}(B_\varepsilon)$.

The equalities (7.2.3) and (7.2.4) are equivalent to the Cauchy problem for the system of ordinary differential equations:

$$\begin{cases} \dfrac{1}{\varepsilon^\beta}\dfrac{d\gamma_l^{(N)}(t)}{dt} + \nu\sum_{j=1}^N a_{lj}\gamma_j^{(N)}(t) + \sum_{i,j=1}^N a_{lij}\gamma_i^{(N)}(t)\gamma_j^{(N)}(t) + \\ \qquad + \sum_{j=1}^N b_{lj}^{(1)}(t)\gamma_j^{(N)}(t) + \sum_{j=1}^N b_{lj}^{(2)}(t)\gamma_j^{(N)}(t) = f_l(t), \\ \qquad\qquad \gamma_l^{(N)}(0) = \gamma_{0k}^{(N)}, \end{cases} \tag{7.2.5}$$

where $l = 1, .., N$:

$$a_{lj} = \int_{B_\varepsilon} \nabla\boldsymbol{\psi}_l \cdot \nabla\boldsymbol{\psi}_j dx = \delta_{lj}\lambda_l, \quad a_{lij} = -\int_{B_\varepsilon} (\boldsymbol{\psi}_i \cdot \nabla)\boldsymbol{\psi}_l \cdot \boldsymbol{\psi}_j dx,$$

$$b_{lj}^{(1)}(t) = -\int_{B_\varepsilon} (\mathbf{g}(x,t)\cdot\nabla)\boldsymbol{\psi}_l \cdot \boldsymbol{\psi}_j dx, \quad b_{lj}^{(2)}(t) = -\int_{B_\varepsilon} (\boldsymbol{\psi}_j \cdot \nabla)\boldsymbol{\psi}_l \cdot \mathbf{g}(x,t) dx,$$

$$f_l(t) = \int_{B_\varepsilon} \mathbf{f}\cdot\boldsymbol{\psi}_l dx.$$

Multiplying the Eqs. (7.2.3) by $\gamma_l^{(N)}(t)$ and summing over $l = 1, \ldots, N$, we obtain:

$$\frac{1}{2\varepsilon^\beta}\frac{d}{dt}\int_{B_\varepsilon} |\mathbf{u}^{(N)}|^2 dx + \nu\int_{B_\varepsilon} |\nabla\mathbf{u}^{(N)}|^2 dx$$

$$= \int_{B_\varepsilon} (\mathbf{u}^{(N)}\cdot\nabla)\mathbf{u}^{(N)}\cdot\mathbf{g}dx + \int_{B_\varepsilon} \mathbf{f}\cdot\mathbf{u}^{(N)}dx.$$

Using (7.1.4) and the Poincaré inequality (2.3.9), we deduce:

$$\left|\int_{B_\varepsilon} (\mathbf{u}^{(N)}\cdot\nabla)\mathbf{u}^{(N)}\cdot\mathbf{g}dx\right| \le \|\mathbf{u}^{(N)}\|_{L^2(B_\varepsilon)}\|\nabla\mathbf{u}^{(N)}\|_{L^2(B_\varepsilon)}\|\mathbf{g}\|_{L^\infty(B_\varepsilon)}$$

$$\le c_1\varepsilon\|\nabla\mathbf{u}^{(N)}\|^2_{L^2(B_\varepsilon)}.$$

Therefore, if $c_1\varepsilon \le \dfrac{\nu}{4}$, then:

$$\frac{1}{2\varepsilon^\beta}\frac{d}{dt}\int_{B_\varepsilon} |\mathbf{u}^{(N)}(\cdot,t)|^2 dx + \nu\int_{B_\varepsilon} |\nabla\mathbf{u}^{(N)}(\cdot,t)|^2 dx$$

$$\le c_1\varepsilon\|\nabla\mathbf{u}^{(N)}(\cdot,t)\|^2_{L^2(B_\varepsilon)} + c\varepsilon\|\mathbf{f}\|_{L^2(B_\varepsilon)}\|\nabla\mathbf{u}^{(N)}(\cdot,t)\|_{L^2(B_\varepsilon)}$$

$$\le \frac{\nu}{2}\|\nabla\mathbf{u}^{(N)}(\cdot,t)\|^2_{L^2(B_\varepsilon)} + c\varepsilon^2\|\mathbf{f}(\cdot,t)\|^2_{L^2(B_\varepsilon)},$$

and we establish:

$$\frac{1}{2\varepsilon^\beta}\frac{d}{dt}\|\mathbf{u}^{(N)}(\cdot,t)\|^2_{L^2(B_\varepsilon)} + \frac{\nu}{2}\|\nabla\mathbf{u}^{(N)}(\cdot,t)\|^2_{L^2(B_\varepsilon)}$$
$$\leq c\varepsilon^2\|\mathbf{f}(\cdot,t)\|^2_{L^2(B_\varepsilon)} \equiv c\varepsilon^2 A_1(t). \tag{7.2.6}$$

Multiplying this relation by $2\varepsilon^\beta$ and using the Poincaré inequality yields:

$$\frac{d}{dt}\|\mathbf{u}^{(N)}(\cdot,t)\|^2_{L^2(B_\varepsilon)} + c_*\varepsilon^{\beta-2}\|\mathbf{u}^{(N)}(\cdot,t)\|^2_{L^2(B_\varepsilon)} \leq c_2\varepsilon^{\beta+2}A_1(t).$$

This inequality, with the omitted second term in the left-hand side, is multiplied by $e^{c_*\varepsilon^{\beta-2}t}$ and integrated over the interval $(0,\tau)$, where $\tau\in(0,2\pi)$:

$$\|\mathbf{u}^{(N)}(\cdot,\tau)\|^2_{L^2(B_\varepsilon)} \leq \|\mathbf{u}_0^{(N)}\|^2_{L^2(B_\varepsilon)}e^{-c_*\tau\varepsilon^{\beta-2}} + c_2\varepsilon^{\beta+2}\int\limits_0^\tau A_1(t)\cdot e^{c_*\varepsilon^{\beta-2}(t-\tau)}dt$$
$$\leq \|\mathbf{u}_0^{(N)}\|^2_{L^2(B_\varepsilon)} + c_2\varepsilon^{\beta+2}\int\limits_0^{2\pi} A_1(t)dt. \tag{7.2.7}$$

In particular:

$$\|\mathbf{u}^{(N)}(\cdot,2\pi)\|^2_{L^2(B_\varepsilon)} \leq \|\mathbf{u}_0^{(N)}\|^2_{L^2(B_\varepsilon)}e^{-c_*2\pi\varepsilon^{\beta-2}} + c_2\varepsilon^{\beta+2}\int\limits_0^{2\pi} A_1(t)dt. \tag{7.2.8}$$

From estimate (7.2.7), it follows the unique solvability of the Cauchy problem (7.2.5) over the entire interval $(0,2\pi)$.

Using the orthogonality of the basis $\{\boldsymbol{\psi}_1,\ldots,\boldsymbol{\psi}_N\}$, we have:

$$\|\mathbf{u}^{(N)}(\cdot,t)\|^2_{L^2(B_\varepsilon)} = \sum_{k=1}^N|\gamma_k^{(N)}(t)|^2. \tag{7.2.9}$$

Let us demonstrate the existence of a 2π-periodic Galerkin approximation. We follow the proof of G. Prouse [183, 184]. Denote:

$$r_0 = \sqrt{\frac{c_2\varepsilon^{\beta+2}}{1-e^{-c_*2\pi\varepsilon^{\beta-2}}}\int\limits_0^{2\pi} A_1(t)dt.}$$

From estimate (7.2.8), we obtain that for every N, the map $M:\mathbf{u}^{(N)}(0)\mapsto\mathbf{u}^{(N)}(2\pi)$ maps the ball $\mathbb{B}_{r_0}\subset\mathbb{R}^N$ of radius r_0 into itself.

Now let us show the continuity of M in $\mathbb{B}_{r_0}$. Let $\{\gamma_{01}^{(N,1)}, \dots, \gamma_{0N}^{(N,1)}\}$ and $\{\gamma_{01}^{(N,2)}, \dots, \gamma_{0N}^{(N,2)}\}$ be two sets of initial data belonging to $\mathbb{B}_{r_0}$ and $\{\gamma_1^{(N,1)}(t), \dots, \gamma_N^{(N,1)}(t)\}$ and $\{\gamma_1^{(N,2)}(t), \dots, \gamma_N^{(N,2)}(t)\}$ be two solutions of problem (7.2.5) corresponding to these initial data. Set $\mathbf{u}^{(N,j)}(x,t) = \sum\limits_{k=1}^{N} \gamma_k^{(N,j)}(t)\boldsymbol{\psi}_k(x)$ for $j = 1, 2$. Then the difference:

$$\mathbf{w}^{(N)}(x,t) = \mathbf{u}^{(N,1)}(x,t) - \mathbf{u}^{(N,2)}(x,t) = \sum_{k=1}^{N} \theta_k^{(N)}(t)\boldsymbol{\psi}_k(x),$$

where $\theta_k^{(N)}(t) = \gamma_k^{(N,1)}(t) - \gamma_k^{(N,2)}(t)$, satisfies the identities:

$$\int\limits_{B_\varepsilon} \Big(\frac{1}{\varepsilon^\beta}\mathbf{w}_t^{(N)} \cdot \boldsymbol{\psi}_l + \nu\nabla\mathbf{w}^{(N)} \cdot \nabla\boldsymbol{\psi}_l\Big)dx = \int\limits_{B_\varepsilon} \Big((\mathbf{u}^{(N,1)} \cdot \nabla))\boldsymbol{\psi}_l \cdot \mathbf{w}^{(N)}$$
$$+(\mathbf{w}^{(N)} \cdot \nabla))\boldsymbol{\psi}_l \cdot \mathbf{u}^{(N,2)}dx + (\mathbf{g} \cdot \nabla)\boldsymbol{\psi}_l \cdot \mathbf{w}^{(N)} + (\mathbf{w}^{(N)} \cdot \nabla)\boldsymbol{\psi}_l \cdot \mathbf{g}\Big)dx, \tag{7.2.10}$$

$l = 1, .., N$, and the initial conditions:

$$\theta_l^{(N)}(0) = \gamma_l^{(N,1)}(0) - \gamma_l^{(N,2)}(0) = \gamma_{0l}^{(N,1)} - \gamma_{0l}^{(N,2)} = \theta_{0l}^{(N)}, \quad l = 1, \dots, N. \tag{7.2.11}$$

The relations (7.2.10) and (7.2.11) can be written in the form:

$$\begin{cases} \dfrac{1}{\varepsilon^\beta}\dfrac{d\theta_l^{(N)}(t)}{dt} + \nu\lambda_l\theta_l^{(N)}(t) + \sum\limits_{i,j=1}^{N} a_{lij}\big[\theta_i^{(N)}(t)\gamma_j^{(N,2)}(t) + \theta_j^{(N)}(t)\gamma_i^{(N,1)}(t)\big] \\ \qquad\qquad + \sum\limits_{j=1}^{N} b_{lj}^{(1)}(t)\theta_j^{(N)}(t) + \sum\limits_{j=1}^{N} b_{lj}^{(2)}(t)\theta_j^{(N)}(t) = 0, \\ \qquad\qquad \theta_l^{(N)}(0) = \theta_{0l}^{(N)}, \end{cases} \tag{7.2.12}$$

$l = 1, .., N$. Here λ_l are the eigenvalues of the Stokes operator. Multiplying equations (7.2.12), respectively, by $\theta_1^{(N)}(t), \dots, \theta_N^{(N)}(t)$ and summing the obtained equalities, due to (7.2.9), we obtain:

$$\frac{1}{2\varepsilon^\beta}\frac{d}{dt}\|\mathbf{w}^{(N)}(\cdot,t)\|_{L^2(B_\varepsilon)}^2 + \nu\lambda_1\|\mathbf{w}^{(N)}(\cdot,t)\|_{L^2(B_\varepsilon)}^2$$
$$+ \sum_{i,j=1}^{N} a_{lij}\theta_i^{(N)}(t)\theta_j^{(N)}(t)\gamma_j^{(N,2)}(t) + \sum_{j=1}^{N} b_{lj}^{(2)}(t)\theta_j^{(N)}(t)\theta_i^{(N)}(t) \le 0. \tag{7.2.13}$$

Here we have used the facts that $a_{lij} = -a_{lji}$, $\lambda_1 \leq \lambda_l$, and:

$$\sum_{i,j=1}^{N} b_{lj}^{(1)}(t)\theta_j^{(N)}(t)\theta_i^{(N)}(t) = -\int_{B_\varepsilon} (\mathbf{g}(x,t)\cdot\nabla)\mathbf{w}^{(N)}\cdot\mathbf{w}^{(N)}dx = 0.$$

Since the initial data lies in $\mathbb{B}_{r_0}$, from (7.2.7), we get that:

$$\max_{t\in[0,T]} \|\mathbf{u}^{(N,2)}(\cdot,t)\|^2_{L^2(B_\varepsilon)} = \max_{t\in[0,T]} \sum_{i=1}^{N} |\gamma_i^{(N,2)}(t)|^2$$

$$\leq \sum_{i=1}^{N} |\gamma_{0i}^{(N,2)}|^2 + c\varepsilon^{\beta+2}\int_0^{2\pi} A_1(t)dt \leq 2r_0^2.$$

From this inequality and (7.1.4), we obtain:

$$\Big|\sum_{i,j=1}^{N} a_{lij}\theta_i^{(N)}(t)\theta_j^{(N)}(t)\gamma_j^{(N,2)}(t)\Big|$$

$$\leq N \max_{l,i,j=1,\ldots,N} |a_{lij}| \max_{t\in[0,2\pi]} \|\mathbf{u}^{(N,2)}(\cdot,t)\|_{L^2(B_\varepsilon)} \sum_{i,j=1}^{N} |\theta_i^{(N)}(t)||\theta_j^{(N)}(t)|$$

$$\leq N^2 \max_{l,i,j=1,\ldots,N} |a_{lij}| \max_{t\in[0,2\pi]} \|\mathbf{u}^{(N,2)}(\cdot,t)\|_{L^2(B_\varepsilon)} \sum_{i=1}^{N} |\theta_i^{(N)}(t)|^2$$

$$\leq c_*\|\mathbf{w}^{(N)}(\cdot,t)\|^2_{L^2(B_\varepsilon)},$$

and:

$$\Big|\sum_{j=1}^{N} b_{lj}^{(2)}(t)\theta_j^{(N)}(t)\theta_i^{(N)}(t)\Big| \leq N \sup_{(x,t)\in B_\varepsilon\times(0,2\pi)} |\mathbf{g}(x,t)| \sum_{i=1}^{N} |\theta_i^{(N)}(t)|^2$$

$$\leq c_{**}\|\mathbf{w}^{(N)}(\cdot,t)\|^2_{L^2(B_\varepsilon)}.$$

Now, from (7.2.13), we deduce that:

$$\frac{1}{\varepsilon^\beta}\frac{d}{dt}\|\mathbf{w}^{(N)}(\cdot,t)\|_{L^2(B_\varepsilon)} \leq (c_0 - \nu\lambda_1)\|\mathbf{w}^{(N)}(\cdot,t)\|_{L^2(B_\varepsilon)},$$

where $c_0 = c_* + c_{**}$, and, thus:

$$\|\mathbf{w}^{(N)}(\cdot,t)\|_{L^2(B_\varepsilon)} \le \|\mathbf{w}_0^{(N)}\|_{L^2(B_\varepsilon)} e^{\varepsilon^\beta(c_0-\nu\lambda_1)t}. \tag{7.2.14}$$

The last inequality means that the operator M admits the estimate:

$$|M(\gamma_{01}^{(N,1)},\dots,\gamma_{0N}^{(N,1)})-M(\gamma_{01}^{(N,2)},\dots,\gamma_{0N}^{(N,2)})| = \sum_{i=1}^{N} |\gamma_i^{(N,1)}(2\pi)-\gamma_i^{(N,2)}(2\pi)|$$

$$\le e^{\varepsilon^\beta(c_0-\nu\lambda_1)2\pi} \sum_{i=1}^{N} |\gamma_{0i}^{(N,1)} - \gamma_{0i}^{(N,2)}|$$

and, hence, M is continuous. Therefore, by Brouwer's fixed point theorem (see Theorem 2.2.9), the map M has a fixed point in $\mathbb{B}_{r_0}$, which ensures for each fixed N the existence of 2π-periodic Galerkin approximations.

2. We now establish a series of preliminary estimates for the periodic-in-time Galerkin approximations $\mathbf{u}^{(N)}$.

2.1. By integrating the inequality (7.2.6) and considering the periodicity condition $\mathbf{u}^{(N)}(x,0) = \mathbf{u}^{(N)}(x,2\pi)$, we can deduce the following result:

$$\int_0^{2\pi}\int_{B_\varepsilon} |\nabla \mathbf{u}^{(N)}(x,t)|^2 dxdt \le c\varepsilon^2 \int_0^{2\pi} A_1(t)dt. \tag{7.2.15}$$

2.2. Next, we multiply (7.2.6) by t and integrate it over the interval $(0,2\pi)$. After some straightforward manipulation, this yields the inequality:

$$\begin{aligned} &\frac{\pi}{\varepsilon^\beta}\int_{B_\varepsilon} |\mathbf{u}^{(N)}(x,2\pi)|^2\,dx + \nu\int_0^{2\pi} t\int_{B_\varepsilon} |\nabla \mathbf{u}^{(N)}(x,t)|^2\,dxdt \\ &\le 2\pi^2\varepsilon^2 \int_0^{2\pi}\int_{B_\varepsilon} |\mathbf{f}(x,t)|^2\,dxdt + \int_0^{2\pi}\int_{B_\varepsilon} |\mathbf{u}^{(N)}(x,t)|^2\,dxdt \\ &\le 2\pi^2\varepsilon^2 \int_0^{2\pi}\int_{B_\varepsilon} |\mathbf{f}(x,t)|^2\,dxdt + c\varepsilon^2\int_0^{2\pi}\int_{B_\varepsilon} |\nabla \mathbf{u}^{(N)}(x,t)|^2\,dxdt \\ &\overset{(7.2.15)}{\le} c\varepsilon^2 \int_0^{2\pi}\int_{B_\varepsilon} |\mathbf{f}(x,t)|^2\,dxdt. \end{aligned} \tag{7.2.16}$$

Utilizing the periodicity condition $\mathbf{u}^{(N)}(x,0) = \mathbf{u}^{(N)}(x,2\pi)$, we get from (7.2.16) the inequality:

$$\int_{B_\varepsilon} |\mathbf{u}^{(N)}(x,0)|^2\,dx \le c\varepsilon^{2+\beta} \int_0^{2\pi}\int_{B_\varepsilon} |\mathbf{f}(x,t)|^2\,dxdt. \tag{7.2.17}$$

Applying mean value theorem for Lebesgue integrals (see Theorem 2.3.6) to the function $\|\nabla\mathbf{u}^{(N)}(\cdot,t)\|^2_{L^2(B_\varepsilon)}$, we deduce, as a consequence of inequality (7.2.15), the existence of a point $t_* \in (0, 2\pi)$ for which the following inequality holds:

$$\|\nabla\mathbf{u}^{(N)}(\cdot,t_*)\|^2_{L^2(B_\varepsilon)} \le c\varepsilon^2 \int_0^{2\pi} A_1(t)dt. \tag{7.2.18}$$

Continuing, by integrating inequality (7.2.6) over the interval $(0,\tau)$, we obtain:

$$\frac{1}{2\varepsilon^\beta}\int_{B_\varepsilon} |\mathbf{u}^{(N)}(x,\tau)|^2\,dx + \frac{\nu}{2}\int_0^{\tau}\int_{B_\varepsilon} |\nabla\mathbf{u}^{(N)}(x,t)|^2\,dxdt$$
$$\le \frac{1}{2\varepsilon^\beta}\int_{B_\varepsilon} |\mathbf{u}^{(N}(x,0)|^2\,dx + c\varepsilon^2\int_0^{\tau}\int_{B_\varepsilon} |\mathbf{f}(x,t)|^2\,dxdt \le c\varepsilon^2\int_0^{\tau}\int_{B_\varepsilon} |\mathbf{f}(x,t)|^2\,dxdt.$$

As the constant in the last inequality is independent of τ, we deduce:

$$\max_{t\in[0,2\pi]} \|\mathbf{u}^{(N)}(\cdot,t)\|^2_{L^2(B_\varepsilon)} \le c\varepsilon^{2+\beta}\int_0^{2\pi}\int_{B_\varepsilon} |\mathbf{f}(x,t)|^2\,dxdt. \tag{7.2.19}$$

2.3. Moving forward, by multiplying (7.2.3) by $\lambda_k\gamma_k^{(N)}(t)$ and summing from $k=1$ to $k=N$, we arrive at the following expression:

$$\int_{B_\varepsilon}\Big(\frac{1}{\varepsilon^\beta}\sum_{k=1}^{N}\mathbf{u}_t^{(N)}\cdot\boldsymbol{\psi}_l\lambda_k\gamma_k^{(N)}(t) - \nu\sum_{k=1}^{N}\Delta\mathbf{u}^{(N)}\cdot\boldsymbol{\psi}_l\lambda_k\gamma_k^{(N)}(t)$$
$$-\sum_{k=1}^{N}((\mathbf{u}^{(N)}+\mathbf{g})\cdot\nabla)\lambda_k\gamma_k^{(N)}(t)\boldsymbol{\psi}_l\cdot\mathbf{u}^{(N)} - \sum_{k=1}^{N}(\mathbf{u}^{(N)}\cdot\nabla)\lambda_k\gamma_k^{(N)}(t)\boldsymbol{\psi}_l\cdot\mathbf{g}\Big)dx$$
$$= \int_{B_\varepsilon}\sum_{k=1}^{N}\mathbf{f}\cdot\boldsymbol{\psi}_l\lambda_k\gamma_k^{(N)}(t)dx,\ l=1,\ldots,N.$$

Utilizing the properties of the Stokes operator (see Sects. 4.2 and 4.3 of Chap. 4), the last equality can be rewritten as follows:

$$\int\limits_{B_\varepsilon} \Big(\frac{1}{\varepsilon^\beta}\mathbf{u}_t^{(N)} \cdot \widetilde{\Delta}\mathbf{u}^{(N)} - \nu\Delta\mathbf{u}^{(N)} \cdot \widetilde{\Delta}\mathbf{u}^{(N)} + ((\mathbf{u}^{(N)} + \mathbf{g}) \cdot \nabla)\mathbf{u}^{(N)} \cdot \widetilde{\Delta}\mathbf{u}^{(N)}$$

$$+(\mathbf{u}^{(N)} \cdot \nabla) \cdot \mathbf{g} \cdot \widetilde{\Delta}\mathbf{u}^{(N)}\Big)dx = \int\limits_{B_\varepsilon} \mathbf{f} \cdot \widetilde{\Delta}\mathbf{u}^{(N)} dx.$$

This can be equivalently expressed as (see the properties of the Stokes operator $\widetilde{\Delta}$):

$$\frac{\nu}{2\varepsilon^\beta}\frac{d}{dt}\int\limits_{B_\varepsilon} |\nabla\mathbf{u}^{(N)}|^2 dx + \int\limits_{B_\varepsilon} |\widetilde{\Delta}\mathbf{u}^{(N)}|^2 dx = -\int\limits_{B_\varepsilon} ((\mathbf{u}^{(N)} + \mathbf{g}) \cdot \nabla)\mathbf{u}^{(N)} \cdot \widetilde{\Delta}\mathbf{u}^{(N)} dx$$

$$-\int\limits_{B_\varepsilon} (\mathbf{u}^{(N)} \cdot \nabla)\mathbf{g} \cdot \widetilde{\Delta}\mathbf{u}^{(N)} dx + \int\limits_{B_\varepsilon} \mathbf{f} \cdot \widetilde{\Delta}\mathbf{u}^{(N)} dx = \sum_{i=1}^{3} J_i. \tag{7.2.20}$$

We proceed to estimate the right-hand side of (7.2.20). Utilizing the inequality (2.3.37), we can bound the terms J_3 and J_2 as follows:

$$\begin{aligned}
|J_3| &\le \delta \int\limits_{B_\varepsilon} |\widetilde{\Delta}\mathbf{u}^{(N)}|^2 dx + c_\delta A_1(t),\\
|J_2| &\le \|\mathbf{u}^{(N)}\|_{L^4(B_\varepsilon)} \|\nabla\mathbf{g}\|_{L^4(B_\varepsilon)} \|\widetilde{\Delta}\mathbf{u}^{(N)}\|_{L^2(B_\varepsilon)}\\
&\le c\varepsilon^{\frac{1}{2}} \|\nabla\mathbf{u}^{(N)}\|_{L^2(B_\varepsilon)} \|\nabla\mathbf{g}\|_{L^4(B_\varepsilon)} \|\widetilde{\Delta}\mathbf{u}^{(N)}\|_{L^2(B_\varepsilon)}\\
&\le c_\delta\varepsilon \|\nabla\mathbf{u}^{(N)}\|^2_{L^2(B_\varepsilon)} \|\nabla\mathbf{g}\|^2_{L^4(B_\varepsilon)} + \delta\|\widetilde{\Delta}\mathbf{u}^{(N)}\|^2_{L^2(B_\varepsilon)}.
\end{aligned} \tag{7.2.21}$$

Using inequalities (2.3.39) and (7.1.4), we find:

$$\|\nabla\mathbf{g}\|^2_{L^4(B_\varepsilon)} \le c\varepsilon^{-1}\|\nabla\mathbf{g}\|_{L^2(B_\varepsilon)}\Big(\|\nabla\mathbf{g}\|^2_{L^2(B_\varepsilon)} + \varepsilon^2\|\nabla^2\mathbf{g}\|^2_{L^2(B_\varepsilon)}\Big)^{1/2} \le c\varepsilon^{-2}.$$

Thus, we can write:

$$|J_2| \le c_\delta\varepsilon^{-1}\|\nabla\mathbf{u}^{(N)}\|^2_{L^2(B_\varepsilon)} + \delta\|\widetilde{\Delta}\mathbf{u}^{(N)}\|^2_{L^2(B_\varepsilon)}. \tag{7.2.22}$$

Let us pass to J_1. Using (7.1.4), we have:

$$\Big|\int\limits_{B_\varepsilon}(\mathbf{g}\cdot\nabla)\mathbf{u}^{(N)}\cdot\widetilde{\Delta}\mathbf{u}^{(N)}dx\Big| \le \|\mathbf{g}\|_{L^\infty(B_\varepsilon)}\|\nabla\mathbf{u}^{(N)}\|_{L^2(B_\varepsilon)}\|\widetilde{\Delta}\mathbf{u}^{(N)}\|_{L^2(B_\varepsilon)}$$
$$\le c_\delta\|\nabla\mathbf{u}^{(N)}\|^2_{L^2(B_\varepsilon)} + \delta\|\widetilde{\Delta}\mathbf{u}^{(N)}\|^2_{L^2(B_\varepsilon)}. \tag{7.2.23}$$

Finally, applying (2.3.37) and (2.3.39), we arrive at:

$$\Big|\int\limits_{B_\varepsilon}(\mathbf{u}^{(N)}\cdot\nabla)\mathbf{u}^{(N)}\cdot\widetilde{\Delta}\mathbf{u}^{(N)}dx\Big| \le \|\mathbf{u}^{(N)}\|_{L^4(B_\varepsilon)}\|\nabla\mathbf{u}^{(N)}\|_{L^4(B_\varepsilon)}\|\widetilde{\Delta}\mathbf{u}^{(N)}\|_{L^2(B_\varepsilon)}$$
$$\le c_\delta\varepsilon^{-1}\|\mathbf{u}^{(N)}\|_{L^2(B_\varepsilon)}\|\nabla\mathbf{u}^{(N)}\|^2_{L^2(B_\varepsilon)}\Big(\|\nabla\mathbf{u}^{(N)}\|^2_{L^2(B_\varepsilon)} + \varepsilon^2\|\nabla^2\mathbf{u}^{(N)}\|^2_{L^2(B_\varepsilon)}\Big)^{1/2}$$
$$+\frac{\delta}{2}\|\widetilde{\Delta}\mathbf{u}^{(N)}\|^2_{L^2(B_\varepsilon)} \le c_\delta\varepsilon^{-1}\|\mathbf{u}^{(N)}\|_{L^2(B_\varepsilon)}\|\nabla\mathbf{u}^{(N)}\|^3_{L^2(B_\varepsilon)}$$
$$+c_\delta\|\mathbf{u}^{(N)}\|_{L^2(B_\varepsilon)}\|\nabla\mathbf{u}^{(N)}\|^2_{L^2(B_\varepsilon)}\|\nabla^2\mathbf{u}^{(N)}\|_{L^2(B_\varepsilon)} + \frac{\delta}{2}\|\widetilde{\Delta}\mathbf{u}^{(N)}\|^2_{L^2(B_\varepsilon)}.$$
$$\le c_\delta(1+\|\mathbf{u}^{(N)}\|^2_{L^2(B_\varepsilon)})\|\nabla\mathbf{u}^{(N)}\|^4_{L^2(B_\varepsilon)} + \delta\|\widetilde{\Delta}\mathbf{u}^{(N)}\|^2_{L^2(B_\varepsilon)}. \tag{7.2.24}$$

Substituting the estimates (7.2.21)–(7.2.24) into (7.2.20), using (7.2.19) and taking $\delta = \frac{1}{8}$ gives:

$$\frac{\nu}{\varepsilon^\beta}\frac{d}{dt}\int\limits_{B_\varepsilon}|\nabla\mathbf{u}^{(N)}|^2dx + \int\limits_{B_\varepsilon}|\widetilde{\Delta}\mathbf{u}^{(N)}|^2dx \le C_1\big(1+\varepsilon^{2+\beta}\int\limits_0^{2\pi}A_1(t)dt\big)\|\nabla\mathbf{u}^{(N)}\|^4_{L^2(B_\varepsilon)}$$
$$+C_3\big(1+\varepsilon^{-1}\big)\|\nabla\mathbf{u}^{(N)}\|^2_{L^2(B_\varepsilon)} + C_2A_1(t). \tag{7.2.25}$$

Let us define $Y(t) = \int\limits_{B_\varepsilon}|\nabla\mathbf{u}^{(N)}(x,t)|^2dx,\quad d_1 = C_1\big(1+\varepsilon^{2+\beta}\int\limits_0^{2\pi}A_1(t)dt\big).$

Then, we can rewrite (7.2.25) as:

$$Y'(t) \le \frac{\varepsilon^\beta}{\nu}d_1\|\nabla\mathbf{u}^{(N)}\|^2_{L^2(B_\varepsilon)}Y(t) + C_3\frac{\varepsilon^\beta}{\nu}(1+\varepsilon^{-1})\|\nabla\mathbf{u}^{(N)}\|^2_{L^2(B_\varepsilon)} + \frac{\varepsilon^\beta}{\nu}C_2A_1(t).$$

Hence, multiplying the above inequality by $e^{-\frac{\varepsilon^\beta}{\nu}\int\limits_{t_*}^{t} d_1\|\nabla\mathbf{u}^{(N)}(\cdot,\tau)\|^2_{L^2(B_\varepsilon)}d\tau}$, where $t > t_*$, we obtain:

$$\left(Y(t)e^{-\frac{\varepsilon^\beta}{\nu}\int\limits_{t_*}^{t} d_1\|\nabla\mathbf{u}^{(N)}(\cdot,\tau)\|^2_{L^2(B_\varepsilon)}\,d\tau}\right)'$$
$$\le C_4\varepsilon^\beta\Big((1+\varepsilon^{-1})\|\nabla\mathbf{u}^{(N)}\|^2_{L^2(B_\varepsilon)}+A_1(t)\Big)e^{-\frac{\varepsilon^\beta}{\nu}\int\limits_{t_*}^{t} d_1\|\nabla\mathbf{u}^{(N)}(\cdot,\tau)\|^2_{L^2(B_\varepsilon)}\,d\tau}$$
$$\le C_4\varepsilon^\beta\Big((1+\varepsilon^{-1})\|\nabla\mathbf{u}^{(N)}\|^2_{L^2(B_\varepsilon)}+A_1(t)\Big).$$

Integrating the last inequality by t over the interval (t_*,τ), where $\tau\in(t_*,2\pi+t_*)$, and using (7.2.15), (7.2.19), and (7.2.18) furnishes:

$$\begin{aligned}Y(\tau)\le cY(t_*)+c\varepsilon^\beta(1+\varepsilon^{-1})\int\limits_{t_*}^{\tau}\|\nabla\mathbf{u}^{(N)}(\cdot,t)\|^2_{L^2(B_\varepsilon)}dt+\varepsilon^\beta\int\limits_{t_*}^{\tau}A_1(t)\,dt\\ \le c\varepsilon^\beta\int\limits_0^{2\pi}A_1(t)dt.\end{aligned}\tag{7.2.26}$$

Thus, we find:

$$\sup_{t\in[0,2\pi]}\|\nabla\mathbf{u}^{(N)}(\,\cdot\,,t)\|^2_{L^2(B_\varepsilon)}\le c\varepsilon^\beta\int\limits_0^{2\pi}A_1(t)dt.\tag{7.2.27}$$

2.4. Substituting (7.2.27) into (7.2.25) and integrating over $(0,2\pi)$ imply:

$$\begin{aligned}\int\limits_0^{2\pi}\int\limits_{B_\varepsilon}|\widetilde{\Delta}\mathbf{u}^{(N)}|^2dxdt\le c\int\limits_0^{2\pi}A_1(t)dt\\ +c\varepsilon^\beta\Big(1+\varepsilon^{2+\beta}\int\limits_0^{2\pi}A_1(t)dt\Big)\int\limits_0^{2\pi}A_1(t)dt\int\limits_0^{2\pi}\|\nabla\mathbf{u}^{(N)}\|^2_{L^2(B_\varepsilon)}dt\\ +c(1+\varepsilon^{-1})\int\limits_0^{2\pi}\|\nabla\mathbf{u}^{(N)}\|^2_{L^2(B_\varepsilon)}dt\le C\int\limits_0^{2\pi}A_1(t)dt.\end{aligned}\tag{7.2.28}$$

Thus:

$$\int\limits_0^{2\pi}\int\limits_{B_\varepsilon}|\widetilde{\Delta}\mathbf{u}^{(N)}|^2dxdt\le c\int\limits_0^{2\pi}A_1(t)dt\tag{7.2.29}$$

2.5. Let us estimate the norm of $\mathbf{u}_t^{(N)}$. We multiply equalities (7.2.3) by $\frac{d}{dt}\gamma_k^{(N)}$ and sum up the resulting relations by k from 1 to N:

$$\frac{1}{\varepsilon^\beta}\int\limits_{B_\varepsilon}|\mathbf{u}_t^{(N)}|^2dx+\frac{\nu}{2}\frac{d}{dt}\int\limits_{B_\varepsilon}|\nabla\mathbf{u}^{(N)}|^2dx$$

$$=-\int\limits_{B_\varepsilon}((\mathbf{u}^{(N)}+\mathbf{g})\cdot\nabla)\mathbf{u}^{(N)}\cdot\mathbf{u}_t dx-\int\limits_{B_\varepsilon}(\mathbf{u}^{(N)}\cdot\nabla)\mathbf{g}\cdot\mathbf{u}_t^{(N)}dx+\int\limits_{B_\varepsilon}\mathbf{f}\cdot\mathbf{u}_t^{(N)}dx$$

$$\leq(\|\mathbf{u}^{(N)}\|_{L^\infty(B_\varepsilon)}+\|\mathbf{g}\|_{L^\infty(B_\varepsilon)})\|\nabla\mathbf{u}^{(N)}\|_{L^2(B_\varepsilon)}\|\mathbf{u}_t^{(N)}\|_{L^2(B_\varepsilon)}$$

$$+\|\mathbf{u}^{(N)}\|_{L^\infty(B_\varepsilon)}\|\nabla\mathbf{g}\|_{L^2(B_\varepsilon)}\|\mathbf{u}_t^{(N)}\|_{L^2(B_\varepsilon)}+\|\mathbf{f}\|_{L^2(B_\varepsilon)}\|\mathbf{u}_t^{(N)}\|_{L^2(B_\varepsilon)}$$

$$\leq\delta\|\mathbf{u}_t^{(N)}\|^2+c_\delta\|\mathbf{u}^{(N)}\|^2_{L^\infty(B_\varepsilon)}\|\nabla\mathbf{u}^{(N)}\|^2_{L^2(B_\varepsilon)}+c_\delta\|\mathbf{g}\|^2_{L^\infty(B_\varepsilon)}\|\nabla\mathbf{u}^{(N)}\|^2_{L^2(B_\varepsilon)}$$

$$+c_\delta\|\mathbf{u}^{(N)}\|^2_{L^\infty(B_\varepsilon)}\|\nabla\mathbf{g}\|^2_{L^2(B_\varepsilon)}+c_\delta A_1(t).$$

By selecting a sufficiently small value of δ and integrating over the interval $[0,2\pi]$, we derive the following inequality:

$$\Big(\frac{1}{\varepsilon^\beta}-\frac{1}{2}\Big)\int\limits_0^{2\pi}\int\limits_{B_\varepsilon}|\mathbf{u}_t^{(N)}|^2dxdt\leq c\int\limits_0^{2\pi}A_1(t)dt$$

$$+c\int\limits_0^{2\pi}\|\mathbf{u}^{(N)}\|^2_{L^\infty(B_\varepsilon)}\|\nabla\mathbf{u}^{(N)}\|^2_{L^2(B_\varepsilon)}dt$$

$$+c\int\limits_0^{2\pi}\|\mathbf{g}\|^2_{L^\infty(B_\varepsilon)}\|\nabla\mathbf{u}^{(N)}\|^2_{L^2(B_\varepsilon)}dt+c\int\limits_0^{2\pi}\|\nabla\mathbf{g}\|^2_{L^2(B_\varepsilon)}\|\mathbf{u}^{(N)}\|^2_{L^\infty(B_\varepsilon)}dt$$

$$\stackrel{(4.3.15)}{\leq}c\int\limits_0^{2\pi}A_1(t)dt+c\varepsilon\sup_{t\in[0,2\pi]}\|\nabla\mathbf{u}^{(N)}\|^2_{L^2(B_\varepsilon)}\int\limits_0^{2\pi}\|\nabla^2\mathbf{u}^{(N)}\|^2_{L^2(B_\varepsilon)}dt$$

$$+c\sup_{t\in[0,2\pi]}\|\mathbf{g}\|^2_{L^\infty(B_\varepsilon)}\int\limits_0^{2\pi}\|\nabla\mathbf{u}^{(N)}\|^2_{L^2(B_\varepsilon)}dt$$

$$+c\varepsilon\sup_{t\in[0,2\pi]}\|\nabla\mathbf{g}\|^2_{L^2(B_\varepsilon)}\int\limits_0^{2\pi}\|\nabla^2\mathbf{u}^{(N)}\|^2_{L^2(B_\varepsilon)}dt$$

$$\overset{(4.3.13)}{\leq} c \int_0^{2\pi} A_1(t)dt + c\varepsilon \sup_{t\in[0,2\pi]} \|\nabla \mathbf{u}^{(N)}\|^2_{L^2(B_\varepsilon)} \int_0^{2\pi} \|\widetilde{\Delta}\mathbf{u}^{(N)}\|^2_{L^2(B_\varepsilon)} dt$$

$$+c \sup_{t\in[0,2\pi]} \|\mathbf{g}\|^2_{L^\infty(B_\varepsilon)} \int_0^{2\pi} \|\nabla \mathbf{u}^{(N)}\|^2_{L^2(B_\varepsilon)} dt$$

$$+c\varepsilon \sup_{t\in[0,2\pi]} \|\nabla \mathbf{g}\|^2_{L^2(B_\varepsilon)} \int_0^{2\pi} \|\widetilde{\Delta}\mathbf{u}^{(N)}\|^2_{L^2(B_\varepsilon)} dt$$

$$\overset{(7.1.4),\,(7.2.15),\,(7.2.27),\,(7.2.28)}{\leq} C \int_0^{2\pi} A_1(t)dt.$$

Thus:

$$\int_0^{2\pi}\int_{B_\varepsilon} |\mathbf{u}_t^{(N)}|^2 dxdt \leq c\varepsilon^\beta \int_0^{2\pi} A_1(t)dt. \tag{7.2.30}$$

3. The validity of estimates (7.2.15), (7.2.29), and (7.2.30) ensures the convergence of the Galerkin approximation and guarantees the existence of the solution $\mathbf{u}$. The demonstration of this assertion closely parallels the proof establishing the convergence of Galerkin approximations to the solution of the initial boundary value problem for the Navier–Stokes equations, as presented in Theorem 6.2.1. With this, the proof is concluded. □

Theorem 7.2.2 *For sufficiently small values of ε, the solution to problem* (7.1.6) *for $n = 2$ is unique.*

Proof

Let us assume the existence of two solutions $\mathbf{u}_1$ and $\mathbf{u}_2$ to problem (7.1.6) that satisfy the conditions outlined in Theorem 7.2.1. By subtracting the integral identity (7.1.6) for $\mathbf{u}_2$ from that of $\mathbf{u}_1$, we obtain the following identity for the difference $\mathbf{w} = \mathbf{u}_1 - \mathbf{u}_2$:

$$\int_0^{2\pi}\int_{B_\varepsilon} \Big(\frac{1}{\varepsilon^\beta}\mathbf{w}_t \cdot \boldsymbol{\eta} + \nu\nabla\mathbf{w}\cdot\nabla\boldsymbol{\eta} - (\mathbf{u}_1\cdot\nabla)\boldsymbol{\eta}\cdot\mathbf{w} - (\mathbf{w}\cdot\nabla)\boldsymbol{\eta}\cdot\mathbf{u}_2$$

$$-(\mathbf{g}\cdot\nabla)\boldsymbol{\eta}\cdot\mathbf{w} - (\mathbf{w}\cdot\nabla)\boldsymbol{\eta}\cdot\mathbf{g}\Big)\,dxdt = 0,$$

where $\boldsymbol{\eta} \in \overset{\circ}{W}{}^{(1,0),2}_{\partial B_\varepsilon}(B_\varepsilon \times (0, 2\pi))$, $\eta(\cdot, t) \in \overset{\circ}{H}{}^{1,2}(B_\varepsilon)$. Choosing $\boldsymbol{\eta} = \mathbf{w}$ leads to:

$$\int_0^{2\pi} \frac{1}{2\varepsilon^\beta} \frac{d}{dt} \int_{B_\varepsilon} |\mathbf{w}|^2\, dx + \nu \int_{B_\varepsilon} |\nabla \mathbf{w}|^2\, dxdt$$

$$= \int_0^{2\pi}\int_{B_\varepsilon} (\mathbf{w}\cdot\nabla)\mathbf{w}\cdot\mathbf{u}_2\, dxdt + \int_0^{2\pi}\int_{B_\varepsilon} (\mathbf{w}\cdot\nabla)\mathbf{w}\cdot\mathbf{g}\, dxdt.$$

The first integral in the above equality is zero due to the periodicity of $\mathbf{w}$. Utilizing inequalities and estimates from earlier in the proof, we obtain:

$$\nu \int_0^{2\pi}\int_{B_\varepsilon} |\nabla \mathbf{w}|^2\, dxdt \leq \int_0^{2\pi} \|\mathbf{w}\|_{L^4(B_\varepsilon)} \|\nabla \mathbf{w}\|_{L^2(B_\varepsilon)} \Big(\|\mathbf{u}_2\|_{L^4(B_\varepsilon)} + \|\mathbf{g}\|_{L^4(B_\varepsilon)}\Big) dt$$

$$\overset{(2.3.39),\,(2.3.37)}{\leq} c\varepsilon \int_0^{2\pi} \|\nabla \mathbf{w}\|^2_{L^2(B_\varepsilon)} \Big(\|\nabla \mathbf{u}_2\|_{L^2(B_\varepsilon)} + \|\nabla \mathbf{g}\|_{L^2(B_\varepsilon)}\Big) dt$$

$$\leq c\varepsilon \left(\sup_{t\in[0,2\pi]} \|\nabla \mathbf{u}_2\|_{L^2(B_\varepsilon)} + \sup_{t\in[0,2\pi]} \|\nabla \mathbf{g}\|_{L^2(B_\varepsilon)} \right) \int_0^{2\pi} \|\nabla \mathbf{w}\|^2_{L^2(B_\varepsilon)} dt$$

$$\leq c\varepsilon \left(\left(\varepsilon^\beta \int_0^{2\pi} A_1(t)\, dt \right)^{1/2} + \varepsilon^{-1/2} \right) \int_0^{2\pi} \|\nabla \mathbf{w}\|^2_{L^2(B_\varepsilon)} dt.$$

For sufficiently small ε (i.e., $c\varepsilon\Big(\Big(\varepsilon^\beta \int_0^{2\pi} A_1(t)\, dt\Big)^{1/2} + \varepsilon^{-1/2}\Big) < \nu$), this implies

$$\int_0^{2\pi}\int_{B_\varepsilon} |\nabla \mathbf{w}|^2\, dxdt = 0,$$

which leads to $\mathbf{u}_1 = \mathbf{u}_2$. Thus, the uniqueness of the solution is established. □

Theorem 7.2.3 *Assuming the conditions stated in Theorem 7.2.1 are met, there exists a time-periodic function $p \in L^2(0, 2\pi; L^2(B_\varepsilon))$ such that the pair $(\mathbf{u}, p)$ satisfies the integral identity:*

$$\int_{B_\varepsilon} \Big(\mathbf{u}_t \cdot \eta + \nu \nabla \mathbf{u} \cdot \nabla \eta - \big((\mathbf{u}+\mathbf{g})\cdot\nabla\big)\eta\cdot\mathbf{u}$$

$$- (\mathbf{u}\cdot\nabla)\eta\cdot\mathbf{g}\Big) dx = \int_{B_\varepsilon} p\, \mathrm{div}\eta\, dx + \int_{B_\varepsilon} \mathbf{f}\cdot\eta\, dx \tag{7.2.31}$$

for all $\boldsymbol{\eta} \in \overset{\circ}{W}{}^{1,2}(B_\varepsilon)$. *The subsequent estimate is valid:*

$$\int_0^{2\pi} \|p(\cdot,t)\|^2_{L^2(B_\varepsilon)} dt \le c\varepsilon^{-1} \int_0^{2\pi} A_1(t)dt. \tag{7.2.32}$$

Furthermore, there exists $\nabla p \in L^2(B_\varepsilon)$, *and the following inequality holds:*

$$\int_0^{2\pi} \|\nabla p(\cdot,t)\|^2_{L^2(B_\varepsilon)} dt \le c \int_0^{2\pi} A_1(t)dt. \tag{7.2.33}$$

The system (6.1.5) *is satisfied almost everywhere in* $B_\varepsilon \times (0,T)$.

Proof
The proof of this theorem closely follows the proof of Theorem 6.2.3 with evident modifications.

□

Remark 7.2.1 The constants in the estimates of Theorems 7.2.1–7.2.3 may depend on $\|\mathbf{f}\|_{L^2(B_\varepsilon\times(0,2\pi))}$, yet they are independent of ε and remain bounded as $\|\mathbf{f}\|_{L^2(B_\varepsilon\times(0,2\pi))}$ tends to zero.

7.2.2 Solvability of the Problem: Case n = 3

This section is devoted to establishing the existence of a unique solution for the problem (7.1.6) in three dimensions.

Theorem 7.2.4 *Let* $B_\varepsilon \subset \mathbb{R}^3$, $\partial B_\varepsilon \in C^2$, *and* $\mathbf{f}$ *lies in* $L^2(\Omega)$. *Suppose that the extended function* $\mathbf{g} \in C^{[\frac{J+1}{2}]+2}(0,2\pi; W^{2,2}(B_\varepsilon))$ *and fulfills conditions* (7.1.2), (7.1.3), *and* (7.1.4). *There exists a constant* c_0 *independent of* ε *such that, if*

$$A_1(t) = \|\mathbf{f}(\cdot,t)\|^2_{L^2(B_\varepsilon)} \le c_0, \tag{7.2.34}$$

then, for sufficiently small ε, *there exists a unique solution for problem* (7.1.6). *Moreover, the subsequent estimate holds:*

$$\sup_{t\in[0,2\pi]} \|\mathbf{u}(\cdot,t)\|^2_{L^2(B_\varepsilon)} + \varepsilon^\beta \int_0^{2\pi} \|\nabla\mathbf{u}(\cdot,t)\|^2_{L^2(B_\varepsilon)} dt \le c\varepsilon^{2+\beta} \int_0^{2\pi} A_1(t)dt, \tag{7.2.35}$$

where $\beta = 0$ *or* $\beta = 2$. *Additionally, if* $\beta = 0$ *and* c_0 *in* (7.2.34) *is suitably small, or if* $\beta = 2$, *then also holds the estimate:*

$$\sup_{t\in[0,2\pi]} \|\nabla \mathbf{u}(\cdot,t)\|^2_{L^2(B_\varepsilon)} + \int_0^{2\pi}\int_{B_\varepsilon} |\mathbf{u}_t(x,t)|^2 dxdt$$
$$+\varepsilon^\beta \int_0^{2\pi} \|\nabla^2 \mathbf{u}(\cdot,t)\|^2_{L^2(B_\varepsilon)} dt \le c\varepsilon^\beta \int_0^{2\pi} A_1(t)dt. \tag{7.2.36}$$

A function $p \in L^2(0,2\pi; L^2(B_\varepsilon))$ exists such that the pair $(\mathbf{u}, p)$ satisfies the integral identity:

$$\int_{B_\varepsilon} \Big(\mathbf{u}_t \cdot \boldsymbol{\eta} + \nu \nabla \mathbf{u} \cdot \nabla \boldsymbol{\eta} - \big((\mathbf{u}+\mathbf{g})\cdot\nabla\big)\boldsymbol{\eta}\cdot\mathbf{u}$$
$$-\big(\mathbf{u}\cdot\nabla\big)\boldsymbol{\eta}\cdot\mathbf{g}\Big)dx = \int_{B_\varepsilon} p\,\mathrm{div}\boldsymbol{\eta} dx + \int_{B_\varepsilon} \mathbf{f}\cdot\boldsymbol{\eta} dx \tag{7.2.37}$$

for all $\boldsymbol{\eta} \in \overset{\circ}{W}{}^{1,2}(B_\varepsilon)$, and p satisfies the following estimate:

$$\int_0^{2\pi} \|p(\cdot,t)\|^2_{L^2(B_\varepsilon)} dt \le c\varepsilon^{-1} \int_0^{2\pi} A_1(t)dt \tag{7.2.38}$$

Moreover, there exists $\nabla p \in L^2(B_\varepsilon \times (0,2\pi))$ such that:

$$\int_0^{2\pi} \|\nabla p(\cdot,t)\|^2_{L^2(B_\varepsilon)} dt \le c \int_0^{2\pi} A_1(t)dt. \tag{7.2.39}$$

The system (7.1.5) *holds almost everywhere in $B_\varepsilon \times (0,T)$.*

Proof
Akin to the two-dimensional case, we employ Galerkin approximations. Mirroring the two-dimensional procedure, we establish the existence of time-periodic Galerkin approximations denoted as $\mathbf{u}^{(N)}(x,t)$. These approximations satisfy the following equalities:

$$\int_{B_\varepsilon} \Big(\frac{1}{\varepsilon^\beta}\mathbf{u}_t^{(N)} \cdot \boldsymbol{\psi}_l + \nu\nabla\mathbf{u}^{(N)}\cdot\nabla\boldsymbol{\psi}_l - \Big((\mathbf{u}^{(N)}+\mathbf{g})\cdot\nabla\Big)\boldsymbol{\psi}_l\cdot\mathbf{u}^{(N)}$$
$$-(\mathbf{u}^{(N)}\cdot\nabla)\boldsymbol{\psi}_l\cdot\mathbf{g}\Big)dx = \int_{B_\varepsilon} \mathbf{f}\cdot\boldsymbol{\psi}_l dx, \quad l=1,..,N. \tag{7.2.40}$$

Moreover, the subsequent estimates hold:

$$\frac{1}{\varepsilon^{\beta}} \sup_{t\in[0,2\pi]} \|\mathbf{u}^{(N)}(\cdot,t)\|^2_{L^2(B_\varepsilon)} + \int_0^{2\pi}\int_{B_\varepsilon} |\nabla \mathbf{u}^{(N)}(x,t)|^2 dxdt \le c\varepsilon^2 \int_0^{2\pi} A_1(t)dt;$$

$$\|\mathbf{u}^{(N)}(\cdot,0)\|^2_{L^2(B_\varepsilon)} \le c\varepsilon^{2+\beta} \int_0^{2\pi} A_1(t)dt. \tag{7.2.41}$$

Furthermore, using the mean value theorem for Lebesgue integrals (Theorem 2.3.6), we obtain for some point $t_* \in [0, 2\pi]$ the estimate (7.2.18):

$$\|\nabla \mathbf{u}^{(N)}(\cdot,t_*)\|^2_{L^2(B_\varepsilon)} \le c_1\varepsilon^2 \int_0^{2\pi} A_1(t)dt. \tag{7.2.42}$$

Given the established estimates of the norms $\|\mathbf{u}^{(N)}(\cdot,t_*)\|^2_{L^2(B_\varepsilon)}$ and $\|\nabla\mathbf{u}^{(N)}(\cdot,t_*)\|^2_{L^2(B_\varepsilon)}$, we can consider problem (7.1.6) as an initial boundary value problem (6.1.6) on the interval $[t_*, 2\pi + t_*]$ with the initial value $\mathbf{u}^{(N)}(\cdot,t_*) \in \overset{\circ}{W}{}^{1,2}(B_\varepsilon)$. Replicating now the reasoning from Theorem 6.2.4 (see (6.2.36)–(6.2.42)), we consequently prove the estimate:

$$\frac{z'(t)}{1+z(t)^2} \le c_2 z(t) + c_3 A_1(t), \tag{7.2.43}$$

where $z(t) = \int_{B_\varepsilon} |\nabla \mathbf{u}(x,t)|^2\, dx$. Integrating inequality (7.2.43) over the interval $[t_*, t]$, where $t \in (t^*, 2\pi + t_*]$, and utilizing estimates (7.2.41) and (7.2.42), we obtain:

$$\begin{aligned}\arctan z(t) &\le \arctan z(t_*) + c_2 \int_{t_*}^{2\pi+t_*} z(t)\, dt + c_3 \int_{t_*}^{2\pi+t_*} A_1(t)\, dt \\ &\le \arctan(\|\nabla\mathbf{u}^{(N)}(\cdot,t_*)\|^2_{L^2(B_\varepsilon)}) + c_4 \int_{t_*}^{2\pi+t_*} A_1(t)\, dt \\ &\le \arctan(c_1\varepsilon^2 \int_0^{2\pi} A_1(t)\, dt) + c_4 \int_0^{2\pi} A_1(t)\, dt = \gamma_0.\end{aligned} \tag{7.2.44}$$

For sufficiently small $\int\limits_0^{2\pi} A_1(t)\,dt$, from (7.2.44), it follows that $\arctan z(t) \leq \gamma_0 < \pi/2$. Then inequality (7.2.44) implies the estimate:

$$z(t) = \int\limits_{B_\varepsilon} |\nabla \mathbf{u}^{(N)}(x,t)|^2\,dx \leq \tan\gamma_0 \leq c_5.$$

Hence, the relation (7.2.43) implies:

$$\frac{1}{1+c_5} z'(t) \leq \frac{z'(t)}{1+z(t)^2} \leq c_2 z(t) + c_3 A_1(t),$$

which further simplifies to:

$$z'(t) \leq c_6 z(t) + c_7 A_1(t).$$

By applying Gronwall's lemma on interval $[t^*, 2\pi + t^*]$, we obtain the estimate (in virtue of (7.2.42) and 2π periodicity of $\mathbf{u}^{(N)}(x,t)$):

$$z(t) \leq c_8 \left(z(t^*) + \int\limits_{t^*}^{2\pi+t^*} A_1(t)\,dt \right) \leq c_9 \int\limits_0^{2\pi} A_1(t)\,dt.$$

Thus, we arrive at:

$$\sup_{t\in[0,2\pi]} \|\nabla \mathbf{u}^{(N)}(\cdot,t)\|^2_{L^2(B_\varepsilon)} \leq c \int\limits_0^{2\pi} A_1(t)\,dt. \tag{7.2.45}$$

Now repeating the reasonings of the last part of Theorem 6.2.4 (see the proof of inequalities (6.2.46)–(6.2.48)) and taking into account the periodicity of $\mathbf{u}^{(N)}(x,t)$, we obtain the estimate (7.2.36) for $\mathbf{u}^{(N)}(x,t)$, and then, in a standard way, we prove the existence of the solution $\mathbf{u}(x,t)$.

The uniqueness proof parallels that of the two-dimensional counterpart (see Theorem 7.2.2). However, instead of inequality (2.3.37), we employ inequality (2.3.38).

Finally, the existence of the pressure p, along with the corresponding estimates (7.2.38) and (7.2.39), follows an identical approach to the analogous estimates presented in Theorem 6.2.6. □

7.3 Asymptotic Expansion

In this section, we outline the procedure for constructing an asymptotic expansion of the solution to problem (7.1.1) in the case where $\beta = 2$. It is noteworthy that the case $\beta = 0$ is completely analogous to the asymptotic expansion elaborated in Sect. 6.3 of Chap. 6. The only distinctive aspect lies in the fact that all participating functions are time-periodic with the period 2π (instead of being equal to zero in some neighborhood of $t = 0$).

Initially, we tackle the time-periodic problem on the graph, aiming to derive the macroscopic pressure as a function periodically dependent on time. This pressure function is linear with respect to the longitudinal variable $x_n^{(e)}$ along each edge. Additionally, this solution complies with Kirchhoff-type junction conditions at the nodes. This problem on the graph is the time-periodic analogue of the problem considered in Sect. 6.3 of Chap. 6.

The solution to this problem on the graph furnishes us with the Poiseuille-type velocity within each cylinder $\Pi_\varepsilon^{(e)}$. This velocity solely depends on the transverse spatial variable $x^{(e)'}$ of the tube. By introducing cutoff functions ζ, equal to one in the main middle part of the cylinder and vanishing in some $O(\varepsilon)$ neighborhood of the nodes, we apply these functions to both the Poiseuille velocity and pressure within each cylinder. This multiplication yields a residual component on the right-hand side of the Navier–Stokes equations, possessing finite support that spans an $O(\varepsilon)$ region surrounding the nodes. In response, we formulate boundary layer correctors tailored to counterbalance this residual. These correctors are solutions to the Stokes equations in the dilated bundles of cylinders extended by outlets to infinity. In this context, the asymptotic expansion of the velocity is constructed in the form:

$$\begin{aligned}\mathbf{u}^{(J)}(x,t) &= \sum_{i=1}^{M} \zeta\Big(\frac{x_n^{(e_i)}}{3r\varepsilon}\Big)\zeta\Big(\frac{|e_i| - x_n^{(e_i)}}{3r\varepsilon}\Big)\sum_{j=0}^{J} \varepsilon^j \mathbf{V}_j^{(e_i)}(y^{(e_i)'},t) \\ &+ \sum_{l=1}^{N}\Big(1 - \zeta\Big(\frac{|x - O_l|}{|e|_{\min}}\Big)\Big)\sum_{j=-1}^{J} \varepsilon^j \mathbf{V}_j^{[BLO_l]}(y,t),\end{aligned} \tag{7.3.1}$$

Here, $y = \dfrac{x^{(e)}}{\varepsilon}$, and $\zeta(\tau)$ stands as a smooth cutoff function, independent of ε, satisfying $\zeta(\tau) = 0$ for $\tau \le 1/3$ and $\zeta(\tau) = 1$ for $\tau \ge 2/3$, and $0 \le \zeta(\tau) \le 1$. The parameter $|e|_{\min}$ signifies the minimum edge length, while $r = 3\max\{\operatorname{diam}\sigma_1, \ldots, \operatorname{diam}\sigma_M\} + 1$. The functions $\mathbf{V}_j^{(e_i)}(y^{(e_i)'},t)$ represent the time-periodic Poiseuille-type velocities, and the boundary layer terms $\mathbf{V}_j^{[BLO_l]}(y,t)$ vanish exponentially as $|y|$ tends to infinity.

The asymptotic expansion of the pressure takes on a similar form:

$$p^{(J)}(x,t)=\sum_{i=1}^{M}\zeta\Big(\frac{x_n^{(e_i)}}{3r\varepsilon}\Big)\zeta\Big(\frac{|e_i|-x_n^{(e_i)}}{3r\varepsilon}\Big)\sum_{j=0}^{J}\varepsilon^{j-2}(-s_j^{(e_i)}(t)x_n^{(e_i)}+a_j^{(e_i)}(t))$$
$$+\sum_{l=1}^{N}\Big(1-\zeta\Big(\frac{|x-O_l|}{|e|_{\min}}\Big)\Big)\sum_{j=-1}^{J}\varepsilon^{j-1}p_j^{[BLO_l]}(y,t). \tag{7.3.2}$$

7.3.1 Time-Periodic Problem on the Graph

Let $L^{(e)}$ be the operator that relates the pressure gradient s with the time-periodic flux $\mathcal{H}$ in an infinite cylindrical pipe with cross section $\sigma^{(e)}$. Specifically, consider the Dirichlet boundary value problem for the following periodic-in-time heat equation in the domain $\sigma^{(e)}$: for a given time-periodic function $\mathcal{S} \in L^2(0,2\pi)$, find $\mathcal{V} \in L^2(0,2\pi;\mathring{W}^{1,2}(\sigma^{(e)}))$ such that $\dfrac{\partial \mathcal{V}}{\partial t} \in L^2(0,2\pi;L^2(\sigma^{(e)}))$ and:

$$\begin{cases} \dfrac{\partial \mathcal{V}}{\partial t}(y^{(e)\prime},t)-\nu\Delta_{y^{(e)\prime}}\mathcal{V}(y^{(e)\prime},t)=\mathcal{S}(t), & y^{(e)\prime}\in\sigma^{(e)}, t\in(0,2\pi),\\ \mathcal{V}(y^{(e)\prime},t)=0, & y^{(e)}\in\partial\sigma^{(e)}, t\in(0,2\pi),\\ \mathcal{V}(y^{(e)\prime},t)=\mathcal{V}(y^{(e)\prime},t+2\pi), & y^{(e)}\in\partial\sigma^{(e)}, t\in[0,2\pi]. \end{cases} \tag{7.3.3}$$

Denote:

$$L^{(e)}\mathcal{S}(t)=\int_{\sigma^{(e)}}\mathcal{V}(y^{(e)\prime},t)dy^{(e)\prime}.$$

The integral $L^{(e)}\mathcal{S}(t)=\int_{\sigma^{(e)}}\mathcal{V}(y^{(e)\prime},t)dy^{(e)\prime}$ defines the bounded linear operator $L^{(e)}$, which acts from $L^2(0,2\pi)$ to $W^{1,2}(0,2\pi)$. Due to results of Sect. 3.3 of Chap. 3, $L^{(e)}$ possesses a bounded inverse $(L^{(e)})^{-1} : L^2(0,2\pi)\longrightarrow W^{1,2}(0,2\pi)$.

Moreover, there exists a unique couple $(\mathcal{V},\mathcal{S})$ satisfying (7.3.3) and the flux condition:

$$\int_{\sigma^{(e)}}\mathcal{V}(y^{(e)\prime},t)dy^{(e)\prime}=\mathcal{H}(t).$$

Additionally, $\mathcal{V}\in L^2(0,2\pi;\ \mathring{W}^{1,2}(\sigma^{(e)}))$, $\dfrac{\partial\mathcal{V}}{\partial t}\in L^2(0,2\pi;\ L^2(\sigma^{(e)}))$, and $\mathcal{S}\in L^2(0,2\pi)$, along with the following estimates:

$$\|\mathcal{V}\|_{L^2(0,2\pi;\,W^{1,2}(\sigma^{(e)}))} + \|\frac{\partial \mathcal{V}}{\partial t}\|_{L^2(0,2\pi;\,L^2(\sigma^{(e)}))} + \|\mathcal{S}\|_{L^2(0,2\pi)} \le c\|\mathcal{H}\|_{W^{1,2}(0,2\pi)}, \tag{7.3.4}$$

and:

$$C^{-1}\|\mathcal{Q}\|_{L^2(0,2\pi)} \le \|L^{(e)}\mathcal{Q}\|_{W^{1,2}(0,2\pi)} \le C\|\mathcal{Q}\|_{L^2(0,2\pi)} \quad \forall \mathcal{Q} \in L^2(0,2\pi). \tag{7.3.5}$$

In these expressions, C is a positive constant depending upon $\sigma^{(e)}$. Lastly, let $\langle f\rangle = \frac{1}{2\pi}\int\limits_0^{2\pi} f(t)dt$ denote the average value of function f over the interval $[0, 2\pi]$.

Lemma 7.3.1 *The following formula:*

$$\langle L^{(e)}\mathcal{S}\rangle = \kappa^{(e)}\langle \mathcal{S}\rangle,$$

is valid, where $\kappa^{(e)} = \int_{\sigma^{(e)}} U_0^{(e)}(y^{(e)\prime})dy^{(e)\prime} > 0$, *and* $U_0^{(e)}$ *is the unique solution to the problem:*

$$\begin{cases} -\nu\Delta'_{y^{(e)\prime}} U_0^{(e)}(y^{(e)\prime}) = 1, \; y^{(e)\prime} \in \sigma^{(e)}, \\ \qquad\qquad U_0^{(e)}(y^{(e)\prime}) = 0, \; y^{(e)\prime} \in \partial\sigma^{(e)}. \end{cases}$$

Proof

Integrating (7.3.3) by t over the interval $[0, 2\pi]$ and using the time periodicity of the solution, we get:

$$\begin{cases} -\nu\Delta'_{y^{(e)\prime}}\langle\mathcal{V}\rangle(y^{(e)\prime})(y^{(e)\prime}) = \langle\mathcal{S}\rangle, \; y^{(e)\prime} \in \sigma^{(e)}, \\ \qquad\qquad \langle\mathcal{V}\rangle(y^{(e)\prime})(y^{(e)\prime}) = 0, \quad y^{(e)\prime} \in \partial\sigma^{(e)}, \end{cases}$$

and, hence, $\langle\mathcal{V}\rangle(y^{(e)\prime}) = \langle\mathcal{S}\rangle U_0^{(e)}(y^{(e)\prime})$. Then:

$$\langle L^{(e)}\mathcal{S}\rangle = \int\limits_{\sigma^{(e)}} \langle\mathcal{V}\rangle(y^{(e)\prime})dy^{(e)\prime} = \langle\mathcal{S}\rangle \int\limits_{\sigma^{(e)}} U_0^{(e)}(y^{(e)\prime})dy^{(e)\prime} = \kappa^{(e)}\langle\mathcal{S}\rangle.$$

□

In particular, we have:

Lemma 7.3.2 *(i)* $\langle L^{(e)}\mathcal{S}\rangle = 0$ *if and only if* $\langle\mathcal{S}\rangle = 0$.
(ii) For any constant C, there holds the equality: $L^{(e)}(C) = \kappa^{(e)}C$.

Next, let us introduce time-periodic functions $\Psi_l \in W^{1,2}(0, 2\pi)$, $l = 1, \dots, N$, and $\mathcal{F}^{(e_i)} \in W^{1,2}(0, 2\pi; L^2(\mathcal{B}))$, $i = 1, \dots, M$. We consider the following problem: find a time-periodic function $p \in L^2(0, 2\pi; W^{1,2}(\mathcal{B}))$ satisfying the following equations on the edges and junction conditions at the nodes and vertices:

$$\begin{cases} -\dfrac{\partial}{\partial x_n^{(e)}}\left(L^{(e)}\dfrac{\partial p}{\partial x_n^{(e)}}(x_n^{(e)},t)\right) = \mathcal{F}^{(e)}(x_n^{(e)},t),\ x_n^{(e)} \in (0,|e|), \\ \qquad\qquad\qquad\qquad\qquad\qquad\qquad \forall e = e_j,\ j = 1,\dots,M, \\ -\sum\limits_{e:O_l\in e}\left(L^{(e)}\dfrac{\partial p}{\partial x_n^{(e)}}\right)(0,t) = \Psi_l(t), \qquad l = 1,\dots,N_1, \\ \quad -\left(L^{(e)}\dfrac{\partial p^{(e)}}{\partial x_n^{(e)}}\right)(0,t) = \Psi_l(t), \qquad l = N_1+1,\dots,N. \end{cases} \tag{7.3.6}$$

The equation $(7.3.6)_1$ holds on each edge e of $\mathcal{B}$, condition $(7.3.6)_2$ holds at each node O_l with the sum taken over all edges having O_l as an endpoint (i.e., over all edges of $\mathcal{B}_l$), and condition $(7.3.6)_3$ is a Neumann-type boundary condition, which holds at each vertex O_l. In conditions $(7.3.6)_2$ and $(7.3.6)_3$, the local coordinate system has the origin at O_l. This problem constitutes a one-dimensional model for the macroscopic pressure defined on the graph $\mathcal{B}$; the right-hand sides represent the sources. The pressure is expected to be a continuous function on the graph since it belongs to $W^{1,2}(\mathcal{B})$; however, at the nodes, one can consider some given jumps of the pressure. This generalized setting can be reduced to (7.3.6) through a change of the unknown function and the right-hand sides.

Let us prove that problem (7.3.6) admits a unique solution vanishing at the vertex O_N, if the compatibility condition:

$$\sum_{i=1}^{M}\int_0^{|e_i|}\mathcal{F}^{(e_i)}dx_n^{(e_i)} + \sum_{l=1}^{N}\Psi_l(t) = 0, \tag{7.3.7}$$

holds for almost all $t \in [0, 2\pi]$.

By differentiating relations (7.3.6) with respect to time t, multiplying by $\psi \in L^2(0, 2\pi; W^{1,2}(\mathcal{B}))$, and integrating by parts over $\mathcal{B} \times (0, 2\pi)$, we obtain the following variational formulation:

Definition 7.3.1 Find a time-periodic function $p \in L^2(0, 2\pi; W^{1,2}(\mathcal{B}))$ that satisfies the integral identity:

$$\alpha(p,\psi) = \beta(\psi), \tag{7.3.8}$$

where the test functions $\psi \in L^2(0, 2\pi; W^{1,2}(\mathcal{B}))$ are time-periodic. The functionals $\alpha(p, \psi)$ and $\beta(\psi)$ are defined as follows:

$$\alpha(p,\psi) = \sum_{i=1}^{M}\int_0^{2\pi}\int_0^{|e_i|}\frac{\partial(L^{(e_i)}p)_t}{\partial x_n^{(e_i)}}\frac{\partial\psi}{\partial x_n^{(e_i)}}dx_n^{(e_i)}dt,$$

$$\beta(\psi) = \sum_{i=1}^{M} \int_0^{2\pi} \int_0^{|e_i|} (\mathcal{F}^{(e_i)})_t \psi dx_n^{(e_i)} dt + \sum_{l=1}^{N} \int_0^{2\pi} (\Psi_l)_t(t)\psi(O_l, t)dt.$$

First, we proceed to study an auxiliary problem.

Lemma 7.3.3 *Assume that the compatibility condition* (7.3.7) *holds. Then, for any given time-periodic functions* $\Psi_l \in W^{1,2}(0, 2\pi)$, $l = 1, \ldots, N$, *and* $\mathcal{F}^{(e_i)}, \mathcal{F}_t^{(e_i)} \in L^2(0, 2\pi; L^2(\mathcal{B}))$, $i = 1, \ldots, M$, *there exists a unique time-periodic function* $p_0 \in L^2(0, 2\pi; W^{1,2}(\mathcal{B}))$ *satisfying the integral identity* (7.3.8), *vanishing at the vertex* O_N, *and having a zero mean value in time:* $p_0(O_N, t) = 0$, $\langle p_0 \rangle = 0$. *Any solution* p *to problem* (7.3.8) *can be expressed as:*

$$p(x, t) = p_0(x, t) + \varphi(x) + \psi(t),$$

where $\varphi \in W^{1,2}(\mathcal{B})$ *and* $\psi \in L^2(0, 2\pi)$.

Proof
We proceed with a Lax–Milgram lemma argument, combined with inequality (7.3.5). Consider the subspaces $W^{1,2}_{O_N}(\mathcal{B})$ of functions in $W^{1,2}(\mathcal{B})$ that vanish at the vertex O_N and the subspace $\widehat{L}^2(0, 2\pi; W^{1,2}_{O_N}(\mathcal{B}))$ of the space $L^2(0, 2\pi; W^{1,2}_{O_N}(\mathcal{B}))$ of elements ψ that have a zero mean value in time.

We verify using (7.3.5) that $\alpha(\cdot, \cdot)$ is a continuous bilinear form with respect to the norm of $\widehat{L}^2(0, T; W^{1,2}_{O_N}(\mathcal{B}))$ and that $\beta(\cdot)$ is a bounded linear functional:

$$|\alpha(p, \psi)| \le \sqrt{\sum_{i=1}^{M} \int_0^{2\pi} \int_0^{|e_i|} \left| \frac{\partial (L^{(e_i)} p)_t}{\partial x_n^{(e_i)}} \right|^2 dx_n^{(e_i)} dt} \sqrt{\sum_{i=1}^{M} \int_0^{2\pi} \int_0^{|e_i|} \left| \frac{\partial \psi}{\partial x_n^{(e_i)}} \right|^2 dx_n^{(e_i)} dt}$$

$$\le \max_{1 \le j \le M} C_{e_j} \sqrt{\sum_{i=1}^{M} \int_0^{2\pi} \int_0^{|e_i|} \left| \frac{\partial p}{\partial x_n^{(e_i)}} \right|^2 dx_n^{(e_i)} dt}$$

$$\times \sqrt{\sum_{i=1}^{M} \int_0^{2\pi} \int_0^{|e_i|} \left| \frac{\partial \psi}{\partial x_n^{(e_i)}} \right|^2 dx_n^{(e_i)} dt.}$$

The boundedness of β follows from the Poincaré inequality in the space $W^{1,2}_{O_N}(\mathcal{B})$ and the inclusion of the space $W^{1,2}_{O_N}(\mathcal{B})$ into $C(\mathcal{B})$.

The coerciveness of α stems from estimates that hold for any edge e and any time-periodic solution $(\mathcal{V}(y^{(e)\prime}, t), \mathcal{S})$ of (7.3.3):

$$\int_0^{2\pi} (L^{(e)}\mathcal{S})_t \mathcal{S} dt = \int_0^{2\pi}\int_{\sigma^{(e)}} (\mathcal{V}(y^{(e)\prime}, t))_t \mathcal{S} dy^{(e)\prime} dt$$

$$= \int_0^{2\pi}\int_{\sigma^{(e)}} \left\{ \left|(\mathcal{V}(y^{(e)\prime}, t))_t)\right|^2 + \frac{\nu}{2}\frac{\partial}{\partial t}\left|\nabla_{y^{(e)\prime}}\mathcal{V}(y^{(e)\prime}, t)\right|^2 \right\} dy^{(e)\prime} dt$$

$$= \int_0^{2\pi}\int_{\sigma^{(e)}} \left|(\mathcal{V}(y^{(e)\prime}, t))_t\right|^2 dy^{(e)\prime} dt$$

$$\geq \frac{1}{|\sigma^{(e)}|}\int_0^{2\pi}\left|\int_{\sigma^{(e)}} ((\mathcal{V}(y^{(e)\prime}, t))_t dy^{(e)\prime}\right|^2 dt = \frac{1}{|\sigma^{(e)}|}\int_0^{2\pi}\left|(L^{(e)}\mathcal{S})_t\right|^2 dt.$$

It is important to note that $\langle L^{(e)}\mathcal{S}\rangle = 0$. Consequently, using the Poincaré inequality and (7.3.5), we obtain:

$$\frac{1}{|\sigma^{(e)}|}\int_0^{2\pi}\left|(L^{(e)}\mathcal{S})_t\right|^2 dt \geq \frac{1}{2|\sigma^{(e)}|}\int_0^{2\pi}\left|(L^{(e)}\mathcal{S})_t\right|^2 dt + \frac{1}{4\pi^2|\sigma^{(e)}|}\int_0^{2\pi}\left|L^{(e)}\mathcal{S}\right|^2 dt$$

$$\geq \frac{1}{2}\frac{C_e^{-2}}{|\sigma^{(e)}|}\int_0^{2\pi}\|p(\cdot, t)\|^2_{W^{1,2}(\mathcal{B})} dt.$$

Hence, we have:

$$\alpha(p, p) \geq \min\left(\frac{1}{2}, \frac{1}{2\pi^2}\right)\frac{\min\limits_{i=1,\dots,M} C_{e_i}^{-2}}{\max\limits_{i=1,\dots,M} |\sigma^{(e_i)}|}\int_0^{2\pi}\|p(\cdot, t)\|^2_{W^{1,2}(\mathcal{B})} dt. \tag{7.3.9}$$

This concludes the proof of the existence and uniqueness of the time-periodic solution p_0 in the subspace $\widehat{L}^2(0, 2\pi; W^{1,2}_{O_N}(\mathcal{B}))$ based on the Lax–Milgram lemma.

We also demonstrate that (7.3.8) remains valid for any test function from $\widehat{L}^2(0, 2\pi; W^{1,2}(\mathcal{B}))$, without requiring the test functions to be zero at the vertex O_N. If $\psi \in \widehat{L}^2(0, 2\pi; W^{1,2}(\mathcal{B}))$, then (7.3.8) holds for the function $\psi(x, t) - \psi(O_N, t)$. Notably, $\alpha(p, \psi(O_N, t)) = 0$, and, due to the compatibility condition (7.3.7), $\beta(\psi(O_N, t)) = 0$. If a test function belongs to $L^2(0, 2\pi; W^{1,2}(\mathcal{B}))$ (i.e., with a nonzero mean value), we can take in (7.3.8) the function $\psi(x_n^{(e)}, t) - \langle\psi\rangle(x_n^{(e)})$. Due to periodicity in time, $\alpha(p, \eta) = 0$ and $\beta(\eta) = 0$ for any function $\eta(x_n^{(e)})$ independent of t (this follows by integrating by parts in the variable t). Hence, (7.3.8) is valid for every $\psi \in L^2(0, 2\pi; W^{1,2}(\mathcal{B}))$.

Consider two solutions p_1 and p_2 of problem (7.3.8). Then, the difference p_1-p_2 can be expressed as:

$$p_1(x,t) - p_2(x,t) = P(x,t) + \langle p_1 - p_2\rangle(x) \\ + (p_1(O_N,t) - p_2(O_N,t) - \langle p_1(O_N,t) - p_2(O_N,t)\rangle),$$

where P is a solution to the homogeneous problem (7.3.8), vanishing at the vertex O_N and having a zero mean value in time. Since such a solution is unique, we conclude that $P = 0$. Therefore, any solution of problem (7.3.8) takes the form $p(x,t) = p_0(x,t) + \phi(x) + \varphi(t)$, with $p_0 \in \widehat{L}^2(0, 2\pi;\ W^{1,2}_{O_N}(\mathcal{B}))$, $\phi \in W^{1,2}(\mathcal{B})$, and $\varphi \in L^2(0, 2\pi)$. Here $\phi = \langle p_1-p_2\rangle$, $\varphi = p_1(O_N,\cdot)-p_2(O_N,\cdot)-\langle p_1(O_N,\cdot)-p_2(O_N,\cdot)\rangle$. □

The proved lemma establishes the existence of a solution for the variational formulation (7.3.8) of problem (7.3.6), which has been differentiated with respect to time. Our next objective is to revert to the original formulation of problem (7.3.6) prior to the temporal differentiation and subsequently demonstrate the existence of a solution, along with its regularity concerning the spatial variable x.

Theorem 7.3.1 *Assuming the compatibility condition* (7.3.7) *holds, problem* (7.3.6) *possesses a time-periodic solution p that lies in the space* $L^2(0, 2\pi;\ W^{1,2}_{O_N}(\mathcal{B}))$. *This solution is unique in this space.*

Proof

1. Consider an arbitrary edge e_i and take a test function with a compact support contained in this edge. The integral identity (7.3.8) then simplifies to:

$$\int_0^{2\pi}\int_0^{|e_i|} \frac{\partial (L^{(e_i)}p)_t}{\partial x_n^{(e_i)}} \frac{\partial \psi}{\partial x_n^{(e_i)}} dx_n^{(e_i)} dt = \int_0^{2\pi}\int_0^{|e_i|} (\mathcal{F}^{(e_i)})_t \psi dx_n^{(e_i)} dt.$$

This implies that for almost all t and any smooth test function ψ dependent only on x and having the support belonging to e_i, the following relation holds:

$$\int_0^{|e_i|} \frac{\partial (L^{(e_i)}p)_t}{\partial x_n^{(e_i)}} \frac{\partial \psi}{\partial x_n^{(e_i)}} dx_n^{(e_i)} dt = \int_0^{|e_i|} (\mathcal{F}^{(e_i)})_t \psi dx_n^{(e_i)} dt.$$

This last identity corresponds to the definition of a weak derivative for functions from $W^{1,2}(e_i)$ and means that $-(\mathcal{F}^{(e_i)})_t$ is a weak derivative of $\dfrac{\partial (L^{(e_i)}p)_t}{\partial x_n^{(e_i)}}$, so that Eq. $(7.3.6)_1$ differentiated with respect to the time variable t

holds for almost all x and t. In a similar way taking test functions having support in a ball $B(O_i, \varepsilon)$ of the radius ε and the center in O_i and letting $\varepsilon \to 0$, we get boundary conditions $(7.3.6)_2$ and $(7.3.6)_3$ differentiated with respect to the time variable for almost all t.

2. Integrating these relations over time, we deduce that the function p_0, from Lemma 7.3.3, satisfies problem (7.3.6) with modified right-hand sides in Eq. $(7.3.6)_1$ and Kirchhoff-type boundary conditions $(7.3.6)_2$ and $(7.3.6)_3$, i.e., with modified forms of $\mathcal{F}^{(e_i)}(x_n^{(e_i)}, t)$ and $\Psi_l(t)$. This leads to the following right-hand sides:

$$\int_0^t (\mathcal{F}^{(e)}(x_n^{(e)}, \theta))_\theta d\theta - \frac{\partial}{\partial x_n^{(e)}}\Big(L^{(e)}\frac{\partial p_0}{\partial x_n^{(e)}}(x_n^{(e)}, 0)\Big)$$

in Eq. $(7.3.6)_1$ and:

$$\int_0^t (\Psi_l(\theta))_\theta d\theta - \sum_{e:O_l\in e}\Big(L^{(e)}\frac{\partial p_0}{\partial x_n^{(e)}}\Big)(0,0)$$

in Kirchhoff-type boundary conditions $(7.3.6)_2$ and $(7.3.6)_3$, respectively, i.e., with:

$$\mathcal{F}^{(e)}(x_n^{(e)}, t) - \mathcal{F}^{(e)}(x_n^{(e)}, 0) - \frac{\partial}{\partial x_n^{(e)}}\Big(L^{(e)}\frac{\partial p_0}{\partial x_n^{(e)}}(x_n^{(e)}, 0)\Big)$$

and:

$$\Psi_l(t) - \Psi_l(0) - \sum_{e:O_l\in e}\Big(L^{(e)}\frac{\partial p_0}{\partial x_n^{(e)}}\Big)(0,0).$$

3. Problem (7.3.6) with the right-hand sides:

$$\mathcal{F}^{(e)}(x_n^{(e)}, 0) + \frac{\partial}{\partial x_n^{(e)}}\Big(L^{(e)}\frac{\partial p_0}{\partial x_n^{(e)}}(x_n^{(e)}, 0)\Big)$$

and:

$$\Psi_l(0) + \sum_{e:O_l\in e}\Big(L^{(e)}\frac{\partial p}{\partial x_n^{(e)}}\Big)(0,0)$$

is equivalent to the following system:

$$\begin{cases} -\dfrac{\partial}{\partial x_n^{(e)}}\Big(\kappa^{(e)}\dfrac{\partial q_0}{\partial x_n^{(e)}}(x_n^{(e)})\Big) = \mathcal{F}^{(e)}(x_n^{(e)},0) \\ \qquad\qquad + \dfrac{\partial}{\partial x_n^{(e)}}\Big(L^{(e)}\dfrac{\partial p_0}{\partial x_n^{(e)}}(x_n^{(e)},0)\Big),\ x_n^{(e)} \in (0,|e|), \\ -\sum\limits_{e:O_l\in e}\Big(\kappa^{(e)}\dfrac{\partial q_0}{\partial x_n^{(e)}}\Big)(0) = \Psi_l(0) + \sum\limits_{e:O_l\in e}\Big(L^{(e)}\dfrac{\partial p_0}{\partial x_n^{(e)}}\Big)(0,0), l = 1,\dots,N_1, \\ -\Big(\kappa^{(e)}\dfrac{\partial q_0^{(e)}}{\partial x_n^{(e)}}\Big)(0) = \Psi_l(0) + \Big(L^{(e)}\dfrac{\partial p_0}{\partial x_n^{(e)}}\Big)(0,0), l = N_1+1,\dots,N. \end{cases} \tag{7.3.10}$$

This problem admits a unique solution (independent of time), $q_0 \in W^{1,2}_{O_N}(\mathcal{B})$ (see Theorem 5.1.1), and the sum $p_0 + q_0$ satisfies all equations of problem (7.3.6). Therefore, the existence of a solution to (7.3.6) has been established.

4. Uniqueness (up to an additive function of time) can be demonstrated by considering solutions p_1 and p_2 to problem (7.3.10). Their difference $p_1 - p_2$ satisfies the homogeneous problem (7.3.8). Assume that the difference $p_1 - p_2$ vanishes at the point O_n. Then, due to coercivity of the bilinear form α (see (7.3.9)), the norm $\|p_1 - p_2\|_{W^{1,2}(\mathcal{B})} = 0$, and so, $p_1 = p_2$.

□

Remark 7.3.1 It is noteworthy that condition (7.3.7) both suffices and is necessary for the existence of a 2π-periodic time solution for problem (7.3.6). This is demonstrated by integrating equation $(7.3.6)_1$ over $\mathcal{B}$, leading to the derivation of (7.3.7).

Removing the requirement $p(O_N, t) = 0$, we get a solution that is unique up to an additive function φ, dependent solely on the time variable t.

7.3.2 Construction of the Asymptotic Expansion

The construction of the asymptotic solution is carried out through induction with respect to the parameter j. At the initial step $j = 0$, we address the following problem on the graph: find a function $p_0 \in L^2(0, 2\pi; W^{1,2}(\mathcal{B}))$ that satisfies the equations:

$$\begin{cases} -\dfrac{\partial}{\partial x_n^{(e)}}\Big(L^{(e)}\dfrac{\partial p_0}{\partial x_n^{(e)}}(x_n^{(e)},t)\Big) = 0, & x_n^{(e)} \in (0,|e|), \forall e = e_j,\ j = 1,\dots,M, \\ -\sum\limits_{e:O_l\in e}\Big(L^{(e)}\dfrac{\partial p_0}{\partial x_n^{(e)}}\Big)(0,t) = 0, & l = 1,\dots,N_1, \\ -\Big(L^{(e)}\dfrac{\partial p_0}{\partial x_n^{(e)}}\Big)(0,t) = \Psi_l(t),\ l = N_1+1,\dots,N. \end{cases} \tag{7.3.11}$$

Here, $\Psi_l(t) = -\int_{\gamma^l} \mathbf{g}_l \cdot \mathbf{n} dS$. The solution $p_0^{(e)}$ is an affine function:

$$p_0^{(e)}(x_n^{(e)}, t) = -s_0^{(e)}(t)x_n^{(e)} + a_0^{(e)}(t).$$

For each edge e_i, the Poiseuille-type velocity $\mathbf{V}_0^{(e_i)}(y^{(e_i)'}, t)$ is defined such that, in local coordinates, its last (normal) component is the solution of problem (7.3.3) with $\mathcal{S}(t) = s_0^{(e)}(t)$, while the tangential components are equal to zero.

The next step involves determining the boundary layer correctors $(\mathbf{V}_0^{[BLO_l]}, P_0^{[BLO_l]})$ as solutions of the time-periodic Stokes equations in the scaled domain Ω_l (see Chaps. 5 and 6). The boundary layer pressure of rank -1 is defined as:

$$P_{-1}^{[BLO_l]}(y, t) = -\Big(\sum_{e:O_l \in e} \zeta\Big(\frac{y_n^{(e)}}{3r}\Big) - 1\Big) p_0(O_l, t),$$

where p_0 is a continuous function on $\mathcal{B}$ without jumps at the nodes, thus making $p_0(O_l, t)$ well-defined. The boundary layer velocity of rank -1 is zero: $\mathbf{V}_{-1}^{[BLO_l]}(y, t) = 0$.

The boundary layer terms $(\mathbf{V}_0^{[BLO_l]}, P_0^{[BLO_l]})$ are determined as the solution to the time-periodic Stokes problem in the unbounded domain Ω_l:

$$\begin{cases} \dfrac{\partial}{\partial t}\mathbf{V}_0^{[BLO_l]} - \nu\Delta_y \mathbf{V}_0^{[BLO_l]} + \nabla_y P_0^{[BLO_l]} = \\ = \displaystyle\sum_{e:O_l \in e} \Big(\zeta\Big(\frac{y_n^{(e)}}{3r}\Big)\frac{\partial}{\partial t} V_0^{(e)}(y^{(e)'}, t) + \nu \frac{\partial^2}{\partial y_n^{(e)2}}\Big(\zeta\Big(\frac{y_n^{(e)}}{3r}\Big)\Big)\mathbf{V}_0^{(e)}(y^{(e)'}, t) \\ \quad + \nabla_y\Big(\zeta\Big(\frac{y_n^{(e)}}{3r}\Big)y_n^{(e)}\Big)s_0^{(e)}(t) - \nabla_y\Big(\zeta\Big(\frac{y_n^{(e)}}{3r}\Big)\Big)a_1^{(e)}(t), \ \ y \in \Omega_l, \ t \in [0, 2\pi], \\ \operatorname{div}_y \mathbf{V}_0^{[BLO_l]} = -\displaystyle\sum_{e:O_l \in e} \frac{\partial}{\partial y_n^{(e)}}\zeta\Big(\frac{y_n^{(e)}}{3r}\Big)V_{0,n}^{(e)}(y^{(e)'}, t), \ \ y \in \Omega_l, \ t \in [0, 2\pi], \\ \mathbf{V}_0^{[BLO_l]}(y, t) = 0, \ y \in \partial\Omega_l, \ t \in [0, 2\pi], \\ \mathbf{V}_0^{[BLO_l]}(y, t) = \mathbf{V}_0^{[BLO_l]}(y, t + 2\pi), \ y \in \partial\Omega_l, \ t \in [0, 2\pi], \end{cases}$$

where the local coordinates have the origin at O_l and $a_1^{(e)}(t)$ is an unknown function. This problem is split into two independent ones: first, we solve it without the term containing $\widehat{a}_1^{(e)}(t)$ in the right-hand side for $(\mathbf{V}_0^{[BLO_l]}, \hat{P}_0^{[BLO_l]})$:

$$\begin{cases} \dfrac{\partial}{\partial t}\mathbf{V}_0^{[BLO_l]} - \nu\Delta_y\mathbf{V}_0^{[BLO_l]} + \nabla_y\hat{P}_0^{[BLO_l]} = \\ = \displaystyle\sum_{e:O_l\in e}\Big(\zeta\Big(\frac{y_n^{(e)}}{3r}\Big)\frac{\partial}{\partial t}V_0^{(e)}(y^{(e)'},t) + \nu\frac{\partial^2}{\partial y_n^{(e)2}}\Big(\zeta\Big(\frac{y_n^{(e)}}{3r}\Big)\Big)\mathbf{V}_0^{(e)}(y^{(e)'},t) + \\ \qquad +\nabla_y\Big(\zeta\Big(\frac{y_n^{(e)}}{3r}\Big)y_n^{(e)}\Big)s_0^{(e)}(t)\Big), \ y\in\Omega_l,\ t\in[0,2\pi], \\ \operatorname{div}_y\mathbf{V}_0^{[BLO_l]} = -\displaystyle\sum_{e:O_l\in e}\frac{\partial}{\partial y_n^{(e)}}\zeta\Big(\frac{y_n^{(e)}}{3r}\Big)V_{0,n}^{(e)}(y^{(e)'},t), \ y\in\Omega_l,\ t\in[0,2\pi], \\ \mathbf{V}_0^{[BLO_l]}(y,t) = 0, \ y\in\partial\Omega_l,\ t\in[0,2\pi], \\ \mathbf{V}_0^{[BLO_l]}(y,t) = \mathbf{V}_0^{[BLO_l]}(y,t+2\pi), \ y\in\Omega_l,\ t\in[0,2\pi], \end{cases}$$

and find a solution $\mathbf{V}_0^{[BLO_l]}$ that decays exponentially as $|y|\to\infty$, while $\hat{P}_0^{[BLO_l]}$ at each outlet $\Pi_\infty^{(e)}$ tends to a constant $\widehat{a}_1^{[BLO_l,e]}(t)$, except for the outlet corresponding to the selected edge e_s where it tends to zero (this is possible because the pressure is defined up to an additive constant); see Sects. 4.4 and 4.6 of Chap. 4.

Then we solve the following problem on the graph: find a periodically time-dependent function $p_1^{(e)}\in L^2(0,2\pi;W^{1,2}(e))$ that satisfies the equations:

$$\begin{cases} -\dfrac{\partial}{\partial x_n^{(e)}}\Big(L^{(e)}\dfrac{\partial p_1^{(e)}}{\partial x_n^{(e)}}(x_n^{(e)},t)\Big) = 0, & x_n^{(e)}\in(0,|e|),\ \forall e=e_j, \\ & j=1,\dots,M, \\ -\displaystyle\sum_{e:O_l\in e}\Big(L^{(e)}\frac{\partial p_1^{(e)}}{\partial x_n^{(e)}}\Big)(0,t) = 0, & l=1,\dots,N_1, \\ -\Big(L^{(e)}\dfrac{\partial p_1^{(e)}}{\partial x_n^{(e)}}\Big)(0,t) = 0, & l=N_1+1,\dots,N, \\ p_1^{(e)}(0,t)-p_1^{(e_s)}(0,t) = \widehat{a}_1^{[BLO_l,e]}(t),\ \forall e\subset\mathcal{B}_l,\ e\neq e_s. \end{cases}$$

Here, e_s is a selected edge of the bundle. This problem has a unique solution p_1 up to an additive function of t, and $p_1^{(e)}(x_n^{(e)},t) = -s_1^{(e)}(t)x_n^{(e)} + a_1^{(e)}(t)$. Finally, we define:

$$P_0^{[BLO_l]} = \hat{P}_0^{[BLO_l]} - \Big(\sum_{e:O_l\in e,e\neq e_s}\zeta\Big(\frac{y_n^{(e)}}{3r}\Big)\Big)\widehat{a}_1^{[BLO_l,e]}(t)$$

$$-\Big(\sum_{e:O_l\in e}\zeta\Big(\frac{y_n^{(e)}}{3r}\Big)-1\Big)p_1^{(e_s)}(0,t).$$

In a similar manner, if O_l is a vertex, the boundary layer problem takes the form:

$$\begin{cases} \frac{\partial}{\partial t}\mathbf{V}_0^{[BLO_l]} - \nu\Delta_y \mathbf{V}_0^{[BLO_l]} + \nabla_y \hat{P}_0^{[BLO_l]} = \\ = \zeta\Big(\frac{y_n^{(e)}}{3r}\Big)\frac{\partial}{\partial t}V_0^{(e)}(y^{(e)'},t) + \nu\frac{\partial^2}{\partial y_n^{(e)2}}\Big(\zeta\Big(\frac{y_n^{(e)}}{3r}\Big)\Big)\mathbf{V}_0^{(e)}(y^{(e)'},t) \\ \qquad +\nabla_y\Big(\zeta\Big(\frac{y_n^{(e)}}{3r}\Big)y_n^{(e)}\Big)s_0^{(e)}(t), \qquad y\in\Omega_l,\ t\in[0,2\pi], \\ \mathrm{div}_y\mathbf{V}_0^{[BLO_l]} = -\frac{\partial}{\partial y_n^{(e)}}\zeta\Big(\frac{y_n^{(e)}}{3r}\Big)V_{0,n}^{(e)}(y^{(e)'},t), \qquad y\in\Omega_l,\ t\in[0,2\pi], \\ \mathbf{V}_0^{[BLO_l]}(y,t) = 0,\ y\in\partial\Omega_l\setminus\gamma^l,\ t\in[0,2\pi], \\ \mathbf{V}_0^{[BLO_l]}(y,t) = \mathbf{g}^l(y,t), \qquad y\in\gamma^l,\ t\in[0,2\pi], \\ \mathbf{V}_0^{[BLO_l]}(y,t) = \mathbf{V}_0^{[BLO_l]}(y,t+2\pi), \qquad y\in\Omega_l,\ t\in[0,2\pi]. \end{cases}$$

Due to condition $(7.3.11)_3$, we have:

$$\int\limits_{\Omega_l}\frac{\partial}{\partial y_n^{(e)}}\zeta\Big(\frac{y_n^{(e)}}{3r}\Big)V_{0,n}^{(e)}(y^{(e)'},t)dy + \int\limits_{\gamma^l}\mathbf{g}^l\cdot\mathbf{n}dS = 0.$$

This compatibility condition ensures the unique existence of a solution $\big(\mathbf{V}_0^{[BLO_l]}, \widehat{P}_0^{[BLO_l]}\big)$ that exponentially tends to zero at infinity (see Sect. 4.6 of Chap. 4).

Assuming that all terms in the asymptotic expansion corresponding to rank $j-1$ are known, and the pressure on the graph $p_j^{(e)}$ is known as well, the transition from rank $j-1$ to rank j can now be described.

Step 1. Given the known pressure on the graph $p_j^{(e)}$, we define functions $s_j^{(e)}(t)$ and $a_j^{(e)}(t)$ for each edge e, such that:

$$p_j^{(e)}(x_n^{(e)},t) = -s_j^{(e)}(t)x_n^{(e)} + a_j^{(e)}(t).$$

Additionally, we define the Poiseuille-type velocity vector $\mathbf{V}_j^{(e)}(y^{(e)'},t)$ for each edge e, where the last component (normal component) corresponds to the solution of problem (7.3.3) with $\mathcal{S}(t) = s_j^{(e)}(t)$, while the tangential components are equal to zero.

Step 2. The boundary layer solution is represented by the periodic-in-time pair $(\mathbf{V}_j^{[BLO_l]}, P_j^{[BLO_l]})$, which satisfies the following problem:

$$\begin{cases} \dfrac{\partial \mathbf{V}_j^{[BLO_l]}}{\partial t} - \nu \Delta_y \mathbf{V}_j^{[BLO_l]} + \nabla_y P_j^{[BLO_l]} = \\ \qquad = \mathbf{f}_j^{[REGO_l]}(y^{(e)'}, t) + \mathbf{f}_j^{[BLO_l]}(y, t), & y \in \Omega_l,\ t \in [0, 2\pi], \\ \qquad \mathrm{div}_y \mathbf{V}_j^{[BLO_l]} = h_j^{[REGO_l]}(y^{(e)'}, t), & y \in \Omega_l,\ t \in [0, 2\pi], \\ \qquad \mathbf{V}_j^{[BLO_l]}(y, t) = 0, & y \in \partial\Omega_l,\ t \in [0, 2\pi], \\ \qquad \mathbf{V}_j^{[BLO_l]}(y, t) = \mathbf{V}_j^{[BLO_l]}(y, t + 2\pi), & y \in \Omega_l,\ t \in [0, 2\pi], \end{cases}$$

where:

$$h_j^{[REGO_l]}(y^{(e)'}, t) = -\sum_{e: O_l \in e} \frac{\partial}{\partial y_n^{(e)}} \zeta\Big(\frac{y_n^{(e)}}{3r}\Big) \mathbf{V}_{j,n}^{(e)}(y^{(e)'}, t),$$

$$\begin{aligned} \mathbf{f}_j^{[REGO_l]}(y^{(e)'}, t) = \sum_{e: O_l \in e} \Big[& \nu \Big(\mathbf{V}_j^{(e)}(y^{(e)'}, t) \frac{\partial^2}{\partial y_n^{(e)2}} \Big(\zeta\Big(\frac{y_n^{(e)}}{3r}\Big)\Big) \\ & - \zeta\Big(\frac{y_n^{(e)}}{3r}\Big) \sum_{k=0}^{j-1} (\mathbf{V}_k^{(e)}(y^{(e)'}, t) \cdot \nabla_y) \\ & \times \Big\{\zeta\Big(\frac{y_n^{(e)}}{3r}\Big) \mathbf{V}_{j-k-1}^{(e)}(y^{(e)'}, t)\Big\} \\ & + s_j^{(e)}(t) y_n^{(e)} \cdot \nabla_y \Big(\zeta\Big(\frac{y_n^{(e)}}{3r}\Big)\Big) - a_{j+1}^{(e)}(t) \cdot \nabla_y \Big(\zeta\Big(\frac{y_n^{(e)}}{3r}\Big)\Big)\Big], \end{aligned}$$

and:

$$\begin{aligned} \mathbf{f}_j^{[BLO_l]}(y, t) = \sum_{e: O_l \in e} \Big[& -\sum_{k=0}^{j-1} (\mathbf{V}_k^{[BLO_l]}(y, t) \cdot \nabla_y) \Big\{ \zeta\Big(\frac{y_n^{(e)}}{3r}\Big) \mathbf{V}_{j-k-1}^{(e)}(y^{(e)'}, t)\Big\} \\ & - \sum_{k=0}^{j-1} (\mathbf{V}_k^{[BLO_l]}(y, t) \cdot \nabla_y) \Big\{ \mathbf{V}_{j-k-1}^{[BLO_l]}(y, t)\Big\} \\ & - \zeta\Big(\frac{y_n^{(e)}}{3r}\Big) \sum_{k=0}^{j-1} (\mathbf{V}_k^{(e)}(y^{(e)'}, t) \cdot \nabla_y) \Big\{ \mathbf{V}_{j-k-1}^{[BLO_l]}(y, t)\Big\}\Big], \end{aligned}$$

with $l = 1, \ldots, N$ and $j = -1, \ldots, J$.

This problem is solved in two steps: first, we find the pair $(\mathbf{V}_j^{[BLO_l]}, \hat{P}_j^{[BLO_l]})$ by solving the same problem without the last term in the definition of $\mathbf{f}_j^{[REGO_l]}$. This has a unique solution $\mathbf{V}_j^{[BLO_l]}(\cdot, t) \in \mathring{W}^{1,2}(\Omega_l)$ and $\widehat{P}_j^{[BLO_l]} \in L^2_{loc}(\Omega_l)$, with t being a parameter, if and only if:

$$\int\limits_{\Omega_l} h_j^{[REGO_l]}(y,t)dy = 0,$$

which can be simplified to:

$$\sum_{e:O_l\in e}\int\limits_{\sigma^{(e)}} \mathbf{V}_j^{(e)}(y^{(e)\prime},t)\cdot \mathbf{n}dy^{(e)\prime} = 0,$$

i.e.:

$$\sum_{e:O_l\in e} L^{(e)}s_j^{(e)}(t) = 0,$$

or:

$$-\sum_{e:O_l\in e}\left(L^{(e)}\frac{\partial p_j^{(e)}}{\partial x_n^{(e)}}\right) = 0,$$

for $l = 1,\ldots,N$. It is satisfied because $p_j^{(e)}$ is a solution to the problem on the graph with this junction condition (by the inductive hypothesis). When $\mathbf{V}^{[BLO_l]}$ exponentially tends to zero as $|y| \to \infty$, the corresponding pressure function $\hat{P}_j^{[BLO_l]}$ stabilizes at infinity outlets to some constants $\widehat{a}_j^{[BLO_l,e]}(t)$, which can differ for different outlets. As the pressure function is defined up to a dependent only on time additive function, we can set the limit equal to zero for the outlet corresponding to the selected edge e_s. We define $\varphi_{l,j+1}^{(e)}(t) = \widehat{a}_j^{[BLO_l,e]}(t)$.

Similarly, in every vertex $O_l,\ l = N_1+1,\ldots,N$, we get for $(\mathbf{V}_j^{[BLO_l]},\ P_j^{[BLO_l]})$ the Stokes problem in Ω_l, which is the same as in the case of nodes O_l with only one difference: there is no summing over $e : O_i \in e$ in the right-hand sides of the equations.

Step 3. Solve the following problem:

$$\begin{cases} -\dfrac{\partial}{\partial x_n^{(e)}}\left(L^{(e)}\dfrac{\partial p_{j+1}^{(e)}}{\partial x_n^{(e)}}(x_n^{(e)},t)\right) = 0, & x_n^{(e)} \in (0,|e|),\ \forall e = e_j, \\ & j = 1,\ldots,M, \\ -\displaystyle\sum_{e:O_l\in e}\left(L^{(e)}\frac{\partial p_{j+1}^{(e)}}{\partial x_n^{(e)}}\right)(0,t) = 0, & l = 1,\ldots,N_1, \\ -\left(L^{(e)}\dfrac{\partial p_{j+1}^{(e)}}{\partial x_n^{(e)}}\right)(0,t) = 0, & l = N_1+1,\ldots,N, \\ p_{j+1}^{(e)}(0,t) - p_{j+1}^{(e_s)}(0,t) = \varphi_{l,j+1}^{(e)}(t),\ \forall e \subset \mathcal{B}_l, e \neq e_s. \end{cases}$$

The local coordinates $x^{(e)}$ are chosen so that all of them have the same origin O_l.

Step 4. Finally, we find the pressure $P_j^{[BLO_l]}(y,t)$ in the boundary layer problem:

$$P_0^{[BLO_l]} = \widehat{P}_0^{[BLO_l]} - \Big(\sum_{e:O_l\in e, e\neq e_s} \zeta\Big(\frac{y_n^{(e)}}{3r}\Big)\Big)\widehat{a}_1^{[BLO_l,e]}(t) - \Big(\sum_{e:O_l\in e} \zeta\Big(\frac{y_n^{(e)}}{3r}\Big) - 1\Big) p_{j+1}^{(e_s)}(0,t).$$

For $j = J$, the last sum is absent. This final step completes the transition from j to $j+1$.

7.4 Residual for the Time-Periodic Setting

As before, we add the divergence corrector $\mathbf{w}^{(J)}$ to the constructed solution $\mathbf{u}^{(J)}$ such that the sum $\mathbf{u}^{(J)} + \mathbf{w}^{(J)}$ is divergence-free, with the same estimates for $\mathbf{w}^{(J)}$ as in previous chapter. The formula for the residual $\mathbf{f}^{(J)}$ for $\beta = 0$ is:

$$\|\mathbf{f}^{(J)}\|_{L^2(B_\varepsilon\times(0,2\pi))} = O(\varepsilon^{J-2}).$$

The remaining part of the justification and the estimates for the asymptotic approximation and the MAPDD solution are the same as in Chap. 6, Sect. 6.3.

For $\beta = 2$, as before, we have to construct first the divergence corrector. Namely, we have:

$$\mathrm{div}\mathbf{v}^{(J)} = h^{(J)}(x,t),$$

where

$$h^{(J)}(x,t) = -\sum_{l=1}^{N} \nabla\zeta\Big(\frac{|x-O_l|}{|e|_{min}}\Big)\cdot \mathbf{V}^{[BLO_l,J]}\Big(\frac{|x-O_l|}{\varepsilon}, t\Big).$$

So, we construct $\mathbf{w}^{(J)} \in L^2(0,2\pi; W^{2,2}(B_\varepsilon)\cap \mathring{W}^{1,2}(B_\varepsilon))$ with $\mathbf{w}_t^{(J)} \in L^2(0,2\pi; \mathring{W}^{1,2}(B_\varepsilon))$ such that:

$$\mathrm{div}\mathbf{w}^{(J)} = -h^{(J)}(y,t).$$

Obviously, $h^{(J)} \in L^2(0,2\pi; \mathring{W}^{1,2}(B_\varepsilon))$, $h_t^{(J)} \in L^2(0,2\pi; L^2(B_\varepsilon))$. Since the support of the function $\nabla\zeta\big(\frac{|x-O_l|}{|e|_{min}}\big)$ belongs to the middle third of every cylinder, there hold the relations:

$$\|h^{(J)}\|_{L^2(0,2\pi;W^{2,2}(B_\varepsilon))} = O(e^{-c_1/\varepsilon}), \quad \left\|\frac{\partial h^{(J)}}{\partial t}\right\|_{L^2(0,2\pi;L^2(B_\varepsilon))} = O(e^{-c_1/\varepsilon}),$$

for some $c_1 > 0$.

Taking into consideration that the support of the divergence corrector lies in the union of middle thirds of the cylinders composing B_ε and that the integral of $h^{(J)}(y,t)$ in each middle third of cylinders vanishes, we conclude that such function exists and the following estimates:

$$\|\mathbf{w}^{(J)}\|_{L^2(0,2\pi;W^{2,2}(B_\varepsilon))} \le \varepsilon^{-2}c\|h^{(J)}\|_{L^2(0,2\pi;W^{1,2}(B_\varepsilon))}, \tag{7.4.1}$$

$$\|\mathbf{w}_t^{(J)}\|_{L^2(0,2\pi;W^{1,2}(B_\varepsilon))} \le \varepsilon^{-1}c\|h_t^{(J)}\|_{L^2(0,2\pi;L^2(B_\varepsilon))} \tag{7.4.2}$$

hold. Here the constant c is independent of t and ε.

Let us define $\widetilde{\mathbf{u}}^{(J)} = \mathbf{v}^{(J)} + \mathbf{w}^{(J)}$. Consequently, $\mathrm{div}\widetilde{\mathbf{u}}^{(J)}$ equals zero, and $\widetilde{\mathbf{u}}^{(J)}$ fulfills the boundary conditions.

Let us compute the residual. Denote, $\mathcal{L}(\mathbf{u}, p) = \varepsilon^{-2}\mathbf{u}_t - \nu\Delta\mathbf{u} + (\mathbf{u}\cdot\nabla)\mathbf{u} + \nabla p$. Let us calculate $\mathcal{L}(\widetilde{\mathbf{u}}^{(J)}, p^{(J)})$ in a half-bundle $HB_{O_l}, l = 1, \dots, N$:

$$-\mathbf{f}_1^{(J)}(x) = \mathcal{L}(\tilde{\mathbf{u}}^{(J)}, p^{(J)})$$

$$= \sum_{J+1\le j\le 2J} \varepsilon^{j-2}\Bigg(\sum_{p+r=j-1=j-1} \Big(\mathbf{V}_p^{(e)}(y_n^{(e)\prime}, t)\zeta\Big(\frac{y_n^{(e)}}{3r}\Big)\cdot\nabla_y\Big)\mathbf{V}_r^{(e)}(y_n^{(e)\prime}, t)\zeta\Big(\frac{y_n^{(e)}}{3r}\Big)$$

$$+ \sum_{p+r=j-1} \zeta\Big(\frac{y_n^{(e)}}{3r}\Big)\big(\mathbf{V}_{(p)}^{(e)}(y^{(e)\prime}, t)\cdot\nabla_y\big)\mathbf{V}_r^{[BLO_l]}(y,t)$$

$$+ \sum_{p+r=j-1} (\mathbf{V}_p^{[BLO_l]}(y,t)\cdot\nabla_y)\Big(\zeta\Big(\frac{y_n^{(e)}}{3r}\Big)\mathbf{V}_{(r)}^{(e)}(y^{(e)\prime}, t)\Big)$$

$$+ \sum_{p+r=j-1} (\mathbf{V}_p^{[BLO_l]}(y,t)\cdot\nabla_y)\mathbf{V}_r^{[BLO_l]}(y,t)\Bigg)$$

$$-\varepsilon^{-2}\mathbf{V}_t^{[BLO_l,J]}(y,t)\zeta(\frac{|x-O_l|}{|e|_{min}})\chi(x) + \nu\Delta\Big(\mathbf{V}^{[BLO_l,J]}(y,t)\zeta(\frac{|x-O_l|}{|e|_{min}})\Big)\chi(x)$$

$$+ \Big((\mathbf{V}^{[BLO_l,J]}(y,t)(1-\zeta(\frac{|x-O_l|}{|e|_{min}})\cdot\nabla))(\mathbf{V}^{[BLO_l,J]}(y,t)(1-\zeta(\frac{|x-O_l|}{|e|_{min}}))\Big)$$

$$-\Big(\mathbf{V}^{[BLO_l,J]}(y,t)\zeta(\frac{|x-O_l|}{|e|_{\min}})\cdot\nabla\Big)\mathbf{V}^{[e,J]}(y^{(e)\prime}, t)$$

$$- \Big(\mathbf{V}^{[e,J]}(y^{(e)\prime}, t)\cdot\nabla\Big)\Big(\mathbf{V}^{[BLO_l,J]}(y,t)\zeta\big(\frac{|x-O_l|}{|e|_{\min}}\big)\Big)$$

$$- \big(\mathbf{V}^{[BLO_l,J]}(y,t)\cdot\nabla)\mathbf{V}^{[BLO_l,J]}(y,t)\Big)\chi(x)$$

$$+\varepsilon^{-2}\mathbf{w}_t^{(J)} - \nu\Delta\mathbf{w}^{(J)} + (\mathbf{w}^{(J)}\cdot\nabla)\mathbf{w}^{(J)} + (\mathbf{v}^{(J)}\cdot\nabla)\mathbf{w}^{(J)} + (\mathbf{w}^{(J)}\cdot\nabla)\mathbf{v}^{(J)}$$

$$+\varepsilon^{J-2}\sum_{e:O_l\in e}\widehat{a}_J^{[BLO_l,e]}(t)\nabla_y\zeta\Big(\frac{y_n^{(e)}}{3r}\Big)$$

$$-\nabla\big(\zeta(\frac{|x-O_l|}{|e|_{min}})P^{[BLO_l,J]}(y,t)\big)\chi(x).$$

Here, as before, $y = \dfrac{x-O_l}{\varepsilon}$; $y^{(e)} = \dfrac{x^{(e)}}{\varepsilon}$; $\chi = \chi_{\operatorname{supp}\zeta\left(\frac{|x-O_l|}{|e|_{min}}\right)}$ is the characteristic function of the set $\operatorname{supp}\zeta\left(\frac{|x-O_l|}{|e|_{min}}\right)$. As before, the terms of the sums $\sum\limits_{e:O_l\in e}$ are extended by zero out of cylinders $\Pi_\varepsilon^{(e)}$.

From the derived formulas, it becomes evident that:

$$\|\mathbf{f}_1^{(J)}\|_{L^2(B_\varepsilon\times(0,2\pi))} = \|\mathcal{L}(\tilde{\mathbf{u}}^{(J)}, p^{(J)})\|_{L^2(B_\varepsilon\times(0,2\pi))} = O(\varepsilon^{J-2}). \tag{7.4.3}$$

7.5 Error Estimate for the Time-Periodic Setting

As in previous chapters, we define the extension of the boundary value $\mathbf{g}$ as the derived asymptotic expansion $\widetilde{\mathbf{u}}^{[J]}$. By subsequently applying Theorems 7.2.1, 7.2.3, and 7.2.4, leveraging corresponding estimates, and arguing as before (see Chap. 6), we arrive at the ensuing result.

Theorem 7.5.1 *If* $\mathbf{g} \in C^{[\frac{J+1}{2}]+3}([0,2\pi]; W^{3/2,2}(\partial B_\varepsilon))$, *then the following estimates hold:*

$$\sup_{t\in[0,2\pi]}\|\mathbf{v}(\cdot,t)-\widetilde{\mathbf{u}}^{(J)}(\cdot,t)\|_{L^2(B_\varepsilon)} + \|\nabla\mathbf{v}-\nabla\widetilde{\mathbf{u}}^{(J)}\|_{L^2(0,2\pi;L^2(B_\varepsilon))} = O(\varepsilon^{J+\frac{n-1}{2}}), \tag{7.5.1}$$

$$\sup_{t\in[0,2\pi]}\|\nabla\mathbf{v}(\cdot,t)-\nabla\widetilde{\mathbf{u}}^{(J)}(\cdot,t)\|_{L^2(B_\varepsilon)} + \|\mathbf{v}_t-\widetilde{\mathbf{u}}_t^{(J)}\|_{L^2(0,2\pi;L^2(B_\varepsilon))}$$
$$+\|\nabla^2\mathbf{v}-\nabla^2\widetilde{\mathbf{u}}^{(J)}\|_{L^2(0,2\pi;L^2(B_\varepsilon))} = O(\varepsilon^{J+\frac{n-3}{2}}). \tag{7.5.2}$$

If $\mathbf{g} \in C^{[\frac{J+1}{2}]+n+3}([0, 2\pi]; W^{3/2,2}(\partial B_\varepsilon))$, *then there exists the pressure* p *such that the following estimate holds:*

$$\|p - p^{(J)}\|_{L^2(0,2\pi;L^2(B_\varepsilon))} = O(\varepsilon^{J+\frac{n-3}{2}}). \tag{7.5.3}$$

Remark 7.5.1 As in the previous chapters, we modify the asymptotic expansion in a way that it belongs to the space $\mathcal{H}_{\text{dec}}(B_\varepsilon)$ for almost all $t \in [0, 2\pi]$, changing the argument $\frac{|x-O_l|}{|e|_{\min}}$ of the cutoff function ζ by $\frac{|x-O_l|}{\delta}$, where $\delta = C_J\varepsilon|\ln\varepsilon||e|_{\min}$, and the constant C_J is chosen so that:

$$\begin{aligned}\widetilde{E}_{\delta/\varepsilon} &= \|\mathbf{V}^{[BLO_l,J]}\|_{W^{(2,1),2}(\Omega_{l,\delta/\varepsilon}\times(0,2\pi))} + \|\nabla P^{[BLO_l,J]}\|_{L^2(\Omega_{l,\delta/\varepsilon}\times(0,2\pi))}\\ &= O(\varepsilon^{J+(n-1)/2+2}).\end{aligned} \tag{7.5.4}$$

Here, as before, $\Omega_{l,R} = \Omega_l \cap \{|y| > R\}$.

Indeed, by Theorem 4.6.3, $\mathbf{V}^{[BLO_l,J]}(y,t)$ and $P^{[BLO_l,J]}(y,t)$ tend to zero exponentially as $|y| \to \infty$, leading to the inequality:

$$\widetilde{E}_R(t) \le c_1 e^{-c_2 R}, \quad c_1, c_2 > 0.$$

Consequently, the estimate $\widetilde{E}_{\delta/\varepsilon} = O(\varepsilon^{J+(n-1)/2+2})$ holds for δ such that:

$$c_2\delta/\varepsilon \ge (J + (n-1)/2 + 2)|\ln\varepsilon|,$$

implying $\delta = C_J\varepsilon|\ln\varepsilon|$.

7.6 Method of Asymptotic Partial Decomposition of the Domain

In this section, we examine the application of the method of asymptotic partial decomposition of the domain (MAPDD) within a time-periodic framework. Similar to Sect. 6, we establish the existence of the MAPDD velocity $\mathbf{u}_d$, which belongs to the function space $L^2(0, T; \mathcal{H}_{0,\text{dec}}(B_\varepsilon, \delta)) \cap L^\infty(0, T; L^2(B_\varepsilon))$. This definition corresponds to the concept of Leray–Hopf solutions for the Navier–Stokes equations.

Definition 7.6.1 Find a time-periodic vector function $\mathbf{v}_d \in L^2(0, 2\pi; \mathcal{H}_{\text{dec}}(B_\varepsilon, \delta)) \bigcap L^\infty(0, 2\pi; L^2(B_\varepsilon))$, such that $\mathbf{v}_d(x,t) = \mathbf{g}(x,t)$ on ∂B_ε and for all time-periodic test functions $\boldsymbol{\eta} \in W^{1,2}(0, 2\pi; \mathcal{H}_{0,\text{dec}}(B_\varepsilon, \delta))$, the following integral identity is valid:

$$\int_0^{2\pi}\int_{B_\varepsilon}\Big(-\varepsilon^{-\beta}\mathbf{v}_d\cdot\frac{\partial\boldsymbol{\eta}}{\partial t}+\nu\nabla\mathbf{v}_d\cdot\nabla\boldsymbol{\eta}+(\mathbf{v}_d\cdot\nabla)\mathbf{v}_d\cdot\boldsymbol{\eta}\Big)\,dxdt=0. \tag{7.6.1}$$

Introduce $\mathbf{u}_d = \mathbf{v}_d - \widetilde{\mathbf{u}}^{(J)}$, where $\widetilde{\mathbf{u}}^{(J)}$ is the asymptotic expansion with the modified cutoff function ζ (see Remark 7.5.1). Then this formulation is equivalent to the following one.

Definition 7.6.2 Find a time-periodic vector-valued function $\mathbf{u}_d \in L^2(0, 2\pi; \mathcal{H}_{0,\mathrm{dec}}(B_\varepsilon, \delta)) \bigcap L^\infty(0, 2\pi; L^2(B_\varepsilon))$ such that for all time-periodic test functions $\boldsymbol{\eta} \in W^{1,2}(0, 2\pi; \mathcal{H}_{0,\mathrm{dec}}(B_\varepsilon, \delta))$, the following integral identity is valid:

$$\int_0^{2\pi}\int_{B_\varepsilon}\Big(-\varepsilon^{-\beta}\mathbf{u}_d\cdot\frac{\partial\boldsymbol{\eta}}{\partial t}+\nu\nabla\mathbf{u}_d\cdot\nabla\boldsymbol{\eta}-((\mathbf{u}_d+\widetilde{\mathbf{u}}^{(J)})\cdot\nabla)\boldsymbol{\eta}\cdot\mathbf{u}_d$$
$$-(\mathbf{u}_d\cdot\nabla)\boldsymbol{\eta}\cdot\widetilde{\mathbf{u}}^{(J)}\Big)\,dxdt=\int_0^{2\pi}\int_{B_e}\mathbf{f}^{(J)}\cdot\boldsymbol{\eta}\,dxdt. \tag{7.6.2}$$

Theorem 7.6.1 *Suppose $B_\varepsilon \subset \mathbb{R}^n$, $n = 2, 3$. For sufficiently small ε, there exists a solution $\mathbf{u}_d \in L^2(0, 2\pi; \mathcal{H}_{0,\mathrm{dec}}(B_\varepsilon, \delta)) \cap L^\infty(0, 2\pi; L^2(B_\varepsilon))$ that satisfies the integral identity* (7.6.2)*. The following estimate holds:*

$$\sup_{t\in[0,2\pi]}\|\mathbf{u}_d(\cdot,t)\|^2_{L^2(B_\varepsilon)}+\varepsilon^\beta\int_0^{2\pi}\|\nabla\mathbf{u}_d(\cdot,\tau)\|^2_{L^2(B_\varepsilon)}d\tau$$
$$\le c\varepsilon^{2+\beta}\int_0^{2\pi}\|\mathbf{f}^{(J)}(\cdot,t)\|^2_{L^2(B_\varepsilon)}\,dt. \tag{7.6.3}$$

Proof
The proof of the theorem combines ideas from the proofs of Theorem 7.2.1 and Theorem 6.10.1. We establish the solvability of the problem (7.6.2) using the Galerkin method. As certain portions of the proof reiterate arguments from the aforementioned theorems, we shall omit them and refer to the corresponding prior proofs.

The Galerkin approximations for the solution of problem (7.6.2) satisfy the following relations:

$$\int\limits_{B_\varepsilon} \Big(\frac{1}{\varepsilon^\beta} \frac{\partial \mathbf{u}_d^{(N)}}{\partial t} \cdot \boldsymbol{\xi}_l + \nu \nabla \mathbf{u}_d^{(N)} \cdot \nabla \boldsymbol{\xi}_l - \big(\mathbf{u}_d^{(N)} + \widetilde{\mathbf{u}}^{(J)}\big) \cdot \nabla\big) \boldsymbol{\xi}_l \cdot \mathbf{u}_d^{(N)}$$
$$- \big(\mathbf{u}_d^{(N)} \cdot \nabla\big) \boldsymbol{\xi}_l \cdot \widetilde{\mathbf{u}}^{(J)} \Big) dx = \int\limits_{B_\varepsilon} \mathbf{f}^{(J)} \cdot \boldsymbol{\xi}_l dx, \tag{7.6.4}$$
$$\gamma_l^{(N)}(0) = \gamma_l^{(N)}(2\pi),$$

where $l = 1, .., N$, $\mathbf{u}_d^{(N)}(x, t) = \sum\limits_{k=1}^{N} \gamma_k^{(N)}(t) \boldsymbol{\xi}_k(x)$, and $\{\boldsymbol{\xi}_k\}_{k=1}^{\infty}$ forms a basis in the space $\mathcal{H}_{0,\mathrm{dec}}(B_\varepsilon, \delta)$. Without loss of generality, we assume that $\{\boldsymbol{\xi}_k\}_{k=1}^{\infty}$ is orthonormal in $L^2(B_\varepsilon)$.

The existence of time-periodic Galerkin approximations $\mathbf{u}_d^{(N)}$ and the subsequent estimate:

$$\max_{t\in[0,2\pi]} \|\mathbf{u}_d^{(N)}(\cdot, t)\|^2_{L^2(B_\varepsilon)} + \varepsilon^\beta \int\limits_0^{2\pi} \int\limits_{B_\varepsilon} |\nabla \mathbf{u}_d^{(N)}(x, t)|^2 dx dt$$
$$\leq c\varepsilon^{2+\beta} \int\limits_0^{2\pi} \|\mathbf{f}^{(J)}(\cdot, t)\|^2_{L^2(B_\varepsilon)} dt \tag{7.6.5}$$

can be established by repeating the arguments from the first part of Theorem 7.2.1.

The estimate (7.6.5) for the Galerkin approximation ensures the existence of a subsequence $\mathbf{u}_d^{(N_l)}$ such that $\mathbf{u}_d^{(N_l)}$ and $\nabla \mathbf{u}_d^{(N_l)}$ converge weakly in $L^2(0, 2\pi; L^2(B_\varepsilon))$ to $\mathbf{u}_d$ and $\nabla \mathbf{u}_d$, respectively. Furthermore, by following the approach in the book by O.A. Ladyzhenskaya [102] (see Chapter VI.7), it can be shown that the sequence $\mathbf{u}_d^{(N_l)}$ converges strongly in $L^2(0, 2\pi; L^2(B_\varepsilon))$. This information suffices to pass to the limit in the integral identity, resulting in $\mathbf{u}_d$ being a weak solution of problem (7.6.2). Clearly, $\mathbf{u}_d$ satisfies the estimate (7.6.3). □

The MAPDD is justified by the following theorem.

Theorem 7.6.2 *Assume that* $\mathbf{g} \in C^{[\frac{J+1}{2}]+3}(\mathbb{R}; W^{2,2}(B_\varepsilon))$, *and let* δ *be* $C_{J+2}\varepsilon|\ln \varepsilon|$, *where the constant* C_{J+2} *satisfies* (7.5.4) *with* J *replaced by* $J+2$. *Then there exists a solution* $\mathbf{v}_d$ *to problem* (7.6.1) *satisfying the following error estimate:*

$$\sup_{t\in[0,2\pi]} \|\mathbf{v}_d - \mathbf{v}\|_{L^2(B_\varepsilon)} + \|\nabla \mathbf{v}_d - \mathbf{v}\|_{L^2(B_\varepsilon)\times(0,2\pi)} = O(\varepsilon^{J+(n-1)/2}). \tag{7.6.6}$$

Proof
Denote $\mathbf{u}_d = \mathbf{v}_d - \widetilde{\mathbf{u}}^{(J')}$, where $J' = J + 2$. Taking into account estimate (7.5.1) and the estimate of Theorem 7.6.1 with J replaced by J' (as in Theorem 6.10.1) and applying the triangle inequality, we get estimate (7.6.6). □

Chapter 8
Bibliographical Remarks

Let us describe briefly the state of the art of the works on partial differential equations set in thin domains. Such domains depend on the small parameter ε, which is the ratio of sizes in different directions. In particular, the thin tube structures belong to this class of domains. The studying of thin structures is motivated by the applications to modeling of industrial installations (rods, bars, frames) as well as by the biological applications (networks of blood vessels). The vast spectrum of the research papers was devoted to the solid mechanics of rods, bars, plates, and rod structures and their dimension reduction based on the asymptotic analysis for the equations of the elasticity theory (see [4, 9, 10, 12, 46–48, 58, 70–72, 77, 97, 112, 124, 128–130, 139–142, 146, 147, 157, 190, 204, 209]; for continuous models for periodic networks of bars, see [135, 139]). Thin structures in fluid mechanics were related to the lubrication problem [14, 33] and to the modeling of viscous flows in thin tubes and tube structures, unions of thin tubes. In the last case, the blood circulation system is a big union B_ε of thin cylindrical "vessels," and ε is the ratio of the radius and the height of the cylinders. This geometrical approximation for such flows has been considered first in [144] where an asymptotic analysis was developed for the steady Stokes and Navier–Stokes equations set in a tube structure with the no-slip boundary condition on the lateral part of the boundary. In order to simplify the solution of equations in thin structures, the method of asymptotic partial decomposition of the domain (MAPDD) has been introduced in [143, 145, 146]. This method combines the description of different dimensions in one hybrid model: it provides the one-dimensional reduction in the main part of the thin structure, at some distance from the junction area, and it keeps the $3D$ (or $2D$) description within the junction area [28, 143, 145]. The asymptotic analysis gives the appropriate interface conditions between these $3D$ and $1D$ descriptions. In particular, for the steady Stokes or Navier–Stokes equations, the above $1D$ models are reduced to some Poiseuille-type flows [28, 145]. Important results on the flows in thin tube structures concern the generalization of asymptotic methods for the

G. Panasenko, K. Pileckas, *Multiscale Analysis of Viscous Flows in Thin Tube Structures*, Advances in Mathematical Fluid Mechanics,
https://doi.org/10.1007/978-3-031-54630-3_8

non-steady Stokes and Navier–Stokes equations. The asymptotic expansion of the solution to the non-steady Navier–Stokes equations was constructed in the set of papers [148–152]. In particular, for the nonstationary Navier–Stokes equations, the $1D$ models are reduced to some non-steady Poiseuille-type flows. It turns out that the keystone of the construction of asymptotic approximations of the solutions of the Stokes and Navier–Stokes equations in thin tube structures is the one-dimensional problem on the graph. This problem relates the Poiseuille-type approximations in the tubes and describes the distribution of the fluxes in the structure. It was introduced and studied in [149] for stationary and nonstationary initial boundary value problems and [153] for the time-periodic setting. Note that the researchers in physiology used one-dimensional approaches for a long time ([125, 126, 181]). The asymptotic analysis and MAPDD for the Stokes and Navier–Stokes equations in the tube structures are in the spotlight of the present book. The auxiliary theorems for the justification of these results use the theory of inverse problem describing the time-dependent Poiseuille flow; see [40, 41, 68, 86, 90, 173, 175–178, 202], etc. for the initial boundary value problem and [16, 67, 86] for the time-periodic case. Chapter III of this book is based on papers [67, 68, 86, 148, 175–177]. Further, the time-periodic setting in thin tube structures for the Navier–Stokes equations was considered in papers [83, 153]. Note that the idea of coupling of models of different dimension was discussed in the engineering literature before the rigorous mathematical studies (see, e.g., [195]), but the question of mathematically consistent junction conditions was an open question.

Evidently, the present monograph cannot cover all topics related to the viscous flows in thin tube structures. Let us describe this adjacent research area. First of all, the Newtonian rheology (Navier–Stokes equations) may be replaced by a non-Newtonian one. There exist several models of non-Newtonian behavior of a viscous fluid (see [50, 69, 92, 179, 187, 188], [93]) as well as close to such rheology micropolar fluid (see [3, 60, 115]). In particular, these rheologies were used for the blood flow description in [1, 63, 185, 186] [69]. Let us mention some papers studying such flows in networks of vessels or thin tube structures: [18, 56, 57] (micropolar fluids) and [156, 166–168] (stationary Navier–Stokes equations with shear rate-dependent viscosity). The new method of partial asymptotic reduction for non-Newtonian setting was introduced in [155]. Other non-Newtonian models in thin structures were considered in [38] (Bingham flow) and [118] (power law). The flow with variable viscosity was considered in [45]. The asymptotic and numerical analysis of the coupling of the Stokes equations and the diffusion–convection equations in thin tube structures was considered in [44]. Asymptotic analysis of flows in thin tube structures was further developed in [96, 121], while the fractal graph structures were considered for modeling of the blood vessel network in [98].

Another important generalization is related to the boundary conditions and, in particular, the case of elastic wall of a tube (the so-called fluid–structure interaction (FSI); see [74, 76] for the general mathematical theory). The dimension reduction of the linearized FSI models for the thin tubes was studied in [30, 39, 158–162, 165]. Numerical experiments for FSI (fluid-structure interaction) dimension reduction in tube structures can be found in [87]. Such models were used for the respiratory track modeling (see [11, 75]).

Curvilinear channels or vessels were considered in [24, 57].

Numerical testing of the MAPDD was performed in [7, 25, 26, 62, 127, 164]. Special numerical methods for the MAPDD were developed in [42, 43, 49, 62, 164].

Note that the asymptotic analysis of thin tube structures uses as the tools the matching method and the boundary layer method (see [81, 82]) as well as the Γ convergence method (see [35, 54]).

Finally, let us cite some general monographs on theoretical fluid dynamics [13, 66, 102, 109, 194, 203].

Bibliography

1. I. Abdullah, N. Amin, A micropolar fluid model of blood flow through a tapered artery with a stenosis. Math. Methods Appl. Sci. **33**, 1910–1923 (2010)
2. R. Adams, J. Fournier, *Sobolev Spaces*. Pure and Applied Mathematics Series, 2nd edn. (Elsevier, Amsterdam, 2003)
3. E.L. Aero, E.V. Kuvshinsky, Main equations of theory of elastic media with rotational interaction of particles. Phys. Solids (Fizika Tverdogo Tela) **2**(7), 1399–1409 (1960) (in Russian)
4. I. Aganović, Z. Tutek, A justification of the one-dimensional model of an elastic beam. Math. Meth. Appl. Sci. **8**, 1–14 (1986)
5. S. Agmon, *Lectures on Elliptic Boundary Value Problems* (AMS Chelsea, Providence, 2010)
6. S. Agmon, A. Douglis, L. Nirenberg, Estimates near the boundary for solutions of elliptic partial differential equations satisfying general boundary conditions. I. Comm. Pure Appl. Math. **12**, 623–727 (1959); II. Comm. Pure Appl. Math. **17**, 35–92 (1964)
7. H. Amar, D. Givoli, Mixed-dimensional modeling of time-dependent wave problems using the Panasenko construction. J. Theor. Comput. Acoust. **26**(3), Article No. 1850034 (2018)
8. M.S. Alnaes, J. Blechta, J. Hake, A. Johansson, B. Kehlet, A. Logg, C. Richardson, J. Ring, M. E. Rognes, G.N. Wells, The FEniCS project version 1.5. Arch. Num. Softw. **3**(100), 9–23 (2015)
9. I. Babuska, L. Li, Hierarchic modelling of plates. Comput. Struct. **40**, 419–430 (1991)
10. I. Babuska, C. Schwab, A posteriori error estimation for the hierarchic models of elliptic boundary value problems on thin domains. SIAM J. Numer. Anal. **33**, 221–246 (1996)
11. L. Baffico, C. Grandmont, B. Maury, Multiscale modeling of the respiratory tract. Math. Models Methods Appl. Sci. **20**(1), 59–93 (2010)
12. N.S. Bakhvalov, G.P. Panasenko, *Homogenisation: Averaging Processes in Periodic Media* (Nauka, Moscow, 1984) (in Russian); English Transl.: (Kluwer, Dordrecht, 1989)
13. G.K. Batchelor, *An Introduction to Fluid Dynamics* (Cambridge University Press, Cambridge, 2002)
14. G. Bayada, M. Chambat, The transition between the Stokes equations and the Reynolds equation: a mathematical proof. Appl. Math. Optim. **14**(1), 73–93 (1986)
15. R. Becker, D. Capatina, R. Luce, D. Trujillo, Finite elements formulation of general boundary conditions for incompressible flows. Compt. Methods. Appl. Mech. Eng. **295**, 240–267 (2015)

G. Panasenko, K. Pileckas, *Multiscale Analysis of Viscous Flows in Thin Tube Structures*, Advances in Mathematical Fluid Mechanics,
https://doi.org/10.1007/978-3-031-54630-3

16. H. Beirao da Veiga, On time-periodic solutions of the Navier–Stokes equations in an unbounded cylindrical domain. Leray's problem for periodic flows. Arch. Rational Mech. Anal. **178**, 301–325 (2005)
17. M. Beneš, P. Kučera, Solutions of the Navier-Stokes equations with various types of boundary conditions. Arch. Math. **98**, 487–497 (2012)
18. M. Beneš, I. Pažanin, Effective flow of incompressible micropolar fluid through a system of thin pipes. Acta Appl. Math. **143**(1), 29–43 (2016)
19. M. Beneš, I. Pažanin, M. Radulović, B. Rukavina, Nonzero boundary conditions for the unsteady micropolar pipe flow: well-posedness and asymptotics. Appl. Math. Comput. **427**, Article No. 127184 (2022)
20. C. Bèque, C. Conca, F. Murat, P. Pironneau, A nouveau sur les équations de Stokes et de Navier–Stokes avec des conditions aux limites sur la pression. C.R. Acad. Sci. Paris, Sér I **304**(1), 23–28 (1987)
21. C. Bèque, C. Conca, F. Murat, P. Pironneau, Les équations de Stokes et de Navier–Stokes avec des conditions aux limites sur la pression, in *Nonlinear Partial Differential Equations and Their Applications. Collège de France Seminar*. Pitman Research Notes in Mathematics Series, vol. IX, 181 304 (1988), pp. 179–264
22. L.M. Bernard, Non-standard Stokes and Navier-Stokes problems: existence and regularity in stationary case. Math. Meth. Applied. Sci. **25**(8), 627–661 (2002)
23. L.M. Bernard, Time-dependent Stokes and Navier-Stokes problems: existence problems with boundary conditions involving pressure, existence and regularity. Nonlinear Anal. Real World Appl. **4**(5), 805–839 (2003)
24. F. Berntsson, A. Ghosh, V.A. Kozlov, S.A. Nazarov, A one-dimensional model of blood flow through a curvilinear artery. Appl. Math. Model. **63**, 633–643 (2018)
25. C. Bertoglio, C. Conca, D. Nolte, G. Panasenko, K. Pileckas, Junction of models of different dimension for flows in tube structures by Womersley-type interface conditions. SIAM J. Appl. Math. **79**(3), 959–985 (2019)
26. C. Bertoglio, D. Nolte, G. Panasenko, K. Pileckas, Reconstruction of the pressure in the method of asymptotic partial decomposition for the flows in tube structures. SIAM J. Appl. Math. **81**(5), 2083–2110 (2021)
27. O.V. Besov, V.P. Il'in, S.M. Nikol'skii, *Integral Representations of Functions, and Embedding Theorems* (Nauka, Moscow, 1996) (in Russian)
28. F. Blanc, O. Gipouloux, G. Panasenko, A.M. Zine, Asymptotic analysis and partial asymptotic decomposition of the domain for Stokes equation in tube structure. Math. Models Methods Appl. Sci. **9**(9), 1351–1378 (1999)
29. S. Blazy, S. Nazarov, M. Specovius-Neugebauer, Artificial boundary conditions of pressure type for viscous flows in a system of pipes. J. Math. Fluid Mech. **9**(1), 1–33 (2007)
30. L. Bociu, S. Čanić, B. Muha, J.T. Webster, Multilayered poroelasticity interacting with Stokes flow. SIAM J. Math. Anal. **53**, 6243–6279 (2021)
31. M.E. Bogovskii, Solutions of some problems of vector analysis related to operators div and grad. Proc. Semin. S.L. Sobolev **1**, 5–40 (1980) (in Russian)
32. D. Bothe, T.K. Kashiwabara, M. Köhne, Strong well-posedness for a class of dynamic outflow boundary conditions for incompressible Newtonian flow. J. Evol. Equ. **17**(1), 131–171 (2017)
33. M. Boukrouche, L. Paoli, F. Ziane, Unsteady micropolar fluid flow in a thin domain with Tresca fluid-solid interface law. Comput. Math. Appl. **77**, 2917–2932 (2019)
34. M. Braack, P.P. Mucha, Directional do-nothing condition for the Navier-Stokes equations. J. Comput. Math. **32**(5), 507–521 (2014)
35. A. Braides, *Γ-Convergence for Beginners* (Oxford University Press, Oxford, 2002)
36. H. Brezis, *Analyse Fonctionnelle: Théorie et Applications* (Masson, Paris, 1983)
37. C.-H. Bruneau, P. Fabrie, New efficient boundary conditions for incompressible Navier-Stokes equations: well-posedness result. RAIRO Model. Math. Anal. Num. **30**(7), 815–840 (1996)
38. R. Bunoiu, A. Gaudiello, A. Leopardi, Asymptotic analysis of a Bingham fluid in a thin T-like shaped structure. J. Math. Pures Appl. **123**, 148–166 (2019)

39. S. Čanić, A. Mikelić, Effective equations describing the flow of a viscous incompressible fluid through a long elastic tube. C.R. Acad. Sci. Paris, Série IIb **330**, 661–666 (2002)
40. J.R. Cannon, The solution of the heat equation subject to specification of energy. Q. Appl. Math. **21**, 155–160 (1963)
41. J.R. Cannon, *The One-Dimensional Heat Equation* (Addison-Wesley, Menoro Park, 1984)
42. E. Canon, F. Chardard, G. Panasenko, O. Štikonienė, Numerical solution of viscous flows in network of thin tubes: equations on the graph. J. Comp. Phys. **435**, Article No. 110262 (2021)
43. E. Canon, F. Chardard, G. Panasenko, O. Štikonienė, Asymptotics and discretization of a weakly singular kernel: application to viscous flows in a network of thin tubes. J. Comp. Phys. **491**, Article No. 112327 (2023)
44. G. Cardone, G. Panasenko, I. Sirakov, Asymptotic analysis and numerical modelling of mass transport in tubular structure. Math. Models Methods Appl. Sci. **20**(3), 397–421 (2010)
45. G. Cardone, R. Fares, G. Panasenko, Asymptotic expansion of the solution of the steady Stokes equation with variable viscosity in a two-dimensional tube structure. J. Math. Phys. **53**, Article No. 103702 (2012)
46. P.G. Ciarlet, *Plates and Junctions in Elastic Multi-structures. An Asymptotic Analysis* (Masson, Paris, 1990)
47. P.G. Ciarlet, *Mathematical Elasticity. II. Theory of Plates* (North-Holland, Amsterdam, 1997)
48. P.G. Ciarlet, P. Destuynder, A justification of the two-dimensional plate model. J. Mécanique **18**, 315–344 (1979)
49. R. Čiegis, G. Panasenko, K. Pileckas, V. Šumskas, ADI scheme for partially dimension reduced heat conduction models. Comput. Math. Appl. **80**(5), 1275–1286 (2020)
50. D. Cioranescu, V. Girault, K.R. Rajagopal, *Mechanics and Mathematics of Fluids of the Differential Type* (Springer International Publishing, Switzerland, 2016)
51. C. Conca, *Problèmes Mathématiques en Couplage Fluide-Structure* (Eyrolles, Paris, 1994)
52. C. Conca, F. Murat, O. Pironneau, The Stokes and Navier-Stokes equations with boundary conditions involving the pressure. Japn. J. Math. **20**(2), 279–318 (1994)
53. C. Conca, C. Pares, O. Pironneau, M. Thiriet, Navier-Stokes equations with imposed pressure and velocity fluxes. Int. J. Num. Methods Fluids **20**(4), 267–287 (1995)
54. G. Dal Maso, *An Introduction to Γ-Convergence* (Birkhäuser, Basel, 1998)
55. T. Dobroserdova, F. Liang, G. Panasenko, Y.V. Vassilevski, Multiscale models of blood flow in the compliant aortic bifurcation. Appl. Math. Lett. **93**, 98–104 (2019)
56. D. Dupuy, G. Panasenko, R. Stavre, Asymptotic methods for micropolar fluids in a tube structure. Math. Models Methods Appl. Sci. **5**(14), 735–758 (2004)
57. D. Dupuy, G. Panasenko, R. Stavre, Asymptotic solution for a micropolar flow in a curvilinear channel. Zeitschrift fur Angewandte Mathematik und Mechanik (ZAMM) **88**(10), 793–807 (2008)
58. M.G. Dzhavadov, Asymptotics of solution of boundary value problem for an elliptic equation stated in thin domains. Diff. Equ. **4**(10), 1901–1909 (1968) (in Russian)
59. A.C. Egloffe, *Study of Some Inverse Problems for the Stokes System. Application to the Lungs.* Thèse de doctorat, l'Université Pierre et Marie Curie - Paris VI, 2012
60. A.C. Eringen, Theory of micropolar fluids. J. Math. Mech. **16**, 1–18 (1966)
61. L.C. Evans, *Partial Differential Equations*. Graduate Studies in Mathematics, vol. 19, 2nd edn. (American Mathematical Society, Providence, 2010)
62. F. Fontvieille, G.P. Panasenko, J. Pousin, Asymptotic decomposition of a singular perturbation problem with unbounded energy. C.R. Mecanique **330**, 507–512 (2002)
63. L. Formaggia, D. Lamponi, A. Quarteroni, One dimensional models for blood flow in arteries. J. Engrg. Math. **47**, 251–276 (2003)
64. J. Fouchet-Incaux, Artificial boundaries and formulations for the incompressible Navier-Stokes equations: applications to air and blood flows. SeMA **64**, 1–40 (2014)
65. A. Friedman, *Partial Differential Equations* (Holt, Rinehart & Winston, New York, 1969)
66. G.P. Galdi, *An Introduction to the Mathematical Theory of the Navier-Stokes Equations. Steady-State Problems* (Springer, New York, 2011)

67. G.P. Galdi, A.M. Robertson, The relation between flow rate and axial pressure gradient for time-periodic Poiseuille flow in a pipe. J. Math. Fluid Mech. **7**(Suppl. No. 2), 215–223 (2004)
68. G.P. Galdi, K. Pileckas, A. Silvestre, On the unsteady Poiseuille flow in a pipe. Zeitschrift fur angew. Mathematik und Physik **58**(6), 994–1007 (2007)
69. P. Galdi, R. Rannacher, A.M. Robertson, S. Turek, *Hemodynamical Flows, Modeling, Analysis and Simulation, Oberwolfach Seminars*, vol. 37 (Birkhäuser, Basel, 2008)
70. A.L. Goldenveizer, The approximation of the plate bending theory by the asymptotic analysis of elasticity theory. Pr. Math. Mech. **26**(4), 668–686 (1962)
71. A.L. Goldenveizer, The approximation of the shell theory by the asymptotic analysis of elasticity theory. Pr. Math. Mech. **27**(4), 593–608 (1963)
72. A.L. Goldenveizer, The principles of reducing three-dimensional problems of elasticity to two-dimensional problems of the theory of plates and shells, in *Proceedings, Eleventh International Congress of Theoretical and Applied Mechanics*, ed. by Görtler, H. (Springer, Berlin, 1964), pp. 306–311
73. L. Grafakos, *Modern Fourier Analysis* (Springer, New York, 2009)
74. C. Grandmont, F. Vergnet, Existence for a quasi-static interaction problem between a viscous fluid and an active structure. J. Math. Fluid Mech. **23**, Article No. 45 (2021)
75. C. Grandmont, B. Maury, A. Soualah, Multiscale modelling of the respiratory track: a theoretical framework, in *Mathematical and Numerical Modelling of the Human Lung. ESAIM Proceedings*, vol. 23 (2008), pp. 10–29
76. P.M. Gresho, Some current CFD issues relevant to the incompressible Navier-Stokes equations. Comput. Methods Appl. Mech. Eng. **87**, 201–252 (1991)
77. G. Griso, *Analyse Asymptotique de Structures Réticulées*, Thèse de l'Université Pierre et Marie Curie, 1996
78. F. Harary, *Graph Theory* (Addison-Wesley, Reading, 1971)
79. J.G. Heywood, On uniqueness questions in the theory of viscous flow. Acta. Math. **136**, 61–102 (1976)
80. J.G. Heywood, R. Rannacher, S. Turek, Artificial boundaries and flux and pressure conditions for the incompressible Navier-Stokes equations. Int. J. Num. Methods Fluids **22**, 325–352 (1996)
81. A.M. Il'in, Difference scheme for a differential equation with small parameter in the higher derivative. Mat. Zametki **6**(2), 237–248 (1969) (in Russian)
82. A.M. Il'in, *Matching of Asymptotic Expansions of Solutions of Boundary Value Problems* (Americal Mathematucal Society, Providence, 1992)
83. R. Juodagalvytė, G. Panasenko, K. Pileckas, Time periodic Navier-Stokes equations in a thin tube structure. Boundary Value Probl. **28** (2020). https://doi.org/10.1186/s13661-020-01334-3
84. L.V. Kantorovich, G.P. Akilov, *Functional Analysis*, 2nd edn. (Pergamon Press, Oxford, 1982)
85. L.V. Kapitanskii, K. Pileckas, Certain problems of vector analysis. Zapiski Nauchn. Sem. LOMI, **138**, 65–85 (1984) (in Russian); English Transl.: Sov. Math. **32**, 469–483 (1986)
86. K. Kaulakytė, N. Kozulinas, K. Pileckas, Time-periodic Poiseuille-type solution with minimally regular flow rate. Nonlinear Analy. Modell. Control **26**(5), 947–968 (2021)
87. K. Kaulakytė, N. Kozulinas, G. Panasenko, K. Pileckas, V. Šumskas, Poiseuille-type approximations for axisymmetric flow in a thin tube with thin stiff elastic wall. Mathematics **11**, Article No. 2106 (2023)
88. T. Kim, Regularity of solutions to the Navier-Stokes equations with a nonstandard boundary condition. Acta Math. Appl. Sin. Engl. Ser. **31**(3), 707–718 (2015)
89. N. Kinash, J. Janno, Inverse problems for a generalized subdiffusion equation with final overdetermination. Math. Model. Anal. **24**, 236–262 (2019)
90. N. Kinash, J. Janno, An inverse problem for a generalized fractional derivative with an application in reconstruction of time- and space-dependent sources in fractional diffusion and wave equations. Mathematics **7**(12), Article No. 1138 (2019)

91. A.A. Kiselev, O.A. Ladyzhenskaya, On the existence and uniqueness of the solution of the nonstationary problem for a viscous, incompressible fluid. Izv. Akad. Nauk SSSR Ser. Mat. **21**(5), 655–680 (1957)
92. N. Klovienė, K. Pileckas, Non-stationary Poiseuille-type solutions for the second-grade fluid flow. Lithuanian Math. Journal **52**(2), 155–171 (2012)
93. N. Klovienė, K. Pileckas, The second grade fluid flow problem in an infinite pipe. Asympt. Analy. **83**(3), 237–262 (2013)
94. A.N. Kolmogorov, S.V. Fomin, *Elements of the Theory of Functions and Functional Analysis* (Dover Publications, Mineola, 1999)
95. M.V. Korobkov, K. Pileckas, R. Russo, Solvability in a finite pipe of steady-state Navier-Stokes equations with boundary conditions involving Bernoulli pressure. Calc. Var. and Partial Diff. Equ. **59**(32), 1–22 (2020)
96. V.A. Kozlov, S.A. Nazarov, One-dimensional model for flows in the junction of thin channels, in particular, arterial trees. Math. Sb. **208**(8), 56–105 (2017)
97. V. Kozlov, V. Maz'ya, A. Movchan, *Asymptotic Analysis of Fields in Multi-Structures*. Oxford Mathematical Monographs (Oxford Science Publications, The Clarendon Press, Oxford University Press, New York, 1999)
98. V. Kozlov, S. Nazarov, G. Zavorokhin, A fractal graph model of capillary type systems. Complex Var. Elliptic Equ. **63**(7–8), 1044–1068 (2018)
99. S. Kračmar, J. Neustupa, Modeling of the unsteady flow through a channel with an artificial outflow condition by the Navier-Stokes variational inequality. Math. Nachrichten **291**, 1801–1814 (2018)
100. P. Kučera, Basic properties of solutions of the non-steady Navier-Stokes equations with mixed boundary conditions in a bounded domain. Ann. Univ. Ferrara Sez. VII Sci Mat. **55**(2), 289–308 (2009)
101. P. Kučera, Z. Skalak, Solutions to the Navier-Stokes equations with mixed boundary conditions. Acta Appl. Math. **54**(3), 275–288 (1998)
102. O.A. Ladyzhenskaia, *The Mathematical Theory of Viscous Incompressible Flow* (Nauka, Moscow, 1970) (in Russian); (English Edition: Gordon and Breach Science Publishers, New York, 1969)
103. O.A. Ladyzhenskaya, *The Boundary Value Problems of Mathematical Physics* (Springer, New York, 1985)
104. O.A. Ladyzhenskaya, V.A. Solonnikov, On some problems of vector analysis and generalized formulations of boundary value problems for the Navier–Stokes equations. Zapiski Nauchn. Sem. LOMI **59**, 81–116 (1976) (in Russian); English Transl.: J. Soviet Math. **10**(2) 257–286 (1978)
105. O.A. Ladyzhenskaya, V.A. Solonnikov, Determination of the solutions of boundary value problems for stationary Stokes and Navier-Stokes equations having an unbounded Dirichlet integral. Zapiski Nauchn. Sem. LOMI **96**, 117–160 (1980) (in Russian); English Transl.: J. Sov. Math. **21**(5), 728–761 (1983)
106. O.A. Ladyzhenskaya, N.N. Ural'ceva, *Linear and Quasi-Linear Elliptic Equations* (Academic, New York, 1968)
107. O.A. Ladyzhenskaya, V.A. Solonnikov, N.N. Ural'ceva, *Linear and Quasilinear Equations of Parabolic Type, Translation of Mathematical Monographs*, vol. 23 (AMS, Providence, 1968)
108. P.K. Lamm, Surveys on solution methods for inverse problems, in *A Survey of Regularization Methods for First-Kind Volterra Equations*, ed. by D. Colton, H.W. Engl, A. Louis, J.R. McLaughlin, W. Rundell (Springer, Vienna, 2000), pp. 53–82
109. L.D. Landau, E.M. Lifshitz, *Fluid Mechanics* (Pergamon Press, Oxford, 1987)
110. E.M. Landis, *Second Order Elliptic and Parabolic Equations* (Nauka, Moscow, 1971)
111. E.M. Landis, G. Panasenko, A theorem on the asymptotic of solutions of elliptic equations with coefficients periodic in all variables except one. Doklady AN SSSR **235**(6), 1253–1255 (1977) (in Russian); English Transl.: Soviet Math. Dokl. **18**(4), 1140–1143 (1977)
112. H. Le Dret, Modelling of the junction between two rods. J. Math. Pures Appl. **68**, 365–397 (1989)

113. J.L. Lions, *Quelques Méthodes de Résolution des Problèmes aux Limites Non Linéaires* (Dunod, Paris, 1969)
114. R. López Pouse, Mean value theorem for Lebesgue integrals. Real Analy. Exchange **37**(1), 439–450 (2011/2012)
115. G. Lukaszewicz, *Micropolar Fluids: Theory and Applications, Birkhäuser* (Basel, Berlin 1999)
116. E.N. Marieb, K. Hoehn, *Human Anatomy and Physiology* (Pearson Education, Singapore, 2018)
117. E. Marušić-Paloka, Steady flow of a non-Newtonian fluid in unbounded channels and pipes. Math. Models Methods Appl. Sci. **10**(9), 1425–1445 (2000). https://doi.org/10.1142/S0218202500000689
118. E. Marušić-Paloka, I. Pažanin, A note on Kirchhoff junction rule for power-law fluids. Zeitschrift für Naturforschung A **70**(9), 695–702 (2015). https://doi.org/10.1515/zna-2015-0148
119. H.N. Mayrovitz, Skin capillary metrics and hemodynamics in the hairless mouse. Microvasc. Res. **43**(1), 46–59 (1992)
120. V.G. Maz'ya, *Sobolev Spaces* (Springer, Berlin, 1985)
121. V.G. Maz'ya, A. Slutskii, Asymptotic analysis of the Navier-Stokes system in a plane domain with thin channels. Asymptotic Anal. **23**(1), 59–89 (2000)
122. V.G. Maz'ya, S. Nazarov, B. Plamenevskij, *Asymptotic Theory of Elliptic Boundary Value Problems in Singularly Perturbed Domains*, vol. 2 (Birkhuser-Verlag, Basel, 2000)
123. R.K. Miller, A. Feldstein, Smoothness of solutions of Volterra integral equations with weakly singular kernels. SIAM J. Math. Anal. **2**, 242–258 (1971)
124. F. Murat, A. Sili, Comportement asymptotique des solutions du système de l' élasticité linéarisé anisotrope hétérogène dans des cylindres minces. Comptes Rendus Acad. Sci. Paris, Série I **328**(1), 179–184 (1999)
125. C. Murray, The physiological principle of minimum work applied to the angle of branching of arteries. J. Gen. Physiol. **9**(6), 835–841 (1926)
126. C. Murray, A relationship between circumference and weight in trees and its bearing on branching angles. J. Gen. Physiol. **10**(5), 725–729 (1927)
127. A. Nachit, G. Panasenko, A.M. Zine, Asymptotic partial domain decomposition in thin tube structures: numerical experiments. Int. J. Multiscale Comput. Eng. **11**(5), 407–441 (2013). https://doi.org/10.1615/IntJMultCompEng
128. S.A. Nazarov, Structure of solutions of elliptic boundary value problems in thin domains. Vestnik Leningrad Univer. **7**, 65–68 (1982) (in Russian)
129. S.A. Nazarov, Asymptotic solution of the Navier-Stokes problem on the flow in thin layer fluid. Siberian Math. J. **31**, 296–307 (1990)
130. S.A. Nazarov, *Asymptotic Theory of Thin Plates and Rods* (Nauchnaya Kniga, Novosibirsk, 2002) (in Russian)
131. S.A. Nazarov, K. Pileckas, Asymptotic conditions at infinity for the Stokes and Navier-Stokes problems in domains with cylindrical outlets to infinity. Quad. Mat. Adv. Fluid Dyn. **4**, 141–245 (1999)
132. S.A. Nazarov, B.A. Plamenevsky, *Elliptic Problems in Domains with Piecewise Smooth Boundary* (Walter de Gruyter, Berlin, 1994)
133. S.A. Nazarov, M. Specovius-Neugebauer, Approximation of unbounded domains by bounded domains. Boundary value problems for Lamé operator. St. Petersburg Math. J. **8**(5), 879–912 (1997)
134. J.J. Nieto, Impulsive resonance periodic problems of first order. Appl. Math. Lett. **15**, 489–493 (2002)
135. A. Noor, M.S. Anderson, W.H. Greene, Continuum models for beam- and platelike lattice structures. AIAA J. **16**, 1219–1228 (1978)
136. B. Nowakowski, G. Stróhmer, In-flow and out-flow problem for the Stokes system. J. Math. Fluid Mech. **22**, Article No. 58 (2020)
137. O.A. Oleinik, G.A. Yosifian, On the asymptotic behaviour at infinity of solutions in linear elasticity. Arch. Rat. Mech. Anal. **78**, 29–53 (1982)

138. O.A. Oleinik, A.S. Shamaev, G.A. Yosif'yan, *Mathematical Problems in Elasticity and Homogenization* (Elsevier, Amsterdam, 1992); Russian Version: *Mathematical Problems in the Theory of Strongly Heterogeneous Elastic Media* (Moscow University Publishers, Moscow, 1990)
139. G.P. Panasenko, The principle of splitting of an averaged operator for a nonlinear system of equations in periodic and random skeletal structures. USSR Dokl. **263**(1), 35–40 (1982) (in Russian); English Transl.: Soviet Math. Dokl. **25**(2), 290–295 (1982)
140. G.P. Panasenko, Asymptotic solutions of the elasticity theory system of equations for lattice and skeletal structures. Math. Sb. **183**(1), 89–113 (1992) (in Russian); English Transl.: Sb. Math. **75**(1), 85–110 (1993)
141. G.P. Panasenko, Asymptotic analysis of bar systems. I. Russ. J. Math. Phys. **2**(3), 325–352 (1994)
142. G.P. Panasenko, Asymptotic analysis of bar systems. II. Russ. J. Math. Phys. **4**(1), 87–116 (1996)
143. G. Panasenko, Method of asymptotic partial decomposition of domain. Math. Models Methods Appl. Sci. **8**(1), 139–156 (1998)
144. G. Panasenko, Asymptotic expansion of the solution of Navier-Stokes equation in a tube structure. C.R. Acad. Sci. Paris **326**, Série IIb, 867–872 (1998)
145. G. Panasenko, Partial asymptotic decomposition of domain: Navier-Stokes equation in tube structure. C.R. Acad. Sci. Paris **326**, Série IIb, 893–898 (1998)
146. G. Panasenko, *Multi-scale Modeling for Structures and Composites* (Springer, Dordrecht, 2005)
147. G. Panasenko, *Introduction to Multiscale Mathematical Modeling* (World Scientific, New Jersey, 2022)
148. G. Panasenko, K. Pileckas, Asymptotic analysis of the nonsteady viscous flow with a given flow rate in a thin pipe. Appl. Analy. **91**(3), 559–574 (2012)
149. G. Panasenko, K. Pileckas, Flows in a tube structure: equation on the graph. J. Math. Phys. **55**, Article No. 081505 (2014). https://doi.org/10.1063/1.4891249
150. G. Panasenko, K. Pileckas, Divergence equation in thin-tube structure. Appl. Analy. **94**(7), 1450–1459 (2015). https://doi.org/10.1080/00036811.2014.933476
151. G. Panasenko, K. Pileckas, Asymptotic analysis of the non-steady Navier-Stokes equations in a tube structure. I. The case without boundary layer-in-time. Nonlinear Analy. Theory Methods Appl. **122**, 125–168 (2015). https://doi.org/10.1016/j.na.2015.03.008
152. G. Panasenko, K. Pileckas, Asymptotic analysis of the non-steady Navier-Stokes equations in a tube structure. II. General case. Nonlinear Analy. Theory Methods Appl. **125**, 582–607 (2015). https://doi.org/10.1016/j.na.2015.05.018
153. G. Panasenko, K. Pileckas, Periodic in time flow in thin structures: equations on the graph. J. Math. Analy. Appl. **490**, Article No. 124335 (2020)
154. G. Panasenko, K. Pileckas, Pressure boundary conditions for viscous flows in thin tube structures: Stokes equations with locally distributed Brinkman term. Math. Modell. Nat. Phenomena **18**, Article No. 17 (2023)
155. G. Panasenko, K. Pileckas, Partial asymptotic dimension reduction for steady state non-Newtonian flow with strain rate dependent viscosity in thin tube structure. J. Math. Mech. Fluids **25**(1), Article No. 11 (2023)
156. G. Panasenko, K. Pileckas, Nonstationary Poiseuille flow of a non-Newtonian fluid with the shear rate-dependent viscosity. Adv. Nonlinear Analy. **12**, Article No. 12023 (2023)
157. G.P. Panasenko, M.V. Reztsov, Averaging the three-dimensional elasticity problem in non-homogeneous plates. Dokl. AN SSSR **294**(5), 1061–1065 (1987) (in Russian); English Transl.: Dokl. Math. **35**(3), 630–636 (1987)
158. G. Panasenko, R. Stavre, Asymptotic analysis of a periodic flow in a thin channel with visco-elastic wall. J. Math. Pures Appl. **85**(4), 558–579 (2006)
159. G. Panasenko, R. Stavre, Asymptotic analysis of a non-periodic flow in a thin channel with visco-elastic wall. Netw. Heterog. Media **3**(3), 651–673 (2008)

160. G. Panasenko, R. Stavre, Well-posedness and asymptotic expansion of solution of Stokes equation set in a thin cylindrical elastic tube, in *Around the Research of Vladimir Maz'ya II, Partial Differential Equations*, ed. by A. Laptev (Springer, New York, 2010), pp. 275–301
161. G.P. Panasenko, R. Stavre, Viscous fluid-thin elastic plate interaction: asymptotic analysis with respect to the rigidity and density of the plate. Appl. Math. Optim. **81**, 141–194 (2020)
162. G.Panasenko, R. Stavre, Three dimensional asymptotic analysis of an axisymmetric flow in a thin tube with thin stiff elastic wall. J. Math. Fluid Mech. **22**, Article No. 20 (2020)
163. G.Panasenko, B. Vernescu, Non-Newtonian flows in domains with non-compact boundaries. Nonlinear Analy. Theory Methods Appl. **183**, 214–229 (2019). https://doi.org/10.1016/j.na.2019.01.015
164. G. Panasenko, M.C. Viallon, Method of asymptotic partial decomposition with discontinuous junctions. Comput. Math. Appl. **105**, 75–93 (2022)
165. G. Panasenko, Y. Sirakov, R. Stavre, Asymptotic and numerical modelling of a flow in a thin channel with visco-elastic wall. Int. J. Multiscale Comput. Eng. **5**(6), 473–482 (2007)
166. G. Panasenko, K. Pileckas, B. Vernescu, Steady state non-Newtonian flow with strain rate dependent viscosity in domains with cylindrical outlets to infinity. Nonlinear Analy. Model. Controle **26**(6), 1166–1199 (2021)
167. G. Panasenko, K. Pileckas, B. Vernescu, Steady state non-Newtonian flow in thin tube structure: equation on the graph. St. Petersburg. Math. J. **33**, 327–340 (2022)
168. G. Panasenko, K. Pileckas, B.Vernescu, Steady state non-Newtonian flow with strain rate dependent viscosity in thin tube structure with no-slop boundary condition. Math. Modell. Nat. Phenomena **17**, Article No. 18 (2022)
169. N.S. Papageorgiou, P. Winkert, *Applied Nonlinear Functional Analysis: An Introduction* (Walter De Gruyter, Berlin, 2018)
170. K. Pileckas, Three-dimensional solenoidal vectors. Zapiski Nauchn. Sem. LOMI **96**, 237–239 (1980) (in Russian); English Transl.: J. Sov. Math. **21**, 821–823 (1983)
171. K. Pileckas, Existence of solutions for the Navier–Stokes equations having an infinite dissipation of energy, in a class of domains with noncompact boundaries. Zapiski Nauchn. Sem. LOMI **110**, 180–202 (1981) (in Russian); English Transl.: J. Sov. Math. **25**(1), 932–947 (1984)
172. K. Pileckas, On spaces of solenoidal vectors. Trudy Mat. Inst. Steklov **159**, 137–149 (1983) (in Russian); English Transl.: Proc. Math. Inst. Steklov **159**(2), 141–154 (1984)
173. K. Pileckas, On the behavior of the nonstationary Poiseuille solution as $t \to \infty$. Siberian Math. J. **46**, 890–900 (2005)
174. K. Pileckas, On the nonstationary linearized Navier-Stokes problem in domains with cylindrical outlets to infinity. Math. Ann. **332**, 395–419 (2005)
175. K. Pileckas, Existence of solutions with the prescribed flux of the Navier-Stokes system in an infinite cylinder. J. Math. Fluid Mech. **8**(4), 542–563 (2006)
176. K. Pileckas, Navier–Stokes system in domains with cylindrical outlets to infinity. Leray's problem, in *Handbook of Mathematical Fluid Dynamics*, ed. by S. Friedlander, D. Serre, vol. 4, Ch. 8 (Elsevier, Amsterdam, 2007), pp. 445–647
177. K. Pileckas, R. Ciegis, Existence of nonstationary Poiseuille type solutions under minimal regularity assumptions. Zeitschrift für angewandte Mathematik und Physik (ZAMP) **71**(6), Article No. 192, 1–17 (2020)
178. K. Pileckas, V. Keblikas, On the existence of the nonstationary Poiseuille solution. Siberian Math. J. **46**(3), 514–526 (2005)
179. K. Pileckas, A. Sequeira, J.H. Videman, Steady flows of viscoelastic fluids in domains with outlets to infinity. J. Math. Fluid Mech. **2**, 185–218 (2000)
180. O. Pironneau, Conditions aux limites sur la pression pour les équations de Stokes et de Navier–Stokes. C.R. Acad. Sci. Paris Sér. I Math. **303**(9), 403–406 (1986)
181. J.L. Poiseuille, Recherches experimentales sur le mouvement des liquides dans les tubes de tres-petits diametres. I. Influence de la pression sur la quantité de liquide qui traverse les tubes de très petits diamrètres. C.R. Acad. Sci. **11**(24), 961–967 (1840)

182. A.I. Prilepko, D.G. Orlovsky, I.A. Vasin, *Methods for Solving Inverse Problems in Mathematical Physics* (Dekker, New York, 1999)
183. G. Prouse, Soluzioni quasi-periodiche dell'equazione differenziale di Navier-Stokes in due dimensioni. Rend. Semin. Mat. Univ. Padova **33**, 186–212 (1963)
184. G. Prouse, Soluzioni periodiche dell'equazione di Navier–Stokes. Atti Accad. Naz. Lincei Rend. Cl. Sci. Fis. Mat. Natur. **35**, 443–447 (1963)
185. A. Quarteroni, L. Formaggia, Mathematical modelling and numerical simulation of the cardiovascular system, in *Modelling of Living Systems*, ed. by N. Ayache. Handbook of Numerical Analysis Series, ed. by P.G. Ciarlet, J.L. Lions (Elsevier, Amsterdam, 2002)
186. A. Quarteroni, A. Veneziani, Analysis of a geometrical multiscale model based on the coupling of ODEs and PDEs for blood flow simulations. Multiscale Model. Simul. **1**(2), 173–195 (2003)
187. K. Rajagopal, A note on unsteady unidirectional flows of a non-Newtonian fluid. Int. J. Nonlinear Mech. **17**(5), 369–373 (1982)
188. K. Rajagopal, A. Gupta, On a class of exact solutions to the equations of motion of a 2D grade fluids. J. Eng. Sci. **19**(7), 1009–1014 (1981)
189. W. Rudin, *Functional Analysis, International Series in Pure and Applied Mathematics* (McGraw-Hill Higher Education, New York, 1991)
190. J. Sanchez-Hubert, E. Sanchez-Palencia, *Coques Elastiques Minces: Propriétés Asymptotiques* (Masson, Paris, 1997)
191. R.L. Sani, P.M. Gresho, Résumé and remarks on the open boundary condition. Int. J. Num. Methos Fluid **18**, 983–1008 (1994)
192. R.L. Sani, J. Shen, O. Pironneau, P.M. Gresho, Pressure boundary condition for the time-dependent incompressible Navier-Stokes equations. Int. J. Numer. Meth. Fluids **50**, 673–682 (2006)
193. W.W. Schmaedeke, Approximate solutions for the Volterra equations of the first kind. J. Math. Anal. Appl. **23**, 604–613 (1968)
194. G. Seregin, *Lecture Notes On Regularity Theory for the Navier-Stokes Equations* (World Scientific Publishing, New Jersey, 2015)
195. S.M. Shugrin, Coupling of one-dimensional and two-dimensional models of water flow. Water Resour. **5**, 5–15 (1987) (in Russian)
196. V.A. Solonnikov, Estimates for solutions of a non-stationary linearized system of Navier–Stokes system. Trudy Mat. Inst. Steklov **70**, 213–317 (1964) (in Russian); English Transl.: Proc. Math. Inst. Steklov **70** (1964)
197. V.A. Solonnikov, On boundary value problems for linear parabolic systems of general form. Trudy Mat. Inst. Steklov **83**, 3–163 (1965) (in Russian); English Transl.: Proc. Math. Inst. Steklov **83**, 1–184 (1965)
198. V.A. Solonnikov, On the boundary value problems for systems elliptic in the sence of A. Douglis, L. Nirenberg, I. Izv. Akad. Nauk SSSR, Ser. Mat. **28**, 665–706 (1964); II. Trudy Mat. Inst. Steklov **92**, 233–297 (1966) (in Russian); English Transl.: I. Amer. Math. Soc. Transl. **56**(2), 192–232 (1966); II. Proc. Steklov Inst. Math. **92**, 269–333 (1966)
199. V.A. Solonnikov, Stokes and Navier–Stokes equations in domains with noncompact boundaries. Nonlinear Partial Diff. Equ. Appl. Pitmann Notes Math. Collège de France Séminar **3**, 240–349 (1983)
200. V.A. Solonnikov, K. Pileckas, Certain spaces of solenoidal vectors and the solvability of the boundary value problem for the Navier–Stokes system of equations in domains with noncompact boundaries. Zapiski Nauchn. Sem. LOMI **73**, 136–151 (1977) (in Russian); English Transl.: J. Sov. Math. **34**(6), 2101–2111 (1986)
201. M. Specovius-Neugebauer, Approximation of the Stokes Dirichlet problem in domains with cylindrical outlets. SIAM J. Math. Anal. **30**(3), 645–677 (1999)
202. S.S. Sritharan, On the acceleration of viscous fluid through an unbounded channel. J. Math. Anal. Appl. **168**, 255–283 (1992)
203. R. Temam, *Navier-Stokes Equations* (North-Holland Pub. Company, Amsterdam, 1977)

204. L. Trabucho, J.M. Viaño, Derivation of generalized models for linear elastic beams by asymptotic expansion methods, in *Applications of Multiple Scalings in Mechanics*, ed. by P.G. Ciarlet, E. Sanchez-Palencia (Masson, Paris, 1987), pp. 302–315
205. V.A. Trenogin, *The Functional Analysis* (Nauka, Moscow, 1980); French Transl.: V.A. Trenoguine, *Analyse Fonctionnelle* (Éditions Mir, Moscow, 1985)
206. F.G. Tricomi, *Integral Equations* (Intersience, New York, 1957)
207. O. Vejvoda, L. Hermann, V. Lovicar, M. Sova, I. Straškraba, M. Štědrý, *Partial Differential Equations: Time-Periodic Solutions* (Martinus Nijhoff Publishers, The Hague, 1981)
208. J.R. Womersley, Method for the calculation of velocity, rate of flow and viscous drag in arteries when the pressure gradient is known. J. Physiol. **127**(3), 553–563 (1955)
209. V.V. Zhikov, S.E. Pastukhova, Homogenization for elasticity problems on periodic networks of critical thickness. Sb. Math. **194**(5), 697–732 (2003)
210. J. Málek, J. Nečas, M. Rokyta, M. Růžička, *Weak and Measure – valued Solutions of Evolutionary PDEs* (Chapman and Hall, London – Weinheim – New York – Tokyo – Melbourne – Madras, 1996)

Index

G. Panasenko, K. Pileckas, *Multiscale Analysis of Viscous Flows in Thin Tube Structures*, Advances in Mathematical Fluid Mechanics,
https://doi.org/10.1007/978-3-031-54630-3